THE AVR MICROCONTROLLER AND EMBEDDED SYSTEMS

Using Assembly and C

Muhammad Ali Mazidi
Sarmad Naimi
Sepehr Naimi

Prentice Hall

Boston   Columbus   Indianapolis   New York   San Francisco   Upper Saddle River
Amsterdam   Cape Town   Dubai   London   Madrid   Milan   Munich   Paris   Montreal   Toronto
Delhi   Mexico City   Sao Paulo   Sydney   Hong Kong   Seoul   Singapore   Taipei   Tokyo

Editor in Chief: Vernon Anthony
Acquisitions Editor: Wyatt Morris
Editorial Assistant: Chris Reed
Director of Marketing: David Gesell
Marketing Manager: Kara Clark
Senior Managing Coordinator: Alicia Wozniak
Marketing Assistant: Les Roberts
Senior Managing Editor: JoEllen Gohr

Project Manager: Rex Davidson
Senior Operations Supervisor: Pat Tonneman
Operations Specialist: Laura Weaver
Art Director: Dianne Ernsberger
Cover Designer: Jeff Vanik
Cover Art: Antonis Papantoniou, Fotolia.com
Text Font: Times Roman

Library of Congress Cataloging in Publication Data

Mazidi, Muhammad Ali.
   The AVR microcontroller and embedded systems: using Assembly and C /
  Muhammad Ali Mazidi, Sarmad Naimi, Sepehr Naimi.
      p. cm.
   ISBN-13: 978-0-13-800331-9 (alk. paper)
   ISBN-10: 0-13-800331-9 (alk. paper)
 1. Atmel AVR microcontroller. 2. Embedded computer systems. 3. Assembler language (Computer program language) 4. C (Computer program language) I. Naimi, Sarmad. II. Naimi, Sepehr. III. Title.
 TJ223.P76M378136 2009
 004.16--dc22
          2009039790

Prentice Hall
is an imprint of

www.pearsonhighered.com

27 18
ISBN 10: 0-13-800331-9
ISBN 13: 978-0-13-800331-9

*This book is dedicated
to the memory of Dr. Kamal Bakhtavar
for all his sacrifices.*
– Muhammad Ali Mazidi

*This book is dedicated
to the memory of Dr. P. Javid
for his inspiring example of dedication to the education of young people.*
– Sarmad Naimi

*This book is dedicated to
Yaran.*
– Sepehr Naimi

Regard man as a mine rich in gems of inestimable value. Education can, alone, cause it to reveal its treasures, and enable mankind to benefit therefrom.

Baha'u'llah

BRIEF CONTENTS

CHAPTERS

0:	Introduction to Computing	1
1:	The AVR Microcontroller: History and Features	39
2:	AVR Architecture and Assembly Language Programming	55
3:	Branch, Call, and Time Delay Loop	107
4:	AVR I/O Port Programming	139
5:	Arithmetic, Logic Instructions, and Programs	161
6:	AVR Advanced Assembly Language Programming	197
7:	AVR Programming in C	255
8:	AVR Hardware Connection, Hex File, and Flash Loaders	289
9:	AVR Timer Programming in Assembly and C	311
10:	AVR Interrupt Programming in Assembly and C	363
11:	AVR Serial Port Programming in Assembly and C	395
12:	LCD and Keyboard Interfacing	429
13:	ADC, DAC, and Sensor Interfacing	463
14:	Relay, Optoisolator, and Stepper Motor Interfacing with AVR	491
15:	Input Capture and Wave Generation in AVR	509
16:	PWM Programming and DC Motor Control in AVR	549
17:	SPI Protocol and MAX7221 Display Interfacing	603
18:	I2C Protocol and DS1307 RTC Interfacing	629

APPENDICES

A:	AVR Instructions Explained	695
B:	Basics of Wire Wrapping	733
C:	IC Interfacing and System Design Issues	737
D:	Flowcharts and Pseudocode	755
E:	AVR Primer for 8051 Programmers	761
F:	ASCII Codes	762
G:	Assemblers, Development Resources, and Suppliers	764
H:	Data Sheets	766

CONTENTS

CHAPTER 0: INTRODUCTION TO COMPUTING   **1**
SECTION 0.1: NUMBERING AND CODING SYSTEMS   2
SECTION 0.2: DIGITAL PRIMER   9
SECTION 0.3: SEMICONDUCTOR MEMORY   13
SECTION 0.4: CPU ARCHITECTURE   29

CHAPTER 1: THE AVR MICROCONTROLLER: HISTORY AND FEATURES   **39**
SECTION 1.1: MICROCONTROLLERS AND EMBEDDED PROCESSORS   40
SECTION 1.2: OVERVIEW OF THE AVR FAMILY   44

CHAPTER 2: AVR ARCHITECTURE AND ASSEMBLY LANGUAGE PROGRAMMING   **55**
SECTION 2.1: THE GENERAL PURPOSE REGISTERS IN THE AVR   56
SECTION 2.2: THE AVR DATA MEMORY   59
SECTION 2.3: USING INSTRUCTIONS WITH THE DATA MEMORY   61
SECTION 2.4: AVR STATUS REGISTER   71
SECTION 2.5: AVR DATA FORMAT AND DIRECTIVES   75
SECTION 2.6: INTRODUCTION TO AVR ASSEMBLY PROGRAMMING   80
SECTION 2.7: ASSEMBLING AN AVR PROGRAM   82
SECTION 2.8: THE PROGRAM COUNTER AND PROGRAM ROM SPACE IN THE AVR   85
SECTION 2.9: RISC ARCHITECTURE IN THE AVR   93
SECTION 2.10: VIEWING REGISTERS AND MEMORY WITH AVR STUDIO IDE   97

CHAPTER 3: BRANCH, CALL, AND TIME DELAY LOOP   **107**
SECTION 3.1: BRANCH INSTRUCTIONS AND LOOPING   108
SECTION 3.2: CALL INSTRUCTIONS AND STACK   118
SECTION 3.3: AVR TIME DELAY AND INSTRUCTION PIPELINE   128

CHAPTER 4: AVR I/O PORT PROGRAMMING   **139**
SECTION 4.1: I/O PORT PROGRAMMING IN AVR   140
SECTION 4.2: I/O BIT MANIPULATION PROGRAMMING   149

CHAPTER 5: ARITHMETIC, LOGIC INSTRUCTIONS, AND PROGRAMS   **161**
SECTION 5.1: ARITHMETIC INSTRUCTIONS   162
SECTION 5.2: SIGNED NUMBER CONCEPTS AND ARITHMETIC OPERATIONS   170
SECTION 5.3: LOGIC AND COMPARE INSTRUCTIONS   176
SECTION 5.4: ROTATE AND SHIFT INSTRUCTIONS AND DATA SERIALIZATION   183
SECTION 5.5: BCD AND ASCII CONVERSION   190

CHAPTER 6: AVR ADVANCED ASSEMBLY LANGUAGE PROGRAMMING   **197**
SECTION 6.1: INTRODUCING SOME MORE ASSEMBLER DIRECTIVES   198
SECTION 6.2: REGISTER AND DIRECT ADDRESSING MODES   202
SECTION 6.3: REGISTER INDIRECT ADDRESSING MODE   208

SECTION 6.4: LOOK-UP TABLE AND TABLE PROCESSING   216
SECTION 6.5: BIT-ADDRESSABILITY   226
SECTION 6.6: ACCESSING EEPROM IN AVR   233
SECTION 6.7: CHECKSUM AND ASCII SUBROUTINES   238
SECTION 6.8: MACROS   244

CHAPTER 7: AVR PROGRAMMING IN C   **255**
SECTION 7.1: DATA TYPES AND TIME DELAYS IN C   256
SECTION 7.2: I/O PROGRAMMING IN C   263
SECTION 7.3: LOGIC OPERATIONS IN C   265
SECTION 7.4: DATA CONVERSION PROGRAMS IN C   275
SECTION 7.5: DATA SERIALIZATION IN C   280
SECTION 7.6: MEMORY ALLOCATION IN C   282

CHAPTER 8: AVR HARDWARE CONNECTION, HEX FILE, AND
FLASH LOADERS   **289**
SECTION 8.1: ATMEGA32 PIN CONNECTION   290
SECTION 8.2: AVR FUSE BITS   294
SECTION 8.3: EXPLAINING THE HEX FILE FOR AVR   300
SECTION 8.4: AVR PROGRAMMING AND TRAINER BOARD   305

CHAPTER 9: AVR TIMER PROGRAMMING IN ASSEMBLY AND C   **311**
SECTION 9.1: PROGRAMMING TIMERS 0, 1, AND 2   313
SECTION 9.2: COUNTER PROGRAMMING   348
SECTION 9.3: PROGRAMMING TIMERS IN C   353

CHAPTER 10: AVR INTERRUPT PROGRAMMING IN ASSEMBLY
AND C   **363**
SECTION 10.1: AVR INTERRUPTS   364
SECTION 10.2: PROGRAMMING TIMER INTERRUPTS   369
SECTION 10.3: PROGRAMMING EXTERNAL HARDWARE
INTERRUPTS   376
SECTION 10.4: INTERRUPT PRIORITY IN THE AVR   381
SECTION 10.5: INTERRUPT PROGRAMMING IN C   385

CHAPTER 11: AVR SERIAL PORT PROGRAMMING IN ASSEMBLY
AND C   **395**
SECTION 11.1: BASICS OF SERIAL COMMUNICATION   396
SECTION 11.2: ATMEGA32 CONNECTION TO RS232   403
SECTION 11.3: AVR SERIAL PORT PROGRAMMING IN ASSEMBLY   405
SECTION 11.4: AVR SERIAL PORT PROGRAMMING IN C   419
SECTION 11.5: AVR SERIAL PORT PROGRAMMING IN ASSEMBLY
AND C USING INTERRUPTS   422

CHAPTER 12: LCD AND KEYBOARD INTERFACING   **429**
SECTION 12.1: LCD INTERFACING   430
SECTION 12.2: KEYBOARD INTERFACING   452

CHAPTER 13: ADC, DAC, AND SENSOR INTERFACING   **463**
SECTION 13.1: ADC CHARACTERISTICS   464
SECTION 13.2: ADC PROGRAMMING IN THE AVR   469

SECTION 13.3: SENSOR INTERFACING AND SIGNAL
    CONDITIONING   480
SECTION 13.4: DAC INTERFACING   484

**CHAPTER 14: RELAY, OPTOISOLATOR, AND STEPPER MOTOR
INTERFACING WITH AVR   491**
SECTION 14.1: RELAYS AND OPTOISOLATORS   492
SECTION 14.2: STEPPER MOTOR INTERFACING   498

CHAPTER 15: INPUT CAPTURE AND WAVE GENERATION IN AVR   509
SECTION 15.1: WAVE GENERATION USING 8-BIT TIMERS   510
SECTION 15.2: WAVE GENERATION USING TIMER1   520
SECTION 15.3: INPUT CAPTURE PROGRAMMING   531
SECTION 15.4: C PROGRAMMING   539

**CHAPTER 16: PWM PROGRAMMING AND DC MOTOR CONTROL
IN AVR   549**
SECTION 16.1: DC MOTOR INTERFACING AND PWM   550
SECTION 16.2: PWM MODES IN 8-BIT TIMERS   560
SECTION 16.3: PWM MODES IN TIMER1   574
SECTION 16.4: DC MOTOR CONTROL USING PWM   597

CHAPTER 17: SPI PROTOCOL AND MAX7221 DISPLAY INTERFACING   603
SECTION 17.1: SPI BUS PROTOCOL   604
SECTION 17.2: SPI PROGRAMMING IN AVR   609
SECTION 17.3: MAX7221 INTERFACING AND PROGRAMMING   615

CHAPTER 18: I2C PROTOCOL AND DS1307 RTC INTERFACING   629
SECTION 18.1: I2C BUS PROTOCOL   630
SECTION 18.2: TWI (I2C) IN THE AVR   638
SECTION 18.3: AVR TWI PROGRAMMING IN ASSEMBLY AND C   642
SECTION 18.4: DS1307 RTC INTERFACING AND PROGRAMMING   654
SECTION 18.5: TWI PROGRAMMING WITH CHECKING STATUS
    REGISTER   668

APPENDIX A: AVR INSTRUCTIONS EXPLAINED   695
SECTION A.1: INSTRUCTION SUMMARY   696
SECTION A.2: AVR INSTRUCTIONS FORMAT   700
SECTION A.3: AVR REGISTER SUMMARY   732

APPENDIX B: BASICS OF WIRE WRAPPING   733

APPENDIX C: IC INTERFACING AND SYSTEM DESIGN ISSUES   737
SECTION C.1: OVERVIEW OF IC TECHNOLOGY   738
SECTION C.2: AVR I/O PORT STRUCTURE AND INTERFACING   744
SECTION C.3: SYSTEM DESIGN ISSUES   750

APPENDIX D: FLOWCHARTS AND PSEUDOCODE   755

APPENDIX E: AVR PRIMER FOR 8051 PROGRAMMERS   761

APPENDIX F: ASCII CODES   **762**

**APPENDIX G: ASSEMBLERS, DEVELOPMENT RESOURCES, AND
   SUPPLIERS**   **764**

APPENDIX H: DATA SHEETS   **766**

INDEX   **771**

PREFACE

Products using microprocessors generally fall into two categories. The first category uses high-performance microprocessors such as the Pentium in applications where system performance is critical. We have an entire book dedicated to this topic, *The x86 PC: Assembly Language, Design, and Interfacing*, published by Prentice Hall. In the second category of applications, performance is secondary; issues of cost, space, power, and rapid development are more critical than raw processing power. The microprocessor for this category is often called a *microcontroller*.

This book is for the second category of applications. The AVR is a widely used microcontroller. This book is intended for use in college-level courses teaching microcontrollers and embedded systems. It not only establishes a foundation of Assembly language programming, but also provides a comprehensive treatment of AVR interfacing for engineering students. From this background, the design and interfacing of microcontroller-based embedded systems can be explored. This book can also be used by practicing technicians, hardware engineers, computer scientists, and hobbyists.

Prerequisites

Readers should have had an introductory digital course. Knowledge of Assembly language would be helpful, but is not necessary. Although this book is written for those with no background in Assembly language programming, students with prior Assembly language experience will be able to gain a mastery of AVR architecture very rapidly and start on their projects right away. For the AVR C programming sections of the book, a basic knowledge of C programming is required. We use the AVR Studio compiler IDE from Atmel throughout the book. The AVR Studio compiler is available for free from the Atmel website (www.atmel.com). We encourage you to use the AVR Studio or some other IDE to simulate and run the programs in this book.

Overview

A systematic, step-by-step approach is used to cover various aspects of AVR C and Assembly language programming and interfacing. Many examples and sample programs are given to clarify the concepts and provide students with an opportunity to learn by doing. Review questions are provided at the end of each section to reinforce the main points of the section.

Chapter 0 covers number systems (binary, decimal, and hex), and provides an introduction to basic logic gates and computer memory. This chapter is designed especially for students, such as mechanical engineering students, who have not taken a digital logic course or those who need to refresh their memory on these topics.

Chapter 1 discusses the history of the AVR and features of the members such as ATmega32. It also provides a list of various members of the AVR family.

Chapter 2 discusses the internal architecture of the AVR and explains the use of a AVR assembler to create ready-to-run programs. It also explores the program counter and the flag register.

In Chapter 3 the topics of loop, jump, and call instructions are discussed, with many programming examples.

Chapter 4 is dedicated to the discussion of I/O ports. This allows students who are working on a project to start experimenting with AVR I/O interfacing and start the hardware project as soon as possible.

Chapter 5 is dedicated to arithmetic, logic instructions, and programs.

Chapter 6 covers the AVR advanced addressing modes and explains how to access the data stored in the look-up table, as well as how to use EEPROM to store data and how to do macros.

The C programming of the AVR is covered in Chapter 7. We use the WinAVR compiler for this and throughout the book. The WinAVR is available for free from the winavr.sourceforge.net website.

In Chapter 8 we discuss the hardware connection of the AVR chip.

Chapter 9 describes the AVR timers and how to use them as event counters.

Chapter 10 provides a detailed discussion of AVR interrupts with many examples on how to write interrupt handler programs.

Chapter 11 is dedicated to serial data communication of the AVR and its interfacing to the RS232. It also shows AVR communication with COM ports of the x86 IBM PC and compatible computers.

Chapter 12 shows AVR interfacing with real-world devices such as LCDs and keyboards.

Chapter 13 shows AVR interfacing with real-world devices such as DAC chips, ADC chips, and sensors.

Chapter 14 covers the basic interfacing of the AVR chip to relays, optoisolators, and stepper motors.

In Chapter 15 we cover how to use AVR timers to generate waves and explain how to capture waves to measure period and duty cycle.

Chapter 16 shows PWM and basic interfacing to DC motors.

Chapter 17 covers the SPI bus protocol and describes how to interface 7-segment displays using MAX7221.

Finally, Chapter 18 shows how to connect and program the DS1307 real-time clock chip using the TWI (I2C) bus protocol.

The appendices have been designed to provide all reference material required for the topics covered in the book. Appendix A describes each AVR instruction in detail, with examples. Appendix A also provides the clock count for instructions and AVR I/O registers. Appendix B describes the basics of wire wrapping. Appendix C examines IC interfacing and logic families, as well as AVR I/O port interfacing and fan-out. Make sure you study this section before connecting the AVR to an external device. In Appendix D, the use of flowcharts and pseudocode is explored. Appendix E is for students familiar with 8051 architectures who need to make a rapid transition to AVR architecture. Appendix F provides the table of ASCII characters. Appendix G lists resources for assembler shareware, AVR trainers, and electronics parts. Appendix H contains data sheets for the AVR chip.

Lab Manual

The lab manual covers some very basic labs and can be found at the **www.MicroDigitalEd.com** website. The more advanced and rigorous lab assignments are left up to the instructors depending on the course objectives, class level, and whether the course is graduate or undergraduate. The support materials for this text and other books by the authors can be found on this website, too.

Solutions Manual/PowerPoint® Slides

The end-of-chapter problems cover some very basic concepts. The more challenging and rigorous homework assignments are left up to the instructors depending on the course objectives, class level, and whether the course is graduate or undergraduate. The solutions manual and PowerPoint® slides for the drawings are available online for instructors only.

Online Instructor Resources

To access supplementary materials online, instructors need to request an instructor access code. Go to **www.prenhall.com**, click the **Instructor Resource Center** link, and then click **Register Today** for an instructor access code. Within 48 hours after registering you will receive a confirming e-mail including an instructor access code. Once you have received your code, go to the site and log on for full instructions on downloading the materials you wish to use.

Acknowledgments

This book is the result of the dedication and encouragement of many individuals. Our sincere and heartfelt appreciation goes to all of them.

Thanks to the reviewers of this edition:
Orod Haghighi Ara, BIHE University;
Arona Kosari, BIHE University;
Anahita Omidvar, BIHE University;
Vahid Mokhtari, BIHE University;
Farshid Hoori, BIHE University;
Navid HajatDoost, BIHE University;
Hootan Rahmanian, BIHE University;
Farzad Sabeti, BIHE University;
Moshtagh Samandari, BIHE University.

Numerous students found errors or made suggestions in improving this book. We would like to thank all of them for their enthusiasm and support. Those students are: Arash Noori, Soroush Taefi, Golriz Nourani, Mozhdeh Amiri, Negar Ziaee Nasrabadi, and Maryam NouhNezhad, all from the computer engineering department of BIHE.

Finally, we would like to thank the people at Prentice Hall, in particular our editor, Wyatt Morris, who continues to support and encourage our writing, and our project manager, Rex Davidson, who made the book a reality. We were lucky to get the best copy editors in the world, Janice Mazidi and Bret Workman. Thank you both for your fantastic job, as usual.

We enjoyed writing this book, and hope you enjoy reading it and using it for your courses and projects. Please let us know if you have any suggestions or find any errors.

Assemblers/Compilers

The AVR Studio can be downloaded from the following website:
http://www.Atmel.com

The WinAVR C compiler for AVR can be downloaded from the following website:
http://winavr.sourceforge.net

The tutorials for all the above assemblers/compilers and AVR Trainer boards can be found on the following website:
http://www.MicroDigitalEd.com

Trademark Information and Acknowledgments

All the figures, tables, and instructions related to the AVR family of microcontrollers used in this textbook belong to Atmel Corporation. Copyright of Atmel Corporation, Inc. 2009, used by permission.

Instruction mnemonics listed in Appendix A are from Atmel Corporation. Copyright of Atmel Corporation, Inc. 2009, used by permission.

The AVR data sheets listed in Appendix H are from Atmel Semiconductor. Copyright of Atmel Semiconductor, Inc. 2009, used by permission.

ABOUT THE AUTHORS

Muhammad Ali Mazidi went to Tabriz University and holds Master's degrees from both Southern Methodist University and the University of Texas at Dallas. He is currently a.b.d. on his Ph.D. in the Electrical Engineering Department of Southern Methodist University. He is co-author of some widely used textbooks, including *The x86 PC, The 8051 Microcontroller and Embedded Systems, The PIC Microcontroller and Embedded Systems,* and *The HCS12 Microcontroller and Embedded Systems,* also available from Prentice Hall. He teaches microprocessor-based system design at DeVry University in Dallas, Texas. He is the founder of MicroDigitalEd.com.

Sarmad Naimi graduated from the Computer Engineering department of BIHE university and is currently working on his Master's degree. His areas of interest include FPGA, RTOS, and real-time embedded systems.

Sepehr Naimi graduated from the Computer Engineering department of BIHE university and is currently working on his Master's degree. His areas of interest include high-performance microcontrollers, RTOS, and real-time embedded systems.

The authors can be contacted at the following e-mail addresses if you have any comments or suggestions, or if you find any errors.

mdebooks@yahoo.com
mmazidi@microdigitaled.com
SarmadNaimi@gmail.com
Sepehr.Naimi@gmail.com

CHAPTER 0

INTRODUCTION
TO COMPUTING

OBJECTIVES

Upon completion of this chapter, you will be able to:

>> Convert any number from base 2, base 10, or base 16 to any of the other two bases
>> Describe the logical operations AND, OR, NOT, XOR, NAND, and NOR
>> Use logic gates to diagram simple circuits
>> Explain the difference between a bit, a nibble, a byte, and a word
>> Give precise mathematical definitions of the terms *kilobyte*, *megabyte*, *gigabyte*, and *terabyte*
>> Describe the purpose of the major components of a computer system
>> Contrast and compare various types of semiconductor memories in terms of their capacity, organization, and access time
>> Describe the relationship between the number of memory locations on a chip, the number of data pins, and the chip's memory capacity
>> Contrast and compare PROM, EPROM, UV-EPROM, EEPROM, Flash memory EPROM, and mask ROM memories
>> Contrast and compare SRAM, NV-RAM, and DRAM memories
>> List the steps a CPU follows in memory address decoding
>> List the three types of buses found in computers and describe the purpose of each type of bus
>> Describe the role of the CPU in computer systems
>> List the major components of the CPU and describe the purpose of each
>> Understand the RISC and Harvard architectures

To understand the software and hardware of a microcontroller-based system, one must first master some very basic concepts underlying computer architecture. In this chapter (which in the tradition of digital computers is called Chapter 0), the fundamentals of numbering and coding systems are presented in Section 0.1. In Section 0.2, an overview of logic gates is given. The semiconductor memory and memory interfacing are discussed in Section 0.3. In Section 0.4, CPUs and Harvard and von Neumann architectures are discussed. Finally, in the last section we give a brief history of RISC architecture. Although some readers may have an adequate background in many of the topics of this chapter, it is recommended that the material be reviewed, however briefly.

SECTION 0.1: NUMBERING AND CODING SYSTEMS

Whereas human beings use base 10 (*decimal*) arithmetic, computers use the base 2 (*binary*) system. In this section we explain how to convert from the decimal system to the binary system, and vice versa. The convenient representation of binary numbers, called *hexadecimal,* also is covered. Finally, the binary format of the alphanumeric code, called *ASCII*, is explored.

Decimal and binary number systems

Although there has been speculation that the origin of the base 10 system is the fact that human beings have 10 fingers, there is absolutely no speculation about the reason behind the use of the binary system in computers. The binary system is used in computers because 1 and 0 represent the two voltage levels of on and off. Whereas in base 10 there are 10 distinct symbols, 0, 1, 2, ..., 9, in base 2 there are only two, 0 and 1, with which to generate numbers. Base 10 contains digits 0 through 9; binary contains digits 0 and 1 only. These two binary digits, 0 and 1, are commonly referred to as *bits*.

Converting from decimal to binary

One method of converting from decimal to binary is to divide the decimal number by 2 repeatedly, keeping track of the remainders. This process continues until the quotient becomes zero. The remainders are then written in reverse order to obtain the binary number. This is demonstrated in Example 0-1.

Example 0-1

Convert 25_{10} to binary.

Solution:

	Quotient	Remainder	
25/2 =	12	1	LSB (least significant bit)
12/2 =	6	0	
6/2 =	3	0	
3/2 =	1	1	
1/2 =	0	1	MSB (most significant bit)

Therefore, $25_{10} = 11001_2$.

Converting from binary to decimal

To convert from binary to decimal, it is important to understand the concept of weight associated with each digit position. First, as an analogy, recall the weight of numbers in the base 10 system, as shown in the diagram. By the same token, each digit position of a number in base 2 has a weight associated with it:

$$
\begin{aligned}
740683_{10} &= \\
3 \times 10^0 &= 3 \\
8 \times 10^1 &= 80 \\
6 \times 10^2 &= 600 \\
0 \times 10^3 &= 0000 \\
4 \times 10^4 &= 40000 \\
7 \times 10^5 &= \underline{700000} \\
&740683
\end{aligned}
$$

$$110101_2 =$$

					Decimal	Binary
1×2^0	=	1×1	=		1	1
0×2^1	=	0×2	=		0	00
1×2^2	=	1×4	=		4	100
0×2^3	=	0×8	=		0	0000
1×2^4	=	1×16	=		16	10000
1×2^5	=	1×32	=		$\underline{32}$	$\underline{100000}$
					53	110101

Knowing the weight of each bit in a binary number makes it simple to add them together to get its decimal equivalent, as shown in Example 0-2.

Example 0-2

Convert 11001_2 to decimal.

Solution:

Weight:	16	8	4	2	1
Digits:	1	1	0	0	1
Sum:	16 +	8 +	0 +	0 +	$1 = 25_{10}$

Knowing the weight associated with each binary bit position allows one to convert a decimal number to binary directly instead of going through the process of repeated division. This is shown in Example 0-3.

Example 0-3

Use the concept of weight to convert 39_{10} to binary.

Solution:

Weight:	32	16	8	4	2	1
	1	0	0	1	1	1
	32 +	0 +	0 +	4 +	2 +	$1 = 39$

Therefore, $39_{10} = 100111_2$.

Hexadecimal system

Base 16, or the *hexadecimal* system as it is called in computer literature, is used as a convenient representation of binary numbers. For example, it is much easier for a human being to represent a string of 0s and 1s such as 100010010110 as its hexadecimal equivalent of 896H. The binary system has 2 digits, 0 and 1. The base 10 system has 10 digits, 0 through 9. The hexadecimal (base 16) system has 16 digits. In base 16, the first 10 digits, 0 to 9, are the same as in decimal, and for the remaining six digits, the letters A, B, C, D, E, and F are used. Table 0-1 shows the equivalent binary, decimal, and hexadecimal representations for 0 to 15.

Table 0-1: Base 16 Number System

Decimal	Binary	Hex
0	0000	0
1	0001	1
2	0010	2
3	0011	3
4	0100	4
5	0101	5
6	0110	6
7	0111	7
8	1000	8
9	1001	9
10	1010	A
11	1011	B
12	1100	C
13	1101	D
14	1110	E
15	1111	F

Converting between binary and hex

To represent a binary number as its equivalent hexadecimal number, start from the right and group 4 bits at a time, replacing each 4-bit binary number with its hex equivalent shown in Table 0-1. To convert from hex to binary, each hex digit is replaced with its 4-bit binary equivalent. See Examples 0-4 and 0-5.

Example 0-4

Represent binary 100111110101 in hex.

Solution:
First the number is grouped into sets of 4 bits: 1001 1111 0101.
Then each group of 4 bits is replaced with its hex equivalent:

1001	1111	0101
9	F	5

Therefore, 100111110101_2 = 9F5 hexadecimal.

Example 0-5

Convert hex 29B to binary.

Solution:

	2	9	B
29B =	0010	1001	1011

Dropping the leading zeros gives 1010011011.

Converting from decimal to hex

Converting from decimal to hex could be approached in two ways:
1. Convert to binary first and then convert to hex. Example 0-6 shows this method of converting decimal to hex.
2. Convert directly from decimal to hex by repeated division, keeping track of the remainders. Experimenting with this method is left to the reader.

Example 0-6

(a) Convert 45_{10} to hex.

32	16	8	4	2	1	First, convert to binary.
1	0	1	1	0	1	$32 + 8 + 4 + 1 = 45$

$$45_{10} = 0010\ 1101_2 = 2D \text{ hex}$$

(b) Convert 629_{10} to hex.

512	256	128	64	32	16	8	4	2	1
1	0	0	1	1	1	0	1	0	1

$$629_{10} = (512 + 64 + 32 + 16 + 4 + 1) = 0010\ 0111\ 0101_2 = 275 \text{ hex}$$

(c) Convert 1714_{10} to hex.

1024	512	256	128	64	32	16	8	4	2	1
1	1	0	1	0	1	1	0	0	1	0

$$1714_{10} = (1024 + 512 + 128 + 32 + 16 + 2) = 0110\ 1011\ 0010_2 = 6B2 \text{ hex}$$

Converting from hex to decimal

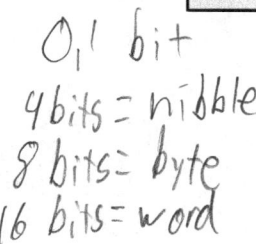

Conversion from hex to decimal can also be approached in two ways:

1. Convert from hex to binary and then to decimal. Example 0-7 demonstrates this method of converting from hex to decimal.
2. Convert directly from hex to decimal by summing the weight of all digits.

Example 0-7

Convert the following hexadecimal numbers to decimal.

(a) $6B2_{16} = 0110\ 1011\ 0010_2$

1024	512	256	128	64	32	16	8	4	2	1
1	1	0	1	0	1	1	0	0	1	0

$$1024 + 512 + 128 + 32 + 16 + 2 = 1714_{10}$$

(b) $9F2D_{16} = 1001\ 1111\ 0010\ 1101_2$

32768	16384	8192	4096	2048	1024	512	256	128	64	32	16	8	4	2	1
1	0	0	1	1	1	1	1	0	0	1	0	1	1	0	1

$$32768 + 4096 + 2048 + 1024 + 512 + 256 + 32 + 8 + 4 + 1 = 40,749_{10}$$

Table 0-2: Counting in Bases

Decimal	Binary	Hex
0	00000	0
1	00001	1
2	00010	2
3	00011	3
4	00100	4
5	00101	5
6	00110	6
7	00111	7
8	01000	8
9	01001	9
10	01010	A
11	01011	B
12	01100	C
13	01101	D
14	01110	E
15	01111	F
16	10000	10
17	10001	11
18	10010	12
19	10011	13
20	10100	14
21	10101	15
22	10110	16
23	10111	17
24	11000	18
25	11001	19
26	11010	1A
27	11011	1B
28	11100	1C
29	11101	1D
30	11110	1E
31	11111	1F

Counting in bases 10, 2, and 16

To show the relationship between all three bases, in Table 0-2 we show the sequence of numbers from 0 to 31 in decimal, along with the equivalent binary and hex numbers. Notice in each base that when one more is added to the highest digit, that digit becomes zero and a 1 is carried to the next-highest digit position. For example, in decimal, $9 + 1 = 0$ with a carry to the next-highest position. In binary, $1 + 1 = 0$ with a carry; similarly, in hex, $F + 1 = 0$ with a carry.

Addition of binary and hex numbers

The addition of binary numbers is a very straightforward process. Table 0-3 shows the addition of two bits. The discussion of subtraction of binary numbers is bypassed since all computers use the addition process to implement subtraction. Although computers have adder circuitry, there is no separate circuitry for subtractors. Instead, adders are used in conjunction with *2's complement* circuitry to perform subtraction. In other words, to implement "$x - y$", the computer takes the 2's complement of y and adds it to x. The concept of 2's complement is reviewed next. Example 0-8 shows the addition of binary numbers.

Table 0-3: Binary Addition

A + B	Carry	Sum
0 + 0	0	0
0 + 1	0	1
1 + 0	0	1
1 + 1	1	0

Example 0-8

Add the following binary numbers. Check against their decimal equivalents.

Solution:

	Binary	Decimal
	1101	13
+	1001	9
	10110	22

2's complement

To get the 2's complement of a binary number, invert all the bits and then add 1 to the result. Inverting the bits is simply a matter of changing all 0s to 1s and 1s to 0s. This is called the *1's complement*. See Example 0-9.

Example 0-9

Take the 2's complement of 10011101.

Solution:

	10011101	binary number
	01100010	1's complement
+	1	
	01100011	2's complement

Addition and subtraction of hex numbers

In studying issues related to software and hardware of computers, it is often necessary to add or subtract hex numbers. Mastery of these techniques is essential. Hex addition and subtraction are discussed separately below.

Addition of hex numbers

This section describes the process of adding hex numbers. Starting with the least significant digits, the digits are added together. If the result is less than 16, write that digit as the sum for that position. If it is greater than 16, subtract 16 from it to get the digit and carry 1 to the next digit. The best way to explain this is by example, as shown in Example 0-10.

Example 0-10

Perform hex addition: 23D9 + 94BE.

Solution:

	23D9	LSD: 9 + 14 = 23	23 − 16 = 7 with a carry
+	94BE	1 + 13 + 11 = 25	25 − 16 = 9 with a carry
	B897	1 + 3 + 4 = 8	
		MSD: 2 + 9 = B	

Subtraction of hex numbers

In subtracting two hex numbers, if the second digit is greater than the first, borrow 16 from the preceding digit. See Example 0-11.

Example 0-11

Perform hex subtraction: 59F − 2B8.

Solution:

	59F	LSD: 8 from 15 = 7
−	2B8	11 from 25 (9 + 16) = 14 (E)
	2E7	2 from 4 (5 − 1) = 2

ASCII code

The discussion so far has revolved around the representation of number systems. Because all information in the computer must be represented by 0s and 1s, binary patterns must be assigned to letters and other characters. In the 1960s a standard representation called *ASCII* (American Standard Code for Information Interchange) was estab-

Hex	Symbol	Hex	Symbol
41	A	61	a
42	B	62	b
43	C	63	c
44	D	64	d
...	...	...	...
59	Y	79	y
5A	Z	7A	z

Figure 0-1. Selected ASCII Codes

lished. The ASCII (pronounced "ask-E") code assigns binary patterns for numbers 0 to 9, all the letters of the English alphabet, both uppercase (capital) and lowercase, and many control codes and punctuation marks. The great advantage of this system is that it is used by most computers, so that information can be shared among computers. The ASCII system uses a total of 7 bits to represent each code. For example, 100 0001 is assigned to the uppercase letter "A" and 110 0001 is for the lowercase "a". Often, a zero is placed in the most-significant bit position to make it an 8-bit code. Figure 0-1 shows selected ASCII codes. A complete list of ASCII codes is given in Appendix F. The use of ASCII is not only standard for keyboards used in the United States and many other countries but also provides a standard for printing and displaying characters by output devices such as printers and monitors.

Notice that the pattern of ASCII codes was designed to allow for easy manipulation of ASCII data. For example, digits 0 through 9 are represented by ASCII codes 30 through 39. This enables a program to easily convert ASCII to decimal by masking off the "3" in the upper nibble. Also notice that there is a relationship between the uppercase and lowercase letters. The uppercase letters are represented by ASCII codes 41 through 5A while lowercase letters are represented by codes 61 through 7A. Looking at the binary code, the only bit that is different between the uppercase "A" and lowercase "a" is bit 5. Therefore, conversion between uppercase and lowercase is as simple as changing bit 5 of the ASCII code.

Review Questions

1. Why do computers use the binary number system instead of the decimal system?
2. Convert 34_{10} to binary and hex.
3. Convert 110101_2 to hex and decimal.
4. Perform binary addition: 101100 + 101.
5. Convert 101100_2 to its 2's complement representation.
6. Add 36BH + F6H.
7. Subtract 36BH – F6H.
8. Write "80x86 CPUs" in its ASCII code (in hex form).

SECTION 0.2: DIGITAL PRIMER

This section gives an overview of digital logic and design. First, we cover binary logic operations, then we show gates that perform these functions. Next, logic gates are put together to form simple digital circuits. Finally, we cover some logic devices commonly found in microcontroller interfacing.

Binary logic

As mentioned earlier, computers use the binary number system because the two voltage levels can be represented as the two digits 0 and 1. Signals in digital electronics have two distinct voltage levels. For example, a system may define 0 V as logic 0 and +5 V as logic 1. Figure 0-2 shows this system with the built-in tolerances for variations in the voltage. A valid digital signal in this example should be within either of the two shaded areas.

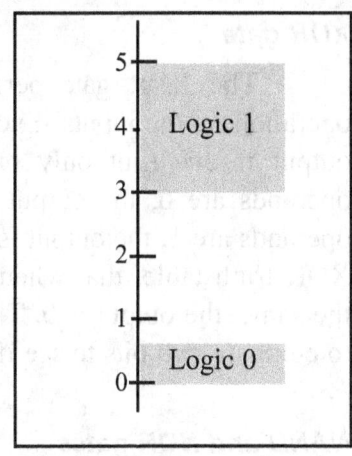

Figure 0-2. Binary Signals

Logic gates

Binary logic gates are simple circuits that take one or more input signals and send out one output signal. Several of these gates are defined below.

AND gate

The AND gate takes two or more inputs and performs a logic AND on them. See the truth table and diagram of the AND gate. Notice that if both inputs to the AND gate are 1, the output will be 1. Any other combination of inputs will give a 0 output. The example shows two inputs, *x* and *y*. Multiple outputs are also possible for logic gates. In the case of AND, if all inputs are 1, the output is 1. If any input is 0, the output is 0.

OR gate

The OR logic function will output a 1 if one or more inputs is 1. If all inputs are 0, then and only then will the output be 0.

Tri-state buffer

A buffer gate does not change the logic level of the input. It is used to isolate or amplify the signal.

Logical AND Function

Inputs	Output
X Y	X AND Y
0 0	0
0 1	0
1 0	0
1 1	1

X $-$)$-$ X AND Y
Y $-$

Logical OR Function

Inputs	Output
X Y	X OR Y
0 0	0
0 1	1
1 0	1
1 1	1

X $-$)$-$ X OR Y
Y $-$

Buffer

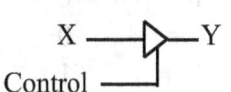

X ——▷— Y

Control ——

Inverter

The inverter, also called NOT, outputs the value opposite to that input to the gate. That is, a 1 input will give a 0 output, while a 0 input will give a 1 output.

XOR gate

The XOR gate performs an exclusive-OR operation on the inputs. Exclusive-OR produces a 1 output if one (but only one) input is 1. If both operands are 0, the output is 0. Likewise, if both operands are 1, the output is also 0. Notice from the XOR truth table, that whenever the two inputs are the same, the output is 0. This function can be used to compare two bits to see if they are the same.

NAND and NOR gates

The NAND gate functions like an AND gate with an inverter on the output. It produces a 0 output when all inputs are 1; otherwise, it produces a 1 output. The NOR gate functions like an OR gate with an inverter on the output. It produces a 1 if all inputs are 0; otherwise, it produces a 0. NAND and NOR gates are used extensively in digital design because they are easy and inexpensive to fabricate. Any circuit that can be designed with AND, OR, XOR, and INVERTER gates can be implemented using only NAND and NOR gates. A simple example of this is given below. Notice in NAND, that if any input is 0, the output is 1. Notice in NOR, that if any input is 1, the output is 0.

Logic design using gates

Next we will show a simple logic design to add two binary digits. If we add two binary digits there are four possible outcomes:

	Carry	Sum
$0 + 0 =$	0	0
$0 + 1 =$	0	1
$1 + 0 =$	0	1
$1 + 1 =$	1	0

Logical Inverter

Input	Output
X	NOT X
0	1
1	0

X ——▷o— NOT X

Logical XOR Function

Inputs	Output
X Y	X XOR Y
0 0	0
0 1	1
1 0	1
1 1	0

X, Y ——)D)— X XOR Y

Logical NAND Function

Inputs	Output
X Y	X NAND Y
0 0	1
0 1	1
1 0	1
1 1	0

X, Y ——D)o—X NAND Y

Logical NOR Function

Inputs	Output
X Y	X NOR Y
0 0	1
0 1	0
1 0	0
1 1	0

X, Y ——)Do— X NOR Y

Notice that when we add 1 + 1 we get 0 with a carry to the next higher place. We will need to determine the sum and the carry for this design. Notice that the sum column above matches the output for the XOR function, and that the carry column matches the output for the AND function. Figure 0-3(a) shows a simple adder implemented with XOR and AND gates. Figure 0-3(b) shows the same logic circuit implemented with AND and OR gates and inverters.

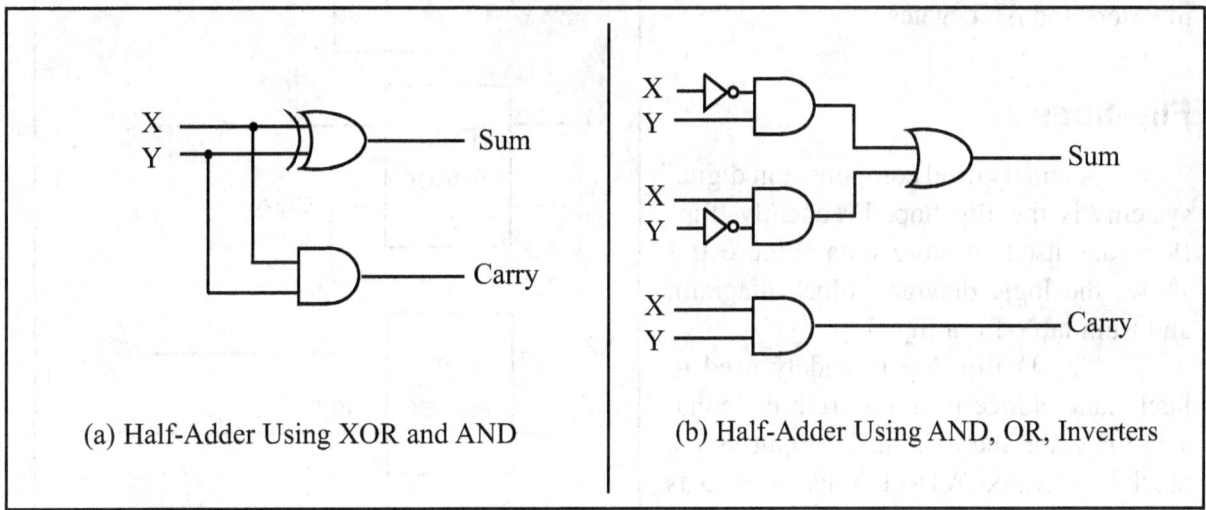

(a) Half-Adder Using XOR and AND   (b) Half-Adder Using AND, OR, Inverters

Figure 0-3. Two Implementations of a Half-Adder

Figure 0-4 shows a block diagram of a half-adder. Two half-adders can be combined to form an adder that can add three input digits. This is called a full-adder. Figure 0-5 shows the logic diagram of a full-adder, along with a block diagram that masks the details of the circuit. Figure 0-6 shows a 3-bit adder using three full-adders.

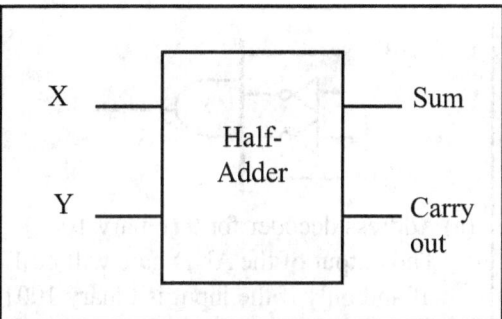

Figure 0-4. Block Diagram of a Half-Adder

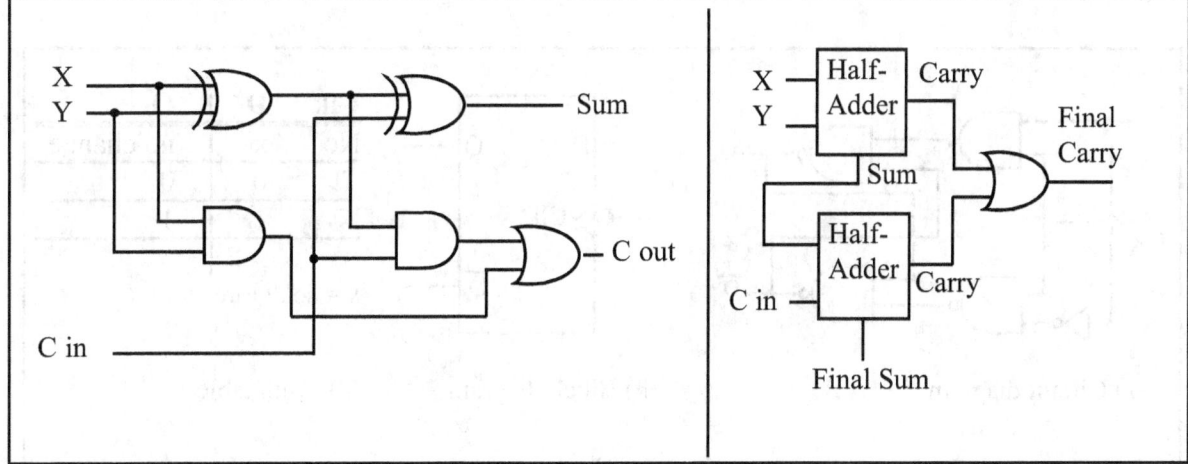

Figure 0-5. Full-Adder Built from a Half-Adder

Decoders

Another example of the application of logic gates is the decoder. Decoders are widely used for address decoding in computer design. Figure 0-7 shows decoders for 9 (1001 binary) and 5 (0101) using inverters and AND gates.

Flip-flops

A widely used component in digital systems is the flip-flop. Frequently, flip-flops are used to store data. Figure 0-8 shows the logic diagram, block diagram, and truth table for a flip-flop.

The D flip-flop is widely used to latch data. Notice from the truth table that a D-FF grabs the data at the input as the clock is activated. A D-FF holds the data as long as the power is on.

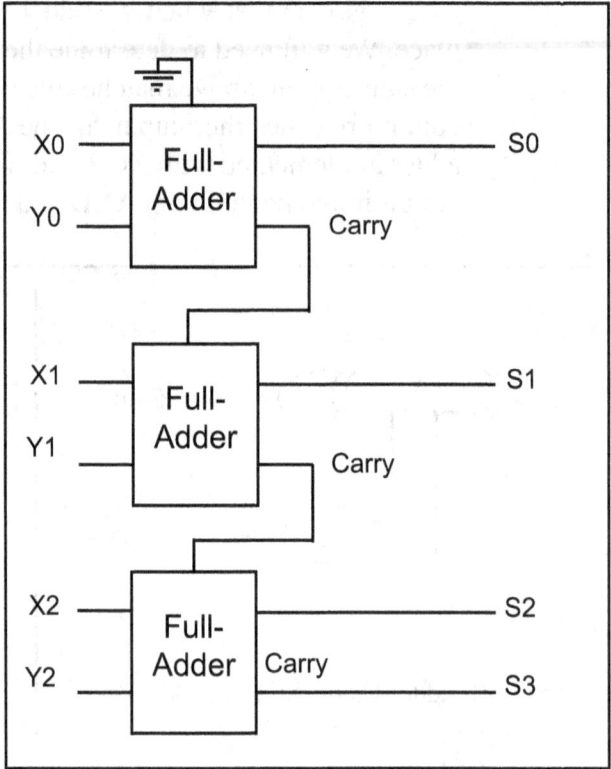

Figure 0-6. 3-Bit Adder Using Three Full-Adders

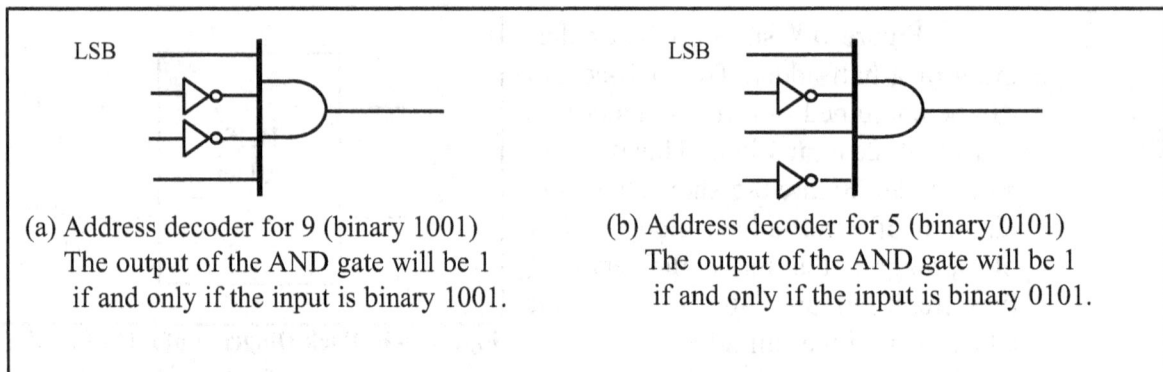

(a) Address decoder for 9 (binary 1001)
The output of the AND gate will be 1
if and only if the input is binary 1001.

(b) Address decoder for 5 (binary 0101)
The output of the AND gate will be 1
if and only if the input is binary 0101.

Figure 0-7. Address Decoders

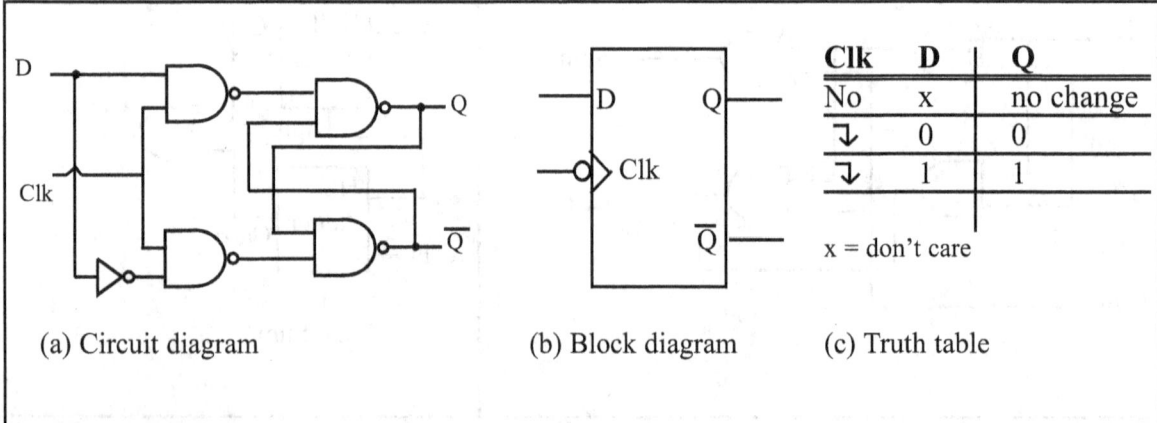

(a) Circuit diagram   (b) Block diagram   (c) Truth table

Clk	D	Q
No	x	no change
⤓	0	0
⤓	1	1

x = don't care

Figure 0-8. D Flip-Flops

Review Questions

1. The logical operation _____ gives a 1 output when all inputs are 1.
2. The logical operation _____ gives a 1 output when one or more of its inputs is 1.
3. The logical operation _____ is often used to compare two inputs to determine whether they have the same value.
4. A _____ gate does not change the logic level of the input.
5. Name a common use for flip-flops.
6. An address _____ is used to identify a predetermined binary address.

SECTION 0.3: SEMICONDUCTOR MEMORY

In this section we discuss various types of semiconductor memories and their characteristics such as capacity, organization, and access time. We will also show how the memory is connected to CPU. Before we embark on the subject of memory, it will be helpful to give an overview of computer organization and review some widely used terminology in computer literature.

Some important terminology

Recall from the discussion above that a *bit* is a binary digit that can have the value 0 or 1. A *byte* is defined as 8 bits. A *nibble* is half a byte, or 4 bits. A *word* is

```
Bit                          0
Nibble                    0000
Byte                 0000 0000
Word 0000 0000 0000 0000
```

two bytes, or 16 bits. The display is intended to show the relative size of these units. Of course, they could all be composed of any combination of zeros and ones.

A *kilobyte* is 2^{10} bytes, which is 1024 bytes. The abbreviation K is often used to represent kilobytes. A *megabyte*, or *meg* as some call it, is 2^{20} bytes. That is a little over 1 million bytes; it is exactly 1,048,576 bytes. Moving rapidly up the scale in size, a *gigabyte* is 2^{30} bytes (over 1 billion), and a *terabyte* is 2^{40} bytes (over 1 trillion). As an example of how some of these terms are used, suppose that a given computer has 16 megabytes of memory. That would be 16×2^{20}, or $2^4 \times 2^{20}$, which is 2^{24}. Therefore 16 megabytes is 2^{24} bytes.

Two types of memory commonly used in microcomputers are *RAM*, which stands for "random access memory" (sometimes called *read/write memory*), and *ROM*, which stands for "read-only memory." RAM is used by the computer for temporary storage of programs that it is running. That data is lost when the computer is turned off. For this reason, RAM is sometimes called *volatile memory*. ROM contains programs and information essential to operation of the computer. The information in ROM is permanent, cannot be changed by the user, and is not lost when the power is turned off. Therefore, it is called *nonvolatile memory*.

Internal organization of computers

The internal working of every computer can be broken down into three parts: CPU (central processing unit), memory, and I/O (input/output) devices. Figure 0-9 shows a block diagram of the internal organization of a computer.

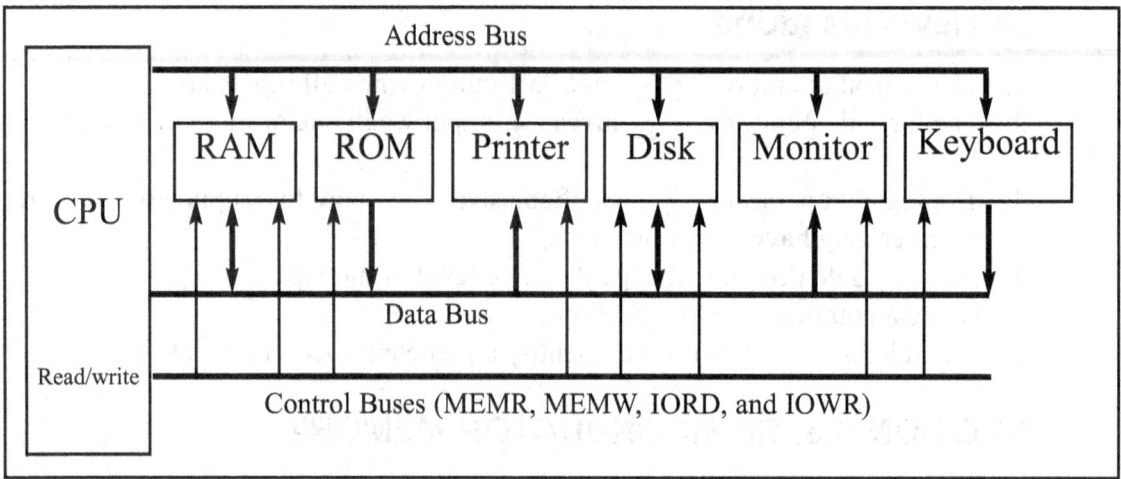

Figure 0-9. Internal Organization of a Computer

The function of the CPU is to execute (process) information stored in memory. The function of I/O devices such as the keyboard and video monitor is to provide a means of communicating with the CPU. The CPU is connected to memory and I/O through strips of wire called a *bus*. The bus inside a computer allows carrying information from place to place just as a street allows cars to carry people from place to place. In every computer there are three types of buses: address bus, data bus, and control bus.

For a device (memory or I/O) to be recognized by the CPU, it must be assigned an address. The address assigned to a given device must be unique; no two devices are allowed to have the same address. The CPU puts the address (in binary, of course) on the address bus, and the decoding circuitry finds the device. Then the CPU uses the data bus either to get data from that device or to send data to it. The control buses are used to provide read or write signals to the device to indicate if the CPU is asking for information or sending information. Of the three buses, the address bus and data bus determine the capability of a given CPU.

More about the data bus

Because data buses are used to carry information in and out of a CPU, the more data buses available, the better the CPU. If one thinks of data buses as highway lanes, it is clear that more lanes provide a better pathway between the CPU and its external devices (such as printers, RAM, ROM, etc.; see Figure 0-9). By the same token, that increase in the number of lanes increases the cost of construction. More data buses mean a more expensive CPU and computer. The average size of data buses in CPUs varies between 8 and 64 bits. Early personal computers such as Apple 2 used an 8-bit data bus, while supercomputers such as Cray used a 64-bit data bus. Data buses are bidirectional, because the CPU must use them either to receive or to send data. The processing power of a computer is related to the size of its buses, because an 8-bit bus can send out 1 byte a time, but a 16-bit bus can send out 2 bytes at a time, which is twice as fast.

More about the address bus

Because the address bus is used to identify the devices and memory connected to the CPU, the more address buses available, the larger the number of devices that can be addressed. In other words, the number of address buses for a CPU determines the number of locations with which it can communicate. The number of locations is always equal to 2^x, where x is the number of address lines, regardless of the size of the data bus. For example, a CPU with 16 address lines can provide a total of 65,536 (2^{16}) or 64K of addressable memory. Each location can have a maximum of 1 byte of data. This is because all general-purpose microprocessor CPUs are what is called *byte addressable*. As another example, the IBM PC AT uses a CPU with 24 address lines and 16 data lines. Thus, the total accessible memory is 16 megabytes (2^{24} = 16 megabytes). In this example there would be 2^{24} locations, and because each location is one byte, there would be 16 megabytes of memory. The address bus is a *unidirectional* bus, which means that the CPU uses the address bus only to send out addresses. To summarize: The total number of memory locations addressable by a given CPU is always equal to 2^x where x is the number of address bits, regardless of the size of the data bus.

CPU and its relation to RAM and ROM

For the CPU to process information, the data must be stored in RAM or ROM. The function of ROM in computers is to provide information that is fixed and permanent. This is information such as tables for character patterns to be displayed on the video monitor, or programs that are essential to the working of the computer, such as programs for testing and finding the total amount of RAM installed on the system, or for displaying information on the video monitor. In contrast, RAM stores temporary information that can change with time, such as various versions of the operating system and application packages such as word processing or tax calculation packages. These programs are loaded from the hard drive into RAM to be processed by the CPU. The CPU cannot get the information from the disk directly because the disk is too slow. In other words, the CPU first seeks the information to be processed from RAM (or ROM). Only if the data is not there does the CPU seek it from a mass storage device such as a disk, and then it transfers the information to RAM. For this reason, RAM and ROM are sometimes referred to as *primary memory* and disks are called *secondary memory*. Next, we discuss various types of semiconductor memories and their characteristics such as capacity, organization, and access time.

Memory capacity

The number of bits that a semiconductor memory chip can store is called chip *capacity*. It can be in units of Kbits (kilobits), Mbits (megabits), and so on. This must be distinguished from the storage capacity of computer systems. While the memory capacity of a memory IC chip is always given in bits, the memory capacity of a computer system is given in bytes. For example, an article in a technical journal may state that the 128M chip has become popular. In that case, it is understood, although it is not mentioned, that 128M means 128 megabits since the article is referring to an IC memory chip. However, if an advertisement states that a computer comes with 128M memory, it is understood that 128M means 128 megabytes since it is referring to a computer system.

Memory organization

Memory chips are organized into a number of locations within the IC. Each location can hold 1 bit, 4 bits, 8 bits, or even 16 bits, depending on how it is designed internally. The number of bits that each location within the memory chip can hold is always equal to the number of data pins on the chip. How many locations exist inside a memory chip? That depends on the number of address pins. The number of locations within a memory IC always equals 2 to the power of the number of address pins. Therefore, the total number of bits that a memory chip can store is equal to the number of locations times the number of data bits per location. To summarize:

1. A memory chip contains 2^x locations, where x is the number of address pins.
2. Each location contains y bits, where y is the number of data pins on the chip.
3. The entire chip will contain $2^x \times y$ bits, where x is the number of address pins and y is the number of data pins on the chip.

Speed

One of the most important characteristics of a memory chip is the speed at which its data can be accessed. To access the data, the address is presented to the address pins, the READ pin is activated, and after a certain amount of time has elapsed, the data shows up at the data pins. The shorter this elapsed time, the better, and consequently, the more expensive the memory chip. The speed of the memory chip is commonly referred to as its *access time*. The access time of memory chips varies from a few nanoseconds to hundreds of nanoseconds, depending on the IC technology used in the design and fabrication process.

The three important memory characteristics of capacity, organization, and access time will be explored extensively in this chapter. Table 0-4 serves as a reference for the calculation of memory organization. Examples 0-12 and 0-13 demonstrate these concepts.

Table 0-4: Powers of 2

x	2^x
10	1K
11	2K
12	4K
13	8K
14	16K
15	32K
16	64K
17	128K
18	256K
19	512K
20	1M
21	2M
22	4M
23	8M
24	16M
25	32M
26	64M
27	128M

ROM (read-only memory)

ROM is a type of memory that does not lose its contents when the power is turned off. For this reason, ROM is also called *nonvolatile* memory. There are different types of read-only memory, such as PROM, EPROM, EEPROM, Flash EPROM, and mask ROM. Each is explained next.

PROM (programmable ROM) and OTP

PROM refers to the kind of ROM that the user can burn information into. In other words, PROM is a user-programmable memory. For every bit of the PROM, there exists a fuse. PROM is programmed by blowing the fuses. If the information burned into PROM is wrong, that PROM must be discarded since its internal fuses are blown permanently. For this reason, PROM is also referred to as

Example 0-12

A given memory chip has 12 address pins and 4 data pins. Find:
(a) the organization, and (b) the capacity.

Solution:

(a) This memory chip has 4,096 locations ($2^{12} = 4{,}096$), and each location can hold 4 bits of data. This gives an organization of 4,096 × 4, often represented as 4K × 4.
(b) The capacity is equal to 16K bits since there is a total of 4K locations and each location can hold 4 bits of data.

Example 0-13

A 512K memory chip has 8 pins for data. Find:
(a) the organization, and (b) the number of address pins for this memory chip.

Solution:

(a) A memory chip with 8 data pins means that each location within the chip can hold 8 bits of data. To find the number of locations within this memory chip, divide the capacity by the number of data pins. 512K/8 = 64K; therefore, the organization for this memory chip is 64K × 8.
(b) The chip has 16 address lines since $2^{16} = 64K$.

OTP (one-time programmable). Programming ROM, also called *burning* ROM, requires special equipment called a ROM burner or ROM programmer.

EPROM (erasable programmable ROM) and UV-EPROM

EPROM was invented to allow making changes in the contents of PROM after it is burned. In EPROM, one can program the memory chip and erase it thousands of times. This is especially necessary during development of the prototype of a microprocessor-based project. A widely used EPROM is called UV-EPROM, where UV stands for ultraviolet. The only problem with UV-EPROM is that erasing its contents can take up to 20 minutes. All UV-EPROM chips have a window through which the programmer can shine ultraviolet (UV) radiation to erase the chip's contents. For this reason, EPROM is also referred to as UV-erasable EPROM or simply UV-EPROM. Figure 0-10 shows the pins for UV-EPROM chips.

To program a UV-EPROM chip, the following steps must be taken:
1. Its contents must be erased. To erase a chip, remove it from its socket on the system board and place it in EPROM erasure equipment to expose it to UV radiation for 15–20 minutes.
2. Program the chip. To program a UV-EPROM chip, place it in the ROM burner (programmer). To burn code or data into EPROM, the ROM burner uses 12.5 volts or higher, depending on the EPROM type. This voltage is referred

to as V_{PP} in the UV-EPROM data sheet.

3. Place the chip back into its socket on the system board.

As can be seen from the above steps, not only is there an EPROM programmer (burner), but there is also separate EPROM erasure equipment. The main problem, and indeed the major disadvantage of UV-EPROM, is that it cannot be erased and programmed while it is in the system board. To provide a solution to this problem, EEPROM was invented.

Notice the patterns of the IC numbers in Table 0-5. For example, part number 27128-25 refers to UV-EPROM that has a capacity of 128K bits and access time of 250 nanoseconds. The capacity of the memory chip is indicated in the part number and the access time is given with a zero dropped. See Example 0-14. In part numbers, C refers to CMOS technology. Notice that 27XX always refers to UV-EPROM chips. For a comprehensive list of available memory chips see the JAMECO (jameco.com) or JDR (jdr.com) catalogs.

27256	27128	2732A	2716		2764			2716	2732A	27128	27256
Vpp	Vpp			Vpp □ 1		28 □ Vcc				Vcc	Vcc
A12	A12			A12 □ 2		27 □ $\overline{PGM}$				$\overline{PGM}$	A14
A7	A7	A7	A7	A7 □ 3		26 □ N.C.		Vcc	Vcc	A13	A13
A6	A6	A6	A6	A6 □ 4		25 □ A8		A8	A8	A8	A8
A5	A5	A5	A5	A5 □ 5		24 □ A9		A9	A9	A9	A9
A4	A4	A4	A4	A4 □ 6		23 □ A11		Vpp	A11	Vpp	Vpp
A3	A3	A3	A3	A3 □ 7		22 □ $\overline{OE}$		$\overline{OE}$	$\overline{OE}$/Vpp	$\overline{OE}$	$\overline{OE}$
A2	A2	A2	A2	A2 □ 8		21 □ A10		A10	A10	A10	A10
A1	A1	A1	A1	A1 □ 9		20 □ $\overline{CE}$		$\overline{CE}$	$\overline{CE}$	$\overline{CE}$	$\overline{CE}$
A0	A0	A0	A0	A0 □ 10		19 □ O7		O7	O7	O7	O7
O0	O0	O0	O0	O0 □ 11		18 □ O6		O6	O6	O6	O6
O1	O1	O1	O1	O1 □ 12		17 □ O5		O5	O5	O5	O5
O2	O2	O2	O2	O2 □ 13		16 □ O4		O4	O4	O4	O4
GND	GND	GND	GND	GND □ 14		15 □ O3		O3	O3	O3	O3

Figure 0-10. Pin Configurations for 27xx ROM Family

Example 0-14

For ROM chip 27128, find the number of data and address pins.

Solution:

The 27128 has a capacity of 128K bits. It has 16K × 8 organization (all ROMs have 8 data pins), which indicates that there are 8 pins for data and 14 pins for address ($2^{14} = 16K$).

Table 0-5: Some UV-EPROM Chips

Part #	Capacity	Org.	Access	Pins	V$_{PP}$
2716	16K	2K × 8	450 ns	24	25 V
2732	32K	4K × 8	450 ns	24	25 V
2732A-20	32K	4K × 8	200 ns	24	21 V
27C32-1	32K	4K × 8	450 ns	24	12.5 V CMOS
2764-20	64K	8K × 8	200 ns	28	21 V
2764A-20	64K	8K × 8	200 ns	28	12.5 V
27C64-12	64K	8K × 8	120 ns	28	12.5 V CMOS
27128-25	128K	16K × 8	250 ns	28	21 V
27C128-12	128K	16K × 8	120 ns	28	12.5 V CMOS
27256-25	256K	32K × 8	250 ns	28	12.5 V
27C256-15	256K	32K × 8	150 ns	28	12.5 V CMOS
27512-25	512K	64K × 8	250 ns	28	12.5 V
27C512-15	512K	64K × 8	150 ns	28	12.5 V CMOS
27C010-15	1024K	128K × 8	150 ns	32	12.5 V CMOS
27C020-15	2048K	256K × 8	150 ns	32	12.5 V CMOS
27C040-15	4096K	512K × 8	150 ns	32	12.5 V CMOS

EEPROM (electrically erasable programmable ROM)

EEPROM has several advantages over EPROM, such as the fact that its method of erasure is electrical and therefore instant, as opposed to the 20-minute erasure time required for UV-EPROM. In addition, in EEPROM one can select which byte to be erased, in contrast to UV-EPROM, in which the entire contents of ROM are erased. However, the main advantage of EEPROM is that one can program and erase its contents while it is still in the system board. It does not require physical removal of the memory chip from its socket. In other words, unlike UV-EPROM, EEPROM does not require an external erasure and programming device. To utilize EEPROM fully, the designer must incorporate the circuitry to program the EEPROM into the system board. In general, the cost per bit for EEPROM is much higher than for UV-EPROM.

Flash memory EPROM

Since the early 1990s, Flash EPROM has become a popular user-programmable memory chip, and for good reasons. First, the erasure of the entire contents takes less than a second, or one might say in a flash, hence its name, Flash memory. In addition, the erasure method is electrical, and for this reason it is sometimes referred to as Flash EEPROM. To avoid confusion, it is commonly called Flash memory. The major difference between EEPROM and Flash memory is that when Flash memory's contents are erased, the entire device is erased, in contrast to EEPROM, where one can erase a desired byte. Although in many Flash memories recently made available the contents are divided into blocks and the erasure can be done block by block, unlike EEPROM, Flash memory has no byte erasure option. Because Flash memory can be programmed while it is in its socket on the system board, it is widely used to upgrade the BIOS ROM of the PC. Some designers believe that Flash memory will replace the hard disk as a mass storage medium.

Table 0-6: Some EEPROM and Flash Chips

EEPROMs

Part No.	Capacity	Org.	Speed	Pins	V_{PP}
2816A-25	16K	2K × 8	250 ns	24	5 V
2864A	64K	8K × 8	250 ns	28	5 V
28C64A-25	64K	8K × 8	250 ns	28	5 V CMOS
28C256-15	256K	32K × 8	150 ns	28	5 V
28C256-25	256K	32K × 8	250 ns	28	5 V CMOS

Flash

Part No.	Capacity	Org.	Speed	Pins	V_{PP}
28F256-20	256K	32K × 8	200 ns	32	12 V CMOS
28F010-15	1024K	128K × 8	150 ns	32	12 V CMOS
28F020-15	2048K	256K × 8	150 ns	32	12 V CMOS

This would increase the performance of the computer tremendously, since Flash memory is semiconductor memory with access time in the range of 100 ns compared with disk access time in the range of tens of milliseconds. For this to happen, Flash memory's program/erase cycles must become infinite, just like hard disks. Program/erase cycle refers to the number of times that a chip can be erased and reprogrammed before it becomes unusable. At this time, the program/erase cycle is 100,000 for Flash and EEPROM, 1000 for UV-EPROM, and infinite for RAM and disks. See Table 0-6 for some sample chips.

Mask ROM

Mask ROM refers to a kind of ROM in which the contents are programmed by the IC manufacturer. In other words, it is not a user-programmable ROM. The term *mask* is used in IC fabrication. Since the process is costly, mask ROM is used when the needed volume is high (hundreds of thousands) and it is absolutely certain that the contents will not change. It is common practice to use UV-EPROM or Flash for the development phase of a project, and only after the code/data have been finalized is the mask version of the product ordered. The main advantage of mask ROM is its cost, since it is significantly cheaper than other kinds of ROM, but if an error is found in the data/code, the entire batch must be thrown away. It must be noted that all ROM memories have 8 bits for data pins; therefore, the organization is ×8.

RAM (random access memory)

RAM memory is called *volatile* memory since cutting off the power to the IC results in the loss of data. Sometimes RAM is also referred to as RAWM (read and write memory), in contrast to ROM, which cannot be written to. There are three types of RAM: static RAM (SRAM), NV-RAM (nonvolatile RAM), and dynamic RAM (DRAM). Each is explained separately.

SRAM (static RAM)

Storage cells in static RAM memory are made of flip-flops and therefore do not require refreshing in order to keep their data. This is in contrast to DRAM, discussed below. The problem with the use of flip-flops for storage cells is that each cell requires at least 6 transistors to build, and the cell holds only 1 bit of data. In recent years, the cells have been made of 4 transistors, which still is too many. The use of 4-transistor cells plus the use of CMOS technology has given birth to a high-capacity SRAM, but its capacity is far below DRAM. Figure 0-11 shows the pin diagram for an SRAM chip.

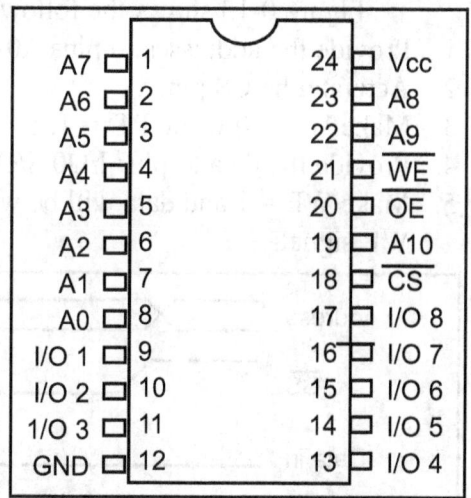

Figure 0-11. 2K × 8 SRAM Pins

The following is a description of the 6116 SRAM pins.

A0–A10 are for address inputs, where 11 address lines gives 2^{11} = 2K.
WE (write enable) is for writing data into SRAM (active low).
OE (output enable) is for reading data out of SRAM (active low)
CS (chip select) is used to select the memory chip.
I/O0–I/O7 are for data I/O, where 8-bit data lines give an organization of 2K × 8.

The functional diagram for the 6116 SRAM is given in Figure 0-12.

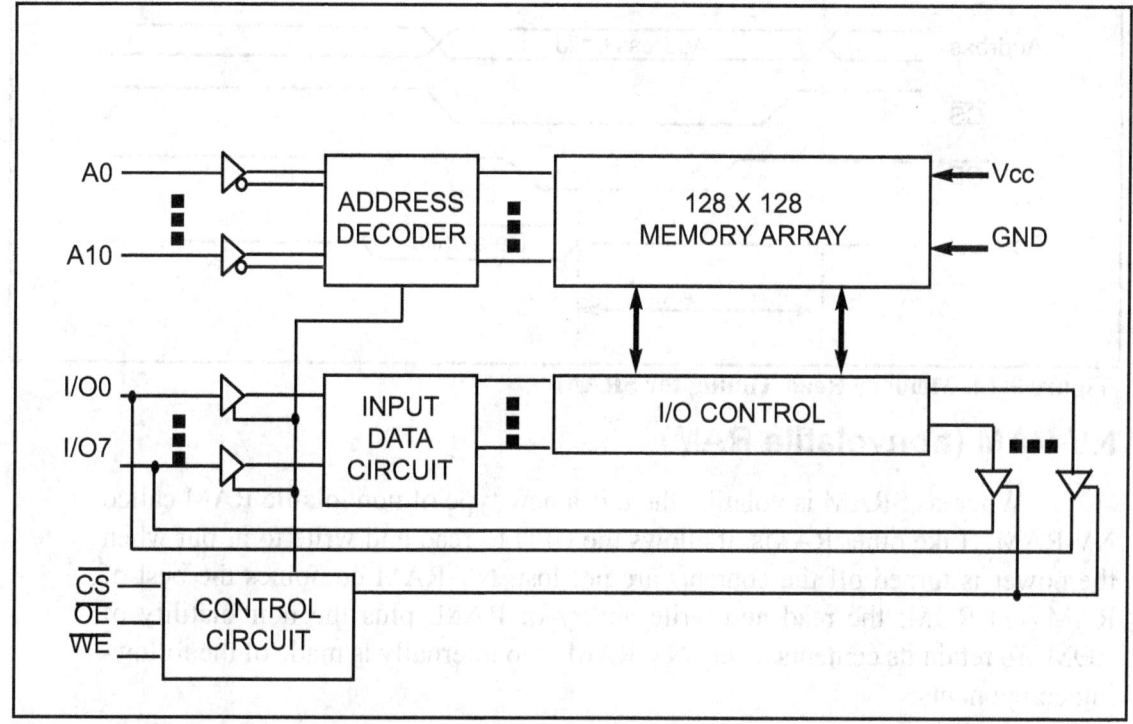

Figure 0-12. Functional Block Diagram for 6116 SRAM

Figure 0-13 shows the following steps to write data into SRAM.

1. Provide the addresses to pins A0–A10.
2. Activate the CS pin.
3. Make WE = 0 while RD = 1.
4. Provide the data to pins I/O0–I/O7.
5. Make WE = 1 and data will be written into SRAM on the positive edge of the WE signal.

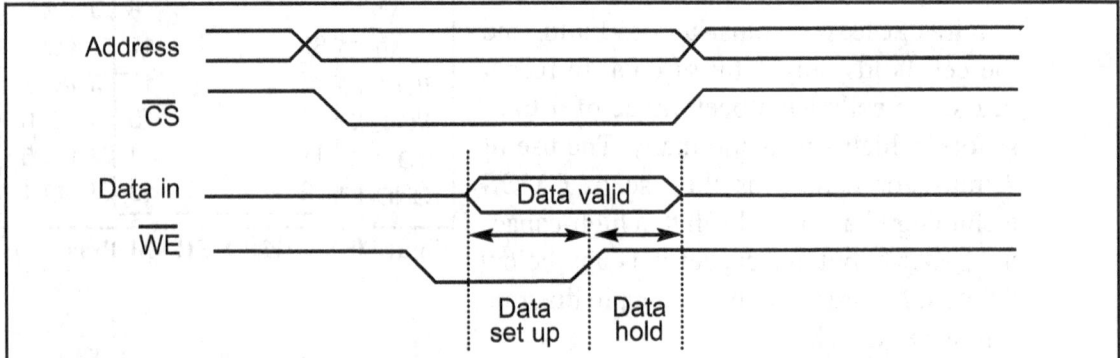

Figure 0-13. Memory Write Timing for SRAM

The following are steps to read data from SRAM. See Figure 0-14.

1. Provide the addresses to pins A0–A10. This is the start of the access time (t_{AA}).
2. Activate the CS pin.
3. While WE = 1, a high-to-low pulse on the OE pin will read the data out of the chip.

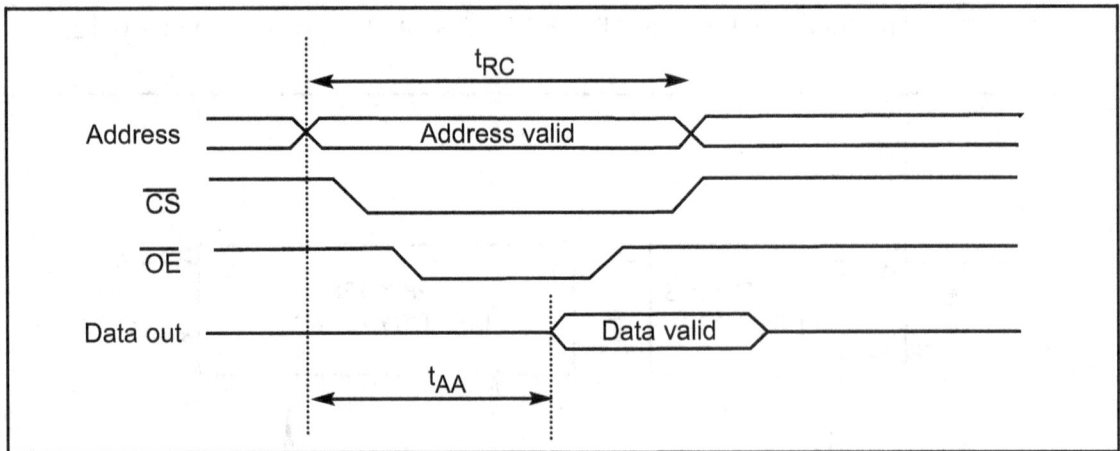

Figure 0-14. Memory Read Timing for SRAM

NV-RAM (nonvolatile RAM)

Whereas SRAM is volatile, there is a new type of nonvolatile RAM called NV-RAM. Like other RAMs, it allows the CPU to read and write to it, but when the power is turned off the contents are not lost. NV-RAM combines the best of RAM and ROM: the read and write ability of RAM, plus the nonvolatility of ROM. To retain its contents, every NV-RAM chip internally is made of the following components:

1. It uses extremely power-efficient (very low-power consumption) SRAM cells built out of CMOS.

Table 0-7: Some SRAM and NV-RAM Chips

SRAM

Part No.	Capacity	Org.	Speed	Pins	V_{PP}
6116P-1	16K	2K × 8	100 ns	24	CMOS
6116P-2	16K	2K × 8	120 ns	24	CMOS
6116P-3	16K	2K × 8	150 ns	24	CMOS
6116LP-1	16K	2K × 8	100 ns	24	Low-power CMOS
6116LP-2	16K	2K × 8	120 ns	24	Low-power CMOS
6116LP-3	16K	2K × 8	150 ns	24	Low-power CMOS
6264P-10	64K	8K × 8	100 ns	28	CMOS
6264LP-70	64K	8K × 8	70 ns	28	Low-power CMOS
6264LP-12	64K	8K × 8	120 ns	28	Low-power CMOS
62256LP-10	256K	32K × 8	100 ns	28	Low-power CMOS
62256LP-12	256K	32K × 8	120 ns	28	Low-power CMOS

NV-RAM from Dallas Semiconductor

Part No.	Capacity	Org.	Speed	Pins	V_{PP}
DS1220Y-150	16K	2K × 8	150 ns	24	
DS1225AB-150	64K	8K × 8	150 ns	28	
DS1230Y-85	256K	32K × 8	85 ns	28	

2. It uses an internal lithium battery as a backup energy source.
3. It uses an intelligent control circuitry. The main job of this control circuitry is to monitor the V_{CC} pin constantly to detect loss of the external power supply. If the power to the V_{CC} pin falls below out-of-tolerance conditions, the control circuitry switches automatically to its internal power source, the lithium battery. The internal lithium power source is used to retain the NV-RAM contents only when the external power source is off.

It must be emphasized that all three of the components above are incorporated into a single IC chip, and for this reason nonvolatile RAM is a very expensive type of RAM as far as cost per bit is concerned. Offsetting the cost, however, is the fact that it can retain its contents up to ten years after the power has been turned off and allows one to read and write in exactly the same way as SRAM. Table 0-7 shows some examples of SRAM and NV-RAM parts.

DRAM (dynamic RAM)

Since the early days of the computer, the need for huge, inexpensive read/write memory has been a major preoccupation of computer designers. In 1970, Intel Corporation introduced the first dynamic RAM (random access memory). Its density (capacity) was 1024 bits and it used a capacitor to store each bit. Using a capacitor to store data cuts down the number of transistors needed to build the cell; however, it requires constant refreshing due to leakage. This is in contrast to SRAM (static RAM), whose individual cells are made of flip-flops. Since each bit in SRAM uses a single flip-flop, and each flip-flop requires six transistors,

SRAM has much larger memory cells and consequently lower density. The use of capacitors as storage cells in DRAM results in much smaller net memory cell size.

The advantages and disadvantages of DRAM memory can be summarized as follows. The major advantages are high density (capacity), cheaper cost per bit, and lower power consumption per bit. The disadvantage is that it must be refreshed periodically because the capacitor cell loses its charge; furthermore, while DRAM is being refreshed, the data cannot be accessed. This is in contrast to SRAM's flip-flops, which retain data as long as the power is on, do not need to be refreshed, and whose contents can be accessed at any time. Since 1970, the capacity of DRAM has exploded. After the 1K-bit (1024) chip came the 4K-bit in 1973, and then the 16K chip in 1976. The 1980s saw the introduction of 64K, 256K, and finally 1M and 4M memory chips. The 1990s saw 16M, 64M, 256M, and the beginning of 1G-bit DRAM chips. In the 2000s, 2G-bit chips are standard, and as the fabrication process gets smaller, larger memory chips will be rolling off the manufacturing line. Keep in mind that when talking about IC memory chips, the capacity is always assumed to be in bits. Therefore, a 1M chip means a 1-megabit chip and a 256K chip means a 256K-bit memory chip. However, when talking about the memory of a computer system, it is always assumed to be in bytes.

Packaging issue in DRAM

In DRAM there is a problem of packing a large number of cells into a single chip with the normal number of pins assigned to addresses. For example, a 64K-bit chip (64K × 1) must have 16 address lines and 1 data line, requiring 16 pins to send in the address if the conventional method is used. This is in addition to V_{CC} power, ground, and read/write control pins. Using the conventional method of data access, the large number of pins defeats the purpose of high density and small packaging, so dearly cherished by IC designers. Therefore, to reduce the number of pins needed for addresses, multiplexing/demultiplexing is used. The method used is to split the address in half and send in each half of the address through the same pins, thereby requiring fewer address pins. Internally, the DRAM structure is divided into a square of rows and columns. The first half of the address is called the row and the second half is called the column. For example, in the case of DRAM of 64K × 1 organization, the first half of the address is sent in through the 8 pins A0–A7, and by activating RAS (row address strobe), the internal latches inside DRAM grab the first half of the address. After that, the second half of the address is sent in through the same pins, and by activating CAS (column address strobe), the internal latches inside DRAM latch the second half of the address. This results in using 8 pins for addresses plus RAS and CAS, for a total of 10 pins, instead of the 16 pins that would be required without multiplexing. To access a bit of data from DRAM, both row and column addresses must be provided. For this concept to work, there must be a 2-by-1 multiplexer outside the DRAM circuitry and a demultiplexer inside every DRAM chip. Due to the complexities associated with DRAM interfacing (RAS, CAS, the need for multiplexer and refreshing circuitry), some DRAM controllers are designed to make DRAM interfacing much easier. However, many small microcontroller-based projects that do not require much RAM (usually less than 64K bytes) use SRAM of types EEPROM and NV-RAM, instead of DRAM.

DRAM organization

In the discussion of ROM, we noted that all of these chips have 8 pins for data. This is not the case for DRAM memory chips, which can have ×1, ×4, ×8, or ×16 organizations. See Example 0-15 and Table 0-8.

In memory chips, the data pins are also called I/O. In some DRAMs there are separate D_{in} and D_{out} pins. Figure 0-15 shows a 256K × 1 DRAM chip with pins A0–A8 for address, RAS and CAS, WE (write enable), and data in and data out, as well as power and ground.

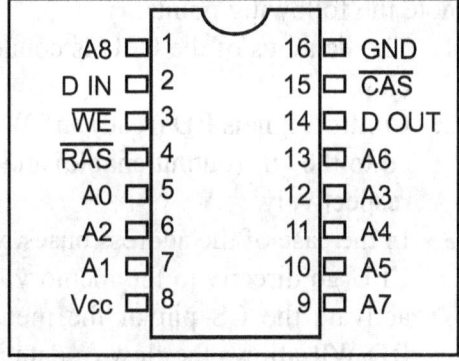

Figure 0-15. 256K × 1 DRAM

Example 0-15

Discuss the number of pins set aside for addresses in each of the following memory chips.   (a) 16K × 4 DRAM   (b) 16K × 4 SRAM
Solution:

Since $2^{14} = 16K$:

(a) For DRAM we have 7 pins (A0–A6) for the address pins and 2 pins for RAS and CAS.
(b) For SRAM we have 14 pins for address and no pins for RAS and CAS since they are associated only with DRAM. In both cases we have 4 pins for the data bus.

Table 0-8: Some DRAMs

Part No.	Speed	Capacity	Org.	Pins
4164-15	150 ns	64K	64K × 1	16
41464-8	80 ns	256K	64K × 4	18
41256-15	150 ns	256K	256K × 1	16
41256-6	60 ns	256K	256K × 1	16
414256-10	100 ns	1M	256K × 4	20
511000P-8	80 ns	1M	1M × 1	18
514100-7	70 ns	4M	4M × 1	20

Memory address decoding

Next we discuss address decoding. The CPU provides the address of the data desired, but it is the job of the decoding circuitry to locate the selected memory block. To explore the concept of decoding circuitry, we look at various methods used in decoding the addresses. In this discussion we use SRAM or ROM for the sake of simplicity.

Memory chips have one or more pins called CS (chip select), which must be activated for the memory's contents to be accessed. Sometimes the chip select is also referred to as chip enable (CE). In connecting a memory chip to the CPU,

---

note the following points.

1. The data bus of the CPU is connected directly to the data pins of the memory chip.
2. Control signals RD (read) and WR (memory write) from the CPU are connected to the OE (output enable) and WE (write enable) pins of the memory chip, respectively.
3. In the case of the address buses, while the lower bits of the addresses from the CPU go directly to the memory chip address pins, the upper ones are used to activate the CS pin of the memory chip. It is the CS pin that along with RD/WR allows the flow of data in or out of the memory chip. No data can be written into or read from the memory chip unless CS is activated.

As can be seen from the data sheets of SRAM and ROM, the CS input of a memory chip is normally active low and is activated by the output of the memory decoder. Normally memories are divided into blocks, and the output of the decoder selects a given memory block. There are three ways to generate a memory block selector: (a) using simple logic gates, (b) using the 74LS138, or (c) using programmable logics such as CPLD and FPGA. Each method is described below.

Simple logic gate address decoder

The simplest method of constructing decoding circuitry is the use of a NAND gate. The output of a NAND gate is active low, and the CS pin is also active low, which makes them a perfect match. In cases where the CS input is active high, an AND gate must be used. Using a combination of NAND gates and inverters, one can decode any address range. An example of this is shown in Figure 0-16, which shows that A15–A12 must be 0011 in order to select the chip. This results in the assignment of addresses 3000H to 3FFFH to this memory chip.

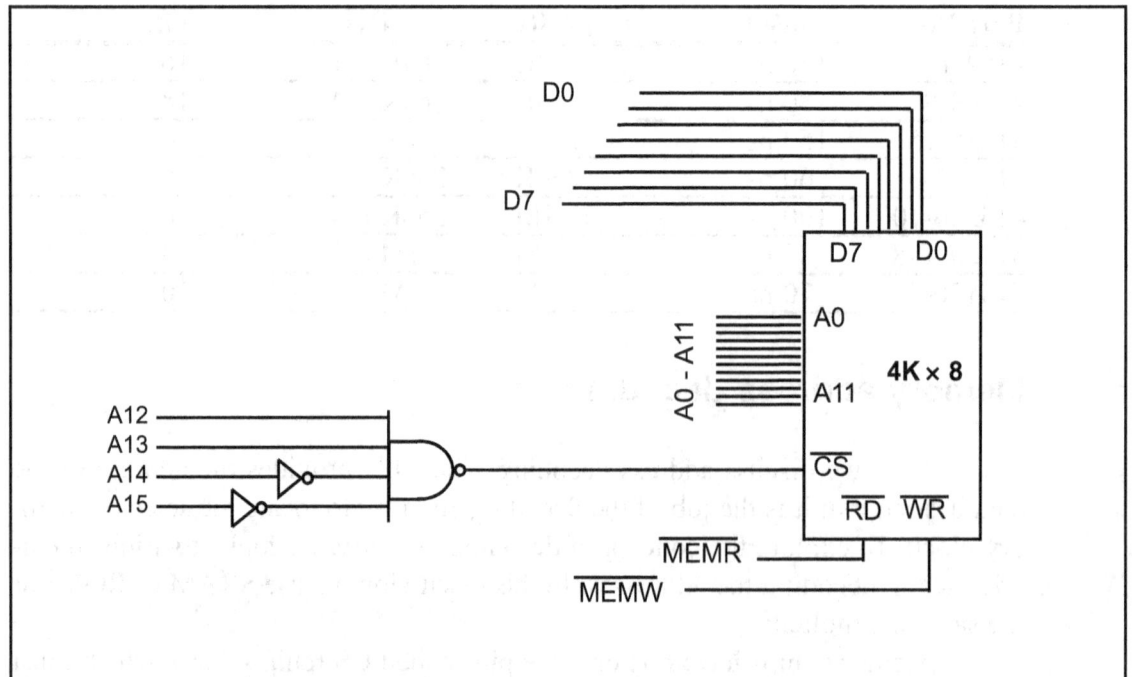

Figure 0-16. Logic Gate as Decoder

Using the 74LS138 3-8 decoder

This used to be one of the most widely used address decoders. The 3 inputs A, B, and C generate 8 active-low outputs Y0–Y7. See Figure 0-17. Each Y output is connected to CS of a memory chip, allowing control of 8 memory blocks by a single 74LS138. In the 74LS138, where A, B, and C select which output is activated, there are three additional inputs, G2A, G2B, and G1. G2A and G2B are both active low, and G1 is active high. If any one of the inputs G1, G2A, or G2B is not connected to an address signal (sometimes they are connected to a control signal), they must be activated permanently by either V_{CC} or ground, depending on the activation level. Example 0-16 shows the design and the address range calculation for the 74LS138 decoder.

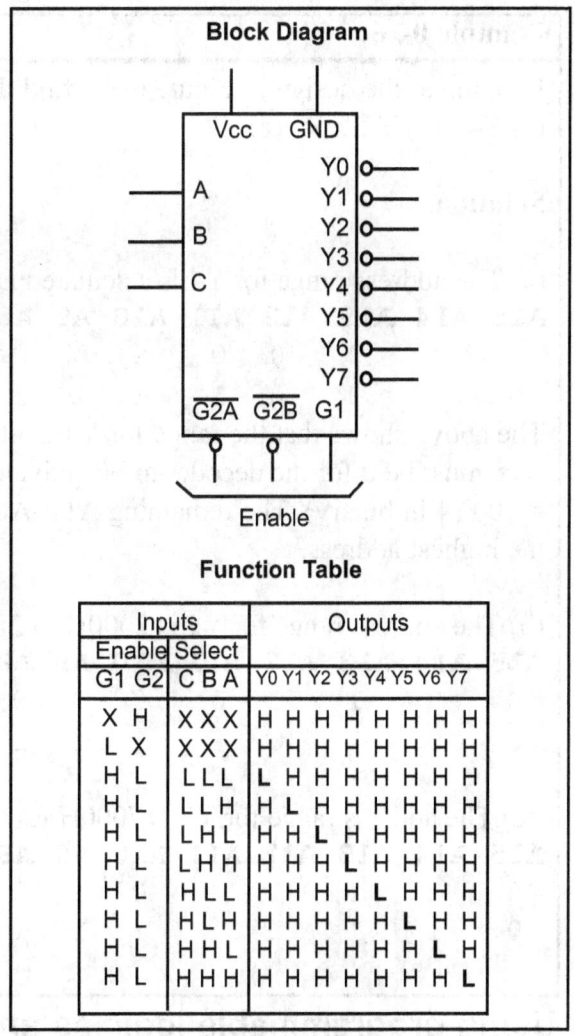

Block Diagram

Function Table

Inputs		Outputs
Enable	Select	
G1 G2	C B A	Y0 Y1 Y2 Y3 Y4 Y5 Y6 Y7
X H	X X X	H H H H H H H H
L X	X X X	H H H H H H H H
H L	L L L	**L** H H H H H H H
H L	L L H	H **L** H H H H H H
H L	L H L	H H **L** H H H H H
H L	L H H	H H H **L** H H H H
H L	H L L	H H H H **L** H H H
H L	H L H	H H H H H **L** H H
H L	H H L	H H H H H H **L** H
H L	H H H	H H H H H H H **L**

Figure 0-17. 74LS138 Decoder

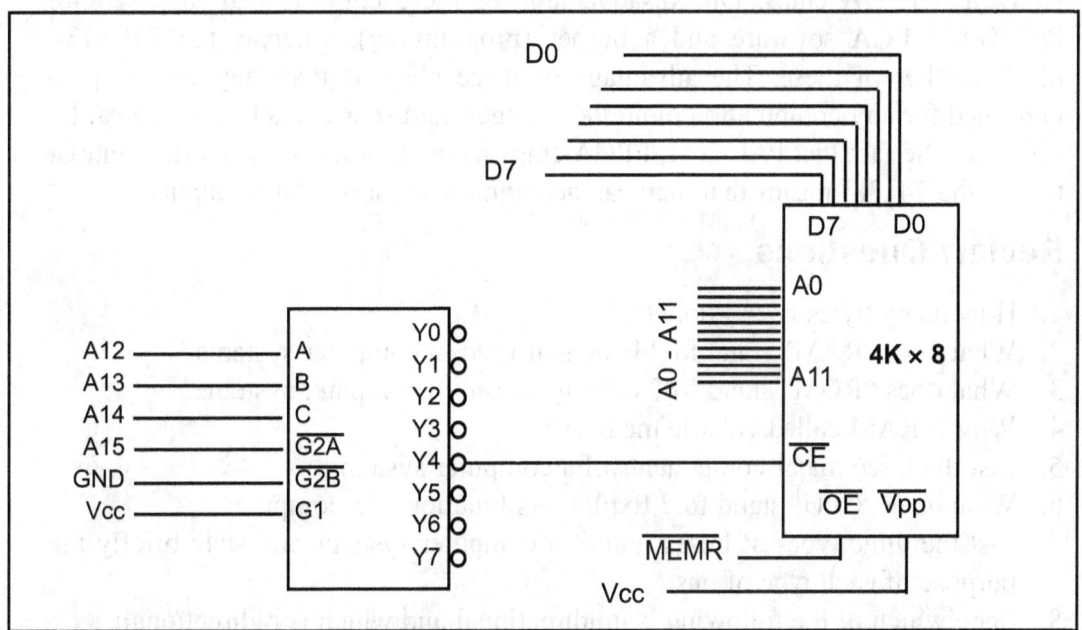

Figure 0-18. Using 74LS138 as Decoder

CHAPTER 0: INTRODUCTION TO COMPUTING   27

Example 0-16

Looking at the design in Figure 0-18, find the address range for the following:
(a) Y4, (b) Y2, and (c) Y7.

Solution:

(a) The address range for Y4 is calculated as follows.

A15	A14	A13	A12	A11	A10	A9	A8	A7	A6	A5	A4	A3	A2	A1	A0
0	1	0	0	0	0	0	0	0	0	0	0	0	0	0	0
0	1	0	0	1	1	1	1	1	1	1	1	1	1	1	1

The above shows that the range for Y4 is 4000H to 4FFFH. In Figure 0-18, notice that A15 must be 0 for the decoder to be activated. Y4 will be selected when A14 A13 A12 = 100 (4 in binary). The remaining A11–A0 will be 0 for the lowest address and 1 for the highest address.

(b) The address range for Y2 is 2000H to 2FFFH.

A15	A14	A13	A12	A11	A10	A9	A8	A7	A6	A5	A4	A3	A2	A1	A0
0	0	1	0	0	0	0	0	0	0	0	0	0	0	0	0
0	0	1	0	1	1	1	1	1	1	1	1	1	1	1	1

(c) The address range for Y7 is 7000H to 7FFFH.

A15	A14	A13	A12	A11	A10	A9	A8	A7	A6	A5	A4	A3	A2	A1	A0
0	1	1	1	0	0	0	0	0	0	0	0	0	0	0	0
0	1	1	1	1	1	1	1	1	1	1	1	1	1	1	1

Using programmable logic as an address decoder

Other widely used decoders are programmable logic chips such as PAL, GAL, and FPGA chips. One disadvantage of these chips is that they require PAL/GAL/FPGA software and a burner (programmer), whereas the 74LS138 needs neither of these. The advantage of these chips is that they can be programmed for any combination of address ranges, and so are much more versatile. This plus the fact that PAL/GAL/FPGA chips have 10 or more inputs (in contrast to 6 in the 74138) means that they can accommodate more address inputs.

Review Questions

1. How many bytes is 24 kilobytes?
2. What does "RAM" stand for? How is it used in computer systems?
3. What does "ROM" stand for? How is it used in computer systems?
4. Why is RAM called volatile memory?
5. List the three major components of a computer system.
6. What does "CPU" stand for? Explain its function in a computer.
7. List the three types of buses found in computer systems and state briefly the purpose of each type of bus.
8. State which of the following is unidirectional and which is bidirectional:
   (a) data bus (b) address bus

9. If an address bus for a given computer has 16 lines, what is the maximum amount of memory it can access?

10. The speed of semiconductor memory is in the range of
    (a) microseconds (b) milliseconds
    (c) nanoseconds   (d) picoseconds

11. Find the organization and chip capacity for each ROM with the indicated number of address and data pins.
    (a) 14 address, 8 data   (b) 16 address, 8 data (c) 12 address, 8 data

12. Find the organization and chip capacity for each RAM with the indicated number of address and data pins.
    (a) 11 address, 1 data SRAM   (b) 13 address, 4 data SRAM
    (c) 17 address, 8 data SRAM   (d) 8 address, 4 data DRAM
    (e) 9 address, 1 data DRAM   (f) 9 address, 4 data DRAM

13. Find the capacity and number of pins set aside for address and data for memory chips with the following organizations.
    (a) 16K × 4 SRAM   (b) 32K × 8 EPROM   (c) 1M × 1 DRAM
    (d) 256K × 4 SRAM   (e) 64K × 8 EEPROM   (f) 1M × 4 DRAM

14. Which of the following is (are) volatile memory?
    (a) EEPROM   (b) SRAM   (c) DRAM   (d) NV-RAM

15. A given memory block uses addresses 4000H–7FFFH. How many kilobytes is this memory block?

16. The 74138 is a(n) _____ by _____ decoder.

17. In the 74138 give the status of G2A and G2B for the chip to be enabled.

18. In the 74138 give the status of G1 for the chip to be enabled.

19. In Example 0-16, what is the range of addresses assigned to Y5?

SECTION 0.4: CPU ARCHITECTURE

In this section we will examine the inside of a CPU. Then, we will compare the Harvard and von Neumann architectures.

Inside CPU

A program stored in memory provides instructions to the CPU to perform an action. See Figure 0-19. The action can simply be adding data such as payroll data or controlling a machine such as a robot. The function of the CPU is to fetch these instructions from memory and execute them. To perform the actions of fetch and execute, all CPUs are equipped with resources such as the following:

1. Foremost among the resources at the disposal of the CPU are a number of *registers*. The CPU uses registers to store information temporarily. The information could be two values to be processed, or the address of the value needed to be fetched from memory. Registers inside the CPU can be 8-bit, 16-bit, 32-bit, or even 64-bit registers, depending on the CPU. In general, the more and bigger the registers, the better the CPU. The disadvantage of more and bigger registers is the increased cost of such a CPU.

2. The CPU also has what is called the *ALU* (arithmetic/logic unit). The ALU section of the CPU is responsible for performing arithmetic functions such as add,

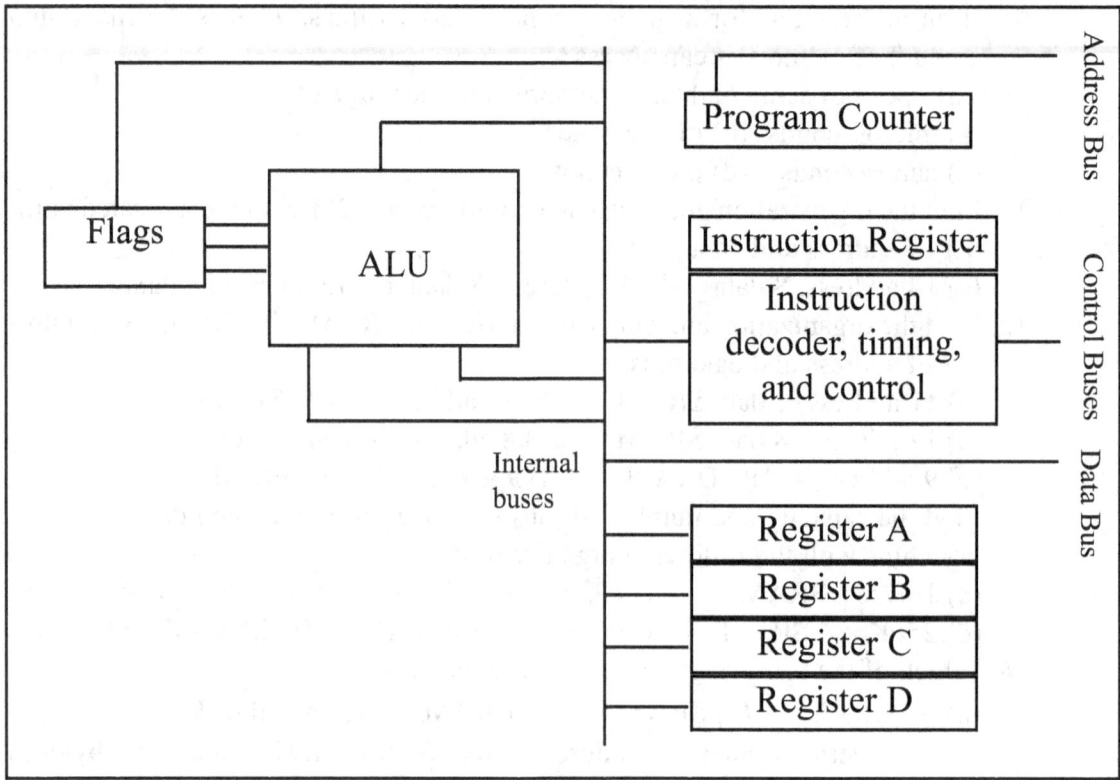

Figure 0-19. Internal Block Diagram of a CPU

subtract, multiply, and divide, and logic functions such as AND, OR, and NOT.

3. Every CPU has what is called a *program counter*. The function of the program counter is to point to the address of the next instruction to be executed. As each instruction is executed, the program counter is incremented to point to the address of the next instruction to be executed. The contents of the program counter are placed on the address bus to find and fetch the desired instruction. In the IBM PC, the program counter is a register called IP, or the instruction pointer.

4. The function of the *instruction decoder* is to interpret the instruction fetched into the CPU. One can think of the instruction decoder as a kind of dictionary, storing the meaning of each instruction and what steps the CPU should take upon receiving a given instruction. Just as a dictionary requires more pages the more words it defines, a CPU capable of understanding more instructions requires more transistors to design.

Internal working of CPUs

To demonstrate some of the concepts discussed above, a step-by-step analysis of the process a CPU would go through to add three numbers is given next. Assume that an imaginary CPU has registers called A, B, C, and D. It has an 8-bit data bus and a 16-bit address bus. Therefore, the CPU can access memory from addresses 0000 to FFFFH (for a total of 10000H locations). The action to be performed by the CPU is to put hexadecimal value 21 into register A, and then add to register A the values 42H and 12H. Assume that the code for the CPU to move a value to register A is 1011 0000 (B0H) and the code for adding a value to register A is 0000 0100 (04H). The necessary steps and code to perform these operations are as follows.

```
Action                            Code    Data
Move value 21H into register A     B0H     21H
Add value 42H to register A        04H     42H
Add value 12H to register A        04H     12H
```

If the program to perform the actions listed above is stored in memory locations starting at 1400H, the following would represent the contents for each memory address location:

```
Memory address   Contents of memory address
1400             (B0)code for moving a value to register A
1401             (21)value to be moved
1402             (04)code for adding a value to register A
1403             (42)value to be added
1404             (04)code for adding a value to register A
1405             (12)value to be added
1406             (F4)code for halt
```

The actions performed by the CPU to run the program above would be as follows:

1. The CPU's program counter can have a value between 0000 and FFFFH. The program counter must be set to the value 1400H, indicating the address of the first instruction code to be executed. After the program counter has been loaded with the address of the first instruction, the CPU is ready to execute.

2. The CPU puts 1400H on the address bus and sends it out. The memory circuitry finds the location while the CPU activates the READ signal, indicating to memory that it wants the byte at location 1400H. This causes the contents of memory location 1400H, which is B0, to be put on the data bus and brought into the CPU.

3. The CPU decodes the instruction B0 with the help of its instruction decoder dictionary. When it finds the definition for that instruction it knows it must bring the byte in the next memory location into register A of the CPU. Therefore, it commands its controller circuitry to do exactly that. When it brings in value 21H from memory location 1401, it makes sure that the doors of all registers are closed except register A. Therefore, when value 21H comes into the CPU it will go directly into register A. After completing one instruction, the program counter points to the address of the next instruction to be executed, which in this case is 1402H. Address 1402 is sent out on the address bus to fetch the next instruction.

4. From memory location 1402H the CPU fetches code 04H. After decoding, the CPU knows that it must add the byte sitting at the next address (1403) to the contents of register A. After the CPU brings the value (in this case, 42H) into register A, it provides the contents of register A along with this value to the ALU to perform the addition. It then takes the result of the addition from the ALU's output and puts it into register A. Meanwhile the program counter becomes 1404, the address of the next instruction.

---

5. Address 1404H is put on the address bus and the code is fetched into the CPU, decoded, and executed. This code again is adding a value to register A. The program counter is updated to 1406H.
6. Finally, the contents of address 1406 are fetched in and executed. This HALT instruction tells the CPU to stop incrementing the program counter and asking for the next instruction. Without the HALT, the CPU would continue updating the program counter and fetching instructions.

Now suppose that address 1403H contained value 04 instead of 42H. How would the CPU distinguish between data 04 to be added and code 04? Remember that code 04 for this CPU means "move the next value into register A." Therefore, the CPU will not try to decode the next value. It simply moves the contents of the following memory location into register A, regardless of its value.

Harvard and von Neumann architectures

Every microprocessor must have memory space to store program (code) and data. While code provides instructions to the CPU, the data provides the information to be processed. The CPU uses buses (wire traces) to access the code ROM and data RAM memory spaces. The early computers used the same bus for accessing both the code and data. Such an architecture is commonly referred to as *von Neumann (Princeton) architecture*. That means for von Neumann computers, the process of accessing the code or data could cause them to get in each other's way and slow down the processing speed of the CPU, because each had to wait for the other to finish fetching. To speed up the process of program execution, some CPUs use what is called *Harvard architecture*. In Harvard architecture, we have separate buses for the code and data memory. See Figure 0-20. That means that we need four sets of buses: (1) a set of data buses for carrying data into and out of the CPU, (2) a set of address buses for accessing the data, (3) a set of data buses for carrying code into the CPU, and (4) an address bus for accessing the code. See Figure 0-20. This is easy to implement inside an IC chip such as a microcontroller where both ROM code and data RAM are internal (on-chip) and distances are on the micron and millimeter scale. But implementing Harvard architecture for systems such as x86 IBM PC-type computers is very expensive because the RAM and ROM that hold code and data are external to the CPU. Separate wire traces for data and code on the motherboard will make the board large and expensive. For example, for a Pentium microprocessor with a 64-bit data bus and a 32-bit address bus we will need about 100 wire traces on the motherboard if it is von Neumann architecture (96 for address and data, plus a few others for control signals of read and write and so on). But the number of wire traces will double to 200 if we use Harvard architecture. Harvard architecture will also necessitate a large number of pins coming out of the microprocessor itself. For this reason you do not see Harvard architecture implemented in the world of PCs and workstations. This is also the reason that microcontrollers such as AVR use Harvard architecture internally, but they still use von Neumann architecture if they need external memory for code and data space. The von Neumann architecture was developed at Princeton University, while the Harvard architecture was the work of Harvard University.

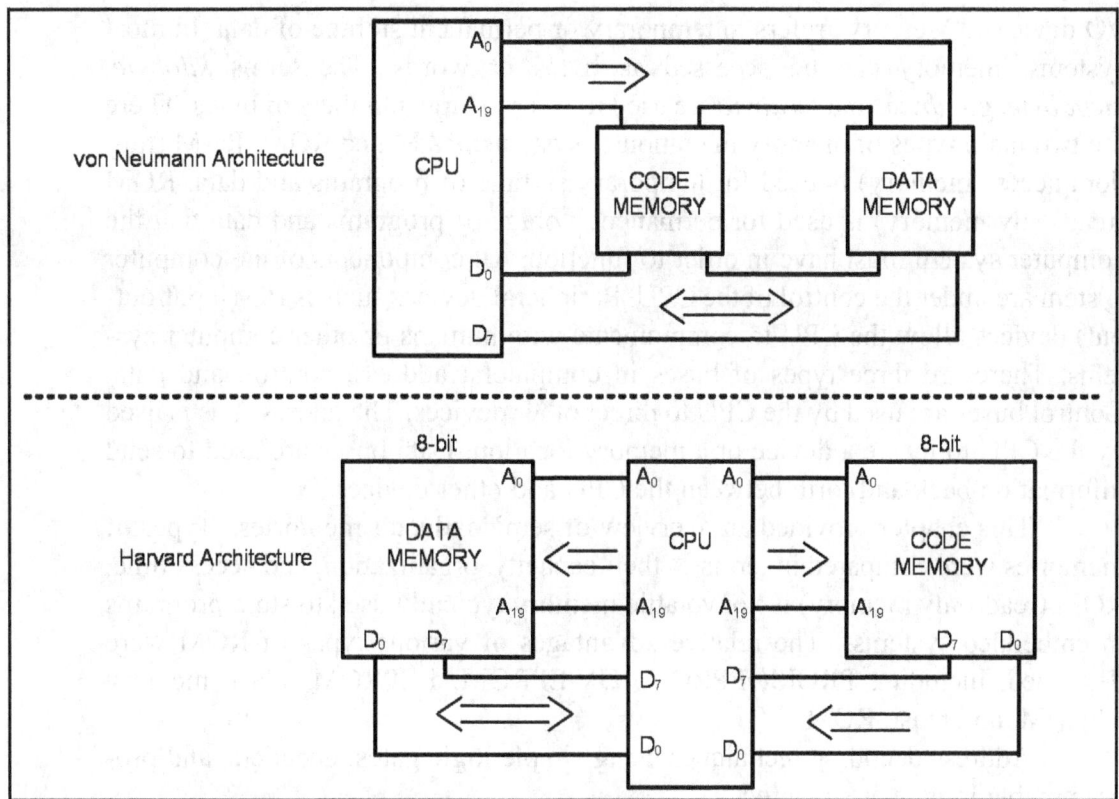

Figure 0-20. von Neumann vs. Harvard Architecture

Review Questions

1. What does "ALU" stand for? What is its purpose?
2. How are registers used in computer systems?
3. What is the purpose of the program counter?
4. What is the purpose of the instruction decoder?
5. True or false. Harvard architecture uses the same address and data buses to fetch both code and data.

SUMMARY

The binary number system represents all numbers with a combination of the two binary digits, 0 and 1. The use of binary systems is necessary in digital computers because only two states can be represented: on or off. Any binary number can be coded directly into its hexadecimal equivalent for the convenience of humans. Converting from binary/hex to decimal, and vice versa, is a straightforward process that becomes easy with practice. ASCII code is a binary code used to represent alphanumeric data internally in the computer. It is frequently used in peripheral devices for input and/or output.

The AND, OR, and inverter logic gates are the basic building blocks of simple circuits. NAND, NOR, and XOR gates are also used to implement circuit design. Diagrams of half-adders and full-adders were given as examples of the use of logic gates for circuit design. Decoders are used to detect certain addresses. Flip-flops are used to latch in data until other circuits are ready for it.

The major components of any computer system are the CPU, memory, and

I/O devices. "Memory" refers to temporary or permanent storage of data. In most systems, memory can be accessed as bytes or words. The terms *kilobyte*, *megabyte*, *gigabyte*, and *terabyte* are used to refer to large numbers of bytes. There are two main types of memory in computer systems: RAM and ROM. RAM (random access memory) is used for temporary storage of programs and data. ROM (read-only memory) is used for permanent storage of programs and data that the computer system must have in order to function. All components of the computer system are under the control of the CPU. Peripheral devices such as I/O (input/output) devices allow the CPU to communicate with humans or other computer systems. There are three types of buses in computers: address, control, and data. Control buses are used by the CPU to direct other devices. The address bus is used by the CPU to locate a device or a memory location. Data buses are used to send information back and forth between the CPU and other devices.

This chapter provided an overview of semiconductor memories. Types of memories were compared in terms of their capacity, organization, and access time. ROM (read-only memory) is nonvolatile memory typically used to store programs in embedded systems. The relative advantages of various types of ROM were described, including PROM, EPROM, UV-EPROM, EEPROM, Flash memory EPROM, and mask ROM.

Address decoding techniques using simple logic gates, decoders, and programmable logic were covered.

The computer organization and the internals of the CPU were also covered.

PROBLEMS

SECTION 0.1: NUMBERING AND CODING SYSTEMS

1. Convert the following decimal numbers to binary:
   (a) 12   (b) 123   (c) 63   (d) 128   (e) 1000
2. Convert the following binary numbers to decimal:
   (a) 100100   (b) 1000001   (c) 11101   (d) 1010   (e) 00100010
3. Convert the values in Problem 2 to hexadecimal.
4. Convert the following hex numbers to binary and decimal:
   (a) 2B9H   (b) F44H   (c) 912H   (d) 2BH   (e) FFFFH
5. Convert the values in Problem 1 to hex.
6. Find the 2's complement of the following binary numbers:
   (a) 1001010   (b) 111001   (c) 10000010   (d) 111110001
7. Add the following hex values:
   (a) 2CH + 3FH   (b) F34H + 5D6H   (c) 20000H + 12FFH
   (d) FFFFH + 2222H
8. Perform hex subtraction for the following:
   (a) 24FH – 129H (b) FE9H – 5CCH   (c) 2FFFFH – FFFFFH
   (d) 9FF25H – 4DD99H
9. Show the ASCII codes for numbers 0, 1, 2, 3, ..., 9 in both hex and binary.
10. Show the ASCII code (in hex) for the following strings:
      "U.S.A. is a country" CR,LF
      "in North America" CR,LF
      (CR is carriage return, LF is line feed)

11. Draw a 3-input OR gate using a 2-input OR gate.
12. Show the truth table for a 3-input OR gate.
13. Draw a 3-input AND gate using a 2-input AND gate.
14. Show the truth table for a 3-input AND gate.
15. Design a 3-input XOR gate with a 2-input XOR gate. Show the truth table for a 3-input XOR.
16. List the truth table for a 3-input NAND.
17. List the truth table for a 3-input NOR.
18. Show the decoder for binary 1100.
19. Show the decoder for binary 11011.
20. List the truth table for a D-FF.

SECTION 0.3: SEMICONDUCTOR MEMORY

21. Answer the following:
    (a) How many nibbles are 16 bits?
    (b) How many bytes are 32 bits?
    (c) If a word is defined as 16 bits, how many words is a 64-bit data item?
    (d) What is the exact value (in decimal) of 1 meg?
    (e) How many kilobytes is 1 meg?
    (f) What is the exact value (in decimal) of 1 gigabyte?
    (g) How many kilobytes is 1 gigabyte?
    (h) How many megs is 1 gigabyte?
    (i) If a given computer has a total of 8 megabytes of memory, how many bytes (in decimal) is this? How many kilobytes is this?
22. A given mass storage device such as a hard disk can store 2 gigabytes of information. Assuming that each page of text has 25 rows and each row has 80 columns of ASCII characters (each character = 1 byte), approximately how many pages of information can this disk store?
23. In a given byte-addressable computer, memory locations 10000H to 9FFFFH are available for user programs. The first location is 10000H and the last location is 9FFFFH. Calculate the following:
    (a) The total number of bytes available (in decimal)
    (b) The total number of kilobytes (in decimal)
24. A given computer has a 32-bit data bus. What is the largest number that can be carried into the CPU at a time?
25. Below are listed several computers with their data bus widths. For each computer, list the maximum value that can be brought into the CPU at a time (in both hex and decimal).
    (a) Apple 2 with an 8-bit data bus
    (b) x86 PC with a 16-bit data bus
    (c) x86 PC with a 32-bit data bus
    (d) Cray supercomputer with a 64-bit data bus
26. Find the total amount of memory, in the units requested, for each of the following CPUs, given the size of the address buses:

(a) 16-bit address bus (in K)
(b) 24-bit address bus (in megs)
(c) 32-bit address bus (in megabytes and gigabytes)
(d) 48-bit address bus (in megabytes, gigabytes, and terabytes)
27. Of the data bus and address bus, which is unidirectional and which is bidirectional?
28. What is the difference in capacity between a 4M memory chip and 4M of computer memory?
29. True or false. The more address pins, the more memory locations are inside the chip. (Assume that the number of data pins is fixed.)
30. True or false. The more data pins, the more each location inside the chip will hold.
31. True or false. The more data pins, the higher the capacity of the memory chip.
32. True or false. The more data pins and address pins, the greater the capacity of the memory chip.
33. The speed of a memory chip is referred to as its _____.
34. True or false. The price of memory chips varies according to capacity and speed.
35. The main advantage of EEPROM over UV-EPROM is _____.
36. True or false. SRAM has a larger cell size than DRAM.
37. Which of the following, EPROM, DRAM, or SRAM, must be refreshed periodically?
38. Which memory is used for PC cache?
39. Which of the following, SRAM, UV-EPROM, NV-RAM, or DRAM, is volatile memory?
40. RAS and CAS are associated with which type of memory?
    (a) EPROM (b) SRAM (c) DRAM (d) all of the above
41. Which type of memory needs an external multiplexer?
    (a) EPROM (b) SRAM (c) DRAM (d) all of the above
42. Find the organization and capacity of memory chips with the following pins.
    (a) EEPROM A0–A14, D0–D7   (b) UV-EPROM A0–A12, D0–D7
    (c) SRAM A0–A11, D0–D7   (d) SRAM A0–A12, D0–D7
    (e) DRAM A0–A10, D0   (f) SRAM A0–A12, D0
    (g) EEPROM A0–A11, D0–D7   (h) UV-EPROM A0–A10, D0–D7
    (i) DRAM A0–A8, D0–D3   (j) DRAM A0–A7, D0–D7

43. Find the capacity, address, and data pins for the following memory organizations.
    (a) 16K × 8 ROM   (b) 32K × 8 ROM
    (c) 64K × 8 SRAM   (d) 256K × 8 EEPROM
    (e) 64K × 8 ROM   (f) 64K × 4 DRAM
    (g) 1M × 8 SRAM   (h) 4M × 4 DRAM
    (i) 64K × 8 NV-RAM

44. Find the address range of the memory design in the diagram.
45. Using NAND gates and inverters, design decoding circuitry for the address range 2000H–2FFFH.
46. Find the address range for Y0, Y3, and Y6 of the 74LS138 for the diagrammed

design.

47. Using the 74138, design the memory decoding circuitry in which the memory block controlled by Y0 is in the range 0000H to 1FFFH. Indicate the size of the memory block controlled by each Y.

48. Find the address range for Y3, Y6, and Y7 in Problem 47.

49. Using the 74138, design memory decoding circuitry in which the memory block controlled by Y0 is in the 0000H to 3FFFH space. Indicate the size of the memory block controlled by each Y.

50. Find the address range for Y1, Y2, and Y3 in Problem 49.

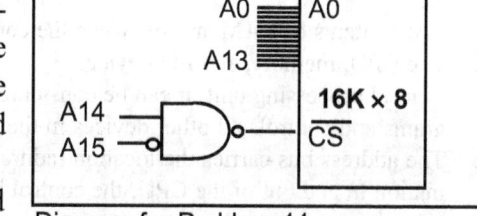

Diagram for Problem 44

SECTION 0.4: CPU AND HARVARD ARCHITEC- TURE

Diagram for Problem 46

51. Which register of the CPU holds the address of the instruction to be fetched?
52. Which section of the CPU is responsible for performing addition?
53. List the three bus types present in every CPU.

ANSWERS TO REVIEW QUESTIONS

SECTION 0.1: NUMBERING AND CODING SYSTEMS

1. Computers use the binary system because each bit can have one of two voltage levels: on and off.
2. $34_{10} = 100010_2 = 22_{16}$
3. $110101_2 = 35_{16} = 53_{10}$
4. 1110001
5. 010100
6. 461
7. 275
8. 38 30 78 38 36 20 43 50 55 73

SECTION 0.2: DIGITAL PRIMER

1. AND
2. OR
3. XOR
4. Buffer
5. Storing data
6. Decoder

SECTION 0.3: SEMICONDUCTOR MEMORY

1. 24,576
2. Random access memory; it is used for temporary storage of programs that the CPU is run-

ning, such as the operating system, word processing programs, etc.

3. Read-only memory; it is used for permanent programs such as those that control the keyboard, etc.

4. The contents of RAM are lost when the computer is powered off.

5. The CPU, memory, and I/O devices

6. Central processing unit; it can be considered the "brain" of the computer; it executes the programs and controls all other devices in the computer.

7. The address bus carries the location (address) needed by the CPU; the data bus carries information in and out of the CPU; the control bus is used by the CPU to send signals controlling I/O devices.

8. (a) bidirectional   (b) unidirectional

9. 64K, or 65,536 bytes

10. c

11. (a) 16K × 8, 128K bits   (b) 64K × 8, 512K   (c) 4K × 8, 32K

12. (a) 2K × 1, 2K bits   (b) 8K × 4, 32K   (c) 128K × 8, 1M
    (d) 64K × 4, 256K   (e) 256K × 1, 256K   (f) 256K × 4, 1M

13. (a) 64K bits, 14 address, and 4 data   (b) 256K, 15 address, and 8 data
    (c) 1M, 10 address, and 1 data   (d) 1M, 18 address, and 4 data
    (e) 512K, 16 address, and 8 data   (f) 4M, 10 address, and 4 data

14. b, c

15. 16K bytes

16. 3, 8

17. Both must be low.

18. G1 must be high.

19. 5000H–5FFFH

SECTION 0.4: CPU ARCHITECTURE

1. Arithmetic/logic unit; it performs all arithmetic and logic operations.

2. They are used for temporary storage of information.

3. It holds the address of the next instruction to be executed.

4. It tells the CPU what actions to perform for each instruction.

5. False

CHAPTER 1

THE AVR MICROCONTROLLER: HISTORY AND FEATURES

OBJECTIVES

Upon completion of this chapter, you will be able to:

>> Compare and contrast microprocessors and microcontrollers
>> Describe the advantages of microcontrollers for some applications
>> Explain the concept of embedded systems
>> Discuss criteria for considering a microcontroller
>> Explain the variations of speed, packaging, memory, and cost per unit and how these affect choosing a microcontroller
>> Compare and contrast the various members of the AVR family
>> Compare the AVR with microcontrollers offered by other manufacturers

This chapter begins with a discussion of the role and importance of micro-controllers in everyday life. In Section 1.1 we also discuss criteria to consider in choosing a microcontroller, as well as the use of microcontrollers in the embedded market. Section 1.2 covers various members of the AVR family and their features. In addition, we provide a brief discussion of alternatives to the AVR chip such as the 8051, PIC, and 68HC11 microcontrollers.

SECTION 1.1: MICROCONTROLLERS AND EMBEDDED PROCESSORS

In this section we discuss the need for microcontrollers and contrast them with general-purpose microprocessors such as the Pentium and other x86 micro-processors. We also look at the role of microcontrollers in the embedded market. In addition, we provide some criteria on how to choose a microcontroller.

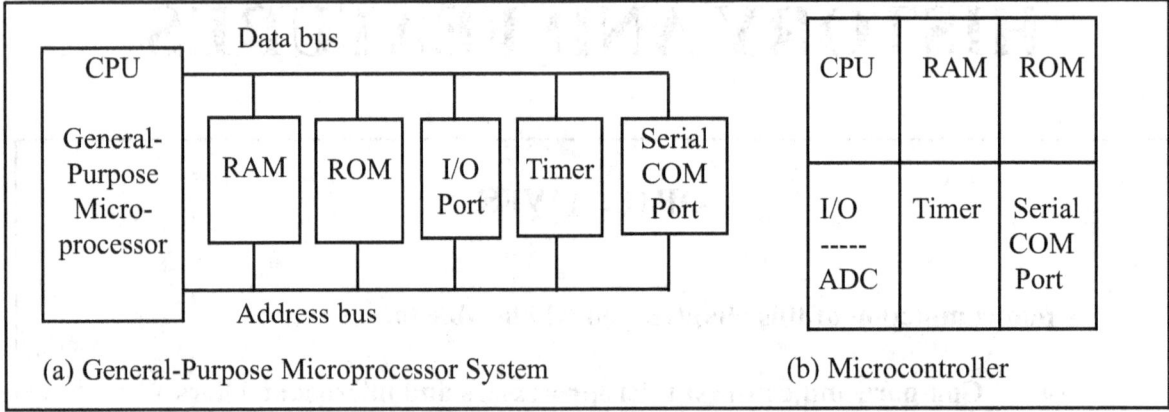

Figure 1-1. Microprocessor System Contrasted with Microcontroller System

Microcontroller versus general-purpose microprocessor

What is the difference between a microprocessor and a microcontroller? By microprocessor is meant the general-purpose microprocessors such as Intel's x86 family (8086, 80286, 80386, 80486, and the Pentium) or Motorola's PowerPC family. These microprocessors contain no RAM, no ROM, and no I/O ports on the chip itself. For this reason, they are commonly referred to as *general-purpose microprocessors*. See Figure 1-1.

A system designer using a general-purpose microprocessor such as the Pentium or the PowerPC must add RAM, ROM, I/O ports, and timers externally to make them functional. Although the addition of external RAM, ROM, and I/O ports makes these systems bulkier and much more expensive, they have the advantage of versatility, enabling the designer to decide on the amount of RAM, ROM, and I/O ports needed to fit the task at hand. This is not the case with microcon-trollers. A microcontroller has a CPU (a microprocessor) in addition to a fixed amount of RAM, ROM, I/O ports, and a timer all on a single chip. In other words, the processor, RAM, ROM, I/O ports, and timer are all embedded together on one chip; therefore, the designer cannot add any external memory, I/O, or timer to it. The fixed amount of on-chip ROM, RAM, and number of I/O ports in microcon-trollers makes them ideal for many applications in which cost and space are criti-

Home

Appliances
Intercom
Telephones
Security systems
Garage door openers
Answering machines
Fax machines
Home computers
TVs
Cable TV tuner
VCR
Camcorder
Remote controls
Video games
Cellular phones
Musical instruments
Sewing machines
Lighting control
Paging
Camera
Pinball machines
Toys
Exercise equipment

Office

Telephones
Computers
Security systems
Fax machine
Microwave
Copier
Laser printer
Color printer
Paging

Auto

Trip computer
Engine control
Air bag
ABS
Instrumentation
Security system
Transmission control
Entertainment
Climate control
Cellular phone
Keyless entry

Table 1-1: Some Embedded Products Using Microcontrollers

cal. In many applications, for example, a TV remote control, there is no need for the computing power of a 486 or even an 8086 microprocessor. In many applications, the space used, the power consumed, and the price per unit are much more critical considerations than the computing power. These applications most often require some I/O operations to read signals and turn on and off certain bits. For this reason some call these processors IBP, "itty-bitty processors." (See "Good Things in Small Packages Are Generating Big Product Opportunities" by Rick Grehan, BYTE magazine, September 1994 (http://www.byte.com) for an excellent discussion of microcontrollers.)

It is interesting to note that many microcontroller manufacturers have gone as far as integrating an ADC (analog-to-digital converter) and other peripherals into the microcontroller.

Microcontrollers for embedded systems

In the literature discussing microprocessors, we often see the term *embedded system*. Microprocessors and microcontrollers are widely used in embedded system products. An embedded system is controlled by its own internal microprocessor (or microcontroller) as opposed to an external controller. Typically, in an embedded system, the microcontroller's ROM is burned with a purpose for specific functions needed for the system. A printer is an example of an embedded system because the processor inside it performs one task only; namely, getting the data and printing it. Contrast this with a Pentium-based PC (or any x86 PC), which can be used for any number of applications such as word processor, print server, bank teller terminal, video game player, network server, or Internet terminal. A PC can also load and run software for a variety of applications. Of course, the reason a PC can perform myriad tasks is that it has RAM memory and an operating system that loads the application software into RAM and lets the CPU run it. In an embedded system, typically only one application software is burned into ROM. An x86 PC contains or is connected to various embedded products such as the keyboard, printer, modem, disk controller, sound card, CD-ROM driver, mouse, and so on. Each one of these peripherals has a microcontroller inside it that performs only one task. For example, inside every mouse a microcontroller performs the task of finding the mouse's position and sending it to the PC. Table 1-1 lists some embedded products.

x86 PC embedded applications

Although microcontrollers are the preferred choice for many embedded systems, sometimes a microcontroller is inadequate for the task. For this reason, in recent years many manufacturers of general-purpose microprocessors such as Intel, Freescale

Semiconductor (formerly Motorola), and AMD (Advanced Micro Devices, Inc.) have targeted their microprocessors for the high end of the embedded market. Intel and AMD push their x86 processors for both the embedded and desktop PC markets. In the early 1990s, Apple computer began using the PowerPC microprocessors (604, 603, 620, etc.) in place of the 680x0 for the Macintosh. In 2007 Apple switched to the x86 CPU for use in the Mac computers. The PowerPC microprocessor is a joint venture between IBM and Freescale, and is targeted for the high end of the embedded market. It must be noted that when a company targets a general-purpose microprocessor for the embedded market it optimizes the processor used for embedded systems. For this reason these processors are often called *high-end embedded processors*. Another chip widely used in the high end of the embedded system design is the ARM (Advanced RISC Machine) microprocessor. Very often the terms *embedded processor* and *microcontroller* are used interchangeably.

One of the most critical needs of an embedded system is to decrease power consumption and space. This can be achieved by integrating more functions into the CPU chip. All the embedded processors based on the x86 and PowerPC 6xx have low power consumption in addition to some forms of I/O, COM port, and ROM, all on a single chip. In high-performance embedded processors, the trend is to integrate more and more functions on the CPU chip and let the designer decide which features to use. This trend is invading PC system design as well. Normally, in designing the PC motherboard we need a CPU plus a chipset containing I/O, a cache controller, a Flash ROM containing BIOS, and finally a secondary cache memory. New designs are emerging in industry. For example, many companies have a chip that contains the entire CPU and all the supporting logic and memory, except for DRAM. In other words, we have the entire computer on a single chip.

Currently, because of Linux and Windows standardization, many embedded systems use x86 PCs. In many cases, using x86 PCs for the high-end embedded applications not only saves money but also shortens development time because a vast library of software already exists for the Linux and Windows platforms. The fact that Windows and Linux are widely used and well-understood platforms means that developing a Windows-based or Linux-based embedded product reduces the cost and shortens the development time considerably.

Choosing a microcontroller

There are five major 8-bit microcontrollers. They are: Freescale Semiconductor's (formerly Motorola) 68HC08/68HC11, Intel's 8051, Atmel's AVR, Zilog's Z8, and PIC from Microchip Technology. Each of the above microcontrollers has a unique instruction set and register set; therefore, they are not compatible with each other. Programs written for one will not run on the others. There are also 16-bit and 32-bit microcontrollers made by various chip makers. With all these different microcontrollers, what criteria do designers consider in choosing one? Three criteria in choosing microcontrollers are as follows: (1) meeting the computing needs of the task at hand efficiently and cost effectively; (2) availability of software and hardware development tools such as compilers, assemblers, debuggers, and emulators; and (3) wide availability and reliable sources of the microcontroller. Next, we elaborate on each of the above criteria.

Criteria for choosing a microcontroller

1. The first and foremost criterion in choosing a microcontroller is that it must meet the task at hand efficiently and cost effectively. In analyzing the needs of a microcontroller-based project, we must first see whether an 8-bit, 16-bit, or 32-bit microcontroller can best handle the computing needs of the task most effectively. Among other considerations in this category are:

   (a) Speed. What is the highest speed that the microcontroller supports?

   (b) Packaging. Does it come in a DIP (dual inline package) or a QFP (quad flat package), or some other packaging format? This is important in terms of space, assembling, and prototyping the end product.

   (c) Power consumption. This is especially critical for battery-powered products.

   (d) The amount of RAM and ROM on the chip.

   (e) The number of I/O pins and the timer on the chip.

   (f) Ease of upgrade to higher-performance or lower-power-consumption versions.

   (g) Cost per unit. This is important in terms of the final cost of the product in which a microcontroller is used. For example, some microcontrollers cost 50 cents per unit when purchased 100,000 units at a time.

2. The second criterion in choosing a microcontroller is how easy it is to develop products around it. Key considerations include the availability of an assembler, a debugger, a code-efficient C language compiler, an emulator, technical support, and both in-house and outside expertise. In many cases, third-party vendor (i.e., a supplier other than the chip manufacturer) support for the chip is as good as, if not better than, support from the chip manufacturer.

3. The third criterion in choosing a microcontroller is its ready availability in needed quantities both now and in the future. For some designers this is even more important than the first two criteria. Currently, of the leading 8-bit microcontrollers, the 8051 family has the largest number of diversified (multiple source) suppliers. (Supplier means a producer besides the originator of the microcontroller.) In the case of the 8051, which was originated by Intel, many companies also currently produce the 8051.

Notice that Freescale Semiconductor (Motorola), Atmel, Zilog, and Microchip Technology have all dedicated massive resources to ensure wide and timely availability of their products because their products are stable, mature, and single sourced. In recent years, companies have begun to sell *Field-Programmable Gate Array* (FPGA) and *Application-Specific Integrated Circuit* (ASIC) libraries for the different microcontrollers.

Mechatronics and microcontrollers

The microcontroller is playing a major role in an emerging field called *mechatronics*. Here is an excellent summary of what the field of mechatronics is all about, taken from the website of Newcastle University (http://mechatronics2004.newcastle.edu.au/mech2004), which holds a major conference every year on this subject:

"Many technical processes and products in the area of mechanical and electrical engineering show an increasing integration of mechanics with electronics and information processing. This integration is between the components (hardware) and the information-driven functions (software), resulting in integrated systems called mechatronic systems.

The development of mechatronic systems involves finding an optimal balance between the basic mechanical structure, sensor and actuator implementation, automatic digital information processing and overall control, and this synergy results in innovative solutions. The practice of mechatronics requires multidisciplinary expertise across a range of disciplines, such as: mechanical engineering, electronics, information technology, and decision making theories."

Review Questions

1. True or false. Microcontrollers are normally less expensive than microprocessors.
2. When comparing a system board based on a microcontroller and a general-purpose microprocessor, which one is cheaper?
3. A microcontroller normally has which of the following devices on-chip?
   (a) RAM   (b) ROM   (c) I/O   (d) all of the above
4. A general-purpose microprocessor normally needs which of the following devices to be attached to it?
   (a) RAM   (b) ROM   (c) I/O   (d) all of the above
5. An embedded system is also called a dedicated system. Why?
6. What does the term *embedded system* mean?
7. Why does having multiple sources of a given product matter?

SECTION 1.2: OVERVIEW OF THE AVR FAMILY

In this section, we first look at the AVR microcontrollers and their features and then examine the different families of AVR in more detail.

A brief history of the AVR microcontroller

The basic architecture of AVR was designed by two students of Norwegian Institute of Technology (NTH), Alf-Egil Bogen and Vegard Wollan, and then was bought and developed by Atmel in 1996.

You may ask what AVR stands for; AVR can have different meanings for different people! Atmel says that it is nothing more than a product name, but it might stand for Advanced Virtual RISC, or Alf and Vegard RISC (the names of the AVR designers).

There are many kinds of AVR microcontroller with different properties. Except for AVR32, which is a 32-bit microcontroller, AVRs are all 8-bit microprocessors, meaning that the CPU can work on only 8 bits of data at a time. Data larger than 8 bits has to be broken into 8-bit pieces to be processed by the CPU. One of the problems with the AVR microcontrollers is that they are not all 100% compatible in terms of software when going from one family to another family. To run programs written for the ATtiny25 on a ATmega64, we must recompile the program and possibly change some register locations before loading it into the ATmega64. AVRs are generally classified into four broad groups: Mega, Tiny,

Special purpose, and Classic. In this book we cover the Mega family because these microcontrollers are widely used. Also, we will focus on ATmega32 since it is powerful, widely available, and comes in DIP packages, which makes it ideal for educational purposes. For those who have mastered the Mega family, understanding the other families is very easy and straightforward. The following is a brief description of the AVR microcontroller.

AVR features

The AVR is an 8-bit RISC single-chip microcontroller with Harvard architecture that comes with some standard features such as on-chip program (code) ROM, data RAM, data EEPROM, timers and I/O ports. See Figure 1-2. Most AVRs have some additional features like ADC, PWM, and different kinds of serial interface such as USART, SPI, I2C (TWI), CAN, USB, and so on. See Figures 1-3 and 1-4. Due to the importance of these peripherals, we have dedicated an entire chapter to many of them. The details of the RAM/ROM memory and I/O features of the Mega are given in the next few chapters.

AVR microcontroller program ROM

In microcontrollers, the ROM is used to store programs and for that reason it is called *program* or *code ROM*. Although the AVR has 8M (megabytes) of program (code) ROM space, not all family members come with that much ROM installed. The program ROM size can vary from 1K to 256K at the time of this writing, depending on the family member. The AVR was one of the first microcontrollers to use on-chip Flash memory for program storage. The Flash memory is

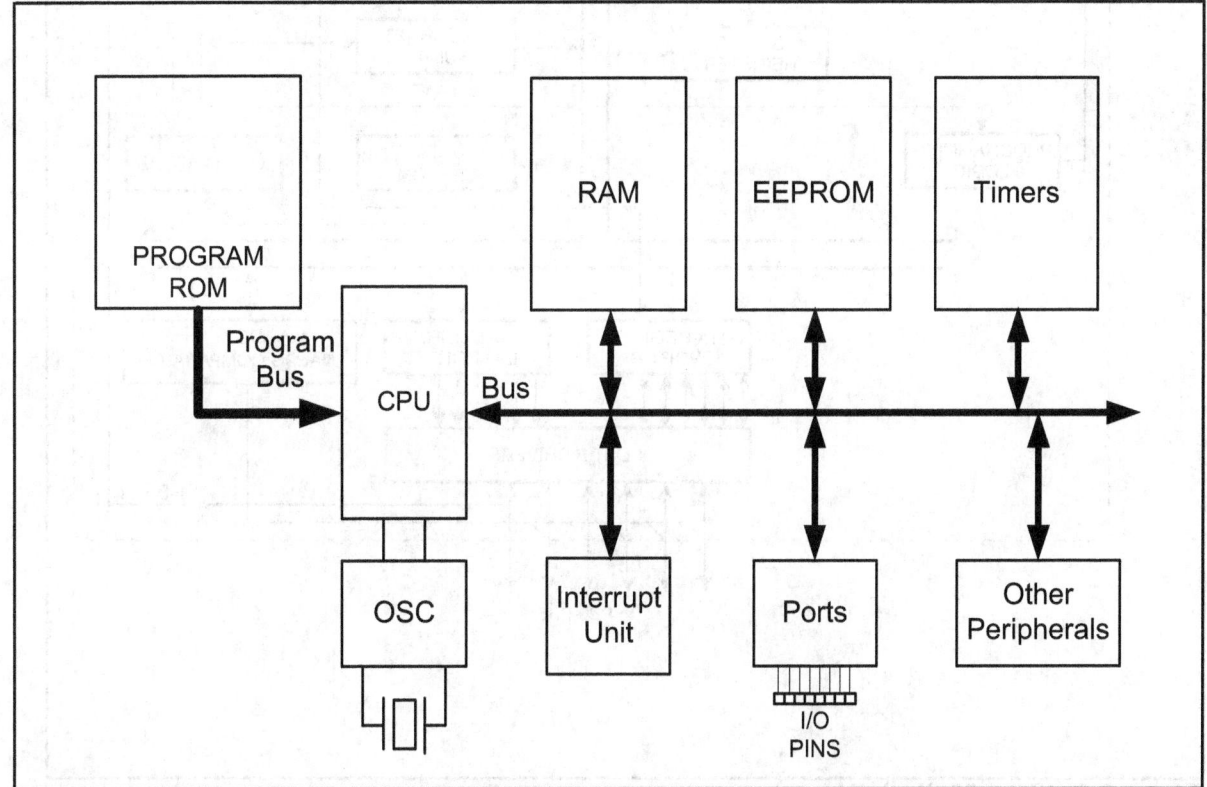

Figure 1-2. Simplified View of an AVR Microcontroller

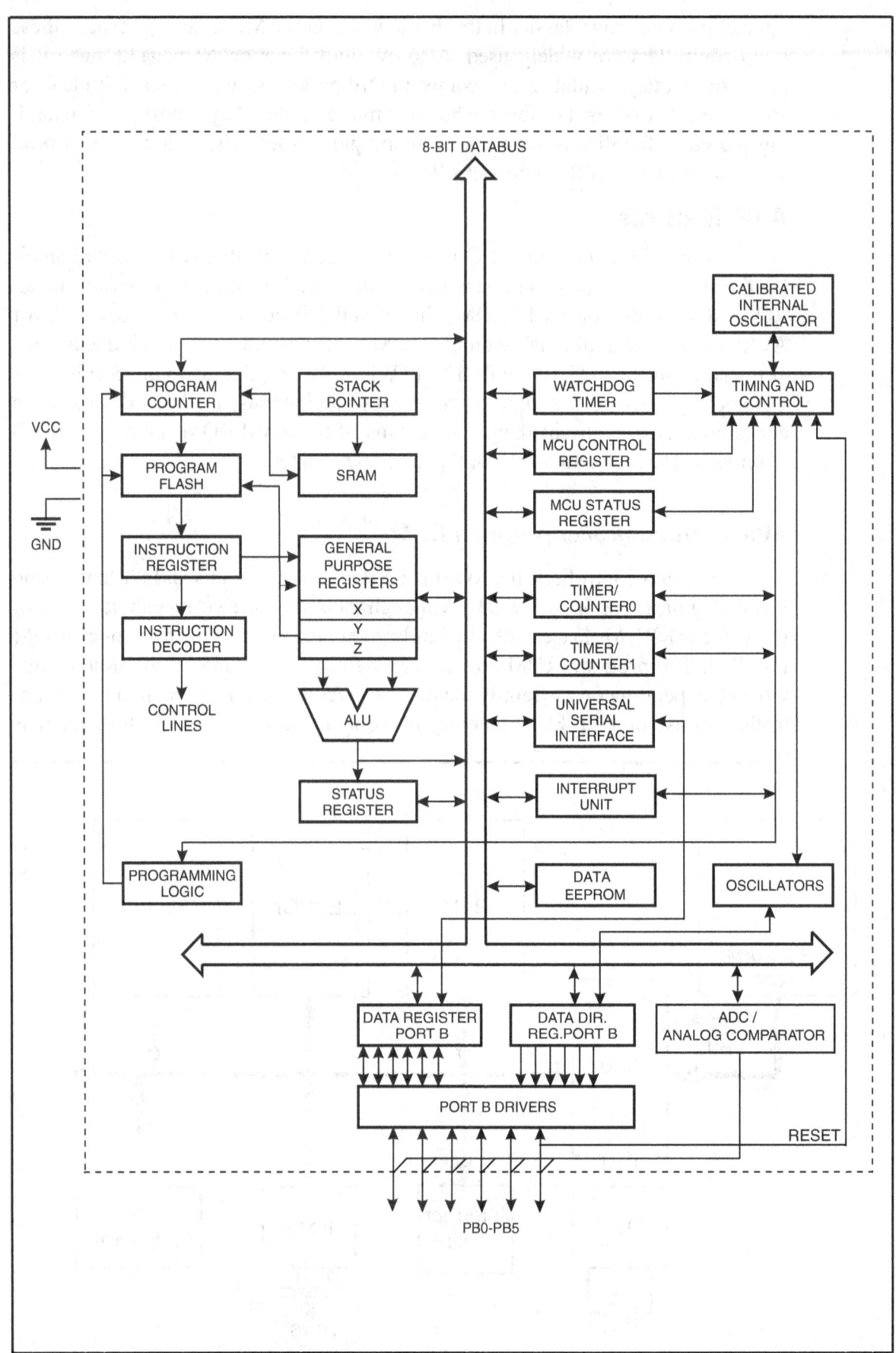

Figure 1-3. ATtiny25 Block Diagram

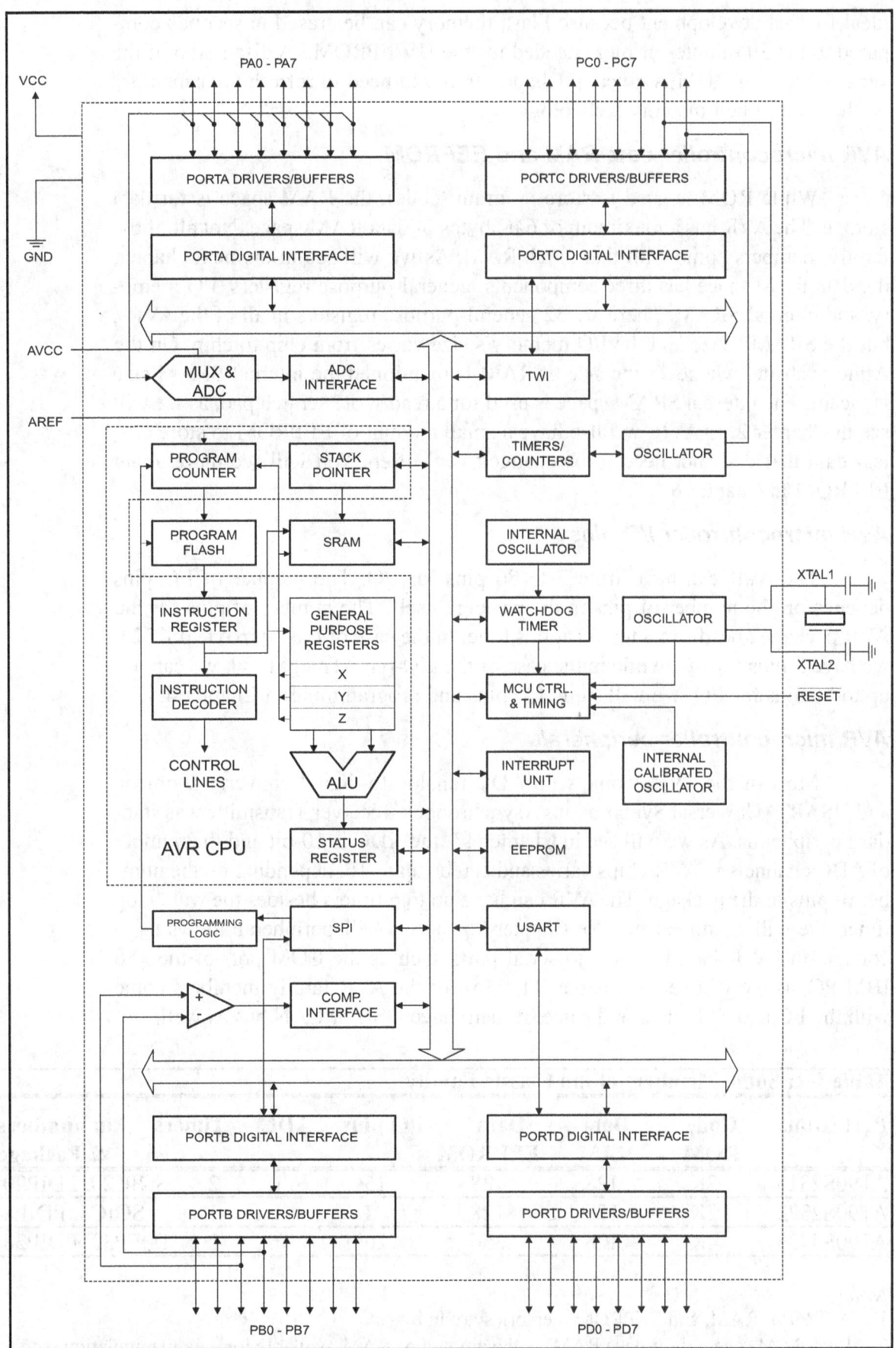

Figure 1-4. ATmega32 Block Diagram

ideal for fast development because Flash memory can be erased in seconds compared to the 20 minutes or more needed for the UV-EPROM. A discussion of the various types of ROM is given in Chapter 0, if you need to refresh your memory on these important memory technologies.

AVR microcontroller data RAM and EEPROM

While ROM is used to store program (code), the RAM space is for data storage. The AVR has a maximum of 64K bytes of data RAM space. Not all of the family members come with that much RAM. As we will see in the next chapter, the data RAM space has three components: general-purpose registers, I/O memory, and internal SRAM. There are 32 general-purpose registers in all of the AVRs, but the SRAM's size and the I/O memory's size varies from chip to chip. On the Atmel website, whenever the size of RAM is mentioned the internal SRAM size is meant. The internal SRAM space is used for a read/write scratch pad, as we will see in Chapter 2. In AVR, we also have a small amount of EEPROM to store critical data that does not need to be changed very often. You will see more about EEPROM in Chapter 6.

AVR microcontroller I/O pins

The AVR can have from 3 to 86 pins for I/O. The number of I/O pins depends on the number of pins in the package itself. The number of pins for the AVR package goes from 8 to 100 at this time. In the case of the 8-pin AT90S2323, we have 3 pins for I/O, while in the case of the 100-pin ATmega1280, we can use up to 86 pins for I/O. We will study I/O pins and programming in Chapter 4.

AVR microcontroller peripherals

Most of the AVRs come with ADC (analog-to-digital converter), timers, and USART (Universal Synchronous Asynchronous Receiver Transmitter) as standard peripherals. As we will see in Chapter 13, the ADC is 10-bit and the number of ADC channels in AVR chips varies and can be up to 16, depending on the number of pins in the package. The AVR can have up to 6 timers besides the watchdog timer. We will examine timers in Chapter 9. The USART peripheral allows us to connect the AVR-based system to serial ports such as the COM port of the x86 IBM PC, as we will see in Chapter 11. Most of the AVR family members come with the I²C and SPI buses and some of them have USB or CAN bus as well.

Table 1-2: Some Members of the Classic Family							
Part Num.	Code ROM	Data RAM	Data EEPROM	I/O pins	ADC	Timers	Pin numbers & Package
AT90S2313	2K	128	128	15	0	2	SOIC20, PDIP20
AT90S2323	2K	128	128	3	0	1	SOIC8, PDIP8
AT90S4433	4K	128	256	20	6	2	TQFP32, PDIP28

Notes:
1. All ROM, RAM, and EEPROM memories are in bytes.
2. Data RAM (general-purpose RAM) is the amount of RAM available for data manipulation (scratch pad) in addition to the register space.

AVR family overview

AVR can be classified into four groups: Classic, Mega, Tiny, and special purpose.

Classic AVR (AT90Sxxxx)

This is the original AVR chip, which has been replaced by newer AVR chips. Table 1-2 shows some members of the Classic AVR that are not recommended for new designs.

Mega AVR (ATmegaxxxx)

These are powerful microcontrollers with more than 120 instructions and lots of different peripheral capabilities, which can be used in different designs. See Table 1-3. Some of their characteristics are as follows:
- Program memory: 4K to 256K bytes
- Package: 28 to 100 pins
- Extensive peripheral set
- Extended instruction set: They have rich instruction sets.

Table 1-3: Some Members of the ATmega Family

Part Num.	Code ROM	Data RAM	Data EEPROM	I/O pins	ADC	Timers	Pin numbers & Package
ATmega8	8K	1K	0.5K	23	8	3	TQFP32, PDIP28
ATmega16	16K	1K	0.5K	32	8	3	TQFP44, PDIP40
ATmega32	32K	2K	1K	32	8	3	TQFP44, PDIP40
ATmega64	64K	4K	2K	54	8	4	TQFP64, MLF64
ATmega1280	128K	8K	4K	86	16	6	TQFP100, CBGA

Notes:
1. All ROM, RAM, and EEPROM memories are in bytes.
2. Data RAM (general-purpose RAM) is the amount of RAM available for data manipulation (scratch pad) in addition to the register space.
3. All the above chips have USART for serial data transfer.

Tiny AVR (ATtinyxxxx)

As its name indicates, the microcontrollers in this group have less instructions and smaller packages in comparison to mega family. You can design systems with low costs and power consumptions using the Tiny AVRs. See Table 1-4. Some of their characteristics are as follows:
- Program memory: 1K to 8K bytes
- Package: 8 to 28 pins
- Limited peripheral set
- Limited instruction set: The instruction sets are limited. For example, some of them do not have the multiply instruction.

Table 1-4: Some Members of the Tiny Family							
Part Num.	Code ROM	Data RAM	Data EEPROM	I/O pins	ADC	Timers	Pin numbers & Package
ATtiny13	1K	64	64	6	4	1	SOIC8, PDIP8
ATtiny25	2K	128	128	6	4	2	SOIC8, PDIP8
ATtiny44	4K	256	256	12	8	2	SOIC14, PDIP14
ATtiny84	8K	512	512	12	8	2	SOIC14, PDIP14

Special purpose AVR

The ICs of this group can be considered as a subset of other groups, but their special capabilities are made for designing specific applications. Some of the special capabilities are: USB controller, CAN controller, LCD controller, Zigbee, Ethernet controller, FPGA, and advanced PWM. See Table 1-5.

Table 1-5: Some Members of the Special Purpose Family							
Part Num.	Code ROM	Data RAM	Data EEPROM	Max I/O pins	Special Capabilities	Timers	Pin numbers & Package
AT90CAN128	128K	4K	4K	53	CAN	4	LQFP64
AT90USB1287	128K	8K	4K	48	USB Host	4	TQFP64
AT90PWM216	16K	1K	0.5K	19	Advanced PWM	2	SOIC24
ATmega169	16K	1K	0.5K	54	LCD	3	TQFP64, MLF64

AVR product number scheme

All of the product numbers start with AT, which stands for Atmel. Now, look at the number located at the end of the product number, from left to right, and find the biggest number that is a power of 2. This number most probably shows the amount of the microcontroller's ROM. For example, in ATmega**128**0 the biggest power of 2 that we can find is 128; so it has 128K bytes of ROM. In ATtiny**44**, the amount of memory is 4K, and so on. Although this rule has a few exceptions such as AT90PWM216, which has 16K of ROM instead of 2K, it works in most of the cases.

Other microcontrollers

There are many other popular 8-bit microcontrollers besides the AVR chip. Among them are the 8051, HCS08, PIC, and Z8. The AVR is made by Atmel Corp, as seen in Table 1-6. Microchip produces the PIC family. Freescale (formerly Motorola) makes the HCS08 and many of its variations. Zilog produces the Z8 microcontroller. The 8051 family is made by Intel and a number of other companies. To contrast the ATmega32 with the 8052 chip and PIC, examine Table 1-7.

For a comprehensive treatment of the 8051, HCS12, and PIC microcontrollers, see "The 8051 Microcontroller and Embedded Systems," "HCS12 Microcontroller and Embedded Systems," and "PIC Microcontroller and Embedded Systems" by Mazidi, et al.

Table 1-6: Some of the Companies that Produce Widely Used 8-bit Microcontrollers

Company	Web Site	Architecture
Atmel	http://www.atmel.com	AVR and 8051
Microchip	http://www.microchip.com	PIC16xxx/18xxx
Intel	http://www.intel.com/design/mcs51	8051
Philips/Signetics	http://www.semiconductors.philips.com	8051
Zilog	http://www.zilog.com	Z8 and Z80
Dallas Semi/Maxim	http://www.maxim-ic.com	8051
Freescale Semi	http://www.freescale.com	68HC11/HCS08

See http://www.microcontroller.com for a complete list.

Table 1-7: Comparison of 8051, PIC18 Family, and AVR (40-pin package)

Feature	8052	PIC18F452	ATmega32
Program ROM	8K	32K	32K
Data RAM (maximum space)	256 bytes	2K	2K
EEPROM	0 bytes	256 bytes	1K
Timers	3	4	3
I/O pins	32	35	32

Review Questions

1. Name three features of the AVR.
2. The AVR is a(n) _____-bit microprocessor.
3. Name the different groups of the AVR chips.
4. Which group of AVR has smaller packages?
5. Give the size of RAM in each of the following:
   (a) ATmega32   (b) ATtiny25
6. Give the size of the on-chip program ROM in each of the following:
   (a) ATtiny84   (b) ATmega32   (c) ATtiny25

See the following websites for AVR microcontrollers and AVR trainers:

http://www.Atmel.com

http://www.MicroDigitalEd.com

http://www.digilentinc.com

SUMMARY

This chapter discussed the role and importance of microcontrollers in everyday life. Microprocessors and microcontrollers were contrasted and compared. We discussed the use of microcontrollers in the embedded market. We also discussed criteria to consider in choosing a microcontroller such as speed, memory, I/O, packaging, and cost per unit. The second section of this chapter described various families of the AVR, such as Mega and Tiny, and their features. In addition, we discussed some of the most common AVR microcontrollers such as the ATmega32 and ATtiny25.

PROBLEMS

SECTION 1.1: MICROCONTROLLERS AND EMBEDDED PROCESSORS

1. True or False. A general-purpose microprocessor has on-chip ROM.
2. True or False. Generally, a microcontroller has on-chip ROM.
3. True or False. A microcontroller has on-chip I/O ports.
4. True or False. A microcontroller has a fixed amount of RAM on the chip.
5. What components are usually put together with the microcontroller onto a single chip?
6. Intel's Pentium chips used in Windows PCs need external _____ and _____ chips to store data and code.
7. List three embedded products attached to a PC.
8. Why would someone want to use an x86 as an embedded processor?
9. Give the name and the manufacturer of some of the most widely used 8-bit microcontrollers.
10. In Question 9, which one has the most manufacture sources?
11. In a battery-based embedded product, what is the most important factor in choosing a microcontroller?
12. In an embedded controller with on-chip ROM, why does the size of the ROM matter?
13. In choosing a microcontroller, how important is it to have multiple sources for that chip?
14. What does the term "third-party support" mean?
15. Suppose that a microcontroller architecture has both 8-bit and 16-bit versions. Which of the following statements is true?
    (a) The 8-bit software will run on the 16-bit system.
    (b) The 16-bit software will run on the 8-bit system.

SECTION 1.2: OVERVIEW OF THE AVR FAMILY

16. What is the advantage of Flash memory over the other kinds of ROM?
17. The ATmega32 has ____ pins for I/O.
18. The ATmega32 has _____ bytes of on-chip program ROM.
19. The ATtiny44 has _____ bytes of on-chip data RAM.
20. The ATtiny44 has _____ ADCs.

---

21. The ATmega64 has _____ bytes of on-chip data RAM.
22. The ATmega1280 has ____ on-chip timer(s).
23. The ATmega32 has ____ bytes of on-chip data RAM.
24. Check the Atmel website to see if there is a RAMless version of the AVR. Give the part number if there is one.
25. Check the Atmel website to see if there is a ROMless version of the AVR. Give the part number if there is one.
26. Check the Atmel website to find three members of the AVR family that have USB controllers.
27. Check the Atmel website to find two members of the AVR family that have CAN controllers.
28. Give the amount of program ROM and data RAM for the following chips:
    (a) ATmega32   (b) ATtiny44   (c) ATtiny84   (d) 90CAN128
29. What are the main differences between the ATmega16 and the ATmega32?
30. The ATmega16 has _____ bytes of data EEPROM.

ANSWERS TO REVIEW QUESTIONS

SECTION 1.1: MICROCONTROLLERS AND EMBEDDED PROCESSORS

1. True
2. A microcontroller-based system
3. (d)
4. (d)
5. It is dedicated because it is does only one type of job.
6. Embedded system means that the application and the processor are combined into a single system.
7. Having multiple sources for a given part means you are not hostage to one supplier. More importantly, competition among suppliers brings about lower cost for that product.

SECTION 1.2: OVERVIEW OF THE AVR FAMILY

1. 64K of RAM space, 8M of on-chip ROM space, a large number of I/O pins, ADC, and different serial protocols such as SPI, USART, I2C, etc.
2. 8
3. Tiny, Mega, Classic, and special purpose
4. Tiny
5. (a) 2K bytes
    (b) 128 bytes
6. (a) 8K bytes   (b) 32K bytes   (c) 2K bytes

CHAPTER 2

AVR ARCHITECTURE AND ASSEMBLY LANGUAGE PROGRAMMING

OBJECTIVES

Upon completion of this chapter, you will be able to:

>> List the registers of the AVR microcontroller
>> Examine the data memory of the AVR microcontroller
>> Perform simple operations, such as ADD and load, and access internal RAM memory in the AVR microcontroller
>> Explain the purpose of the status register
>> Discuss data RAM memory space allocation in the AVR microcontroller
>> Code simple AVR Assembly language instructions
>> Describe AVR data types and directives
>> Assemble and run a AVR program using AVR Studio
>> Describe the sequence of events that occur upon AVR power-up
>> Examine programs in AVR ROM code
>> Detail the execution of AVR Assembly language instructions
>> Understand the RISC and Harvard architectures of the AVR microcontroller
>> Examine the AVR's registers and data RAM using the AVR Studio simulator

CPUs use registers to store data temporarily. To program in Assembly language, we must understand the registers and architecture of a given CPU and the role they play in processing data. In Section 2.1 we look at the general purpose registers (GPRs) of the AVR. We demonstrate the use of GPRs with simple instructions such as LDI and ADD. Allocation of RAM memory inside the AVR and the addressing mode of the AVR are discussed in Sections 2.2 and 2.3. In Section 2.4 we discuss the status register's flag bits and how they are affected by arithmetic instructions. In Section 2.5 we look at some widely used Assembly language directives, pseudocode, and data types related to the AVR. In Section 2.6 we examine Assembly language and machine language programming and define terms such as mnemonics, opcode, operand, and so on. The process of assembling and creating a ready-to-run program for the AVR is discussed in Section 2.7. Step-by-step execution of an AVR program and the role of the program counter are examined in Section 2.8. The merits of RISC architecture are examined in Section 2.9. Section 2.10 discusses the AVR Studio.

SECTION 2.1: THE GENERAL PURPOSE REGISTERS IN THE AVR

CPUs use many registers to store data temporarily. To program in Assembly language, we must understand the registers and architecture of a given CPU and the role they play in processing data. In this section we look at the general purpose registers (GPRs) of the AVR and we demonstrate the use of GPRs with simple instructions such as LDI and ADD.

AVR microcontrollers have many registers for arithmetic and logic operations. In the CPU, registers are used to store information temporarily. That information could be a byte of data to be processed, or an address pointing to the data to be fetched. The vast majority of AVR registers are 8-bit registers. In the AVR there is only one data type: 8-bit. The 8 bits of a register are shown in the diagram below. These range from the MSB (most-significant bit) D7

D7	D6	D5	D4	D3	D2	D1	D0

R0
R1
R2
⋮
R14
R15
R16
R17
R18
⋮
R30
R31

to the LSB (least-significant bit) D0. With an 8-bit data type, any data larger than 8 bits must be broken into 8-bit chunks before it is processed.

In AVR there are 32 general purpose registers. They are R0–R31 and are located in the lowest location of memory address. See Figure 2-1. All of these registers are 8 bits.

The general purpose registers in AVR are the same as the accumulator in other microprocessors. They can be used by all arithmetic and logic instructions. To understand the use of the general purpose registers, we will show it in the context of two simple instructions: LDI and ADD.

Figure 2-1. GPRs

LDI instruction

Simply stated, the LDI instruction copies 8-bit data into the general purpose registers. It has the following format:

```
LDI Rd,K     ;load Rd (destination) with Immediate value K
             ;d must be between 16 and 31
```

K is an 8-bit value that can be 0–255 in decimal, or 00–FF in hex, and Rd is R16 to R31 (any of the upper 16 general purpose registers). The I in LDI stands for "immediate." If we see the word "immediate" in any instruction, we are dealing with a value that must be provided right there with the instruction. The following instruction loads the R20 register with a value of 0x25 (25 in hex).

```
LDI R20,0x25              ;load R20 with 0x25 (R20 = 0x25)
```

The following instruction loads the R31 register with the value 0x87 (87 in hex).

```
LDI R31,0x87              ;load 0x87 into R31   (R31 = 0x87)
```

The following instruction loads R25 with the value 0x15 (15 in hex and 21 in decimal).

```
LDI R25,0x79              ;load 0x79 into R25  (R25 = 0x79)
```

> **Note:** We cannot load values into registers R0 to R15 using the LDI instruction. For example, the following instruction is not valid:
> ```
> LDI R5,0x99              ;invalid instruction
> ```

Notice the position of the source and destination operands. As you can see, the LDI loads the right operand into the left operand. In other words, the destination comes first.

To write a comment in Assembly language we use ';'. It is the same as '//' in C language, which causes the remainder of the line of code to be ignored. For instance, in the above examples the expressions mentioned after ';' just explain the functionality of the instructions to you, and do not have any effects on the execution of the instructions.

When programming the GPRs of the AVR microcontroller with an immediate value, the following points should be noted:

1. If we want to present a number in hex, we put a dollar sign ($) or a 0x in front of it. If we put nothing in front of a number, it is in decimal. For example, in "LDI R16,50", R16 is loaded with 50 in decimal, whereas in "LDI R16,0x50", R16 is loaded with 50 in hex.

2. If values 0 to F are moved into an 8-bit register such as GPRs, the rest of the bits are assumed to be all zeros. For example, in "LDI R16,0x5" the result will be R16 = 0x05; that is, R16 = 00000101 in binary.

3. Moving a value larger than 255 (FF in hex) into the GPRs will cause an error.
   ```
   LDI    R17, 0x7F2  ;ILLEGAL $7F2 > 8 bits ($FF)
   ```

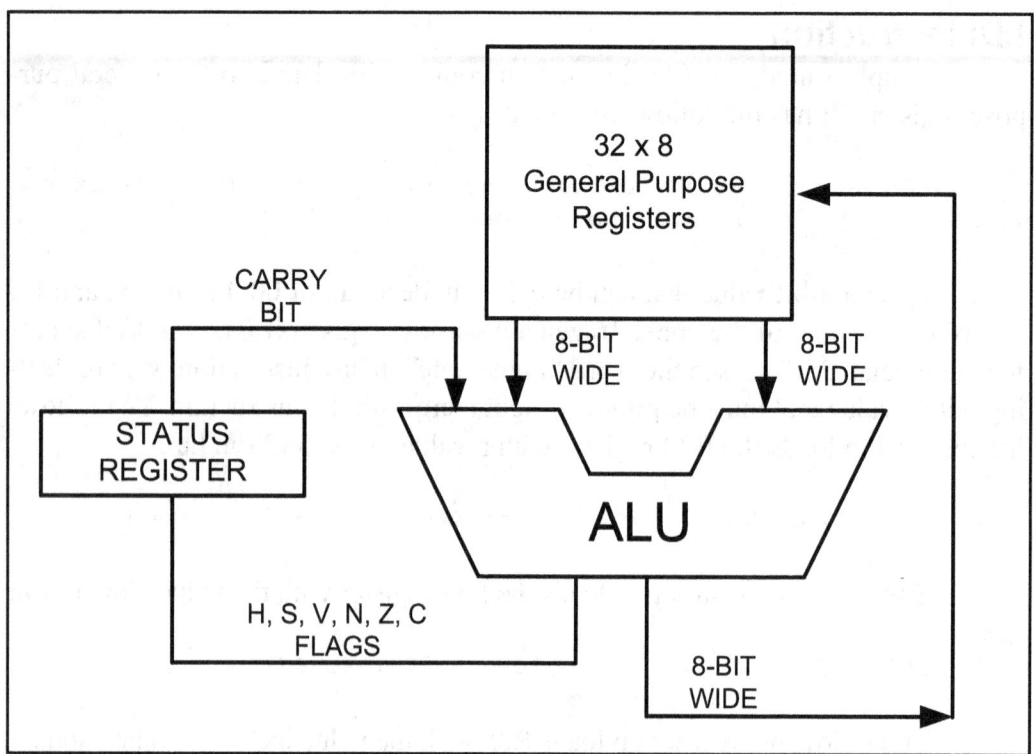

Figure 2-2. AVR General Purpose Registers and ALU

ADD instruction

The ADD instruction has the following format:

```
ADD   Rd,Rr  ;ADD Rr to Rd and store the result in Rd
```

The ADD instruction tells the CPU to add the value of Rr to Rd and put the result back into the Rd register. To add two numbers such as 0x25 and 0x34, one can do the following:

```
LDI R16,0x25        ;load 0x25 into R16
LDI R17,0x34        ;load 0x34 into R17
ADD R16,R17         ;add value R17 to R16  (R16 = R16 + R17)
```

Executing the above lines results in R16 = 0x59 (0x25 + 0x34 = 0x59)

Figure 2-2 shows the general purpose registers (GPRs) and the ALU in AVR. The affect of arithmetic and logic operations on the status register will be discussed in Section 2.4.

Review Questions

1. Write instructions to move the value 0x34 into the R29 register.
2. Write instructions to add the values 0x16 and 0xCD. Place the result in the R19 register.
3. True or false. No value can be moved directly into the GPRs.
4. What is the largest hex value that can be moved into an 8-bit register? What is the decimal equivalent of that hex value?
5. The vast majority of registers in the AVR are _____-bit.

SECTION 2.2: THE AVR DATA MEMORY

In AVR microcontrollers there are two kinds of memory space: code memory space and data memory space. Our program is stored in code memory space, whereas the data memory stores data. We will examine the code memory space in Section 2.8. In this section, we will discuss the data memory space. The data memory is composed of three parts: GPRs (general purpose registers), I/O memory, and internal data SRAM. See Figure 2-3.

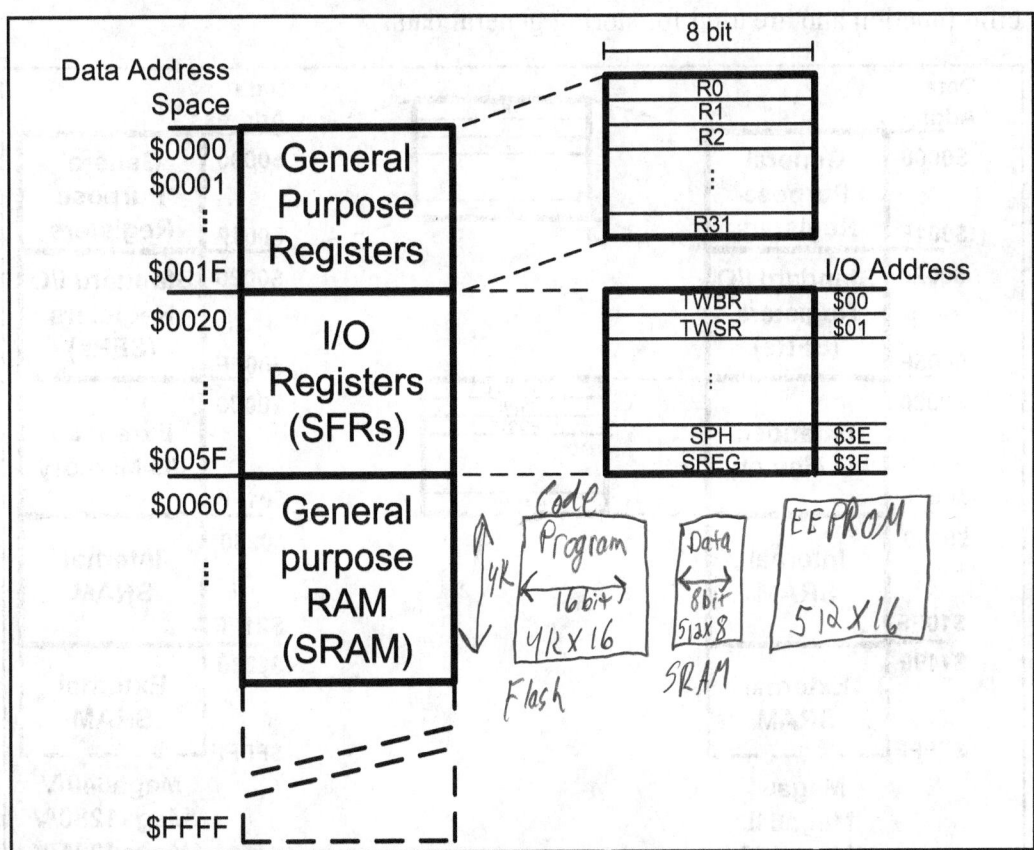

Figure 2-3. The Data Memory for AVRs with No Extended I/O Memory

GPRs (general purpose registers)

As we discussed in the last section, the GPRs use 32 bytes of data memory space. They always take the address location $00–$1F in the data memory space, regardless of the AVR chip number. See Figure 2-3.

I/O memory (SFRs)

The I/O memory is dedicated to specific functions such as status register, timers, serial communication, I/O ports, ADC, and so on. The function of each I/O memory location is fixed by the CPU designer at the time of design because it is used for control of the microcontroller or peripherals. The AVR I/O memory is made of 8-bit registers. The number of locations in the data memory set aside for I/O memory depends on the pin numbers and peripheral functions supported by that chip, although the number can vary from chip to chip even among members of the same family. However, all of the AVRs have at least 64 bytes of I/O memory locations. This 64-byte section is called *standard I/O memory*. In AVRs with more than 32 I/O pins (e.g., ATmega64, ATmega128, and ATmega256) there is also an extended I/O memory, which contains the registers for controlling the extra ports and the extra peripherals. See Figures 2-3 and 2-4. In other microcontrollers the I/O registers are called *SFRs (special function registers)* since each one is dedicated to a specific function. In contrast to SFRs, the GPRs do not have any specific function and are used for storing general data.

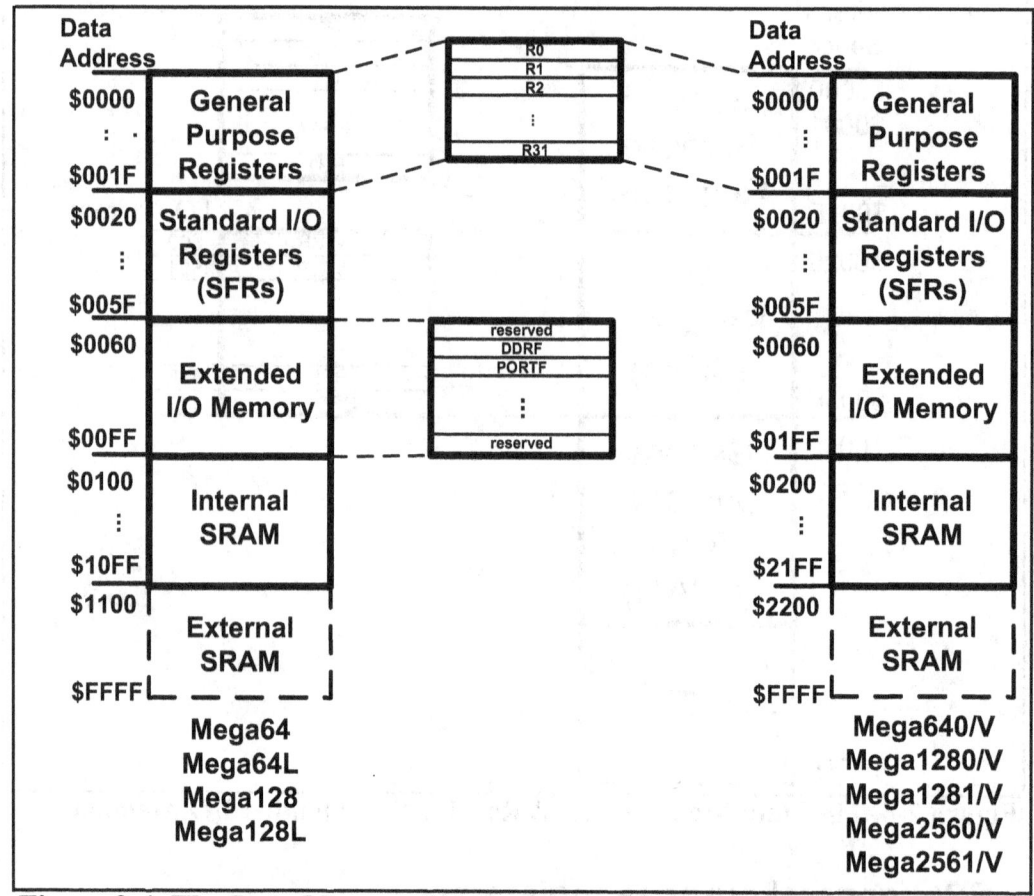

Figure 2-4. The Data Memory for the AVRs with Extended I/O Memory

Internal data SRAM

Internal data SRAM is widely used for storing data and parameters by AVR programmers and C compilers. Generally, this is called *scratch pad*. Each location of the SRAM can be accessed directly by its address. We will use these locations in future chapters to store data brought into the CPU via I/O and serial ports. Each location is 8 bits wide and can be used to store any data we want as long as it is 8-bit. Again, the size of SRAM can vary from chip to chip, even among members of the same family. See Table 2-1 for a comparison of the data memories of various AVR chips. Also, see Figure 2-4.

SRAM vs. EEPROM in AVR chips

The AVR has an EEPROM memory that is used for storing data. As you saw in Chapter 0, EEPROM does not lose its data when power is off, whereas SRAM does. So, the EEPROM is used for storing data that should rarely be changed and should not be lost when the power is off (e.g., options and settings); whereas the SRAM is used for storing data and parameters that are changed frequently. The three parts of the data memory (GPRs, SFRs, and the internal SRAM) are made of SRAM. The EEPROM memory of AVR chips is covered in Chapter 6.

In AVR datasheets, EEPROM refers to the EEPROM's size, and SRAM is the internal SRAM size. By adding the sizes of GPR, SFRs (I/O registers), and SRAMs we get the data memory size. See Table 2-1.

Table 2-1: Data Memory Size for AVR Chips

	Data Memory (Bytes)	=	I/O Registers (Bytes)	+	SRAM (Bytes)	+	General Purpose Register
ATtiny25	224		64		128		32
ATtiny85	608		64		512		32
ATmega8	1120		64		1024		32
ATmega16	1120		64		1024		32
ATmega32	2144		64		2048		32
ATmega128	4352		64+160		4096		32
ATmega2560	8704		64+416		8192		32

Extracted from http://www.atmel.com

Review Questions

1. True or false. The I/O registers are used for storing data.
2. The GPRs together with I/O registers and SRAM are called_____.
3. The I/O registers in AVR are _____-bit.
4. The data memory space in AVR is divided into _____ parts.
5. The data memory space in AVR can be a maximum of _____ bytes.
6. The standard I/O memory space in AVR is _____ bytes.

SECTION 2.3: USING INSTRUCTIONS WITH THE DATA MEMORY

The instructions we have used so far worked with the immediate (constant) value of K and the GPRs. They also used the GPRs as their destination. We saw simple examples of using LDI and ADD earlier in Section 2.1. The AVR allows direct access to other locations in the data memory. In this section we show the instructions accessing various locations of the data memory. This is one of the most important sections in the book for mastering the topic of AVR Assembly language programming.

LDS instruction (LoaD direct from data Space)

```
LDS    Rd, K   ;load Rd with the contents of location K (0 ≤ d ≤ 31)
               ;K is an address between $0000 to $FFFF
```

The LDS instruction tells the CPU to load (copy) one byte from an address in the data memory to the GPR. After this instruction is executed, the GPR will have the same value as the location in the data memory. The location in the data memory can be in any part of the data space; it can be one of the I/O registers, a location in the internal SRAM, or a GPR. For example, the "LDS R20,0x1" instruction will copy the contents of location 1 (in hex) into R20. As you can see in Figure 2-3, location 1 of the data memory is in the GPR part, and it is the address of R1. So, the instruction copies R1 to R20.

The following instruction loads R5 with the contents of location 0x200. As you can see in Figure 2-3, 0x200 is located in the internal SRAM:

```
LDS R5,0x200 ;load R5 with the contents of location $200
```

The following program adds the contents of location 0x300 to location 0x302. To do so, first it loads R0 with the contents of location 0x300 and R1 with the contents of location 0x302, then adds R0 to R1:

```
LDS    R0, 0x300    ;R0 = the contents of location 0x300
LDS    R1, 0x302    ;R1 = the contents of location 0x302
ADD    R1, R0       ;add R0 to R1
```

You can see the execution of "LDS R0,0x300" and "LDS R1,0x302" instructions in Figure 2-5. Figure 2-6 shows the contents of R0, R1 and locations 300 and 302 of data memory before and after the execution of each of the instructions, assuming that locations $300 and $302 contain a and β, respectively.

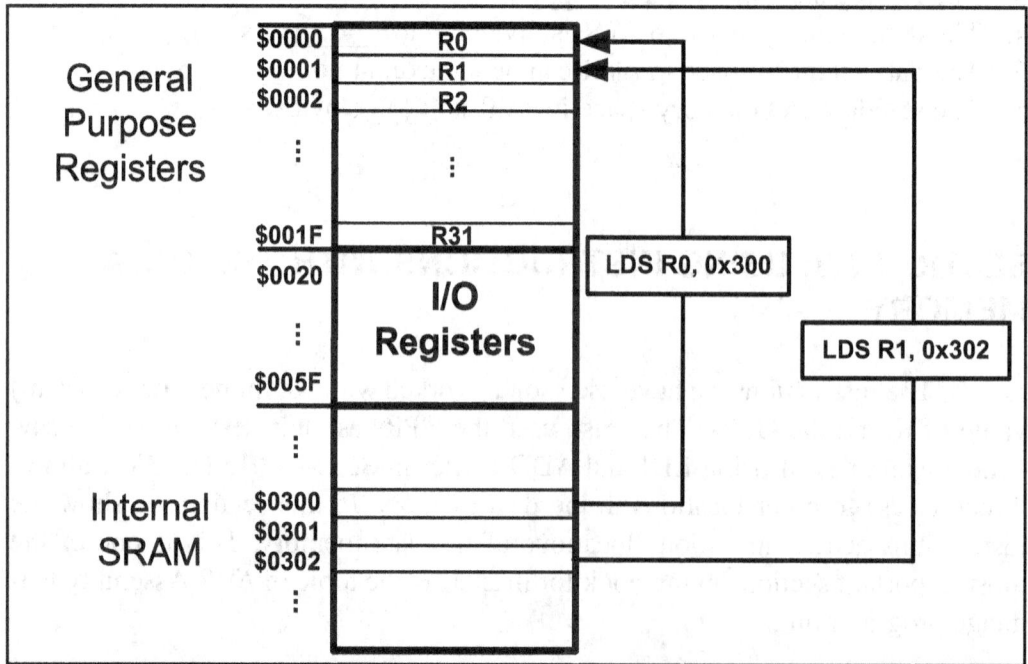

Figure 2-5. Execution of "LDS R0,0x300" and "LDS R1,0x302" Instructions

	R0	R1	Loc $300	Loc $302
Before LDS R0,0x300	?	?	α	β
After LDS R0,0x300	α	?	α	β
After LDS R1,0x302	α	β	α	β
After ADD R0, R1	α + β	β	α	β

Figure 2-6. The Contents of R0, R1, and Locations $300 and $302

STS instruction (STore direct to data Space)

```
STS    K, Rr ;store register into location K
              ;K is an address between $0000 to $FFFF
```

The STS instruction tells the CPU to store (copy) the contents of the GPR to an address location in the data memory space. After this instruction is executed, the location in the data space will have the same value as the GPR. The location can be in any part of the data memory space; it can be one of the I/O registers, a location in the SRAM, or a GPR. For example, the "STS 0x1,R10" instruction will copy the contents of R10 into location 1. As you can see in Figure 2-3, location 1 of the data memory is in the GPR part, and it is the address of R1. So, the instruction copies R10 to R1.

The following instruction stores the contents of R25 to location 0x230. As you can see in Figure 2-3, 0x230 is located in the internal SRAM:

```
STS 0x230, R25    ;store R25 to data space location 0x230
```

The following program first loads the R16 register with value 0x55, then moves this value around to I/O registers of ports B, C, and D. As shown in Figure 2-7, the addresses of PORTB, PORTC, and PORTD are 0x38, 0x35, and 0x32, respectively:

```
LDI    R16, 0x55  ;R16 = 55 (in hex)
STS    0x38, R16   ;copy R16 to Port B (PORTB = 0x55)
STS    0x35, R16   ;copy R16 to Port C (PORTC = 0x55)
STS    0x32, R16   ;copy R16 to Port D (PORTD = 0x55)
```

As we saw in Figure 2-3, PORTB, PORTC, and PORTD are part of the special function registers in the I/O memory. They can be connected to the I/O pins of the AVR microcontroller as we will see in Chapter 4. We can also store the contents of a GPR into any location in the SRAM region of the data space. The following program will put 0x99 into locations 0x200–0x203 of the SRAM region in the data memory:

Address	Data
$200	0x99
$201	0x99
$202	0x99
$203	0x99

```
LDI    R20, 0x99   ;R20 = 0x99
STS    0x200, R20  ;store R20 in loc 0x200
STS    0x201, R20  ;store R20 in loc 0x201
STS    0x202, R20
STS    0x203, R20  ;see the Mem. contents->
```

Notice that you cannot copy (store) an immediate value directly into the SRAM location in the AVR. This must be done via the GPRs.

The following program adds the contents of location 0x220 to location 0x221, and stores the result in location 0x221:

```
LDS    R30, 0x220   ;load R30 with the contents of location 0x220
LDS    R31, 0x221   ;load R31 with the contents of location 0x221
ADD    R31, R30     ;add R30 to R31
STS    0x221, R31   ;store R31 to data space location 0x221
```

See Examples 2-1 and 2-2.

IN instruction (IN from I/O location)

```
IN     Rd, A ;load an I/O location to the GPR (0 ≤ d ≤31),(0 ≤ A ≤ 63)
```

The IN instruction tells the CPU to load one byte from an I/O register to the GPR. After this instruction is executed, the GPR will have the same value as the I/O register. For example, the "IN R20,0x16" instruction will copy the contents of location 16 (in hex) of the I/O memory into R20. As you can see in Figure 2-7, each location in I/O memory has two addresses: I/O address and data memory address. Each location in the data memory has a unique address called the *data memory address*. Each I/O register has a relative address in comparison to the beginning of the I/O memory; this address is called the *I/O address*. See Figure 2-3. You see the list of I/O registers in Figure 2-7.

Address		Name	Address		Name	Address		Name
Mem.	I/O		Mem.	I/O		Mem.	I/O	
$20	$00	TWBR	$36	$16	PINB	$4B	$2B	OCR1AH
$21	$01	TWSR	$37	$17	DDRB	$4C	$2C	TCNT1L
$22	$02	TWAR	$38	$18	PORTB	$4D	$2D	TCNT1H
$23	$03	TWDR	$39	$19	PINA	$4E	$2E	TCCR1B
$24	$04	ADCL	$3A	$1A	DDRA	$4F	$2F	TCCR1A
$25	$05	ADCH	$3B	$1B	PORTA	$50	$30	SFIOR
$26	$06	ADCSRA	$3C	$1C	EECR	$51	$31	OCDR
$27	$07	ADMUX	$3D	$1D	EEDR			OSCCAL
$28	$08	ACSR	$3E	$1E	EEARL	$52	$32	TCNT0
$29	$09	UBRRL	$3F	$1F	EEARH	$53	$33	TCCR0
$2A	$0A	UCSRB	$40	$20	UBRRC	$54	$34	MCUCSR
$2B	$0B	UCSRA			UBRRH	$55	$35	MCUCR
$2C	$0C	UDR	$41	$21	WDTCR	$56	$36	TWCR
$2D	$0D	SPCR	$42	$22	ASSR	$57	$37	SPMCR
$2E	$0E	SPSR	$43	$23	OCR2	$58	$38	TIFR
$2F	$0F	SPDR	$44	$24	TCNT2	$59	$39	TIMSK
$30	$10	PIND	$45	$25	TCCR2	$5A	$3A	GIFR
$31	$11	DDRD	$46	$26	ICR1L	$5B	$3B	GICR
$32	$12	PORTD	$47	$27	ICR1H	$5C	$3C	OCR0
$33	$13	PINC	$48	$28	OCR1BL	$5D	$3D	SPL
$34	$14	DDRC	$49	$29	OCR1BH	$5E	$3E	SPH
$35	$15	PORTC	$4A	$2A	OCR1AL	$5F	$3F	SREG

Note: Although memory address $20-$5F is set aside for I/O registers (SFR) we can access them as I/O locations with addresses starting at $00.

Figure 2-7. I/O Registers of the ATmega32 and Their Data Memory Address Locations

Example 2-1

State the contents of RAM locations $212 to $216 after the following program is executed:

```
LDI    R16, 0x99    ;load R16 with value 0x99
STS    0x212, R16
LDI    R16, 0x85    ;load R16 with value 0x85
STS    0x213, R16
LDI    R16, 0x3F    ;load R16 with value 0x3F
STS    0x214, R16
LDI    R16, 0x63    ;load R16 with value 0x63
STS    0x215, R16
LDI    R16, 0x12    ;load R16 with value 0x12
STS    0x216, R16
```

Solution:

After the execution of STS 0x212, R16 data memory location $212 has value 0x99; after the execution of STS 0x213, R16 data memory location $213 has value 0x85; after the execution of STS 0x214, R16 data memory location $214 has value 0x3F; after the execution of STS 0x215, R16 data memory location $215 has value 0x63; and so on, as shown in the chart.

Address	Data
$212	0x99
$213	0x85
$214	0x3F
$215	0x63
$216	0x12

Example 2-2

State the contents of R20, R21, and data memory location 0x120 after the following program:

```
LDI    R20, 5       ;load R20 with 5
LDI    R21, 2       ;load R21 with 2
ADD    R20, R21     ;add R21 to R20
ADD    R20, R21     ;add R21 to R20
STS    0x120, R20   ;store in location 0x120 the contents of R20
```

Solution:

The program loads R20 with value 5. Then it loads R21 with value 2. Then it adds the R21 register to R20 twice. At the end, it stores the result in location 0x120 of data memory.

Location	Data		Location	Data		Location	Data		Location	Data		Location	Data
R20	5		R20	5		R20	7		R20	9		R20	9
R21			R21	2		R21	2		R21	2		R21	2
0x120			0x120			0x120			0x120			0x120	9

After	After	After	After	After
LDI R20, 5	LDI R21, 2	ADD R20, R21	ADD R20, R21	STS 0x120, R20

In the IN instruction, the I/O registers are referred to by their I/O addresses. For example, the "IN R20,0x16" instruction will copy the contents of location $16 of the I/O memory (whose data memory address is 0x36) into R20. As shown in Figure 2-7, I/O address 0x16 belongs to PINB, so the instruction copies the contents of PINB to R20.

The following instruction loads R19 with the contents of location 0x10 of the I/O memory:

```
IN R19,0x10        ;load R19 with location $10 (R19 = PIND)
```

To work with the I/O registers more easily, we can use their names instead of their I/O addresses. For example, the following instruction loads R19 with the contents of PIND:

```
IN R19,PIND        ;load R19 with PIND
```

Notice that to be able to use the names of the I/O addresses instead of the I/O addresses we should include the proper header files, as discussed in Section 2.5. The details of I/O ports are discussed in Chapter 4.

The following program adds the contents of PIND to PINB, and stores the result in location 0x300 of the data memory:

```
IN     R1,PIND      ;load R1 with PIND
IN     R2,PINB      ;load R2 with PINB
ADD    R1, R2       ;R1 = R1 + R2
STS    0x300, R1    ;store R1 to data space location $300
```

IN vs. LDS

As we mentioned earlier, we can use the LDS instruction to copy the contents of a memory location to a GPR. This means that we can load an I/O register into a GPR, using the LDS instruction. So, what is the advantage of using the IN instruction for reading the contents of I/O registers over using the LDS instruction? The IN instruction has the following advantages:

1. The CPU executes the IN instruction faster than LDS. As you will see in Chapter 3, the IN instruction lasts 1 machine cycle, whereas LDS lasts 2 machine cycles.
2. The IN is a 2-byte instruction, whereas LDS is a 4-byte instruction. This means that the IN instruction occupies less code memory.
3. When we use the IN instruction, we can use the names of the I/O registers instead of their addresses.
4. The IN instruction is available in all of the AVRs, whereas LDS is not implemented in some of the AVRs.

Notice that in using the IN instruction we can access only the standard I/O memory, while we can access all parts of the data memory using the LDS instruction.

OUT instruction (OUT to I/O location)

```
OUT A, Rr ;store register to I/O location (0 ≤ r ≤ 31),(0 ≤A ≤ 63)
```

The OUT instruction tells the CPU to store the GPR to the I/O register. After the instruction is executed, the I/O register will have the same value as the GPR. For example, the "OUT PORTD,R10" instruction will copy the contents of R10 into PORTD (location 12 of the I/O memory).

Notice that in the OUT instruction, the I/O registers are referred to by their I/O addresses (like the IN instruction).

The following program copies 0xE6 to the SPL register:

```
LDI    R20,0xE6    ;load R20 with 0xE6
OUT    SPL, R20    ;out R20 to SPL
```

We must remember that we cannot copy an immediate value to an I/O register nor to an SRAM location.

The following program copies PIND to PORTA:
```
IN     R0, PIND    ;load R20 with the contents of I/O reg PIND
OUT    PORTA, R0   ;out R20 to PORTA
```

In Example 2-3 we use JMP to repeat an action indefinitely. JMP is similar to "goto" in the C language. We will study looping in Chapter 3.

Example 2-3

Write a program to get data from the PINB and send it to the I/O register of PORT C continuously.

Solution:

```
AGAIN:IN    R16, PINB    ;bring data from PortB into R16
      OUT   PORTC,R16    ;send it to Port C
      JMP   AGAIN        ;keep doing it forever
```

MOV instruction

The MOV instruction is used to copy data among the GPR registers of R0–R31. It has the following format:

```
MOV    Rd,Rr           ;Rd = Rr (copy Rr to Rd)
                       ;Rd and Rr can be any of the GPRs
```

For example, the following instruction copies the contents of R20 to R10:

```
MOV    R10,R20         ;R10 = R20
```

For instance, if R20 contains 60, after execution of the above instruction both R20 and R10 will contain 60.

More ALU instructions involving the GPRs

The following program adds 0x19 to the contents of location 0x220 and stores the result in location 0x221:

```
LDI    R20, 0x19    ;load R20 with 0x19
LDS    R21, 0x220   ;load R21 with the contents of location 0x220
ADD    R21, R20      ;R21 = R21 + R20
STS    0x221, R21   ;store R21 to location 0x221
```

INC instruction

```
INC    Rd      ;increment the contents of Rd by one (0 ≤ d ≤ 31)
```

The INC instruction increments the contents of Rd by 1. For example, the following instruction adds 1 to the contents of R2:

```
INC    R2               ;R2 = R2 + 1
```

The following program increments the contents of data memory location 0x430 by 1:

```
LDS    R20, 0x430   ;R20 = contents of location 0x430
INC    R20          ;R20 = R20 + 1
STS    0x430, R20   ;store R20 to location 0x430
```

SUB instruction

The SUB instruction has the following format:

```
SUB   Rd,Rr          ;Rd = Rd - Rr
```

The SUB instruction tells the CPU to subtract the value of Rr from Rd and put the result back into the Rd register. To subtract 0x25 from 0x34, one can do the following:

```
LDI    R20, 0x34   ;R20 = 0x34
LDI    R21, 0x25   ;R20 = 0x25
SUB    R20, R21    ;R20 = R20 - R21
```

The following program subtracts 5 from the contents of location 0x300 and stores the result in location 0x320:

```
LDS    R0, 0x300    ;R0 = contents of location 0x300
LDI    R16, 0x5     ;R16 = 0x5
SUB    R0, R16      ;R0 = R0 - R16
STS    0x320,R0     ;store the contents of R0 to location 0x320
```

The following program decrements the contents of R10, by 1:

```
LDI    R16, 0x1     ;load 1 to R16
SUB    R10, R16     ;R10 = R10 - R16
```

Table 2-2: ALU Instructions Using Two GPRs

Instruction		
ADD	Rd, Rr	ADD Rd and Rr
ADC	Rd, Rr	ADD Rd and Rr with Carry
AND	Rd, Rr	AND Rd with Rr
EOR	Rd, Rr	Exclusive OR Rd with Rr
OR	Rd, Rr	OR Rd with Rr
SBC	Rd, Rr	Subtract Rr from Rd with carry
SUB	Rd, Rr	Subtract Rr from Rd without carry

Rd and Rr can be any of the GPRs. See Chapter 5 for examples of the instructions in Table 2-2.

DEC instruction

The DEC instruction has the following format:

```
DEC   Rd              ;Rd = Rd - 1
```

The DEC instruction decrements (subtracts 1 from) the contents of Rd and puts the result back into the Rd register. For example, the following instruction subtracts 1 from the contents of R10:

```
DEC   R10             ;R10 = R10 - 1
```

In the following program, we put the value 3 into R30. Then the value in R30 is decremented.

```
LDI   R30, 3          ;R30 = 3
DEC   R30             ;R30 has 2
DEC   R30             ;R30 has 1
DEC   R30             ;R30 has 0
```

In the next chapter we will use the DEC instruction for looping.

Table 2-3: Some Instructions Using a GPR as Operand

Instruction		
CLR	Rd	Clear Register Rd
INC	Rd	Increment Rd
DEC	Rd	Decrement Rd
COM	Rd	One's Complement Rd
NEG	Rd	Negative (two's complement) Rd
ROL	Rd	Rotate left Rd through carry
ROR	Rd	Rotate right Rd through carry
LSL	Rd	Logical Shift Left Rd
LSR	Rd	Logical Shift Right Rd
ASR	Rd	Arithmetic Shift Right Rd
SWAP	Rd	Swap nibbles in Rd

Chapters 3 through 6 will show how to use the instructions in Table 2-3.

COM instruction

The "COM Rd" instruction complements (inverts) the contents of Rd and places the result back into the Rd register. In the following program, we put 0x55 into R16 and then send it to the SFR location of PORTB. Then the content of R16 is complemented, which becomes AA in hex. The 01010101 (0x55) is inverted and becomes 10101010 (0xAA).

```
LDI   R16,0x55    ;R16 = 0x55
OUT   PORTB, R16  ;copy R16 to Port B SFR (PB = 0x55)
COM   R16         ;complement R16         (R16 = 0xAA)
OUT   PORTB, R16  ;copy R16 to Port B SFR (PB = 0xAA)
```

Examine Example 2-4.

Example 2-4

Write a simple program to toggle the I/O register of PORT B continuously forever.

Solution:

```
      LDI   R20, 0x55   ;R20 = 0x55
      OUT   PORTB, R20  ;move R20 to Port B SFR (PB = 0x55)
L1:   COM   R20         ;complement R20
      OUT   PORTB, R20  ;move R20 to Port B SFR
      JMP   L1          ;repeat forever (see Chapter 3 for JMP)
```

The above concepts are important and must be understood since there are a large number of instructions with these formats.

Regarding Tables 2-2 and 2-3 the following points must be noted:

1. The instructions in Table 2-2 operate on two GPR registers of source (Rr) and destination (rd) and then place the result in the destination register (Rd)
2. The instructions in Table 2-3 operate on a single GPR register and place the result in the same register.

Review Questions

1. True or false. No value can be loaded directly into internal SRAM.
2. Write instructions to load value 0x95 into the *SPL* I/O register.
3. Write instructions to add 2 to the contents of R18.
4. Write instructions to add the values 0x16 and 0xCD. Place the result in location 0x400 of the data memory.
5. What is the largest hex value that can be moved into a location in the data memory? What is the decimal equivalent of the hex value?
6. "ADD R16, R3" puts the result in _____ .
7. What does "OUT OCR0, R23" do?
8. What is wrong with "STS OCR0, R23"? What does it do?

SECTION 2.4: AVR STATUS REGISTER

Like all other microprocessors, the AVR has a flag register to indicate arithmetic conditions such as the carry bit. The flag register in the AVR is called the *status register (SReg)*. In this section, we discuss various bits of this register and provide some examples of how it is altered. Chapters 3 and 5 show how the flag bits of the status register are used.

AVR status register

The status register is an 8-bit register. It is also referred to as the *flag register*. See Figure 2-8 for the bits of the status register. The bits C, Z, N, V, S, and H are called *conditional flags*, meaning that they indicate some conditions that result after an instruction is executed. Each of the conditional flags can be used to perform a conditional branch (jump), as we will see in Chapters 3 and 5.

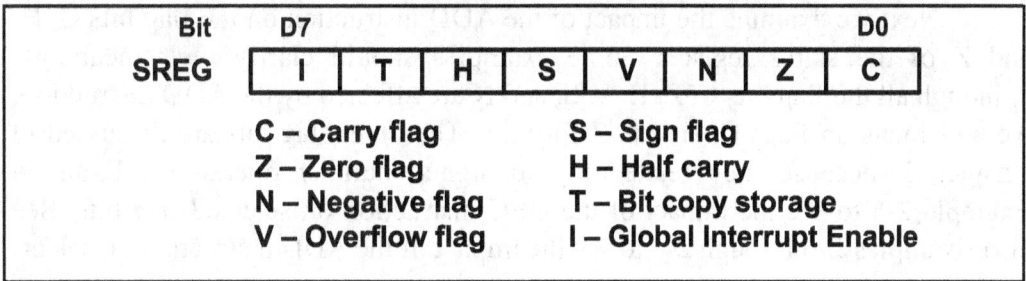

Figure 2-8. Bits of Status Register (SREG)

The following is a brief explanation of the flag bits of the status register. The impact of instructions on this register is then discussed.

C, the carry flag

This flag is set whenever there is a carry out from the D7 bit. This flag bit is affected after an 8-bit addition or subtraction. Chapter 5 shows how the carry flag is used.

Z, the zero flag

The zero flag reflects the result of an arithmetic or logic operation. If the result is zero, then Z = 1. Therefore, Z = 0 if the result is not zero. See Chapter 3 to see how we use the Z flag for looping.

N, the negative flag

Binary representation of signed numbers uses D7 as the sign bit. The negative flag reflects the result of an arithmetic operation. If the D7 bit of the result is zero, then N = 0 and the result is positive. If the D7 bit is one, then N = 1 and the result is negative. The negative and V flag bits are used for the signed number arithmetic operations and are discussed in Chapter 5.

V, the overflow flag

This flag is set whenever the result of a signed number operation is too large, causing the high-order bit to overflow into the sign bit. In general, the carry flag is used to detect errors in unsigned arithmetic operations while the overflow flag is used to detect errors in signed arithmetic operations. The V and N flag bits are used for signed number arithmetic operations and are discussed in Chapter 5.

S, the Sign bit

This flag is the result of Exclusive-ORing of N and V flags. See Chapter 5 for more information.

H, Half carry flag

If there is a carry from D3 to D4 during an ADD or SUB operation, this bit is set; otherwise, it is cleared. This flag bit is used by instructions that perform BCD (binary coded decimal) arithmetic. In some microprocessors this is called the AC flag (Auxiliary Carry flag). See Chapter 5 for more information.

The T flag bit is discussed in Chapter 6 while Chapter 10 covers the I flag.

ADD instruction and the status register

Next we examine the impact of the ADD instruction on the flag bits C, H, and Z of the status register. Some examples should clarify their meanings. Although all the flag bits C, Z, H, V, S, and N are affected by the ADD instruction, we will focus on flags C, H, and Z for now. The other flag bits are discussed in Chapter 5, because they relate only to signed number operations. Examine Example 2-5 to see the impact of the DEC instruction on selected flag bits. See also Examples 2-6 through 2-8 to see the impact of the ADD instruction on selected flag bits.

Example 2-5

Show the status of the Z flag during the execution of the following program:

```
    LDI    R20,4        ;R20 = 4
    DEC    R20          ;R20 = R20 - 1
    DEC    R20          ;R20 = R20 - 1
    DEC    R20          ;R20 = R20 - 1
    DEC    R20          ;R20 = R20 - 1
```

Solution:

The Z flag is one when the result is zero. Otherwise, it is cleared (zero). Thus:

After	Value of R20	The Z flag
LDI R20, 4	4	0
DEC R20	3	0
DEC R20	2	0
DEC R20	1	0
DEC R20	0	1

Example 2-6

Show the status of the C, H, and Z flags after the addition of 0x38 and 0x2F in the following instructions:

```
LDI    R16, 0x38
LDI    R17, 0x2F
ADD    R16, R17    ;add R17 to R16
```

Solution:

$$
\begin{array}{rl}
\$38 & \quad 0011\ 1000 \\
+\ \underline{\$2F} & \quad \underline{0010\ 1111} \\
\$67 & \quad 0110\ 0111 \qquad R16 = 0x67
\end{array}
$$

C = 0 because there is no carry beyond the D7 bit.
H = 1 because there is a carry from the D3 to the D4 bit.
Z = 0 because the R16 (the result) has a value other than 0 after the addition.

Example 2-7

Show the status of the C, H, and Z flags after the addition of 0x9C and 0x64 in the following instructions:

```
LDI    R20, 0x9C
LDI    R21, 0x64
ADD    R20, R21    ;add R21 to R20
```

Solution:

$$
\begin{array}{rl}
\$9C & \quad 1001\ 1100 \\
+\ \underline{\$64} & \quad \underline{0110\ 0100} \\
\$100 & \quad 0000\ 0000 \qquad R20 = 00
\end{array}
$$

C = 1 because there is a carry beyond the D7 bit.
H = 1 because there is a carry from the D3 to the D4 bit.
Z = 1 because the R20 (the result) has value 0 in it after the addition.

Example 2-8

Show the status of the C, H, and Z flags after the addition of 0x88 and 0x93 in the following instructions:

```
LDI    R20, 0x88
LDI    R21, 0x93
ADD    R20, R21    ;add R21 to R20
```

Solution:

$$
\begin{array}{rl}
\$88 & \quad 1000\ 1000 \\
+\ \$\underline{93} & \quad \underline{1001\ 0011} \\
\$11B & \quad 0001\ 1011 \qquad R20 = 0x1B
\end{array}
$$

C = 1 because there is a carry beyond the D7 bit.
H = 0 because there is no carry from the D3 to the D4 bit.
Z = 0 because the R20 has a value other than 0 after the addition.

Not all instructions affect the flags

Some instructions affect all the six flag bits C, H, Z, S, V, and N (e.g., ADD). But some instructions affect no flag bits at all. The load instructions are in this category. And some instructions affect only some of the flag bits. The logic instructions (e.g., AND) are in this category.

Table 2-4 shows the instructions and the flag bits affected by them. Appendix A provides a complete list of all the instructions and their associated flag bits.

Table 2-4: Instructions That Affect Flag Bits

Instruction	C	Z	N	V	S	H
ADD	X	X	X	X	X	X
ADC	X	X	X	X	X	X
ADIW	X	X	X	X	X	
AND		X	X	X	X	
ANDI		X	X	X	X	
CBR		X	X	X	X	
CLR		X	X	X	X	
COM	X	X	X	X	X	
DEC		X	X	X	X	
EOR		X	X	X	X	
FMUL	X	X				
INC		X	X	X	X	
LSL	X	X	X	X		X
LSR	X	X	X	X		
OR		X	X	X	X	
ORI		X	X	X	X	
ROL	X	X	X	X		X
ROR	X	X	X	X		
SEN			1			
SEZ		1				
SUB	X	X	X	X	X	X
SUBI	X	X	X	X	X	X
TST		X	X	X	X	

Note: X can be 0 or 1. (See Chapter 5 for how to use these instructions.)

Flag bits and decision making

There are instructions that will make a conditional jump (branch) based on the status of the flag bits. Table 2-5 provides some of these instructions. Chapter 3 discusses the conditional branch instructions and how they are used.

Table 2-5: AVR Branch (Jump) Instructions Using Flag Bits

Instruction	Action
BRLO	Branch if C = 1
BRSH	Branch if C = 0
BREQ	Branch if Z = 1
BRNE	Branch if Z = 0
BRMI	Branch if N = 1
BRPL	Branch if N = 0
BRVS	Branch if V = 1
BRVC	Branch if V = 0

Review Questions

1. The flag register in the AVR is called the _____.
2. What is the size of the flag register in the AVR?
3. Find the C, Z, and H flag bits for the following code:

```
LDI    R20, 0x9F
LDI    R21, 0x61
ADD    R20, R21
```

4. Find the C, Z, and H flag bits for the following code:

```
LDI    R17, 0x82
LDI    R23, 0x22
ADD    R17, R23
```

5. Find the C, Z, and H flag bits for the following code:

```
LDI    R20, 0x67
LDI    R21, 0x99
ADD    R20, R21
```

SECTION 2.5: AVR DATA FORMAT AND DIRECTIVES

In this section we look at some widely used data formats and directives supported by the AVR assembler.

AVR data type

The AVR microcontroller has only one data type. It is 8 bits, and the size of each register is also 8 bits. It is the job of the programmer to break down data larger than 8 bits (00 to 0xFF, or 0 to 255 in decimal) to be processed by the CPU. For examples of how to process data larger than 8 bits, see Chapter 5. The data types used by the AVR can be positive or negative. A discussion of signed numbers is given in Chapter 5 also. The bit-addressable data is discussed in Chapters 4 and 6.

Data format representation

There are four ways to represent a byte of data in the AVR assembler. The numbers can be in hex, binary, decimal, or ASCII formats. The following are examples of how each works.

Hex numbers

There are two ways to show hex numbers:

1. Put 0x (or 0X) in front of the number like this: LDI R16, 0x99
2. Put $ in front of the number, like this: LDI R22, $99

We use both of these methods in this book, because many application notes out there use one of them and we need to get used to them.

Here are a few lines of code that use the hex format:

```
LDI    R28,$75      ;R28 = 0x75
SUBI   R28,0x11     ;R28 = 0x75 - 0x11 = 0x64
SUBI   R28,0X20     ;R28 = 0x64 - 0x20 = 0x44
ANDI   R28,0xF      ;R28 = 0x44 - 0x0F = 0x35
```

Binary numbers

There is only one way to represent binary numbers in an AVR assembler. It is as follows:

```
LDI R16,0b10011001      ;R16 = 10011001 or 99 in hex
```

The uppercase B will also work. Here are some examples of how to use it:

```
LDI    R23,0b00100101   ;R23 = $25
SUBI   R23,0B00010001   ;R23 = $25 - $11 = $14
```

Decimal numbers

To indicate decimal numbers in an AVR assembler we simply use the decimal (e.g., 12) and nothing before or after it. Here are some examples of how to use it:

```
LDI  R17, 12  ;R17 = 00001100 or 0C in hex
SUBI R17, 2   ;R17 = 12 - 2 = 10 where 10 is equal to 0x0A
```

ASCII characters

To represent ASCII data in an AVR assembler we use single quotes as follows:

```
LDI R23,'2'  ;R23 = 00110010 or 32 in hex (See Appendix F)
```

This is the same as other assemblers such as the 8051 and x86. Here are some more examples:

```
LDI   R20,'9';R20 = 0x39, which is hex number for ASCII '9'
SUBI R20,'1';R20 = 0x39 - 0x31 = 0x8
            ;(31 hex is for ASCII '1')
```

To represent a string, double quotes are used; and for defining ASCII strings (more than one character), we use the .DB (define byte) directive. We will see .DB usage in Chapter 6.

Assembler directives

While instructions tell the CPU what to do, directives (also called *pseudo-instructions*) give directions to the assembler. For example, the LDI and ADD instructions are commands to the CPU, but .EQU, .DEVICE, and .ORG are directives to the assembler. The following sections present some more widely used directives of the AVR and how they are used. The directives help us develop our program easier and make our program legible (more readable).

.EQU (equate)

This is used to define a constant value or a fixed address. The .EQU directive does not set aside storage for a data item, but associates a constant number with a data or an address label so that when the label appears in the program, its constant will be substituted for the label. The following uses .EQU for the counter constant, and then the constant is used to load the R21 register:

```
.EQU  COUNT = 0x25
...   ....
      LDI   R21, COUNT        ;R21 = 0x25
```

When executing the above instruction "LDI R21, COUNT", the register R21 will be loaded with the value 25H. What is the advantage of using .EQU? Assume that a constant (a fixed value) is used throughout the program, and the programmer wants to change its value everywhere. By the use of .EQU, the programmer can change it once and the assembler will change all of its occurrences throughout the program. This allows the programmer to avoid searching the entire program trying to find every occurrence.

We mentioned earlier that we can use the names of the I/O registers instead of their addresses (e.g., we can write "OUT PORTA,R20" instead of "OUT 0x1B,R20"). This is done with the help of the .EQU directive. In include files such as M32DEF.INC the I/O register names are associated with their addresses using the .EQU directive. For example, in M32DEF.INC the following pseudo-instruction exists, which associates 0x1B (the address of PORTB) with the PORTB.

```
.EQU  PORTB = 0x1B
```

.SET

This directive is used to define a constant value or a fixed address. In this regard, the .SET and .EQU directives are identical. The only difference is that the value assigned by the .SET directive may be reassigned later.

Using .EQU for fixed data assignment

To get more practice using .EQU to assign fixed data, examine the following:

```
         ;in hexadecimal
.EQU DATA1 = 0x39       ;one way to define hex value
.EQU DATA2 = $39        ;another way to define hex value

         ;in binary
```

```
.EQU DATA3 = 0b00110101    ;binary (35 in hex)

                ;in decimal
.EQU DATA4 = 39            ;decimal numbers (27 in hex)

                ;in ASCII
.EQU DATA5 = '2'           ;ASCII characters
```

We use .DB to allocate code ROM memory locations for fixed data such as ASCII strings. See Chapter 6 for more examples.

Using .EQU for SFR address assignment

.EQU is also widely used to assign SFR addresses. Examine the following code:

```
.EQU COUNTER = 0x00    ;counter value 00
.EQU PORTB = 0x18      ;SFR Port B address
LDI  R16, COUNTER      ;R16 = 0x00
OUT  PORTB, R16        ;Port B (loc 0x18) now has 00 too
```

Using .EQU for RAM address assignment

Another common usage of .EQU is for the address assignment of the internal SRAM. Examine the following rewrite of an earlier example using .EQU:

```
.EQU  SUM = 0x120      ;assign RAM loc to SUM
LDI   R20, 5           ;load R20 with 5
LDI   R21, 2           ;load R21 with 2
ADD   R20, R21         ;R20 = R20 + R21
ADD   R20, R21         ;R20 = R20 + R21
STS   SUM, R20         ;store the result in loc 0x120
```

This is especially helpful when the address needs to be changed in order to use a different AVR chip for a given project. It is much easier to refer to a name than a number when accessing RAM address locations.

.ORG (origin)

The .ORG directive is used to indicate the beginning of the address. It can be used for both code and data.

.INCLUDE directive

The .include directive tells the AVR assembler to add the contents of a file to our program (like the #include directive in C language). In Table 2-6, you see the files that you must include whenever you want to use any of the AVRs.

For example, when you want to use ATmega32, you must write the following instruction at the beginning of your program:

```
.INCLUDE "M32DEF.INC"
```

Table 2-6: Some of the Common AVRs and Their Include Files		
ATMEGA	**ATTINY**	**Special Purpose**
ATmega8 m8def.inc	ATtiny11 tn11def.inc	AT90CAN32 can32def.inc
ATmega16 m16def.inc	ATtiny12 tn12def.inc	AT90CAN64 can64def.inc
ATmega32 m32def.inc	ATtiny22 tn22def.inc	AT90PWM2 pwm2def.inc
ATmega64 m64def.inc	ATtiny44 tn44def.inc	AT90PWM3 pwm3def.inc
ATmega128 m128def.inc	ATtiny85 tn85def.inc	AT90USB646 usb646def.inc
ATmega256 m256def.inc		
ATmega2560 m2560def.inc		

Rules for labels in Assembly language

By choosing label names that are meaningful, a programmer can make a program much easier to read and maintain. There are several rules that names must follow. First, each label name must be unique. The names used for labels in Assembly language programming consist of alphabetic letters in both uppercase and lowercase, the digits 0 through 9, and the special characters question mark (?), period (.), at (@), underline (_), and dollar sign ($). The first character of the label must be an alphabetic character. In other words, it cannot be a number. Every assembler has some reserved words that must not be used as labels in the program. Foremost among the reserved words are the mnemonics for the instructions. For example, "LDI" and "ADD" are reserved because they are instruction mnemonics. In addition to the mnemonics there are some other reserved words. Check your assembler for the list of reserved words.

Review Questions

1. Give two ways for hex data representation in the AVR assembler.
2. Show how to represent decimal 99 in formats of (a) hex, (b) decimal, and (c) binary in the AVR assembler.
3. What is the advantage in using the .EQU directive to define a constant value?
4. Show the hex number value used by the following directives:
   (a) `.EQU ASC_DATA = '4'`   (b) `.EQU MY_DATA=0B00011111`
5. Give the value in R22 for the following:
   ```
   .EQU   MYCOUNT = 15
   LDI    R22, MYCOUNT
   ```
6. Give the value in data memory location 0x200 for the following:
   ```
   .EQU   MYCOUNT = 0x95
   .EQU   MYREG = 0x200
   LDI    R22, MYCOUNT
   STS    MYREG, R22
   ```
7. Give the value in data memory 0x63 for the following:
   ```
   .EQU MYDATA = 12
   .EQU MYREG = 0x63
   .EQU FACTOR = 0x10
   LDI R19, MYDATA
   ADD R19, FACTOR
   STS MYREG, R19
   ```

In this section we discuss Assembly language format and define some widely used terminology associated with Assembly language programming.

While the CPU can work only in binary, it can do so at a very high speed. It is quite tedious and slow for humans, however, to deal with 0s and 1s in order to program the computer. A program that consists of 0s and 1s is called *machine language*. In the early days of the computer, programmers coded programs in machine language. Although the hexadecimal system was used as a more efficient way to represent binary numbers, the process of working in machine code was still cumbersome for humans. Eventually, Assembly languages were developed, which provided mnemonics for the machine code instructions, plus other features that made programming faster and less prone to error. The term *mnemonic* is frequently used in computer science and engineering literature to refer to codes and abbreviations that are relatively easy to remember. Assembly language programs must be translated into machine code by a program called an *assembler*. Assembly language is referred to as a *low-level language* because it deals directly with the internal structure of the CPU. To program in Assembly language, the programmer must know all the registers of the CPU and the size of each, as well as other details.

Today, one can use many different programming languages, such as BASIC, Pascal, C, C++, Java, and numerous others. These languages are called *high-level languages* because the programmer does not have to be concerned with the internal details of the CPU. Whereas an *assembler* is used to translate an Assembly language program into machine code (sometimes also called *object code* or opcode for operation code), high-level languages are translated into machine code by a program called a *compiler*. For instance, to write a program in C, one must use a C compiler to translate the program into machine language. Next we look at AVR Assembly language format.

Structure of Assembly language

An Assembly language program consists of, among other things, a series of lines of Assembly language instructions. An Assembly language instruction consists of a mnemonic, optionally followed by one or two operands. The operands are the data items being manipulated, and the mnemonics are the commands to the CPU, telling it what to do with those items.

An Assembly language program (see Program 2-1) is a series of statements, or lines, which are either Assembly language instructions such as ADD and LDI, or statements called *directives*. While instructions tell the CPU what to do, directives (also called *pseudo-instructions*) give directions to the assembler. For example, in Program 2-1, while the LDI and ADD instructions are commands to the CPU, .ORG and .EQU are directives to the assembler. The directive .ORG tells the assembler to place the opcode at memory location 0, while .EQU introduces a new expression equal to a known expression.

An Assembly language instruction consists of four fields:

```
[ label:]     mnemonic   [ operands]   [ ;comment]
```

```
        ;AVR Assembly Language Program To Add Some Data.
        ;store SUM in SRAM location 0x300.

        .EQU  SUM   = 0x300      ;SRAM loc $300 for SUM

        .ORG 00                  ;start at address 0
        LDI R16, 0x25            ;R16 = 0x25
        LDI R17, $34             ;R17 = 0x34
        LDI R18, 0b00110001      ;R18 = 0x31
        ADD R16, R17             ;add R17 to R16
        ADD R16, R18             ;add R18 to R16
        LDI R17, 11              ;R17 = 0x0B
        ADD R16, R17             ;add R17 to R16
        STS SUM, R16             ;save the SUM in loc $300
HERE:   JMP HERE                 ;stay here forever
```

Program 2-1: Sample of an Assembly Language Program

Brackets indicate that a field is optional and not all lines have them. Brackets should not be typed in. Regarding the above format, the following points should be noted:

1. The label field allows the program to refer to a line of code by name. The label field cannot exceed a certain number of characters. Check your assembler for the rule.
2. The Assembly language mnemonic (instruction) and operand(s) fields together perform the real work of the program and accomplish the tasks for which the program was written. In Assembly language statements such as

```
    LDI    R23, $55
    ADD    R23, R19
    SUBI   R23, $67
```

ADD and LDI are the mnemonics that produce opcodes; the "$55" and "$67" are the operands. Instead of a mnemonic and an operand, these two fields could contain assembler pseudo-instructions, or directives. Remember that directives do not generate any machine code (opcode) and are used only by the assembler, as opposed to instructions that are translated into machine code (opcode) for the CPU to execute. In Program 2-1 the commands .ORG (origin) and .EQU are examples of directives. More of these pseudo-instructions are discussed in future chapters.
3. The comment field begins with a semicolon comment indicator ";". Comments may be at the end of a line or on a line by themselves. The assembler ignores comments, but they are indispensable to programmers. Although comments are optional, it is recommended that they be used to describe the program in a way that makes it easier for someone else to read and understand.
4. Notice the label "HERE" in the label field in Program 2-1. In the JMP the AVR is told to stay in this loop indefinitely. If your system has a monitor program you do not need this line and should delete it from your program. In Section 2.7 we will see how to create a ready-to-run program.

Review Questions

1. What is the purpose of pseudo-instructions?
2. _____ are translated by the assembler into machine code, whereas _____ are not.
3. True or false. Assembly language is a high-level language.
4. Which of the following instructions produces opcode? List all that do.
   (a) `LDI R16,0x25` (b) `ADD R23,R19` (c) `.ORG 0x500` (d) `JMP HERE`
5. Pseudo-instructions are also called _____.
6. True or false. Assembler directives are not used by the CPU itself. They are simply a guide to the assembler.
7. In Question 4, which one is an assembler directive?

SECTION 2.7: ASSEMBLING AN AVR PROGRAM

Now that the basic form of an Assembly language program has been given, the next question is: How it is created, assembled, and made ready to run? The steps to create an executable Assembly language program (Figure 2-9) are outlined as follows:

1. First we use a text editor to type in a program similar to Program 2-1. In the case of the AVR microcontrollers, we use the AVRStudio IDE, which has a text editor, assembler, simulator, and much more all in one software package. It is an excellent development software that supports all the AVR chips and is free. Many editors or word processors are also available that can be used to create or edit the program. A widely used editor is the Notepad in Windows, which comes with all Microsoft operating systems. Notice that the editor must be able

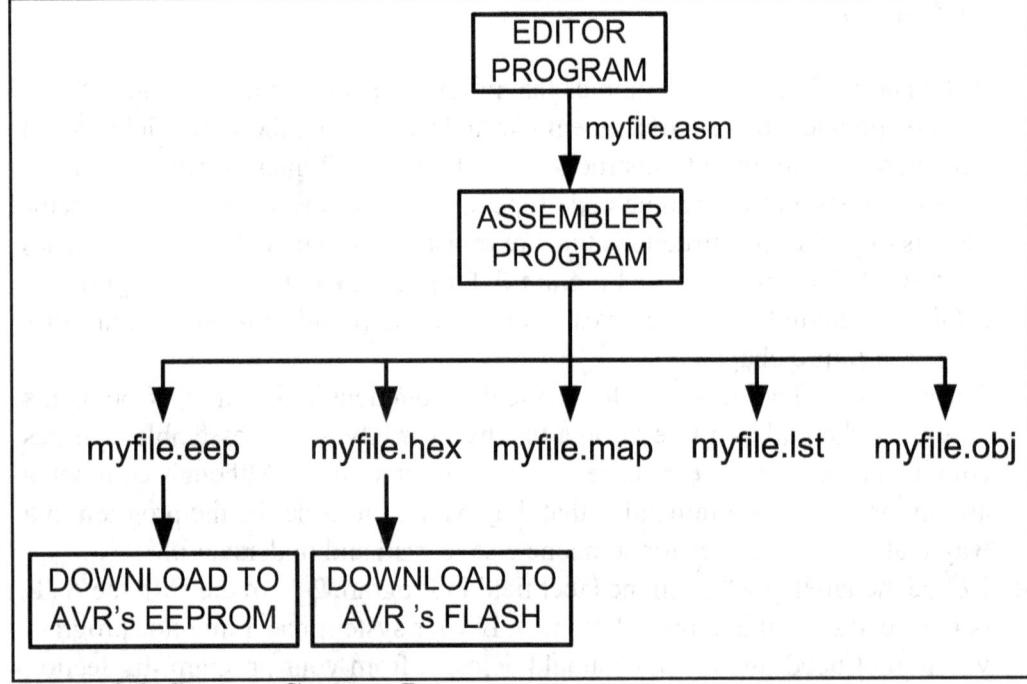

Figure 2-9. Steps to Create a Program

```
AVRASM: AVR macro assembler 2.1.2 (build 99 Nov  4 2005 09:35:05)
Copyright (C) 1995-2005 ATMEL Corporation

F:\AVR\Sample\Sample.asm(7): error: Invalid register
F:\AVR\Sample\Sample.asm(8): error: Operand(s) out of range in 'ldi r17,0x3432'
F:\AVR\Sample\Sample.asm(9): error: Undefined symbol: R38
F:\AVR\Sample\Sample.asm(9): error: Invalid register
F:\AVR\Sample\Sample.asm(16): No EEPROM data, deleting F:\AVR\Sample\Sample.eep

Assembly failed, 4 errors, 0 warnings
```

Figure 2-10. Sample of an AVR Error

to produce an ASCII file. For assemblers, the file names follow the usual DOS conventions, but the source file has the extension "asm". The "asm" extension for the source file is used by an assembler in the next step.

2. The "asm" source file containing the program code created in step 1 is fed to the AVR assembler. The assembler produces an object file, a hex file, an eeprom file, a list file, and a map file. The object file has the extension "obj", the hex file extension is "hex", the list file extension is "lst", the map file extension is "map", and the eeprom file has the extension "eep". After a successful link, the hex file is ready to be burned into the AVR's program ROM and is downloaded into the AVR Trainer. We can write the eeprom file into the AVR's EEPROM to initialize the EEPROM. See Chapter 8 for more details.

More about asm and object files

The asm file is also called the *source* file and must have the "asm" extension. As mentioned earlier, this file is created with a text editor such as Windows Notepad. Many assemblers come with a text editor. The assembler converts the asm file's Assembly language instructions into machine language and provides the obj (object) file. The object file, as mentioned earlier, has an "obj" as its extension. The object file is used as input to a simulator or an emulator.

Before we can assemble a program to create a ready-to-run program, we must make sure that it is error free. The AVR Studio IDE provides us error messages and we examine them to see the nature of syntax errors. The assembler will not assemble the program until all the syntax errors are fixed. A sample of an error message is shown in Figure 2-10.

"lst" and "map" files

The map file shows the labels defined in the program together with their values. Examine Figure 2-11. It shows the Map file of Program 2-1.

The lst (list) file, which is optional, is very useful to the programmer. The

```
AVRASM ver. 2.1.2  F:\AVR\Sample\Sample.asm Sun Apr 06 23:39:32 2008



EQU   SUM          00000300
CSEG  HERE         00000009
```

Figure 2-11. Map File of Program 2-1

```
AVRASM ver.  2.1.2   F:\AVR\Sample\Sample.asm Tue Mar 11 11:28:34 2008

                     ;store SUM in SRAM location 0x300.
                        .DEVICE ATMega32
                        .EQU  SUM   = 0x300       ;SRAM loc $300 for SUM

                        .ORG 00                   ;start at address 0
000000 e205            LDI R16, 0x25             ;R16 = 0x25
000001 e314            LDI R17, $34              ;R17 = 0x34
000002 e321            LDI R18, 0b00110001       ;R18 = 0x31
000003 0f01            ADD R16, R17              ;add R17 to R16
000004 0f02            ADD R16, R18              ;add R18 to R16
000005 e01b            LDI R17, 11               ;R17 = 0x0B
000006 0f01            ADD R16, R17              ;add R17 to R16
000007 9300 0300       STS SUM, R16              ;save the SUM in loc $300
000009 940c 0009 HERE: JMP HERE                  ;stay here forever


RESOURCE USE INFORMATION
------------------------

...
Memory use summary [bytes]:
Segment    Begin     End        Code    Data    Used    Size    Use%
-------------------------------------------------------------------
[.cseg]  0x000000 0x000016       22       0      22 unknown      -
[.dseg]  0x000060 0x000060        0       0       0 unknown      -
[.eseg]  0x000000 0x000000        0       0       0 unknown      -

Assembly complete, 0 errors, 0 warnings
```

Figure 2-12. List File of Program 2-1

list shows the binary and source code; it also shows which instructions are used in the source code, and the amount of memory the program uses. See Figure 2-12.

Many assemblers assume that the list file is not wanted unless you indicate that you want to produce it. These files can be accessed by a text editor such as Notepad and displayed on the monitor, or sent to the printer to get a hard copy. The programmer uses the list and map files to ensure correct system design.

There are many different AVR assemblers available for free nowadays. If you use the Windows operating system, AVR Studio can be a good choice. It has a nice environment and provides great help.

Review Questions

1. True or false. The AVR Studio IDE and Windows Notepad text editor both produce an ASCII file.
2. True or false. The extension for the source file is "asm".
3. Which of the following files can be produced by a text editor?
   (a) myprog.asm (b) myprog.obj (c) myprog.hex (d) myprog.lst
4. Which of the following files is produced by an assembler?
   (a) myprog.asm (b) myprog.obj (c) myprog.hex (d) myprog.lst

SECTION 2.8: THE PROGRAM COUNTER AND PROGRAM ROM SPACE IN THE AVR

In this section we discuss the role of the program counter (PC) in executing a program and show how the code is fetched from ROM and executed. We will also discuss the program (code) ROM space for various AVR family members. Finally, we examine the Harvard architecture of the AVR.

Program counter in the AVR

The most important register in the AVR microcontroller is the PC (program counter). The program counter is used by the CPU to point to the address of the next instruction to be executed. As the CPU fetches the opcode from the program ROM, the program counter is incremented automatically to point to the next instruction. The wider the program counter, the more memory locations a CPU can access. That means that a 14-bit program counter can access a maximum of 16K ($2^{14} = 16K$) program memory locations.

In AVR microcontrollers each Flash memory location is 2 bytes wide. For example, in ATmega32, whose Flash is 32K bytes, the Flash is organized as $16K \times 16$, and its program counter is 14 bits wide ($2^{14} = 16K$ memory locations). The ATmega64 has a 15-bit program counter, so its Flash has 32K locations ($2^{15} = 32K$), with each location containing 2 bytes ($32K \times 2$ bytes = 64K bytes).

In the case of a 16-bit program counter, the code space is 64K ($2^{16} = 64K$), which occupies the 0000–$FFFF address range. The program counter in the AVR family can be up to 22 bits wide. This means that the AVR family can access program addresses 000000 to $3FFFFF, a total of 4M locations. Because each Flash location is 2 bytes wide, the AVR can have a maximum of 8M bytes of code. However, at the time of this writing, none of the members of the AVR family have the entire 8M bytes of on-chip ROM installed. See Table 2-7.

Table 2-7: AVR On-chip ROM Size and Address Space

	On-chip Code ROM (Bytes)	Code Address Range (Hex)	ROM Organization
ATtiny25	2K	00000–003FF	1K × 2 bytes
ATmega8	8K	00000–00FFF	4K × 2 bytes
ATmega32	32K	00000–03FFF	16K × 2 bytes
ATmega64	64K	00000–07FFF	32K × 2 bytes
ATmega128	128K	00000–0FFFF	64K × 2 bytes
ATmega256	256K	00000–1FFFF	128K × 2 bytes

ROM memory map in the AVR family

As we just discussed, some family members have only a few kilobytes of on-chip ROM and some, such as the ATmega128, have 128K of ROM. The point to remember is that no member of the AVR family can access more than 4M words of opcode because the program counter in the AVR can be a maximum of 22 bits wide (000000 to $3FFFFF address range). It must be noted that while the first

Example 2-9

Find the ROM memory address of each of the following AVR chips:
(a) ATtiny25 with 2 KB
(b) ATmega16 with 16 KB
(c) ATmega64 with 64 KB

Solution:

(a) With 2K bytes of on-chip ROM memory, we have 2048 bytes (2 × 1024 = 2048). As each address location in AVR is 2 bytes, its Flash has 1024 locations (2048 / 2 = 1024). This maps to address locations of 0000 to $03FF. Notice that 0 is always the first location.

(b) With 16K bytes of on-chip ROM memory, we have 16,384 bytes (16 × 1024 = 16,384), and 8192 locations (16384 / 2 = 8192), which gives 0000–$1FFF.

(c) With 64K we have 65,535 bytes (64 × 1024 = 65,535), and 32,768 locations. Converting 32,768 to hex, we get $8000; therefore, the memory space is 0000 to $7FFF.

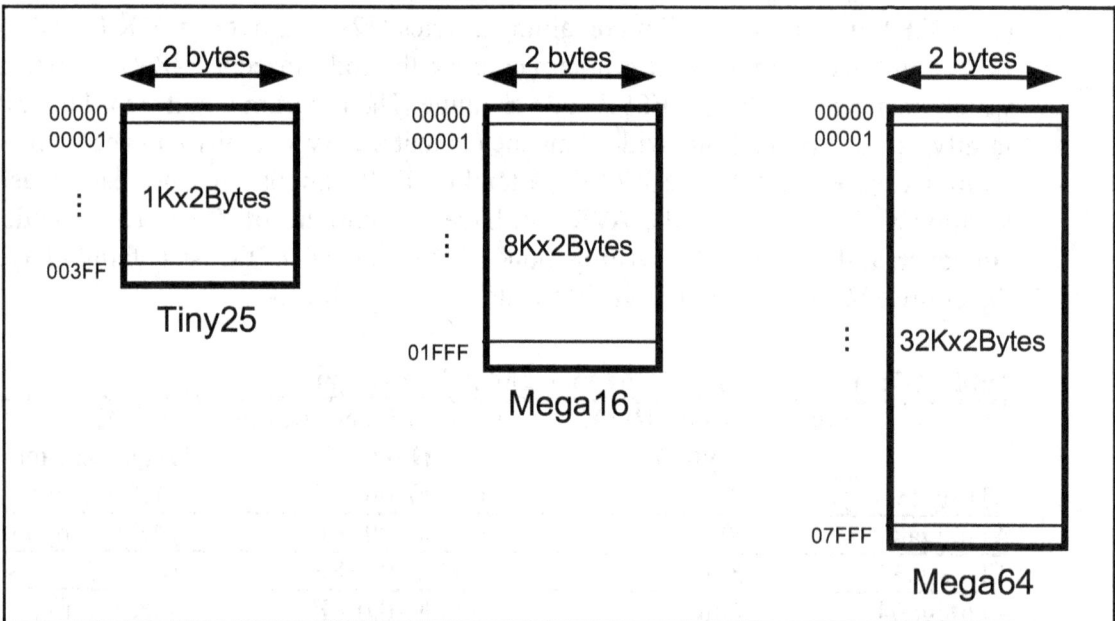

Figure 2-13. AVR On-Chip Program (code) ROM Address Range

location of program ROM inside the AVR has the address of 000000, the last location can be different depending on the size of the ROM on the chip. (See Figure 2-13.) Among the AVR family members, the ATmega8 has 8K of on-chip ROM. This 8K ROM memory is organized as 4K × 2 bytes and has memory addresses of 00000 to $00FFF. Therefore, the first location of on-chip ROM of this AVR has an address of 00000 and the last location has the address of $00FFF. Look at Example 2-9 to see how this is computed.

Where the AVR wakes up when it is powered up

One question that we must ask about any microcontroller (or microprocessor) is: At what address does the CPU wake up when power is applied? Each microprocessor is different. In the case of the AVR microcontrollers (that is, all members regardless of the family and variation), the microcontroller wakes up at memory address 0000 when it is powered up. By powering up we mean applying V_{CC} to the RESET pin as discussed in Chapter 8. In other words, when the AVR is powered up, the PC (program counter) has the value of 00000 in it. This means that it expects the first opcode to be stored at ROM address $00000. For this reason, in the AVR system, the first opcode must be burned into memory location $00000 of program ROM because this is where it looks for the first instruction when it is booted. We achieve this by using the .ORG statement in the source program as shown earlier. Next we discuss the step-by-step action of the program counter in fetching and executing a sample program.

Placing code in program ROM

To get a better understanding of the role of the program counter in fetching and executing a program, we examine the action of the program counter as each instruction is fetched and executed. First, we examine once more the list file of the sample program and show how the code is placed into the Flash ROM of the AVR chip. As we can see, the opcode and operand for each instruction are listed on the left side of the list file.

After the program is burned into ROM of an AVR family member such as ATmega32 or ATtiny11, the opcode and operand are placed in ROM memory locations starting at 0000 as shown in the Program 2-1 list file.

The list shows that address 0000 contains E205, which is the opcode for moving a value into R16, and the operand (in this case 0x25) to be moved to R16. Therefore, the instruction "LDI R16,0x25" has a machine code of "E205", where E is the opcode and 205 is the operand. See Figures 2-14 and 2-15. Similarly, the machine code "E314" is located in ROM memory location 0001 and represents the opcode and the operands for the instruction "LDI R17,$34". In

1 1 1 0	k k k k	d d d d	k k k k

LDI Rd, k $16 \leq d \leq 31, \ 0 \leq K \leq 255$

Figure 2-14. The Machine Code for Instruction "LDI Rd, k" in Binary

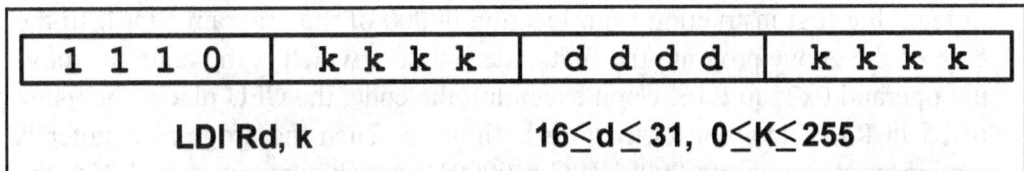

E k_1 d k_0	E 2 0 5
LDI Rd, k_1k_0	LDI R16, 0x25
E 3 1 4	E 3 2 1
LDI R17, 0x34	LDI R18, 0x31

Figure 2-15. The Machine Code for Instruction "LDI Rd, k" in Hex

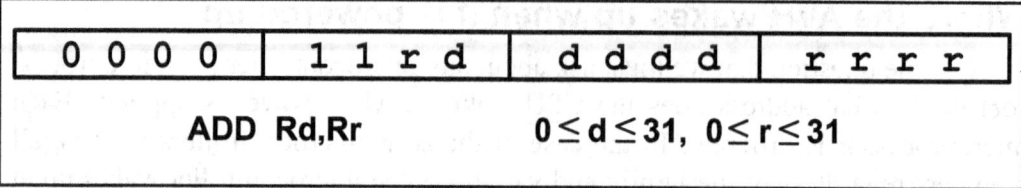

| 0 0 0 0 | 1 1 r d | d d d d | r r r r |

ADD Rd,Rr $0 \leq d \leq 31, \; 0 \leq r \leq 31$

Figure 2-16. The Machine Code for Instruction "ADD Rd,Rr" in Binary

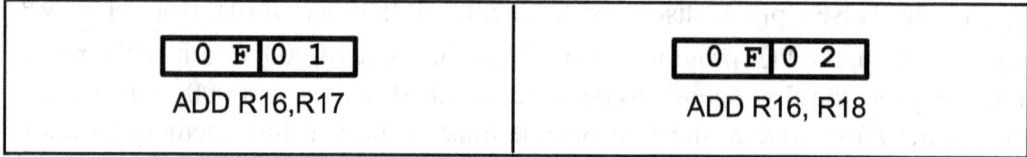

| 0 F 0 1 | | 0 F 0 2 |
| ADD R16,R17 | | ADD R16, R18 |

Figure 2-17. The Machine Code for Instruction "ADD Rd,Rr" in Hex

the same way, machine code "E321" is located in memory location 0002 and represents the opcode and the operand for the instruction "LDI R18,0b00110001". The memory location 0003 has the machine code of 0F01, which is the opcode and the operands for the instruction "ADD R16,R17". Similarly, the machine code "0F02" is located in memory location 0004 and represents the opcode and the operands for the instruction "ADD R16, R18". See Figures 2-16 and 2-17. The memory location 0005 has the opcode and operand for the "LDI R17,11" instruction. The memory location 0006 has the opcode and operand for the "ADD R16,R17" instruction. The opcode for instruction "STS SUM,R16" is located at address 00007 and its address of 0x300 at address 00008. The opcode for "JMP HERE" and its target address are located in locations 00009 and 0000A. While all the instructions in this program are 2-byte instructions, the JMP and STS instructions are 4-byte instructions. The reasons are explained at the end of this section.

Executing a program instruction by instruction

Assuming that the above program is burned into the ROM of an AVR chip, the following is a step-by-step description of the action of the AVR upon applying power to it:

1. When the AVR is powered up, the PC (program counter) has 00000 and starts to fetch the first instruction from location 00000 of the program ROM. In the case of the above program the first code is E205, which is the code for moving operand 0x25 to R16. Upon executing the code, the CPU places the value of 25 in R16. Now one instruction is finished. Then the program counter is incremented to point to 00001 (PC = 00001), which contains code E314, the machine code for the instruction "LDI R17,0x34".

2. Upon executing the machine code E314, the value 0x34 is loaded to R17. Then the program counter is incremented to 0002.

3. ROM location 0002 has the machine code for instruction "LDI R18, 0x31". This instruction is executed and now PC = 0003.

4. This process goes on until all the instructions up to "ADD R16,R17" are fetched and executed. Notice that all the above instructions are 2-byte instructions; that is, each one takes two bytes of ROM (one word).

5. Now PC = 0007 points to the next instruction, which is "STS SUM,R16". This is a 2-word (4-byte) instruction. It takes addresses of 07 and 08. When the

instruction is executed, the content of R16 is stored into memory location 0x300. After the execution of this instruction, PC = 0009.

6. Now PC = 0009 points to the next instruction, which is "JMP HERE". This is a 2-word (4-byte) instruction. It takes addresses of 09 and 0A. After the execution of this instruction, PC = 0009. This keeps the program in an infinite loop. The fact that the program counter points at the next instruction to be executed explains why some microprocessors (notably the x86) call the program counter the *instruction pointer*.

ROM width in the AVR

As we have seen so far in this section, each location of the address space holds two bytes (a word). If we have 16 address lines, this will give us 2^{16} locations, which is 64K of memory location with an address map of 0000–FFFFH. To bring in more information (code or data) into the CPU, AVR increased the width of the data bus to 16 bits. In other words, the AVR is word-addressable. In contrast, the 8051 CPU is byte addressable. In a sense, the data bus is like traffic lanes on the highway where each lane is 8 bits wide. The more lanes, the more information we can bring into the CPU for processing. For the AVR, the internal data bus between the code ROM and the CPU is 16 bits wide, as shown in Figure 2-18. Therefore, the 64K ROM space is shown as 32K × 16 using a 16-bit word

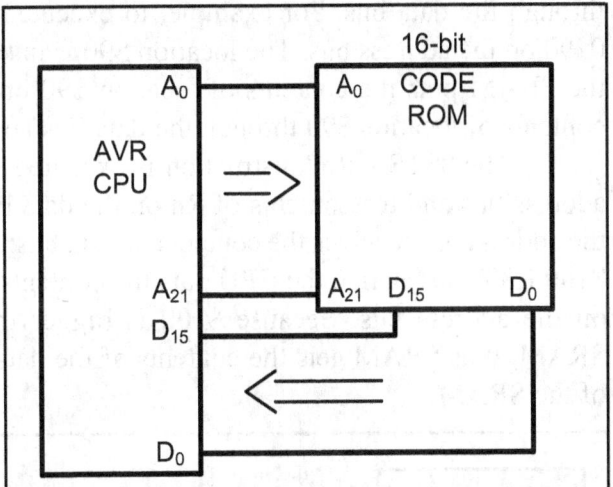

Figure 2-18. Program ROM Width for the AVR

data size. The same rule applies to the entire program address space of AVR, which is 8M, organized as 4M × 16. The widening of the data path between the program ROM and the CPU is another way in which the AVR designers increased the processing power of the AVR family. Another reason to make the code ROM 16 bits wide is to match it with the instruction width of the AVR because the vast majority of the instructions are 2-byte instructions. This way, the CPU brings in an instruction from ROM every time it makes a trip to the program ROM. That will make instruction fetch a single cycle, as we will see in the next chapter when instruction timing is discussed.

The AVR designers have made all instructions either 2-byte or 4-byte; there are no 1-byte or 3-byte instructions, as is the case with the x86 and 8051 chips. This is part of the RISC architectural philosophy, which we will study in the next section. It must also be noted that the data memory SRAM in the AVR microcontroller is still 8-bit, and it is byte-addressable.

Harvard architecture in the AVR

As we mentioned in Chapter 0, AVR uses Harvard architecture, which means that there are separate buses for the code and the data memory. See Figure 2-19. The Program Bus provides access to the Program Flash ROM whereas the Data Bus is used for bringing data to the CPU.

As we can see in Figure 2-19, in the Program Bus, the data bus is 16 bits wide and the address bus is as wide as the PC register to enable the CPU to address the entire Program Flash ROM.

In the Data Bus, the data bus is 8 bits wide. As a result, the CPU can access one byte of data at a time. The address bus is 16 bits wide. Thus the data memory space can be up to 64K bytes.

In Sections 2-2 and 2-3, you learned about data memory space and how to use the STS and LDS instructions. When the CPU wants to execute the "LDS Rn,k" instruction, it puts k on the address bus of the Data Bus, and receives data through the data bus. For example, to execute "LDS R20, 0x90", the CPU puts 0x90 on the address bus. The location $90 is in the SRAM (see Figure 2-4). Thus, the SRAM puts the contents of location $90 on the data bus. The CPU gets the contents of location $90 through the data bus and puts it in R20.

The "STS k,Rn" instruction is executed similarly. The CPU puts k on the address bus and the contents of Rn on the data bus. The unit whose address is on the address bus receives the contents of data bus. For example, to execute the "STS $100,R30" instruction the CPU puts the contents of R30 on the data bus and $100 on the address bus. Because $100 is bigger than $60, the address belongs to SRAM; thus SRAM gets the contents of the data bus and puts it in location $100 of the SRAM.

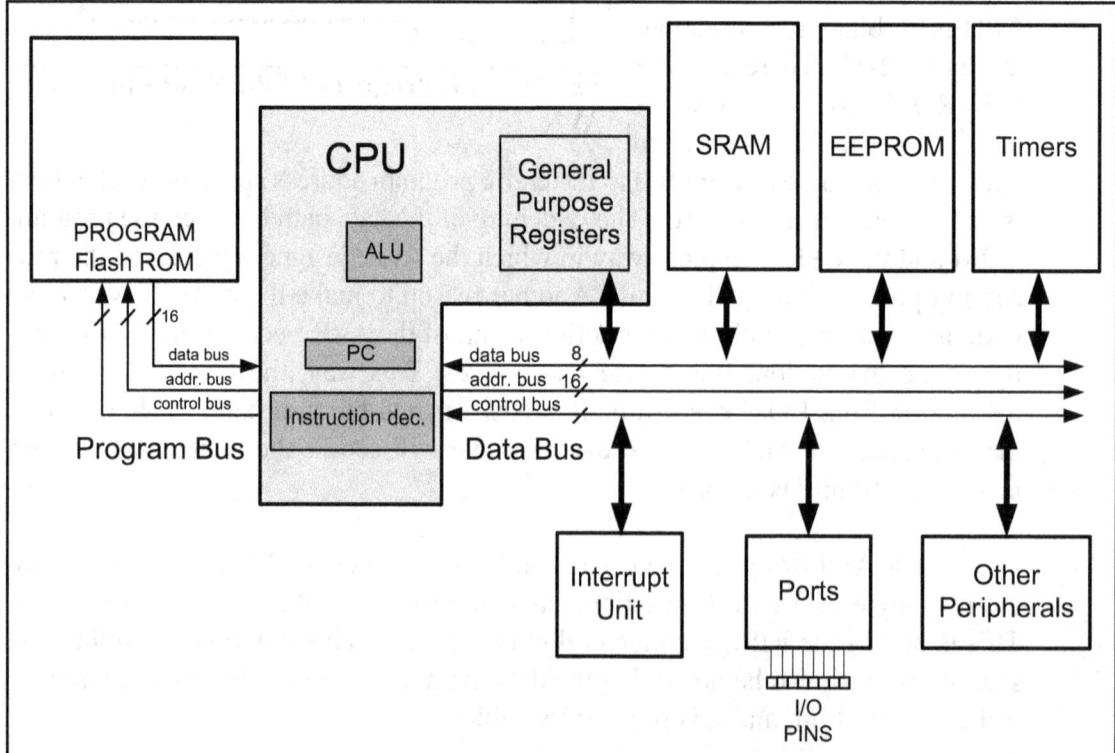

Figure 2-19. Harvard Architecture in the AVR

Examine the placing of the code in the AVR ROM, shown in Figure 2-20. The low byte goes to the low memory location, and the high byte goes to the high memory address. This convention is called little endian to contrast it with big endian. The origin of the terms *big endian* and *little endian* is from an argument in a Gulliver's Travels story over how an egg should be opened: from the big end or the little end. In the big endian method, the high byte goes to the low address, whereas in the little endian method, the high byte goes to the high address and the low byte to the low address. All Intel microprocessors and many microcontrollers use the little endian convention. Freescale (formerly Motorola) microprocessors, along with some mainframes, use big endian. The difference might seem as trivial as whether to break an egg from the big end or the little end, but it is a nuisance in converting software from one camp to be run on a computer of the other camp. Some microprocessors, such as the PowerPC from IBM/Freescale, let the software designer choose little endian or big endian convention.

Address	High byte	Low byte
00000	E2	05
00001	E3	14
00002	E3	21
00003	0F	01
00004	0F	02
00005	E0	1B
00006	0F	01
00007	93	00
00008	03	00
00009	94	0C
0000A	00	09

Figure 2-20. AVR Program ROM Contents for Program 2-1 List File

Instruction size of the AVR

Recall that the AVR instructions are either 2-byte or 4-byte. Almost all the instructions in the AVR are 2-byte instructions. The exceptions are STS, JMP, and a few others. Next we explore the instruction size and formation for a few of the instructions we have used in this chapter. This should give you some insight into the instructions of the AVR.

LDI instruction formation

The LDI is a 2-byte (16-bit) instruction. Of the 16 bits, the first 4 bits are set aside for the opcode, the second and the fourth 4 bits are used for the value of 00 to $FF, and the third 4 bits present the destination register. This is shown below.

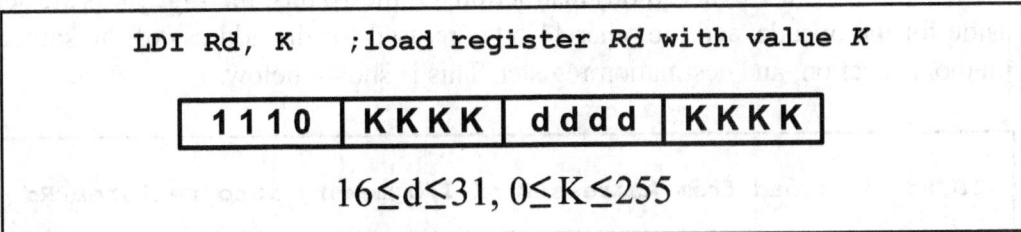

```
    LDI Rd, K    ;load register Rd with value K

    1110 | KKKK | dddd | KKKK

       16≤d≤31, 0≤K≤255
```

ADD instruction formation

The ADD is a 2-byte (16-bit) instruction. Of the 16 bits, the first 6 bits are set aside for the opcode, and the other 10 bits represent the source and the destination registers. This is shown below.

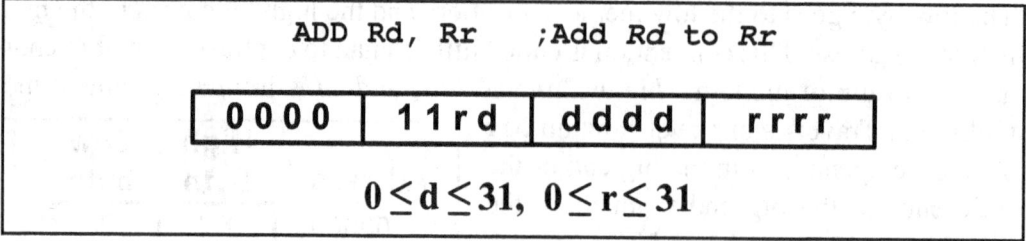

STS instruction formation

The STS is a 4-byte (32-bit) instruction. Of the 32 bits, the first 16 bits are set aside for the opcode and the address of the source, and the other 16 bits are used for the address of the destination. This is shown below.

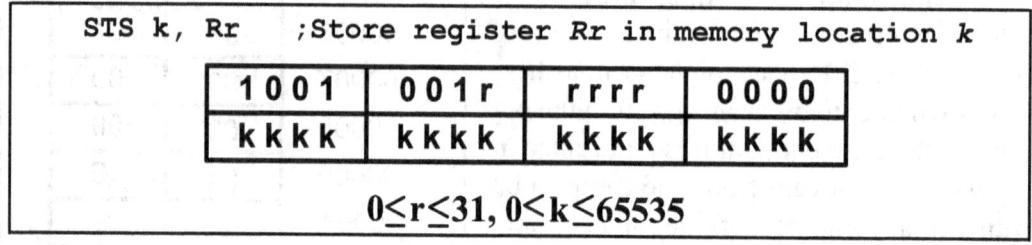

LDS instruction formation

The LDS is a 4-byte (32-bit) instruction. Of the 32 bits, the first 16 bits are set aside for the opcode and the destination register, and the other 16 bits are used for the address of the source memory location. This is shown below.

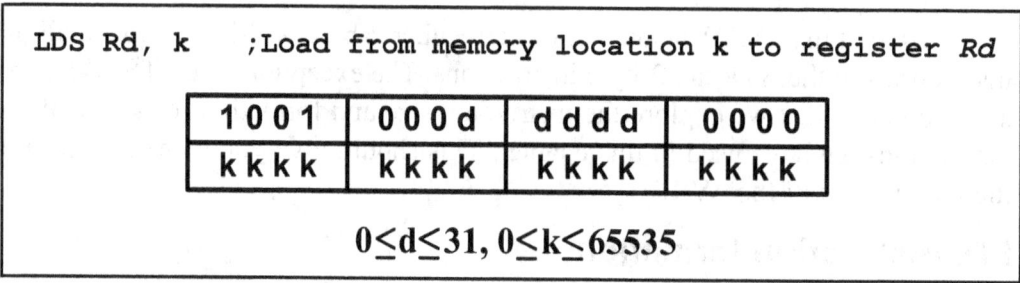

IN instruction formation

The IN is a 2-byte (16-bit) instruction. Of the 16 bits, the first 5 bits are set aside for the opcode, and the other 11 bits are used for the address of the source memory location, and destination register. This is shown below.

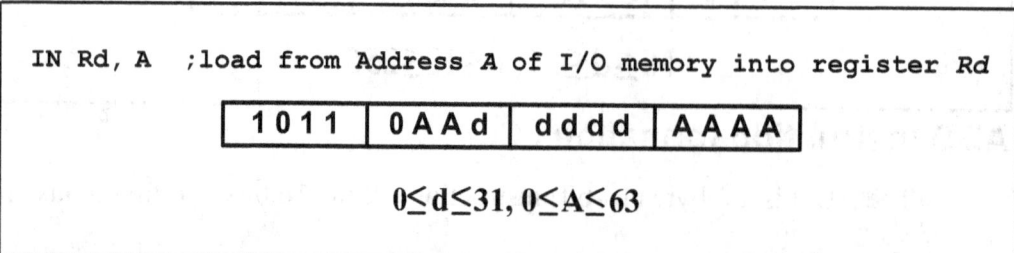

OUT instruction formation

The OUT is a 2-byte (16-bit) instruction. Of the 16 bits, the first 5 bits are set aside for the opcode, and the other 11 bits are used for the address of the source memory location and destination register. This is shown below.

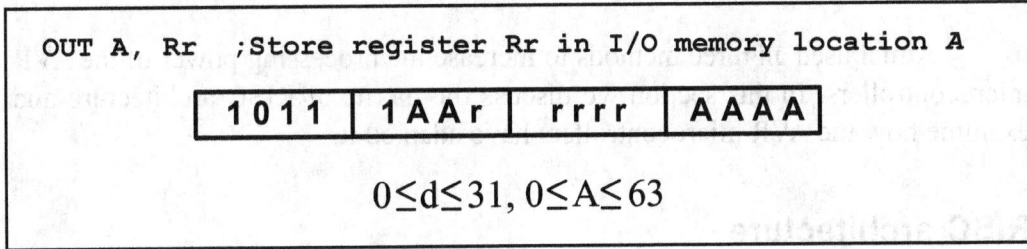

```
OUT A, Rr   ;Store register Rr in I/O memory location A

       1011  | 1AAr | rrrr | AAAA

           0≤d≤31, 0≤A≤63
```

JMP instruction formation

The JMP is a 4-byte (32-bit) instruction. Of the 32 bits, only 10 bits are set aside for the opcode, and the rest (22 bits) are used for the target address of the JMP. This is shown below.

The 22-bit address gives us 4M of address space; so, it can address all of the ROM space.

```
          JMP k     ;Jump to address k

       1001  | 010k | kkkk | 110k
       kkkk  | kkkk | kkkk | kkkk

                 0≤k≤4M
```

Review Questions

1. In the AVR, the program counter is, at most, _____ bits wide.
2. True or false. Every member of the AVR family, regardless of the program ROM size, wakes up at memory $0000 when it is powered up.
3. At what ROM location do we store the first opcode of an AVR program?
4. The instruction "LDI R20,0x44" is a ____-byte instruction.
5. The instruction "JMP label" is a ____-byte instruction.
6. True or false. All the instructions in the AVR are 2- or 4-byte instructions.

SECTION 2.9: RISC ARCHITECTURE IN THE AVR

There are three ways available to microprocessor designers to increase the processing power of the CPU:

1. Increase the clock frequency of the chip. One drawback of this method is that the higher the frequency, the more power and heat dissipation. Power and heat dissipation is especially a problem for hand-held devices.
2. Use Harvard architecture by increasing the number of buses to bring more information (code and data) into the CPU to be processed. While in the case of

x86 and other general purpose microprocessors this architecture is very expensive and unrealistic, in today's microcontrollers this is not a problem. As we saw in the last section, the AVR has Harvard architecture.

3. Change the internal architecture of the CPU and use what is called RISC architecture.

Atmel used all three methods to increase the processing power of the AVR microcontrollers. In this section we discuss the merits of RISC architecture and examine how the AVR microcontrollers have adapted it.

RISC architecture

In the early 1980s, a controversy broke out in the computer design community, but unlike most controversies, it did not go away. Since the 1960s, in all mainframes and minicomputers, designers put as many instructions as they could think of into the CPU. Some of these instructions performed complex tasks. An example is adding data memory locations and storing the sum into memory. Naturally, microprocessor designers followed the lead of minicomputer and mainframe designers. Because these microprocessors used such a large number of instructions, many of which performed highly complex activities, they came to be known as CISC (complex instruction set computer) processors. According to several studies in the 1970s, many of these complex instructions etched into CPUs were never used by programmers and compilers. The huge cost of implementing a large number of instructions (some of them complex) into the microprocessor, plus the fact that a good portion of the transistors on the chip are used by the instruction decoder, made some designers think of simplifying and reducing the number of instructions. As this concept developed, the resulting processors came to be known as RISC (reduced instruction set computer).

Features of RISC

The following are some of the features of RISC as implemented by the AVR microcontroller.

Feature 1

RISC processors have a fixed instruction size. In a CISC microcontroller such as the 8051, instructions can be 1, 2, or even 3 bytes. For example, look at the following instructions in the 8051:

```
CLR   C                        ;clear Carry flag, a 1-byte instruction
ADD   Accumulator, #mybyte     ;a 2-byte instruction
LJMP  target_address           ;a 3-byte instruction
```

This variable instruction size makes the task of the instruction decoder very difficult because the size of the incoming instruction is never known. In a RISC architecture, the size of all instructions is fixed. Therefore, the CPU can decode the instructions quickly. This is like a bricklayer working with bricks of the same size as opposed to using bricks of variable sizes. Of course, it is much more efficient to use bricks of the same size. In the last section we saw how the AVR uses 2-byte instructions with very few 4-byte instructions.

Feature 2

One of the major characteristics of RISC architecture is a large number of registers. All RISC architectures have at least 32 registers. Of these 32 registers, only a few are assigned to a dedicated function. One advantage of a large number of registers is that it avoids the need for a large stack to store parameters. Although a stack can be implemented on a RISC processor, it is not as essential as in CISC because so many registers are available. In the AVR microcontrollers the use of 32 general purpose registers satisfies this RISC feature. The stack for the AVR is covered in the next chapter.

Feature 3

RISC processors have a small instruction set. RISC processors have only basic instructions such as ADD, SUB, MUL, LOAD, STORE, AND, OR, EOR, CALL, JUMP, and so on. The limited number of instructions is one of the criticisms leveled at the RISC processor because it makes the job of Assembly language programmers much more tedious and difficult compared to CISC Assembly language programming. This is one reason that RISC is used more commonly in high-level language environments such as the C programming language rather than Assembly language environments. It is interesting to note that some defenders of CISC have called it "complete instruction set computer" instead of "complex instruction set computer" because it has a complete set of every kind of instruction. How many of these instructions are used and how often is another matter. The limited number of instructions in RISC leads to programs that are large. Although these programs can use more memory, this is not a problem because memory is cheap. Before the advent of semiconductor memory in the 1960s, however, CISC designers had to pack as much action as possible into a single instruction to get the maximum bang for their buck. In the ATmega we have around 130 instructions. We will examine more of the instruction set for the AVR in future chapters.

Feature 4

At this point, one might ask, with all the difficulties associated with RISC programming, what is the gain? The most important characteristic of the RISC processor is that more than 95% of instructions are executed with only one clock cycle, in contrast to CISC instructions. Even some of the 5% of the RISC instructions that are executed with two clock cycles can be executed with one clock cycle by juggling instructions around (code scheduling). Code scheduling is most often the job of the compiler. We will examine the instruction cycle time and pipelining of the AVR in Chapter 3.

Feature 5

RISC processors have separate buses for data and code. In all the x86 processors, like all other CISC computers, there is one set of buses for the address (e.g., A0–A24 in the 80286) and another set of buses for data (e.g., D0–D15 in the 80286) carrying opcodes and operands in and out of the CPU. To access any section of memory, regardless of whether it contains code or data operands, the same

---

address bus and data bus are used. In RISC processors, there are four sets of buses: (1) a set of data buses for carrying data (operands) in and out of the CPU, (2) a set of address buses for accessing the data, (3) a set of buses to carry the opcodes, and (4) a set of address buses to access the opcodes. The use of separate buses for code and data operands is commonly referred to as Harvard architecture. We examined the Harvard architecture of the AVR in the previous section.

Feature 6

Because CISC has such a large number of instructions, each with so many different addressing modes, microinstructions (microcode) are used to implement them. The implementation of microinstructions inside the CPU employs more than 40–60% of transistors in many CISC processors. RISC instructions, however, due to the small set of instructions, are implemented using the hardwire method. Hardwiring of RISC instructions takes no more than 10% of the transistors.

Feature 7

RISC uses load/store architecture. In CISC microprocessors, data can be manipulated while it is still in memory. For example, in instructions such as "ADD Reg, Memory", the microprocessor must bring the contents of the external memory location into the CPU, add it to the contents of the register, then move the result back to the external memory location. The problem is there might be a delay in accessing the data from external memory. Then the whole process would be stalled, preventing other instructions from proceeding in the pipeline. In RISC, designers did away with these kinds of instructions. In RISC, instructions can only load from external memory into registers or store registers into external memory locations. There is no direct way of doing arithmetic and logic operations between a register and the contents of external memory locations. All these instructions must be performed by first bringing both operands into the registers inside the CPU, then performing the arithmetic or logic operation, and then sending the result back to memory. This idea was first implemented by the Cray 1 supercomputer in 1976 and is commonly referred to as load/store architecture. In the last section, we saw that the arithmetic and logic operations are between the data memory (internal) locations, but none involves a ROM location. For example, there is no "ADD ROM-Loc" instruction in AVR.

In concluding this discussion of RISC processors, it is interesting to note that RISC technology was explored by the scientists at IBM in the mid-1970s, but it was David Patterson of the University of California at Berkeley who in 1980 brought the merits of RISC concepts to the attention of computer scientists. It must also be noted that in recent years CISC processors such as the Pentium have used some RISC features in their design. This was the only way they could enhance the processing power of the x86 processors and stay competitive. Of course, they had to use lots of transistors to do the job, because they had to deal with all the CISC instructions of the x86 processors and the legacy software of DOS/Windows.

Review Questions

1. What do RISC and CISC stand for?
2. True or false. The CISC architecture executes the vast majority of its instructions in 2, 3, or more clock cycles, while RISC executes them in one clock.
3. RISC processors normally have a _____ (large, small) number of general-purpose registers.
4. True or false. Instructions such as "ADD R16, ROMmemory" do not exist in RISC microcontrollers such as the AVR.
5. How many instructions does the ATmega have?
6. True or false. While CISC instructions are of variable sizes, RISC instructions are all the same size.
7. Which of the following operations do not exist for the ADD instruction in RISC?
   (a) register to register (b) immediate to register (c) memory to memory
8. True or false. Harvard architecture uses the same address and data buses to fetch both code and data.

SECTION 2.10: VIEWING REGISTERS AND MEMORY WITH AVR STUDIO IDE

The AVR microcontroller has great tools and support systems, many of them free or inexpensive. AVR Studio is an assembler and simulator provided for free by Atmel Corporation and can be downloaded from the www.atmel.com website. See http://www.MicroDigitalEd.com for tutorials on how to use the AVR Studio assembler and simulator.

Many assemblers and C compilers come with a simulator. Simulators allow us to view the contents of registers and memory after executing each instruction (single-stepping). It is strongly recommended to use a simulator to single-step some of the programs in this chapter and future chapters. Single-stepping a program with a simulator gives us a deeper understanding of microcontroller architecture, in addition to the fact that we can use it to find the errors in our programs.

Figures 2-21 through 2-23 show screenshots for AVR simulators from AVR Studio.

See the following website for a tutorial on using AVR Studio:

http://www.MicroDigitalEd.com

```
Memory                                                                    ×
Program    ▼   8/16   abc.   Address: 0x00            Cols: Auto ▼
000000 05 E2 14 E3 21 E3 01 0F 02 0F 1B E0 01 0F 00 93  .â.ã!ã.....à...`
000008 00 03 0C 94 09 00 FF FF FF FF FF FF FF FF FF FF  ...".ÿÿÿÿÿÿÿÿÿÿ
000010 FF FF FF FF FF FF FF FF FF FF FF FF FF FF FF FF  ÿÿÿÿÿÿÿÿÿÿÿÿÿÿÿÿ
000018 FF FF FF FF FF FF FF FF FF FF FF FF FF FF FF FF  ÿÿÿÿÿÿÿÿÿÿÿÿÿÿÿÿ
000020 FF FF FF FF FF FF FF FF FF FF FF FF FF FF FF FF  ÿÿÿÿÿÿÿÿÿÿÿÿÿÿÿÿ
000028 FF FF FF FF FF FF FF FF FF FF FF FF FF FF FF FF  ÿÿÿÿÿÿÿÿÿÿÿÿÿÿÿÿ
000030 FF FF FF FF FF FF FF FF FF FF FF FF FF FF FF FF  ÿÿÿÿÿÿÿÿÿÿÿÿÿÿÿÿ
```

Figure 2-21. Data Memory Window in AVR Studio IDE

```
Disassembler                                            _ □ ×
   ---- program2_1.asm -------------------------------------
   7:            LDI R16, 0x25          ;R16 = 0x25
⇨|+00000000:   E205        LDI       R16,0x25       Load immediate
   8:            LDI R17, $34           ;R17 = 0x34
  +00000001:    E314        LDI       R17,0x34       Load immediate
   9:            LDI R18, 0b00110001    ;R18 = 0x31
  +00000002:    E321        LDI       R18,0x31       Load immediate
  10:            ADD R16, R17           ;add R17 to R16
  +00000003:    0F01        ADD       R16,R17        Add without carry
  11:            ADD R16, R18           ;add R18 to R16
  +00000004:    0F02        ADD       R16,R18        Add without carry
  12:            LDI R17, 11            ;R17 = 0x0B
  +00000005:    E01B        LDI       R17,0x0B       Load immediate
```

Figure 2-22. Program ROM (Disassembler) Window in AVR Studio IDE

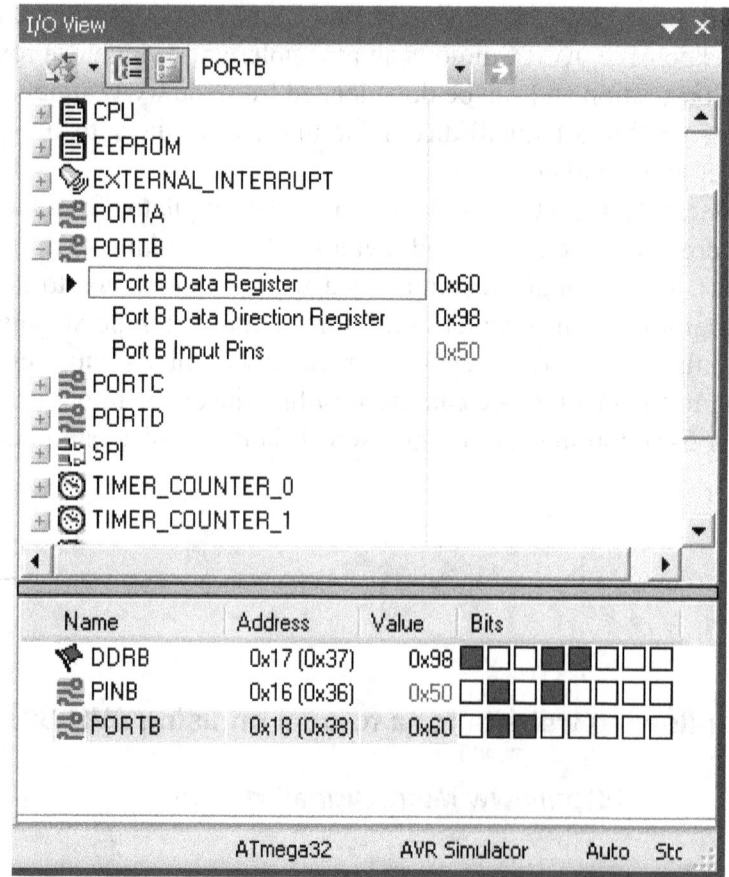

Figure 2-23. I/O View Window in AVR Studio IDE

SUMMARY

This chapter began with an exploration of the major registers of the AVR, including general purpose registers, I/O registers, and internal SRAM, and the program counter. The use of these registers was demonstrated in the context of programming examples. The process of creating an Assembly language program was described from writing the source file, to assembling it, linking, and executing the program. The PC (program counter) register always points to the next instruction to be executed. The way the AVR uses program Flash ROM space was explored because AVR Assembly language programmers must be aware of where programs are placed in ROM, and how much memory is available.

An Assembly language program is composed of a series of statements that are either instructions or pseudo-instructions, also called *directives*. Instructions are translated by the assembler into machine code. Pseudo-instructions are not translated into machine code: They direct the assembler in how to translate instructions into machine code. Some pseudo-instructions, called *data directives*, are used to define data. Data is allocated in byte-size increments. The data can be in binary, hex, decimal, or ASCII formats.

Flags are useful to programmers because they indicate certain conditions, such as carry or zero, that result from execution of instructions. The concepts of the RISC and Harvard architectures were also explored.

The RISC architecture allows the design of much more powerful microcontrollers. It has a simple instruction set and uses a large number of registers. Harvard architecture allows us to bring more code and data to the CPU faster. The use of a wider data bus in the AVR allows us to fetch an instruction every cycle because the AVR instructions are typically 2 bytes.

PROBLEMS

SECTION 2.1: THE GENERAL PURPOSE REGISTERS IN THE AVR

1. AVR is a(n) _____-bit microcontroller.
2. The general purpose registers are _____ bits wide.
3. The value in LDI is _____ bits wide.
4. The largest number that can be loaded into the GPRs is _____ in hex.
5. What is the result of the following code and where is it kept?
   ```
   LDI        R20,$15
   LDI        R21,$13
   ADD        R20,R21
   ```

6. Which of the following is (are) illegal?
   (a) LDI R20,500   (b) LDI   R23,50   (c) LDI   R1,00
   (d) LDI R16,$255   (e) LDI   R42,$25   (f) LDI R23,0xF5
   (g) LDI 123,0x50

7. Which of the following is (are) illegal?
   (a) ADD R20,R11   (b) ADD R16,R1   (c) ADD R52,R16

8. What is the result of the following code and where is it kept?
```
LDI   R19,$25
ADD   R19,$1F
```
9. What is the result of the following code and where is it kept?
```
LDI   R21,0x15
ADD   R21,0xEA
```
10. True or false. We have 32 general purpose registers in the AVR.

SECTION 2.2: THE AVR DATA MEMORY

11. AVR data memory consists of _____ (Flash ROM, internal SRAM).
12. True or false. The special function register in AVR is called the I/O register.
13. True or false. The I/O registers are part of the data memory space.
14. True or false. The general-purpose registers are not part of the data memory space.
15. True or false. The data memory is the same size in all members of AVR.
16. If we add the I/O registers, internal RAM, and general purpose register sizes together we should get the total space for the _____.
17. Find the data memory size for the following AVR chips:
    (a) ATmega32   (b) ATmega16   (c) ATtiny44
18. What is the difference between the EEPROM and data RAM space in the AVR?
19. Can we have an AVR chip with no EEPROM?
20. Can we have an AVR chip with no data memory?
21. What is the address range for the internal RAM?
22. What is the maximum number of bytes that the AVR can have for the data memory?

SECTION 2.3: USING INSTRUCTIONS WITH THE DATA MEMORY

23. Show a simple code to load values $30 and $97 into locations $105 and $106, respectively.
24. Show a simple code to load the value $55 into locations $300–$308.
25. Show a simple code to load the value $5F into the PORTB I/O register.
26. True or false. We cannot load immediate values into the internal RAM directly.
27. Show a simple code to (a) load the value $11 into locations $100–$105, and (b) add the values together and place the result in R20 as they are added.
28. Repeat Problem 27, except place the result in location $105 after the addition is done.
29. Show a simple code to (a) load the value $15 into location $67, and (b) add it to R19 five times and place the result in R19 as the values are added. R19 should be zero before the addition starts.
30. Repeat Problem 29, except place the result in location $67.
31. Write a simple code to complement the contents of location $68 and place the result in R27.
32. Write a simple code to copy data from location $68 to PORTC using R19.

33. The status register is a(n) _____ -bit register.
34. Which bits of the status register are used for the C and H flag bits, respectively?
35. Which bits of the status register are used for the V and N flag bits, respectively?
36. In the ADD instruction, when is C raised?
37. In the ADD instruction, when is H raised?
38. What is the status of the C and Z flags after the following code?

```
LDI  R20,0xFF
LDI  R21,1
ADD  R20,R21
```

39. Find the C flag value after each of the following codes:

```
(a) LDI  R20,0x54    (b) LDI  R23,0      (c) LDI  R30,0xFF
    LDI  R25,0xC4        LDI  R16,0xFF        LDI  R18,0x05
    ADD  R20,R25        ADD  R23,R16        ADD  R30,R18
```

40. Write a simple program in which the value 0x55 is added 5 times.

SECTION 2.5: AVR DATA FORMAT AND DIRECTIVES

41. State the value (in hex) used for each of the following data:

```
.EQU  MYDAT_1  =  55
.EQU  MYDAT_2  =  98
.EQU  MYDAT_3  =  'G'
.EQU  MYDAT_4  =  0x50
.EQU  MYDAT_5  =  200
.EQU  MYDAT_6  =  'A'
.EQU  MYDAT_7  =  0xAA
.EQU  MYDAT_8  =  255
.EQU  MYDAT_9  =  0B10010000
.EQU  MYDAT_10  =  0b01111110
.EQU  MYDAT_11  =  10
.EQU  MYDAT_12  =  15
```

42. State the value (in hex) for each of the following data:

```
.EQU  DAT_1  =  22
.EQU  DAT_2  =  $56
.EQU  DAT_3  =  0b10011001
.EQU  DAT_4  =  32
.EQU  DAT_5  =  0xF6
.EQU  DAT_6  =  0B11111011
```

43. Show a simple code to (a) load the value $11 into locations $60–$65, and (b) add them together and place the result in R29 as the values are added. Use .EQU to assign the names TEMP0–TEMP5 to locations $60–$65.

44. Assembly language is a _____ (low, high)-level language while C is a _____ (low, high)-level language.
45. Of C and Assembly language, which is more efficient in terms of code generation (i.e., the amount of ROM space it uses)?
46. Which program produces the obj file?
47. True or false. The source file has the extension "asm".
48. True or false. The source code file can be a non-ASCII file.
49. True or false. Every source file must have .ORG and .EQU directives.
50. Do the .ORG and .SET directives produce opcodes?
51. Why are the directives also called pseudocode?
52. True or false. The .ORG directive appears in the ".lst" file.
53. The file with the _____ extension is downloaded into AVR Flash ROM.
54. Give three file extensions produced by AVR Studio.

SECTION 2.8: THE PROGRAM COUNTER AND PROGRAM ROM SPACE IN THE AVR

55. Every AVR family member wakes up at address _____ when it is powered up.
56. A programmer puts the first opcode at address $100. What happens when the microcontroller is powered up?
57. Find the number of bytes each of the following instructions takes:
    (a) `LDI R19,0x5`   (b) `LDI R30,$9F`   (c) `ADD R20,R21`
    (d) `ADD R22,R20`   (e) `LDI R18,0x41`   (f) `LDI R28,20`
    (g) `ADD R1,R3`   (h) `JMP`
58. Write a program to (a) place each of your 5-digit ID numbers into a RAM locations starting at address 0x100, (b) add each digit to R19 and store the sum in RAM location 0x306, and (c) use the program listing to show the ROM memory addresses and their contents.
59. Find the address of the last location of on-chip program ROM for each of the following:
    (a) AVR with 32 KB   (b) AVR with 8 KB
    (c) AVR with 64 KB   (d) AVR with 16 KB
    (f) AVR with 128 KB
60. Show the lowest and highest values (in hex) that the ATmega32 program counter can take.
61. A given AVR has $7FFF as the address of the last location of its on-chip ROM. What is the size of on-chip ROM for this AVR?
62. Repeat Question 61 for $3FF.
63. Find the on-chip program ROM size in K for the AVR chip with the following address ranges:
    (a) $0000–$1FFF   (b) $0000–$3FFF
    (c) $0000–$7FFF   (d) $0000–$FFFF
    (e) $0000–$1FFFF   (f) $00000–$3FFFF
    (g) $00000–$FFF   (h) $00000–$1FF

64. Find the on-chip program ROM size in K for the AVR chips with the following address ranges:
   (a) $00000–$3FF   (b) $00000–$7FF
   (c) $00000–$7FFFF   (d) $00000–$FFFFF
   (e) $00000–$1FFFFF   (f) $00000–$3FFFFF
   (g) $00000–$5FFF   (h) $00000–$BFFFF

(Some of the above might not be in production yet.)

65. How wide is the program ROM in the AVR chip?
66. How wide is the data bus between the CPU and the program ROM in the AVR chip?
67. In instruction "LDI R21,K" explain why the K value cannot be larger than 255 decimal.
68. $0C01 is the machine code for the _____ (LDI, STS, JMP, ADD) instruction.
69. In "STS memLocation,R22", explain what the size of the instruction is and how it allows one to cover the entire range of the data memory in the AVR chip.
70. In "LDS Rd, memLocation", explain what the size of the instruction is and how it allows one to cover the entire range of the data memory in the AVR chip.
71. Explain how the instruction "JMP target-addr" is able to cover the entire 4M address space of the AVR chip.

SECTION 2.9: RISC ARCHITECTURE IN THE AVR

72. What do RISC and CISC stand for?
73. In _____ (RISC, CISC) architecture we can have 1-, 2-, 3-, or 4-byte instructions.
74. In _____ (RISC, CISC) architecture instructions are fixed in size.
75. In _____ (RISC, CISC) architecture instructions are mostly executed in one or two cycles.
76. In _____ (RISC, CISC) architecture we can have an instruction to ADD a register to external memory.
77. True or false. Most instructions in CISC are executed in one or two cycles.

ANSWERS TO REVIEW QUESTIONS

SECTION 2.1: THE GENERAL PURPOSE REGISTERS IN THE AVR

1. LDI R29,0x34
2. LDI R18,0x16
    LDI R19,0xCD
    ADD R19,R18
3. False
4. FF hex and 255 in decimal
5. 8

SECTION 2.2: THE AVR DATA MEMORY

1. False
2. Data memory
3. 8
4. 3
5. 64K
6. 64

SECTION 2.3: USING INSTRUCTIONS WITH THE DATA MEMORY

1. True
2. LDI R20,0x95
   OUT SPL,R20
3. LDI R19,2
   ADD R18,R19
4. LDI R20,0x16
   LDI R21,0xCD
   ADD R20,R21
5. FF in hex or 255 in decimal
6. R16
7. It copies the contents of R23 into the OCR0 I/O register.
8. The OCR0 presents the I/O address of the OCR0 register ($3C); but we should use the data memory address while using the STS instruction. The instruction copies the contents of R23 into the location with the data memory address of $3C (the EECR I/O register).

SECTION 2.4: AVR STATUS REGISTER

1. SREG
2. 8 bits
3.

Hex	binary	
9F	1001 1111	
+ 61	+ 0110 0001	
100	10000 0000	This leads to C = 1, H = 1, and Z = 1.

4.

Hex	binary	
82	1000 0010	
+ 22	+ 0010 0010	
A4	1010 0100	This leads to C = 0, H = 0, and Z = 0.

5.

Hex	binary	
67	0110 0111	
+ 99	+ 1001 1001	
100	10000 0000	This leads to C = 1, H = 1, and Z = 1.

SECTION 2.5: AVR DATA FORMAT AND DIRECTIVES

1. .EQU DATA1 = 0x9F
   .EQU DATA2 = $9F

2. .EQU DATA1 = 0x99
   .EQU DATA2 = 99
   .EQU DATA3 = 0b10011001

3. If the value is to be changed later, it can be done once in one place instead of at every occurrence.

4. (a) $34   (b) $1F
5. 15 in decimal (0x0F in hex)
6. Value of location 0x200 = (0x95)
7. $0C + $10 = $1C will be in data memory location $63.

SECTION 2.6: INTRODUCTION TO AVR ASSEMBLY PROGRAMMING

1. The real work is performed by instructions such as LDI and ADD. Pseudo-instructions, also called assembly directives, instruct the assembler in doing its job.
2. The instruction mnemonics, pseudo-instructions
3. False
4. All except (c)
5. Assembler directives
6. True
7. (c)

SECTION 2.7: ASSEMBLING AN AVR PROGRAM

1. True
2. True
3. (a)
4. (b), (c), and (d)

SECTION 2.8: THE PROGRAM COUNTER AND PROGRAM ROM SPACE IN THE AVR

1. 22
2. True
3. 0000H
4. 2
5. 4
6. True

SECTION 2.9: RISC ARCHITECTURE IN THE AVR

1. RISC is reduced instruction set computer; CISC stands for complex instruction set computer.
2. True
3. Large
4. True
5. Around 130
6. True
7. (c)
8. False

CHAPTER 3

BRANCH, CALL, AND TIME DELAY LOOP

OBJECTIVES

Upon completion of this chapter, you will be able to:

>>   Code AVR Assembly language instructions to create loops
>>   Code AVR Assembly language conditional branch instructions
>>   Explain conditions that determine each conditional branch instruction
>>   Code JMP (long jump) instructions for unconditional jumps
>>   Calculate target addresses for conditional branch instructions
>>   Code AVR subroutines
>>   Describe the stack and its use in subroutines
>>   Discuss pipelining in the AVR
>>   Discuss crystal frequency versus instruction cycle time in the AVR
>>   Code AVR programs to generate a time delay

In the sequence of instructions to be executed, it is often necessary to transfer program control to a different location. There are many instructions in AVR to achieve this. This chapter covers the control transfer instructions available in AVR Assembly language. In Section 3.1, we discuss instructions used for looping, as well as instructions for conditional and unconditional branches (jumps). In Section 3.2, we examine the stack and the CALL instruction. In Section 3.3, instruction pipelining of the AVR is examined. Instruction timing and time delay subroutines are also discussed in Section 3.3.

SECTION 3.1: BRANCH INSTRUCTIONS AND LOOPING

In this section we first discuss how to perform a looping action in AVR and then the branch (jump) instructions, both conditional and unconditional.

Looping in AVR

Repeating a sequence of instructions or an operation a certain number of times is called a *loop*. The loop is one of most widely used programming techniques. In the AVR, there are several ways to repeat an operation many times. One way is to repeat the operation over and over until it is finished, as shown below:

```
LDI  R16,0        ;R16 = 0
LDI  R17,3        ;R17 = 3
ADD  R16,R17      ;add value 3 to R16 (R16 = 0x03)
ADD  R16,R17      ;add value 3 to R16 (R16 = 0x06)
ADD  R16,R17      ;add value 3 to R16 (R16 = 0x09)
ADD  R16,R17      ;add value 3 to R16 (R16 = 0x0C)
ADD  R16,R17      ;add value 3 to R16 (R16 = 0x0F)
ADD  R16,R17      ;add value 3 to R16 (R16 = 0x12)
```

In the above program, we add 3 to R16 six times. That makes $6 \times 3 = 18 = 0x12$. One problem with the above program is that too much code space would be needed to increase the number of repetitions to 50 or 100. A much better way is to use a loop. Next, we describe the method to do a loop in AVR.

Using BRNE instruction for looping

The BRNE (branch if not equal) instruction uses the zero flag in the status register. The BRNE instruction is used as follows:

```
BACK: ........    ;start of the loop
      ........    ;body of the loop
      ........    ;body of the loop
      DEC Rn      ;decrement Rn, Z = 1 if Rn = 0
      BRNE BACK   ;branch to BACK if Z = 0
```

In the last two instructions, the Rn (e.g., R16 or R17) is decremented; if it is not zero, it branches (jumps) back to the target address referred to by the label. Prior to the start of the loop, the Rn is loaded with the counter value for the number of repetitions. Notice that the BRNE instruction refers to the Z flag of the status register affected by the previous instruction, DEC. This is shown in Example 3-1.

In the program in Example 3-1, register R16 is used as a counter. The counter is first set to 10. In each iteration, the DEC instruction decrements the R16 and sets the flag bits accordingly. If R16 is not zero (Z = 0), it jumps to the target address associated with the label "AGAIN". This looping action continues until R16 becomes zero. After R16 becomes zero (Z = 1), it falls through the loop and executes the instruction immediately below it, in this case "OUT PORTB,R20". See Figure 3-1.

Example 3-1

Write a program to (a) clear R20, then (b) add 3 to R20 ten times, and (c) send the sum to PORTB. Use the zero flag and BRNE.

Solution:

```
;this program adds value 3 to the R20 ten times
.INCLUDE "M32DEF.INC"
      LDI   R16, 10      ;R16 = 10 (decimal) for counter
      LDI   R20, 0       ;R20 = 0
      LDI   R21, 3       ;R21 = 3
AGAIN:ADD   R20, R21     ;add 03 to R20 (R20 = sum)
      DEC   R16          ;decrement R16 (counter)
      BRNE  AGAIN        ;repeat until COUNT = 0
      OUT   PORTB,R20    ;send sum to PORTB
```

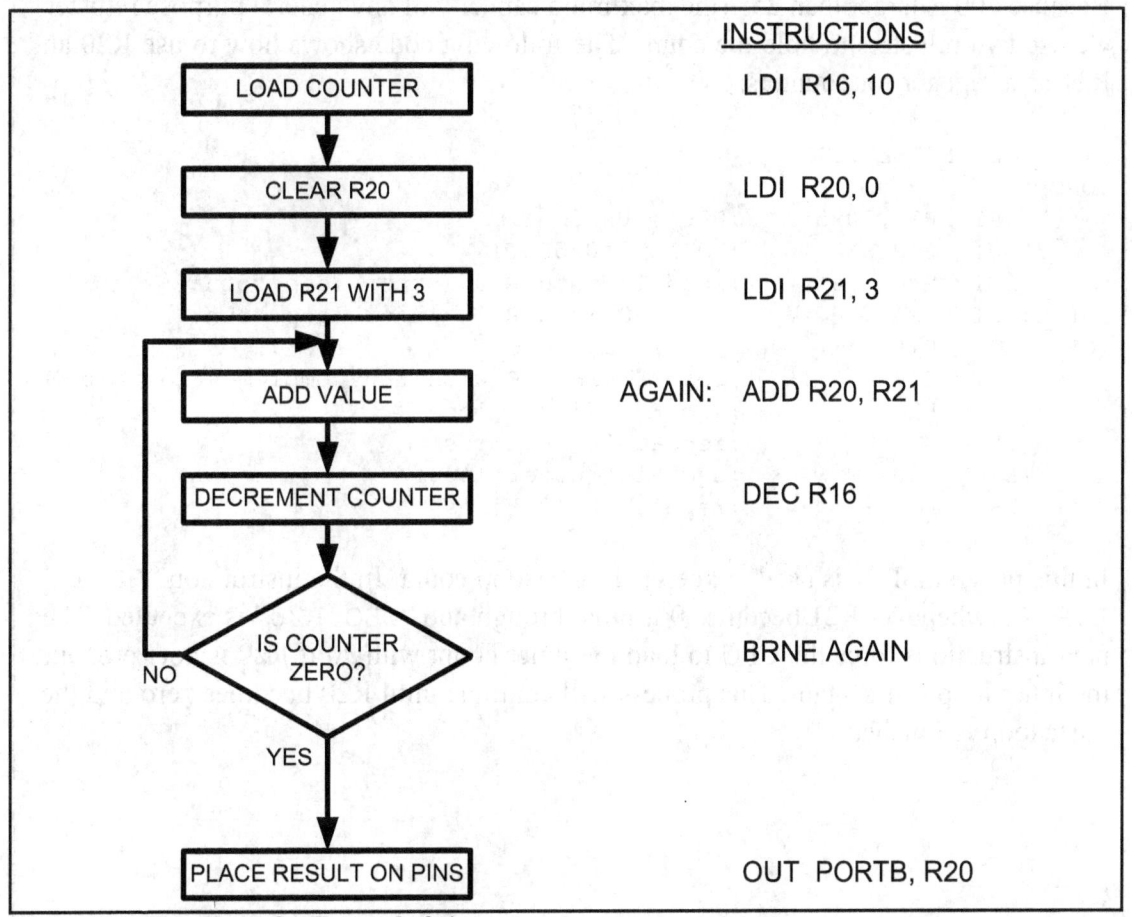

Figure 3-1. Flowchart for Example 3-1

Example 3-2

What is the maximum number of times that the loop in Example 3-1 can be repeated?

Solution:

Because location R16 is an 8-bit register, it can hold a maximum of 0xFF (255 decimal); therefore, the loop can be repeated a maximum of 255 times. Example 3-3 shows how to solve this limitation.

Loop inside a loop

As shown in Example 3-2, the maximum count is 255. What happens if we want to repeat an action more times than 255? To do that, we use a loop inside a loop, which is called a *nested loop*. In a nested loop, we use two registers to hold the count. See Example 3-3.

Example 3-3

Write a program to (a) load the PORTB register with the value 0x55, and (b) complement Port B 700 times.

Solution:

Because 700 is larger than 255 (the maximum capacity of any general purpose register), we use two registers to hold the count. The following code shows how to use R20 and R21 as a register for counters.

```
.INCLUDE "M32DEF.INC"
.ORG 0
        LDI   R16, 0x55     ;R16 = 0x55
        OUT   PORTB, R16     ;PORTB = 0x55
        LDI   R20, 10        ;load 10 into R20 (outer loop count)
LOP_1:LDI   R21, 70        ;load 70 into R21 (inner loop count)
LOP_2:COM   R16            ;complement R16
        OUT   PORTB, R16     ;load PORTB SFR with the complemented value
        DEC   R21            ;dec R21 (inner loop)
        BRNE  LOP_2          ;repeat it 70 times
        DEC   R20            ;dec R20 (outer loop)
        BRNE  LOP_1          ;repeat it 10 times
```

In this program, R21 is used to keep the inner loop count. In the instruction "BRNE LOP_2", whenever R21 becomes 0 it falls through and "DEC R20" is executed. The next instructions force the CPU to load the inner count with 70 if R20 is not zero, and the inner loop starts again. This process will continue until R20 becomes zero and the outer loop is finished.

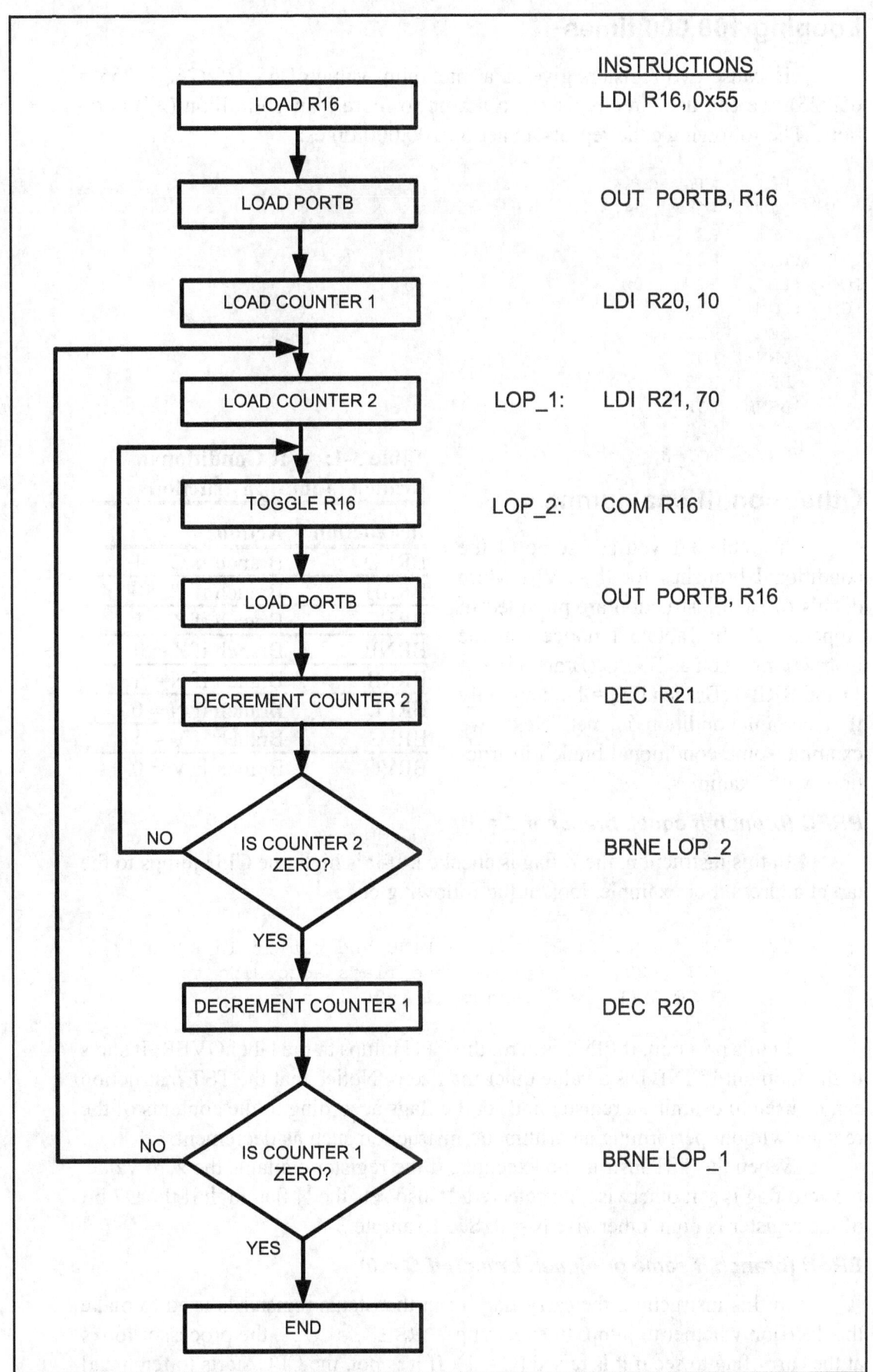

LOAD R16	LDI R16, 0x55
LOAD PORTB	OUT PORTB, R16
LOAD COUNTER 1	LDI R20, 10
LOAD COUNTER 2	LOP_1: LDI R21, 70
TOGGLE R16	LOP_2: COM R16
LOAD PORTB	OUT PORTB, R16
DECREMENT COUNTER 2	DEC R21
IS COUNTER 2 ZERO? NO ... YES	BRNE LOP_2
DECREMENT COUNTER 1	DEC R20
IS COUNTER 1 ZERO? NO ... YES	BRNE LOP_1
END	

Figure 3-2. Flowchart for Example 3-3

CHAPTER 3: BRANCH, CALL, AND TIME DELAY LOOP   111

Looping 100,000 times

Because two registers give us a maximum value of 65,025 ($255 \times 255 = 65,025$), we can use three registers to get up to more than 16 million (2^{24}) iterations. The following code repeats an action 100,000 times:

```
        LDI    R16, 0x55
        OUT    PORTB, R16
        LDI    R23, 10
LOP_3:LDI    R22, 100
LOP_2:LDI    R21, 100
LOP_1:COM    R16
        DEC    R21
        BRNE   LOP_1
        DEC    R22
        BRNE   LOP_2
        DEC    R23
        BRNE   LOP_3
```

Other conditional jumps

In Table 3-1 you see some of the conditional branches for the AVR. More details of each instruction are provided in Appendix A. In Table 3-1 notice that the instructions, such as BREQ (Branch if Z = 1) and BRLO (Branch if C = 1), jump only if a certain condition is met. Next, we examine some conditional branch instructions with examples.

Table 3-1: AVR Conditional Branch (Jump) Instructions

Instruction	Action
BRLO	Branch if C = 1
BRSH	Branch if C = 0
BREQ	Branch if Z = 1
BRNE	Branch if Z = 0
BRMI	Branch if N = 1
BRPL	Branch if N = 0
BRVS	Branch if V = 1
BRVC	Branch if V = 0

BREQ (branch if equal, branch if Z = 1)

In this instruction, the Z flag is checked. If it is high, the CPU jumps to the target address. For example, look at the following code.

```
OVER: IN    R20, PINB    ;read PINB and put it in R20
       TST   R20          ;set the flags according to R20
       BREQ  OVER         ;jump if R20 is zero
```

In this program, if PINB is zero, the CPU jumps to the label OVER. It stays in the loop until PINB has a value other than zero. Notice that the TST instruction can be used to examine a register and set the flags according to the contents of the register without performing an arithmetic instruction such as decrement.

When the TST instruction executes, if the register contains the zero value, the zero flag is set; otherwise, it is cleared. It also sets the N flag high if the D7 bit of the register is high, otherwise N = 0. See Example 3-4.

BRSH (branch if same or higher, branch if C = 0)

In this instruction, the carry flag bit in the Status register is used to make the decision whether to jump. In executing "BRSH label", the processor looks at the carry flag to see if it is raised (C = 1). If it is not, the CPU starts to fetch and execute instructions from the address of the label. If C = 1, it will not branch but

Example 3-4

Write a program to determine if RAM location 0x200 contains the value 0. If so, put 0x55 into it.

Solution:

```
      .EQU   MYLOC=0x200
      LDS    R30, MYLOC
      TST    R30                ;set the flag
                                ;(Z=1 if R30 has zero value)
      BRNE   NEXT               ;branch if R30 is not zero (Z=0)
      LDI    R30, 0x55          ;put 0x55 if R30 has zero value
      STS    MYLOC,R30          ;and store a copy to loc $200
NEXT: ...
```

will execute the next instruction below BRSH. Study Example 3-5 to see how BRSH is used to add numbers together when the sum is higher than $FF. Note that there is also a "BRLO label" instruction. In the BRLO instruction, if C = 1 the CPU jumps to the target address. We will give more examples of these instructions in the context of some applications in Chapter 5.

The other conditional branch instructions in Table 3-1 are discussed in Chapter 5 when arithmetic operations with signed numbers are discussed.

Example 3-5

Find the sum of the values 0x79, 0xF5, and 0xE2. Put the sum into R20 (low byte) and R21 (high byte).

Solution:

```
.INCLUDE "M32DEF.INC"
.ORG 0
      LDI    R21, 0       ;clear high byte (R21 = 0)
      LDI    R20, 0       ;clear low byte (R20 = 0)
      LDI    R16, 0x79
      ADD    R20, R16     ;R20 = 0 + 0x79 = 0x79, C = 0
      BRSH   N_1          ;if C = 0, add next number
      INC    R21          ;C = 1, increment (now high byte = 0)
N_1:  LDI    R16, 0xF5
      ADD    R20, R16     ;R20 = 0x79 + 0xF5 = 0x6E and C = 1
      BRSH   N_2          ;branch if C = 0
      INC    R21          ;C = 1, increment (now high byte = 1)
N_2:  LDI    R16, 0xE2
      ADD    R20, R16     ;R20 = 0x6E + 0xE2 = 0x50 and C = 1
      BRSH   OVER         ;branch if C = 0
      INC    R21          ;C = 1, increment (now high byte = 2)
OVER:                     ;now low byte = 0x50, and high byte = 02
```

	R21 (high byte)	R20 (low byte)
At first	$0	$00
Before LDI R16,0xF5	$0	$79
Before LDI R16,0xE2	$1	$6E
At the end	$2	$50

All conditional branches are short jumps

It must be noted that all conditional jumps are short jumps, meaning that the address of the target must be within 64 bytes of the program counter (PC). This concept is discussed next.

Example 3-6

Using the following list file, verify the jump forward address calculation.

```
LINE    ADDRESS      Machine Mnemonic Operand
3:     +00000000:   E050   LDI    R21, 0      ;clear high byte (R21 = 0)
4:     +00000001:   E040   LDI    R20, 0      ;clear low byte (R20 = 0)
5:     +00000002:   E709   LDI    R16, 0x79
6:     +00000003:   0F40   ADD    R20, R16    ;R20 = 0 + 0x79 = 0x79, C = 0
7:     +00000004:   F408   BRSH   N_1         ;if C = 0, add next number
8:     +00000005:   9543   INC    R21   ;C = 1, increment (now high byte = 0)
9:     +00000006:   EF05 N_1:LDI    R16, 0xF5
10:    +00000007:   0F40   ADD    R20, R16    ;R20 = $79 + $F5 = 6E and C = 1
11:    +00000008:   F408   BRSH   N_2         ;branch if C = 0
12:    +00000009:   9553   INC    R21   ;C = 1, increment (now high byte = 1)
@0000000A: n_2
13:    +0000000A:   EE02 N_2:LDI    R16, 0xE2
14:    +0000000B:   0F40   ADD    R20, R16    ;R20 = $6E + $E2 = $50 and C = 1
15:    +0000000C:   F408   BRSH   OVER        ;branch if C = 0
16:    +0000000D:   9553   INC    R21   ;C = 1, increment (now high byte = 2)
@0000000E: over
```

Solution:

First notice that the BRSH instruction jumps forward. The target address for a forward jump is calculated by adding the PC of the following instruction to the second byte of the branch instruction. Recall that each instruction takes 2 bytes. In line 7 the instruction "BRSH N_1" has the machine code of F408. To distinguish the operand and opcode parts, we should compare the machine code with the BRSH instruction format. In the following, you see the format of the BRSH instruction. The bits marked by k are

1111	01kk	kkkk	k000

the operand bits while the remainder are the opcode bits. In this example the machine code's equivalent in binary is 1111 0100 0000 1000. If we compare it with the BRSH format, we see that the operand is 000001 and the opcode is 111101000. The 01 is the relative address, relative to the address of the next instruction INC R21, which is 000005. By adding 000001 to 000005, the target address of the label N_1, which is 000006, is generated. Likewise for line 11, the "BRSH N_2" instruction, and line 000015, the "BRSH OVER" instruction jumps forward because the relative value is positive.

All the conditional jump instructions, whose mnemonics begin with BR (e.g., BRNE and BRIE), have the same instruction format, and the opcode changes from instruction to instruction. So, we can calculate the short branch address for any of them, as we did in this example.

Calculating the short branch address

All conditional branches such as BRSH, BREQ, and BRNE are short branches due to the fact that they are all 2-byte instructions. In these instructions the opcode is 9 bits and the relative address is 7 bits. The target address is relative to the value of the program counter. If the relative address is positive, the jump is forward. If the relative address is negative, then the jump is backwards. The relative address can be a value from –64 to +63. To calculate the target address, the relative address is added to the PC of the next instruction (target address = relative address + PC). See Example 3-6. We do the same thing for the backward branch, although the second byte is negative. That is, we add it to the PC value of the next instruction. See Example 3-7.

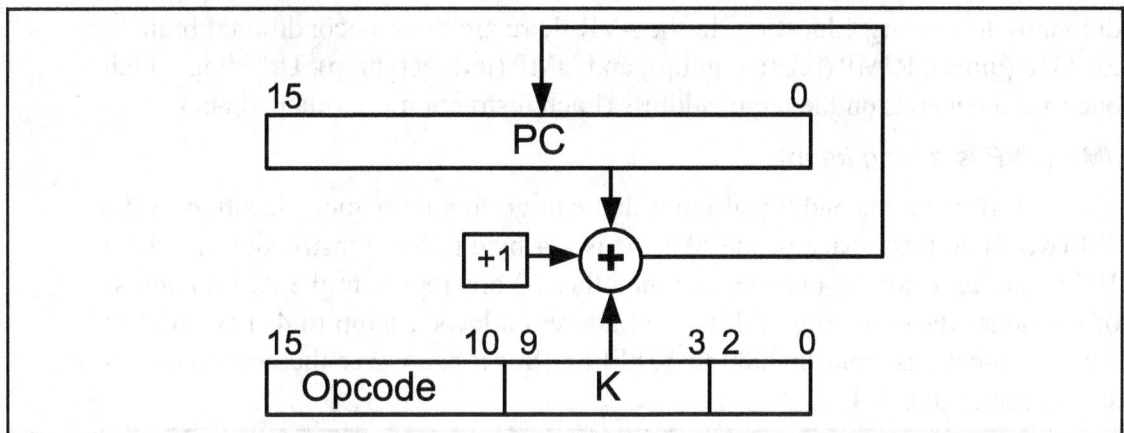

Figure 3-3. Calculating the Target Address in Conditional Branch Instructions

---

Example 3-7

Verify the calculation of backward jumps for the listing of Example 3-1, shown below.

Solution:
```
LINE   ADDRESS     Machine Mnemonic Operand
3:     +00000000:  E00A    LDI       R16, 10    ;R16 = 10 (decimal) for counter
4:     +00000001:  E040    LDI       R20, 0     ;R20 = 0
5:     +00000002:  E053    LDI       R21, 3     ;R21 = 3
6:     +00000003:  0F45 AGAIN:ADD    R20, R21   ;add 03 to R20 (R20 = sum)
7:     +00000004:  950A    DEC       R16        ;decrement R16 (counter)
8:     +00000005:  F7E9    BRNE      AGAIN      ;repeat until COUNT = 0
9:     +00000006:  BB48    OUT       PORTB,R20  ;send sum to PORTB SFR
```

In the program list, "BRNE AGAIN" has machine code F7E9. To specify the operand and opcode, we compare the instruction with the branch instruction format, which you saw in the previous example. Because the binary equivalent of the instruction is 1111 0111 1110 1001, the opcode is 111101001 and the operand (relative address) is 1111101. The 1111101 gives us –3, which means the displacement is –3. When the relative address of –3 is added to 000006, the address of the instruction below, we have –3 + 06 = 03 (the carry is dropped). Notice that 000003 is the address of the label AGAIN. 1111101 is a negative number, which means it will branch backward. For further discussion of the addition of negative numbers, see Chapter 5.

CHAPTER 3: BRANCH, CALL, AND TIME DELAY LOOP

You might ask why we add the relative address to the address of the next instruction. (Why don't we add the relative address to the address of the current instruction?) Before an instruction is executed, it should be fetched. So, the branch instructions are executed after they are fetched. The PC points to the instruction that should be fetched next. So, when the branch instructions are executed, the PC is pointing to the next instruction. That is why we add the relative address to the address of the next instruction. We will discuss the execution of instructions in the last section of this chapter.

Unconditional branch instruction

The unconditional branch is a jump in which control is transferred unconditionally to the target location. In the AVR there are three unconditional branches: JMP (jump), RJMP (relative jump), and IJMP (indirect jump). Deciding which one to use depends on the target address. Each instruction is explained next.

JMP (JMP is a long jump)

JMP is an unconditional jump that can go to any memory location in the 4M (word) address space of the AVR. It is a 4-byte (32-bit) instruction in which 10 bits are used for the opcode, and the other 22 bits represent the 22-bit address of the target location. The 22-bit target address allows a jump to 4M (words) of memory locations from 000000 to $3FFFFF. So, it can cover the entire address space. See Figure 3-4.

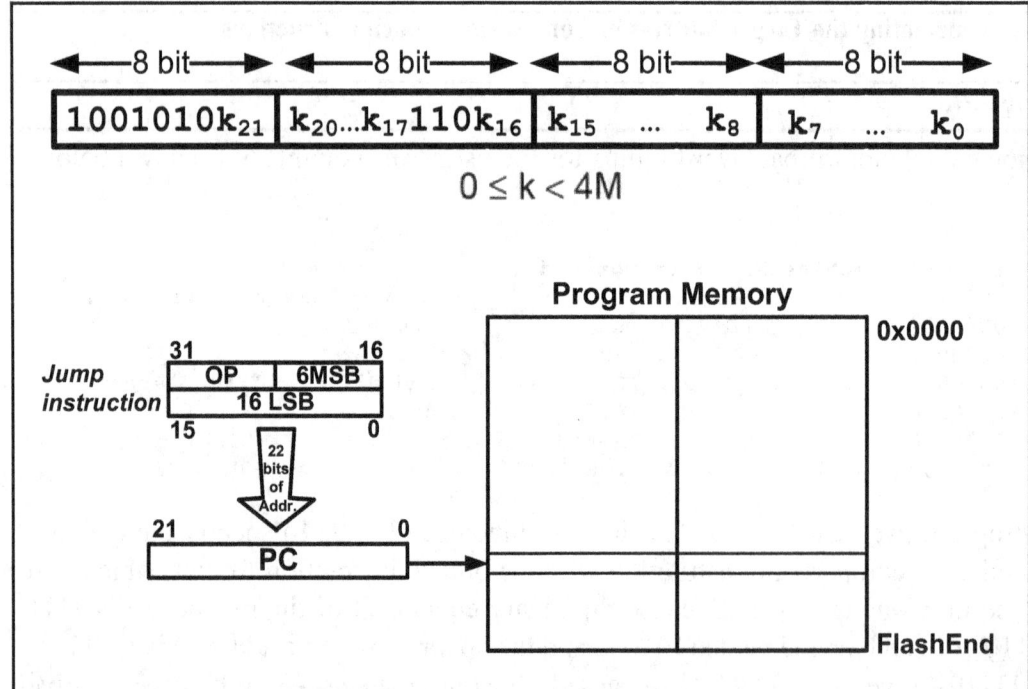

Figure 3-4. JMP Instruction

Remember that although the AVR can have ROM space of 8M bytes, not all AVR family members have that much on-chip program ROM. Some AVR family members have only 4K–32K of on-chip ROM for program space; consequently, every byte is precious. For this reason there is also an RJMP (relative jump) instruction, which is a 2-byte instruction as opposed to the 4-byte JMP instruction. This can save some bytes of memory in many applications where ROM memory space is in short supply. RJMP is discussed next.

RJMP (relative jump)

In this 2-byte (16-bit) instruction, the first 4 bits are the opcode and the rest (lower 12 bits) is the relative address of the target location. The relative address range of 000 – $FFF is divided into forward and backward jumps; that is, within –2048 to +2047 words of memory relative to the address of the current PC (program counter). If the jump is forward, then the relative address is positive. If the jump is backward, then the relative address is negative. In this regard, RJMP is like the conditional branch instructions except that 12 bits are used for the offset address instead of 7. This is shown in detail in Figure 3-5.

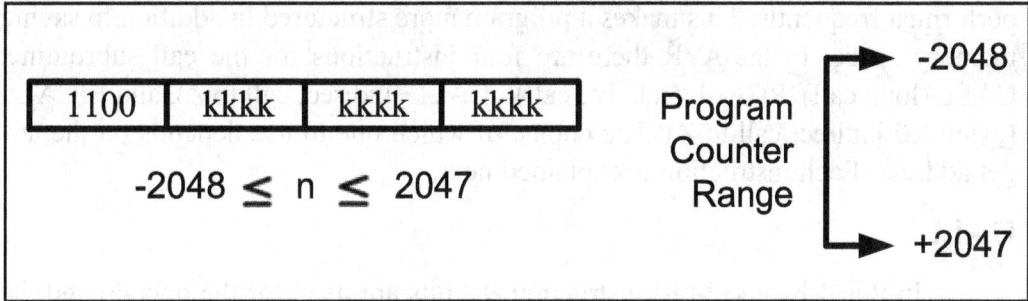

Figure 3-5. RJMP (Relative Jump) Instruction Address Range

Notice that this is a 2-byte instruction, and is preferred over the JMP because it takes less ROM space.

IJMP (indirect jump)

IJMP is a 2-byte instruction. When the instruction executes, the PC is loaded with the contents of the Z register, so it jumps to the address pointed to by the Z register. As you will see in Chapter 6, Z is a 2-byte register, so IJMP can jump within the lowest 64K words of the program memory.

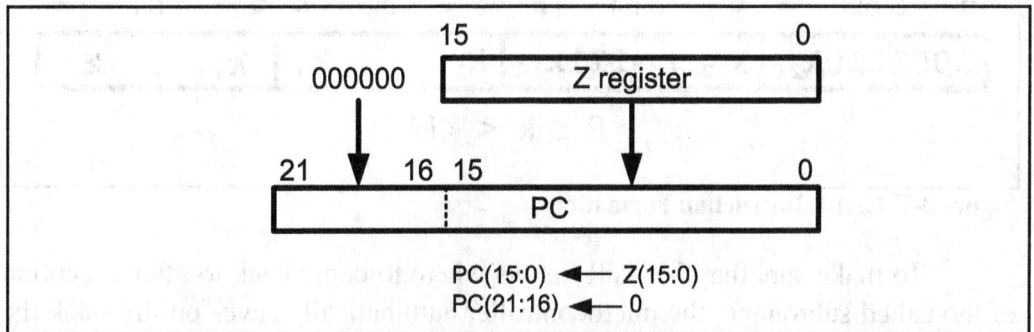

Figure 3-6. IJMP (Indirect Jump) Instruction Target Address

In the other jump instructions, the target address is static, which means that in a specific condition they jump to a fixed point. But IJMP has a dynamic target point, and we can dynamically change the target address by changing the Z register's contents through the program.

Review Questions

1. The mnemonic BRNE stands for _____.
2. True or false. "BRNE BACK" makes its decision based on the last instruction affecting the Z flag.
3. "BRNE HERE" is a ___ -byte instruction.
4. In "BREQ NEXT", which register's content is checked to see if it is zero?
5. JMP is a(n) ___ -byte instruction.

SECTION 3.2: CALL INSTRUCTIONS AND STACK

Another control transfer instruction is the CALL instruction, which is used to call a subroutine. Subroutines are often used to perform tasks that need to be performed frequently. This makes a program more structured in addition to saving memory space. In the AVR there are four instructions for the call subroutine: CALL (long call) RCALL (relative call), ICALL (indirect call to Z), and EICALL (extended indirect call to Z). The choice of which one to use depends on the target address. Each instruction is explained next.

CALL

In this 4-byte (32-bit) instruction, 10 bits are used for the opcode and the other 22 bits, k21–k0, are used for the address of the target subroutine, just as in the JMP instruction. Therefore, CALL can be used to call subroutines located anywhere within the 4M address space of 000000–$3FFFFF for the AVR, as shown in Figure 3-7.

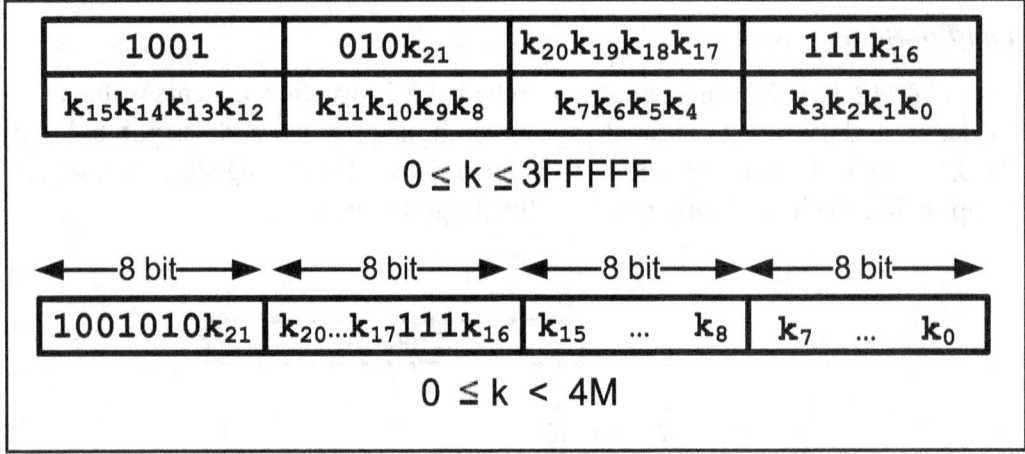

Figure 3-7. CALL Instruction Formation

To make sure that the AVR knows where to come back to after execution of the called subroutine, the microcontroller automatically saves on the stack the address of the instruction immediately below the CALL. When a subroutine is called, control is transferred to that subroutine, and the processor saves the PC (program counter) of the next instruction on the stack and begins to fetch instructions from the new location. After finishing execution of the subroutine, the RET instruction transfers control back to the caller. Every subroutine needs RET as the last instruction.

Stack and stack pointer in AVR

The stack is a section of RAM used by the CPU to store information temporarily. This information could be data or an address. The CPU needs this storage area because there are only a limited number of registers.

How stacks are accessed in the AVR

If the stack is a section of RAM, there must be a register inside the CPU to point to it. The register used to access the stack is called the SP (stack pointer) register.

In I/O memory space, there are two registers named SPL (the low byte of the SP) and SPH (the high byte of the SP). The SP is implemented as two registers. The SPH register presents the high byte of the SP while the SPL register presents the lower byte.

Figure 3-8. SP (Stack Pointer) in AVR

The stack pointer must be wide enough to address all the RAM. So, in the AVRs with more than 256 bytes of memory the SP is made of two 8-bit registers (SPL and SPH), while in the AVRs with less than 256 bytes the SP is made of only SPL, as an 8-bit register can address 256 bytes of memory.

The storing of CPU information such as the program counter on the stack is called a PUSH, and the loading of stack contents back into a CPU register is called a POP. In other words, a register is pushed onto the stack to save it and popped off the stack to retrieve it. The following describes each process.

Pushing onto the stack

The stack pointer (SP) points to the top of the stack (TOS). As we push data onto the stack, the data are saved where the SP points to, and the SP is decremented by one. Notice that this is the same as with many other microprocessors, notably x86 processors, in which the SP is decremented when data is pushed onto the stack.

To push a register onto stack we use the PUSH instruction.

```
PUSH Rr    ;Rr can be any of the general purpose registers (R0-R31)
```

For example, to store the value of R10 we can write the following instruction:

```
PUSH R10    ;store R10 onto the stack, and decrement SP
```

Popping from the stack

Popping the contents of the stack back into a given register is the opposite process of pushing. When the POP instruction is executed, the SP is incremented and the top location of the stack is copied back to the register. That means the stack is LIFO (Last-In-First-Out) memory.

To retrieve a byte of data from stack we can use the POP instruction.

```
POP Rr    ;Rr can be any of the general purpose registers (R0-R31)
```

For example, the following instruction pops from the top of stack and copies to R10:

```
POP R16   ;increment SP, and then load the top of stack to R10
```

Example 3-8

This example shows the stack and stack pointer and the registers used after the execution of each instruction.

```
.INCLUDE "M32DEF.INC"
    .ORG 0
    ;initialize the SP to point to the last location of RAM (RAMEND)
    LDI   R16, HIGH(RAMEND)          ;load SPH
    OUT   SPH, R16
    LDI   R16, LOW(RAMEND)           ;load SPL
    OUT   SPL, R16

    LDI   R31, 0
    LDI   R20, 0x21
    LDI   R22, 0x66

    PUSH  R20
    PUSH  R22

    LDI   R20, 0
    LDI   R22, 0

    POP   R22
    POP   R31
```

Solution:

After the execution of	Contents of some of the registers				Stack
	R20	R22	R31	SP	
OUT SPL,R16	$0	$0	0	$085F	85D / 85E / 85F — SP
LDI R22, 0x66	$21	$66	0	$085F	85D / 85E / 85F — SP
PUSH R20	$21	$66	0	$085E	85D / 85E — SP / 85F 21
PUSH R22	$21	$66	0	$085D	85D — SP / 85E 66 / 85F 21
LDI R22, 0	$0	$0	0	$085D	85D — SP / 85E 66 / 85F 21
POP R22	$0	$66	0	$085E	85D / 85E — SP / 85F 21
POP R31	$0	$66	$21	$085F	85D / 85E / 85F — SP

Initializing the stack pointer

When the AVR is powered up, the SP register contains the value 0, which is the address of R0. Therefore, we must initialize the SP at the beginning of the program so that it points to somewhere in the internal SRAM. In AVR, the stack grows from higher memory location to lower memory location (when we push onto the stack, the SP decrements). So, it is common to initialize the SP to the uppermost memory location.

Different AVRs have different amounts of RAM. In the AVR assembler RAMEND represents the address of the last RAM location. So, if we want to initialize the SP so that it points to the last memory location, we can simply load RAMEND into the SP. Notice that SP is made of two registers, SPH and SPL. So, we load the high byte of RAMEND into SPH, and the low byte of RAMEND into the SPL.

Example 3-8 shows how to initialize the SP and use the PUSH and POP instructions. In the example you can see how the stack changes when the PUSH and POP instructions are executed.

For more information about RAMEND, LOW, and HIGH, see Section 6-1.

CALL instruction and the role of the stack

When a subroutine is called, the processor first saves the address of the instruction just below the CALL instruction on the stack, and then transfers control to that subroutine. This is how the CPU knows where to resume when it returns from the called subroutine.

For the AVRs whose program counter is not longer than 16 bits (e.g., ATmega128, ATmega32), the value of the program counter is broken into 2 bytes. The higher byte is pushed onto the stack first, and then the lower byte is pushed.

For the AVRs whose program counters are longer than 16 bits but shorter than 24 bits, the value of the program counter is broken up into 3 bytes. The highest byte is pushed first, then the middle byte is pushed, and finally the lowest byte is pushed. So, in both cases, the higher bytes are pushed first.

RET instruction and the role of the stack

When the RET instruction at the end of the subroutine is executed, the top location of the stack is copied back to the program counter and the stack pointer is incremented. When the CALL instruction is executed, the address of the instruction below the CALL instruction is pushed onto the stack; so, when the execution of the function finishes and RET is executed, the address of the instruction below the CALL is loaded into the PC, and the instruction below the CALL instruction is executed.

To understand the role of the stack in call instruction and returning, examine the contents of the stack and stack pointer (see Examples 3-9 and Example 3-10). The following points should be noted for the program in Example 3-9:

1. Notice the DELAY subroutine. After the first "CALL DELAY" is executed, the address of the instruction right below it, "LDI R16, 0xAA", is pushed onto

the stack, and the AVR starts to execute instructions at address 0x300.

2. In the DELAY subroutine, the counter R20 is set to 255 (R20 = 0xFF); therefore, the loop is repeated 255 times. When R20 becomes 0, control falls to the RET instruction, which pops the address from the top of the stack into the program counter and resumes executing the instructions after the CALL.

Example 3-9

Toggle all the bits of Port B by sending to it the values $55 and $AA continuously. Put a time delay between each issuing of data to Port B.

Solution:

```
.INCLUDE "M32DEF.INC"
.ORG 0
        LDI R16,HIGH(RAMEND)     ;load SPH
        OUT SPH,R16
        LDI R16,LOW(RAMEND)      ;load SPL
        OUT SPL,R16

BACK:
        LDI   R16,0x55      ;load R16 with 0x55
        OUT   PORTB,R16     ;send 55H to port B
        CALL  DELAY         ;time delay
        LDI   R16,0xAA      ;load R16 with 0xAA
        OUT   PORTB,R16     ;send 0xAA to port B
        CALL  DELAY         ;time delay
        RJMP  BACK          ;keep doing this indefinitely
;------ this is the delay subroutine
        .ORG 0x300          ;put time delay at address 0x300
DELAY:
        LDI   R20,0xFF      ;R20 = 255,the counter
AGAIN:
        NOP                 ;no operation wastes clock cycles
        NOP
        DEC   R20
        BRNE  AGAIN         ;repeat until R20 becomes 0
        RET                 ;return to caller
```

The amount of time delay in Example 3-9 depends on the frequency of the AVR. How to calculate the exact time will be explained in the last section of this chapter.

The upper limit of the stack

As mentioned earlier, we can define the stack anywhere in the general purpose memory. So, in the AVR the stack can be as big as its RAM. Note that we must not define the stack in the register memory, nor in the I/O memory. So, the SP must be set to point above 0x60.

In AVR, the stack is used for calls and interrupts. We must remember that upon calling a subroutine, the stack keeps track of where the CPU should return after completing the subroutine. For this reason, we must be very careful when manipulating the stack contents.

Example 3-10

Analyze the stack for the CALL instructions in the following program.

```
            .INCLUDE "M32DEF.INC"
            .ORG 0
+00000000:      LDI R16,HIGH(RAMEND)    ;initialize SP
+00000001:      OUT SPH,R16
+00000002:      LDI R16,LOW(RAMEND)
+00000003:      OUT SPL,R16

            BACK:
+00000004:      LDI   R16,0x55      ;load R16 with 0x55
+00000005:      OUT   PORTB,R16     ;send 55H to port B
+00000006:      CALL DELAY          ;time delay
+00000008:      LDI   R16,0xAA      ;load R16 with 0xAA
+00000009:      OUT   PORTB,R16     ;send 0xAA to port B
+0000000A:      CALL DELAY          ;time delay
+0000000C:      RJMP BACK           ;keep doing this indefinitely
            ;-----this is the delay subroutine
            .ORG 0x300        ;put time delay at address 0x300
            DELAY:
+00000300:      LDI   R20,0xFF      ;R20 = 255, the counter
            AGAIN:
+00000301:      NOP                 ;no operation wastes clock cycles
+00000302:      NOP
+00000303:      DEC   R20
+00000304:      BRNE AGAIN          ;repeat until R20 becomes 0
+00000305:      RET                 ;return to caller
```

Solution:

When the first CALL is executed, the address of the instruction "LDI R16,0xAA" is saved (pushed) onto the stack. The last instruction of the called subroutine must be a RET instruction, which directs the CPU to pop the contents of the top location of the stack into the PC and resume executing at address 0008. The diagrams show the stack frame after the CALL and RET instructions.

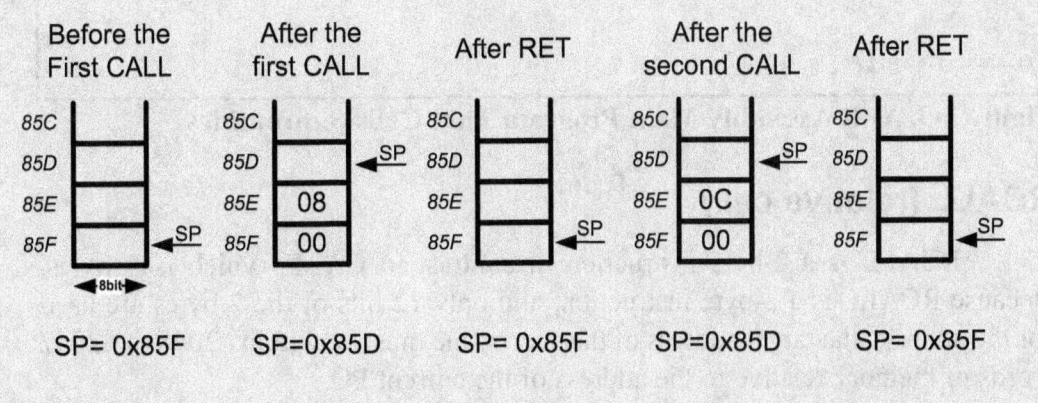

Calling many subroutines from the main program

In Assembly language programming, it is common to have one main program and many subroutines that are called from the main program. See Figure 3-9. This allows you to make each subroutine into a separate module. Each module can be tested separately and then brought together with the main program. More importantly, in a large program the modules can be assigned to different programmers in order to shorten development time.

It needs to be emphasized that in using CALL, the target address of the subroutine can be anywhere within the 4M (word) memory space of the AVR. See Example 3-11. This is not the case for the RCALL instruction, which is explained next.

```
.INCLUDE "M32DEF.INC"    ;Modify for your chip

;MAIN program calling subroutines
            .ORG  0
MAIN:       CALL  SUBR_1
            CALL  SUBR_2
            CALL  SUBR_3
            CALL  SUBR_4
HERE:       RJMP  HERE          ;stay here
;————————end of MAIN
;
SUBR_1:     ....
            ....
            RET
;————————end of subroutine 1
;
SUBR_2:     ....
            ....
            RET
;————————end of subroutine 2

SUBR_3:     ....
            ....
            RET
;————————end of subroutine 3

SUBR_4:     ....
            ....
            RET
;————————end of subroutine 4
```

Figure 3-9. AVR Assembly Main Program That Calls Subroutines

RCALL (relative call)

RCALL is a 2-byte instruction in contrast to CALL, which is 4 bytes. Because RCALL is a 2-byte instruction, and only 12 bits of the 2 bytes are used for the address, the target address of the subroutine must be within –2048 to +2047 words of memory relative to the address of the current PC.

Example 3-11

Write a program to count up from 00 to $FF and send the count to Port B. Use one CALL subroutine for sending the data to Port B and another one for time delay. Put a time delay between each issuing of data to Port B.

Solution:

```
                    .INCLUDE "M32DEF.INC"

                    .DEF COUNT=R20
                    .ORG 0
+00000000:          LDI  R16,HIGH(RAMEND)
+00000001:          OUT  SPH,R16
+00000002:          LDI  R16,LOW(RAMEND)
+00000003:          OUT  SPL,R16            ;initialize stack pointer

+00000004:          LDI  COUNT,0            ;count = 0
            BACK:
+00000005:          CALL DISPLAY
+00000007:          RJMP BACK

                    ;-----Increment and put it in PORTB
            DISPLAY:
+00000008:             INC  COUNT           ;increment count
+00000009:             OUT  PORTB, COUNT    ;send it to PORTB
+0000000A:             CALL DELAY
+0000000C:             RET                  ;return to caller

                    ;-----This is the delay subroutine
                    .ORG 0x300              ;put time delay at address 0x300
            DELAY:
+00000300:          LDI  R16,0xFF           ;R16 = 255
+00000301:  AGAIN:
+00000302:          NOP
+00000303:          NOP
+00000304:          NOP
+00000305:          DEC  R16
+00000306:          BRNE AGAIN              ;repeat until R16 becomes 0
+00000307:          RET                     ;return to caller
```

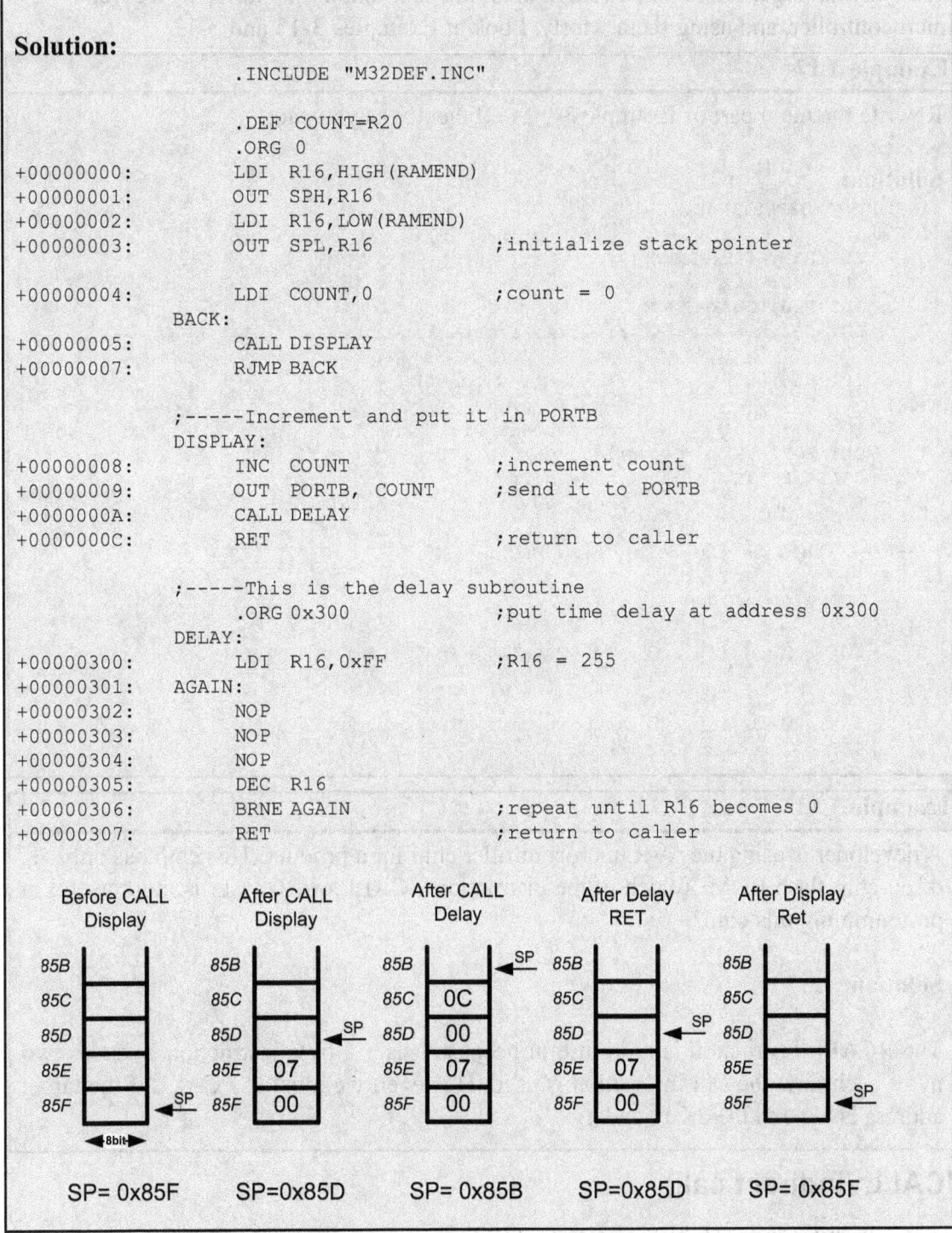

There is no difference between RCALL and CALL in terms of saving the program counter on the stack or the function of the RET instruction. The only difference is that the target address for CALL can be anywhere within the 4M address space of the AVR while the target address of RCALL must be within a 4K range.

In many variations of the AVR marketed by Atmel Corporation, on-chip ROM is as low as 4K. In such cases, the use of RCALL instead of CALL can save a number of bytes of program ROM space.

Of course, in addition to using compact instructions, we can program efficiently by having a detailed knowledge of all the instructions supported by a given microcontroller, and using them wisely. Look at Examples 3-12 and 3-13.

Example 3-12

Rewrite the main part of Example 3-9 as efficiently as you can.

Solution:
```
.INCLUDE "M32DEF.INC"
.ORG 0
      LDI  R16,HIGH(RAMEND)
      OUT  SPH,R16
      LDI  R16,LOW(RAMEND)
      OUT  SPL,R16            ;initialize stack pointer

      LDI  R16,0x55           ;load R16 with 55H
BACK:
      COM  R16                ;complement R16
      OUT  PORTB,R16          ;load port B SFR
      RCALL DELAY             ;time delay
      RJMP BACK               ;keep doing this indefinitely

;---------this is the delay subroutine
DELAY:
      LDI  R20,0xFF
AGAIN:
      NOP                     ;no operation wastes clock cycles
      NOP
      DEC  R20
      BRNE AGAIN              ;repeat until R20 becomes 0
      RET
```

Example 3-13

A developer is using the AVR microcontroller chip for a product. This chip has only 4K of on-chip flash ROM. Which of the instructions, CALL or RCALL, is more useful in programming this chip?

Solution:

The RCALL instruction is more useful because it is a 2-byte instruction. It saves two bytes each time the call instruction is used. However, we must use CALL if the target address is beyond the 2K boundary.

ICALL (indirect call)

In this 2-byte (16-bit) instruction, the Z register specifies the target address. When the instruction is executed, the address of the next instruction is pushed into the stack (like CALL and RCALL) and the program counter is loaded with the contents of the Z register. So, the Z register should contain the address of a function when the ICALL instruction is executed. Because the Z register is 16 bits wide, the ICALL instruction can call the subroutines that are within the lowest 64K words of the program memory. (The target address calculation in ICALL is the same as for the IJMP instruction.) See Figure 3-10.

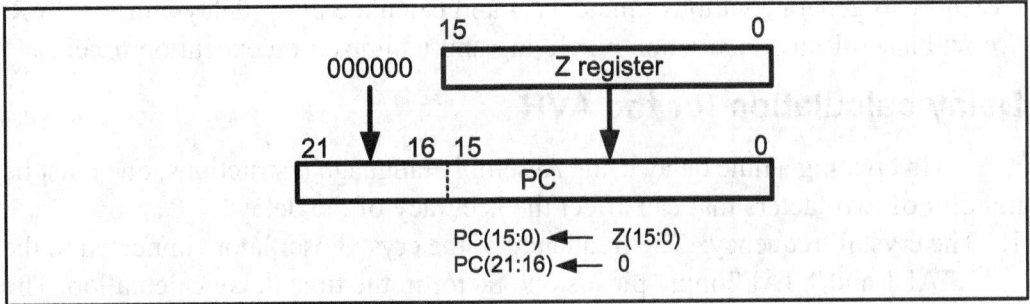

Figure 3-10. ICALL Instruction

In the AVRs with more than 64K words of program memory, the EICALL (extended indirect call) instruction is available. The EICALL loads the Z register into the lower 16 bits of the PC and the EIND register into the upper 6 bits of the PC. Notice that EIND is a part of I/O memory. See Figure 3-11.

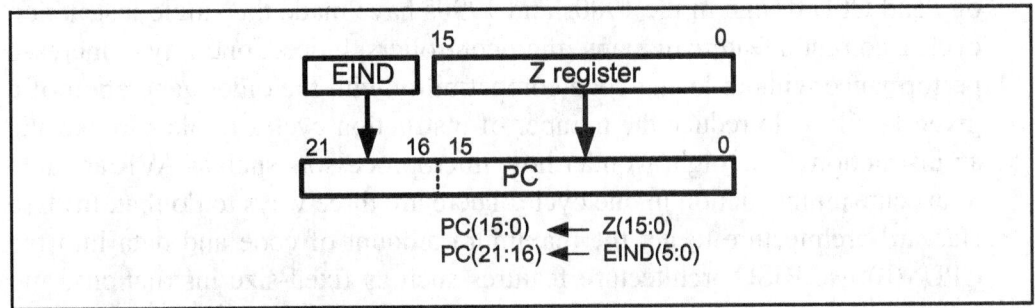

Figure 3-11. EICALL Instruction

The ICALL and EICALL instructions can be used to implement pointer to function.

Review Questions

1. True or false. In the AVR, control can be transferred anywhere within the 4M of code space by using the CALL instruction.
2. The CALL instruction is a(n) ___ -byte instruction.
3. True or false. In the AVR, control can be transferred anywhere within the 4M of code space by using the RCALL instruction.
4. With each CALL instruction, the stack pointer register, SP, is _____ (incremented, decremented).
5. With each RET instruction, the SP is _____ (incremented, decremented).
6. On power-up, the SP points to address _____ .
7. How deep is the size of the stack in the AVR?
8. The RCALL instruction is a(n) ___ -byte instruction.
9. _____ (RCALL, CALL) takes more ROM space.

SECTION 3.3: AVR TIME DELAY AND INSTRUCTION PIPELINE

In the last section we used the DELAY subroutine. In this section we discuss how to generate various time delays and calculate exact delays for the AVR. We will also discuss instruction pipelining and its impact on execution time.

Delay calculation for the AVR

In creating a time delay using Assembly language instructions, one must be mindful of two factors that can affect the accuracy of the delay:

1. The crystal frequency: The frequency of the crystal oscillator connected to the XTAL1 and XTAL2 input pins is one factor in the time delay calculation. The duration of the clock period for the instruction cycle is a function of this crystal frequency.
2. The AVR design: Since the 1970s, both the field of IC technology and the architectural design of microprocessors have seen great advancements. Due to the limitations of IC technology and limited CPU design experience for many years, the instruction cycle duration was longer. Advances in both IC technology and CPU design in the 1980s and 1990s have made the single instruction cycle a common feature of many microcontrollers. Indeed, one way to increase performance without losing code compatibility with the older generation of a given family is to reduce the number of instruction cycles it takes to execute an instruction. One might wonder how microprocessors such as AVR are able to execute an instruction in one cycle. There are three ways to do that: (a) Use Harvard architecture to get the maximum amount of code and data into the CPU, (b) use RISC architecture features such as fixed-size instructions, and finally (c) use pipelining to overlap fetching and execution of instructions. We examined the Harvard and RISC architectures in Chapter 2. Next, we discuss pipelining.

Pipelining

In early microprocessors such as the 8085, the CPU could either fetch or execute at a given time. In other words, the CPU had to fetch an instruction from memory, then execute it; and then fetch the next instruction, execute it, and so on. The idea of pipelining in its simplest form is to allow the CPU to fetch and execute at the same time, as shown in Figure 3-12. (An instruction fetches while the

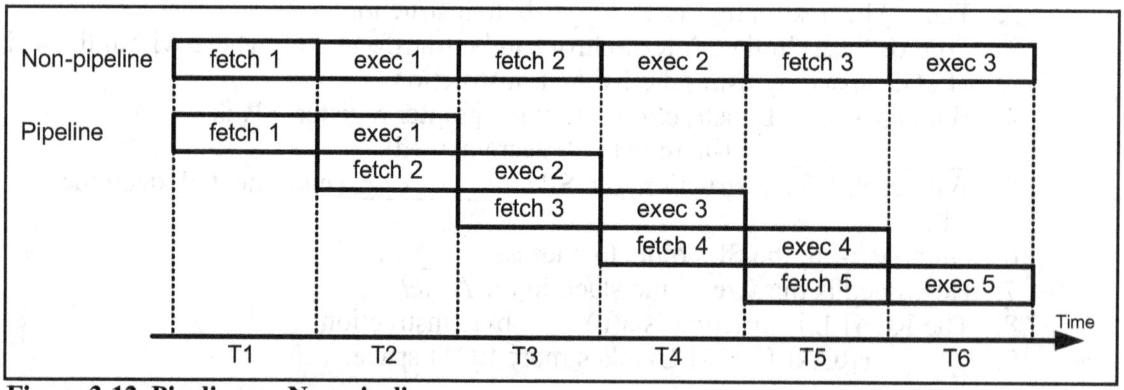

Figure 3-12. Pipeline vs. Non-pipeline

128

previous instruction executes.)

We can use a pipeline to speed up execution of instructions. In pipelining, the process of executing instructions is split into small steps that are all executed in parallel. In this way, the execution of many instructions is overlapped. One limitation of pipelining is that the speed of execution is limited to the slowest stage of the pipeline. Compare this to making pizza. You can split the process of making pizza into many stages, such as flattening the dough, putting on the toppings, and baking, but the process is limited to the slowest stage, baking, no matter how fast the rest of the stages are performed. What happens if we use two or three ovens for baking pizzas to speed up the process? This may work for making pizza but not for executing programs, because in the execution of instructions we must make sure that the sequence of instructions is kept intact and that there is no out-of-step execution.

Stage 1	Stage 2	Stage 3
READ OPERANDS	PROCESS	WRITE BACK

Figure 3-13. Single Cycle ALU Operation

AVR multistage execution pipeline

As shown in Figure 3-13, in the AVR, each instruction is executed in 3 stages: operand fetch, ALU operation execution, and result write back.

In step 1, the operand is fetched. In step 2, the operation is performed; for example, the adding of the two numbers is done. In step 3, the result is written into the destination register. In reality, one can construct the AVR pipeline for three instructions, as is shown in Figure 3-14.

It should be noted that in many computer architecture books, the process stage is referred to as *execute* and write back is called *write*.

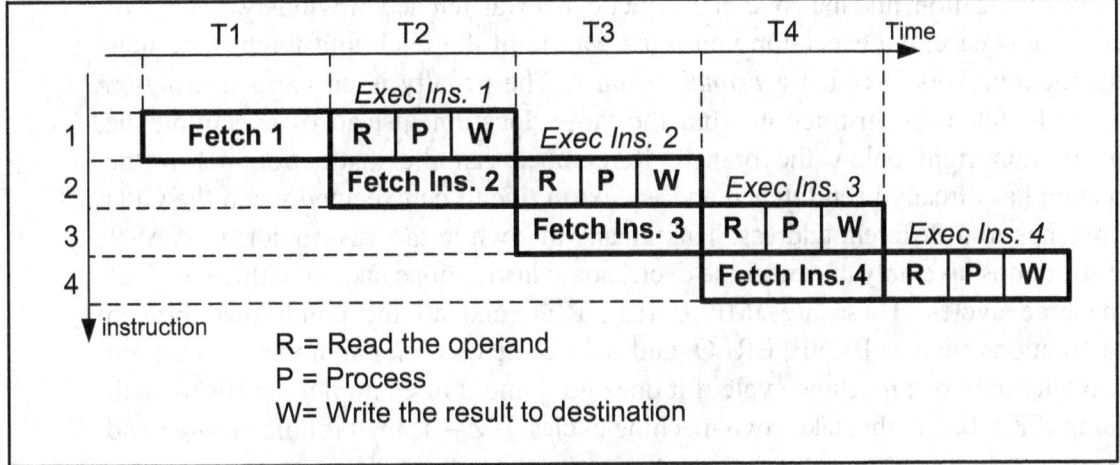

Figure 3-14. Pipeline Activity for Both Fetch and Execute

Instruction cycle time for the AVR

It takes a certain amount of time for the CPU to execute an instruction. This time is referred to as *machine cycles*. Because all the instructions in the AVR are either 1-word (2-byte) or 2-word (4-byte), most instructions take no more than one or two machine cycles to execute. (Notice, however, that some instructions such as JMP and CALL could take up to three or four machine cycles.) Appendix A provides a list of AVR instructions and their cycles. In the AVR family, the length of the machine cycle depends on the frequency of the oscillator connected to the AVR system. The crystal oscillator, along with on-chip circuitry, provide the clock source for the AVR CPU (see Chapter 8). In the AVR, one machine cycle consists of one oscillator period, which means that with each oscillator clock, one machine cycle passes. Therefore, to calculate the machine cycle for the AVR, we take the inverse of the crystal frequency, as shown in Example 3-14.

Example 3-14

The following shows the crystal frequency for four different AVR-based systems. Find the period of the instruction cycle in each case.

(a) 8 MHz   (b) 16 MHz   (c) 10 MHz   (d) 1 MHz

Solution:

(a) instruction cycle is 1/8 MHz = 0.125 μs (microsecond) = 125 ns (nanosecond)
(b) instruction cycle = 1/16 MHz = 0.0625 μs = 62.5 ns (nanosecond)
(c) instruction cycle = 1/10 MHz = 0.1 μs = 100 ns
(d) instruction cycle = 1/1 MHz = 1 μs

Branch penalty

The overlapping of fetch and execution of the instruction is widely used in today's microcontrollers such as AVR. For the concept of pipelining to work, we need a buffer or queue in which an instruction is prefetched and ready to be executed. In some circumstances, the CPU must flush out the queue. For example, when a branch instruction is executed, the CPU starts to fetch codes from the new memory location, and the code in the queue that was fetched previously is discarded. In this case, the execution unit must wait until the fetch unit fetches the new instruction. This is called a *branch penalty*. The penalty is an extra instruction cycle to fetch the instruction from the target location instead of executing the instruction right below the branch. Remember that the instruction below the branch has already been fetched and is next in line to be executed when the CPU branches to a different address. This means that while the vast majority of AVR instructions take only one machine cycle, some instructions take two, three, or four machine cycles. These are JMP, CALL, RET, and all the conditional branch instructions such as BRNE, BRLO, and so on. The conditional branch instruction can take only one machine cycle if it does not jump. For example, the BRNE will jump if Z = 0, and that takes two machine cycles. If Z = 1, then it falls through and it takes only one machine cycle. See Examples 3-15 and 3-16.

Example 3-15

For an AVR system of 1 MHz, find how long it takes to execute each of the following instructions:

(a)	LDI	(b)	DEC	(c)	LD
(d)	ADD	(e)	NOP	(f)	JMP
(g)	CALL	(h)	BRNE	(i)	.DEF

Solution:

The machine cycle for a system of 1 MHz is 1 μs, as shown in Example 3-14. Appendix A shows instruction cycles for each of the above instructions. Therefore, we have:

Instruction		Instruction cycles	Time to execute
(a)	LDI	1	1 × 1 μs = 1 μs
(b)	DEC	1	1 × 1 μs = 1 μs
(c)	OUT	1	1 × 1 μs = 1 μs
(d)	ADD	1	1 × 1 μs = 1 μs
(e)	NOP	1	1 × 1 μs = 1 μs
(f)	JMP	3	3 × 1 μs = 2 μs
(g)	CALL	4	4 × 1 μs = 4 μs
(h)	BRNE	2/1	(2 μs taken, 1 μs if it falls through)
(i)	.DEF	0	(directive instructions do not produce machine instructions)

Example 3-16

Find the size of the delay of the code snippet below if the crystal frequency is 10 MHz:

Solution:

From Appendix A, we have the following machine cycles for each instruction of the DELAY subroutine:

```
                         Instruction Cycles
        .DEF COUNT = R20            0
DELAY:  LDI        COUNT, 0xFF      1
AGAIN:  NOP                         1
        NOP                         1
        DEC        COUNT            1
        BRNE       AGAIN            2/1
        RET                         4
```

Therefore, we have a time delay of $[1 + ((1 + 1 + 1 + 2) \times 255) + 4] \times 0.1$ μs $= 128.0$ μs. Notice that BRNE takes two instruction cycles if it jumps back, and takes only one when falling through the loop. That means the above number should be 127.9 μs.

Delay calculation for AVR

As seen in the last section, a delay subroutine consists of two parts: (1) setting a counter, and (2) a loop. Most of the time delay is performed by the body of the loop, as shown in Examples 3-17 and 3-18.

Very often we calculate the time delay based on the instructions inside the loop and ignore the clock cycles associated with the instructions outside the loop.

In Example 3-16, the largest value the R20 register can take is 255. One way to increase the delay is to use NOP instructions in the loop. NOP, which stands for "no operation," simply wastes time, but takes 2 bytes of program ROM space,

---

Example 3-17

Find the size of the delay in the following program if the crystal frequency is 1 MHz:

```
.INCLUDE "M32DEF.INC"
.ORG 0
        LDI R16,HIGH(RAMEND)   ;initialize SP
        OUT SPH,R16
        LDI R16,LOW(RAMEND)
        OUT SPL,R16
BACK:
        LDI R16,0x55           ;load R16 with 0x55
        OUT PORTB,R16          ;send 55H to port B
        RCALL DELAY            ;time delay
        LDI R16,0xAA           ;load R16 with 0xAA
        OUT PORTB,R16          ;send 0xAA to port B
        RCALL DELAY            ;time delay
        RJMP BACK              ;keep doing this indefinitely
;-----this is the delay subroutine
        .ORG 0x300             ;put time delay at address 0x300
DELAY:  LDI R20,0xFF           ;R20 = 255,the counter
AGAIN:
        NOP                    ;no operation wastes clock cycles
        NOP
        DEC R20
        BRNE AGAIN             ;repeat until R20 becomes 0
        RET                    ;return to caller
```

Solution:

From Appendix A, we have the following machine cycles for each instruction of the DELAY subroutine:

	Instruction	Cycles
DELAY:	LDI R20,0xFF	1
AGAIN:	NOP	1
	NOP	1
	DEC R20	1
	BRNE AGAIN	2/1
	RET	4

Therefore, we have a time delay of $[1 + (255 \times 5) - 1 + 4] \times 1 \ \mu s = 1279 \ \mu s$.

132

which is too heavy a price to pay for just one instruction cycle. A better way is to use a nested loop.

Loop inside a loop delay

Another way to get a large delay is to use a loop inside a loop, which is also called a *nested loop*. See Example 3-18. Compare that with Example 3-19 to see the disadvantage of using many NOPs. Also see Example 3-20.

From these discussions we conclude that the use of instructions in generating time delay is not the most reliable method. To get more accurate time delay we

---

Example 3-18

For an instruction cycle of 1 μs (a) find the time delay in the following subroutine, and (b) find the amount of ROM it takes.

```
                              Instruction Cycles
DELAY:     LDI      R16,200      1
AGAIN:     LDI      R17,250      1
HERE:      NOP                   1
           NOP                   1
           DEC      R17          1
           BRNE     HERE         2/1
           DEC      R16          1
           BRNE     AGAIN        2/1
           RET                   4
```

Solution:

(a)
For the HERE loop, we have $[(5 \times 250) - 1] \times 1$ μs = 1249 μs. (We should subtract 1 for the times BRNE HERE falls through.) The AGAIN loop repeats the HERE loop 200 times; therefore, we have 200×1249 μs = 249,800 μs, if we do not include the overhead. However, the following instructions of the outer loop add to the delay:

```
AGAIN:     LDI      R17,250      1
           .....
           DEC      R16          1
           BRNE     AGAIN        2/1
```

The above instructions at the beginning and end of the AGAIN loop add $[(4 \times 200) - 1] \times 1$ μs = 799 μs to the time delay. As a result we have 249,800 + 799 = 250,599 μs for the total time delay associated with the above DELAY subroutine. Notice that this calculation is an approximation because we have ignored the "LDI R16,200" instruction and the last instruction, RET, in the subroutine.

(b)
There are 9 instructions in the above DELAY program, and all the instructions are 2-byte instructions. That means that the loop delay takes 22 bytes of ROM code space.

Example 3-19

Find the time delay for the following subroutine, assuming a crystal frequency of 1 MHz. Discuss the disadvantage of this over Example 3-18.

```
                            Machine Cycles

DELAY:  LDI     R16, 200          1
AGAIN:  NOP                       1
        NOP                       1
        NOP                       1
        NOP                       1
        NOP                       1
        NOP                       1
        NOP                       1
        NOP                       1
        NOP                       1
        NOP                       1
        NOP                       1
        NOP                       1
        DEC     R16               1
        BRNE    AGAIN             2
        RET                       4
```

Solution:

The time delay inside the AGAIN loop is $[200(13 + 2)] \times 1$ µs $= 3000$ µs. NOP is a 2-byte instruction, even though it does not do anything except to waste cycle time. There are 16 instructions in the above DELAY program, and all the instructions are 2-byte instructions. This means the loop delay takes 32 bytes of ROM code space, and gives us only a 3000 µs delay. That is the reason we use a nested loop instead of NOP instructions to create a time delay. Chapter 9 shows how to use AVR timers to create delays much more efficiently.

use timers, as described in Chapter 9. We can use AVR Studio's simulator to verify delay time and number of cycles used. Meanwhile, to get an accurate time delay for a given AVR microcontroller, we must use an oscilloscope to measure the exact time delay.

Review Questions

1. True or false. In the AVR, the machine cycle lasts 1 clock period of the crystal frequency.
2. The minimum number of machine cycles needed to execute an AVR instruction is _____.
3. For Question 2, what is the maximum number of cycles needed, and for which instructions?
4. Find the machine cycle for a crystal frequency of 12 MHz.
5. Assuming a crystal frequency of 1 MHz, find the time delay associated with the loop section of the following DELAY subroutine:

```
        DELAY:      LDI         R20, 100
        HERE:       NOP
```

Example 3-20

Write a program to toggle all the bits of I/O register PORTB every 1 s. Assume that the crystal frequency is 8 MHz and the system is using an ATmega32.

Solution:

```
.INCLUDE "M32DEF.INC"
.ORG 0
      LDI    R16,HIGH(RAMEND)
      OUT    SPH,R16
      LDI    R16,LOW(RAMEND)
      OUT    SPL,R16

      LDI    R16,0x55            ;load R16 with 0x55
BACK:
      COM    R16                 ;complement PORTB
      OUT    PORTB,R16           ;send it to port B
      CALL   DELAY_1S            ;time delay
      RJMP   BACK                ;keep doing this indefinitely

DELAY_1S:
      LDI    R20,32
L1:   LDI    R21,200
L2:   LDI    R22,250
L3:
      NOP
      NOP
      DEC    R22
      BRNE   L3

      DEC    R21
      BRNE   L2

      DEC    R20
      BRNE   L1
      RET
```

Machine cycle = 1 / 8 MHz = 125 ns
Delay = 32 × 200 × 250 × 5 × 125 ns = 1,000,000,000 ns = 1,000,000 μs = 1 s.
In this calculation, we have not included the overhead associated with the two outer loops. Use the AVR Studio simulator to verify the delay.

```
      NOP
      NOP
      NOP
      NOP
      DEC    R20
      BRNE   HERE
      RET
```

6. True or false. In the AVR, the machine cycle lasts 6 clock periods of the crystal frequency.
7. Find the machine cycle for an AVR if the crystal frequency is 8 MHz.
8. True or false. In the AVR, the instruction fetching and execution are done at the same time.
9. True or false. JMP and RCALL will always take 3 machine cycles.
10. True or false. The BRNE instruction will always take 2 machine cycles.

SUMMARY

The flow of a program proceeds sequentially, from instruction to instruction, unless a control transfer instruction is executed. The various types of control transfer instructions in Assembly language include conditional and unconditional branches, and call instructions.

Looping in AVR Assembly language is performed using an instruction to decrement a counter and to jump to the top of the loop if the counter is not zero. This is accomplished with the BRNE instruction. Other branch instructions jump conditionally, based on the value of the carry flag, the Z flag, or other bits of the status register. Unconditional branches can be long or short, depending on the location of the target address. Special attention must be given to the effect of CALL and RCALL instructions on the stack.

PROBLEMS

SECTION 3.1: BRANCH INSTRUCTIONS AND LOOPING

1. In the AVR, looping action with the "BRNE target" instruction is limited to _____ iterations.
2. If a conditional branch is not taken, what is the next instruction to be executed?
3. In calculating the target address for a branch, a displacement is added to the contents of register _____.
4. The mnemonic RJMP stands for _____ and it is a(n) ___-byte instruction.
5. The JMP instruction is a(n) ___-byte instruction.
6. What is the advantage of using RJMP over JMP?
7. True or false. The target of a BRNE can be anywhere in the 4M word address space.
8. True or false. All AVR branch instructions can branch to anywhere in the 4M word address space.
9. Which of the following instructions is (are) 2-byte instructions.
    (a) BREQ   (b) BRSH   (c) JMP   (d) RJMP
10. Dissect the RJMP instruction, indicating how many bits are used for the operand and the opcode, and indicate how far it can branch.
11. True or false. All conditional branches are 2-byte instructions.
12. Show code for a nested loop to perform an action 1,000 times.
13. Show code for a nested loop to perform an action 100,000 times.
14. Find the number of times the following loop is performed:

```
        LDI        R20, 200
  BACK: LDI        R21, 100
  HERE: DEC        R21
        BRNE       HERE
        DEC        R20
        BRNE       BACK
```

15. The target address of a BRNE is backward if the relative address of the opcode is _____ (negative, positive).

16. The target address of a BRNE is forward if the relative address of the opcode is _____ (negative, positive).

SECTION 3.2: CALL INSTRUCTIONS AND STACK

17. CALL is a(n) ___-byte instruction.
18. RCALL is a(n) ___-byte instruction.
19. True or false. The RCALL target address can be anywhere in the 4M (word) address space.
20. True or false. The CALL target address can be anywhere in the 4M address space.
21. When CALL is executed, how many locations of the stack are used?
22. When RCALL is executed, how many locations of the stack are used?
23. Upon reset, the SP points to location _____.
24. Describe the action associated with the RET instruction.
25. Give the size of the stack in AVR.
26. In AVR, which address is pushed into the stack when a call instruction is executed.

SECTION 3.3: AVR TIME DELAY AND INSTRUCTION PIPELINE

27. Find the oscillator frequency if the machine cycle = 1.25 µs.
28. Find the machine cycle if the crystal frequency is 20 MHz.
29. Find the machine cycle if the crystal frequency is 10 MHz.
30. Find the machine cycle if the crystal frequency is 16 MHz.
31. True or false. The CALL and RCALL instructions take the same amount of time to execute even though one is a 4-byte instruction and the other is a 2-byte instruction.
32. Find the time delay for the delay subroutine shown below if the system has an AVR with a frequency of 8 MHz:

```
        LDI     R16, 200
BACK:   LDI     R18, 100
HERE:   NOP
        DEC     R18
        BRNE    HERE
        DEC     R16
        BRNE    BACK
```

33. Find the time delay for the delay subroutine shown below if the system has an AVR with a frequency of 8 MHz:

```
        LDI     R20, 200
BACK:   LDI     R22, 100
HERE:   NOP
        NOP
        DEC     R22
        BRNE    HERE
        DEC     R20
        BRNE    BACK
```

34. Find the time delay for the delay subroutine shown below if the system has an AVR with a frequency of 4 MHz:

```
                 LDI          R20, 200
       BACK:     LDI          R21, 250
       HERE:     NOP
                 DEC          R21
                 BRNE         HERE
                 DEC          R20
                 BRNE         BACK
```

35. Find the time delay for the delay subroutine shown below if the system has an AVR with a frequency of 10 MHz:

```
                 LDI          R20, 200
       BACK:     LDI          R25, 100
                 NOP
                 NOP
                 NOP
       HERE      DEC          R25
                 BRNE         HERE
                 DEC          R20
                 BRNE         BACK
```

ANSWERS TO REVIEW QUESTIONS

SECTION 3.1: BRANCH INSTRUCTIONS AND LOOPING

1. Branch if not equal
2. True
3. 2
4. Z flag of SREG (status register)
5. 4

SECTION 3.2: CALL INSTRUCTIONS AND STACK

1. True
2. 4
3. False
4. Decremented
5. Incremented
6. 0
7. The AVR's stack can be as big as its RAM memory.
8. 2
9. CALL

SECTION 3.3: AVR TIME DELAY AND INSTRUCTION PIPELINE

1. True
2. 1
3. 4; CALL, RET
4. 12 MHz / 4 = 3 MHz, and MC = 1/3 MHz = 0.333 μs
5. $[100 \times (1 + 1 + 1 + 1 + 1 + 1 + 2)] \times 1\,\mu$s = 800 μs = 0.8 milliseconds
6. False. It takes 4 clocks.
7. Machine cycle = 1 / 8 MHz = 0.125 μs = 125 ns
8. True
9. True
10. False. Only if it branches to the target address.

CHAPTER 4

AVR I/O PORT PROGRAMMING

OBJECTIVES

Upon completion of this chapter, you will be able to:

>> List all the ports of the AVR
>> Describe the dual role of AVR pins
>> Code Assembly language to use the ports for input or output
>> Explain the dual role of Ports A, B, C, and D
>> Code AVR instructions for I/O handling
>> Code I/O bit-manipulation programs for the AVR
>> Explain the bit-addressability of AVR ports

This chapter describes I/O port programming of the AVR with many examples. In Section 4.1, we describe I/O access using byte-size data, and in Section 4.2, bit manipulation of the I/O ports is discussed in detail.

SECTION 4.1: I/O PORT PROGRAMMING IN AVR

In the AVR family, there are many ports for I/O operations, depending on which family member you choose. Examine Figure 4-1 for the ATmega32 40-pin chip. A total of 32 pins are set aside for the four ports PORTA, PORTB, PORTC, and PORTD. The rest of the pins are designated as VCC, GND, XTAL1, XTAL2, RESET, AREF, AGND, and AVCC. They are discussed in Chapter 8.

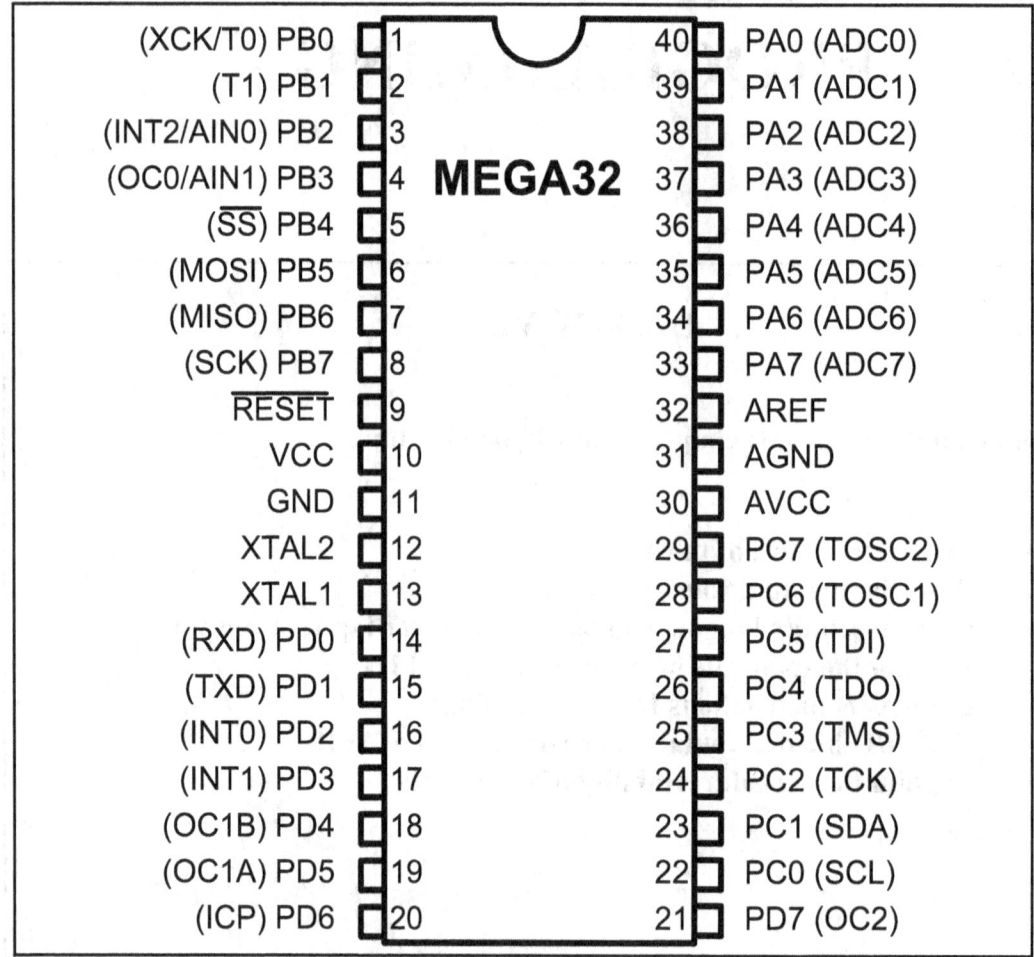

Figure 4-1. ATmega32 Pin Diagram

I/O port pins and their functions

The number of ports in the AVR family varies depending on the number of pins on the chip. The 8-pin AVR has port B only, while the 64-pin version has ports A through F, and the 100-pin AVR has ports A through L, as shown in Table 4-1. The 40-pin AVR has four ports. They are PORTA, PORTB, PORTC, and PORTD. To use any of these ports as an input or output port, it must be programmed, as we will explain throughout this section. In addition to being used for simple I/O, each

Table 4-1: Number of Ports in Some AVR Family Members

Pins	8-pin	28-pin	40-pin	64-pin	100-pin
Chip	ATtiny25/45/85	ATmega8/48/88	ATmega32/16	ATmega64/128	ATmega1280
Port A			X	X	X
Port B	6 bits	X	X	X	X
Port C		7 bits	X	X	X
Port D		X	X	X	X
Port E				X	X
Port F				X	X
Port G				5 bits	6 bits
Port H					X
Port J					X
Port K					X
Port L					X

Note: X indicates that the port is available.

port has some other functions such as ADC, timers, interrupts, and serial communication pins. Figure 4-1 shows alternate functions for the ATmega32 pins. We will study all these alternate functions in future chapters. In this chapter we focus on the simple I/O function of the AVR family. Not all ports have 8 pins. For example, in the ATmega8, Port C has 7 pins. Each port has three I/O registers associated with it, as shown in Table 4-2. They are designated as PORTx, DDRx, and PINx. For example, for Port B we have PORTB, DDRB, and PINB. Notice that DDR stands for Data Direction Register, and PIN stands for Port INput pins. Also notice that each of the I/O registers is 8 bits wide, and each port has a maximum of 8 pins; therefore each bit of the I/O registers affects one of the pins (see Figure 4-2; the content of bit 0 of DDRB represents the direction of the PB0 pin, and so on). Next, we describe how to access the I/O registers associated with the ports.

Table 4-2: Register Addresses for ATmega32 Ports

Port	Address	Usage
PORTA	$3B	output
DDRA	$3A	direction
PINA	$39	input
PORTB	$38	output
DDRB	$37	direction
PINB	$36	input
PORTC	$35	output
DDRC	$34	direction
PINC	$33	input
PORTD	$32	output
DDRD	$31	direction
PIND	$30	input

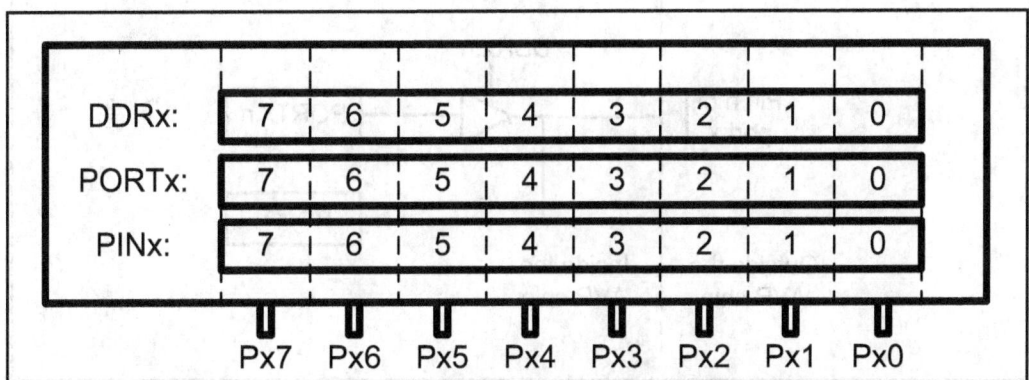

Figure 4-2. Relations Between the Registers and the Pins of AVR

DDRx register role in outputting data

Each of the ports A–D in the ATmega32 can be used for input or output. The DDRx I/O register is used solely for the purpose of making a given port an input or output port. For example, to make a port an output, we write 1s to the DDRx register. In other words, to output data to all of the pins of the Port B, we must first put 0b11111111 into the DDRB register to make all of the pins output.

The following code will toggle all 8 bits of Port B forever with some time delay between "on" and "off" states:

```
        LDI   R16,0xFF    ;R16 = 0xFF = 0b11111111
        OUT   DDRB,R16    ;make Port B an output port (1111 1111)
L1:     LDI   R16,0x55    ;R16 = 0x55 = 0b01010101
        OUT   PORTB,R16   ;put 0x55 on port B pins
        CALL  DELAY
        LDI   R16,0xAA    ;R16 = 0xAA = 0b10101010
        OUT   PORTB,R16   ;put 0xAA on port B pins
        CALL  DELAY
        RJMP  L1
```

It must be noted that unless we set the DDRx bits to one, the data will not go from the port register to the pins of the AVR. This means that if we remove the first two lines of the above code, the 0x55 and 0xAA values will not get to the pins. They will be sitting in the I/O register of Port B inside the CPU.

To see the role of the DDRx register in allowing the data to go from Portx to the pins, examine Figure 4-3. For more information about the internal circuitry of I/O ports, see Appendix C.

DDR register role in inputting data

To make a port an input port, we must first put 0s into the DDRx register for that port, and then bring in (read) the data present at the pins. As an aid for remembering that the port is input when the DDR bits are 0s, imagine a person who has 0 dollars. The person can only get money, not give it. Similarly, when DDR contains 0s, the port gets data.

Notice that upon reset, all ports have the value 0x00 in their DDR registers. This means that all ports are configured as input as we will see next.

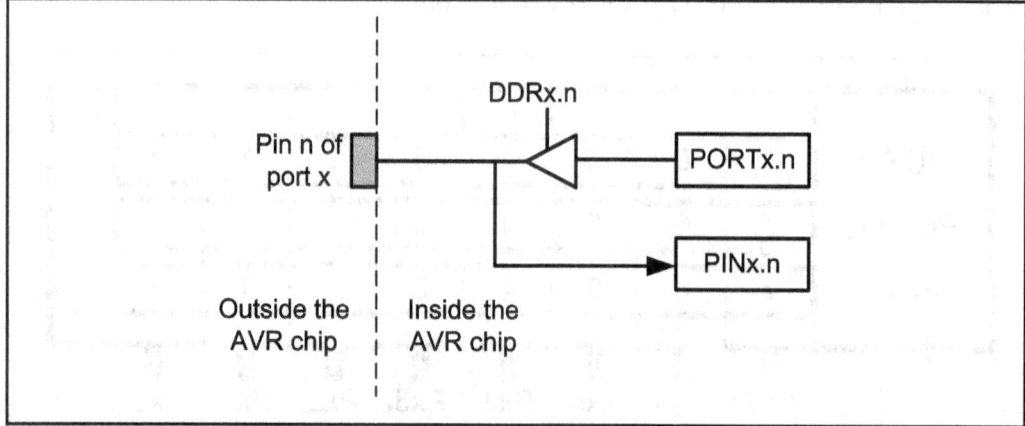

Figure 4-3. The I/O Port in AVR

PIN register role in inputting data

To read the data present at the pins, we should read the PIN register. It must be noted that to bring data into CPU from pins we read the contents of the PINx register, whereas to send data out to pins we use the PORTx register.

PORT register role in inputting data

There is a pull-up resistor for each of the AVR pins. If we put 1s into bits of the PORTx register, the pull-up resistors are activated. In cases in which nothing is connected to the pin or the connected devices have high impedance, the resistor pulls up the pin. See Figure 4-4.

If we put 0s into the bits of the PORTx register, the pull-up resistor is inactive.

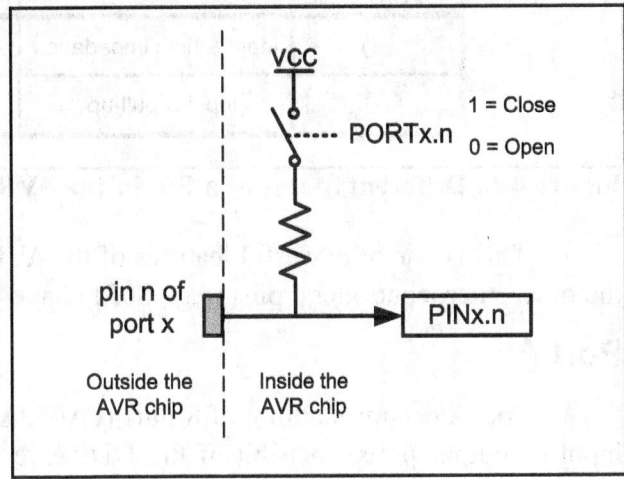

Figure 4-4. The Pull-up Resistor

The following code gets the data present at the pins of port C and sends it to port B indefinitely, after adding the value 5 to it:

```
.INCLUDE "M32DEF.INC"
      LDI    R16,0x00    ;R16 = 00000000 (binary)
      OUT    DDRC,R16    ;make Port C an input port
      LDI    R16,0xFF    ;R16 = 11111111 (binary)
      OUT    DDRB,R16    ;make Port B an output port(1 for Out)
L2:   IN     R16,PINC    ;read data from Port C and put in R16
      LDI    R17,5
      ADD    R16,R17     ;add 5 to it
      OUT    PORTB,R16   ;send it to Port B
      RJMP   L2          ;continue forever
```

If we want to make the pull-up resistors of port C active, we must put 1s into the PORTC register. The program becomes as follows:

```
.INCLUDE    "M32DEF.INC"
      LDI    R16,0xFF    ;R16 = 11111111 (binary)
      OUT    DDRB,R16    ;make Port B an output port
      OUT    PORTC,R16   ;make the pull-up resistors of C active
      LDI    R16,0x00    ;R16 = 00000000 (binary)
      OUT    DDRC,R16    ;Port C an input port (0 for I)
L2:   IN     R16,PINC    ;move data from Port C to R16
      LDI    R17,5
      ADD    R16,R17     ;add some value to it
      OUT    PORTB,R16   ;send it to Port B
      RJMP   L2          ;continue forever
```

Again, it must be noted that unless we clear the DDR bits (by putting 0s there), the data will not be brought into the registers from the pins of Port C. To

---

see the role of the DDRx register in allowing the data to come into the CPU from the pins, examine Figure 4-3.

The pins of the AVR microcontrollers can be in four different states according to the values of PORTx and DDRx, as shown in Figure 4-5.

PORTx \ DDRx	0	1
0	Input & high impedance	Out 0
1	Input & pull-up	Out 1

Figure 4-5. Different States of a Pin in the AVR Microcontroller

This is one of powerful features of the AVR microcontroller, since most of the other microcontrollers' pins (e.g., 8051) have fewer states.

Port A

Port A occupies a total of 8 pins (PA0–PA7). To use the pins of Port A as input or output ports, each bit of the DDRA register must be set to the proper value. For example, the following code will continuously send out to Port A the alternating values of 0x55 and 0xAA:

```
;toggle all bits of PORTA
.INCLUDE    "M32DEF.INC"
      LDI   R16,0xFF    ;R16 = 11111111 (binary)
      OUT   DDRA,R16    ;make Port A an output port
L1:   LDI   R16,0x55    ;R16 = 0x55
      OUT   PORTA,R16   ;put 0x55 on Port A pins
      CALL  DELAY
      LDI   R16,0xAA    ;R16 = 0xAA
      OUT   PORTA,R16   ;put 0xAA on Port A pins
      CALL  DELAY
      RJMP  L1
```

It must be noted that 0x55 (01010101) when complemented becomes 0xAA (10101010).

Port A as input

In order to make all the bits of Port A an input, DDRA must be cleared by writing 0 to all the bits. In the following code, Port A is configured first as an input port by writing all 0s to register DDRA, and then data is received from Port A and saved in a RAM location:

```
.INCLUDE    "M32DEF.INC"
      .EQU  MYTEMP 0x100      ;save it here
      LDI   R16,0x00    ;R16 = 00000000 (binary)
      OUT   DDRA,R16    ;make Port A an input port (0 for In)
      NOP               ;synchronizer delay
      IN    R16,PINA    ;move from pins of Port A to R16
      STS   MYTEMP,R16  ;save it in MYTEMP
```

Synchronizer delay

The input circuit of the AVR has a delay of 1 clock cycle. In other words, the PIN register represents the data that was present at the pins one clock ago. In the above code, when the instruction "IN R16, PINA" is executed, the PINA register contains the data, which was present at the pins one clock before. That is why the NOP is put before the "IN R16, PINA" instruction. (If the NOP is omitted, the read data is the data of the pins when the port was output.)

For more information see Section C-2.

Port B

Port B occupies a total of 8 pins (PB0–PB7). To use the pins of Port B as input or output ports, each bit of the DDRB register must be set to the proper value.

For example, the following code will continuously send out the alternating values of 0x55 and 0xAA to Port B:

```
        ;toggle all bits of PORTB
.INCLUDE    "M32DEF.INC"
        LDI    R16,0xFF    ;R16 = 11111111 (binary)
        OUT    DDRB,R16    ;make Port B an output port (1 for Out)
L1:     LDI    R16,0x55    ;R16 = 0x55
        OUT    PORTB,R16   ;put 0x55 on Port B pins
        CALL   DELAY
        LDI    R16,0xAA    ;R16 = 0xAA
        OUT    PORTB,R16   ;put 0xAA on Port B pins
        CALL   DELAY
        RJMP   L1
```

Port B as input

In order to make all the bits of Port B an input, DDRB must be cleared by writing 0 to all the bits. In the following code, Port B is configured first as an input port by writing all 0s to register DDRB, and then data is received from Port B and saved in some RAM location:

```
.INCLUDE    "M32DEF.INC"
    .EQU   MYTEMP=0x100 ;save it here
    LDI    R16,0x00    ;R16 = 00000000 (binary)
    OUT    DDRB,R16    ;make Port B an input port (0 for In)
    NOP
    IN     R16,PINB    ;move from pins of Port B to R16
    STS    MYTEMP,R16  ;save it in MYTEMP
```

Dual role of Ports A and B

The AVR multiplexes an analog-to-digital converter through Port A to save I/O pins. The alternate functions of the pins for Port A are shown in Table 4-3. We will show how to use Port A's ADC in Chapter 13. Because many projects use an ADC, we usually do not use Port A for simple I/O functions.

The AVR multiplexes some other functions through Port B to save pins.

The alternate functions of the pins for Port B are shown in Table 4-4. We will show how to use the alternate functions of Port B in future chapters.

Table 4-3: Port A Alternate Functions

Bit	Function
PA0	ADC0
PA1	ADC1
PA2	ADC2
PA3	ADC3
PA4	ADC4
PA5	ADC5
PA6	ADC6
PA7	ADC7

Table 4-4: Port B Alternate Functions

Bit	Function
PB0	XCK/T0
PB1	T1
PB2	INT2/AIN0
PB3	OC0/AIN1
PB4	SS
PB5	MOSI
PB6	MISO
PB7	SCK

Port C

Port C occupies a total of 8 pins (PC0–PC7). To use the pins of Port C as input or output ports, each bit of the DDRC register must be set to the proper value. For example, the following code will continuously send out the alternating values of 0x55 and 0xAA to Port C:

```
        ;toggle all bits of PORTB
.INCLUDE    "M32DEF.INC"
        LDI    R16,0xFF    ;R16 = 11111111 (binary)
        OUT    DDRC,R16    ;make Port C an output port (1 for Out)
L1:     LDI    R16,0x55    ;R16 = 0x55
        OUT    PORTC,R16   ;put 0x55 on Port C pins
        CALL   DELAY
        LDI    R16,0xAA    ;R16 = 0xAA
        OUT    PORTC,R16   ;put 0xAA on Port C pins
        CALL   DELAY
        RJMP   L1
```

Port C as input

In order to make all the bits of Port C an input, DDRC must be cleared by writing 0 to all the bits. In the following code, Port C is configured first as an input port by writing all 0s to register DDRC, and then data is received from Port C and saved in a RAM location:

```
.INCLUDE    "M32DEF.INC"
        .EQU   MYTEMP 0x100   ;save it here
        LDI    R16,0x00    ;R16 = 00000000 (binary)
        OUT    DDRC,R16    ;make Port C an input port (0 for In)
        NOP
        IN     R16,PINC    ;move from pins of Port C to R16
        STS    MYTEMP,R16  ;save it in MYTEMP
```

Port D

Port D occupies a total of 8 pins (PD0–PD7). To use the pins of Port D as input or output ports, each bit of the DDRD register must be set to the proper value. For example, the following code will continuously send out to Port D the alternating values of 0x55 and 0xAA:

```
                ;toggle all bits of PORTB
.INCLUDE        "M32DEF.INC"
        LDI     R16,0xFF        ;R16 = 11111111 (binary)
        OUT     DDRD,R16        ;make Port D an output port (1 for Out)
L1:     LDI     R16,0x55        ;R16 = 0x55
        OUT     PORTD,R16       ;put 0x55 on Port D pins
        CALL    DELAY
        LDI     R16,0xAA        ;R16 = 0xAA
        OUT     PORTD,R16       ;put 0xAA on Port D pins
        CALL    DELAY
        RJMP    L1
```

Port D as input

In order to make all the bits of Port D an input, DDRD must be cleared by writing 0 to all the bits. In the following code, Port D is configured first as an input port by writing all 0s to register DDRD, and then data is received from Port D and saved in a RAM location:

```
.INCLUDE        "M32DEF.INC"
.EQU MYTEMP 0x100       ;save it here

        LDI     R16,0x00        ;R16 = 00000000 (binary)
        OUT     DDRD,R16        ;make Port D an input port (0 for In)
        NOP
        IN      R16,PIND        ;move from pins of Port D to R16
        STS     MYTEMP,R16      ;save it in MYTEMP
```

Dual role of Ports C and D

The alternate functions of the pins for Port C are shown in Table 4-5. We will show how to use Port C's alternate functions in future chapters. The alternate functions of the pins for Port D are shown in Table 4-6. We will show how to use Port D's alternate functions in future chapters.

Table 4-5: Port C Alternate Functions

Bit	Function
PC0	SCL
PC1	SDA
PC2	TCK
PC3	TMS
PC4	TDO
PC5	TDI
PC6	TOSC1
PC7	TOSC2

Table 4-6: Port D Alternate Functions

Bit	Function
PD0	PSP0/C1IN+
PD1	PSP1/C1IN-
PD2	PSP2/C2IN+
PD3	PSP3/C2IN-
PD4	PSP4/ECCP1/P1A
PD5	PSP5/P1B
PD6	PSP6/P1C
PD7	PSP7/P1D

Example 4-1

Write a test program for the AVR chip to toggle all the bits of PORTB, PORTC, and PORTD every 1/4 of a second. Assume a crystal frequency of 1 MHz.

Solution:

```
;tested with AVR Studio for the ATmega32 and XTAL = 1 MHz
;to select the XTAL frequency in AVR Studio, press ALT+O
.INCLUDE "M32DEF.INC"
      LDI   R16, HIGH(RAMEND)
      OUT   SPH, R16
      LDI   R16, LOW(RAMEND)
      OUT   SPL, R16    ;initialize stack pointer

      LDI   R16, 0xFF
      OUT   DDRB, R16   ;make Port B an output port
      OUT   DDRC, R16   ;make Port C an output port
      OUT   DDRD, R16   ;make Port D an output port

      LDI   R16, 0x55   ;R16 = 0x55
L3:   OUT   PORTB, R16  ;put 0x55 on Port B pins
      OUT   PORTC, R16  ;put 0x55 on Port C pins
      OUT   PORTD, R16  ;put 0x55 on Port D pins
      CALL  QDELAY      ;quarter of a second delay
      COM   R16         ;complement R16
      RJMP  L3

;---------------1/4 SECOND DELAY
QDELAY:
      LDI   R21, 200
D1:   LDI   R22, 250
D2:   NOP
      NOP
      DEC   R22
      BRNE  D2
      DEC   R21
      BRNE  D1
      RET
```

Calculations:

1 / 1 MHz = 1 µs
Delay = 200 × 250 × 5 MC × 1 µs = 250,000 µs (If we include the overhead, we will have 250,608 µs. See Example 3-18 in the previous chapter.)

Use the AVR Studio simulator to verify the delay size.

Review Questions

1. There are a total of _____ ports in the ATmega32.
2. True or false. All of the ATmega32 ports have 8 pins.
3. True or false. Upon power-up, the I/O pins are configured as output ports.
4. Code a simple program to send 0x99 to Port B and Port C.
5. To make Port B an output port, we must place _____ in register _____.
6. To make Port B an input port, we must place _____ in register _____.
7. True or false. We use a PORTx register to send data out to AVR pins.
8. True or false. We use PINx to bring data into the CPU from AVR pins.

SECTION 4.2: I/O BIT MANIPULATION PROGRAMMING

In this section we further examine the AVR I/O instructions. We pay special attention to I/O bit manipulation because it is a powerful and widely used feature of the AVR family.

I/O ports and bit-addressability

Sometimes we need to access only 1 or 2 bits of the port instead of the entire 8 bits. A powerful feature of AVR I/O ports is their capability to access individual bits of the port without altering the rest of the bits in that port. For all AVR ports, we can access either all 8 bits or any single bit without altering the rest. Table 4-7 lists the single-bit instructions for the AVR. Although the instructions in Table 4-7 can be used for any of the lower 32 I/O registers, I/O port operations use them most often. We will see the use of these instructions throughout future chapters. Table 4-8 shows the lower 32 I/O registers.

Next we describe all these instructions and examine their usage.

Table 4-7: Single-Bit (Bit-Oriented) Instructions for AVR

Instruction		Function
SBI	ioReg,bit	Set Bit in I/O register (set the bit: bit = 1)
CBI	ioReg,bit	Clear Bit in I/O register (clear the bit: bit = 0)
SBIC	ioReg,bit	Skip if Bit in I/O register Cleared (skip next instruction if bit = 0)
SBIS	ioReg,bit	Skip if Bit in I/O register Set (skip next instruction if bit = 1)

Address		Name	Address		Name	Address		Name
Mem.	I/O		Mem.	I/O		Mem.	I/O	
$20	$00	TWBR	$2B	$0B	UCSRA	$36	$16	PINB
$21	$01	TWSR	$2C	$0C	UDR	$37	$17	DDRB
$22	$02	TWAR	$2D	$0D	SPCR	$38	$18	PORTB
$23	$03	TWDR	$2E	$0E	SPSR	$39	$19	PINA
$24	$04	ADCL	$2F	$0F	SPDR	$3A	$1A	DDRA
$25	$05	ADCH	$30	$10	PIND	$3B	$1B	PORTA
$26	$06	ADCSRA	$31	$11	DDRD	$3C	$1C	EECR
$27	$07	ADMUX	$32	$12	PORTD	$3D	$1D	EEDR
$28	$08	ACSR	$33	$13	PINC	$3E	$1E	EEARL
$29	$09	UBRRL	$34	$14	DDRC	$3F	$1F	EEARH
$2A	$0A	UCSRB	$35	$15	PORTC			

Table 4-8: The Lower 32 I/O Registers

SBI (set bit in I/O register)

To set HIGH a single bit of a given I/O register, we use the following syntax:
```
SBI ioReg, bit_num
```
where ioReg can be the lower 32 I/O registers (addresses 0 to 31) and bit_num is the desired bit number from 0 to 7. In Table 4-8 you see the list of the lower 32 I/O registers. Although the bit-oriented instructions can be used for manipulation of bits D0–D7 of the lower 32 I/O registers, they are mostly used for I/O ports. For example the following instruction sets HIGH bit 5 of Port B:
```
SBI PORTB, 5
```
In Figure 4-6, you see the SBI instruction format.

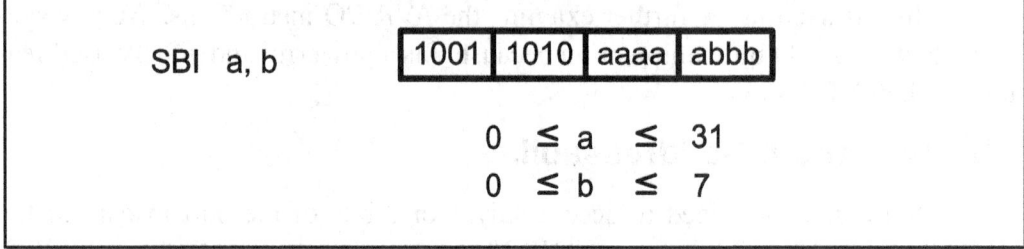

Figure 4-6. SBI (Set Bit) Instruction Format

CBI (Clear Bit in I/O register)

To clear a single bit of a given I/O register, we use the following syntax:
```
CBI ioReg, bit_number
```
For example, the following code toggles pin PB2 continuously:

```
        SBI    DDRB, 2          ;bit = 1, make PB2 an output pin
AGAIN:SBI    PORTB, 2         ;bit set (PB2 = high)
        CALL   DELAY
        CBI    PORTB, 2         ;bit clear (PB2 = low)
        CALL   DELAY
        RJMP   AGAIN
```

Remember that for I/O ports, we must set the appropriate bit in the DDRx register if we want the pin to be output.

Notice that PB2 is the third bit of Port B (the first bit is PB0, the second bit is PB1, etc.). This is shown in Table 4-9. See Example 4-2 for an example of bit manipulation of I/O bits.

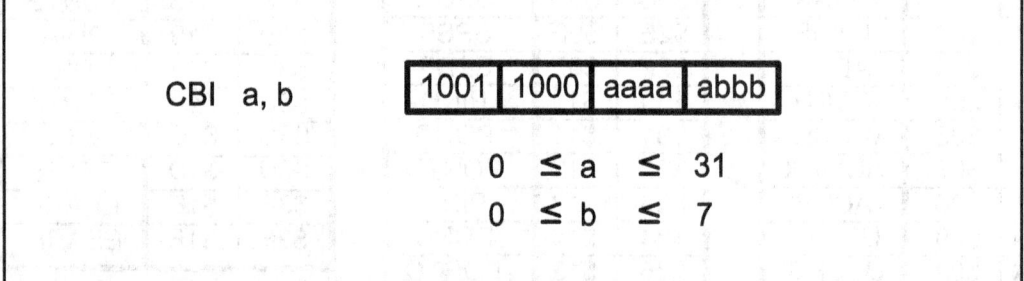

Figure 4-7. CBI (Clear Bit) Instruction Format

Table 4-9: Single-Bit Addressability of Ports for ATmega32/16

PORT	PORTB	PORTC	PORTD	Port Bit
PA0	PB0	PC0	PD0	D0
PA1	PB1	PC1	PD1	D1
PA2	PB2	PC2	PD2	D2
PA3	PB3	PC3	PD3	D3
PA4	PB4	PC4	PD4	D4
PA5	PB5	PC5	PD5	D5
PA6	PB6	PC6	PD6	D6
PA7	PB7	PC7	PD7	D7

Notice in Example 4-2 that unused portions of Port C are undisturbed. This single-bit addressability of I/O ports is one of the most powerful features of the AVR microcontroller.

Example 4-2

An LED is connected to each pin of Port D. Write a program to turn on each LED from pin D0 to pin D7. Call a delay subroutine before turning on the next LED.

Solution:

```
.INCLUDE "M32DEF.INC"
LDI    R20, HIGH(RAMEND)
OUT    SPH, R20
LDI    R20, LOW(RAMEND)
OUT    SPL, R20      ;initialize stack pointer
LDI    R20, 0xFF
OUT    PORTD, R20    ;make PORTD an output port
SBI    PORTD,0       ;set bit PD0
CALL   DELAY         ;delay before next one
SBI    PORTD,1       ;turn on PD1
CALL   DELAY         ;delay before next one
SBI    PORTD,2       ;turn on PD2
CALL   DELAY
SBI    PORTD,3
CALL   DELAY
SBI    PORTD,4
CALL   DELAY
SBI    PORTD,5
CALL   DELAY
SBI    PORTD,6
CALL   DELAY
SBI    PORTD,7
CALL   DELAY
```

Example 4-3

Write the following programs:
(a) Create a square wave of 50% duty cycle on bit 0 of Port C.
(b) Create a square wave of 66% duty cycle on bit 3 of Port C.

Solution:

(a) The 50% duty cycle means that the "on" and "off" states (or the high and low portions of the pulse) have the same length. Therefore, we toggle PC0 with a time delay between each state.

```
.INCLUDE "M32DEF.INC"

        LDI   R20, HIGH(RAMEND)
        OUT   SPH, R20
        LDI   R20, LOW(RAMEND)
        OUT   SPL, R20    ;initialize stack pointer

        SBI   DDRC, 0     ;set bit 0 of DDRC (PC0 = out)
HERE:   SBI   PORTC, 0    ;set to HIGH PC0 (PC0 = 1)
        CALL  DELAY       ;call the delay subroutine
        CBI   PORTC, 0    ;PC0 = 0
        CALL  DELAY
        RJMP  HERE        ;keep doing it
```

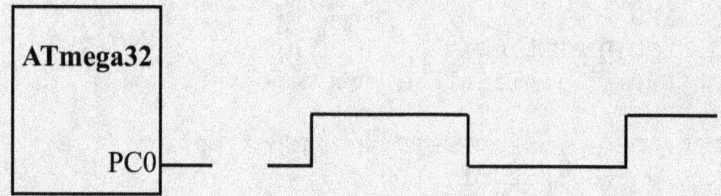

(b) A 66% duty cycle means that the "on" state is twice the "off" state.

```
        ....
        SBI   DDRC, 3     ;set bit 3 of DDRC (PC3 = out)
HERE:   SBI   PORTC, 3    ;set to HIGH PC3 (PC3 = 1)
        CALL  DELAY       ;call the delay subroutine
        CALL  DELAY       ;call the delay subroutine
        CBI   PORTC, 3    ;PC3 = 0
        CALL  DELAY
        RJMP  HERE        ;keep doing it
```

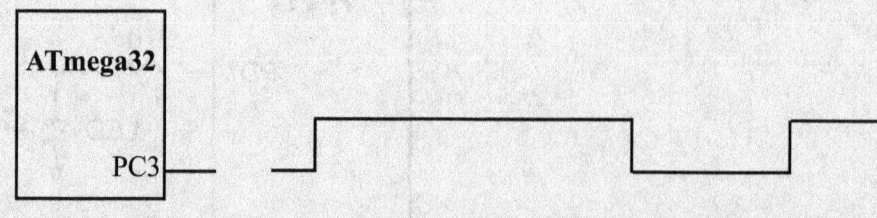

Checking an input pin

To make decisions based on the status of a given bit in the file register, we use the SBIC (Skip if Bit in I/O register Cleared) and SBIS (Skip if Bit in I/O register Set) instructions. These single-bit instructions are widely used for I/O operations. They allow you to monitor a single pin and make a decision depending on whether it is 0 or 1. Again it must be noted that the SBIC and SBIS instructions can be used for any bits of the lower 32 I/O registers, including the I/O ports A, B, C, D, and so on.

SBIS (Skip if Bit in I/O register Set)

To monitor the status of a single bit for HIGH, we use the SBIS instruction. This instruction tests the bit and skips the next instruction if it is HIGH. See Figure 4-8. Example 4-4 shows how it is used.

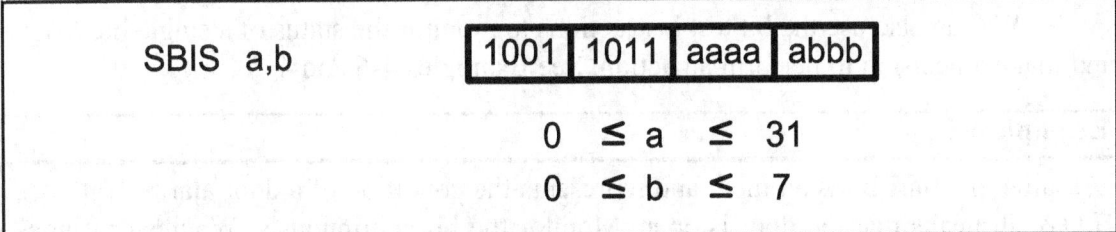

Figure 4-8. SBIS (Skip If Bit in I/O Register Set) Instruction Format

---

Example 4-4

Write a program to perform the following:
(a) Keep monitoring the PB2 bit until it becomes HIGH;
(b) When PB2 becomes HIGH, write the value $45 to Port C, and also send a HIGH-to-LOW pulse to PD3.

Solution:

```
.INCLUDE "M32DEF.INC"

        CBI     DDRB, 2      ;make PB2 an input
        LDI     R16, 0xFF
        OUT     DDRC, R16    ;make Port C an output port
        SBI     DDRD, 3      ;make PD3 an output
AGAIN:  SBIS    PINB, 2      ;skip if Bit PB2 is HIGH
        RJMP    AGAIN        ;keep checking if LOW
        LDI     R16, 0x45
        OUT     PORTC, R16   ;write 0x45 to port C
        SBI     PORTD, 3     ;set bit PD3 (H-to-L)
        CBI     PORTD, 3     ;clear bit PD3
HERE: RJMP    HERE
```

In this program, "SBIS PINB, 2" instruction stays in the loop as long as PB2 is LOW. When PB2 becomes HIGH, it skips the branch instruction to get out of the loop, and writes the value $45 to Port C. It also sends a HIGH-to-LOW pulse to PD3.

SBIC (Skip if Bit in I/O register Cleared)

To monitor the status of a single bit for LOW, we use the SBIC instruction. This instruction tests the bit and skips the instruction right below it if the bit is LOW. See Figure 4-9. Example 4-5 shows how it is used.

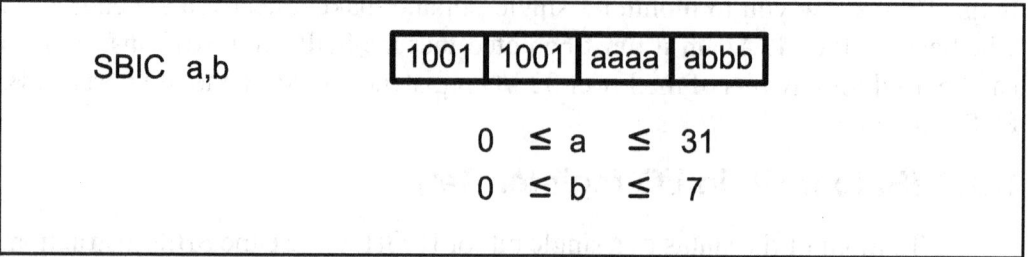

Figure 4-9. SBIC (Skip if Bit in I/O Register Cleared) Instruction Format

Monitoring a single bit

We can also use the bit test instructions to monitor the status of a single bit and make a decision to perform an action. See Examples 4-6 and 4-7.

Example 4-5

Assume that bit PB3 is an input and represents the condition of a door alarm. If it goes LOW, it means that the door is open. Monitor the bit continuously. Whenever it goes LOW, send a HIGH-to-LOW pulse to port PC5 to turn on a buzzer.

Solution:
```
.INCLUDE "M32DEF.INC"

        CBI    DDRB, 3      ;make PB3 an input
        SBI    DDRC, 5      ;make PC5 an output
HERE:   SBIC   PINB, 3      ;keep monitoring PB3 for HIGH
        RJMP   HERE         ;stay in the loop
        SBI    PORTC,5      ;make PC5 HIGH
        CBI    PORTC,5      ;make PC5 LOW for H-to-L
        RJMP   HERE
```

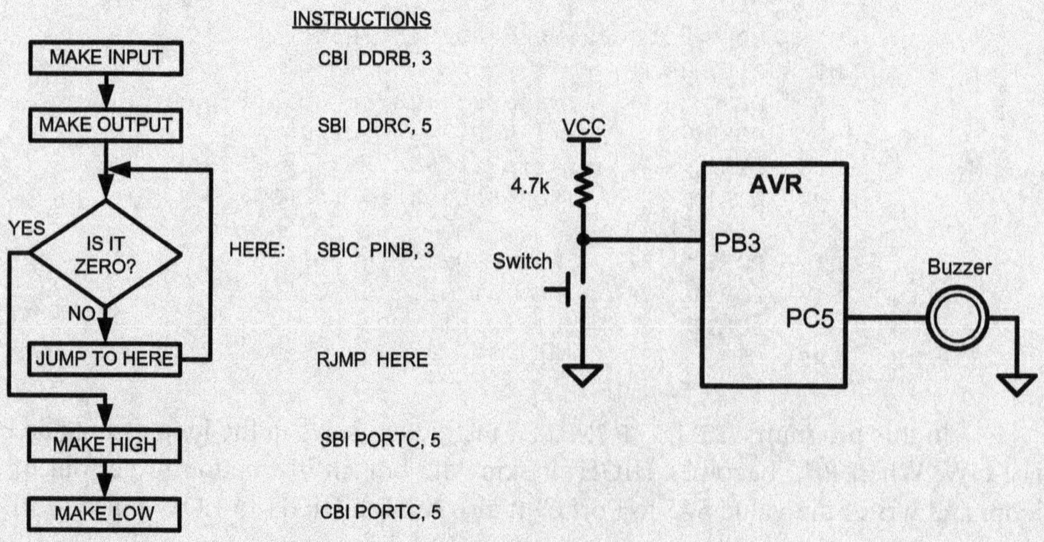

INSTRUCTIONS

MAKE INPUT		CBI DDRB, 3
MAKE OUTPUT		SBI DDRC, 5
IS IT ZERO? (YES / NO)	HERE:	SBIC PINB, 3
JUMP TO HERE		RJMP HERE
MAKE HIGH		SBI PORTC, 5
MAKE LOW		CBI PORTC, 5

VCC
4.7k
Switch
AVR
PB3
PC5
Buzzer

154

Example 4-6

A switch is connected to pin PB2. Write a program to check the status of SW and perform the following:
(a) If SW = 0, send the letter 'N' to PORTD.
(b) If SW = 1, send the letter 'Y' to PORTD.

Solution:

```
        .INCLUDE "M32DEF.INC"

        CBI     DDRB, 2     ;make PB2 an input
        LDI     R16, 0xFF
        OUT     DDRD, R16   ;make PORTD an output port

AGAIN:SBIS    PINB, 2     ;skip next if PB bit is HIGH
        RJMP    OVER        ;SW is LOW
        LDI     R16, 'Y'    ;R16 = 'Y' (ASCII letter Y)
        OUT     PORTD, R16  ;PORTD = 'Y'
        RJMP    AGAIN
OVER:   LDI     R16, 'N'    ;R16 = 'N' (ASCII letter Y)
        OUT     PORTD, R16  ;PORTD = 'N'
        RJMP    AGAIN
```

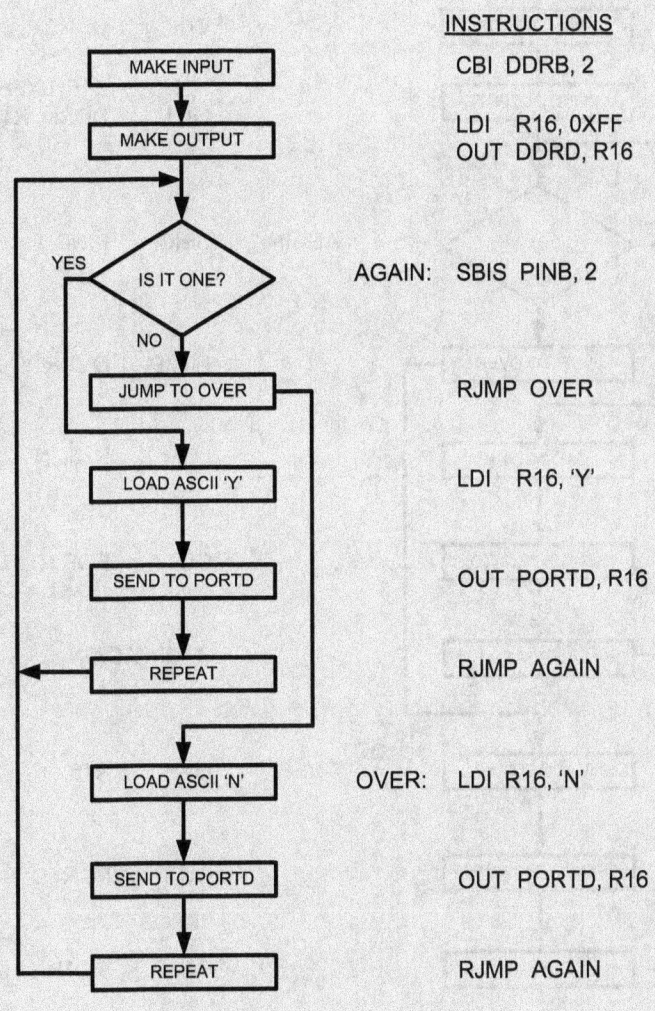

Flowchart	INSTRUCTIONS
MAKE INPUT	CBI DDRB, 2
MAKE OUTPUT	LDI R16, 0XFF OUT DDRD, R16
IS IT ONE? (YES / NO)	AGAIN: SBIS PINB, 2
JUMP TO OVER	RJMP OVER
LOAD ASCII 'Y'	LDI R16, 'Y'
SEND TO PORTD	OUT PORTD, R16
REPEAT	RJMP AGAIN
LOAD ASCII 'N'	OVER: LDI R16, 'N'
SEND TO PORTD	OUT PORTD, R16
REPEAT	RJMP AGAIN

Example 4-7

Rewrite the program of Example 4-6, using the SBIC instruction instead of SBIS.

Solution:

```
.INCLUDE "M32DEF.INC"

        CBI    DDRB, 2      ;make PB2 an input
        LDI    R16, 0xFF
        OUT    DDRD, R16    ;make PORTD an output port

AGAIN:  SBIC   PINB, 2      ;skip next if PB bit is LOW
        RJMP   OVER         ;SW is HIGH
        LDI    R16, 'N'     ;R16 = 'N' (ASCII letter N)
        OUT    PORTD, R16   ;PORTD = 'N'
        RJMP   AGAIN

OVER:   LDI    R16, 'Y'     ;R16 = 'Y' (ASCII letter Y)
        OUT    PORTD, R16   ;PORTD = 'Y'
        RJMP   AGAIN
```

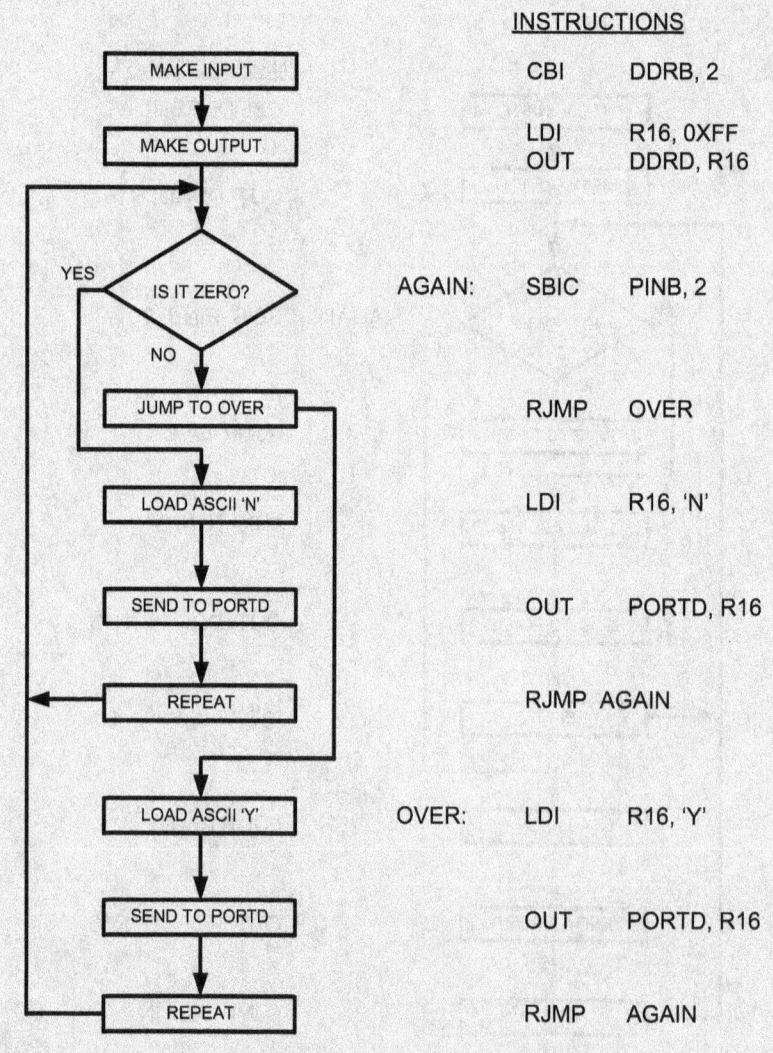

	INSTRUCTIONS	
MAKE INPUT	CBI	DDRB, 2
MAKE OUTPUT	LDI	R16, 0XFF
	OUT	DDRD, R16
IS IT ZERO? (YES/NO)	AGAIN: SBIC	PINB, 2
JUMP TO OVER	RJMP	OVER
LOAD ASCII 'N'	LDI	R16, 'N'
SEND TO PORTD	OUT	PORTD, R16
REPEAT	RJMP AGAIN	
LOAD ASCII 'Y'	OVER: LDI	R16, 'Y'
SEND TO PORTD	OUT	PORTD, R16
REPEAT	RJMP	AGAIN

Reading a single bit

We can also use the bit test instructions to read the status of a single bit and send it to another bit or save it. This is shown in Examples 4-8 and 4-9.

Example 4-8

A switch is connected to pin PB0 and an LED to pin PB7. Write a program to get the status of SW and send it to the LED.

Solution:

```
.INCLUDE "M32DEF.INC"
      CBI   DDRB, 0      ;make PB0 an input
      SBI   DDRB, 7      ;make PB7 an output
AGAIN:SBIC  PINB, 0      ;skip next if PB0 is clear
      RJMP  OVER         ;(JMP is OK too)
      CBI   PORTB, 7
      RJMP  AGAIN        ;we can use JMP too
OVER: SBI   PORTB, 7
      RJMP  AGAIN        ;we can use JMP too
```

Example 4-9

A switch is connected to pin PB0. Write a program to get the status of SW and save it in location 0x200.

Solution:

```
.EQU MYTEMP = 0x200     ;set aside location 0x200
.INCLUDE "M32DEF.INC"
      CBI   DDRB, 0      ;make PB0 an input
AGAIN:SBIC  PINB, 0      ;skip next if PB0 is clear
      RJMP  OVER         ;(JMP is OK too)
      LDI   R16, 0
      STS   MYTEMP,R16   ;save it in MYTEMP
      RJMP  AGAIN        ;we can use JMP too
OVER: LDI   R16,0x1      ;move 1 to R16
      STS   MYTEMP,R16   ;save it in MYTEMP
      RJMP  AGAIN        ;we can use JMP too
```

Review Questions

1. True or false. The instruction "SBI PORTB, 1" makes pin PB1 HIGH while leaving other pins of PORTB unchanged, if bit 1 of the DDR bits is configured for output.
2. Show one way to toggle the pin PB7 continuously using AVR instructions.
3. Write instructions to get the status of PB2 and put it on PB0.
4. Write instructions to toggle both bits of PD7 and PD0 continuously.
5. According to Figure 4-7, what does the machine instruction $9819 do?

SUMMARY

This chapter focused on the I/O ports of the AVR. The four ports of the ATmega32, PORTA, PORTB, PORTC, and PORTD, were explored. These ports can be used for input or output. All the ports have alternate functions. The three registers associated with each port are PORTx, DDRx, and PINx. Their role in I/O manipulation was examined. Then, I/O instructions of the AVR were explained, and numerous examples were given. We also showed the bit-addressability of AVR ports.

CAUTION

We strongly recommend that you study Section C.2 (Appendix C) before connecting any external hardware to your AVR system. Failure to use the right instruction or the right connection to port pins can damage the ports of your AVR chip.

PROBLEMS

SECTION 4.1: I/O PORT PROGRAMMING IN AVR

1. The ATmega32 has a DIP package of _____ pins.
2. In ATmega32, how many pins are assigned to V_{CC} and GND?
3. In the ATmega32, how many pins are designated as I/O port pins?
4. How many pins are designated as PORTA in the 40-pin DIP package and what are their numbers?
5. How many pins are designated as PORTB in the 40-pin DIP package and what are their numbers?
6. How many pins are designated as PORTC in the 40-pin DIP package and what are their numbers?
7. How many pins are designated as PORTD in the 40-pin DIP package and what are their numbers?
8. Upon reset, all the bits of ports are configured as _____ (input, output).
9. Explain the role of DDRx and PORTx in I/O operations.

10. Write a program to get 8-bit data from PORTC and send it to PORTB and PORTD.
11. Write a program to get 8-bit data from PORTD and send it to PORTB and PORTC.
12. Which pins are for RxD and TxD?
13. Give data memory location assigned to DDR registers of Ports A–C for the ATmega32.
14. Write a program to toggle all the bits of PORTB and PORTC continuously
    (a) using 0xAA and 0x55 (b) using the COM instruction.

SECTION 4.2: I/O BIT MANIPULATION PROGRAMMING

15. Which ports of the ATmega32 are bit-addressable?
16. What is the advantage of bit-addressability for AVR ports?
17. Is the instruction "COM PORTB" a valid instruction?
18. Write a program to toggle PB2 and PB5 continuously without disturbing the rest of the bits.
19. Write a program to toggle PD3, PD7, and PC5 continuously without disturbing the rest of the bits.
20. Write a program to monitor bit PC3. When it is HIGH, send 0x55 to PORTD.
21. Write a program to monitor the PB7 bit. When it is LOW, send $55 and $AA to PORTC continuously.
22. Write a program to monitor the PA0 bit. When it is HIGH, send $99 to PORTB. If it is LOW, send $66 to PORTB.
23. Write a program to monitor the PB5 bit. When it is HIGH, make a LOW-to-HIGH-to-LOW pulse on PB3.
24. Write a program to get the status of PC3 and put it on PC4.
25. Create a flowchart and write a program to get the statuses of PD6 and PD7 and put them on PC0 and PC7, respectively.
26. Write a program to monitor the PB5 and PB6 bits. When both of them are HIGH, send $AA to PORTC; otherwise, send $55 to PORTC.
27. Write a program to monitor the PB5 and PB6 bits. When either of them is HIGH, send $AA to PORTC; otherwise, send $55 to PORTC.
28. Referring to Figure 4-8 and Table 4-8, write the machine equivalent of "SBIS PINB,3".
29. Referring to Figure 4-6 and Table 4-8, write the machine equivalent of the "SBI PORTA,2" instruction.

ANSWERS TO REVIEW QUESTIONS

SECTION 4.1: I/O PORT PROGRAMMING IN AVR

1. 4
2. True
3. False
4. ```
   LDI  R16,0xFF
   OUT  DDRB,R16
   OUT  DDRC,R16
   LDI  R16,0x99
   OUT  PORTB,R16
   OUT  PORTC,R16
   ```
5. $FF, DDRB
6. $00, DDRB
7. True
8. True

SECTION 4.2: I/O BIT MANIPULATION PROGRAMMING

1. True
2. ```
           CBI  DDRB,7
   H1:     SBI  PORTB,7
           CBI  PORTB,7
           RJMP H1
   ```

3. ```
           CBI   DDRB,2
           SBI   DDRB,0
   AGAIN:  SBIS  PINB,2
           RJMP  OVER
           SBI   PORTB,0
           RJMP  AGAIN
   OVER:   CBI   PORTB,0
           RJMP  AGAIN
   ```

4. ```
           SBI  DDRD,0
           SBI  DDRD,7
   H2:     SBI  PORTD,0
           SBI  PORTD,7
           CBI  PORTD,0
           CBI  PORTD,7
           RJMP H2
   ```

5. $9819 is 1001 1000 0001 1001 in binary; according to Figure 4-7, this is the CBI instruction, where a = 00011 = 3 and b = 001 = 1. According to Table 4-8, 3 is the I/O address of TWDR; thus, this is the "CBI TWDR,1" instruction and clears bit 1 of the TWDR register.

CHAPTER 5

ARITHMETIC, LOGIC INSTRUCTIONS, AND PROGRAMS

OBJECTIVES

Upon completion of this chapter, you will be able to:

>> Define the range of numbers possible in AVR unsigned data
>> Code addition and subtraction instructions for unsigned data
>> Code AVR multiplication instructions
>> Code AVR programs for division
>> Code AVR Assembly language logic instructions AND, OR, and EX-OR
>> Use AVR logic instructions for bit manipulation
>> Use compare instructions for program control
>> Code conditional branch instructions
>> Code AVR rotate instructions and data serialization
>> Contrast and compare packed and unpacked BCD data
>> Code AVR programs for ASCII and BCD data conversion

This chapter describes AVR arithmetic and logic instructions. Program examples are given to illustrate the application of these instructions. In Section 5.1 we discuss instructions and programs related to addition, subtraction, multiplication, and division of unsigned numbers. Signed numbers are described in Section 5.2. In Section 5.3, we discuss the logic instructions AND, OR, and XOR, as well as the compare instruction. The rotate and shift instructions and data serialization are explained in Section 5.4. In Section 5.5 we introduce BCD and ASCII conversion.

SECTION 5.1: ARITHMETIC INSTRUCTIONS

Unsigned numbers are defined as data in which all the bits are used to represent data and no bits are set aside for the positive or negative sign. This means that the operand can be between 00 and FFH (0 to 255 decimal) for 8-bit data.

Addition of unsigned numbers

In the AVR, the add operation has two general purpose registers as inputs and the result will be stored in the first (left) register. One form of the ADD instruction in the AVR is:

```
ADD   Rd,Rr   ;Rd = Rd + Rr
```

The instruction adds Rr (resource) to Rd (destination) and stores the result in Rd. It could change any of the Z, C, N, V, H or S bits of the status register, depending on the operands involved. The effect of the ADD instruction on N and V is discussed in Section 5.2 because these bits are relevant mainly in signed number operations. Look at Example 5-1.

Notice that none of the AVR addition instructions support direct memory access; that is, we cannot add a memory location to another memory location or register. To add a memory location we should first load it to any of the R0–R31 registers and then use the ADD operation on it. Look at Example 5-2.

Example 5-1

Show how the flag register is affected by the following instructions.

```
        LDI    R21,0xF5    ;R21 = F5H
        LDI    R22,0x0B    ;R22 = 0x0BH
        ADD    R21,R22     ;R21 = R21+R22 = F5+0B = 00 and C = 1
```

Solution:
```
     F5H           1111 0101
  +  0BH        +  0000 1011
    100H           0000 0000
```

After the addition, register R21 contains 00 and the flags are as follows:
C = 1 because there is a carry out from D7.
Z = 1 because the result in destination register (R21) is zero.
H = 1 because there is a carry from D3 to D4.

Example 5-2

Assume that RAM location 400H has the value of 33H. Write a program to find the sum of location 400H of RAM and 55H. At the end of the program, R21 should contain the sum.

Solution:

```
LDS    R2,0x400    ;R2 = 33H (location 0x400 of RAM)
LDI    R21,0x55    ;R21 = 55
ADD    R21,R2      ;R21 = R21 + R2 = 55H + 33H = 88H, C = 0
```

ADC and addition of 16-bit numbers

When adding two 16-bit data operands, we need to be concerned with the propagation of a carry from the lower byte to the higher byte. This is called *multi-byte addition* to distinguish it from the addition of individual bytes. The instruction ADC (ADD with carry) is used on such occasions.

For example, look at the addition of 3CE7H + 3B8DH, as shown next.

```
        1
      3C E7
   +  3B 8D
      78 74
```

When the first byte is added, there is a carry (E7 + 8D = 74, C = 1). The carry is propagated to the higher byte, which results in 3C + 3B + 1 = 78 (all in hex). Example 5-3 shows the above steps in an AVR program.

Example 5-3

Write a program to add two 16-bit numbers. The numbers are 3CE7H and 3B8DH. Assume that R1 = 8D, R2 = 3B, R3 = E7, and R4 = 3C. Place the sum in R3 and R4; R3 should have the lower byte.

Solution:

```
;R1 = 8D
;R2 = 3B
;R3 = E7
;R4 = 3C

ADD    R3,R1    ;R3 = R3 + R1 = E7 + 8D = 74 and C = 1
ADC    R4,R2    ;R4 = R4 + R2 + carry, adding the upper byte
                ;with carry from lower byte
                ;R4 = 3C + 3B + 1 = 78H (all in hex)
```

Notice the use of ADD for the lower byte and ADC for the higher byte.

Subtraction of unsigned numbers

In many microprocessors, there are two different instructions for subtraction: SUB and SUBB (subtract with borrow). In the AVR we have five instructions for subtraction: SUB, SBC, SUBI, SBCI, and SBIW. Figure 5-1 shows a summary of each instruction.

```
SUB    Rd,Rr          ;Rd=Rd-Rr
SBC    Rd,Rr          ;Rd=Rd-Rr-c
SUBI   Rd,K           ;Rd=Rd-K
SBCI   Rd,K           ;Rd=Rd-K-c
SBIW   Rd:Rd+1,K      ;Rd+1:Rd=Rd+1:Rd-K
```

Figure 5-1.

The SBC and SBCI instructions are subtract with borrow. In the AVR, we use the C (carry) flag for the borrow and that is why they are called SBC (SuB with Carry). In this section we will examine some of these commands.

SUB Rd,Rr (Rd = Rd – Rr)

In subtraction, the AVR microcontrollers (indeed, all modern CPUs) use the 2's complement method. Although every CPU contains adder circuitry, it would be too cumbersome (and take too many transistors) to design separate subtractor circuitry. For this reason, the AVR uses adder circuitry to perform the subtraction command. Assuming that the AVR is executing a simple subtract instruction and that C = 0 prior to the execution of the instruction, one can summarize the steps of the hardware of the CPU in executing the SUB instruction for unsigned numbers as follows:

1. Take the 2's complement of the subtrahend (right-hand operand).
2. Add it to the minuend (left-hand operand).
3. Invert the carry.

Example 5-4

Show the steps involved in the following.

```
        LDI    R20, 0x23        ;load 23H into R20
        LDI    R21, 0x3F        ;load 3FH into R21
        SUB    R21, R20         ;R21 <- R21-R20
```

Solution:

```
     R21 = 3F   0011 1111         0011 1111
   - R20 = 23   0010 0011       + 1101 1101  (2's complement)
          1C                    1 0001 1100
                                C = 0, D7 = N = 0 (result is positive)
```

The flags would be set as follows: N = 0, C = 0. (Notice that there is a carry but C = 0. We will discuss this more in the next section.) The programmer must look at the N (or C) flag to determine if the result is positive or negative.

These two steps are performed for every SUB instruction by the internal hardware of the CPU, regardless of the source of the operands, provided that the addressing mode is supported. It is after these two steps that the result is obtained and the flags are set. Example 5-4 illustrates the two steps.

After the execution of the SUB instruction, if $N = 0$ (or $C = 0$), the result is positive; if $N = 1$ (or $C = 1$), the result is negative and the destination has the 2's complement of the result. Normally, the result is left in 2's complement, but the NEG (negate, which is 2's complement) instruction can be used to change it. The other subtraction instructions for subtract are SUBI and SBIW, which subtract an immediate (constant) value from a register. SBIW subtracts an immediate value in the range of 0–63 from a register pair and stores the result in the register pair. Notice that only the last eight registers can be used with SBIW. See Examples 5-5 and 5-6.

Example 5-5

Write a program to subtract 18H from 29H and store the result in R21 (a) without using the SUBI instruction, and (b) using the SUBI instruction.

Solution:

(a)
```
     LDI    R21,0x29    ;R21 = 29H
     LDI    R22,0x18    ;R22 = 18H
     SUB    R21,R22     ;R21 = R21 - R22 = 29 - 18 = 11 H
```

(b)
```
     LDI    R21,0x29    ;R21 = 29H
     SUBI   R21,0x18    ;R21 = R21 - 18 = 29 - 18 = 11 H
```

Example 5-6

Write a program to subtract 18H from 2917H and store the result in R25 and R24.

Solution:

```
     LDI    R25,0x29           ;load the high byte (R25 = 29H)
     LDI    R24,0x17           ;load the low byte (R24 = 17H)
     SBIW   R25:R24,0x18       ;R25:R24 <- R25:R24 - 0x18
                               ;28FF = 2917 - 18
```

Notice that you should use SBIW Rd+1:Rd,K format. If SBIW Rd:Rd+1,K format is used, the assembler will assemble your code as if you had typed SBIW Rd+1:Rd,K Change the third line of the code from SBIW R25:R24,0x18 to SBIW R24:R25,0x18 and examine the result.

CHAPTER 5: ARITHMETIC, LOGIC INSTRUCTIONS, AND PROGRAMS   165

SBC (Rd ← Rd – Rr – C) subtract with borrow (denoted by C)

This instruction is used for multibyte numbers and will take care of the borrow of the lower byte. If the borrow flag is set to one (C = 1) prior to executing the SBC instruction, this operation also subtracts 1 from the result. See Example 5-7.

Example 5-7

Write a program to subtract two 16-bit numbers: 2762H – 1296H. Assume R26 = (62) and R27 = (27). Place the difference in R26 and R27; R26 should have the lower byte.

Solution:

```
;R26 = (62)
;R27 = (27)

LDI   R28,0x96     ;load the low byte (R28 = 96H)
LDI   R29,0x12     ;load the high byte (R29 = 12H)
SUB   R26,R28      ;R26 = R26 - R28 = 62 - 96 = CCH
                   ;C = borrow = 1, N = 1
SBC   R27,R29      ;R27 = R27 - R29 - C
                   ;R27 = 27 - 12 - 1 = 14H
```

After the SUB, R26 has = 62H – 96H = CCH and the carry flag is set to 1, indicating there is a borrow (notice, N = 1). Because C = 1, when SBC is executed R27 has 27H – 12H – 1 = 14H. Therefore, we have 2762H – 1296H = 14CCH.

The C flag in subtraction for AVR

Notice that the AVR is like other CPUs such as the x86 and the 8051 when it comes to the carry flag in subtract operations. In the AVR, after subtract operations, the carry is inverted by the CPU itself and we examine the C flag to see if the result is positive or negative. This means that, after subtract operations, if C = 1, the result is negative, and if C = 0, the result is positive. If you study Example 5-4 again, you will see that there was a carry from MSB, but C = 0. Now you know the reason; it is because the CPU inverts the carry flag after the SUB instruction. Notice that the CPU does not invert the carry flag after the ADD instruction.

Multiplication of unsigned numbers

The AVR has several instructions dedicated to multiplication. Here we will discuss the MUL instruction. Other instructions are similar to MUL but are used for signed numbers. See Table 5-1.

MUL is a byte-by-byte multiply instruction. In byte-by-byte multiplication, operands must be in registers. After multiplication, the 16-bit unsigned product is placed in R1 (high byte) and R0 (low byte). Notice that if any of the operands is selected from R0 or R1 the result will overwrite those registers after multiplication.

Table 5-1: Multiplication Summary

Multiplication	Application	Byte1	Byte2	High byte of result	Low byte of result
MUL Rd, Rr	Unsigned numbers	Rd	Rr	R1	R0
MULS Rd, Rr	Signed numbers	Rd	Rr	R1	R0
MULSU Rd, Rr	Unsigned numbers with signed numbers	Rd	Rr	R1	R0

The following example multiplies 25H by 65H.

```
LDI    R23,0x25     ;load 25H to R23
LDI    R24,0x65     ;load 65H to R24
MUL    R23,R24      ;25H * 65H = E99 where
                    ;R1 = 0EH and R0 = 99H
```

Division of unsigned numbers

AVR has no instruction for divide operation. We can write a program to perform division by repeated subtraction. In dividing a byte by a byte, the numerator is placed in a register and the denominator is subtracted from it repeatedly. The quotient is the number of times we subtracted and the remainder is in the register upon completion. See Program 5-1.

```
.DEF  NUM = R20
.DEF  DENOMINATOR = R21
.DEF  QUOTIENT = R22


      LDI    NUM,95            ;NUM = 95
      LDI    DENOMINATOR,10    ;DENOMINATOR = 10
      CLR    QUOTIENT          ;QUOTIENT = 0

L1:   INC    QUOTIENT
      SUB    NUM, DENOMINATOR
      BRCC   L1                ;branch if C is zero

      DEC    QUOTIENT          ;once too many
      ADD    NUM, DENOMINATOR  ;add back to it


HERE: JMP HERE                 ;stay here forever
```

Program 5-1: Divide Function

An application for division

Sometimes a sensor is connected to an ADC (analog-to-digital converter) and the ADC represents some quantity such as temperature or pressure. The 8-bit ADC provides data in hex in the range of 00–FFH. This hex data must be converted to decimal. We do that by dividing it by 10 repeatedly, saving the remainders, as shown in Examples 5-8 and 5-9.

Example 5-8

Assume that the data memory location 0x315 has value FD (hex). Write a program to convert it to decimal. Save the digits in locations 0x322, 0x323, and 0x324, where the least-significant digit is in location 0x322.

Solution:

```
.EQU HEX_NUM = 0x315

.EQU RMND_L = 0x322
.EQU RMND_M = 0x323
.EQU RMND_H = 0x324


.DEF NUM = R20
.DEF DENOMINATOR = R21
.DEF QUOTIENT = R22


        LDI    R16,0xFD              ;$FD = 253 in decimal
        STS    HEX_NUM,R16           ;store $FD in location 0x315

;====================

        LDS    NUM, HEX_NUM
        LDI    DENOMINATOR,10        ;DENOMINATOR = 10

L1:     INC    QUOTIENT              ;
        SUB    NUM, DENOMINATOR;
        BRCC   L1                    ;if C = 0 go back

        DEC    QUOTIENT              ;once too many
        ADD    NUM, DENOMINATOR      ;add back to it
        STS    RMND_L, NUM           ;store remainder as the 1st digit

        MOV    NUM, QUOTIENT
        LDI    QUOTIENT,0

L2:     INC    QUOTIENT
        SUB    NUM, DENOMINATOR
        BRCC   L2

        DEC    QUOTIENT              ;once too many
        ADD    NUM, DENOMINATOR      ;add back to it
        STS    RMND_M, NUM           ;store remainder as the 2nd digit

        STS    RMND_H, QUOTIENT      ;store quotient as the 3rd digit

HERE:   JMP    HERE                  ;stay here forever
```

To convert a single decimal digit to ASCII format, we OR it with 30H. See Section 5.5.

Example 5-9

Analyze the program in Example 5-8 for a numerator of 253.

Solution:

To convert a binary (hex) value to decimal, we divide it by 10 repeatedly until the quotient is less than 10. After each division the remainder is saved. In the case of an 8-bit binary, such as FDH, we have 253 decimal, as shown below.

```
                Quotient   Remainder
253/10 =        25         3 (low digit)
25/10  =        2          5 (middle digit)
                           2 (high digit)
```

Therefore, we have FDH = 253.

Review Questions

1. In unsigned byte-by-byte multiplication, the product will be placed in register(s) _____ .
2. Is "MUL R2,0x10" a valid AVR instruction? Explain your answer.
3. In AVR, the largest two numbers that can be multiplied are _____ and _____.
4. True or false. The MUL instruction works on R0 and R1 only.
5. The instruction "ADD R20, R21" places the sum in _____.
6. Why is the following ADD instruction illegal? "ADD R1,0x04"
7. Rewrite the instruction above in correct format.
8. The instruction "SUB R1, R2" places the result in _____.
9. Find the value of the C flags in each of the following.
   (a) LDI R21 0x4F   (b) LDI   R21, 0x9C
       LDI R22 0xB1   LDI   R22, 0x63
       ADD R21, R22   ADD   R21, R22
10. Show how the CPU would subtract 05H from 43H.
11. If C = 1, R1 = 95H, and R2 = 4FH prior to the execution of "SBC R1,R2", what will be the contents of R1 and C after the subtraction?

CHAPTER 5: ARITHMETIC, LOGIC INSTRUCTIONS, AND PROGRAMS   169

SECTION 5.2: SIGNED NUMBER CONCEPTS AND ARITHMETIC OPERATIONS

All data items used so far have been unsigned numbers, meaning that the entire 8-bit operand was used for the magnitude. Many applications require signed data. In this section the concept of signed numbers is discussed along with related instructions. If your applications do not involve signed numbers, you can bypass this section.

Concept of signed numbers in computers

In everyday life, numbers are used that could be positive or negative. For example, a temperature of 5 degrees below zero can be represented as –5, and 20 degrees above zero as +20. Computers must be able to accommodate such numbers. To do that, computer scientists have devised the following arrangement for the representation of signed positive and negative numbers: The most significant bit (MSB) is set aside for the sign (+ or –), while the rest of the bits are used for the magnitude. The sign is represented by 0 for positive (+) numbers and 1 for negative (–) numbers. Signed byte representation is discussed below.

Signed 8-bit operands

In signed byte operands, D7 (MSB) is the sign, and D0 to D6 are set aside for the magnitude of the number. If D7 = 0, the operand is positive, and if D7 = 1, it is negative. The N flag in the status register is the D7 bit.

Positive numbers

The range of positive numbers that can be represented by the format shown in Figure 5-2 is 0 to +127. If a positive number is larger than +127, a 16-bit operand must be used.

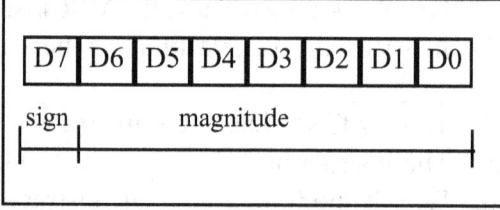

Figure 5-2. 8-Bit Signed Operand

Negative numbers

For negative numbers, D7 is 1; however, the magnitude is represented in its 2's complement. Although the assembler does the conversion, it is still important to understand how the conversion works. To convert to negative number representation (2's complement), follow these steps:

1. Write the magnitude of the number in 8-bit binary (no sign).
2. Invert each bit.
3. Add 1 to it.

Examples 5-10, 5-11, and 5-12 on the next page demonstrate these three steps.

Example 5-10

Show how the AVR would represent –5.

Solution:

Observe the following steps.

```
1.    0000 0101        5 in 8-bit binary
2.    1111 1010        invert each bit
3     1111 1011        add 1 (which becomes FB in hex)
```

Therefore, –5 = FBH, the signed number representation in 2's complement for –5. The D7 = N = 1 indicates that the number is negative.

Example 5-11

Show how the AVR would represent –34H.

Solution:

Observe the following steps.

```
1.    0011 0100        34H given in binary
2.    1100 1011        invert each bit
3     1100 1100        add 1 (which is CC in hex)
```

Therefore, –34 = CCH, the signed number representation in 2's complement for 34H. The D7 = N = 1 indicates that the number is negative.

Example 5-12

Show how the AVR would represent –128.

Solution:

Observe the following steps.

```
1.    1000 0000        128 in 8-bit binary
2.    0111 1111        invert each bit
3     1000 0000        add 1 (which becomes 80 in hex)
```

Therefore, –128 = 80H, the signed number representation in 2's complement for –128. The D7 = N = 1 indicates that the number is negative. Notice that 128 (binary 10000000) in unsigned representation is the same as signed –128 (binary 10000000).

From the examples above, it is clear that the range of byte-sized negative numbers is –1 to –128. The following lists byte-sized signed number ranges:

Decimal	Binary	Hex
-128	1000 0000	80
-127	1000 0001	81
-126	1000 0010	82
...	....	..
-2	1111 1110	FE
-1	1111 1111	FF
0	0000 0000	00
+1	0000 0001	01
+2	0000 0010	02
..	....	..
+127	0111 1111	7F

Overflow problem in signed number operations

When using signed numbers, a serious problem sometimes arises that must be dealt with. This is the overflow problem. The AVR indicates the existence of an error by raising the V (overflow) flag, but it is up to the programmer to take care of the erroneous result. The CPU understands only 0s and 1s and ignores the human convention of positive and negative numbers. What is an overflow? If the result of an operation on signed numbers is too large for the register, an overflow has occurred and the programmer must be notified. Look at Example 5-13.

Example 5-13

Examine the following code and analyze the result, including the N and V flags.

```
    LDI   R20,0X60    ;R20 = 0110 0000 (+70)
    LDI   R21,0x46    ;R21 = 0100 0110 (+96)
    ADD   R20,R21     ;R20 = (+96) + (+70) = 1010 0110
                      ;R20 = A6H = -90 decimal, INVALID!!
```

Solution:

```
    +96     0110 0000
 +  +70     0100 0110
 +  166     1010 0110   N = 1 (negative) and V = 1 Sum = -90
```

According to the CPU, the result is negative (N = 1), which is wrong. The CPU sets V = 1 to indicate the overflow error. Remember that the N flag is the D7 bit. If N = 0, the sum is positive, but if N = 1, the sum is negative.

In Example 5-13, +96 was added to +70 and the result, according to the CPU, was –90. Why? The reason is that the result was larger than what R0 could contain. Like all other 8-bit registers, R0 could only contain values less than or equal to +127. The designers of the CPU created the overflow flag specifically for the purpose of informing the programmer that the result of the signed number operation is erroneous. The N flag is D7 of the result. If N = 0, the sum is positive (+) and if N = 1, then the sum is negative.

When is the V flag set?

In 8-bit signed number operations, V is set to 1 if either of the following two conditions occurs:

1. There is a carry from D6 to D7 but no carry out of D7 (C = 0).
2. There is a carry from D7 out (C = 1) but no carry from D6 to D7.

In other words, the overflow flag is set to 1 if there is a carry from D6 to D7 or from D7 out, but not both. This means that if there is a carry both from D6 to D7 and from D7 out, V = 0. In Example 5-13, because there is only a carry from D6 to D7 and no carry from D7 out, V = 1.

In Example 5-14, because there is only a carry from D7 out and no carry from D6 to D7, V = 1.

Example 5-14

Examine the following code, noting the role of the V and N flags:
```
    LDI    R20,0x80     ;R20 = 1000 0000 (80H = -128)
    LDI    R21,0xFE     ;R21 = 1111 1110 (FEH = -2)
    ADD    R20,R21      ;R20 = (-128) + (-2)
                        ;R20 = 1000000 + 11111110 = 0111 1110,
                        ;N = 0, R0 = 7EH = +126, invalid
```
Solution:
```
  -128          1000 0000
  + - 2         1111 1110
  - 130         0111 1110   N = 0 (positive) and V = 1
```

According to the CPU, the result is +126, which is wrong, and V = 1 indicates that. Notice that the N flag indicates the sign of the corrupted result, not the sign that the real result should have.

Further considerations on the V flag

In the ADD instruction, there are two different conditions. Either the operands have the same sign or the signs of the operands are different. When we ADD two numbers with different signs, the absolute value of the result is smaller than the operands before executing the ADD instruction. So overflow definitely cannot happen after two operands with different signs are added. Overflow is possible only when we ADD two operands with the same sign. In this case the absolute value of the result is larger than the operands before executing the ADD instruction. So it is possible that the result will be too large for the register and cause overflow. If we ADD two numbers with the same sign, the results should have the same sign too. If we add two numbers with the same sign and the result sign is different, we know that overflow has occurred. That is exactly the way that the CPU knows when to set the V flag. In the AVR the equation of the V flag is as follows:

$$V = Rd7 \cdot Rr7 \cdot \overline{R7} + \overline{Rd7} \cdot \overline{Rr7} \cdot R7$$

where Rd7 and Rr7 are the 7th bit of the operands and R7 is the 7th bit of

---

the result. We can extend this concept to the SUB instructions (a − b = a + (−b))

Study Examples 5-15 and 5-16 to understand the overflow flag in signed arithmetic.

Example 5-15

Examine the following code, noting the role of the V and N flags:

```
        LDI     R20,-2              ;R20 = 1111 1110 (R20 = FEH)
        LDI     R21,-5              ;R21 = 1111 1110 (R21 = FBH)
        ADD     R20,R21             ;R20 = (-2) + (-5) = -7 or F9H
                                    ;correct, since V = 0
```
Solution:

```
  -2        1111 1110
+ -5        1111 1011
  - 7       1111 1001   and V = 0 and N = 1. Sum is negative
```

According to the CPU, the result is –7, which is correct, and the V indicates that (V = 0).

Example 5-16

Examine the following code, noting the role of the V and N flags:

```
        LDI     R20,7           ;R20 = 0000 0111
        LDI     R20,18          ;R20 = 0001 0010
        ADD     R20,R21         ;R20 = (+7) + (+18)
                                ;R20 = 00000111 + 00010010 = 0001 1001
                                ;R20 = (+7) + (+18) = +25, N = 0, positive
                                ;and correct, V = 0
```
Solution:

```
  + 7 0000 0111
+ +18 0001 0010
  +25 0001 1001   N = 0 (positive 25) and V = 0
```

According to the CPU, this is +25, which is correct, and V = 0 indicates that.

From Examples 5-14 to 5-16, we conclude that in any signed number addition, V indicates whether the result is valid or not. If V = 1, the result is erroneous; if V = 0, the result is valid. We can state emphatically that in unsigned number addition, the programmer must monitor the status of C (carry flag), and in signed number addition, the V (overflow) flag must be monitored. In the AVR, instructions such as BRCS and BRCC allow the program to branch right after the addition of unsigned numbers according to the value of C flag. There are also the BRVC and the BRVS instructions for the V flag that allow us to correct the signed number error. We also have two branch instructions for the N flag (negative), BRPL and BRMI.

What is the difference between the N and S flags?

As we mentioned before, in signed numbers the N flag represents the D7 bit of the result. If the result is positive, the N flag is zero, and if the result is negative, the N flag is one, which is why it is called the Negative flag.

In operations on signed numbers, overflow is possible. Overflow corrupts the result and negates the sign bit. So if you ADD two positive numbers, in case of overflow, the N flag would be 1 showing that the result is negative! The S flag helps you to know the sign of the real result. It checks the V flag in addition to the D7 bit. If V = 0, it shows that overflow has not occurred and the S flag will be the same as D7 to show the sign of the result. If V = 1, it shows that overflow has occurred and the S flag will be opposite to the D7 to show the sign of the real (not the corrupted) result. See Example 5-17.

Example 5-17

Study Examples 5-13 through 5-16 again and state what the value of the S flag is in each of them and whether the value of the S flag is the same as that of the N flag.

Solution:
Example 5-13: Because two positive numbers are added, the sign of the real result is positive, so S = 0 (for positive). The value of the S flag is not the same as that of the N flag (1) because there is overflow (V = 1).

Example 5-14: Because two negative numbers are added, the sign of the real result is negative, so S = 1 (for negative). The value of the S flag is not the same as that of the N flag (0) because there is overflow (V = 1).

Example 5-15: Because two negative numbers are added, the sign of the real result is negative, so S = 1 (for negative). The value of the S flag is the same as that of the N flag (1) because there is no overflow (V = 0).

Example 5-16: Because two positive numbers are added, the sign of the real result is positive, so S = 0 (for positive). The value of the S flag is the same as that of the N flag (0) because there is overflow (V = 0).

Instructions to create 2's complement

The AVR has a special instruction to make the 2's complement of a number. It is called NEG (negate), which is discussed in the next section.

Review Questions

1. In an 8-bit operand, bit _____ is used for the sign bit.
2. Convert –16H to its 2's complement representation.
3. The range of byte-sized signed operands is _____ to _____.
4. Show +9 and –9 in binary.
5. Explain the difference between a carry and an overflow.

SECTION 5.3: LOGIC AND COMPARE INSTRUCTIONS

Apart from I/O and arithmetic instructions, logic instructions are some of the most widely used instructions. In this section we cover Boolean logic instructions such as AND, OR, Exclusive-OR (XOR), and complement. We will also study the compare instruction.

AND

```
AND Rd,Rr   ;Rd = Rd AND Rr
```

This instruction will perform a logical AND on the two operands and place the result in the left-hand operand. There is also the "ANDI Rd, k" instruction in which the right-hand operand can be a constant value. The AND instruction will affect the Z, S, and N flags. N is D7 of the result, and Z = 1 if the result is zero. The AND instruction is often used to mask (set to 0) certain bits of an operand. See Example 5-18.

Logical AND Function

Inputs		Output
X	Y	X AND Y
0	0	0
0	1	0
1	0	0
1	1	1

X —⊐D— X AND Y
Y —

Example 5-18

Show the results of the following.

```
LDI    R20,0x35    ;R20 = 35H
ANDI   R20,0x0F    ;R20 = R20 AND 0FH (now R20 = 05)
```

Solution:

```
        35H    0011 0101
AND     0FH    0000 1111
        ----------------
        05H    0000 0101           ;35H AND 0FH = 05H, Z = 0, N = 0
```

OR

```
OR Rd,Rr    ;Rd = Rd OR Rr
```

This instruction will perform a logical OR on the two operands and place the result in the left-hand operand. There is also the "ORI Rd, k" instruction in which the right-hand operand can be a constant value. The OR instruction will affect the Z, S, and N flags. N is D7 of the result and Z = 1 if the result is zero. The OR instruction can be used to set certain bits of an operand to 1. See Example 5-19.

Logical OR Function

Inputs		Output
X	Y	X OR Y
0	0	0
0	1	1
1	0	1
1	1	1

X —⊐D— X OR Y
Y —

Example 5-19

(a) Show the results of the following:

```
        LDI     R20,0x04                ;R20 = 04
        ORI     R20,0x30                ;now R20 = 34H
```

(b) Assume that PB2 is used to control an outdoor light, and PB5 to control a light inside a building. Show how to turn "on" the outdoor light and turn "off" the inside one.

Solution:

```
(a)         04H     0000 0100
      OR    30H     0011 0000
                    -------------------
            34H     0011 0100        04 OR 30 = 34H, Z = 0 and N = 0
(b)
      SBI   DDRB, 2         ;bit 2 of Port B is output
      SBI   DDRB, 5         ;bit 5 of Port B is output
      IN    R20, PORTB      ;move PORTB to R20.(Notice that we read
                            ;the value of PORTB instead of PINB
                            ;because we want to know the last value
                            ;of PORTB, not the value of the AVR
                            ;chip pins.)
      ORI   R20, 0b00000100 ;set bit 2 of R20 to one
      ANDI  R20, 0b11011111 ;clear bit 5 of R20 to zero
      OUT   PORTB, R20      ;out R20 to PORTB

HERE: JMP HERE              ;stop here
```

EX-OR

Logical XOR Function

Inputs		Output
A	**B**	**A XOR B**
0	0	0
0	1	1
1	0	1
1	1	0

```
        EOR     Rd,Rs         ;Rd = Rd XOR Rs
```

This instruction will perform a logical EX-OR on the two operands and place the result in the left-hand operand. The EX-OR instruction will affect the Z, S, and N flags. N is D7 of the result and Z = 1 if the result is zero. See Example 5-20.

A —⊐⊐D— A XOR B
B

Example 5-20

Show the results of the following:

```
        LDI     R20, 0x54
        LDI     R21, 0x78
        EOR     R20, R21
```

Solution:

```
            54H     0101 0100
      XOR   78H     0111 1000
                    ---------------
            2CH     0010 1100        54H XOR 78H = 2CH, Z = 0, N = 0
```

EX-OR can also be used to see if two registers have the same value. The "EOR R0, R1" instruction will EX-OR the R0 register and R1, and put the result in R0. If both registers have the same value, 00 is placed in R0 and the Z flag is set (Z = 1). Then, we can use the BREQ or BRNE instruction to make a decision based on the result. See Examples 5-21 and 5-22.

Example 5-21

The EX-OR instruction can be used to test the contents of a register by EX-ORing it with a known value. In the following code, we show how EX-ORing the value 45H with itself will raise the Z flag:

```
OVER:        IN    R20,PINB
             LDI   R21,0x45
             EOR   R20,R21
             BRNE  OVER
```

Solution:

```
    45H          01000101
    45H          01000101
    00           00000000
```

EX-ORing a number with itself sets it to zero with Z = 1. We can use the BREQ instruction to make the decision. EX-ORing with any other number will result in a nonzero value.

Example 5-22

Read and test PORTB to see whether it has the value 45H. If it does, send 99H to PORTC; otherwise, it is cleared.

Solution:

```
        LDI    R20,0xFF     ;R20 = 0xFF
        OUT    DDRC,R20     ;Port C is output
        LDI    R20,0x00     ;R20 = 0
        OUT    DDRB,R20     ;Port B is input
        OUT    PORTC,R20    ;PORTC = 00
        LDI    R21,0x45     ;R21 = 45
HERE:
        IN     R20,PINB     ;get a byte
        EOR    R20,R21      ;EX-OR with 0x45

        BRNE   HERE         ;branch if PORTB has value other than 45
        LDI    R20,0x99     ;R20 = 0x99
        OUT    PORTC,R20    ;PORTC = 99h
EXIT:   JMP    EXIT         ;stop here
```

Another widely used application of EX-OR is to toggle the bits of an operand. The following code demonstrates how to use EX-OR to toggle the bits of an operand.

```
        LDI    R20,0xFF
        EOR    R0, R20      ;EX-OR R0 with 1111 1111 will
                            ;change all the bits of R0 to
                            ;opposite
```

COM (complement)

This instruction complements the contents of a register. The complement action changes the 0s to 1s, and the 1s to 0s. This is also called *1's complement*.

```
LDI    R20,0xAA    ;R20 = 0xAA
COM    R20         ;now R20 = 55H
```

Logical Inverter

Input	Output
X	NOT X
0	1
1	0

X ——▷o— NOT X

NEG (negate)

This instruction takes the 2's complement of a register. See Example 5-23.

Example 5-23

Find the 2's complement of the value 85H. Notice that 85H is –123.
Solution:
```
LDI    R21, 0x85              ;85H = 1000 0101
                              ;1's = 0111 1010
                                            + 1
                              _____
NEG    R21              ;2's comp    0111 1011 = 7BH
```

Compare instructions

```
CP     Rd,Rr
```

The AVR has the CP instruction for the compare operation. The compare instruction is really a subtraction, except that the values of the operands do not change. There is also the "CPI Rd, k" instruction in which the right-hand operand can be a constant value.

The AVR has some conditional branch instructions that can be used after the CP instruction to make decisions based on the result of the CP instruction. In Chapter 3 we used some of them. Next, you will learn some other conditional branches.

Conditional branch instructions

As we studied in Chapter 3, conditional branches alter the flow of control if a condition is true. In the AVR there are at least two conditional jumps for each flag of the status register. Here we will describe eight of the most important conditional jumps. Others are similar but of different flags. Table 5-2 shows the conditional branch instructions that we will describe in this section.

Table 5-2: AVR Compare Instructions

BREQ	Branch if equal	Branch if Z = 1
BRNE	Branch if not equal	Branch if Z = 0
BRSH	Branch if same or higher	Branch if C = 0
BRLO	Branch if lower	Branch if C = 1
BRLT	Branch if less than (signed)	Branch if S = 1
BRGE	Branch if greater than or equal (signed)	Branch if S = 0
BRVS	Branch if Overflow flag set	Branch if V = 1
BRVC	Branch if Overflow flag clear	Branch if V = 0

BREQ and BRNE instructions

```
BREQ  k      ;if (Z = 1) then branch
             ;else continue
```

The BREQ makes decisions based on the Z flag. If Z = 1 the BREQ instruction branches. Notice that after the CP instructions, the Z = 1 means that the operands were equal, and after the DEC instruction it means that the operand is now equal to zero.

The BRNE instruction, like the BREQ, makes decisions based on the Z flag, but it branches when Z = 0. (After the CP instructions, Z = 0 means that the operands were not equal, and after the DEC instruction it means that the operand is not equal to zero.) See Example 5-24.

Notice that the BREQ and BRNE instructions can be used for both signed, and unsigned numbers.

Example 5-24

Write a program to monitor PORTB continuously for the value 63H. It should stop monitoring only if PORTB = 63H.

Solution:
```
      LDI    R20,0x00
      OUT    DDRB,R20      ;PORT B is input
      LDI    R21,0x63
AGAIN:
      IN     R20,PINB
      CP     R20,R21       ;compare with 0x63, Z = 1 if yes
      BRNE   AGAIN         ;go to AGAIN if PORTB is not equal to 0x63
      ....
```

BRSH and BRLO instructions

```
BRSH k       ;if (C = 0) then branch
             ;else continue
```

The BRSH makes decisions based on the C flag. If C = 0 (which, after the CP instructions for unsigned numbers, means that the left-hand operand of the CP instruction was the same as or higher than the right-hand operand) the CPU will jump.

The BRLO instruction, like the BRSH, makes decisions based on the C flag, but it branches when C = 1. (After the CP instructions for unsigned numbers, C = 1 means that the left-hand operand of the CP instruction was lower than the right-hand operand.) See Example 5-25.

Notice that the BRSH and the BRLO instructions can be used to compare unsigned numbers. To compare signed numbers you should use the BRGE and BRLT instructions. We will discuss them in more detail next.

We can use more than one conditional branch instruction to make more complicated decisions. See Example 5-26.

Example 5-25

Write a program to find the greater of the two values 27 and 54, and place it in R20.

Solution:

```
.EQU    VAL_1=27
.EQU    VAL_2=54

        LDI     R20,VAL_1       ;R20 = VAL_1
        LDI     R21,VAL_2       ;R21 = VAL_2
        CP      R21,R20         ;compare R21 and R20
        BRLO    NEXT            ;if R21<R20 (branch if lower) go to NEXT
        LDI     R20,VAL_2       ;R20 = VAL_2
NEXT:
```

Example 5-26

Assume that Port B is an input port connected to a temperature sensor. Write a program to read the temperature and test it for the value 75. According to the test results, place the temperature value into the registers indicated by the following.

If T = 75	then R16 = T	; R17 = 0 ; R18 = 0
If T > 75	then R16 = 0	; R17 = T ; R18 = 0
If T < 75	then R16 = 0	; R17 = 0 ; R18 = T

Solution:

```
            LDI     R20,0x00        ;R20 = 0
            OUT     DDRB,R20        ;Port B = input

            CLR     R16             ;R16 = 0
            CLR     R17             ;R17 = 0
            CLR     R18             ;R18 = 0

            IN      R20,PINB
            CPI     R20,75          ;compare R20 (PORTB) and 75
            BRSH    SAME_HI

                                    ;executes when R20 < 75
            MOV     R18,R20
            RJMP    CNTNU
SAME_HI:                            ;executes when R20 >= 75

            BRNE    HI

            MOV     R16,R20         ;executes when R20 = 75
            RJMP    CNTNU
HI:                                 ;executes when R20 > 75
            MOV     R17,R20
CNTNU:      ....
```

BRGE and BRLT instructions

The BRGE makes decisions based on the S flag. If S = 0 (which, after the CP instruction for signed numbers, means that the left-hand operand of the CP instruction was greater than or equal to the right-hand operand) the BRGE instruction branches in a forward or backward direction relative to program counter.

The BRLT is like the BRGE, but it branches when S = 1. Notice that the BRGE, and the BRLT are used with signed numbers.

BRVS and BRVC instructions

As we mentioned before, the V (overflow) flag must be monitored by the programmer to detect overflow and handle the error. The BRVC and BRVS instructions let you check the value of the V flag and change the flow of the program if overflow has occurred. See Example 5-27.

Example 5-27

Write a program to add two signed numbers. The numbers are in R21 and R22. The program should store the result in R21. If the result is not correct, the program should put 0xAA on PORTA and clear R21.

Solution:

```
        LDI   R21,0xFA      ;R21 = 0xFA
        LDI   R22,0x05      ;R22 = 0x05
        LDI   R23,0xFF      ;R23 = 0xFF
        OUT   DDRA,R23      ;Port A is output
        ADD   R21,R22       ;R21 = R21 + R22
        BRVC  NEXT          ;if V = 0 ( no error) then go to next
        LDI   R23,0xAA      ;R23 = 0xAA
        OUT   PORTA,R23     ;send 0xAA to PORTA
        LDI   R21,0x00      ;clear R21
NEXT: ...
```

Review Questions

1. Find the content of R20 after the following code in each case:
   (a) ```LDI R20,0x37``` (b) ```LDI R20,0x37```   (c) ```LDI R20,0x37```
      ```LDI R21,0xCA```   ```LDI R21,0xCA```   ```LDI R21,0xCA```
      ```AND R20,R21```   ```OR R20,21```   ```EOR```
2. To mask certain bits of R20, we must AND it with _____.
3. To set certain bits of R20 to 1, we must OR it with _____.
4. EX-ORing an operand with itself results in _____.
5. True or false. The CP instruction alters the contents of its operands.
6. Find the contents of register R20 after execution of the following code:
```
        LDI   R20,0
        LDI   R21,0x99
        LDI   R22,0xFF
        OR    R20,R21
        EOR   R20,R22
```

SECTION 5.4: ROTATE AND SHIFT INSTRUCTIONS AND DATA SERIALIZATION

In many applications there is a need to perform a bitwise rotation of an operand. In the AVR the rotation instructions ROL and ROR are designed specifically for that purpose. They allow a program to rotate a register right or left through the carry flag. We explore the rotate instructions next because they are widely used in many different applications.

Rotating through the carry

There are two rotate instructions in the AVR. They involve the carry flag. Each is shown next.

ROR instruction

```
ROR Rd        ;rotate Rd right through carry
```

In the ROR, as bits are rotated from left to right, the carry flag enters the MSB, and the LSB exits to the carry flag. In other words, in ROR the C is moved to the MSB, and the LSB is moved to the C. In

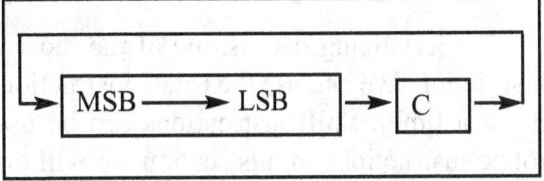

reality, the carry flag acts as if it is part of the register, making it a 9-bit register. Examine the following code.

```
CLC                      ;make C = 0 (carry is 0)
LDI        R20,0x26      ;R20 = 0010 0110
ROR        R20           ;R20 = 0001 0011 C = 0
ROR        R20           ;R20 = 0000 1001 C = 1
ROR        R20           ;R20 = 1000 0100 C = 1
```

ROL instruction

The other rotating instruction is ROL. In ROL, as bits are shifted from right to left, the carry flag enters the LSB, and the MSB exits to the carry flag. In other words, in ROL the C is moved to the LSB, and

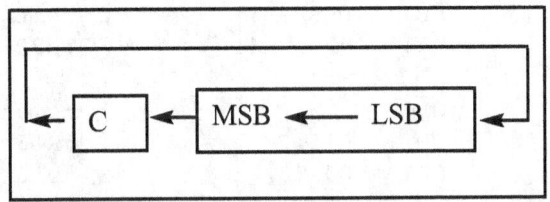

the MSB is moved to the C. See the following code and diagram. Again, the carry flag acts as if it is part of the register, making it a 9-bit register. Examine the following code.

```
SEC                      ;make C = 1
LDI        R20,0x15      ;R20 = 0001 0101
ROL        R20           ;R20 = 0010 1011 C = 0
ROL        R20           ;R20 = 0101 0110 C = 0
ROL        R20           ;R20 = 1010 1100 C = 0
ROL        R20           ;R20 = 0101 1000 C = 1
```

Serializing data

Serializing data is a way of sending a byte of data one bit at a time through a single pin of the microcontroller. There are two ways to transfer a byte of data serially:

1. Using the serial port. In using the serial port, programmers have very limited control over the sequence of data transfer. The details of serial port data transfer are discussed in Chapter 11.
2. The second method of serializing data is to transfer data one bit at a time and control the sequence of data and spaces between them. In many new generations of devices such as LCD, ADC, and ROM, the serial versions are becoming popular because they take up less space on a printed circuit board. Next, we discuss how to use rotate instructions in serializing data.

Serializing a byte of data

Serializing data is one of the most widely used applications of the rotate instruction. We can use the rotate instruction to transfer a byte of data serially (one bit at a time). Shift instructions can be used for the same job. After presenting rotate instructions in this section we will discuss shift instructions in more detail. Example 5-28 shows how to transfer an entire byte of data serially via any AVR pin.

Example 5-28

Write a program to transfer the value 41H serially (one bit at a time) via pin PB1. Put one high at the start and end of the data. Send the LSB first.

Solution:

```
.INCLUDE "M32DEF.INC"

        SBI     DDRB, 1             ;bit 1 of Port B is output
        LDI     R20,0x41            ;R20 = the value to be sent

        CLC                         ;clear carry flag
        LDI     R16, 8              ;R16 = 8
        SBI     PORTB, 1            ;bit 1 of PORTB is 1

AGAIN:
        ROR     R20                 ;rotate right R20 (send LSB to C flag)
        BRCS    ONE                 ;if C = 1 then go to ONE
        CBI     PORTB, 1            ;bit 1 of PORTB is cleared to zero
        JMP     NEXT                ;go to NEXT
ONE:    SBI     PORTB, 1            ;bit 1 of PORTB is set to one
NEXT:
        DEC     R16                 ;decrement R16
        BRNE    AGAIN               ;if R16 is not zero then go to AGAIN
        SBI     PORTB, 1            ;bit 1 of PORTB is set to one

HERE:   JMP     HERE                ;RB1 = high
```

Example 5-29 also shows how to bring in a byte of data serially. We will see how to use these concepts for a serial RTC (real-time clock) chip in Chapter 16. Example 5-30 shows how to scan the bits in a byte.

Example 5-29

Write a program to bring in a byte of data serially via pin RC7 and save it in R20 register. The byte comes in with the LSB first.

Solution:

```
.INCLUDE    "M32DEF.INC"

        CBI    DDRC, 7      ;bit 7 of Port C is input
        LDI    R16, 8       ;R16 = 8
        LDI    R20, 0       ;R20 = 0
AGAIN:
        SBIC   PINC, 7      ;skip the next line if bit 7 of Port C is 0
        SEC                 ;set carry flag to one
        SBIS   PINC, 7      ;skip the next line if bit 7 of Port C is 1
        CLC                 ;clear carry flag to zero
        ROR    R20          ;rotate right R20. move C flag to MSB of R21
        DEC    R16          ;decrement R16
        BRNE   AGAIN        ;if R16 is not zero go to AGAIN

HERE:   JMP    HERE         ;stop here
```

Example 5-30

Write a program that finds the number of 1s in a given byte.

Solution:

```
.INCLUDE    "M32DEF.INC"

        LDI    R20, 0x97
        LDI    R30, 0       ;number of 1s
        LDI    R16, 8       ;number of bits in a byte

AGAIN:
        ROR    R20          ;rotate right R20 and move LSB to C flag
        BRCC   NEXT         ;if C = 0 then go to NEXT
        INC    R30          ;increment R30
NEXT:
        DEC    R16          ;decrement R16
        BRNE   AGAIN        ;if R16 is not zero then go to AGAIN

        ROR    R20          ;one more time to leave R20 unchanged

HERE:   JMP    HERE         ;stop here
```

Shift instructions

There are three shift instructions in the AVR. All of them involve the carry flag. Each is shown next.

LSL instruction

```
LSL Rd        ;logical shift left
```

In LSL, as bits are shifted from right to left, 0 enters the LSB, and the MSB exits to the carry flag. In other words, in LSL, 0 is

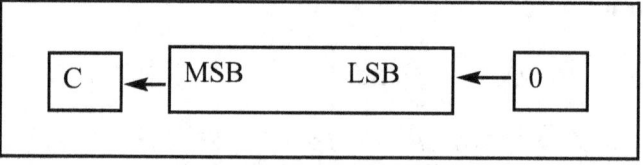

moved to the LSB, and the MSB is moved to the C flag. Notice that this instruction multiplies the content of the register by 2 assuming that after LSL the carry flag is not set. Examine the following code.

```
CLC                    ;make C = 0  (carry is 0 )
LDI        R20,0x26    ;R20 = 0010 0110(38)   c = 0
LSL        R20         ;R20 = 0100 1100(76)   C = 0
LSL        R20         ;R20 = 1001 1000(152)  C = 0
LSL        R20         ;R20 = 0011 0000(48)   C = 1
                       ;as C = 1 and content of R20
                       ;is not multiplied by 2
```

LSR instruction

The second shift instruction is LSR. In LSR, as bits are shifted from left to right, 0 enters the MSB, and the LSB exits to the carry flag. In other words, in

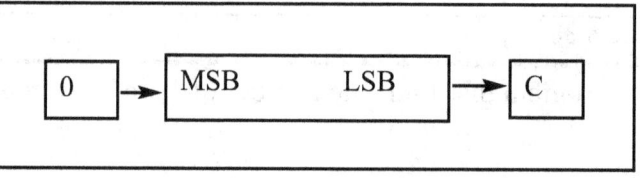

LSR, 0 is moved to the MSB, and the LSB is moved to the C flag. Notice that this instruction divides the content of the register by 2 and the carry flag contains the remainder of the division. Examine the following code.

```
LDI        R20,0x26    ;R20 = 0010 0110 (38)
LSR        R20         ;R20 = 0001 0011 (19) C = 0
LSR        R20         ;R20 = 0000 1001 (9)  C = 1
LSR        R20         ;R20 = 0000 0100 (4)  C = 1
```

LSR cannot be used to divide signed numbers by 2. See Example 5-31.

ASR instruction

The third shift instruction is ASR, which means arithmetic shift right. The ASR instruction can divide signed numbers by two. In ASR, as bits are

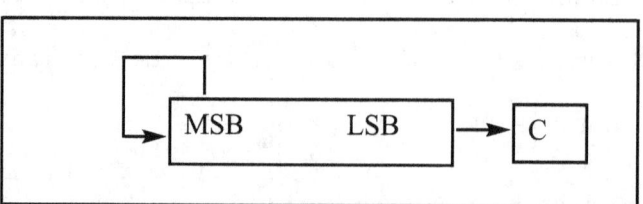

shifted from left to right, the MSB is held constant and the LSB exits to the carry flag. In other words, MSB is not changed but is copied to D6, D6 is moved to D5, D5 is moved to D4, and so on. Examine the following code.

```
LDI     R20,0D60     ;R20 = 1101 0000(-48) c = 0
LSR     R20          ;R20 = 1110 1000(-24) C = 0
LSR     R20          ;R20 = 1111 0100(-12) C = 0
LSR     R20          ;R20 = 1111 1010(-6)  C = 0
LSR     R20          ;R20 = 1111 1101(-3)  C = 0
LSR     R20          ;R20 = 1111 1110(-1)  C = 1
```

Example 5-32 shows how we can use ROR to divide a register by a number that is a power of 2.

Example 5-31

Assume that R20 has the number –6. Show that LSR cannot be used to divide the content of R20 by 2. Why?

Solution:

```
LDI   R20,0xFA    ;R20 = 1111 1010 (-6)
LSR   R20         ;R20 = 0111 1101 (+125)
                  ;-6 divided by 2 is not +125 and
                  ;the answer is not correct
```

Because LSR shifts the sign bit it changes the sign of the number and therefore cannot be used for signed numbers.

Example 5-32

Assume that R20 has the number 48. Show how we can use ROR to divide R20 by 8.

Solution:

```
                  ;to divide a number by 8 we can
                  ;shift it 3 bits to the right. without
                  ;LSR we have to ROR 3 times and
                  ;clear carry flag before
                  ;each rotation

LDI   R20,0x30    ;R20 = 0011 0000 (48)
CLC               ;clear carry flag
ROR   R20         ;R20 = 0001 1000 (24)
CLC               ;clear carry flag
ROR   R20         ;R20 = 0000 1100 (12)
CLC               ;clear carry flag
ROR   R20         ;R20 = 0000 0110 (6)
                  ;48 divided by 8 is 6 and
                  ;the answer is correct
```

SWAP instruction

```
SWAP  Rd      ;swap nibbles
```

Another useful instruction is the SWAP instruction. It works on R0–R31. It swaps the lower nibble and the higher nibble. In other words, the lower 4 bits are put into the higher 4 bits, and the higher 4 bits are put into the lower 4 bits. See the diagrams below.

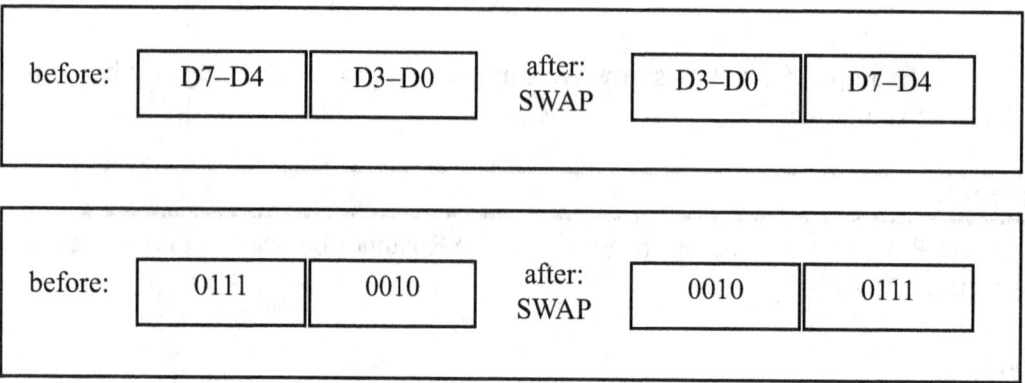

Example 5-33 shows how to exchange nibbles of a byte with and without SWAP instruction.

Example 5-33

(a) Find the contents of the R20 register in the following code .

```
        LDI   R20, 0x72
        SWAP  R20
```

(b) In the absence of a SWAP instruction, how would you exchange the nibbles? Write a simple program to show the process.

Solution:

(a)
```
        LDI   R20, 0x72        ;R20 = 0x72
        SWAP  R20              ;R20 = 0x27
```
(b)
```
        LDI   R20,0x72
        LDI   R16,4
        LDI   R21,0
BEGIN:
        CLC
        ROL   R20
        ROL   R21
        DEC   R16
        BRNE  BEGIN
        OR    R20, R21
HERE:   JMP   HERE
```

188

Review Questions

1. What is the value of R20 after the following code is executed?

```
LDI    R20,0x40
CLC
ROR    R20
ROR    R20
ROR    R20
ROR    R20
```

2. What is the value of R20 after the following code is executed?

```
LDI    R20,0x40
CLC
ROL    R20
ROL    R20
ROL    R20
ROL    R20
```

3. What is the value of R20 after execution of the following code?

```
LDI    R20,0x40
SEC
ROL    R20
SWAP   R20
```

4. What is the value of R20 after execution of the following code?

```
LDI    R20,0x00
SEC
ROL    R20
CLC
ROL    R20
SEC
ROL    R20
CLC
ROL    R20
SEC
ROL    R20
CLC
ROL    R20
SEC
ROL    R20
CLC
ROL    R20
```

5. What is the value of R20 after execution of the last code if you replace ROL with the ROR instruction?

6. How many LSR instructions are needed to divide a number by 32?

CHAPTER 5: ARITHMETIC, LOGIC INSTRUCTIONS, AND PROGRAMS   189

SECTION 5.5: BCD AND ASCII CONVERSION

In this section you will learn about packed and unpacked BCD numbers. We will also show how you can change packed BCD to unpacked BCD and vice versa. Then you will learn to convert ASCII codes to BCD and vice versa.

BCD (binary coded decimal) number system

BCD stands for *binary coded decimal*. BCD is needed because in everyday life we use the digits 0 to 9 for numbers, not binary or hex numbers. Binary representation of 0 to 9 is called BCD (see Figure 5-3). In computer literature, one encounters two terms for BCD numbers: (1) unpacked BCD, and (2) packed BCD. We describe each one next.

Digit	BCD
0	0000
1	0001
2	0010
3	0011
4	0100
5	0101
6	0110
7	0111
8	1000
9	1001

Figure 5-3. BCD Code

Unpacked BCD

In unpacked BCD, the lower 4 bits of the number represent the BCD number, and the rest of the bits are 0. For example, "0000 1001" and "0000 0101" are unpacked BCD for 9 and 5, respectively. Unpacked BCD requires 1 byte of memory, or an 8-bit register, to contain it.

Packed BCD

In packed BCD, a single byte has two BCD numbers in it: one in the lower 4 bits, and one in the upper 4 bits. For example, "0101 1001" is packed BCD for 59H. Only 1 byte of memory is needed to store the packed BCD operands. Thus, one reason to use packed BCD is that it is twice as efficient in storing data.

ASCII numbers

On ASCII keyboards, when the key "0" is activated, "011 0000" (30H) is provided to the computer. Similarly, 31H (011 0001) is provided for key "1", and so on, as shown in Table 5-3.

Table 5-3: ASCII and BCD Codes for Digits 0–9

Key	ASCII (hex)	Binary	BCD (unpacked)
0	30	011 0000	0000 0000
1	31	011 0001	0000 0001
2	32	011 0010	0000 0010
3	33	011 0011	0000 0011
4	34	011 0100	0000 0100
5	35	011 0101	0000 0101
6	36	011 0110	0000 0110
7	37	011 0111	0000 0111
8	38	011 1000	0000 1000
9	39	011 1001	0000 1001

190

It must be noted that BCD numbers are universal, although ASCII is standard in the United States (and many other countries). Because the keyboard, printers, and monitors all use ASCII, how does data get converted from ASCII to BCD, and vice versa? These are the subjects covered next.

Packed BCD to ASCII conversion

In many systems we have what is called a *real-time clock* (RTC). The RTC provides the time of day (hour, minute, second) and the date (year, month, day) continuously, regardless of whether the power is on or off (see Chapter 16). This data, however, is provided in packed BCD. For this data to be displayed on a device such as an LCD, or to be printed by the printer, it must be in ASCII format.

To convert packed BCD to ASCII, you must first convert it to unpacked BCD. Then the unpacked BCD is tagged with 011 0000 (30H). The following demonstrates converting packed BCD to ASCII. See also Example 5-34.

```
Packed BCD        Unpacked BCD          ASCII
29H               02H  & 09H            32H  & 39H
0010 1001         0000 0010 &           0011 0010 &
                  0000 1001             0011 1001
```

Example 5-34

Assume that R20 has packed BCD. Write a program to convert the packed BCD to two ASCII numbers and place them in R21 and R22.

Solution:

```
.INCLUDE    "M32DEF.INC"

    LDI    R20,0x29    ;the packed BCD to be converted is 29

    MOV    R21,R20     ;R21 = R20 = 29H
    ANDI   R21,0x0F    ;mask the upper nibble (R21 = 09H)
    ORI    R21,0x30    ;make it ASCII (R21 = 39H)

    MOV    R22,R20     ;R22 = R20 = 29H
    SWAP   R22         ;swap nibbles (R22 = 92H)
    ANDI   R22,0x0F    ;mask the upper nibble (R22 = 02)
    ORI    R22,0x30    ;make it ASCII (R22 = 32H)

HERE: JMP  HERE
```

ASCII to packed BCD conversion

To convert ASCII to packed BCD, you first convert it to unpacked BCD (to get rid of the 3), and then combine it to make packed BCD. For example, for 4 and 7 the keyboard gives 34 and 37, respectively. The goal is to produce 47H or "0100 0111", which is packed BCD. This process is illustrated next.

```
Key    ASCII    Unpacked BCD    Packed BCD
4      34       00000100
7      37       00000111        01000111 which is 47H
```

```
LDI     R21,'4'     ;load character 4 to R21
LDI     R22,'7'     ;load character 7 to R22
ANDI    R21,0x0F    ;mask upper nibble of R21
SWAP    R21         ;swap nibbles of R21
                    ;to make upper nibble of packed BCD
ANDI    R22,0x0F    ;mask upper nibble of R22
OR      R22,R21     ;join R22 and R21 to make packed BCD
MOV     R20,R22     ;move the result to R20
```

After this conversion, the packed BCD numbers are processed and the result will be in packed BCD format.

Review Questions

1. For the following decimal numbers, give the packed BCD and unpacked BCD representations.
   (a) 15   (b) 99
   (c) 25   (d) 55
2. Show the binary and hex formats for "76" and its BCD version.
3. Does the R20 register contain 54H after the following instruction is executed?
   `LDI R20,54`
4. 67H in BCD when converted to ASCII is ____hex and ____hex.

SUMMARY

This chapter discussed arithmetic instructions for both signed and unsigned data in the AVR. Unsigned data uses all 8 bits of the byte for data, making a range of 0 to 255 decimal. Signed data uses 7 bits for the data and 1 for the sign bit, making a range of –128 to +127 decimal.

In coding arithmetic instructions for the AVR, special attention has to be given to the possibility of a carry or overflow condition.

This chapter defined the logic instructions AND, OR, XOR, and complement. In addition, AVR Assembly language instructions for these functions were described. These functions are often used for bit manipulation purposes.

Compare and conditional branch instructions were described using different examples.

The rotate and swap instructions of the AVR are used in many applications such as serial devices. These instructions were discussed in detail.

Binary coded decimal (BCD) data represents the digits 0 through 9. Both packed and unpacked BCD formats were discussed. This chapter also described BCD and ASCII conversions.

PROBLEMS

SECTION 5.1: ARITHMETIC INSTRUCTIONS

1. Find the C, Z, and H flags for each of the following:

```
(a)    LDI    R20,0x3F    (b)    LDI    R20,0x99
       LDI    R21,0x45           LDI    R21,0x58
       ADD    R20,R21            ADD    R20,R21
(c)    LDI    R20,0xFF    (d)    LDI    R20,0xFF
       CLR    R21                LDI    R21,0x1
       SEC                       ADD    R20,R21
       ADC    R20,R21
```

2. Write a program to add 25 to the day of your birthday and save the result in R20.
3. Write a program to add the following numbers and save the result in R20.
        0x25, 0x19, 0x12
4. Modify Problem 3 to add the result with 0x3D.
5. State the steps that the SUB instruction will go through for each of the following.
    (a) 23H – 12H   (b) 43H – 53H   (c) 99 – 99
6. For Problem 5, write a program to perform each operation.
7. Write a program to add 7F9AH to BC48H and save the result in R20 (low byte) and R21 (high byte).
8. Write a program to subtract 7F9AH from BC48H and save the result in R20 (low byte) and R21 (high byte).
9. Show how to perform 77×34 in the AVR.
10. Show how to perform 64/4 in the AVR.
11. The MUL instruction places the result in registers _____ and_____.

SECTION 5.2: SIGNED NUMBER CONCEPTS AND ARITHMETIC OPERATIONS

12. Show how the following numbers are represented by the assembler:
    (a) –23   (b) +12   (c) –28
    (d) +6FH   (e) –128   (f) +127
13. The memory addresses in computers are _____ (signed, unsigned) numbers.
14. Write a program for each of the following and indicate the status of the V flag for each:

    (a) (+15) + (–12)   (b) (–123) + (–127)
    (c) (+25H) + (+34H)   (d) (–127) + (+127)

15. Explain the difference between the C and V flags and where each one is used.
16. When is the V flag raised? Explain.
17. Which register holds the V flag?
18. How do you detect the V flag in the AVR? How do you detect the C flag?

---

19. Find the contents of register R20 after each of the following instructions:

```
(a)  LDI   R20,0x65      (b)  LDI   R20, 0x70
     LDI   R21,0x76           LDI   R21, 0x6B
     AND   R20,R21            OR    R20, R21


(c)  LDI   R20,0x95      (d)  LDI   R20, 0x5D
     LDI   R21,0xAA           LDI   R21, 0x75
     EOR   R20,R21            AND   R20, R21


(e)  LDI   R20,0x0C5     (f)  LDI   R20, 0x6A
     LDI   R21,0x12           LDI   R21, 0x6E
     OR    R20,R21            EOR   R20, R21


(g)  LDI   R20,0x37
     LDI   R21,0x26
     OR    R20,R21
```

20. Explain how the BRSH instruction works.
21. Does the compare instruction affect the flag bits of the status register?
22. Assume that R20 = 85H. Indicate whether the conditional branch is executed in each of the following cases:

```
(a)   LDI    R21, 0x90      (b)   LDI   R21, 0x70
      CP     R20,R21              CP    R20, R21
      BRLO   NEXT                 BRSH  NEXT
      . . .                       . . .
```

23. For Problem 22, indicate the value in R20 after execution of each program.

SECTION 5.4: ROTATE AND SHIFT INSTRUCTIONS AND DATA SERIALIZATION

24. Find the contents of R20 after each of the following is executed:

```
(a)  LDI   R20, 0x56      (b)  LDI   R20, 0x39
     SWAP  R20                 SEC
     CLC                       ROL   R20
     ROR   R20                 ROL   R0
     ROR   R20


(c)  CLC                  (d)  CP    R20, R20
     LDI   R20, 0x4D           LDI   R20, 0x7A
     SWAP  R20                 ROR   R20
     ROL   R20
     ASR   R20
```

25. Show the code to replace the SWAP instruction:
    (a) using the ROL instruction
    (b) using the ROR instruction
26. Write a program that finds the number of zeros in an 8-bit data item.

27. Write a program that finds the position of the first high in an 8-bit data item. The data is scanned from D0 to D7. Give the result for 68H.

28. Write a program that finds the position of the first high in an 8-bit data item. The data is scanned from D7 to D0. Give the result for 68H.

SECTION 5.5: BCD AND ASCII CONVERSION

29. Write a program to convert the following packed BCD numbers to ASCII. Place the ASCII codes into R20 and R21.

     (a)   0x76
     (b)   0x87

30. For 3 and 2 the keyboard gives 33H and 32H, respectively. Write a program to convert these values to packed BCD and store the result in R20.

ANSWERS TO REVIEW QUESTIONS

SECTION 5.1: ARITHMETIC INSTRUCTIONS

1. R1:R0
2. No. Because immediate addressing mode is not supported.
3. 255, 255.
4. False.
5. R20.
6. We cannot use immediate addressing mode with ADD.
7. "ADI R1,0x04"
8. R1
9. (a) R21 = 00 and C = 1
   (b) R21 = FF and C = 0
10.
```
    43H  0100 0011                        0100 0011
  - 05H  0000 0101  2's complement    +  1111 1011
    3EH                                   0011 1110
```
11. R1 = 95H – 4FH – 1 = 45H, C = 0.

SECTION 5.2: SIGNED NUMBER CONCEPTS AND ARITHMETIC OPERATIONS

1. D7
2. 16H is 00010110 in binary and its 2's complement is 1110 1010 or
   −16H = EA in hex.
3. −128 to +127
4. +9 = 00001001 and −9 = 11110111 or F7 in hex.
5. An overflow is a carry into the sign bit (D7), but the carry is a carry out of register.

SECTION 5.3: LOGIC AND COMPARE INSTRUCTIONS

1. (a) 02 H
   (b) FF H
   (c) FD H
2. Zero
3. One
4. All zeros

---

5. False
6. 66H

SECTION 5.4: ROTATE AND SHIFT INSTRUCTIONS AND DATA SERIALIZATION

1. 04H
2. 02H
3. 18H
4. AAH
5. 55H
6. 5 LSL instructions

SECTION 5.5: BCD AND ASCII CONVERSION

1. (a) 15H = 0001 0101 packed BCD, 0000 0001,0000 0101 unpacked BCD
   (b) 99H = 1001 1001 packed BCD, 0000 1001,0000 1001 unpacked BCD
   (c) 25H = 0010 0101 packed BCD, 0000 0010,0000 0101 unpacked BCD
   (d) 55H = 0101 0101 packed BCD, 0000 1001,0101 0101 unpacked BCD
2. 3736H = 00110111 00110110B
   and in BCD we have 76H = 0111 0110B
3. No. We need to write it as 0x54 (to indicate that it is hex) or 0b01010100. The value 54 is interpreted as 36H by the assembler.
4. 36H, 37H

CHAPTER 6

AVR ADVANCED ASSEMBLY LANGUAGE PROGRAMMING

OBJECTIVES

Upon completion of this chapter, you will be able to:

>> List all the addressing modes of the AVR microcontroller
>> Contrast and compare the addressing modes
>> Code AVR Assembly language instructions using each addressing mode
>> Access the data RAM file register using various addressing modes
>> Code AVR instructions to manipulate a look-up table
>> Access fixed data residing in the program Flash ROM space
>> Discuss how to create macros
>> Explain how to write data to EEPROM memory of the AVR
>> Explain how to read data from EEPROM memory of the AVR
>> Code AVR programs to create and test the checksum byte
>> Code AVR programs for ASCII data conversion

In Section 6.1, you learn some new assembler directives that are used throughout this chapter. In Sections 6.2 through 6.4 we see the different ways in which we can access program and data memories in the AVR.

Section 6.5 explains the bit-addressability of the data memory space. In Section 6.6 we discuss how to access EEPROM in the AVR. Checksum generation and BCD-ASCII conversions are covered in Section 6.7. Macros are examined in Section 6.8.

SECTION 6.1: INTRODUCING SOME MORE ASSEMBLER DIRECTIVES

In Chapter 2, we introduced the assembler directives .ORG, .SET, and .INCLUDE. In this section, you will learn some other useful directives.

Arithmetic and logic expressions with constant values

As you saw in Chapter 2, we can define constant values using .EQU. The AVR Studio IDE supports arithmetic operations between expressions. See Table 6-1. For example, in the following program R24 is loaded with 29, which is the result of the arithmetic expression "((ALFA-BETA)*2)+9".

```
.EQU ALFA = 50
.EQU BETA = 40
LDI R23,ALFA                ;R23 = ALFA = 50
LDI R24,((ALFA-BETA)*2)+9   ;R24 = ((50-40)*2)+9 = 29
```

The AVR Studio IDE supports logic operations between expressions as well. See Table 6-2. For example, in the following program R21 is loaded with 0x14:

```
.EQU C1 = 0x50
.EQU C2 = 0x10
.EQU C3 = 0x04
LDI R21,(C1&C2)|C3  ;R21=(0x10&0x50)|0x04 = 0x10|0x04= 0x14
```

In Table 6-3 you see the shift operators, which are very useful. They shift left and right a constant value. For example, the following instruction loads the R20 register with 0b00001110:

```
LDI R16,0b00000111<<1 ;R16 = 0b00001110
```

One of the uses of shift operators is for initializing the registers. For exam-

Table 6-1: Arithmetic Operators	
Symbol	Action
+	Addition
-	Subtraction
*	Multiplication
/	Division
%	Modulo

Table 6-2: Logic Operators	
Symbol	Action
&	Bitwise AND
|	Bitwise OR
^	Bitwise XOR
~	Bitwise NOT

Table 6-3: Shift Operators

Symbol	Action	Example
<<	Shifts left the left expression by the number of places given by the right expression	LDI R20,0b101<<2 ;R20=0b10100
>>	Shifts right the left expression by the number of places given by the right expression	LDI R20,0b100>>1 ;R20=0b010

```
           Bit    D7                            D0
          SREG  | I | T | H | S | V | N | Z | C |
```

Figure 6-1. Bits of the Status Register

ple, suppose we want to set the Z and C bits of the SREG (Status Register) register and clear the others. Look at Figure 6-1. If we load 0b00000011 to SREG the task will be done:

```
LDI R20, 0b00000011   ;Z = 1, C = 1
OUT SREG,R20
```

In this example, we calculated the 0b00000011 number by looking at Figure 6-1. But imagine you are writing a program and you want to do the same task; you have to open the datasheet or a reference book to see the structure of the SREG register. To make the task simpler, the names of the register bits are defined in the header files of each AVR microcontroller. For example, in M32DEF.INC there are the following lines of code:

```
...
; SREG - Status Register
.equ  SREG_C    = 0   ;carry flag
.equ  SREG_Z    = 1   ;zero flag
.equ  SREG_N    = 2   ;negative flag
.equ  SREG_V    = 3   ;2's complement overflow flag
.equ  SREG_S    = 4   ;sign bit
.equ  SREG_H    = 5   ;half carry flag
.equ  SREG_T    = 6   ;bit copy storage
.equ  SREG_I    = 7   ;global interrupt enable
...
```

So, we can use the names of the bits instead of remembering the structure of the registers or finding them in the datasheet. For example, the following program sets the Z flag of the SREG register and clears the other bits:

```
LDI   R16, 1<<SREG_Z ;R16= 1 << 1 = 0b00000010
OUT   SREG,R16       ;SREG = 0b00000010 (set Z and clear others)
```

As another example, the following program sets the V and S flags of SREG:

```
LDI   R16,(1<<SREG_V)|(1<<SREG_S)      ;R16=0b1000|0b10000=0b11000
OUT   SREG,R16    ;SREG = 0b00011000 (set V and S, clear others)
```

In Example 6-1, you see the usage of the directives in I/O port programming.

Example 6-1

Write codes to set PB2 and PB4 of PORTB to 1 and clear the other pins
(a) without the directives, and
(b) using the directives.

Solution:

(a)
```
    LDI   R20,0x14           ;R20 = 0x14
    OUT   PORTB,R20          ;PORTB = R20
```

To make the code more readable, we can write the number in binary as well:

```
    LDI   R20,0b00010100     ;R20 = 0x14
    OUT   PORTB,R20          ;PORTB = 0x14
```

(b)
```
    LDI   R20,(1<<4)|(1<<2)  ;R20 = (0b10000 | 0b00100) = 0b10100
    OUT   PORTB,R20          ;PORTB = R20
```

As we mentioned before, the names of the register bits are defined in the header files of each AVR microcontroller. PB2 and PB4 are defined equal to 2 and 4, as well. Therefore, we can write the code as shown below:

```
    LDI   R20,(1<<PB4)|(1<<PB2) ;set the PB4 and PB2 bits
    OUT   PORTB,R20                ;PORTB = R20
```

Notice that when the assembler wants to convert a code to machine language it substitutes all of the assembler directives with their equivalent values. Thus, using the directives has no side effects on the performance of our code but rather makes our code more readable. See Examples 6-2 and 6-3.

Example 6-2

What does the AVR assembler do while assembling the following program?
```
.equ  C1 = 2
.equ  C2 = 3
LDI R20,C1|(1<<C2)  ;R20= 2|(1<<3)= 0b00000010|0b00001000= 0b00001010
```

Solution:

.equ is an assembler directive. When assembling ".equ C1 = 2", the assembler assigns value 2 to C1. Similarly, while assembling the ".equ C2 = 3" instruction, it assigns the value 3 to C2.

When the assembler converts the "LDI R20,C1|(1<<C2)" instruction to machine language, it knows the values of C1 and C2. Thus it calculates the value of "C1|(1<<C2)", and then replaces the expression with its value. Therefore, "LDI R20,C1|(1<<C2)" will be converted to "LDI R20,0b00001010". Then the assembler converts the instruction to machine language.

Example 6-3

What does the AVR assembler do while assembling the following program?
```
.INCLUDE "M32DEF.INC"
      LDI   R20,(1<<PB4)|(1<<PB2);set the PB4 and PB2 bits
      OUT   DDRB,R20            ;DDRB = R20
HERE: RJMP  HERE
```

Solution:

Including a header file at the beginning of a program is similar to copying all the contents of the header file to the beginning of the program. Thus, the assembler, first assembles the contents of M32DEF.INC. The header file contains some ".equ" instructions, such as ".equ PB4 = 4". Thus, after reading the header file the assembler learns that PB4 is equal to 4, PB2 is equal to 2, and so on. Thus, when it wants to assemble instructions such as "LDI R20,(1<<PB4)|(1<<PB2)", it knows the values of PB2 and PB4. It calculates the value of "(1<<PB4)|(1<<PB2)" and substitutes it.

It is highly recommended that you take a look at the M32DEF.INC file. The file is located in the following path, if you did not change it while installing the AVR Studio software:
Program Files\Atmel\AVR Tools\AvrAssembler2\Appnotes\m32def.inc

HIGH( ) and LOW( ) functions

The HIGH( ) and LOW( ) functions give the higher and the lower bytes of a 16-bit value. For example, in the following program 0x55 and 0x44 are loaded into R16 and R17, respectively:

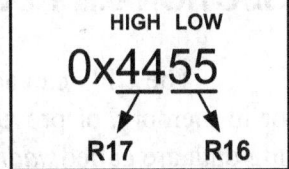

```
      LDI   R16,LOW(0x4455)  ;R16 = 0x55
      LDI   R17,HIGH(0x4455) ;R17 = 0x44
```

In Chapter 2, we used the following instructions to make the stack pointer refer to the last location of the memory:
```
      LDI   R16,HIGH(RAMEND) ;R16 = 0x08 (for ATmega32)
      OUT   SPH,R16          ;SPH = the high byte of address
      LDI   R16,LOW(RAMEND)  ;R16 = 0x5f
      OUT   SPL,R16          ;SPL = the low byte of address
```

But how do the instructions work? In the AVR header files (e.g., M32DEF.INC) RAMEND is defined equal to the address of the last location of the memory. For example, in *M32DEF.INC* there is the following line:

```
      .equ  RAMEND     = 0x085f
```

The HIGH( ) and LOW( ) functions split the RAMEND into two bytes, $08 and $5F. They go to SPH and SPL, respectively.
You can see the list of the different directives available in the AVR by using the help feature of AVR Studio. (Choose the *assembler Help* option from the *Help* menu and then click on the *Expressions* topic.)

Review Questions

1. Indicate the value loaded into the registers in the following program:
```
.EQU CONST1 = 0x10
.EQU CONST2 = 0x91
.EQU CONST3 = 0x14
.EQU ADDR = (0x91 << 1)+1
LDI   R20,CONST1&CONST2
LDI   R21,CONST2|CONST3
LDI   R30,LOW(ADDR)
LDI   R31,HIGH(ADDR)
```

2. What does the following code do?
```
LDI   R16,(1<<SREG_V)|(1<<SREG_Z)
OUT   SREG,R16    ;SREG = 0b00011000
```

3. Using the assembler directives write a program that sets the Z and C flags and clears the other flags.

4. Calculate the values that are loaded into the TCNT1L and TCNT1H I/O registers.
```
LDI   R16,HIGH(15900)
OUT   TCNT1H,R16          ;TCNT1H = HIGH(15900)
LDI   R16,LOW(15900)
OUT   TCNT1L,R16          ;TCNT1L = LOW(15900)
```

SECTION 6.2: REGISTER AND DIRECT ADDRESSING MODES

The CPU can access data in various ways. The data could be in a register, or in memory, or provided as an immediate value. These various ways of accessing data are called *addressing modes*. In Sections 6.2 through 6.6 we discuss AVR addressing modes in the context of some examples.

The various addressing modes of a microprocessor are determined when it is designed, and therefore cannot be changed by the programmer. The AVR provides a total of 13 distinct addressing modes, which can be categorized into the following groups:

1. Single-Register (Immediate)
2. Register
3. Direct
4. Register indirect
5. Flash Direct
6. Flash Indirect

In this section we look at immediate, two-register, and direct addressing modes. In Section 6.3 we cover accessing RAM data memory using the register indirect mode. Section 6.4 explains how to access fixed data and look-up tables stored in program ROM.

Single-register (immediate) addressing mode

In this addressing mode, the operand is a register. See the examples below.

```
NEG    R18              ;negate the contents of R18
COM    R19              ;complement the contents of R19
INC    R20              ;increment R20
DEC    R21              ;decrement R21
ROR    R22              ;rotate right R22
```

In some of the instructions there is also a constant value with the register operand. See the examples below.

```
LDI    R19,0x25         ;load 0x25 into R19
SUBI   R19,0x6          ;subtract 0x6 from R19
ANDI   R19,0b01000000   ;AND R19 with 0x40
```

The constant value is sometimes referred to as *immediate data* since the operand comes immediately after the opcode when the instruction is assembled; and the addressing mode is referred to as *immediate addressing mode* in some microcontrollers. But the AVR datasheet refers to this mode as a subset of the single-register addressing mode. This addressing mode can be used to load data into any of the R16–R31 general purpose registers. The immediate addressing mode is also used for arithmetic and logic instructions. Note that the letter "I" in instructions such as LDI, ANDI, and SUBI means "immediate." See Figures 6-2a and 6-2b.

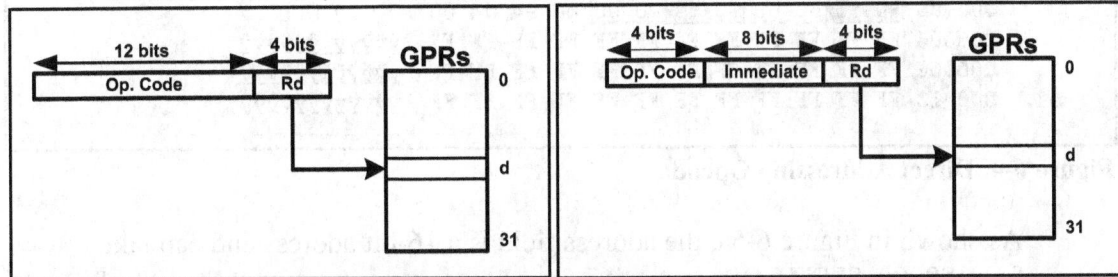

Figure 6-2a. Single-Register Addressing　　**Figure 6-2b. Single-Register (with immediate)**

We can use the .EQU directive to access immediate data, as shown below.
```
.EQU COUNT = 0x30
...    ...
LDI    R16,COUNT            ;R16 = 0x30
```

Two-register addressing mode

Two-register addressing mode involves the use of two registers to hold the data to be manipulated. See Figure 6-3.

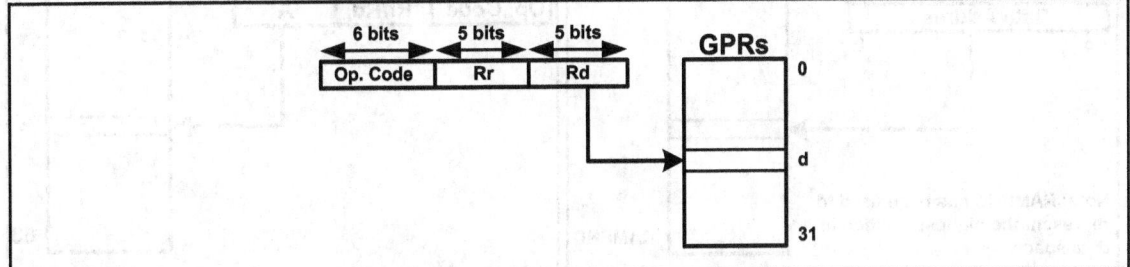

Figure 6-3. Two-Register Addressing

Examples of two-register addressing mode are as follows:

```
ADD    R20,R23            ;add R23 to R20
SUB    R29,R20            ;subtract R20 from R29
AND    R16,R17            ;AND R16 with 0x40
MOV    R23,R19            ;copy the contents of R19 to R23
```

Direct addressing mode

The entire data memory can be accessed using either direct or register indirect addressing modes. The register indirect addressing mode will be discussed in the next section. In direct addressing mode, the operand data is in a RAM memory location whose address is known, and this address is given as a part of the instruction. Contrast this with immediate addressing mode in which the operand data itself is provided with the instruction. Examine the following instructions:

```
LDS    R19,0x560   ;load R19 with the contents of memory loc $560
STS    0x40,R19    ;store R19 to data space location 0x40
```

The two instructions use direct addressing mode. If we dissect the opcode we see that the addresses are embedded in the instruction, as shown in Figure 6-4.

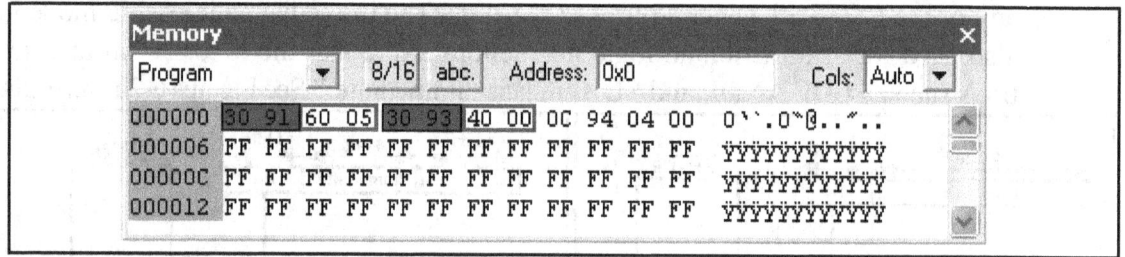

Figure 6-4. Direct Addressing Opcode

As shown in Figure 6-5a, the address field is a 16-bit address and can take values from $0000–$FFFF. Of course, it is much easier to use names instead of addresses in the program, and we have seen many examples of them in the last few chapters. It must be noted that data memory does not support immediate addressing mode. In other words, to move data into internal RAM or to I/O registers, we must first move it to a GPR (R16–R31), and then move it from the GPR to the data memory space using the STS instruction. For example, if we want to store 0x95 in memory location 0x520 we should write the following program, as there is no

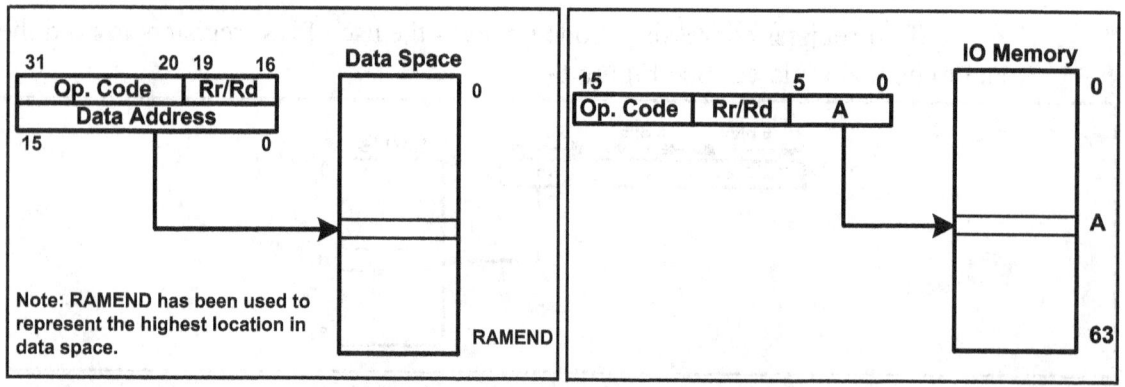

Figure 6-5a. Direct Data Addressing   **Figure 6-5b. I/O Direct Addressing**

instruction for storing immediate values in memory locations:

```
LDI    R19,0x95    ;load 0x95 into R19
STS    0x520,R19   ;store R19 into data location 0x520
```

I/O direct addressing mode

To access the I/O registers there is a special mode called *I/O direct addressing mode*. The I/O direct addressing mode can address only the standard I/O registers. The IN and OUT instructions use this addressing mode. Examine the following instruction, which copies the contents of PINB to PORTC:

```
IN     R18,0x16    ;R18 = contents of location $16 (PINB)
OUT    0x15,R18    ;PORTC (location $15) = R18
```

As shown in Figure 6-5b, the address field is a 6-bit address and can take values from $00 to $3F, which is from 00 to 63 in decimal. So, it can address the entire standard I/O register memory space.

The AVR registers for Ports A, B, and so on are part of the group of registers commonly referred to as *I/O registers*. There are many I/O registers and they are widely used, as we will discuss in future chapters. The I/O registers can be accessed by their names (which is much easier) or by their addresses. For example, PINB has address 0x16, and PORTC the address $15, as shown in Table 6-4. Notice how the following pairs of instructions mean the same thing:

```
OUT 0x15,R19    ;is the same as the next instruction
OUT PORTC,R19   ;which means copy R19 into Port C

IN R26,0x16     ;is the same as the next instruction
IN R26,PINB     ;which means copy PINB into R26
```

Table 6-4: Selected ATmega32 I/O Register Addresses

Symbol	Name	I/O Address	Data Memory Addr.
PIND	Port D input pins	$10	$30
DDRD	Data Direction, Port D	$11	$31
PORTD	Port D data register	$12	$32
PINC	Port C input pins	$13	$33
DDRC	Data Direction, Port C	$14	$34
PORTC	Port C data register	$15	$35
PINB	Port B input pins	$16	$36
DDRB	Data Direction, Port B	$17	$37
PORTB	Port B data register	$18	$38
PINA	Port A input pins	$19	$39
DDRA	Data Direction, Port A	$1A	$3A
PORTA	Port A data register	$1B	$3B
SPL	Stack Pointer, Low byte	$3D	$5D
SPH	Stack Pointer, High byte	$3E	$5E

Table 6-4 lists some of the AVR I/O registers and their addresses. The following points should be noted about the addresses of I/O registers:

1. As shown in Figures 2-3 and 2-4, the addresses between $20 and $5F of the data space have been assigned to standard I/O registers in all of the AVRs. These I/O registers have two addresses: I/O address and data memory address. The I/O address is used when we use the I/O direct addressing mode, while the data memory address is used when we use the direct addressing mode; in other words, the standard I/O registers can be accessed using both the direct addressing and the I/O addressing modes. For example, the following pairs of instructions do the same thing, but the IN and OUT instructions are more efficient, as mentioned in Section 2-3:

```
OUT 0x15,R19 ;PORTC=R19 (0x15 is the I/O addr. of PORTC)
STS 0x35,R19 ;PORTC=R19 (0x35 is the data memory addr. of PORTC)

IN  R19,0x16 ;R19=PINB (0x16 is the I/O addr. of PINB)
LDS R19,0x36 ;R19=PINB (0x36 is the data memory addr. of PINB)
```

2. Some AVRs have less than 64 I/O registers. So, some locations of the standard I/O memory are not used by the I/O registers. The unused locations are reserved and must not be used by the AVR programmer.

3. Some AVRs have more than 64 I/O registers. The extra I/O registers are located above the data memory address $5F. The data memory allocated to the extra I/O registers is called *extended I/O memory*. As shown in Figure 6-2b, in the I/O direct addressing mode, the address field is a 6-bit address and can take values from $00–$3F, which is from 00 to 63 in decimal. So, it can address only the standard I/O register memory, and it cannot be used for addressing the extended I/O memory. For example, the following instruction causes an error, since the I/O address must be between 0 and $3F:

```
OUT 0x65,R19          ;illegal as the address is above $3F
```

To access the extended I/O registers we can use the direct addressing mode. For example, in ATmega128, PORTF has the memory address of 0x62. So, the following instruction stores the contents of R20 in PORTF.

```
STS 0x62,R20          ;PORTF = R20
```

4. The I/O registers can have different addresses in different AVR microcontrollers. For example, the I/O address $2 is assigned to TWAR in the ATmega32, while the same address is assigned to DDRE in ATmega128. This means that in ATmega32, the instruction "OUT 0x2,R20" copies the contents of R20 to TWAR, while the same instruction, in ATmega128, copies the contents of R20 to DDRE. In other words, the same instruction can have different meanings in different AVR microcontrollers. This can cause problems if you want to run programs written for one AVR on another AVR. For example, if you have written a code for ATmega32 and you want to run it on an ATmega128, it might be necessary to change some register locations before loading it into the ATmega128.

The best way to solve this problem is to use the names of the registers instead of their addresses. For example, the instruction "OUT TWAR,R20" has the same meaning on all the AVRs. Therefore, using the names of the registers instead of their addresses makes our code more portable. See Example 6-4.

Example 6-4

Write code to send $55 to Port B. Include
(a) the register name,
(b) the I/O address, and
(c) the data memory address.

Solution:

(a)
```
    LDI    R20,0xFF      ;R20 = 0xFF
    OUT    DDRB,R20      ;DDRB = R20 (Port B output)
    LDI    R20,0x55      ;R20 = $55
    OUT    PORTB,R20     ;Port B = 0x55
```

(b)   From Table 6-4, DDRB I/O address = $17 and PORTB I/O address = $18.

```
    LDI    R20,0xFF      ;R20 = 0xFF
    OUT    0x17,R20      ;DDRB = R20 (Port B output)
    LDI    R20,0x55      ;R20 = $55
    OUT    0x18,R20      ;Port B = 0x55
```

(c)   From Table 6-4, DDRB data memory address = $37 and PORTB data memory address = $38.

```
    LDI    R20,0xFF      ;R20 = 0xFF
    STS    0x37,R20      ;DDRB = R20 (Port B output)
    LDI    R20,0x55      ;R20 = $55
    STS    0x38,R20      ;Port B = 0x55
```

Review Problems

1. Can the programmer of a microcontroller make up new addressing modes?
2. Show the instructions to load 1000 0000 (binary) into register SPL.
3. True or false. In immediate addressing the value comes immediately after the opcode.
4. True or false. We can access the extended I/O registers using the I/O direct addressing mode.
5. True or false. SPL is an I/O register.
6. True or false. Using the names of the registers makes the code more portable.

SECTION 6.3: REGISTER INDIRECT ADDRESSING MODE

We can use direct or register indirect addressing modes to access data stored in the data memory. In the previous section we showed how to use direct addressing mode. The register indirect addressing mode is a very important addressing mode in the AVR. This topic will be discussed thoroughly in this section.

Register indirect addressing mode

In the register indirect addressing mode, a register is used as a pointer to the data memory location. In the AVR, three registers are used for this purpose: X, Y, and Z. These are 16-bit registers allowing access to the entire 65,536 bytes of data memory space in the AVR.

Each of the registers is made by combining two specific GPRs; for example, combining R26 and R27 makes the X register. In this case R26 is the lower byte of X, and R27 is the higher byte. The Y and Z registers are made by combining R29:R28 and R31:R30, respectively. See Figure 6-6. The R26, R27, R28, R29, R30, and R31 GPRs can be referred to as XL, XH, YL,YH, ZL, and ZH, respectively. For example, "LDI XL,0x31" is the same as "LDI R26,0x31" since XL is another name for R26.

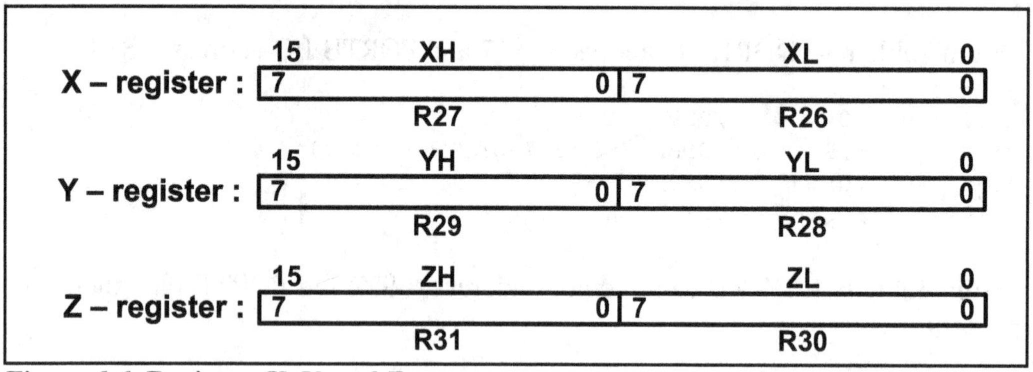

Figure 6-6. Registers X, Y, and Z

The 16-bit registers X, Y, and Z are widely used as pointers. We can use them with the LD instruction to read the value of a location pointed to by these registers. For example, the following instruction reads the value of the location pointed to by the X pointer.

```
LD      R24, X       ;load into R24 from location pointed to by X
```

For instance, the following program loads the contents of location 0x130 into R18:

```
LDI    XL, 0x30     ;load R26 (the low byte of X) with 0x30
LDI    XH, 0x01     ;load R27 (the high byte of X) with 0x1
LD     R18, X       ;copy the contents of location 0x130 to R18
```

The above program loads 0x130 into the X register; this is done by loading 0x30 into R26 (the low byte of X) and 0x1 into R27 (the high byte of X). Then it loads R18 with the contents of the location to which X points. See Figure 6-7.

The ST instruction can be used to write a value to a location to which any of the X, Y, and Z registers points. For example, the following program stores the contents of R23 into location 0x139F:

```
LDI    ZL, 0x9F     ;load 0x9F into the low byte of Z
LDI    ZH, 0x13     ;load 0x13 into the high byte of Z (Z=0x139F)
ST     X, R23       ;store the contents of location 0x139F in R23
```

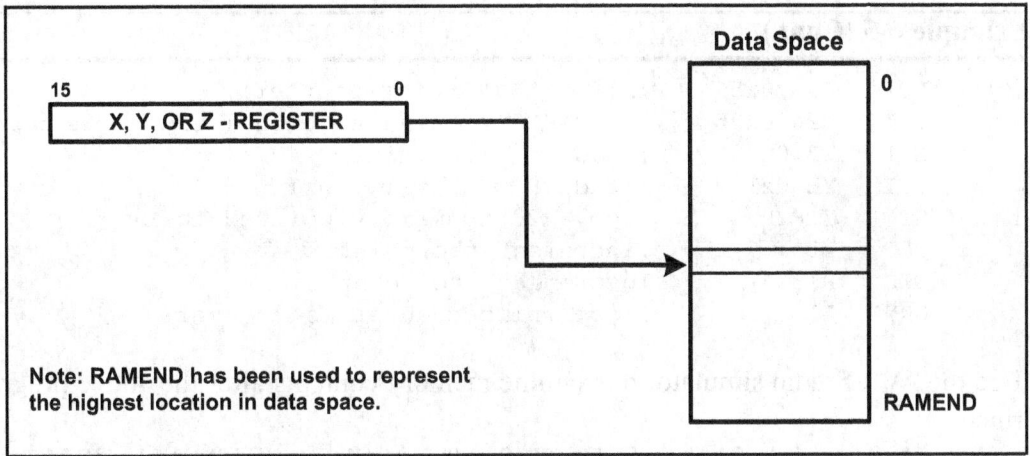

Figure 6-7. Register Indirect Addressing Mode

Advantages of register indirect addressing mode

One of the advantages of register indirect addressing mode is that it makes accessing data dynamic rather than static, as with direct addressing mode. Example 6-5 shows three cases of copying $55 into RAM locations $140 through $144 . Notice in solution (b) that two instructions are repeated numerous times. We can create a loop with those two instructions as shown in solution (c). Solution (c) is the most efficient and is possible only because of the register indirect addressing mode.

Example 6-5

Write a program to copy the value $55 into memory locations $140 through $144 using
(a) direct addressing mode,
(b) register indirect addressing mode without a loop, and
(c) a loop.

Solution:

```
(a)    LDI    R17,0x55        ;load R17 with value 0x55
       STS    0x140,R17       ;copy R17 to memory location 0x140
       STS    0x141,R17       ;copy R17 to memory location 0x141
       STS    0x142,R17       ;copy R17 to memory location 0x142
       STS    0x143,R17       ;copy R17 to memory location 0x143
       STS    0x144,R17       ;copy R17 to memory location 0x144

(b)    LDI    R16,0x55     ;load R16 with value 0x55
       LDI    YL,0x40      ;load R28 with value 0x40 (low byte of addr.)
       LDI    YH,0x1       ;load R29 with value 0x1 (high byte of addr.)
       ST     Y,R16        ;copy R16 to memory location 0x140
       INC    YL           ;increment the low byte of Y
       ST     Y,R16        ;copy R16 to memory location 0x141
       INC    YL           ;increment the pointer
       ST     Y,R16        ;copy R16 to memory location 0x142
       INC    YL           ;increment the pointer
       ST     Y,R16        ;copy R16 to memory location 0x143
       INC    YL           ;increment the pointer
       ST     Y,R16        ;copy R16 to memory location 0x144
```

Example 6-5 (Cont.)

```
(c)     LDI    R16,0x5      ;R16 = 5 (R16 for counter)
        LDI    R20,0x55     ;load R20 with value 0x55 (value to be copied)
        LDI    YL,0x40      ;load YL with value 0x40
        LDI    YH,0x1       ;load YH with value 0x1
L1:     ST     Y,R20        ;copy R20 to memory pointed to by Y
        INC    YL           ;increment the pointer
        DEC    R16          ;decrement the counter
        BRNE   L1           ;loop while counter is not zero
```

Use the AVR Studio simulator to examine memory contents after the above program is run.

$140 = (\$55)   \$141 = (\$55)   \$142 = (\$55)   \$143 = (\$55)   144 = (\$55)$

In Example 6-5, we must use "INC YL" to increment the pointer because there is no such instruction as "INC Y". Looping is not possible in direct addressing mode, and that is the main difference between the direct and register indirect addressing modes. For example, trying to copy a string of data located in consecutive locations of data RAM is much more efficient and dynamic using register indirect addressing mode than using direct addressing mode. See Example 6-6.

Auto-increment and auto-decrement options for pointer registers

Because the pointer registers (X, Y, and Z) are 16-bit registers, they can go from $0000 to $FFFF, which covers the entire 64K memory space of the AVR. Using the "INC ZL" instruction to increment the pointer can cause a problem when an address such as $5FF is incremented. The instruction "INC ZL" will not propagate the carry into the ZH register. The AVR gives us the options of auto-increment and auto-decrement for pointer registers to overcome this problem. The syntax used for the LD instruction in such cases is shown in Table 6-5.

Table 6-5: AVR Auto-Increment/Decrement of Pointer Registers for LD Instruction

Instruction		Function
LD	Rn,X	After loading location pointed to by X, the X stays the same.
LD	Rn,X+	After loading location pointed to by X, the X is incremented.
LD	Rn,-X	The X is decremented, then the location pointed to by X is loaded.
LD	Rn,Y	After loading location pointed to by Y, the Y stays the same.
LD	Rn,Y+	After loading location pointed to by Y, the Y is incremented.
LD	Rn,-Y	The Y is decremented, then the location pointed to by Y is loaded.
LDD	Rn,Y+q	After loading location pointed to by Y+q, the Y stays the same.
LD	Rn,Z	After loading location pointed to by Z, the Z stays the same.
LD	Rn,Z+	After loading location pointed to by Z, the Z is incremented.
LD	Rn,-Z	The Z is decremented, then the location pointed to by Z is loaded.
LDD	Rn,Z+q	After loading location pointed to by Z+q, the Z stays the same.

Note: This table shows the syntax for the LD instruction, but it works for all such instructions. The auto-decrement or auto-increment affects the entire 16 bits of the pointer register and has no effect on the status register. This means that pointer register going from FFFF to 0000 will not raise any flag.

Example 6-6

Assume that RAM locations $90–$94 have a string of ASCII data, as shown below.

    $90 = ('H') $91 = ('E') $92 = ('L') $93 = ('L') $94 = ('O')

Write a program to get each character and send it to Port B one byte at a time. Show the program using:

(a) Direct addressing mode.
(b) Register indirect addressing mode.

Solution:

(a) Using direct addressing mode

```
LDI   R20,0xFF
OUT   DDRB,R20          ;make Port B an output
LDS   R20,0x90          ;R20 = contents of location 0x90
OUT   PORTB,R20         ;PORTB = R20
LDS   R20,0x91          ;R20 = contents of location 0x91
OUT   PORTB,R20         ;PORTB = R20
LDS   R20,0x92          ;R20 = contents of location 0x92
OUT   PORTB,R20         ;PORTB = R20
LDS   R20,0x93          ;R20 = contents of location 0x93
OUT   PORTB,R20         ;PORTB = R20
LDS   R20,0x94          ;R20 = contents of location 0x94
OUT   PORTB,R20         ;PORTB = R20
```

(b) Using register indirect addressing mode

```
        LDI   R16,0x5          ;R16=0x5 (R16 for counter)
        LDI   R20,0xFF
        OUT   DDRB,R20          ;make Port B an output
        LDI   ZL,0x90          ;the low byte of address (ZL = 0x90)
        LDI   ZH,0x0           ;the high byte of address (ZH = 0x0)
L1:     LD    R20,Z            ;read from location pointed to by Z
        INC   ZL               ;increment pointer
        OUT   PORTB,R20         ;send to PortB the contents of R20
        DEC   R16              ;decrement counter
        BRNE  L1               ;if R16 is not zero go to L1
```

When simulating the above program on the AVR Studio, make sure that memory locations $90–$94 have the message "HELLO".

See Figures 6-8 and 6-9. Then, see Examples 6-7 through 6-9.

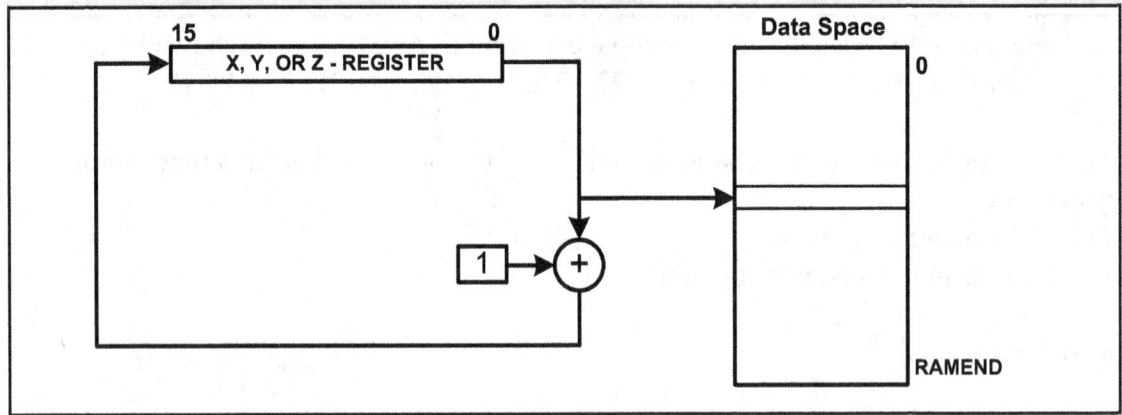

Figure 6-8. Register Indirect Addressing with Post-increment

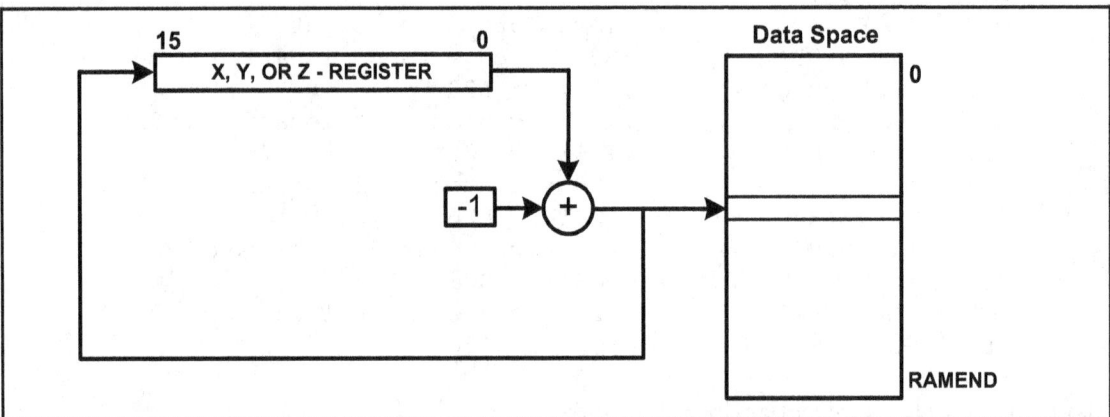

Figure 6-9. Register Indirect Addressing with Pre-decrement

Example 6-7

Write a program to clear 16 memory locations starting at data memory address $60.
Use the following:
(a) INC Rn
(b) Auto-increment

Solution:

```
(a)     LDI     R16, 16         ;R16 = 16 (counter value)
        LDI     XL, 0x60        ;XL = the low byte of address
        LDI     XH, 0x00        ;XH = the high byte of address
        LDI     R20, 0x0        ;R20 = 0
L1:     ST      X, R20          ;clear location X points to
        INC     XL              ;increment pointer
        DEC     R16             ;decrement counter
        BRNE    L1              ;loop until counter = zero

(b)     LDI     R16, 16         ;R16 = 16 (counter value)
        LDI     XL, 0x60        ;the low byte of X = 0x60
        LDI     XH, 0x00        ;the high byte of X = 0
        LDI     R20, 0x0        ;R20 = 0
L1:     ST      X+, R20         ;clear location X points to
        DEC     R16             ;decrement counter
        BRNE    L1              ;loop until counter = zero
```

Example 6-8

Assume that data memory locations $240–$243 have the following hex data. Write a program to add them together and place the result in locations $220 and $221.

$240 = ($7D) $241 = ($EB) $242 = ($C5) $243 = ($5B)

Solution:

```
        .INCLUDE "M32DEF.INC"
        .EQU L_BYTE = 0x220    ;RAM loc for L_Byte
        .EQU H_BYTE = 0x221    ;RAM loc for H_Byte
        LDI    R16,4
        LDI    R20,0
        LDI    R21,0
        LDI    XL, 0x40    ;the low byte of X = 0x40
        LDI    XH, 0x02    ;the high byte of X = 02
L1:     LD     R22, X+     ;read contents of location where X points to
        ADD    R20, R22
        BRCC   L2          ;branch if C = 0
        INC    R21         ;increment R21
L2:     DEC    R16         ;decrement counter
        BRNE   L1          ;loop until counter is zero
        ST     L_BYTE, R20 ;store the low byte of the result in $220
        ST     H_BYTE, R21 ;store the high byte of the result in $221
```

Example 6-9

Write a program to copy a block of 5 bytes of data from data memory locations starting at $130 to RAM locations starting at $60.

Solution:

```
        LDI    R16, 16     ;R16 = 16 (counter value)
        LDI    XL, 0x30    ;the low byte of address
        LDI    XH, 0x01    ;the high byte of address
        LDI    YL, 0x60    ;the low byte of address
        LDI    YH, 0x00    ;the high byte of address
L1:     LD     R20, X+     ;read where X points to
        ST     Y+, R20     ;store R20 where Y points to
        DEC    R16         ;decrement counter
        BRNE   L1          ;loop until counter = zero
```

Before we run the above program.
```
130 = ('H') 131 = ('E') 132 = ('L') 133 = ('L') 134 = ('O')
```

After the program is run, the addresses $60–$64 have the same data as $130–$134.
```
130 = ('H') 131 = ('E') 132 = ('L') 133 = ('L') 134 = ('O')
60 = ('H')  61 = ('E')  62 = ('L')  63 = ('L')  64 = ('O')
```

To see an example of how to use all three pointer registers, study and simulate Example 6-10.

Example 6-10

Two multibyte numbers are stored in locations $130–$133 and $150–$153. Write a program to add the multibyte numbers and save the result in address $160–$163.

```
      $C7659812
+     $2978742A
```

Solution:

```
        .INCLUDE "M32DEF.INC"
        LDI    R16, 4           ;R16 = 4 (counter value)
        LDI    XL, 0x30
        LDI    XH, 0x1          ;load pointer. X = $130
        LDI    YL, 0x50
        LDI    YH, 0x1          ;load pointer. Y = $150
        LDI    ZL, 0x60
        LDI    ZH, 0x1          ;load pointer. Z = $160
        CLC                     ;clear carry
L1:     LD     R18, X+          ;copy memory to R18 and INC X
        LD     R19, Y+          ;copy memory to R19 and INC Y
        ADC    R18,R19          ;R18 = R18 + R19 + carry
        ST     Z+,R18           ;store R18 in memory and INC Z
        DEC    R16              ;decrement R16 (counter)
        BRNE   L1               ;loop until counter = zero
```

Before the addition we have:

MSByte   **LSByte**
```
133 = ($C7) 132 = ($65) 131 = ($98)  130 = ($12)
153 = ($29) 152 = ($78) 151 = ($74)  150 = ($2A)
```

After the addition we have:
```
163 = ($F0) 162 = ($DE) 161 = (0C)  160 = (3C)
```
Notice that we are using the little endian convention of storing a low byte to a low address, and a high byte to a high address. Single-step the program in AVR Studio and examine the pointer registers and memory contents to gain insight into register indirect addressing mode.

Register indirect with displacement

Suppose we want to read a byte that is a few bytes higher than where the Z register points to. To do so we can increment the Z register so that it points to the desired location and then read it. But there is an easier way; we can use the register indirect with displacement. In this addressing mode a fixed number is added to the Z register. For example, if we want to read from the location that is 5 bytes after the location to which Z points, we can write the following instruction:

```
LDD   R20, Z+5    ;load from Z+5 into R20
```

The general format of the instruction is as follows:

```
LDD   Rd, Z+q     ;load from Z+q into Rd
```

where q is a number between 0 to 63, and Rd is any of the general purpose registers. See Figure 6-10.

To store a byte of data in a data memory location using the register indirect

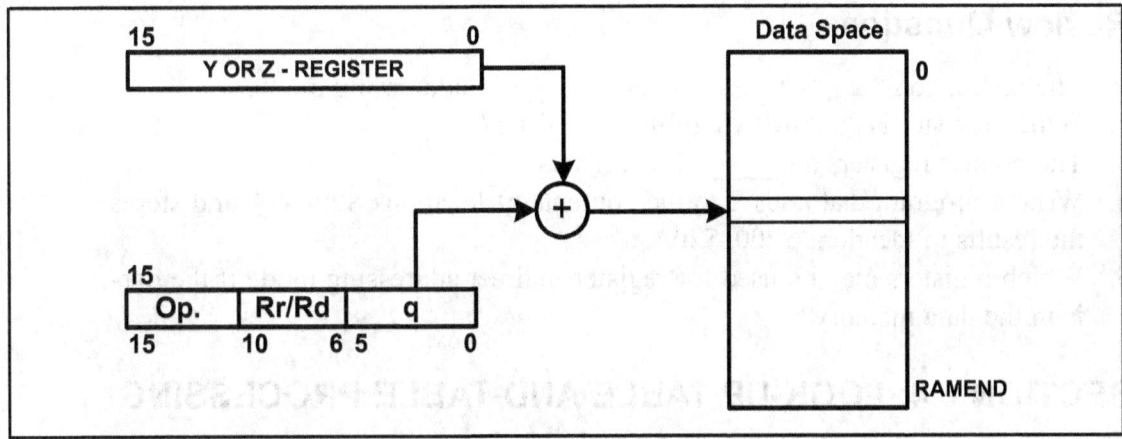

Figure 6-10. Register Indirect with Displacement

with displacement addressing mode we can use STD (Store with Displacement). The instruction is as follows:

```
STD    Z+q,Rr        ;store Rr into location Z+q
```

For example, the following instruction writes the contents of R20 into the location that is five bytes away from where Z points to:

```
STD    Z+5,R20       ;store R20 into location Z+5
```

To see an example of how to use the addressing mode, see Example 6-11.

Example 6-11

Write a function that adds the contents of three continuous locations of data space and stores the result in the first location. The Z register should point to the first location before the function is called.

Solution:

```
.INCLUDE "M32DEF.INC"
       LDI    R16,HIGH(RAMEND)   ;initialize the stack pointer
       OUT    SPH,R16
       LDI    R16,LOW(RAMEND)
       OUT    SPL,R16
       LDI    ZL,0x00            ;initialize the Z register
       LDI    ZH,2
       CALL   ADD3LOC            ;call add3loc
HERE:  JMP    HERE               ;loop forever
ADD3LOC:
       LDI    R21,0              ;R21 = 0
       LD     R20,Z              ;R20 = contents of location  Z
       LDD    R16,Z+1            ;R16 = contents of location Z+1
       ADD    R20,R16            ;R20 = R20 + R16
       BRCC   L1                 ;branch if carry cleared
       INC    R21                ;increment R21 if carry occurred
L1:    LDD    R16,Z+2            ;R16 = contents of location Z+2
       ADD    R20,R16            ;R20 = R20 + R16
       BRCC   L2                 ;branch if carry cleared
       INC    R21                ;increment R21
L2:    ST     Z,R20              ;store R20 into location Z
       STD    Z+1,R21            ;store R21 into location Z+1
       RET
```

Review Questions

1. The instruction "LD R19,0x95" uses _____ addressing mode.
2. Which register is the low byte of the X register?
3. The pointer registers are _____-bit registers.
4. Write a program that adds 2 to the contents of locations $90–$9A and stores the results in locations $200–$20A.
5. Which registers may be used for register indirect addressing mode if the data is in the data memory?

SECTION 6.4: LOOK-UP TABLE AND TABLE PROCESSING

So far, we have seen that the AVR has a maximum of 8M bytes of code (program) space and 64K of data memory space. We can use the code space to store fixed data. In this section we discuss how to access fixed data residing in the program ROM space of the AVR. First we examine how to store fixed data in the program ROM space using the .DB (define byte) directive.

.DB (define byte) and fixed data in program ROM

The .DB data directive is widely used to allocate ROM program (code) memory in byte-sized chunks. In other words, .DB is used to define an 8-bit fixed data. When .DB is used to define fixed data, the numbers can be in decimal, binary, hex, or ASCII formats. The .DB directive is widely used to define ASCII strings.

See Example 6-12. In Example 6-12 notice that each location of program memory is 2 bytes, whereas the .DB directive allocates byte-sized chunks. If we allocate a few bytes of data using the .DB directive, the first byte goes to the low byte of ROM location; the second byte goes to the high byte of ROM location; the third byte goes to the low byte of the next location of program ROM; and so on. In the cases in which we allocate an odd number of ROM locations using .DB, the assembler will automatically make the number of allocated locations even by placing a zero into the high byte of the last location. In other words, even if we allocate a fraction of a program ROM location, the assembler will allocate the whole location and load the unused part of it with zero. In Example 6-12 notice also that we must use single quotes (') for a single character and double quotes (") for a string.

AVR assembly also allows the use of .DW in place of .DB to define values greater than 255 (0xFF) but not larger than 65,535 (0xFFFF). See Example 6-13.

Reading table elements in the AVR

Example 6-12 showed how to place fixed data into program ROM. Now, we need to have a register pointer to point to the data to be fetched from the code space. The Z register can be used for this purpose. For this reason we can call it register indirect flash addressing mode. This is an addressing mode widely used to access data elements located in the program space of the AVR. In AVR terminology, there are two register indirect flash addressing modes: *program memory constant addressing* and *program memory addressing with post-increment*. In the pro-

Example 6-12

Assume that we have burned the following fixed data into the program ROM of an AVR chip. Give the contents of each ROM location starting at $500. See Appendix F for the hex values of the ASCII characters.

```
;MY DATA IN FLASH ROM
     .ORG $500
DATA1: .DB 1,8,5,3
DATA2: .DB 28              ;DECIMAL(1C in hex)
DATA3: .DB 0b00110101      ;BINARY (35 in hex)
DATA4: .DB 0x39            ;HEX
     .ORG 0x510
DATA4: .DB 'Y'             ;single ASCII char
DATA5: .DB '2','0','0','5';ASCII numbers
     .ORG $516
DATA6: .DB "Hello ALI"     ;ASCII string
```

Solution:

DATA1 has four bytes of data. The ".ORG $500" directive causes the assembler to put the first byte of DATA1 in the low byte of location $500. The second byte of DATA1, which is 8, goes to the high byte of location $500; the third byte goes to the low byte of location $501, and the fourth byte goes to the high byte of location $501.

DATA2 will be located after DATA1, in location $502 of memory. As DATA2 has one byte of data and each location of program is 2 bytes wide, the assembler puts zero in the high byte of location $502.

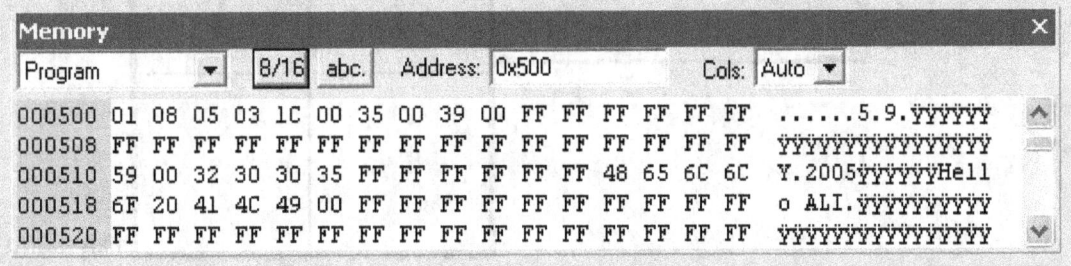

Example 6-13

Give the contents of each ROM location starting at $600.

```
     .ORG $600
DATA1: .DW 0x1234,0x1122
DATA2: .DW 28              ;DECIMAL (001C in hex)
DATA3: .DW 0x2239          ;HEX
```

Solution:

Since AVR is little endian, the low byte of 0x1234, which is 0x34, goes to the low byte of location $600, and its high byte goes to the high byte.

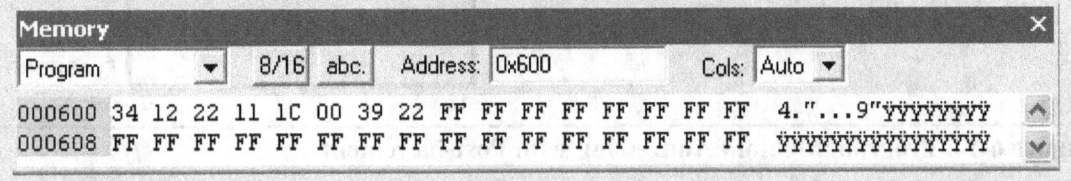

gram memory constant addressing mode, the content of Z does not change when the instruction is executed, which is why it is called *constant addressing*; whereas in the program memory addressing *with post-increment*, the content of Z increments after each execution.

"LPM Rn, Z" uses program memory constant addressing mode, while "LPM Rn,Z+" uses program memory addressing with post-increment. (See Table 6-6.) In Figures 6-11 and 6-12 you see the addressing modes.

There is a group of AVR instructions designed for table processing. Table 6-6 shows the instructions for table reading in the AVR.

The "LPM Rn, Z" instruction loads the byte pointed to by Z into the Rn. As you know, in the AVR, each location of the program memory is 2 bytes. So, we should mention if we want to read the low byte or the high byte. The least significant bit (LSB) of the Z register indicates whether the low byte or the high byte should be read. If LSB = 0, then the low byte will be read; otherwise, the high byte

Table 6-6: AVR Table Read Instructions

Instruction	Function	Description
LPM Rn,Z	Load from Program Memory	After read, Z stays the same
LPM Rn,Z+	Load from Program Memory with post-inc.	Reads and increments Z

Note: The byte of data is read into the Rn register from code space pointed to by Z.

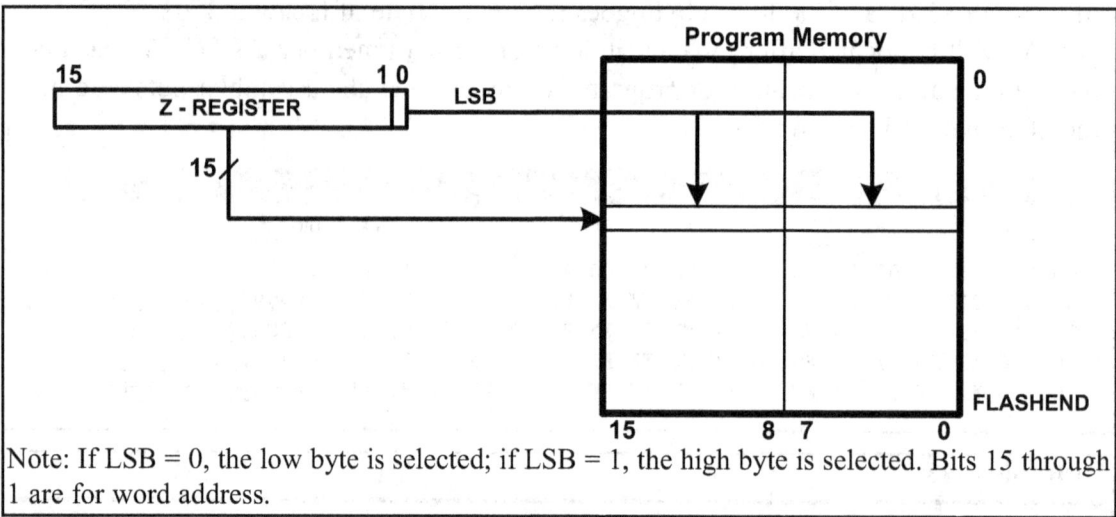

Note: If LSB = 0, the low byte is selected; if LSB = 1, the high byte is selected. Bits 15 through 1 are for word address.

Figure 6-11. Program Memory Constant Addressing

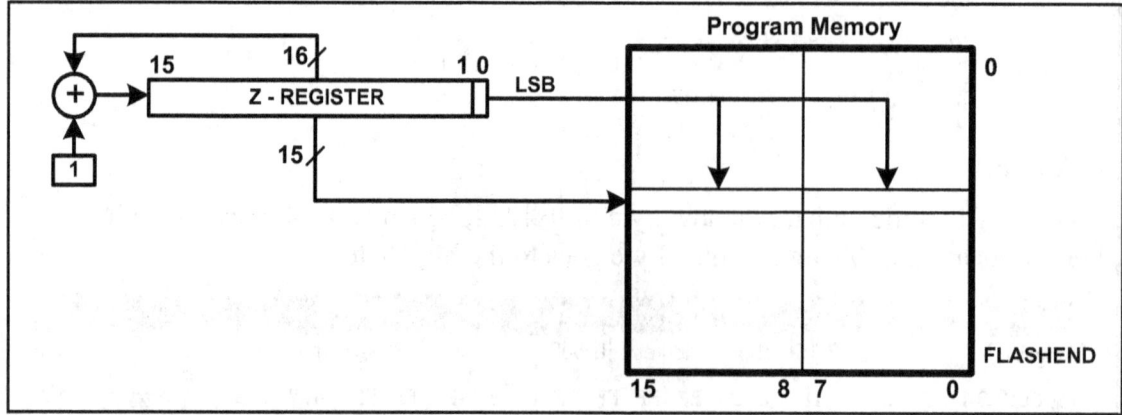

Figure 6-12. Program Memory Addressing with Post-increment

will be read. The other bits of the Z register (bit 1 to bit 15) represent the address of the location that should be read. See Figure 6-11.

Figure 6-13b shows the value that should be loaded into the Z register in order to address each byte of the program memory. For example, to address the low byte of location $0002, we should load the Z register with $0005, as shown below:

```
LDI ZH, 0x00     ;load ZH with 0x00 (the high byte of addr.)
LDI ZL, 0x05     ;load ZL with 0x05 (the low byte of addr.)
LPM R16, Z       ;load R16 with contents of location Z
```

Low	High	Address
0000 0000 0000 0000	0000 0000 0000 0001	0000 0000 0000 0000
0000 0000 0000 0010	0000 0000 0000 0011	0000 0000 0000 0001
0000 0000 0000 0100	0000 0000 0000 0101	0000 0000 0000 0010
0000 0000 0000 0110	0000 0000 0000 0111	0000 0000 0000 0011
0000 0000 0000 1000	0000 0000 0000 1001	0000 0000 0000 0100
0000 0000 0000 1010	0000 0000 0000 1011	0000 0000 0000 0101
⋮	⋮	
1111 1111 1111 1100	1111 1111 1111 1101	0111 1111 1111 1110
1111 1111 1111 1110	1111 1111 1111 1111	0111 1111 1111 1111

Figure 6-13a. Values of Z (in Binary)

Low	High	Address
$0000	$0001	$0000
$0002	$0003	$0001
$0004	$0005	$0002
$0006	$0007	$0003
$0008	$0009	$0004
$000A	$000B	$0005
$FFFC	$FFFD	$7FFE
$FFFE	$FFFF	$7FFF

Figure 6-13b. Values of Z

We can write the code using the HIGH and LOW directives as well:

```
LDI ZH, HIGH(0x0005) ;load ZH with 0x00 (the high byte of addr.)
LDI ZL, LOW (0x0005) ;load ZL with 0x05 (the low byte of addr.)
LPM R16, Z           ;load R16 with contents of location Z
```

As you see in Figure 6-13a, to read the low byte of each location we should shift the address of that location one bit to the left. For instance, to access the low byte of location 0b00000101, we should load Z with 0b000001010. To read the high byte, we shift the address to the left and we set bit 0 to one.

We can shift the address using the << directive as well. For example, the following program reads the low byte of location $100:

```
LDI ZH, HIGH($100<<1) ;load ZH with the high byte of addr.
LDI ZL, LOW ($100<<1) ;load ZL with the low byte of addr.
LPM R16, Z            ;load R16 with contents of location Z
```

If we *OR* a number with 1, its bit 0 will be set. Thus, the following program reads the high byte of location $100.

```
LDI ZH, HIGH(($100<<1)|1)
LDI ZL, LOW (($100<<1)|1)
LPM R16, Z            ;load R16 with contents of location Z
```

See Examples 6-14 and 6-15.

Example 6-14

In this program, assume that the phrase "WORLD PEACE." is burned into ROM locations starting at $500, and the program is burned into ROM locations starting at 0. Analyze how the program works and state where "WORLD PEACE." is stored after this program is run.

```
        .ORG   $0000                  ;burn into ROM starting at 0
        LDI    R20,0xFF
        OUT    DDRB,R20               ;make PB an output
        LDI    ZL,LOW(MYDATA<<1)       ;ZL = 0x00 (low byte of address)
        LDI    ZH,HIGH(MYDATA<<1)      ;ZH = 0x05 (high byte of address)
        LPM    R20,Z
        OUT    PORTB,R20    ;send it to port B
        INC    ZL           ;ZL = 01 pointing to next byte (A01)
        LPM    R20,Z        ;load R20 with 'W' (char pointed to by Z)
        OUT    PORTB,R20    ;send it to port B
        INC    ZL           ;ZL = 02 pointing to next byte (A02)
        LPM    R20,Z        ;load R20 with 'O' (char pointed to by Z)
        OUT    PORTB,R20    ;send it to port B
        INC    ZL           ;ZL = 03 pointing to next byte (A03)
        LPM    R20,Z        ;load R20 with 'R' (char pointed to by Z)
        OUT    PORTB,R20    ;send it to port B
        INC    ZL           ;ZL = 04 pointing to next byte (A04)
        LPM    R20,Z        ;load R20 with 'L' (char pointed to by Z)
        OUT    PORTB,R20    ;send it to port B
        INC    ZL           ;ZL = 05 pointing to next byte (A05)
        LPM    R20,Z        ;load R20 with 'D' (char pointed to by Z)
        OUT    PORTB,R20    ;send it to port B
HERE:   RJMP   HERE         ;stay here forever
        .ORG   $500  ;data is burned into program space starting at $500
MYDATA: .DB "WORLD PEACE."
```

R31 / R30

```
 15              8 7              1 0
Z = 0 0 0 0 1 0 1 0 0 0 0 0 0 0 0 0
     The address of location 0x500    Low
```

Memory — Program — 8/16

000500	57 4F	WO
000501	52 4C	RL
000502	44 20	D
000503	50 45	PE
000504	41 43	AC
000505	45 2E	E.

Solution:

In the above program, ROM locations $500–$505 have the following contents.

```
$500 (Low byte) = ('W')     $500 (High byte) = ('O')
$501 (Low byte) = ('R')     $501 (High byte) = ('L')
$502 (Low byte) = ('D')     $502 (High byte) = (' ')
$503 (Low byte) = ('P')     $503 (High byte) = ('E')
$504 (Low byte) = ('A')     $504 (High byte) = ('C')
$505 (Low byte) = ('E')     $505 (High byte) = ('.')
```

We start with Z = $0A00 (R31:R30 = $A00). The instruction "LPM R20,Z" loads R20 with the contents of the low byte of ROM location $500. Register R20 contains $57, the ASCII value for 'W'. This is loaded to Port B. Next, ZL is incremented to make Z = $A01. The LPM instruction will get the contents of the high byte of ROM location $500, which is character 'O'. After this program is run, we send the ASCII values for the characters 'W', 'O', 'R', 'L', and 'D' to Port B one character at a time. The loop version of this program is given in the next example.

Example 6-15

Assuming that program ROM space starting at $500 contains "WORLD PEACE.". write a program to send all the characters to Port B one byte at a time.

Solution:

```
(a) This method uses a counter


.ORG  $0000                     ;burn into ROM starting at 0
.INCLUDE "M32DEF.INC"
      LDI   R16,11
      LDI   R20,0xFF
      OUT   DDRB,R20             ;make PB an output
      LDI   ZH,HIGH(MYDATA<<1);ZH = high byte of addr.
      LDI   ZL,LOW(MYDATA<<1) ;ZL = low byte of addr.
L1:   LPM   R20,Z
      OUT   PORTB,R20           ;send it to Port B
      INC   ZL                  ;pointing to next byte
      DEC   R16                 ;decrement counter
      BRNE  L1                  ;repeat if counter not zero
HERE: RJMP  HERE                ;stay here forever
;-------------------------------------------------
;data is burned into code (program) space starting at $500
      .ORG  0x500
MYDATA DB   "WORLD PEACE."



(b) This method uses null char for end of string


.ORG  $0000                     ;burn into ROM starting at 0
.INCLUDE "M32DEF.INC"
      LDI   R20,0xFF
      OUT   DDRB,R20             ;make PB an output
      LDI   ZH,HIGH(MYDATA<<1);ZH = high byte of addr.
      LDI   ZL,LOW(MYDATA<<1) ;ZL = low byte of addr.
L1:   LPM   R20,Z               ;bring in next byte
      CPI   R20,0               ;compare R20 with 0
      BREQ  HERE                ;branch if equal
      OUT   PORTB,R20           ;send it to Port B
      INC   ZL                  ;pointing to next byte
      RJMP  L1                  ;repeat
HERE: RJMP  HERE                ;stay here forever
;-------------------------------------------------
;data is burned into code (program) space starting at $500
      .ORG 0x500
MYDATA: .DB "WORLD PEACE",0   ;notice null
```

Memory						✕
Program ▼	8/16	abc.	Address: 500		Cols: Auto ▼	

```
000500  57 4F 52 4C 44 20 50 45 41 43 45 00 FF FF FF FF   WORLD PEACE.ÿÿÿÿ
000508  FF FF FF FF FF FF FF FF FF FF FF FF FF FF FF FF   ÿÿÿÿÿÿÿÿÿÿÿÿÿÿÿÿ
```

Auto-increment option for Z

Using the "INC ZL" instruction to increment the pointer can cause a problem when an address such as $5FF is incremented. The carry will not propagate into ZH. The AVR gives us the option of LPM Rn, Z+ (load program memory with post-increment) as shown in Table 6-6. See Examples 6-16 and 6-17.

Example 6-16

Repeat Example 6-15, using auto-increment.

Solution:

```
.ORG   $0000              ;burn into ROM starting at 0
.INCLUDE "M32DEF.INC"
       LDI    R20,0xFF
       OUT    DDRB,R20            ;make PB an output
       LDI    ZH,HIGH(MYDATA<<1)      ;ZH = high byte of addr.
       LDI    ZL,LOW(MYDATA<<1)       ;ZL = low byte of addr.
L1:    LPM    R20,Z+              ;bring in next byte and inc. Z
       CPI    R20,0        ;compare R20 with 0
       BREQ   HERE         ;branch if equal
       OUT    PORTB,R20    ;send it to Port B
       RJMP   L1           ;repeat
HERE:  RJMP   HERE         ;stay here forever
;data is burned into code (program) space starting at $500
       .ORG   0x500
MYDATA: .DB "WORLD PEACE",0         ;notice null
```

Example 6-17

Assume that ROM space starting at $100 contains the message "The Promise of World Peace". Write a program to bring this message into the CPU one byte at a time and place the bytes in RAM locations starting at $140.

Solution:

```
.EQU  RAM_BUF = 0x140
.ORG $0000                         ;burn into ROM starting at 0
.INCLUDE "M32DEF.INC"
       LDI R20,0xFF
       OUT DDRB,R20                ;make PB an output
       LDI ZH,HIGH(MYDATA<<1)  ;ZH = high byte of addr.
       LDI ZL,LOW(MYDATA<<1)   ;ZL = low byte of addr.
       LDI XH,HIGH(RAM_BUF)    ;XH = $1, high byte of RAM addr.
       LDI XL,LOW(RAM_BUF)     ;XL = $40, low byte of RAM addr.
L1:    LPM    R20,Z+              ;bring in next byte and increment Z
       CPI    R20,0        ;compare R20 with 0
       BREQ HERE           ;branch if end of string
       ST     X+,R20       ;store R20 in RAM and increment X
       RJMP L1             ;repeat
HERE:  RJMP HERE           ;stay here forever
;----------------------message
       .ORG 0x100          ;data burned starting at 0x100
MYDATA: .DB "The Promise of World Peace",0     ;notice null
```

Look-up table

The look-up table is a widely used concept in microcontroller programming. It allows access to elements of a frequently used table with minimum operations. As an example, assume that for a certain application we need $4 + x^2$ values in the range of 0 to 9. We can use a look-up table instead of calculating the values, which takes some time. In the AVR, to get the table element we add the index to the address of the look-up table. This is shown in Examples 6-18 through 6-20.

Example 6-18

Assume that the lower three bits of Port C are connected to three switches. Write a program to send the following ASCII characters to Port D based on the status of the switches.

000	'0'
001	'1'
010	'2'
011	'3'
100	'4'
101	'5'
110	'6'
111	'7'

Solution:

```
.ORG 0
.INCLUDE "M32DEF.INC"
      LDI   R16,0x0
      OUT   DDRC,R16                      ;DDRC = 0x00 (port C as input)
      LDI   R16,0xFF
      OUT   DDRD,R16                      ;DDRD = 0xFF (port D as output)
      LDI   ZH,HIGH(ASCII_TABLE<<1) ;ZH = high byte of addr.
BEGIN:IN    R16,PINC                      ;read from port C into R16
      ANDI  R16,0b00000111                ;mask upper 5 bits
      LDI   ZL,LOW(ASCII_TABLE<<1)  ;ZL = the low byte of addr.
      ADD   ZL,R16                        ;add PINC to the addr
      LPM   R17,Z                         ;get ASCII from look-up table
      OUT   PORTD,R17
      RJMP  BEGIN

;look-up table for ASCII numbers 0-7
.ORG 0x20
ASCII_TABLE:
      .DB '0','1','2','3','4','5','6','7'
```

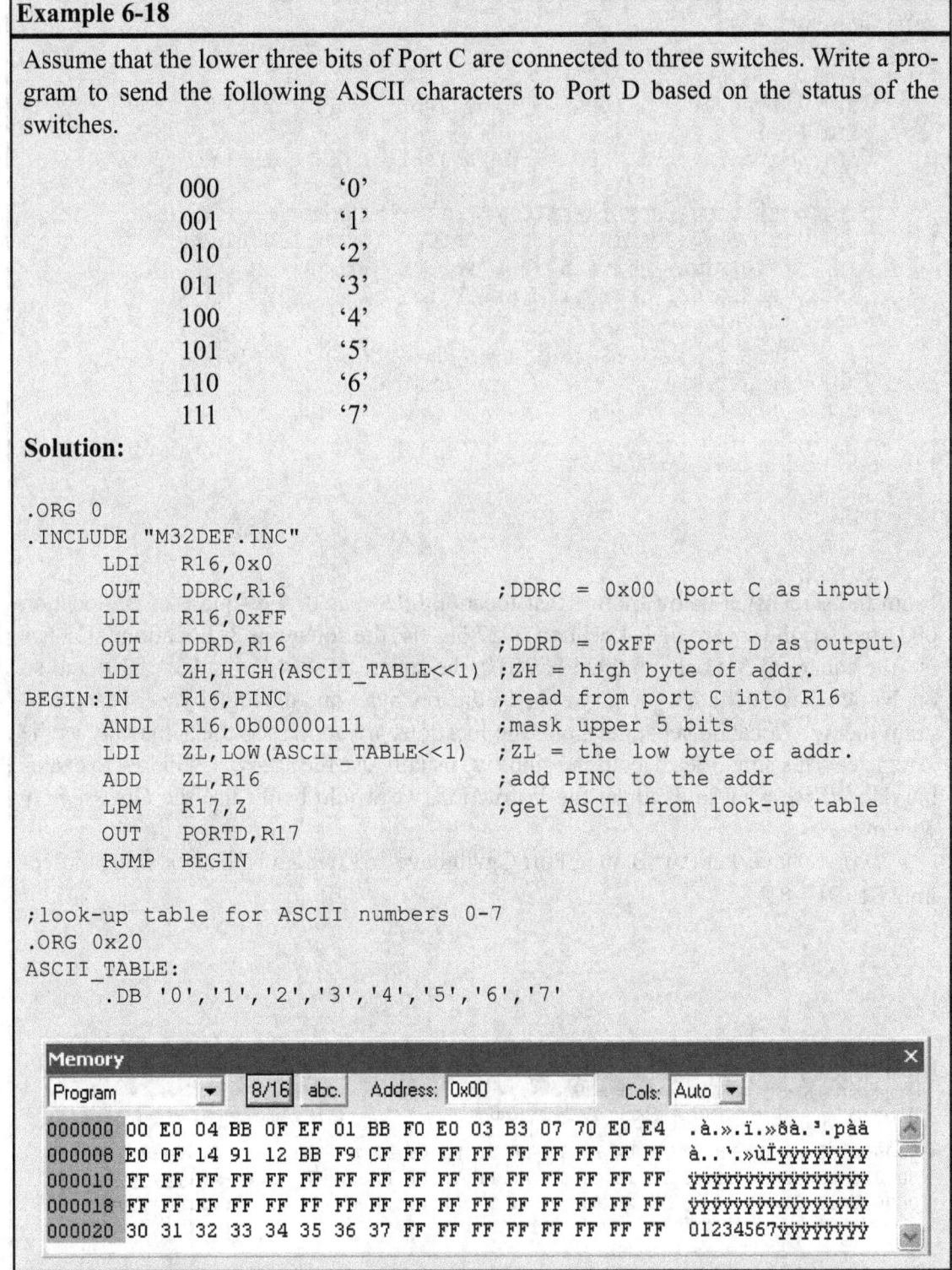

Example 6-19

Write a program to get the x value from Port B and send x^2 to Port C. Assume that PB3–PB0 has the x value of 0–9. Use a look-up table instead of a multiply instruction.

What is the value of Port C if we have 9 at Port B?

Solution:

```
.INCLUDE "M32DEF.INC"
      .ORG 0
      LDI   R16,0x00
      OUT   DDRB,R16     ;DDRB = 0x00 (Port B as input)
      LDI   R16,0xFF
      OUT   DDRC,R16     ;DDRC = 0xFF (Port C as output)

      LDI   ZH,HIGH(XSQR_TABLE<<1)  ;ZH = high byte of addr.
L1:   LDI   ZL,LOW(XSQR_TABLE<<1)   ;ZL = low byte of addr.
      IN    R16,PINB     ;read from Port B into R16
      ANDI  R16,0x0F     ;mask upper bits
      ADD   ZL,R16
      LPM   R18,Z        ;get x2 from the look-up table
      OUT   PORTC,R18
      RJMP  L1

;look-up table for square of numbers 0-9
.ORG 0x10
XSQR_TABLE:
      .DB 0, 1, 4, 9, 16, 25, 36, 49, 64, 81
```

From the screenshot below, notice that location 0020 has 0, the square of 0. Location 0021 has 01, the square of 1. Location 0022 has 04, the square of 2. Location 0023 has 09, the square of 3. Location 0024 has $10, the square of 4 ($4 \times 4 = 16 = \$10$) and so on. Notice that the Memory window shows the low bytes and the high bytes of each program memory location separately, and the locations are addressed the same way as the Z register. This simplifies debugging since we usually use the *Memory* window to examine data. If we want to examine the instruction, we would better use the *Disassembly* window.

   If we have 9 at Port B, then Port C will have $51, which is the hex value of decimal 81 ($9^2 = 81$).

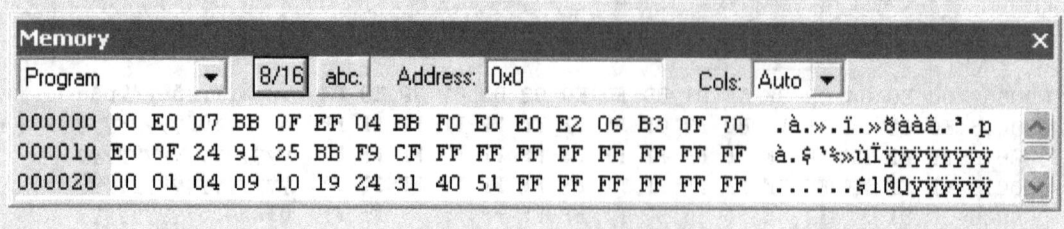

Example 6-20

Write a program to get the x value from Port B and send $x^2 + 2x + 3$ to Port C. Assume PB3–PB0 has the x value of 0–9. Use a look-up table instead of a multiply instruction.

Solution:

```
      .ORG 0
      .INCLUDE "M32DEF.INC"
      LDI R16,0x00
      OUT    DDRB,R16     ;DDRB = 0x00 (Port B as input)
      LDI    R16,0xFF
      OUT    DDRC,R16     ;DDRC = 0xFF (Port C as output)

      LDI    ZH,HIGH(TABLE<<1) ;ZH = high byte of addr.
L1:   LDI    ZL,LOW(TABLE<<1)  ;ZL = low byte of addr.
      IN     R16,PINB     ;read from Port B into R16
      ANDI   R16,0x0F     ;mask upper bits
      ADD    ZL,R16
      LPM    R18,Z        ;get x2 + 2x + 3 from the look-up table
      OUT    PORTC,R18
      RJMP L1

.ORG 0x10
TABLE:
      .DB 3, 6, 11, 18, 27, 38, 51, 66, 83, 102
```

Accessing a look-up table in RAM

The look-up table elements can also be in RAM instead of ROM. Sometimes we need to bring in the elements of the look-up table from RAM because the elements are dynamic and can change. In the AVR, we can do that using the pointers.

Writing table elements in AVR

In AVR we also have the SPM instruction, which allows us to write (store) data into program memory. See the AVR datasheets to see how to write to Flash ROM.

Review Questions

1. The instruction "LPM" uses register _____ as the address pointer.
2. What register holds data once it is read by the LPM Rd,Z instruction?
3. What is the size of Z? How much ROM space does it cover?
4. What register is incremented upon execution of the LPM Rd,Z+ instruction?
5. What is the difference between the LPM and ELPM instructions?
6. When should we make our look-up table in RAM?
7. True or false. We can write into program memory using the SPM instruction.

SECTION 6.5: BIT-ADDRESSABILITY

Many microprocessors such as the 386 or Pentium allow programs to access registers and I/O ports in byte size only. In other words, if you need to check a single bit of an I/O port, you must read the entire byte first and then manipulate the whole byte with some logic instructions to get hold of the desired single bit. This is not the case with the AVR as we saw in Chapter 4. In this section, we provide more programming examples of bit manipulation using the bit-addressable and byte-addressable options of the AVR family.

In Table 6-7, some of the bit-oriented instructions are given. Notice that the bit-oriented instructions use only one addressing mode, the direct addressing mode. In the previous sections of this chapter we showed various addressing modes of byte-addressable space of the AVR, among them register indirect addressing mode for both data RAM and program (code) ROM. Note that there is no register indirect addressing mode for bit-oriented instructions in the AVR, nor are there any bit-oriented instructions for program memory.

Manipulating the bits of general purpose registers

In this part we discuss how to set, clear, or copy the bits of a GPR.

Setting the bits

The SBR (Set Bits in Register) instruction sets the specified bits in the general purpose register. It has the following format:

```
SBR Rd,K     ;set bits in register Rd
```

K is an 8-bit value that can be 00–FF in hex, and Rd is R16 to R31 (any of the 16 general purpose registers). The SBR instruction is just another name for the ORI instruction and it sets any of the bits of the general purpose register whose bit in the K variable is 1. For example, in the following program the SBR instruction sets the bits 2, 5, and 6 regardless of their previous values.

```
LDI R17,0b01011001   ;R17 = 0x59
SBR R17,0b01100100   ;set bits 2, 5, and 6 in register R17
```

When execution of the above instructions is finished, R17 contains 0x7D. Notice that the SBR instruction is a byte-oriented instruction as it manipulates the whole byte at one time.

Clearing the bits

The CBR (Clear Bits in Register) instruction clears the specified bits in the general purpose register. It has the following format:

```
CBR Rd,K     ;clear bits in register Rd
```

K is an 8-bit value that can be 00–FF in hex, and Rd is R16 to R31 (any of the 16 general purpose registers). The CBR instruction clears any of the bits of the general purpose register whose bit in the K variable is 1. For example, in the following program the CBR instruction clears the bits 2, 5, and 6 regardless of their

Table 6-7: Single-Bit (Bit-Oriented) Instructions for AVR

Instruction	Function
SBI A,b	Set Bit b in I/O register
CBI A,b	Clear Bit b in I/O register
SBIC A,b	Skip next instruction if Bit b in I/O register is Cleared
SBIS A,b	Skip next instruction if Bit b in I/O register is Set
BST Rr,b	Bit store from register Rr to T
BLD Rd,b	Bit load from T to Rd
SBRC Rr,b	Skip next instruction if Bit b in Register is Cleared
SBRS Rr,b	Skip next instruction if Bit b in Register is Set
BRBS s,k	Branch if Bit s in status register is Set
BRBC s,k	Branch if Bit s in status register is Cleared

Note: A can be any location of the I/O register.

previous values.

```
LDI R17,0b01011001  ;R17 = 0x59
CBR R17,0b01100100  ;clear bits 2, 5, and 6 in register R17
```

After the execution of the above instructions, R17 contains 0x19.

Copying a bit

As we saw in Chapter 2, one of the bits in the SREG (status register) is named T (temporary), which is used when we want to copy a bit of data from one GPR to another GPR. The BST (Bit Store from register to T) and BLD (Bit Load from T to register) instructions can be used to copy a bit of a register to a specific bit of another register. The "BST Rd,b" instruction stores bit b from Rd to the T flag, while the "BLD Rr,b" instruction copies the T flag to bit b in register Rr.

For example, the following program copies bit 3 from R17 to bit 5 in register R19:

```
BST R17,3   ;store bit 3 from R17 to the T flag
BLD R19,5   ;copy the T flag to bit 5 in R19
```

See Example 6-21.

Example 6-21

A switch is connected to pin PB4. Write a program to get the status of the switch and save it in D0 of internal RAM location 0x200.

Solution:

```
.EQU MYREG = 0x200       ;set aside loc 0x200
     CBI   DDRB,0         ;make PB0 an input
     IN    R17,PINB       ;R17 = PINB
     BST   R17,4          ;T = PINB.4
     LDI   R16,0x00       ;R16 = 0
     BLD   R16,0          ;R16.0 = T
     STS   MYREG,R16      ;copy R16 to location $200
HERE: JMP   HERE
```

Checking a bit

To see if a bit of a general purpose register is set or cleared, we can use the SBRS (Skip next instruction if Bit in Register is Set) and SBRC (Skip next instruction if Bit in register is Cleared) instructions.

The SBRS instruction tests a bit of a register and skips the instruction right below it if the bit is HIGH. The format of the SBRS instruction is as follows:

```
SBRS Rd,b   ;skip next instruction if Bit b in Rd is set
```

For example, in the following program the "LDI   R20,0x55" instruction will not be executed since bit 3 of R17 is set.

```
LDI   R17,0b0001010
SBRS  R17,3   ;skip next instruction if Bit 3 in R17 is set
LDI   R20,0x55
LDI   R30,0x33
```

The SBRC instruction skips the next instruction if a bit of a GPR is cleared. It has the following format:

```
SBRC Rd,b ;skip next instruction if Bit b in Rd is cleared
```

For example, in the following program the "LDI   R20,0x55" instruction will not be executed since bit 2 of R16 is cleared.

```
LDI   R16,0b0001010
SBRC  R16,2 ;skip next instruction if Bit 2 in R16 is cleared
LDI   R20,0x55
LDI   R30,0x33
```

See Example 6-22.

Example 6-22

A switch is connected to pin PC7. Using the SBRS instruction, write a program to check the status of the switch and perform the following:
(a) If switch = 0, send letter 'N' to Port D.
(b) If switch = 1, send letter 'Y' to Port D.

Solution:

```
.INCLUDE "M32DEF.INC"   ;include a file according to the IC you use
      CBI   DDRC,7       ;make PC7 an input
      LDI   R16,0xFF
      OUT   DDRD,R16     ;make Port D an output port
AGAIN:IN    R20,PINC     ;R20 = PINC
      SBRS  R20,7        ;skip next line if Bit PC7 is set
      RJMP  OVER         ;it must be LOW
      LDI   R16,'Y'      ;R16 = 'Y' ASCII letter Y
      OUT   PORTD,R16    ;issue R16 to PD
      RJMP  AGAIN        ;we could use JMP instead
OVER: LDI   R16,'N'      ;R16 = 'N' ASCII letter N
      OUT   PORTD,R16    ;issue R16 to PORTD
      RJMP  AGAIN        ;we can use JMP too
```

228

Manipulating the bits of I/O registers

As we discussed in Chapter 4, we can set and clear the lower 32 I/O registers (addresses 0 to 31) using the SBI (Set bit in I/O register) and CBI (Clear bit in I/O register) instructions. For example, the following two instructions set the PORTA.1 and clear the PORTB.4, respectively:

```
SBI PORTA,1      ;set Bit 1 in PORTA
CBI PORTB,4      ;clear Bit 4 in PORTB
```

See Example 6-23.

Example 6-23

Write a program to toggle PB2 a total of 200 times.

Solution:

```
        LDI    R16,200     ;load the count into R16
        SBI    DDRB,2      ;DDRB.1 = 1, make RB1 an output
AGAIN:  SBI    PORTB,2     ;set bit PB2 (toggle PB2)
        CBI    PORTB,2     ;clear bit PB2 (toggle PB2)
        DEC    R16         ;decrement R16
        BRNE   AGAIN       ;continue until counter is zero
```

In Chapter 4 we mentioned that we can test a bit in the lower 32 I/O registers using the SBIS (Skip if Bit in I/O register is Set) and SBIC (Skip if Bit in I/O register is Cleared) instructions. See Examples 6-24 and 6-25.

Example 6-24

Rewrite the program of Example 6-22 using the SBIC instruction.

Solution:

```
.INCLUDE "M32DEF.INC"    ;include a proper file
        CBI    DDRC,7      ;make PC7 an input
        LDI    R16,0xFF
        OUT    DDRD,R16    ;make Port D an output port
AGAIN:  IN     R20,PINC    ;R20 = PINC
        SBRC   R20,7       ;skip next line if Bit PC7 is cleared
        RJMP   OVER        ;it must be HIGH
        LDI    R16,'N'     ;R16 = 'N' ASCII letter N
        OUT    PORTD,R16   ;issue R16 to PD
        RJMP   AGAIN       ;we could use JMP instead
OVER:   LDI    R16,'Y'     ;R16 = 'Y' ASCII letter Y
        OUT    PORTD,R16   ;issue R16 to PORTD
        RJMP   AGAIN       ;we can use JMP too
```

Example 6-25

Rewrite the program of Example 6-22 using the SBIS instruction.

Solution:

```
.INCLUDE "M32DEF.INC"    ;include a proper file
      CBI   DDRC,7       ;make PC7 an input
      LDI   R16,0xFF
      OUT   DDRD,R16      ;make Port D an output port
AGAIN:SBIS  PINC,7       ;skip next line if Bit PC7 is set
      RJMP  OVER         ;it must be LOW
      LDI   R16,'Y'      ;R16 = 'Y' ASCII letter Y
      OUT   PORTD,R16     ;issue R16 to PD
      RJMP  AGAIN        ;we could use JMP instead
OVER: LDI   R16,'N'      ;R16 = 'N' ASCII letter N
      OUT   PORTD,R16     ;issue R16 to PORTD
      RJMP  AGAIN        ;we can use JMP too
```

Status register bit-addressability

Now let's see how we can use bit-addressability of the status register. As we discussed in Chapter 2, the bits of the status register are used for the flags C, Z, N, V, S, H, T, and I. The status register is shown in Figure 6-14.

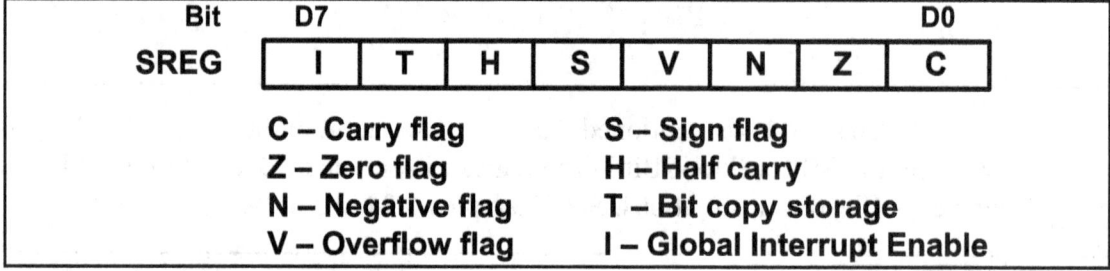

Figure 6-14. Bits of the Status Register

Checking a flag bit

There are some instructions for checking the bits in the status register, as shown in Table 6-8. All of the instructions are derived from two instructions: BRBS (Branch if status flag is Set) and BRBC (Branch if status flag is Cleared). The instructions are as follows:

```
BRBS  s,k        ;branch if status flag bit is set
BRBC  s,k        ;branch if status flag bit is cleared
```

where s is a number between 0 and 7, and represents the bit in the status register, and k is the relative address of the target location to which the instruction branches when the condition is true.

For example, in the following program the LDI instruction is not executed when the carry flag is set:

```
BRBS  0,L1       ;branch if status flag bit 0 is set
LDI   R20,3
L1:
```

Table 6-8: AVR Conditional Branch (Jump) Instructions

Instruction	Action	Instruction	Action
BRCS	Branch if C = 1	BRCC	Branch if C = 0
BRLO	Branch if C = 1	BRSH	Branch if C = 0
BREQ	Branch if Z = 1	BRNE	Branch if Z = 0
BRMI	Branch if N = 1	BRPL	Branch if N = 0
BRVS	Branch if V = 1	BRVC	Branch if V = 0
BRLT	Branch if S = 1	BRGE	Branch if S = 0
BRHS	Branch if H = 1	BRHC	Branch if H = 0
BRTS	Branch if T = 1	BRTC	Branch if T = 0
BRIE	Branch if I = 1	BRID	Branch if I = 0

We can write the same program using the "BRCS L1" instruction as follows:

```
      BRCS   L1          ;branch if carry flag is set
      LDI    R20,3
L1:
```

Since it is hard to memorize the bits of the status register and use the BRBC and BRBS instructions, we can use the instructions in Table 6-8 to simplify checking the bits of the status register.

Manipulating a bit

To set a flag we can use the BSET instruction.

```
      BSET   s           ;flag bit set
```

where *s* is a number between 0 and 7, and represents the bit to be set in the status register.

For example, the following instruction sets the carry flag.

```
      BSET   0           ;set bit 0 (carry flag)
```

As another example, the instruction "BSET 2" sets the N (Negative) flag.

To clear a flag we can use the BCLR (flag bit clear) instruction.

```
      BCLR   s           ;flag bit clear
```

where *s* is a number between 0 and 7, and represents the bit to be cleared in the status register.

For example, the following instruction clears the carry flag.

```
      BCLR   0           ;clear bit 0 (carry flag)
```

As another example, the instruction "BCLR 1" clears the Z (Zero) flag.

A more convenient way is to use the CLZ instruction, as shown in Table 6-9.

Table 6-9: Manipulating the Flags of the Status Register

Instruction	Action		Instruction	Action	
SEC	Set Carry	C = 1	CLC	Clear Carry	C = 0
SEZ	Set Zero	Z = 1	CLZ	Clear Zero	Z = 0
SEN	Set Negative	N = 1	CLN	Clear Negative	N = 0
SEV	Set overflow	V = 1	CLV	Clear overflow	V = 0
SES	Set Sign	S = 1	CLS	Clear Sign	S = 0
SEH	Set Half carry	H = 1	CLH	Clear Half carry	H = 0
SET	Set Temporary	T = 1	CLT	Clear Temporary	T = 0
SEI	Set Interrupt	I = 1	CLI	Clear Interrupt	I = 0

Internal RAM bit-addressability

The internal RAM is not bit-addressable. So, in order to manipulate a bit of the internal RAM location, you should bring it into the general purpose register and then manipulate it, as shown in Examples 6-26 and 6-27.

Example 6-26

Write a program to see if the internal RAM location $195 contains an even value. If so, send it to Port B. If not, make it even and then send it to Port B.

Solution 1:

```
.EQU  MYREG = 0x195          ;set aside loc 0x195
      LDI    R16,0xFF
      OUT    DDRB,R16         ;make Port B an output port
AGAIN:LDS    R16,MYREG
      SBRS   R16,0            ;bit test D0, skip if set
      RJMP   OVER             ;it must be LOW
      CBR    R16,0b00000001   ;clear bit D0 = 0
OVER: OUT    PORTB,R16        ;copy it to Port B
      JMP    AGAIN            ;we can use RJMP too
```

Solution 2:

```
.EQU  MYREG = 0x195          ;set aside loc 0x195
      LDI    R16,0xFF
      OUT    DDRB,R16         ;make Port B an output port
AGAIN:LDS    R16,MYREG
      CBR    R16,0b00000001   ;clear bit D0 = 0
OVER: OUT    PORTB,R16        ;copy it to Port B
      JMP    AGAIN            ;we can use RJMP too
```

Example 6-27

Write a program to see if the internal RAM location $137 contains an even value. If so, write 0x55 into location $200. If not, write 0x63 into location $200.

Solution:

```
.EQU   MYREG = 0x137      ;set aside location 0x137
.EQU   RESULT= 0x200
       LDS   R16,MYREG
       SBRC  R16,0        ;skip if clear Bit D0 of R16 register is clr
       RJMP  OVER         ;it is odd
       LDI   R16,0x55
       STS   RESULT,R16
       RJMP  HERE
OVER:  LDI   R16,0x63
       STS   RESULT,R16
HERE:  RJMP  HERE
```

Review Questions

1. True or false. All registers of the AVR are bit-addressable.
2. True or false. The status register of the AVR is bit-addressable.
3. Indicate which of the following registers are bit-addressable.
    (a) Port A (b) Port B   (c) R19   (d) status register (e) PC register
4. How would you check to see whether bit D1 of R23 is HIGH or LOW?
5. Show how to clear the carry flag.
6. State what each instruction does.
    (a) `SBR R16,0x1`   (b) `CBR R30,0x7`   (c) `BST R19,2`
    (d) `SBI PORTB,4`   (e) `CBI SREG,1`   (f) `CLI`

SECTION 6.6: ACCESSING EEPROM IN AVR

Every member of the AVR microcontrollers has some amount of on-chip EEPROM. In Table 6-10 you can see the amount of EEPROM memory in each member of the ATmega family. As we mentioned in Chapter 0, the data in SRAM will be lost if the power is disconnected. However, we need a place to save our data to protect them against power failure. EEPROM memory can save stored data even when the power is cut off. In this section we will show how to write to EEPROM memory and how to access it.

Table 6-10: Size of EEPROM Memory in ATmega Family

Chip	Bytes	Chip	Bytes	Chip	Bytes
ATmega8	512	ATmega16	512	ATmega32	1024
ATmega64	2048	ATmega128	4096	ATmega256RZ	4096
ATmega640	4096	ATmega1280	4096	ATmega2560	4096

EEPROM registers

There are three I/O registers that are directly related to EEPROM. These are EECR (EEPROM Control Register), EEDR (EEPROM Data Register), and EEARH-EEARL (EEPROM Address Register High-Low). Each of these registers is discussed in detail in this section.

EEPROM Data Register (EEDR)

To write data to EEPROM, you have to write it to the EEDR register and then transfer it to EEPROM. Also, if you want to read from EEPROM you have to read from EEDR. In other words, EEDR is a bridge between EEPROM and CPU.

EEPROM Address Register (EEARH and EEARL)

The EEARH:EEARL registers together make a 16-bit register to address each location in EEPROM memory space. When you want to read from or write to EEPROM, you should load the EEPROM location address in EEARs. As you see in Figure 6-15, only 10 bits of the EEAR registers are used in ATmega32. Because ATmega32 has 1024-byte EEPROM locations, we need 10 bits to address each location in EEPROM space. In ATmega16, 9 bits of the EEAR registers are used because ATmega16 has 512 bytes of EEPROM, and to address 512 bytes we need a 9-bit address.

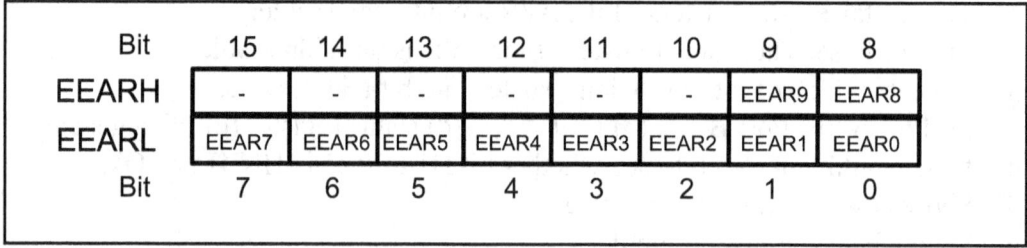

Figure 6-15. EEPROM Address Registers

EEPROM Control Register (EECR)

The EECR register is used to select the kind of operation to perform on. The operation can be start, read, and write. In Figure 6-16 you see the bits of the EECR register. The bits are as follows:

EEPROM Read Enable (EERE): Setting this bit to one will cause a read operation if EEWE is zero. When a read operation starts, one byte of EEPROM will be read into the EEPROM Data Register (EEDR). The EEAR register specifies the address of the desired byte.

EEPROM Write Enable (EEWE) and EEPROM Master Write Enable (EEMWE): When EEMWE is set, setting EEWE within four clock cycles will start a write operation. If EEMWE is zero, setting EEWE to one will have no effect.

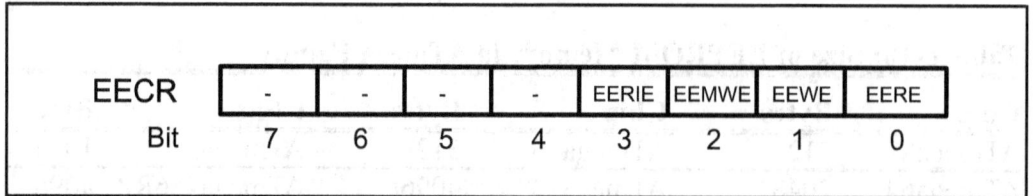

Figure 6-16. EEPROM Control Registers

When you set EEMWE to one, the hardware clears the bit to zero after four clock cycles. This prevents unwanted write operations on EEPROM contents. Notice that you cannot start read or write operations before the last write operation is finished. You can check for this by polling the EEWE bit. If EEWE is zero it means that EEPROM is ready to start a new read or write operation.

EEPROM Ready Interrupt Enable (EERIE): In Chapter 10 you will learn about interrupts in AVR. As you see in Figure 6-16, bits 4 to 7 of EECR are unused at the present time and are reserved.

Programming the AVR to write on EEPROM

To write on EEPROM the following steps should be followed. Notice that steps 2 and 3 are optional, and the order of the steps is not important. Also note that you cannot do anything between step 4 and step 5 because the hardware clears the EEMWE bit to zero after four clock cycles.

1. Wait until EEWE becomes zero.
2. Write new EEPROM address to EEAR (optional).
3. Write new EEPROM data to EEDR (optional).
4. Set the EEMWE bit to one (in EECR register).
5. Within four clock cycles after setting EEMWE, set EEWE to one.

See Example 6-28 to see how we write a byte on EEPROM.

Example 6-28

Write an AVR program to store 'G' into location 0x005F of EEPROM .

Solution:

```
.INCLUDE "M16DEF.INC"

WAIT:                  ;wait for last write to finish
SBIC   EECR,EEWE       ;check EEWE to see if last write is finished
RJMP   WAIT            ;wait more
LDI    R18,0           ;load high byte of address to R18
LDI    R17,0x5F        ;load low byte of address to R17
OUT    EEARH, R18      ;load high byte of address to EEARH
OUT    EEARL, R17      ;load low byte of address to EEARL
LDI    R16,'G'         ;load 'G' to R16
OUT    EEDR,R16        ;load R16 to EEPROM Data Register
SBI    EECR,EEMWE      ;set Master Write Enable to one
SBI    EECR,EEWE       ;set Write Enable to one
```

Run and simulate the code on AVR Studio to see how the content of the EEPROM changes after the last line of code. Enter four NOP instructions before the last line, change the 'G' to 'H', and run the code again. Explain why the code doesn't store 'H' at location 0x005F of EEPROM.

Programming the AVR to read from EEPROM

To read from EEPROM the following steps should be taken. Note that step 2 is optional.

1. Wait until EEWE becomes zero.
2. Write new EEPROM address to EEAR (optional).
3. Set the EERE bit to one.
4. Read EEPROM data from EEDR.

See Example 6-29 to see how we read a byte from EEPROM.

Example 6-29

Write an AVR program to read the content of location 0x005F of EEPROM into PORTB.

Solution:

```
.INCLUDE "M16DEF.INC"
      LDI    R16,0xFF
      OUT    DDRB,R16
WAIT:                         ;wait for last write to finish
      SBIC   EECR,EEWE        ;check EEWE to see if last write is finished
      RJMP   WAIT             ;wait more
      LDI    R18,0            ;load high byte of address to R18
      LDI    R17,0x5F         ;load low byte of address to R17
      OUT    EEARH, R18       ;load high byte of address to EEARH
      OUT    EEARL, R17       ;load low byte of address to EEARL
      SBI    EECR,EERE        ;set Read Enable to one
      IN     R16,EEDR         ;load EEPROM Data Register to R16
      OUT    PORTB,R16        ;out R16 to PORTB
```

Initializing EEPROM

In Section 6-4, you saw how to allocate program memory using the .DB directive. We can also allocate and initialize the EEPROM using the .DB directive. If we write .ESEG before a definition, the variable will be located in the EEPROM, whereas .CSEG before a definition causes the variable to be allocated in the code (program) memory. By default the variables are located in the program memory. For example, the following code allocates locations $10 and $11 of EEPROM for DATA1 and DATA2, and initializes them with $95 and $19, respectively:

```
      .ESEG
      .ORG $10
DATA1:      .DB    $95
DATA2:      .DB    $19
```

The following code allocates DATA1 and DATA3 in program memory and DATA2 in EEPROM:

```
DATA1:      .DB    $10    ;by default it is located in code memory
      .ESEG
DATA2:      .DB    $20    ;it is located in EEPROM
DATA3:      .DB    $35    ;it is located in EEPROM
      .CSEG
DATA4:      .DB    $45    ;it is located in code memory
```

See Example 6-30.

Example 6-30

Write a program that counts how many times a system has been powered up.

Solution:

```
.INCLUDE "M32DEF.INC"
    LDI    R20,HIGH(RAMEND)
    OUT    SPH,R20
    LDI    R20,LOW(RAMEND)
    OUT    SPL,R20              ;initialize stack pointer

    LDI    XH,HIGH(COUNTER)
    LDI    XL,LOW(COUNTER)      ;X points to COUNTER
    CALL   LOAD_FROM_EEPROM     ;load R20 with value of COUNTER
    INC    R20                  ;increment R20
    CALL   STORE_IN_EEPROM      ;store R20 in EEPROM
HERE: RJMP  HERE

;-----Load R20 with contents of location X of EEPROM
LOAD_FROM_EEPROM:
    SBIC   EECR, EEWE
    RJMP   LOAD_FROM_EEPROM     ;wait while EEPROM is busy
    OUT    EEARH,XH
    OUT    EEARL,XL             ;EEAR = X
    SBI    EECR,EERE            ;set Read Enable to one
    IN     R20,EEDR             ;load EEPROM Data Register to r20
    RET

;-----Store R20 into location X of EEPROM
STORE_IN_EEPROM:
    SBIC   EECR, EEWE
    RJMP   STORE_IN_EEPROM      ;wait while EEPROM is busy
    OUT    EEARH,XH
    OUT    EEARL,XL             ;EEAR = X
    OUT    EEDR,R20
    SBI    EECR,EEMWE           ;set Master Write Enable to one
    SBI    EECR,EEWE            ;write EEDR into EEPROM
    RET
;-------EEPROM
.ESEG
.ORG 0
COUNTER:    .DB    0
```

COUNTER is initialized with $0. Then, it is incremented on each power-up.

Review Questions

1. True or false. The AVR EEPROM memory is used for both program code and data.
2. True or false. The ATmega32 has 1,024 bytes of EEPROM memory.
3. True or false. In the AVR, EEPROM contents are lost when power to the chip is cut off.
4. True or false. In the AVR, EEPROM memory is read and write memory.
5. True or false. Every AVR chip comes with 1 KB of EEPROM.

SECTION 6.7: CHECKSUM AND ASCII SUBROUTINES

In this section we look at some widely used subroutines: checksum byte, BCD, and ASCII conversion.

Checksum byte in EEPROM

To ensure the integrity of ROM contents, every system must perform a checksum calculation. The checksum will detect any corruption of the contents of ROM. One cause of ROM corruption is current surge, either when the system is turned on, or during operation. To ensure data integrity in ROM, the checksum process uses what is called a *checksum byte*. The checksum byte is an extra byte that is tagged to the end of a series of bytes of data. To calculate the checksum byte of a series of bytes of data, the following steps can be taken:

1. Add the bytes together and drop the carries.
2. Take the 2's complement of the total sum, and that is the checksum byte, which becomes the last byte of the series.

To perform a checksum operation, add all the bytes, including the checksum byte. The result must be zero. If it is not zero, one or more bytes of data have been changed (corrupted). To clarify these important concepts, see Example 6-31.

Checksum program

The checksum generation and testing program is given in subroutine form. Five subroutines perform the following operations:

1. Retrieve the data from EEPROM.
2. Test the checksum byte for any data error.
3. Initialize variables if the checksum byte is corrupted.
4. Calculate the checksum byte.
5. Store the data in EEPROM.

Each of these subroutines can be used in other applications. Example 6-31 shows how to manually calculate the checksum for a list of values. Also, see Program 6-1.

```
;PROG 6-1: CHECKSUM
.INCLUDE "M32DEF.INC"
.EQU   OPTION_SIZE = 0x4
.EQU   RAM_OPTIONS = 0x100
;------------main program
.ORG   0
       LDI    R16,HIGH(RAMEND)
       OUT    SPH,R16
       LDI    R16,LOW(RAMEND)
       OUT    SPL,R16            ;SP points to RAMEND
       RCALL  LOAD_OPTIONS       ;load options
       RCALL  TEST_CHKSUM        ;test checksum
       TST    R20
       BREQ   L1                 ;if data is not corrupted go to L1
```

Example 6-31

Assume that we have 4 bytes of hexadecimal data: $25, $62, $3F, and $52.

(a) Find the checksum byte.
(b) Perform the checksum operation to ensure data integrity.
(c) If the second byte, $62, has been changed to $22, show how the checksum method detects the error.

Solution:

(a) Find the checksum byte.

```
    $25
  + $62
  + $3F
  + $52
    ─────
    $118      (Dropping the carry of 1, we have $18. Its 2's complement is $E8. Therefore,
              the checksum byte is $E8.)
```

(b) Perform the checksum operation to ensure data integrity.

```
    $25
  + $62
  + $3F
  + $52
  + $E8
    ─────
    $200      (Dropping the carries, we see 00, indicating that the data is not corrupted.)
```

(c) If the second byte, $62, has been changed to $22, show how the checksum method detects the error.

```
    $25
  + $22
  + $3F
  + $52
  + $E8
    ─────
    $1C0      (Dropping the carry, we get $C0, which is not 00. This means that the data
              is corrupted.)
```

```
        RCALL INIT_OPTIONS        ;initialize options
L1:     ;Here you can use the options

        RCALL CAL_CHKSUM          ;calculating checksum
        RCALL STORE_OPTIONS       ;storing options in EEPROM
HERE:   RJMP  HERE

;-----Load R20 with contents of location X of EEPROM
LOAD_FROM_EEPROM:
        SBIC  EECR, EEWE
        RJMP  LOAD_FROM_EEPROM    ;wait while EEPROM is busy
        OUT   EEARH,XH
        OUT   EEARL,XL            ;EEAR = X
        SBI   EECR,EERE           ;set Read Enable to one
        IN    R20,EEDR            ;load EEPROM Data Register to R20
        RET
;-----Store R20 into location X of EEPROM
```

```
STORE_IN_EEPROM:
        SBIC  EECR, EEWE
        RJMP  STORE_IN_EEPROM   ;wait while EEPROM is busy
        OUT   EEARH,XH
        OUT   EEARL,XL          ;EEAR = X
        OUT   EEDR,R20
        SBI   EECR,EEMWE        ;set Master Write Enable to one
        SBI   EECR,EEWE         ;write EEDR into EEPROM
        RET
;-----copying the data from EEPROM to internal SRAM
LOAD_OPTIONS:
        LDI   XL,LOW(E_OPTIONS)
        LDI   XH,HIGH(E_OPTIONS)     ;X points to E_OPTIONS
        LDI   YL,LOW(RAM_OPTIONS)
        LDI   YH,HIGH(RAM_OPTIONS)   ;Y points to RAM_OPTIONS
        LDI   R16,OPTION_SIZE+1      ;COUNTER = OPTION_SIZE+1
LL1:    CALL  LOAD_FROM_EEPROM       ;load R20 with EEPROM loc X
        ST    Y+,R20                 ;store R20 in RAM loc Y
        INC   XL                ;increment XL
        BRNE  LL2               ;if not carry go to LL2
        INC   XH
LL2:    DEC   R16               ;decrement COUNTER
        BRNE  LL1               ;if COUNTER not zero go to LL1
        RET                     ;return
;-------copying data from code ROM to data RAM
INIT_OPTIONS:
        LDI   ZL,LOW(FLASH_OPTIONS<<1);Z points to FLASH_OPTIONS
        LDI   ZH,HIGH(FLASH_OPTIONS<<1)
        LDI   YL,LOW(RAM_OPTIONS)
        LDI   YH,HIGH(RAM_OPTIONS)   ;Y points to RAM_OPTIONS
        LDI   R16,OPTION_SIZE        ;COUNTER = OPTION_SIZE
H1:     LPM   R18,Z+          ;load R18 with program mem. location Z
        ST    Y+,R18            ;store R18 in loc Y of RAM
        DEC   R16               ;decrement COUNTER
        BRNE  H1                ;if COUNTER is not zero go to H1
        RET                     ;return
;-------calculating checksum byte
CAL_CHKSUM:
        LDI   YL,LOW(RAM_OPTIONS)
        LDI   YH,HIGH(RAM_OPTIONS)   ;Y points to RAM_OPTIONS
        LDI   R16,OPTION_SIZE   ;COUNTER = OPTION_SIZE
        LDI   R20,0             ;SUM = 0
CL1:    LD    R17,Y+            ;load R17 with contents of loc Y
        ADD   R20,R17           ;SUM = SUM + R17
        DEC   R16               ;decrement COUNTER
        BRNE  CL1               ;if COUNTER is not zero go to CL1
        NEG   R20               ;two's complement SUM
        ST    Y,R20             ;store checksum in loc Y of RAM
        RET                     ;return
;-------testing checksum byte
TEST_CHKSUM:
        LDI   YL,LOW(RAM_OPTIONS)
        LDI   YH,HIGH(RAM_OPTIONS)    ;Y points to RAM_OPTIONS
        LDI   R16,OPTION_SIZE+1
        LDI   R20,0             ;SUM = 0
TL1:    LD    R17,Y+            ;load R17 with contents of loc Y
        ADD   R20,R17           ;SUM = SUM + R17
```

```
        DEC    R16                     ;decrement COUNTER
        BRNE   TL1                     ;loop while COUNTER is not zero
        RET
;-----copying the data from internal SRAM to EEPROM
STORE_OPTIONS:
        LDI    XL,LOW(E_OPTIONS)
        LDI    XH,HIGH(E_OPTIONS)      ;X points to E_OPTIONS
        LDI    YL,LOW(RAM_OPTIONS)
        LDI    YH,HIGH(RAM_OPTIONS)    ;Y points to RAM_OPTIONS
        LDI    R16,OPTION_SIZE+1       ;COUNTER = OPTION_SIZE+1
SL1:    LD     R20, Y+
        CALL   STORE_IN_EEPROM  ;store R20 in loc X
        INC    XL                      ;increment XL
        BRNE   SL2                     ;if not carry go to SL2
        INC    XH
SL2:    DEC    R16                     ;decrement COUNTER
        BRNE   SL1                     ;loop while COUNTER is not zero
        RET                            ;return
;-------initial values in program ROM
FLASH_OPTIONS: .DB        0x25,0x62,0x3F,0x52
;-------EEPROM
.ESEG
.ORG  $0
E_OPTIONS: .DB     0x25,0x62,0x3F,0x52
```

BCD to ASCII conversion program

Many RTCs (real-time clocks) provide time and date in BCD format. To display the BCD data on an LCD or a PC screen, we need to convert it to ASCII. Program 6-2 (a) transfers packed BCD data from program ROM to data RAM, (b) converts packed BCD to ASCII, and (c) sends the ASCII to port B for display. The displaying of data on an LCD will be shown in Chapter 12. See Chapter 5 for the BCD to ASCII conversion algorithm.

```
;PROG 6-2: CONVERTING PACKED BCD TO ASCII
.INCLUDE "M32DEF.INC"
.EQU RAM_ADDR =    0x80
        LDI    R16,HIGH(RAMEND)
        OUT    SPH,R16
        LDI    R16,LOW(RAMEND)
        OUT    SPL,R16                 ;SP = RAMEND
        CALL   BCD_ASCII_COV
HERE: RJMP   HERE
;-----convert packed BCD to ASCII
BCD_ASCII_COV:
        LDI    ZL,LOW(MYBYTE<<1)
        LDI    ZH,HIGH(MYBYTE<<1)      ;Z = MYBYTE
        LDI    XL,LOW(RAM_ADDR)
        LDI    XH,HIGH(RAM_ADDR)       ;X = RAM_ADDR
        LDI    R16,4                   ;COUNTER = 4
L1:     LPM    R20,Z+                  ;
        MOV    R21,R20                 ;R21 = R20
        ANDI   R21,0x0F                ;mask the upper nibble
        ORI    R21,0x30                ;make it an ASCII
        ST     X+,R21
```

```
        SWAP    R20             ;swap the nibbles
        ANDI    R20,0x0F        ;mask the upper nibble
        ORI     R20,0x30        ;make it an ASCII
        ST      X+,R20
        DEC     R16             ;decrement COUNTER
        BRNE    L1              ;loop while COUNTER is not zero
        RET                     ;return
;-----send ASCII to Port B
SEND_TO_PORTB:
        LDI     XH,HIGH(RAM_ADDR)
        LDI     XL,LOW(RAM_ADDR)    ;X = RAM_ADDR
        LDI     R16,8               ;COUNTER = 8
L2:     LD      R20,X+              ;
        OUT     PORTB,R20           ;PORTB = R20
        DEC     R16                 ;decrement counter
        BRNE    L2                  ;loop while counter is not zero
        RET
MYBYTE: .DB 0x25, 0x67, 0x39, 0x52
```

Binary (hex) to ASCII conversion program

Many ADC (analog-to-digital converter) chips provide output data in binary (hex). To display the data on an LCD or PC screen, we need to convert it to ASCII. The code for the binary-to-ASCII conversion is shown in Program 6-3. Notice that the subroutine gets a byte of 8-bit binary (hex) data from Port B and converts it to decimal digits, and the second subroutine converts the decimal digits to ASCII digits and saves them. We are saving the low digit in the lower address location and the high digit in higher address location. This is referred to as the little-endian convention (i.e., low byte to low location, and high byte to high location). All AVR products use the little-endian convention. For the binary-to-ASCII conversion algorithm see Chapter 5.

```
;PROG 6-3: CONVERTING BINARY TO ASCII
.INCLUDE "M32DEF.INC"
.DEF    NUM = R20
.DEF    DENOMINATOR = R21
.DEF    QUOTIENT = R22
.EQU    RAM_ADDR = 0x200
.EQU    ASCII_RESULT = 0x210
;--------------main program
.ORG 0
        LDI     R18,HIGH(RAMEND)
        OUT     SPH,R18
        LDI     R18,LOW(RAMEND)
        OUT     SPL,R18
        LDI     R16,0x00
        OUT     DDRA,R16
        RCALL   BIN_DEC_CONVRT
        RCALL   DEC_ASCI_CONVRT
END:    RJMP    END

;-----------Converting BIN(HEX) TO DEC (00-FF TO 000-255)
BIN_DEC_CONVRT:
        LDI     XL,LOW(RAM_ADDR)    ;save DEC digits in these locations
        LDI     XH,HIGH(RAM_ADDR)
```

242

```
        IN    NUM,PINA              ;read data from PORT A
        LDI   DENOMINATOR,10
        RCALL DIVIDE                ;QUOTIENT=PINA/10 NUM=PINA%10
        ST    X+,NUM                ;save lower digit
        MOV   NUM,QUOTIENT
        RCALL DIVIDE                ;divide by 10 once more
        ST    X+,NUM                ;save the next digit
        ST    X+,QUOTIENT           ;save the last digit
        RET
DEC_ASCI_CONVRT:
        LDI   XL,LOW(RAM_ADDR)          ;addr. of DEC data
        LDI   XH,HIGH(RAM_ADDR)
        LDI   YL,LOW(ASCII_RESULT)      ;addr. of ASCII data
        LDI   YH,HIGH(ASCII_RESULT)
        LDI   R16,3                     ;count
BACK:   LD    R20,X+                    ;get DEC digit
        ORI   R20,0x30                  ;make it an ASCII digit
        ST    Y+,R20                    ;store it
        DEC   R16                       ;decrement counter
        BRNE  BACK                      ;repeat until the last one
        RET
;----------------------------------------------------------------
DIVIDE:
        LDI   QUOTIENT,0
L1:     INC   QUOTIENT
        SUB   NUM, DENOMINATOR
        BRCC  L1
        DEC   QUOTIENT
        ADD   NUM, DENOMINATOR
        RET
```

We can write a function that directly converts binary to ASCII as shown below:

```
.INCLUDE "M32DEF.INC"
.DEF  NUM = R20
.DEF  DENOMINATOR = R21
.DEF  QUOTIENT = R22
.EQU  ASCII_RESULT = 0x210
;--------------main program
.ORG 0
        LDI   R18,HIGH(RAMEND)
        OUT   SPH,R18
        LDI   R18,LOW(RAMEND)
        OUT   SPL,R18             ;initialize stack pointer
        LDI   R16,0x00
        OUT   DDRA,R16
        RCALL BIN_ASCII_CONVRT
HERE:   RJMP  HERE
;-----------Converting BIN(HEX) TO DEC (00-FF TO 000-255)
BIN_ASCII_CONVRT:
        LDI   XL,LOW(ASCII_RESULT)     ;save results in these loc.
        LDI   XH,HIGH(ASCII_RESULT)
        IN    NUM,PINA                 ;read data from PORT A
        LDI   DENOMINATOR,10
        RCALL DIVIDE                    ;QUOTIENT=PINA/10    NUM=PINA%10
```

```
        ORI    NUM,0x30            ;make it an ASCII digit
        ST     X+,NUM             ;save lower digit
        MOV    NUM,QUOTIENT
        RCALL  DIVIDE             ;divide by 10 once more
        ORI    NUM,0x30            ;make it an ASCII digit
        ST     X+,NUM             ;save the next digit
        ORI    QUOTIENT,0x30       ;make it an ASCII digit
        ST     X+,QUOTIENT        ;save the last digit
        RET
```

SECTION 6.8: MACROS

In this section we explore macros and their use in Assembly language programming. The format and usage of macros are defined and many examples of their applications are examined.

What is a macro and how is it used?

There are applications in Assembly language programming in which a group of instructions performs a task that is used repeatedly. For example, moving data into a RAM location is done repeatedly in the same program. It does not make sense to rewrite this code every time it is needed. Therefore, to reduce the time that it takes to write code and reduce the possibility of errors, the concept of macros was born. Macros allow the programmer to write the task (code to perform a specific job) once only, and to invoke it whenever it is needed.

Macro definition

Every macro definition must have three parts, as follows:

.MACRO   name

......

.ENDMACRO

The .MACRO directive indicates the beginning of the macro definition and the .ENDMACRO directive signals the end. What goes between the .MACRO and .ENDMACRO directives is called the *body* of the macro. The name must be unique and must follow Assembly language naming conventions. A macro can take up to 10 parameters. The parameters can be referred to as @0 to @9 in the body of the macro. After the macro has been written, it can be invoked (or called) by its name, and appropriate values are substituted for parameters.

For example, moving immediate data into I/O register data RAM is a widely used service, but there is no instruction for that. We can use a macro to do the job as shown in the following code:

```
.MACRO     LOADIO
       LDI    R20,@1
       OUT    @0,R20
.ENDMACRO
```

The above is the macro definition. Note that parameters @0 and @1 are mentioned in the body of the macro.

The following are three examples of how to use the above macro:

```
1. LOADIO    PORTA, 0x20        ;send value 0x20 to PORTA

2. .EQU      VAL_1 = 0xFF
   LOADIO    DDRC, VAL_1
3. LOADIO    SPL, 0x55          ;send value $55 to SPL
```

Now examine Program 6-4 to see how to use a macro in a program.

```
;Program 6-4: toggling Port B using macros
;-------------------------------------------
.INCLUDE "M32DEF.INC"
.MACRO LOADIO
     LDI   R20,@1
     OUT   @0,R20
.ENDMACRO

;------------------------------time delay macro
.MACRO DELAY
     LDI @0,@1
BACK:
     NOP
     NOP
     NOP
     NOP
     DEC   @0
     BRNE BACK
.ENDMACRO

;------------------------------program starts
     .ORG  0
     LOADIO  DDRB,0xFF        ;make PORTB output
L1:  LOADIO  PORTB,0x55       ; PORTB = 0x55
     DELAY   R18,0x70         ;delay
     LOADIO  PORTB,0xAA       ; PB = 0xAA
     DELAY   R18,0x70         ;delay
     RJMP    L1
```

.INCLUDE directive

Assume that several macros are used in every program. Must they be rewritten every time? The answer is no, if the concept of the .INCLUDE directive is known. The .INCLUDE directive allows a programmer to write macros and save them in a file, and later bring them into any program file. For example, assume that the following widely used macros were written and then saved under the filename "MYMACRO1.MAC".

Assuming that the LOADIO and DELAY macros are saved on a disk under the filename "MYMACRO1.MAC", the .INCLUDE directive can be used to bring this file into any ".asm" file and then the program can call upon any of the macros as many times as needed. When a file includes all macros, the macros are listed at the beginning of the ".lst" file and, as they are expanded, will be part of the program.

To understand this, see Program 6-5.

```
;Program 6-5: toggling Port B using macros
.INCLUDE "M32DEF.INC"
.INCLUDE "MYMACRO1.MAC" ;get macros from macro file
;--------------------------program starts
        .ORG   0
        LOADIO  DDRB,0xFF
L1:     LOADIO  PORTB,0x55
        DELAY   R18,0x70
        LOADIO  PORTB,0xAA
        DELAY   R18,0x70
        RJMP    L1
```

.LISTMAC directive

When viewing the .lst file with macros, the details of the macros are not displayed. This means that the bodies of the macros are not displayed when they are invoked during the code. But when we are debugging the code we might need to see exactly what instructions are executed. Using the .LISTMAC directive we can turn on the display of the bodies of macros in the list file. For example, examine the following code:

```
.INCLUDE "M32DEF.INC"
.MACRO    LOADIO
        LDI   R20,@1
        OUT   @0,R20
.ENDMACRO

        LOADIO    PORTA,0x20
        LOADIO    DDRA,0x53
HERE:   JMP       HERE
```

The assembler provides the following code in the .lst file:

```
                    .MACRO    LOADIO
                        LDI   R20,@1
                        OUT   @0,R20
                    .ENDMACRO

000000 e240
000001 bb4b         LOADIO    PORTA,0x20
000002 e543
000003 bb4a         LOADIO    DDRA,0x53
000004 940c 0004 HERE:JMP     HERE
```

If we add the .LISTMAC directive to the above code:

```
.MACRO    LOADIO
        LDI   R20,@1
        OUT   @0,R20
.ENDMACRO
.LISTMAC
        LOADIO    PORTA,0x20
```

```
                   LOADIO        DDRA,0x53
HERE:JMP      HERE
```

The assembler expands the macro by providing the following code in the .lst file:

```
                        .MACRO        LOADIO
                        LDI    R20,@1
                        OUT    @0,R20
                   .ENDMACRO
                   .LISTMAC
                   +
000000 e240        +LDI R20 , 0x20
000001 bb4b        +OUT PORTA , R20
                        LOADIO        PORTA,0x20
                   +
000002 e543        +LDI R20 , 0x53
000003 bb4a        +OUT DDRA , R20
                        LOADIO        DDRA,0x53
000004 940c 0004 HERE:JMP      HERE
```

The + indicates that the code is from the macro.

Macros vs. subroutines

Macros and subroutines are useful in writing assembly programs, but each has limitations. Macros increase code size every time they are invoked. For example, if you call a 10-instruction macro 10 times, the code size is increased by 100 instructions; whereas, if you call the same subroutine 10 times, the code size is only that of the subroutine instructions. On the other hand, a function call takes 3 or 4 clocks and the RET instruction takes 4 clocks to get executed. So, using functions adds around 8 clock cycles. The subroutines use stack space as well when called, while the macros do not.

Review Questions

1. Discuss the benefits of macro programming.
2. List the three parts of a macro.
3. Explain and contrast the macro definition and invoking the macro.

SUMMARY

This chapter described the addressing modes of the AVR. Immediate addressing mode uses a constant for the operand. Direct or register indirect addressing modes can be used to access data stored in data memory of the AVR. Register indirect addressing mode uses a register as a pointer to the data. The advantage of this is that it makes addressing dynamic rather than static. Program memory addressing mode is widely used in accessing data elements of look-up table entries located in the program Flash ROM space of the AVR. The AVR allows the reading of fixed data stored in program Flash ROM space, in addition to writing to Flash ROM.

The I/O registers can be accessed by six different addressing modes: I/O direct addressing (by their names or their addresses), direct data addressing, data indirect with displacement, data indirect addressing, data indirect addressing with pre-decrement, and data indirect addressing with post-increment.

We also discussed the bit-addressable locations and showed how to use single-bit instructions to access them directly.

We also explained how to access EEPROM and how to use checksum to make sure data is not corrupted.

Macros were also explored and their advantages were discussed.

PROBLEMS

SECTION 6.1: INTRODUCING SOME MORE ASSEMBLER DIRECTIVES

1. Indicate the value loaded into the registers in the following program:
```
.EQU  C1 = 0x20
.EQU  C2 = 0x6F
.EQU  C3 = 0x14
LDI   R20,(C1&C2)|C3
LDI   R21,C2-(C1+C3)
```

2. Indicate the value loaded into R30, R31, and R20 in the following program:

```
.ORG 0x0
.EQU  DATA_ADDR = (OUR_DATA<<1)
      LDI   R30,LOW(DATA_ADDR)
      LDI   R31,HIGH(DATA_ADDR)
      LPM   R20,Z

.ORG 0x100
OUR_DATA: .DB 20,'A','C'
```

SECTION 6.2: REGISTER AND DIRECT ADDRESSING MODES

3. Which of the following are invalid uses of immediate addressing mode?
    (a) LDI R20,0x24   (b) STS 0x70, 0x30   (c) OUT 0x20,0x42
4. Identify the addressing mode for each of the following:
        (a) OUT PORTB,R20 (b) LDI R20, 0x50   (c) LDS 0x40,R20
        (d) ADD R20,R25   (e) MOV R20,R25
5. Indicate the addresses assigned to each of the following:
        (a) PORTB   (b) PORTC   (c) DDRC
        (d) DDRD   (e) SPL   (f) SPH
        (g) SREG
6. In accessing the I/O registers, we should use _____ addressing mode.
7. What does the following instruction do? "STS 0xF0,R20"
8. What does the following instruction do? "OUT PORTC,R19"

9. The byte addresses assigned to the internal SRAM are _____ to _____ in ATmega32. (Hint: To calculate the address of the last location, add the size of SRAM in ATmega32 to the address of the first location of SRAM and decrease the result by one.)

10. The byte addresses assigned to the SRAM are _____ to _____ in ATmega16.

11. Write a program to add the following data and place the result in RAM location $200: The data values are 6, 9, 2, 5, 7

SECTION 6.3: REGISTER INDIRECT ADDRESSING MODE

12. Which registers are allowed to be used as a pointer for register indirect addressing mode when accessing data RAM? Give their names and show how they are loaded.

13. Write a program to copy $AA into RAM locations $80 to $9F.

14. Write a program to clear RAM locations $90 to $12F.

15. Write a program to copy 10 bytes of data starting at RAM address $80 to RAM locations starting at $90.

16. Write a program to toggle RAM locations $80 to $8F.

SECTION 6.4: LOOK-UP TABLE AND TABLE PROCESSING

17. Compile and state the contents of each ROM location for the following data:
```
          .ORG  0x200
MYDAT_1:   .DB   "Earth"
MYDAT_2:   .DB   "987-65"
MYDAT_3:   .DB   "GABEH 98"
```

18. Compile and state the contents of each ROM location for the following data:
```
          .ORG  0x340
DAT_1: .DB 0x22,0x56, 0b10011001, 32, 0xF6, 0b11111011
```

19. Which register is allowed to be used as a pointer for register indirect addressing mode when accessing data stored in program ROM? Give the name and show how it is loaded.

20. What is the size of the Z register? How much ROM space does the LPM instruction cover?

21. Write a program to read data from the low byte of Flash ROM location 0x200.

22. Write a program to read data from the high byte of Flash ROM location 0x340.

23. Write a program to read the following message from ROM and place it in data RAM starting at 0x60:
```
          .ORG 0x600
MYDATA:    .DB   "1-800-999-9999",0
```

24. Write a program to find y where $y = x^2 + 2x + 5$, and x is between 0 and 9.

25. Write a program to find y where $y = 20x + 5$, and x is between 0 and 9.

26. Write a program to read the following message from ROM and place it in data RAM starting at 40:
```
          .ORG 0x700
MYDATA:    .DB   "The earth is but one country",0
```

27. True or false. In all AVR members we can access the Flash ROM memory.

28. True or false. The ELPM instruction works for all AVR members.

29. Assume that the lower four bits of PORTB are connected to four switches.

Write a program to send the following ASCII characters to a PORTC, based on the status of the switches:

0000	'0'
0001	'1'
0010	'2'
0011	'3'
0100	'4'
0101	'5'
0110	'6'
0111	'7'
1000	'8'
1001	'9'
1010	'A'
1011	'B'
1100	'C'
1101	'D'
1110	'E'
1111	'F'

SECTION 6.5: BIT-ADDRESSABILITY

30. Write a program to generate a square wave with 75% duty cycle on bit PB5.
31. Write a program to generate a square wave with 80% duty cycle on bit PC7.
32. Write a program to monitor PB4. When it goes HIGH, the program will generate a sound (square wave of 50% duty cycle) on pin PB7.
33. Write a program to monitor PC1. When it goes LOW, the program will send the value $55 to PD.
34. What register does the carry flag belong to?
35. What bit address is assigned to the Z flag?
36. Which of the following instructions are valid? If valid, indicate which bit is altered.

(a) SBI PORTB,1   (b) CBI PORTC.3   (c) SBR SREG,1
(d) SBR R20,1   (e) BLD PORTD,0   (f) BST R20,3
(g) CLV R3   (h) CLN

37. "SBI PORTB, 0" is a(n) _____ (valid, invalid) instruction.
38. Which of the I/O ports of PORTB, PORTC, and PORTD are bit-addressable?
39. Which of the general purpose registers are bit-addressable?
40. Give an instruction to clear the carry flag.
41. Show how would you check whether the C flag is HIGH.
42. Show how would you check whether the Z flag is HIGH.
43. Give the bit locations in the status register assigned to the flag bits C, Z, H, and V.
44. True or false. I/O registers are not bit-addressable.
45. Write instructions to save the C flag bit in bit 4 of location 0x60.
46. Write instructions to save the H flag bit in bit 2 of location 0x160.
47. Write instructions to save the Z flag bit in bit 7 of location 0x120.
48. Write instructions to see whether the D0 and D1 bits of register R20 are LOW.

If so, divide register R20 by 4.

49. Write a program to see whether the D7 bit of register R25 is HIGH. If so, send 0xFF to PORTD.

50. Write a program to set HIGH all the bits of the PORTC I/O register using the following methods:

(a) byte addresses (b) bit addresses

51. Write a program to see whether the R24 register is divisible by 8.

SECTION 6.6: ACCESSING EEPROM IN AVR

52. Write a program that writes 0 in EEPROM locations $0 to $30.

53. Write a program to copy 10 bytes of data starting at RAM address $80 to EEPROM locations starting at $10.

54. Write a program to copy 10 bytes of data starting at EEPROM address $10 to RAM locations starting at $80.

55. Write a program that calculates the sum of the values of locations $10 to $20 of EEPROM.

SECTION 6.7: CHECKSUM AND ASCII SUBROUTINES

56. Find the checksum byte for the following ASCII message: "Hello"

57. In each of the following cases perform checksum calculation to see if data is corrupted or not.

(a) Data = $65, $09, and $95; checksum = $23.

(b) Data = $71, $69, $38, and $81; checksum = $6D.

58. True or false. If we add all bytes, including the checksum byte, and the result is $00, there is no error in the data.

59. Write a program to (a) get the data "Hello, my fellow world citizens" from program ROM, (b) calculate the checksum byte, and (c) test the checksum byte for any data error.

60. To display data on LCD or PC monitors, it must be in _____ (binary, BCD, ASCII).

61. Write a program to convert a series of packed BCD numbers to ASCII. Assume that the packed BCD is located in ROM locations starting at $700. Place the ASCII codes in RAM locations starting at $40.

```
            .ORG $700
MYDATA:     .DB    $76, $87, $98, $43
```

62. Write a program to convert a series of ASCII numbers to packed BCD. Assume that the ASCII data is located in ROM locations starting at $300. Place the BCD data in RAM locations starting at $60.

```
            .ORG $300
MYDATA:     .DB    "87675649"
```

63. Write a program to get an 8-bit binary number from PORTD, convert it to ASCII, and save the result in RAM locations $40, $41, and $42. What is the result if PORTD has 1000 1101 binary as input?

SECTION 6.8: MACROS

64. Give two advantages of macros.
65. Which uses more program Flash ROM space: a macro or a subroutine?

ANSWERS TO REVIEW QUESTIONS

SECTION 6.1: INTRODUCING SOME MORE ASSEMBLER DIRECTIVES

1. R20 = 0x10&0x91 = 0x10
    R21 = 0x91|0x14 = 0x95
    Z = ZH:ZL = 0x123
2. It sets the V and Z flags and clears the other flags.
3.  ```
    LDI     R16,(1<<SREG_Z)|(1<<SREG_C)
    OUT     SREG,R16    ;set Z and C, clear others
    ```
4. 15900 is $3E1C in hex. Therefore, TCNT1H is loaded with $3E and $1C is loaded into TCNT1L.

SECTION 6.2: REGISTER AND DIRECT ADDRESSING MODES

1. No
2.  ```
    LDI R20,0b10000000
    OUT SPL,R20
    ```
3. True
4. False
5. True
6. True

SECTION 6.3: REGISTER INDIRECT ADDRESSING MODE

1. Indirect
2. R26
3. 16
4.
```
        .INCLUDE "M32DEF.INC"
        LDI     XL,$90
        LDI     XH,$00
        LDI     YL,$00
        LDI     YH,$2
        LDI     R16,11
        LDI     R22,2
L1:     LD      R20,X+
        ADD     R20,R22
        ST      Y+,R20
        DEC     R16
        BRNE    L1
HERE:   RJMP    HERE
```
5. X, Y, Z

SECTION 6.4: LOOK-UP TABLE AND TABLE PROCESSING

1. Z
2. Rd
3. 16 bits, 32K words

4. Z
5. ELPM can address up to 4M words of Flash memory.
6. When we want to be able to change the look-up table
7. True

SECTION 6.5: BIT-ADDRESSABILITY

1. False
2. True
3. a, b, and d
4. BST R23,1 ;T = R23.1
   BRTS L1   ;branch if T = 1 (branch if R23.1 is high)

   ....
   L1:
5. CLC
6. (a) It sets to HIGH bit 0 of R16.
   (b) It clears bits 0, 1, and 2 of R30.
   (c) It stores bit 2 of R19 to the T flag.
   (d) It sets to HIGH bit 4 of PORTB.
   (e) It clears bit 1 of the status register.
   (f) It clears the I flag of the status register.

SECTION 6.6: ACCESSING EEPROM IN AVR

1. False
2. True
3. False
4. True
5. False

SECTION 6.8: MACROS

1. Macro programming can save the programmer time by allowing a set of frequently repeated instructions to be invoked within the program with a single line. This can also make the code easier to read.
2. The three parts of a macro are the .MACRO directive, the body, and the .ENDMACRO directive.
3. The macro definition is the list of statements the macro will perform. It begins with the .MACRO directive and ends with the .ENDMACRO directive. The macro is invoked whenever it is called from within an Assembly language program. The macro is expanded when the Assembly program replaces the line invoking the macro with the Assembly language code in the body of the macro.

CHAPTER 7

AVR PROGRAMMING IN C

Why program the AVR in C?

Compilers produce hex files that we download into the Flash of the microcontroller. The size of the hex file produced by the compiler is one of the main concerns of microcontroller programmers because microcontrollers have limited on-chip Flash. For example, the Flash space for the ATmega16 is 16K bytes.

How does the choice of programming language affect the compiled program size? While Assembly language produces a hex file that is much smaller than C, programming in Assembly language is often tedious and time consuming. On the other hand, C programming is less time consuming and much easier to write, but the hex file size produced is much larger than if we used Assembly language. The following are some of the major reasons for writing programs in C instead of Assembly:

1. It is easier and less time consuming to write in C than in Assembly.
2. C is easier to modify and update.
3. You can use code available in function libraries.
4. C code is portable to other microcontrollers with little or no modification.

Several third-party companies develop C compilers for the AVR microcontroller. Our goal is not to recommend one over another, but to provide you with the fundamentals of C programming for the AVR. You can use the compiler of your choice for the chapter examples and programs. For this book we have chosen AVR GCC compiler to integrate with AVR Studio. At the time of the writing of this book AVR GCC and AVR Studio are available as a free download from the Web. See http://www.MicroDigitalEd.com for tutorials on AVR Studio and the AVR GCC compiler.

C programming for the AVR is the main topic of this chapter. In Section 7.1, we discuss data types, and time delays. I/O programming is shown in Section 7.2. The logic operations AND, OR, XOR, inverter, and shift are discussed in Section 7.3. Section 7.4 describes ASCII and BCD conversions and checksums. In Section 7.5, data serialization for the AVR is shown. In Section 7.6, memory allocation in C is discussed.

SECTION 7.1: DATA TYPES AND TIME DELAYS IN C

In this section we first discuss C data types for the AVR and then provide code for time delay functions.

C data types for the AVR C

One of the goals of AVR programmers is to create smaller hex files, so it is worthwhile to re-examine C data types. In other words, a good understanding of C data types for the AVR can help programmers to create smaller hex files. In this section we focus on the specific C data types that are most common and widely used in AVR C compilers. Table 7-1 shows data types and sizes, but these may vary from one compiler to another.

Table 7-1: Some Data Types Widely Used by C Compilers

Data Type	Size in Bits	Data Range/Usage
unsigned char	8-bit	0 to 255
char	8-bit	−128 to +127
unsigned int	16-bit	0 to 65,535
int	16-bit	−32,768 to +32,767
unsigned long	32-bit	0 to 4,294,967,295
long	32-bit	−2,147,483,648 to +2,147,483,648
float	32-bit	±1.175e-38 to ±3.402e38
double	32-bit	±1.175e-38 to ±3.402e38

Unsigned char

Because the AVR is an 8-bit microcontroller, the character data type is the most natural choice for many applications. The unsigned char is an 8-bit data type that takes a value in the range of 0–255 (00–FFH). It is one of the most widely used data types for the AVR. In many situations, such as setting a counter value, where there is no need for signed data, we should use the unsigned char instead of the signed char.

In declaring variables, we must pay careful attention to the size of the data and try to use unsigned char instead of int if possible. Because the AVR microcontroller has a limited number of registers and data RAM locations, using int in place of char can lead to the need for more memory space. Such misuse of data types in compilers such as Microsoft Visual C++ for x86 IBM PCs is not a significant issue.

Remember that C compilers use the signed char as the default unless we put the keyword *unsigned* in front of the char (see Example 7-1). We can also use the unsigned char data type for a string of ASCII characters, including extended ASCII characters. Example 7-2 shows a string of ASCII characters. See Example 7-3 for toggling a port 200 times.

Example 7-1

Write an AVR C program to send values 00–FF to Port B.

Solution:

```
#include <avr/io.h>                    //standard AVR header

int main(void)
{
  unsigned char z;
  DDRB = 0xFF;                          //PORTB is output
  for(z = 0; z <= 255; z++)
    PORTB = z;

  return 0;
}
//Notice that the program never exits the for loop because if you
//increment an unsigned char variable when it is 0xFF, it will
//become zero.
```

Example 7-2

Write an AVR C program to send hex values for ASCII characters of 0, 1, 2, 3, 4, 5, A, B, C, and D to Port B.

Solution:

```c
#include <avr/io.h>                    //standard AVR header

int main(void)                         //the code starts from here
{
    unsigned char myList[]= "012345ABCD";
    unsigned char z;
    DDRB = 0xFF;                       //PORTB is output
    for(z=0; z<10; z++)                //repeat 10 times and increment z
        PORTB = myList[z];             //send the character to PORTB

    while(1);                          //needed if running on a trainer
    return 0;
}
```

Example 7-3

Write an AVR C program to toggle all the bits of Port B 200 times.

Solution:

```c
//toggle PB 200 times
#include <avr/io.h>                    //standard AVR header

int main(void)                         //the code starts from here
{
    DDRB = 0xFF;                       //PORTB is output
    PORTB = 0xAA;                      //PORTB is 10101010
    unsigned char z;

    for(z=0; z < 200; z++)             //run the next line 200 times
        PORTB = ~ PORTB;               //toggle PORTB

    while(1);                          //stay here forever
    return 0;
}
```

Signed char

The signed char is an 8-bit data type that uses the most significant bit (D7 of D7–D0) to represent the − or + value. As a result, we have only 7 bits for the magnitude of the signed number, giving us values from −128 to +127. In situations where + and − are needed to represent a given quantity such as temperature, the use of the signed char data type is necessary (see Example 7-4).

Again, notice that if we do not use the keyword *unsigned,* the default is the signed value. For that reason we should stick with the unsigned char unless the data needs to be represented as signed numbers.

258

Example 7-4

Write an AVR C program to send values of –4 to +4 to Port B.

Solution:

```
#include <avr/io.h>                    //standard AVR header

int main(void)
{
  char mynum[] = {-4,-3,-2,-1,0,+1,+2,+3,+4};
  unsigned char z;

  DDRB = 0xFF;                         //PORTB is output

  for(z=0; z<=8; z++)
    PORTB = mynum[z];

  while(1);                            //stay here forever
  return 0;
}
```

Run the above program on your simulator to see how PORTB displays values of FCH, FDH, FEH , FFH, 00H, 01H, 02H, 03H, and 04H (the hex values for –4, –3, –2, –1, 0, 1, etc.). See Chapter 5 for discussion of signed numbers.

Unsigned int

The unsigned int is a 16-bit data type that takes a value in the range of 0 to 65,535 (0000–FFFFH). In the AVR, unsigned int is used to define 16-bit variables such as memory addresses. It is also used to set counter values of more than 256. Because the AVR is an 8-bit microcontroller and the int data type takes two bytes of RAM, we must not use the int data type unless we have to. Because registers and memory accesses are in 8-bit chunks, the misuse of int variables will result in larger hex files, slower execution of program, and more memory usage. Such misuse is not a problem in PCs with 512 megabytes of memory, the 32-bit Pentium's registers and memory accesses, and a bus speed of 133 MHz. For AVR programming, however, do not use signed int in places where unsigned char will do the job. Of course, the compiler will not generate an error for this misuse, but the overhead in hex file size will be noticeable. Also, in situations where there is no need for signed data (such as setting counter values), we should use unsigned int instead of signed int. This gives a much wider range for data declaration. Again, remember that the C compiler uses signed int as the default unless we specify the keyword *unsigned*.

Signed int

Signed int is a 16-bit data type that uses the most significant bit (D15 of D15–D0) to represent the – or + value. As a result, we have only 15 bits for the magnitude of the number, or values from –32,768 to +32,767.

Other data types

The unsigned int is limited to values 0–65,535 (0000–FFFFH). The AVR C compiler supports long data types, if we want values greater than 16-bit. Also, to deal with fractional numbers, most AVR C compilers support float and double data types. See Examples 7-5 and 7-6.

Example 7-5

Write an AVR C program to toggle all bits of Port B 50,000 times.

Solution:
```
#include <avr/io.h>                    //standard AVR header
int main(void)
{
  unsigned int z;
  DDRB = 0xFF;                         //PORTB is output

  for(z=0; z<50000; z++)
  {
    PORTB = 0x55;
    PORTB = 0xAA;
  }

  while(1);                            //stay here forever
  return 0;
}
```

Run the above program on your simulator to see how Port B toggles continuously. Notice that the maximum value for unsigned int is 65,535.

Example 7-6

Write an AVR C program to toggle all bits of Port B 100,000 times.

Solution:
```
//toggle PB 100,00 times
#include <avr/io.h>                    //standard AVR header
int main(void)
{
  unsigned long z;                     //long is used because it should
                                       //store more than 65535.
  DDRB = 0xFF;                         //PORTB is output

  for(z=0; z<100000; z++){
    PORTB = 0x55;
    PORTB = 0xAA;
  }

  while(1);                            //stay here forever
  return 0;
}
```

Time delay

There are three ways to create a time delay in AVR C

1. Using a simple `for` loop
2. Using predefined C functions
3. Using AVR timers

In creating a time delay using a for loop, we must be mindful of two factors that can affect the accuracy of the delay:

1. The crystal frequency connected to the XTAL1–XTAL2 input pins is the most important factor in the time delay calculation. The duration of the clock period for the instruction cycle is a function of this crystal frequency.

2. The second factor that affects the time delay is the compiler used to compile the C program. When we program in Assembly language, we can control the exact instructions and their sequences used in the delay subroutine. In the case of C programs, it is the C compiler that converts the C statements and functions to Assembly language instructions. As a result, different compilers produce different code. In other words, if we compile a given C program with different compilers, each compiler produces different hex code.

For the above reasons, when we use a loop to write time delays for C, we must use the oscilloscope to measure the exact duration. Look at Example 7-7. Notice that most compilers do some code optimization before generating a .hex file. In this process they may omit the delay loop because it does not do anything other than wasting CPU time. In these compilers, you have to set the level of optimization to zero (none). To see how you can set the level of optimization for WinAVR and AVR Studio, refer to www.MicroDigitalEd.com.

Example 7-7

Write an AVR C program to toggle all the bits of Port B continuously with a 100 ms delay. Assume that the system is ATmega 32 with XTAL = 8 MHz.

Solution:

```
#include <avr/io.h>                    //standard AVR header
void delay100ms(void)
{
  unsigned int i;
  for(i=0; i<42150; i++);             //try different numbers on your
}                                     //compiler and examine the result.

int main(void)
{
  DDRB = 0xFF;                         //PORTB is output
  while (1)
  {
    PORTB = 0xAA;
    delay100ms();
    PORTB = 0x55;
    delay100ms();
  }
  return 0;
}
```

Another way of generating time delay is to use predefined functions such as _delay_ms() and _delay_us() defined in delay.h in WinAVR or delay_ms() and delay_us() defined in delay.h in CodeVision. The only drawback of using these functions is the portability problem. Because different compilers do not use the same name for delay functions, you have to change every place in which the delay functions are used, if you want to compile your program on another compiler. To overcome this problem, programmers use macro or wrapper function. Wrapper functions do nothing more than call the predefined delay function. If you use wrapper functions and decide to change your compiler, instead of changing all instances of predefined delay functions, you simply change the wrapper function. Look at Example 7-8. Notice that calling a wrapper function may take some microseconds.

The use of the AVR timer to create time delays will be discussed in Chapter 9.

Example 7-8

Write an AVR C program to toggle all the pins of Port C continuously with a 10 ms delay. Use a predefined delay function in Win AVR.

Solution:

```
#include <util/delay.h>              //delay loop functions
#include <avr/io.h>                  //standard AVR header

int main(void)
{
    void delay_ms(int d)            //delay in d microseconds
    {
        _delay_ms(d);
    }
    DDRB = 0xFF;                     //PORTA is output
    while (1){
        PORTB = 0xFF;
        delay_ms(10);
        PORTB = 0x55;
        delay_ms(10);
    }
    return 0;
}
```

Review Questions

1. Give the magnitude of the unsigned char and signed char data types.
2. Give the magnitude of the unsigned int and signed int data types.
3. If we are declaring a variable for a person's age, we should use the ___ data type.
4. True or false. Using predefined functions of compilers to create a time delay is not recommended if you want your code to be portable to other compilers.
5. Give two factors that can affect the delay size.

SECTION 7.2: I/O PROGRAMMING IN C

As we stated in Chapter 4, all port registers of the AVR are both byte accessible and bit accessible. In this section we look at C programming of the I/O ports for the AVR. We look at both byte and bit I/O programming.

Byte size I/O

To access a PORT register as a byte, we use the PORTx label where x indicates the name of the port. We access the data direction registers in the same way, using DDRx to indicate the data direction of port x. To access a PIN register as a byte, we use the PINx label where x indicates the name of the port. See Examples 7-9, 7-10, and 7-11.

Example 7-9

LEDs are connected to pins of Port B. Write an AVR C program that shows the count from 0 to FFH (0000 0000 to 1111 1111 in binary) on the LEDs.

Solution:

```c
#include <avr/io.h>                    //standard AVR header
int main(void)
{
  DDRB = 0xFF;                         //Port B is output
  while (1)
  {
    PORTB = PORTB + 1;
  }
  return 0;
}
```

Example 7-10

Write an AVR C program to get a byte of data from Port B, and then send it to Port C.

Solution:

```c
#include <avr/io.h>                    //standard AVR header
int main(void)
{
  unsigned char temp;

  DDRB = 0x00;                         //Port B is input
  DDRC = 0xFF;                         //Port C is output

  while(1)
  {
    temp = PINB;
    PORTC = temp;
  }
  return 0;
}
```

Example 7-11

Write an AVR C program to get a byte of data from Port C. If it is less than 100, send it to Port B; otherwise, send it to Port D.

Solution:

```c
#include <avr/io.h>                        //standard AVR header
int main(void)
{
  DDRC = 0;                                //Port C is input
  DDRB = 0xFF;                             //Port B is output
  DDRD = 0xFF;                             //Port D is output
  unsigned char temp;
  while(1)
  {
    temp = PINC;                           //read from PINB
    if ( temp < 100 )
      PORTB = temp;
    else
      PORTD = temp;
  }
  return 0;
}
```

Bit size I/O

The I/O ports of ATmega32 are bit-accessible. But some AVR C compilers do not support this feature, and the others do not have a standard way of using it. For example, the following line of code can be used in CodeVision to set the first pin of Port B to one:

```c
PORTB.0 = 1;
```

but it cannot be used in other compilers such as WinAVR.

To write portable code that can be compiled on different compilers, we must use AND and OR bit-wise operations to access a single bit of a given register.

So, you can access a single bit without disturbing the rest of the byte. In next section you will see how to mask a bit of a byte. You can use masking for both bit-accessible and byte-accessible ports and registers.

Review Questions

1. Write a short program that toggles all bits of Port C.
2. True or false. All bits of Port B are bit addressable.
3. Write a short program that toggles bit 2 of Port C using the functions of your compiler.
4. True or false. To access the data direction register of Port B, we use DDRB.

SECTION 7.3: LOGIC OPERATIONS IN C

One of the most important and powerful features of the C language is its ability to perform bit manipulation. Because many books on C do not cover this important topic, it is appropriate to discuss it in this section. This section describes the action of bit-wise logic operators and provides some examples of how they are used.

Bit-wise operators in C

While every C programmer is familiar with the logical operators AND (&&), OR (||), and NOT (!), many C programmers are less familiar with the bit-wise operators AND (&), OR (|), EX-OR (^), inverter (~), shift right (>>), and shift left (<<). These bit-wise operators are widely used in software engineering for embedded systems and control; consequently, their understanding and mastery are critical in microcontroller-based system design and interfacing. See Table 7-2.

Table 7-2: Bit-wise Logic Operators for C

		AND	OR	EX-OR	Inverter
A	B	A&B	A|B	A^B	Y=~B
0	0	0	0	0	1
0	1	0	1	1	0
1	0	0	1	1	
1	1	1	1	0	

The following shows some examples using the C bit-wise operators:
1. 0x35 & 0x0F = 0x05   /* ANDing */
2. 0x04 | 0x68 = 0x6C   /* ORing   */
3. 0x54 ^ 0x78 = 0x2C   /* XORing */
4. ~0x55 = 0xAA   /* Inverting 55H */

Examples 7-12 through 7-20 show how the bit-wise operators are used in C. Run these programs on your simulator and examine the results.

Example 7-12

Run the following program on your simulator and examine the results.

```
#include <avr/io.h>          //standard AVR header
int main(void)
{
  DDRB = 0xFF;               //make Port B output
  DDRC = 0xFF;               //make Port C output
  DDRD = 0xFF;               //make Port D output
  PORTB = 0x35 & 0x0F;       //ANDing
  PORTC = 0x04 | 0x68;       //ORing
  PORTD = 0x54 ^ 0x78;       //XORing
  PORTB = ~0x55;             //inverting
  while (1);
  return 0;
}
```

Example 7-13

Write an AVR C program to toggle only bit 4 of Port B continuously without disturbing the rest of the pins of Port B.

Solution:

```c
#include <avr/io.h>                    //standard AVR header

int main(void)
{
  DDRB = 0xFF;                         //PORTB is output

  while(1)
  {
    PORTB = PORTB | 0b00010000;    //set bit 4 (5th bit) of PORTB
    PORTB = PORTB & 0b11101111;    //clear bit 4 (5th bit) of PORTB
  }

  return 0;
}
```

Example 7-14

Write an AVR C program to monitor bit 5 of port C. If it is HIGH, send 55H to Port B; otherwise, send AAH to Port B.

Solution:

```c
#include <avr/io.h>               //standard AVR header

int main(void)
{
  DDRB = 0xFF;                //PORTB is output
  DDRC = 0x00;                //PORTC is input
  DDRD = 0xFF;                //PORTB is output

  while(1)
  {
    if (PINC & 0b00100000)    //check bit 5 (6th bit) of PINC
      PORTB = 0x55;
    else
      PORTB = 0xAA;
  }

  return 0;
}
```

Example 7-15

A door sensor is connected to bit 1 of Port B, and an LED is connected to bit 7 of Port C. Write an AVR C program to monitor the door sensor and, when it opens, turn on the LED.

Solution:

```
#include <avr/io.h>                    //standard AVR header

int main(void)
{
  DDRB = DDRB & 0b11111101;            //pin 1 of Port B is input
  DDRC = DDRC | 0b10000000;            //pin 7 of Port C is output

  while(1)
  {
    if (PINB & 0b00000010)             //check pin 1 (2nd pin) of PINB
      PORTC = PORTC | 0b10000000;      //set pin 7 (8th pin) of PORTC
    else
      PORTC = PORTC & 0b01111111;      //clear pin 7 (8th pin) of PORTC
  }
  return 0;
}
```

Example 7-16

The data pins of an LCD are connected to Port B. The information is latched into the LCD whenever its Enable pin goes from HIGH to LOW. The enable pin is connected to pin 5 of Port C (6th pin). Write a C program to send "The Earth is but One Country" to this LCD.

Solution:

```
#include <avr/io.h>                    //standard AVR header

int main(void)
{
  unsigned char message[] = "The Earth is but One Country";
  unsigned char z;

  DDRB = 0xFF;                         //Port B is output
  DDRC = DDRC | 0b00100000;            //pin 5 of Port C is output

  for ( z = 0; z < 28; z++)
  {
    PORTB = message[z];
    PORTC = PORTC | 0b00100000;        //pin LCD_EN of Port C is 1
    PORTC = PORTC & 0b11011111;        //pin LCD_EN of Port C is 0
  }
  while (1);
  return 0;
}
//In Chapter 12 we will study more about LCD interfacing
```

Example 7-17

Write an AVR C program to read pins 1 and 0 of Port B and issue an ASCII character to Port D according to the following table:

	pin1	pin0	
	0	0	send '0' to Port D (notice ASCII '0' is 0x30)
	0	1	send '1' to Port D
	1	0	send '2' to Port D
	1	1	send '3' to Port D

Solution:

```c
#include <avr/io.h>                //standard AVR header

int main(void)
{
  unsigned char z;
  DDRB = 0;                        //make Port B an input
  DDRD = 0xFF;                     //make Port D an output
  while(1)                         //repeat forever
  {
    z = PINB;                      //read PORTB
    z = z & 0b00000011;            //mask the unused bits
    switch(z)                      //make decision
    {
      case(0):
      {
        PORTD = '0';    //issue ASCII 0
        break;
      }
      case(1):
      {
        PORTD = '1';    //issue ASCII 1
        break;
      }
      case(2):
      {
        PORTD = '2';    //issue ASCII 2
        break;
      }
      case(3):
      {
        PORTD = '3';    //issue ASCII 3
        break;
      }
    }
  }
  return 0;
}
```

Example 7-18

Write an AVR C program to monitor bit 7 of Port B. If it is 1, make bit 4 of Port B input; otherwise, change pin 4 of Port B to output.

Solution:

```c
#include <avr/io.h>                          //standard AVR header

int main(void)
{
  DDRB = DDRB & 0b01111111;                  //bit 7 of Port B is input

  while (1)
  {
    if(PINB & 10000000)
      DDRB = DDRB & 0b11101111;              //bit 4 of Port B is input
    else
      DDRB = DDRB | 0b00010000;              //bit 4 of Port B is output
  }

  return 0;
}
```

Example 7-19

Write an AVR C program to get the status of bit 5 of Port B and send it to bit 7 of port C continuously.

Solution:

```c
#include <avr/io.h>                          //standard AVR header

int main(void)
{
  DDRB = DDRB & 0b11011111;                  //bit 5 of Port B is input
  DDRC = DDRC | 0b10000000;                  //bit 7 of Port C is output

  while (1)
  {
    if(PINB & 0b00100000 )
      PORTC = PORTC | 0b10000000;  //set bit 7 of Port C to 1
    else
      PORTC = PORTC & 0b01111111;  //clear bit 7 of Port C to 0
  }
  return 0;
}
```

Example 7-20

Write an AVR C program to toggle all the pins of Port B continuously.
(a) Use the inverting operator.   (b) Use the EX-OR operator.
Solution:

(a)
```
#include <avr/io.h>                    //standard AVR header
int main(void)
{
     DDRB = 0xFF;                      //Port B is output
     PORTB = 0xAA;
     while (1)
          PORTB = ~ PORTB;            //toggle PORTB
     return 0;
}
```

(b)
```
#include <avr/io.h>                    //standard AVR header
int main(void)
{
     DDRB = 0xFF;                      //Port B is output
     PORTB = 0xAA;
     while (1)
          PORTB = PORTB ^ 0xFF;
     return 0;
}
```

```
4:                       {
+00000049:    EF8F      SER      R24              Set Register
+0000004A:    BB87      OUT      0x17,R24         Out to I/O location
6:                 PORTB = 0xAA;
+0000004B:    EA8A      LDI      R24,0xAA         Load immediate
+0000004C:    BB88      OUT      0x18,R24         Out to I/O location
9:                 PORTB =~ PORTB ;
+0000004D:    B388      IN       R24,0x18         In from I/O location
+0000004E:    9580      COM      R24              One's complement
+0000004F:    BB88      OUT      0x18,R24         Out to I/O location
+00000050:    CFFC      RJMP     PC-0x0003        Relative jump
+00000051:    94F8      CLI                       Global Interrupt Disab
+00000052:    CFFF      RJMP     PC-0x0000        Relative jump
```
Disassembly of Example 7-20 Part a

```
4:                       {
+00000049:    EF8F      SER      R24              Set Register
+0000004A:    BB87      OUT      0x17,R24         Out to I/O location
6:                 PORTB = 0xAA;
+0000004B:    EA8A      LDI      R24,0xAA         Load immediate
+0000004C:    BB88      OUT      0x18,R24         Out to I/O location
9:                 PORTB = PORTB ^ 0xFF ;
+0000004D:    B388      IN       R24,0x18         In from I/O location
+0000004E:    9580      COM      R24              One's complement
+0000004F:    BB88      OUT      0x18,R24         Out to I/O location
+00000050:    CFFC      RJMP     PC-0x0003        Relative jump
+00000051:    94F8      CLI                       Global Interrupt Disab
+00000052:    CFFF      RJMP     PC-0x0000        Relative jump
```
Disassembly of Example 7-20 Part b

270

Examine the Assembly output for parts (a) and (b) of Example 7-20. You will notice that the generated codes are the same because they do exactly the same thing.

Compound assignment operators in C

To reduce coding (typing) we can use compound statements for bit-wise operators in C. See Table 7-3 and Example 7-21.

Table 7-3: Compound Assignment Operator in C

Operation	Abbreviated Expression	Equal C Expression
And assignment	a &= b	a = a & b
OR assignment	a |= b	a = a | b

Example 7-21

Using bitwise compound assignment operators
(a) Rewrite Example 7-18   (b) Rewrite Example 7-19

Solution:
(a)

```
#include <avr/io.h>            //standard AVR header
int main(void)
{
  DDRB &= DDRB & 0b11011111;  //bit 5 of Port B is input
  while (1)
  {
    if(PINB & 0b00100000)
      DDRB &= 0b11101111;     //bit 4 of Port B is input
    else
      DDRB |= 0b00010000;     //bit 4 of Port B is output
  }
  return 0;
}
```

(b)

```
#include <avr/io.h>            //standard AVR header
int main(void)
{
  DDRB &= 0b11011111;         //bit 5 of Port B is input
  DDRC |= 0b10000000;         //bit 7 of Port C is output

  while (1)
  {
    if(PINB & 0b00100000)
      PORTC |=  0b10000000;   //set bit 7 of Port C to 1
    else
      PORTC &= 0b01111111;    //clear bit 7 of Port C to 0
  }
  return 0;
}
```

CHAPTER 7: AVR PROGRAMMING IN C

Bit-wise shift operation in C

There are two bit-wise shift operators in C. See Table 7-4.

Table 7-4: Bit-wise Shift Operators for C

Operation	Symbol	Format of Shift Operation
Shift right	>>	data >> number of bits to be shifted right
Shift left	<<	data << number of bits to be shifted left

The following shows some examples of shift operators in C:

1. 0b00010000 >> 3 = 0b00000010   /* shifting right 3 times */
2. 0b00010000 << 3 = 0b10000000   /* shifting left 3 times */
3. 1 << 3 = 0b00001000   /* shifting left 3 times */

Bit-wise shift operation and bit manipulation

Reexamine the last 10 examples. To do bit-wise I/O operation in C, we need numbers like 0b00100000 in which there are seven zeroes and one one. Only the position of the one varies in different programs. To leave the generation of ones and zeros to the compiler and improve the code clarity, we use shift operations. For example, instead of writing "0b00100000" we can write "0b00000001 << 5" or we can write simply "1<<5".

Sometimes we need numbers like 0b11101111. To generate such a number, we do the shifting first and then invert it. For example, to generate 0b11101111 we can write ~ (1<<5). See Example 7-22.

Example 7-22

Write code to generate the following numbers:

(a) A number that has only a one in position D7
(b) A number that has only a one in position D2
(c) A number that has only a one in position D4
(d) A number that has only a zero in position D5
(e) A number that has only a zero in position D3
(f) A number that has only a zero in position D1

Solution:

(a) (1<<7)
(b) (1<<2)
(c) (1<<4)
(d) ~(1<<5)
(e) ~(1<<3)
(f) ~(1<<1)

Examples 7-23 and 7-24 are the same as Examples 7-18 and 7-19, but they use shift operation.

Example 7-23

Write an AVR C program to monitor bit 7 of Port B. If it is 1, make bit 4 of Port B input; else, change pin 4 of Port B to output.

Solution:

```
#include <avr/io.h>                        //standard AVR header

int main(void)
{
   DDRB = DDRB & ~(1<<7);                   //bit 7 of Port B is input

   while (1)
   {
      if(PINB & (1<<7))
         DDRB = DDRB & ~(1<<4);             //bit 4 of Port B is input
      else
         DDRB = DDRB | (1<<4);              //bit 4 of Port B is output
   }

   return 0;
}
```

Example 7-24

Write an AVR C program to get the status of bit 5 of Port B and send it to bit 7 of port C continuously.

Solution:

```
#include <avr/io.h>                        //standard AVR header

int main(void)
{
   DDRB = DDRB & ~(1<<5);                   //bit 5 of Port B is input
   DDRC = DDRC | (1<<7);                    //bit 7 of Port C is output

   while (1)
   {
      if(PINB & (1<<5) )
         PORTC = PORTC | (1<<7);            //set bit 7 of Port C to 1
      else
         PORTC = PORTC & ~(1<<7);           //clear bit 7 of Port C to 0
   }
   return 0;
}
```

CHAPTER 7: AVR PROGRAMMING IN C

As we mentioned before, bit-wise shift operation can be used to increase code clarity. See Example 7-25.

Example 7-25

A door sensor is connected to the port B pin 1, and an LED is connected to port C pin 7. Write an AVR C program to monitor the door sensor and, when it opens, turn on the LED.

Solution:

```c
#include <avr/io.h>                    //standard AVR header
#define LED 7
#define SENSOR 1

int main(void)
{
  DDRB = DDRB & ~(1<<SENSOR);         //SENSOR pin is input
  DDRC = DDRC | (1<< LED);            //LED pin is output

  while(1)
  {
    if (PINB & (1 << SENSOR))         //check SENSOR pin of PINB
      PORTC = PORTC | (1<<LED);       //set LED pin of Port C
    else
      PORTC = PORTC & ~(1<<LED);      //clear LED pin of Port C
  }
  return 0;
}
```

Notice that to generate more complicated numbers, we can OR two simpler numbers. For example, to generate a number that has a one in position D7 and another one in position D4, we can OR a number that has only a one in position D7 with a number that has only a one in position D4. So we can simply write (1<<7)|(1<<4). In future chapters you will see how we use this method.

Review Questions

1. Find the content of PORTB after the following C code in each case:
   (a) PORTB=0x37&0xCA;
   (b) PORTB=0x37|0xCA;
   (c) PORTB=0x37^0xCA;
2. To mask certain bits we must AND them with _____.
3. To set high certain bits we must OR them with _____.
4. EX-ORing a value with itself results in _____.
5. Find the contents of PORTC after execution of the following code:
   ```c
   PORTC = 0;
   PORTC = PORTC | 0x99;
   PORTC = ~PORTC;
   ```
6. Find the contents of PORTC after execution of the following code:
   ```c
   PORTC = ~(0<<3);
   ```

SECTION 7.4: DATA CONVERSION PROGRAMS IN C

Recall that BCD numbers were discussed in Chapters 5 and 6. As stated there, many newer microcontrollers have a real-time clock (RTC) where the time and date are kept even when the power is off. Very often the RTC provides the time and date in packed BCD. To display them, however, we must convert them to ASCII. In this section we show the application of logic and rotate instructions in the conversion of BCD and ASCII.

ASCII numbers

On ASCII keyboards, when the "0" key is activated, "0011 0000" (30H) is provided to the computer. Similarly, 31H (0011 0001) is provided for the "1" key, and so on, as shown in Table 7-5.

Table 7-5: ASCII Code for Digits 0–9

Key	ASCII (hex)	Binary	BCD (unpacked)
0	30	011 0000	0000 0000
1	31	011 0001	0000 0001
2	32	011 0010	0000 0010
3	33	011 0011	0000 0011
4	34	011 0100	0000 0100
5	35	011 0101	0000 0101
6	36	011 0110	0000 0110
7	37	011 0111	0000 0111
8	38	011 1000	0000 1000
9	39	011 1001	0000 1001

Packed BCD to ASCII conversion

The RTC provides the time of day (hour, minute, second) and the date (year, month, day) continuously, regardless of whether the power is on or off. This data is provided in packed BCD. To convert packed BCD to ASCII, you must first convert it to unpacked BCD. Then the unpacked BCD is tagged with 011 0000 (30H). The following demonstrates converting from packed BCD to ASCII. See also Example 7-26.

```
Packed BCD    Unpacked BCD        ASCII
0x29          0x02, 0x09          0x32, 0x39
00101001      00000010,00001001   00110010,00111001
```

ASCII to packed BCD conversion

To convert ASCII to packed BCD, you first convert it to unpacked BCD (to get rid of the 3), and then combine the numbers to make packed BCD. For example, 4 and 7 on the keyboard give 34H and 37H, respectively. The goal is to produce 47H or "0100 0111", which is packed BCD.

Key	ASCII	Unpacked BCD	Packed BCD
4	34	00000100	
7	37	00000111	01000111 or 47H

See Example 7-27.

Example 7-26

Write an AVR C program to convert packed BCD 0x29 to ASCII and display the bytes on PORTB and PORTC.

Solution:

```
#include <avr/io.h>                    //standard AVR header
int main(void)
{
  unsigned char x, y;
  unsigned char mybyte = 0x29;

  DDRB = DDRC = 0xFF;                   //make Ports B and C output
  x = mybyte & 0x0F;                    //mask upper 4 bits
  PORTB = x | 0x30;                     //make it ASCII
  y = mybyte & 0xF0;                    //mask lower 4 bits
  y = y >> 4;                           //shift it to lower 4 bits
  PORTC = y | 0x30;                     //make it ASCII

  return 0;
}
```

Example 7-27

Write an AVR C program to convert ASCII digits of '4' and '7' to packed BCD and display them on PORTB.

Solution:

```
#include <avr/io.h>              //standard AVR header

int main(void)
{
  unsigned char bcdbyte;
  unsigned char w = '4';
  unsigned char z = '7';
  DDRB = 0xFF;                   //make Port B an output
  w = w & 0x0F;                  //mask 3
  w = w << 4;                    //shift left to make upper BCD digit
  z = z & 0x0F;                  //mask 3
  bcdbyte = w | z;               //combine to make packed BCD
  PORTB = bcdbyte;

  return 0;
}
```

Checksum byte in ROM

To ensure the integrity of data, every system must perform the checksum calculation. When you transmit data from one device to another or when you save and restore data to a storage device you should perform the checksum calculation to ensure the integrity of the data. The checksum will detect any corruption of data. To ensure data integrity, the checksum process uses what is called a *checksum byte*. The checksum byte is an extra byte that is tagged to the end of a series of bytes of data. To calculate the checksum byte of a series of bytes of data, the following steps can be taken:

1. Add the bytes together and drop the carries.
2. Take the 2's complement of the total sum. This is the checksum byte, which becomes the last byte of the series.

To perform the checksum operation, add all the bytes, including the checksum byte. The result must be zero. If it is not zero, one or more bytes of data have been changed (corrupted). See Examples 7-28 through 7-30.

Example 7-28

Assume that we have 4 bytes of hexadecimal data: 25H, 62H, 3FH, and 52H.
(a) Find the checksum byte, (b) perform the checksum operation to ensure data integrity, and (c) if the second byte, 62H, has been changed to 22H, show how checksum detects the error.

Solution:

(a)   Find the checksum byte.

```
        25H
    +   62H
    +   3FH
    +   52H
    1  18H  (dropping carry of 1 and taking 2's complement, we get E8H)
```

(b)   Perform the checksum operation to ensure data integrity.

```
        25H
    +   62H
    +   3FH
    +   52H
    +   E8H
    2  00H  (dropping the carries we get 00, which means data is not corrupted)
```

(c)   If the second byte, 62H, has been changed to 22H, show how checksum detects the error.

```
        25H
    +   22H
    +   3FH
    +   52H
    +   E8H
    1  C0H  (dropping the carry, we get C0H, which means data is corrupted)
```

Example 7-29

Write an AVR C program to calculate the checksum byte for the data given in Example 7-28.

Solution:

```c
#include <avr/io.h>                    //standard AVR header
int main(void)
{
    unsigned char mydata[] = { 0x25,0x62,0x3F,0x52};
    unsigned char sum = 0;
    unsigned char x;
    unsigned char chksumbyte;

    DDRA = 0xFF;                       //make Port A output
    DDRB = 0xFF;                       //make Port B output
    DDRC = 0xFF;                       //make Port C output

    for(x=0;  x<4;  x++)
    {
        PORTA = mydata[x];        //issue each byte to PORTA
        sum = sum + mydata[x];    //add them together
        PORTB = sum;              //issue the sum to PORTB
    }
    chksumbyte = ~sum + 1;            //make 2's complement (invert +1)
    PORTC = chksumbyte;              //show the checksum byte
    return 0;
}
```

Example 7-30

Write a C program to perform step (b) of Example 7-28. If the data is good, send ASCII character 'G' to PORTD. Otherwise, send 'B' to PORTD.

Solution:

```c
#include <avr/io.h>                        //standard AVR header
int main(void)
{
    unsigned char mydata[] = { 0x25,0x62,0x3F,0x52,0xE8};
    unsigned char chksum = 0;
    unsigned char x;
    DDRD = 0xFF;                         //make Port D an output
    for(x=0;x<5;x++)
       chksum = chksum + mydata[x];     //add them together
    if(chksum == 0)
       PORTD = 'G';
    else
       PORTD = 'B';
    return 0;
}
```

Change one or two values in the mydata array and simulate the program to see the results.

Binary (hex) to decimal and ASCII conversion in C

The printf function is part of the standard I/O library in C and can do many things including converting data from binary (hex) to decimal, or vice versa. But printf takes a lot of memory space and increases your hex file substantially. For this reason, in systems based on the AVR microcontroller, it is better to know how to write our own conversion function instead of using printf.

One of the most widely used conversions is binary to decimal conversion. In devices such as ADCs (Analog-to-Digital Converters), the data is provided to the microcontroller in binary. In some RTCs, the time and dates are also provided in binary. In order to display binary data, we need to convert it to decimal and then to ASCII. Because the hexadecimal format is a convenient way of representing binary data, we refer to the binary data as hex. The binary data 00–FFH converted to decimal will give us 000 to 255. One way to do that is to divide it by 10 and keep the remainder, as was shown in Chapters 5 and 6. For example, 11111101 or FDH is 253 in decimal. The following is one version of an algorithm for conversion of hex (binary) to decimal:

Hex	Quotient	Remainder
FD/0A	19	3 (low digit) LSD
19/0A	2	5 (middle digit)
		2 (high digit) (MSD)

Example 7-31 shows the C program for the above algorithm.

Example 7-31

Write an AVR C program to convert 11111101 (FD hex) to decimal and display the digits on PORTB, PORTC, and PORTD.

Solution:

```
#include <avr/io.h>                      //standard AVR header
int main(void)
{
    unsigned char x, binbyte, d1, d2, d3;
    DDRB = DDRC = DDRD =0xFF;            //Ports B, C, and D output
    binbyte = 0xFD;                     //binary (hex) byte
    x = binbyte / 10;                   //divide by 10
    d1 = binbyte % 10;                  //find remainder (LSD)
    d2 = x % 10;                        //middle digit
    d3 = x / 10;                        //most-significant digit (MSD)
    PORTB = d1;
    PORTC = d2;
    PORTD = d3;

    return 0;
}
```

Many compilers have some predefined functions to convert data types. In Table 7-6 you can see some of them. To use these functions, the stdlib.h file should be included. Notice that these functions may vary in different compilers.

Table 7-6: Data Type Conversion Functions in C

Function signature	Description of functions
int atoi(char *str)	Converts the string str to integer.
long atol(char *str)	Converts the string str to long.
void itoa(int n, char *str)	Converts the integer n to characters in string str.
void ltoa(int n, char *str)	Converts the long n to characters in string str.
float atof(char *str)	Converts the characters from string str to float.

Review Questions

1. For the following decimal numbers, give the packed BCD and unpacked BCD representations:
   (a) 15   (b) 99
2. Show the binary and hex for "76".
3. 67H in BCD when converted to ASCII is ____H and ____H.
4. Does the following convert unpacked BCD to ASCII?
   ```
   mydata=0x09+0x30;
   ```
5. Why is the use of packed BCD preferable to ASCII?
6. Which takes more memory space to store numbers: packed BCD or ASCII?
7. In Question 6, which is more universal?
8. Find the checksum byte for the following values: 22H, 76H, 5FH, 8CH, 99H.
9. To test data integrity, we add the bytes together, including the checksum byte. The result must be equal to ____ if the data is not corrupted.
10. An ADC provides an output of 0010 0110. How do we display that on the screen?

SECTION 7.5: DATA SERIALIZATION IN C

Serializing data is a way of sending a byte of data one bit at a time through a single pin of a microcontroller. There are two ways to transfer a byte of data serially:

1. Using the serial port. In using the serial port, the programmer has very limited control over the sequence of data transfer. The details of serial port data transfer are discussed in Chapter 11.
2. The second method of serializing data is to transfer data one bit a time and control the sequence of data and spaces between them. In many new generations of devices such as LCD, ADC, and EEPROM, the serial versions are becoming popular because they take up less space on a printed circuit board. Although we can use standards such as I²C, SPI, and CAN, not all devices support such standards. For this reason we need to be familiar with data serialization using the C language.

Examine the next four examples to see how data serialization is done in C.

Example 7-32

Write an AVR C program to send out the value 44H serially one bit at a time via PORTC, pin 3. The LSB should go out first.

Solution:

```c
#include <avr/io.h>
#define serPin 3

int main(void)
{
    unsigned char conbyte = 0x44;
    unsigned char regALSB;
    unsigned char x;
    regALSB = conbyte;
    DDRC |= (1<<serPin);

    for(x=0;x<8;x++)
      {
        if(regALSB & 0x01)
            PORTC |= (1<<serPin);
        else
            PORTC &= ~(1<<serPin);
        regALSB = regALSB >> 1;
      }
    return 0;
}
```

Example 7-33

Write an AVR C program to send out the value 44H serially one bit at a time via PORTC, pin 3. The MSB should go out first.

Solution:

```c
#include <avr/io.h>
#define serPin 3
int main(void)
{
    unsigned char conbyte = 0x44;
    unsigned char regALSB;
    unsigned char x;
    regALSB = conbyte;
    DDRC |= (1<<serPin);
    for(x=0;x<8;x++)
      {
        if(regALSB & 0x80)
            PORTC |= (1<<serPin);
        else
            PORTC &= ~(1<<serPin);
        regALSB = regALSB << 1;
      }
    return 0;
}
```

Example 7-34

Write an AVR C program to bring in a byte of data serially one bit at a time via PORTC, pin 3. The LSB should come in first.

Solution:
```c
//Bringing in data via PC3 (SHIFTING RIGHT)
#include <avr/io.h>            //standard AVR header
#define serPin 3
int main(void)
{
    unsigned char x;
    unsigned char REGA=0;
    DDRC &= ~(1<<serPin);      //serPin as input
    for(x=0; x<8; x++)         //repeat for each bit of REGA
      {
        REGA = REGA >> 1;      //shift REGA to right one bit
        REGA |= (PINC &(1<<serPin)) << (7-serPin); //copy bit serPin
      }                        //of PORTC to MSB of REGA.
    return 0;
}
```

Example 7-35

Write an AVR C program to bring in a byte of data serially one bit at a time via PORTC, pin 3. The MSB should come in first.

Solution:

```c
#include <avr/io.h>                   //standard AVR header
#define serPin 3

int main(void)
{
    unsigned char x;
    unsigned char REGA=0;
    DDRC &= ~(1<<serPin);      //serPin as input
    for(x=0; x<8; x++)         //repeat for each bit of REGA
      {
        REGA = REGA << 1;      //shift REGA to left one bit
        REGA |= (PINC &(1<<serPin))>> serPin; //copy bit serPin of
      }                        //PORT C to LSB of REGA.
    return 0;
}
```

SECTION 7.6: MEMORY ALLOCATION IN C

Using program (code) space for predefined fixed data is a widely used option in the AVR, as we saw in Chapter 6. In that chapter we saw how to use Assembly language programs to access the data stored in ROM. Next, we do the same thing with C language.

282

Flash, RAM, and EEPROM data space in the AVR

In the AVR we have three spaces in which to store data. They are as follows:

1. The 64K bytes of SRAM space with address range 0000–FFFFH. As we have seen in previous chapters, many AVR chips have much less than 64K bytes for the SRAM. We also have seen how we can read (from) or write (into) this RAM space directly or indirectly. We store temporary variables in SRAM since the SRAM is the scratch pad.

2. The 2M words (4M bytes) of code (program) space with addresses of 000000–1FFFFFH. This 2M words of on-chip Flash ROM space is used primarily for storing programs (opcodes) and therefore is directly under control of the program counter (PC). As we have seen in the previous chapters, many AVR chips have much less than 2M words of on-chip program ROM (see Table 7-7). We have also seen how to access the program space for the purpose of data storage (see Chapter 6).

3. EEPROM. As we mentioned before, EEPROM can save data when the power is off. That is why we use EEPROM to save variables that should not be lost when the power is off. For example, the temperature set point of a cooling system should be changed by users and cannot be stored in program space. Also, it should be saved when the power is off, so we place it in EEPROM. Also, when there is not enough code space, we can place permanent variables in EEPROM to save some code space.

Table 7-7: Memory Size for Some ATmega Family Members(Bytes)

	Flash	SRAM	EEPROM
ATmega 8	8K	256	256
ATmega 16	16K	1K	512
ATmega 32	32K	2K	1K
ATmega 64	64K	4K	2K
ATmega 128	128K	8K	4K

In Chapter 6 we saw how to read from or write to EEPROM. In this chapter we will show the same concept using C programming. Notice that different C compilers may have their built-in functions or directives to access each type of memory. In CodeVision, to define a const variable in the Flash memory, you only need to put the Flash directive before it. Also, to define a variable in EEPROM, you can put the eeprom directive in front of it:

```
flash unsigned char mynum[] = "Hello";    //use Flash code space
eeprom unsigned char = 7                   //use EEPROM space
```

To learn how you can use the built-in functions or directives of your compiler, you should consult the manual for your compiler. Also, you can download some examples using different compilers from www.MicroDigitalEd.com.

See www.MicroDigitalEd.com for using Flash data space to store fix data

EEPROM access in C

In Chapter 6 we saw how we can access EEPROM using Assembly language. Next, we do the same thing with C language. Notice that as we mentioned before, most compilers have some built-in functions or directives to make the job of accessing the EEPROM memory easier. See Examples 7-36 and 7-37 to learn how we access EEPROM in C.

Example 7-36

Write an AVR C program to store 'G' into location 0x005F of EEPROM.

Solution:

```
#include <avr/io.h>              //standard AVR header
int main(void)
{
   while(EECR & (1<<EEWE));      //wait for last write to finish
   EEAR = 0x5f;                  //write 0x5F to address register
   EEDR = 'G';                   //write 'G' to data register
   EECR |= (1<<EEMWE);           //write one to EEMWE
   EECR |= (1<<EEWE);            //start EEPROM write
      return 0;
}
```

Example 7-37

Write an AVR C program to read the content of location 0x005F of EEPROM into PORTB.

Solution:

```
#include <avr/io.h>              //standard AVR header
int main(void)
{
   DDRB = 0xFF;                  //make PORTB an output
   while(EECR & (1<<EEWE));      //wait for last write to finish
   EEAR = 0x5f;                  //write 0x5F to address register
   EECR |= (1<<EERE);            //start EEPROM read by writing EERE
   PORTB = EEDR;                 //move data from data register to PORTB
}
```

Review Questions

1. The AVR family has a maximum of ____ of program ROM space.
2. The ATmega128 has ____ of program ROM.
3. True or false. The program (code) ROM space can be used for data storage, but the data space cannot be used for code.
4. True or false. Using the program ROM space for data means the data is fixed and static.
5. If we have a message string with a size of over 1000 bytes, then we use _____ (program ROM, data RAM) to store it.

SUMMARY

This chapter dealt with AVR C programming, specifically I/O programming and time delays in C. We also showed the logic operators AND, OR, XOR, and complement. In addition, some applications for these operators were discussed. This chapter described BCD and ASCII formats and conversions in C. We also discussed how to access EEPROM in C. The widely used technique of data serialization was also discussed.

PROBLEMS

SECTION 7.1: DATA TYPES AND TIME DELAYS IN C

1. Indicate what data type you would use for the following variables:
    (a) temperature
    (b) the number of days in a week
    (c) the number of days in a year
    (d) the number of months in a year
    (e) a counter to track the number of people getting on a bus
    (f) a counter to track the number of people going to a class
    (g) an address of 64K RAM space
    (h) the age of a person
    (i) a string for a message to welcome people to a building
2. Give the hex value that is sent to the port for each of the following C statements:
    (a) PORTB=14;   (b) PORTB=0x18;   (c) PORTB='A';
    (d) PORTB=7;   (e) PORTB=32;   (f) PORTB=0x45;
    (g) PORTB=255;   (h) PORTB=0x0F;
3. Give two factors that can affect time delay in the AVR microcontroller.
4. Of the two factors in Problem 3, which can be set by the system designer?
5. Can the programmer set the number of clock cycles used to execute an instruction? Explain your answer.
6. Explain why various C compilers produce different hex file sizes.

SECTION 7.2: I/O PROGRAMMING IN C

7. What is the difference between PORTC=0x00 and DDRC=0x00?
8. Write a C program to toggle all bits of Port B every 200 ms.
9. Write a C program to toggle bits 1 and 3 or Port B every 200 ms.
10. Write a time delay function for 100 ms.
11. Write a C program to toggle only bit 3 of PORT C every 200 ms.
12. Write a C program to count up Port B from 0–99 continuously.

SECTION 7.3: LOGIC OPERATIONS IN C

13. Indicate the data on the ports for each of the following:
    Note: The operations are independent of each other.
    (a) PORTB=0xF0&0x45;   (b) PORTB=0xF0&0x56;

```
(c)   PORTB=0xF0^0x76;        (d)   PORTC=0xF0&0x90;
(e)   PORTC=0xF0^0x90;        (f)   PORTC=0xF0|0x90;
(g)   PORTC=0xF0&0xFF;        (h)   PORTC=0xF0|0x99;
(i)   PORTC=0xF0^0xEE;        (j)   PORTC=0xF0^0xAA;
```
14. Find the contents of the port after each of the following operations:
```
(a)   PORTB=0x65&0x76;        (b)   PORTB=0x70|0x6B;
(c)   PORTC=0x95^0xAA;        (d)   PORTC=0x5D&0x78;
(e)   PORTC=0xC5|0x12;        (f)   PORTD=0x6A^0x6E;
(g)   PORTB=0x37|0x26;
```
15. Find the port value after each of the following is executed:
```
(a)   PORTB=0x65>>2;          (b)   PORTC=0x39<<2;
(c)   PORTB=0xD4>>3;          (d)   PORTB=0xA7<<2;
```
16. Show the C code to swap 0x95 to make it 0x59.
17. Write a C program that finds the number of zeros in an 8-bit data item.

SECTION 7.4: DATA CONVERSION PROGRAMS IN C

18. Write a C program to convert packed BCD 0x34 to ASCII and display the bytes on PORTB and PORTC.
19. Write a program to convert ASCII digits of '7' and '2' to packed BCD and display them on PORTB.

SECTION 7.5: DATA SERIALIZATION IN C

20. Write a C program to that finds the number of 1s in a given byte.

SECTION 7.6: MEMORY ALLOCATION IN C

21. Indicate what type of memory (data SRAM or code space) you would use for the following variables:
    (a) temperature
    (b) the number of days in a week
    (c) the number of days in a year
    (d) the number of months in a year
22. True or false. When using code space for data, the total size of the array should not exceed 256 bytes.
23. Why do we use the code space for video game characters and shapes?
24. What is the advantage of using code space for data?
25. What is the drawback of using program code space for data?
26. Write a C program to send your first and last names to EEPROM.
27. Indicate what type of memory (data RAM, or code ROM space) you would use for the following variables:
    (a) a counter to track the number of people getting on a bus
    (b) a counter to track the number of people going to a class
    (c) an address of 64K RAM space
    (d) the age of a person
    (e) a string for a message to welcome people to a building

28. Why do we not use the data RAM space for video game characters and shapes?
29. What is the drawback of using RAM data space for fixed data?

30. What is the advantage of using data RAM space for variables?

ANSWERS TO REVIEW QUESTIONS

SECTION 7.1: DATA TYPES AND TIME DELAYS IN C

1. 0 to 255 for unsigned char and –128 to +127 for signed char
2. 0 to 65,535 for unsigned int and –32,768 to +32,767 for signed int
3. Unsigned char
4. True
5. (a) Crystal frequency of the AVR system
   (b) Compiler used for C

SECTION 7.2: I/O PROGRAMMING IN C

1.
```
void main()
{
   DDRC = 0xFF;      //PORTC is output
   PORTC = 0x55;     //PORTC is 0101 0101
   PORTC = 0xAA;     //PORTC is 1010 1010
}
```
2. True
3. In CodeVision, the code can be:
```
void main()
{
   DDRC.2 = 1;
   PORTC.2 = 0;
}
```
4. True

SECTION 7.3: LOGIC OPERATIONS IN C

1. (a) 02H
   (b) FFH
   (c) FDH
2. Zeros
3. One
4. All zeros
5. 66H
6. ~ ((0000 0000) << 3) = ~ (1111 1111) = FFH

SECTION 7.4: DATA CONVERSION PROGRAMS IN C

1. (a) 15H = 0001 0101 packed BCD, 0000 0001,0000 0101 unpacked BCD
   (b) 99H = 1001 1001 packed BCD, 0000 1001,0000 1001 unpacked BCD
2. "76" = 3736H = 00110111 00110110B
3. 36, 37
4. Yes, because mydata = 0x39
5. Space savings
6. ASCII
7. BCD
8. E4H
9. 0
10. First, convert from binary to decimal, then convert to ASCII, and then send the results to the screen and we will see 038.

CHAPTER 7: AVR PROGRAMMING IN C   **287**

SECTION 7.6: PROGRAM ROM ALLOCATION IN C

1. 2M words (4M bytes)
2. 128K bytes
3. True
4. True
5. Program ROM

CHAPTER 8

AVR HARDWARE CONNECTION, HEX FILE, AND FLASH LOADERS

OBJECTIVES

Upon completion of this chapter, you will be able to:

>> Explain the function of the reset pin of the AVR microcontroller
>> Show the hardware connection of the AVR chip
>> Show the use of a crystal oscillator for a clock source
>> Explain how to design an AVR-based system
>> Explain the role of brown-out detection voltage (BOD) in system reset
>> Explain the role of the fuse bytes in an AVR-based system
>> Show the design of the AVR trainer
>> Code a test program in Assembly and C for testing the AVR
>> Show how to download programs into the AVR system using AVRISP
>> Explain the hex file characteristics

This chapter describes the process of physically connecting and testing AVR-based systems. In the first section we describe the functions of ATmega32 pins. The fuse bits of the AVR are explored in Section 8.2. In Section 8.3 we explain the characteristics of hex files that are produced by AVR Studio. In Section 8.4 we discuss the various methods of loading a program into the AVR microcontroller. It also shows the hardware connection for an AVR trainer using the ATmega16 or ATmega32 chips.

SECTION 8.1: ATMEGA32 PIN CONNECTION

The ATmega family members come in different packages, such as DIP (dual in-line package), MLF (Micro Lead Frame Package), and QFP (quad flat package). They all have many pins that are dedicated to various functions such as I/O, ADC, timer, and interrupts. Notice that Atmel provides a 28-pin version of the ATmega family (ATmega8) with a reduced number of I/O ports for less demanding applications. In Chapter 1 you can see members of the ATmega family and their characteristics. Because the vast majority of educational trainers use the 40-pin chip, we will concentrate on that. Figure 8-1 shows the pins for the ATmega32.

Examine Figure 8-1. Notice that of the 40 pins, a total of 32 are set aside for the four Ports A, B, C, and D, with their alternate functions. The rest of the pins are designated as VCC, AVCC, AREF, GND, XTAL1, XTAL2, and RESET. Next, we describe the function of each pin.

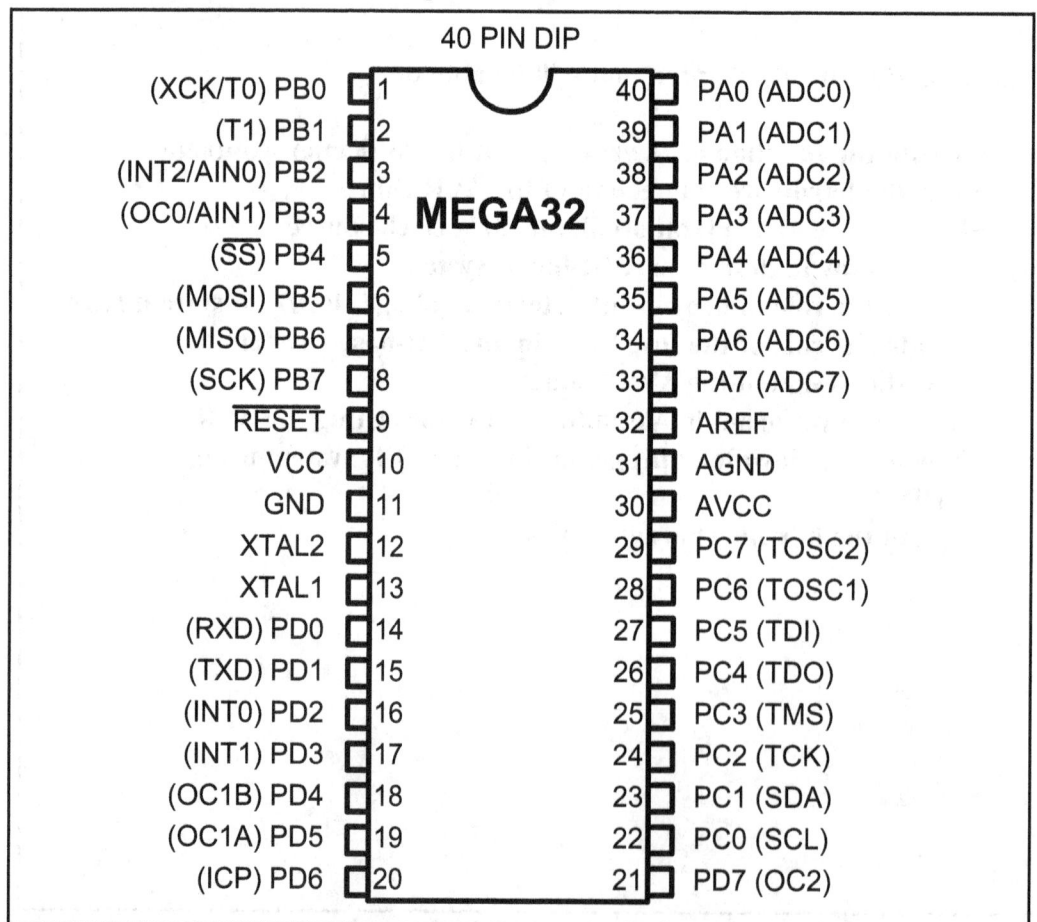

Figure 8-1. ATmega32 Pin Diagram

VCC

This pin provides supply voltage to the chip. The typical voltage source is +5 V. Some AVR family members have lower voltage for VCC pins in order to reduce the noise and power dissipation of the AVR system. For example, ATmega32L operation voltage is 2.7–5.5 V. We can choose other options for the operating voltage level by setting BOD fuse bits. The BOD fuse bits are discussed in the next section.

AVCC

AVCC is the supply voltage pin for Port A and the A/D Converter. It should be externally connected to VCC, even if the ADC is not used. In Chapter 13 you will see how to connect this pin if you want to use ADC.

AREF

AREF is the analog reference pin for ADC. In Chapter 13 we will discuss it further.

GND

Two pins are also used for ground. In chips with 40 pins and more, it is common to have multiple pins for VCC and GND. This will help reduce the noise (ground bounce) in high-frequency systems.

XTAL1 and XTAL2

The ATmega32 has many options for the clock source. Most often a quartz crystal oscillator is connected to input pins XTAL1 and XTAL2. The quartz crystal oscillator connected to the XTAL1 and XTAL2 pins also needs two capacitors. One side of each capacitor is connected to the ground as shown in Figure 8-2. Notice that ATmega32 microcontrollers can have speeds of 0 Hz to 16 MHz.

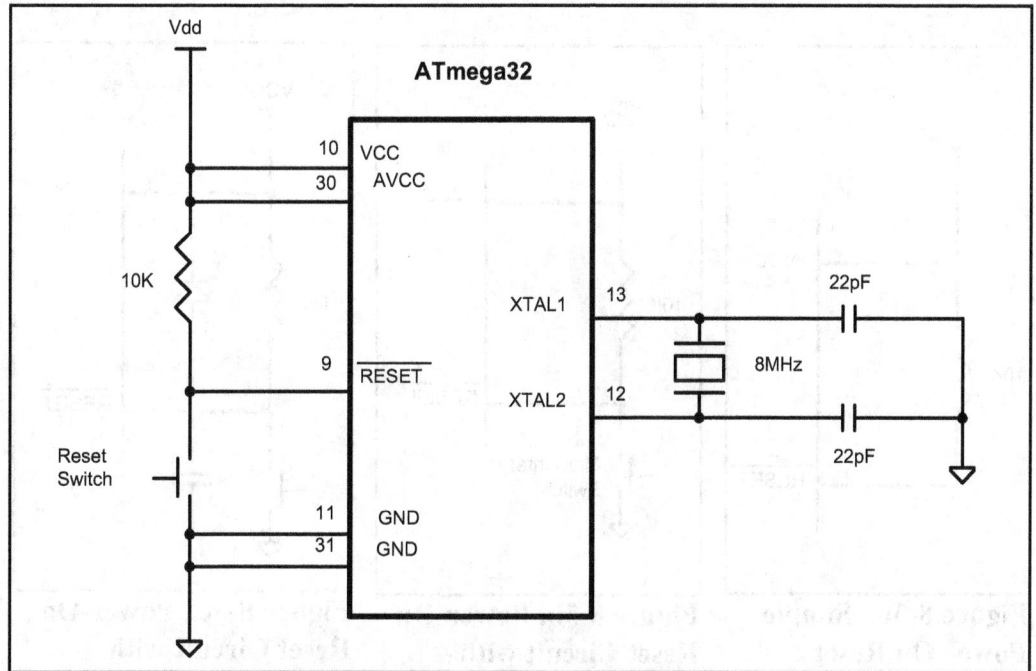

Figure 8-2. Minimum Connection for ATmega32

We can choose options for the clock source and frequency by setting some fuse bits. The fuse bits are discussed in the next section.

RESET

Pin 9 (in the ATmega32, 40-pin DIP) is the RESET pin. It is an input and is active-LOW (normally HIGH). When a LOW pulse is applied to this pin, the microcontroller will reset and terminate all activities. After applying reset, contents of all registers and SRAM locations will be cleared. Notice that after applying reset, all ports will be input because contents of all DDR registers are cleared. The CPU will start executing the program from run location 0x00000 after a brief delay when the RESET pin is forced low and then released.

Table 8-1: RESET Values for Some AVR Registers

Register	Reset Value (hex)
PC	000000
R0–R31	00
DDRx	00
PORTx	00

Figures 8-3a, 8-3b, and 8-3c show three ways of connecting the RESET pin. Figure 8-3b uses a momentary switch for reset circuitry. The most difficult time for any system is during the power-up. The CPU needs both a stable clock source and a stable voltage level to function properly. Some designers put a 10 nF capacitor between the RESET pin and GND to filter the noise during reset and working time. See Figure 3-8c. The diode protects the RESET pin from being powered by the capacitor when the power is off. The AVR chips come with some features that help the reset process. We can choose these features by setting the bits in the fuse bytes. The fuse bits for the reset are discussed in the next section. In addition to the RESET pin there are other sources of reset in the AVR family that will be discussed in future chapters.

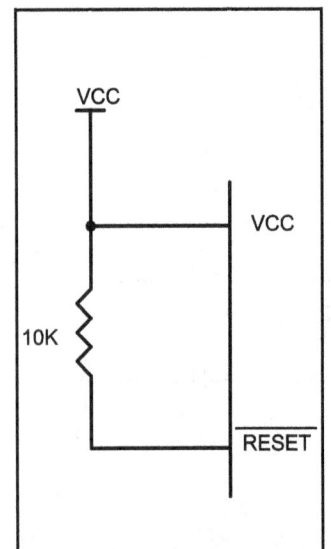

Figure 8-3a. Simple Power-On Reset Circuit

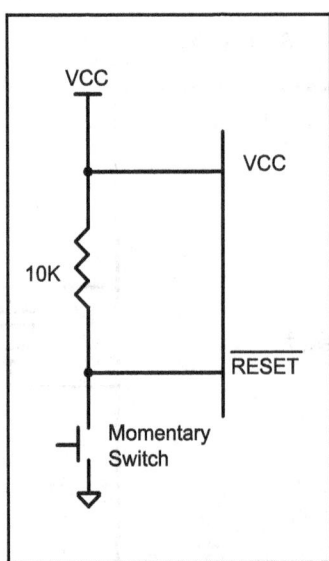

Figure 8-3b. Power-On Reset Circuit with Momentary Switch

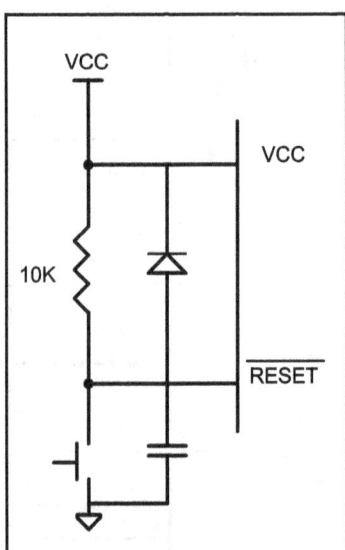

Figure 8-3c. Power-On Reset Circuit with Capacitor and Diode

The number of I/O ports varies among the AVR family members, as we saw in Chapter 4. The following is another look at them for the ATmega32.

Ports A, B, C, D, and E

As shown in Figure 8-1 (and discussed in Chapter 4), the Ports A, B, C, and D use a total of 32 pins. Tables 8-2 through 8-5 provide summaries of features of Ports A–D and their alternative functions. We will study the alternative functions of these pins in future chapters, as we discuss the AVR features.

Table 8-2: Port A Alternate Functions

Bit	Function
PA0	ADC0
PA1	ADC1
PA2	ADC2
PA3	ADC3
PA4	ADC4
PA5	ADC5
PA6	ADC6
PA7	ADC7

Table 8-3: Port B Alternate Functions

Bit	Function
PB0	XCK/T0
PB1	T1
PB2	INT2/AIN0
PB3	OC0/AIN1
PB4	SS
PB5	MOSI
PB6	MISO
PB7	SCK

Table 8-4: Port C Alternate Functions

Bit	Function
PC0	SCL
PC1	SDA
PC2	TCK
PC3	TMS
PC4	TDO
PC5	TDI
PC6	TOSC1
PC7	TOSC2

Table 8-5: Port D Alternate Functions

Bit	Function
PD0	RXD
PD1	TXD
PD2	INT0
PD3	INT1
PD4	OC1B
PD5	OC1A
PD6	ICP
PD7	OC2

Review Questions

1. Which pin is used to reset the ATmega32 chip?
2. Upon power-up, the R0–R31 registers have a value of ____.
3. True or false. Upon power-up, the CPU continues running the code from the line it was running before reset.
4. Reset is an active-_____ (LOW, HIGH) pin.
5. What is the operating voltage of ATmega32L?

SECTION 8.2: AVR FUSE BITS

There are some features of the AVR that we can choose by programming the bits of fuse bytes. These features will reduce system cost by eliminating any need for external components.

ATmega32 has two fuse bytes. Tables 8-6 and 8-7 give a short description of the fuse bytes. Notice that the default values can be different from production to production and time to time. In this section we examine some of the basic fuse bits. The Atmel website (http://www.atmel.com) provides the complete description of fuse bits for the AVR microcontrollers. It must be noted that if a fuse bit is incorrectly programmed, it can cause the system to fail. An example of this is changing the SPIEN bit to 0, which disables SPI programming mode. In this case you will not be able to program the chip any more! Also notice that the fuse bits are '0' if they are programmed and '1' when they are not programmed.

In addition to the fuse bytes in the AVR, there are 4 lock bits to restrict access to the Flash memory. These allow you to protect your code from being copied by others. In the development process it is not recommended to program lock bits because you may decide to read or verify the contents of Flash memory. Lock bits are set when the final product is ready to be delivered to market. In this book we do not discuss lock bits. To study more about lock bits you can read the data sheets for your chip at http://www.atmel.com.

Table 8-6: Fuse Byte (High)

Fuse High Byte	Bit No.	Description	Default Value
OCDEN	7	Enable OCD	1 (unprogrammed)
JTAGEN	6	Enable JTAG	0 (programmed)
SPIEN	5	Enable SPI serial program and data downloading	0 (programmed)
CKOPT	4	Oscillator options	1 (unprogrammed)
EESAVE	3	EEPROM memory is preserved through the chip erase	1 (unprogrammed)
BOOTSZ1	2	Select boot size	0 (programmed)
BOOTSZ0	1	Select boot size	0 (programmed)
BOOTRST	0	Select reset vector	1 (unprogrammed)

Table 8-7: Fuse Byte (Low)

Fuse High Byte	Bit No.	Description	Default Value
BODLEVEL	7	Brown-out detector trigger level	1 (unprogrammed)
BODEN	6	Brown-out detector enable	1 (unprogrammed)
SUT1	5	Select start-up time	1 (unprogrammed)
SUT0	4	Select start-up time	0 (programmed)
CKSEL3	3	Select clock source	0 (programmed)
CKSEL2	2	Select clock source	0 (programmed)
CKSEL1	1	Select clock source	0 (programmed)
CKSEL0	0	Select clock source	1 (unprogrammed)

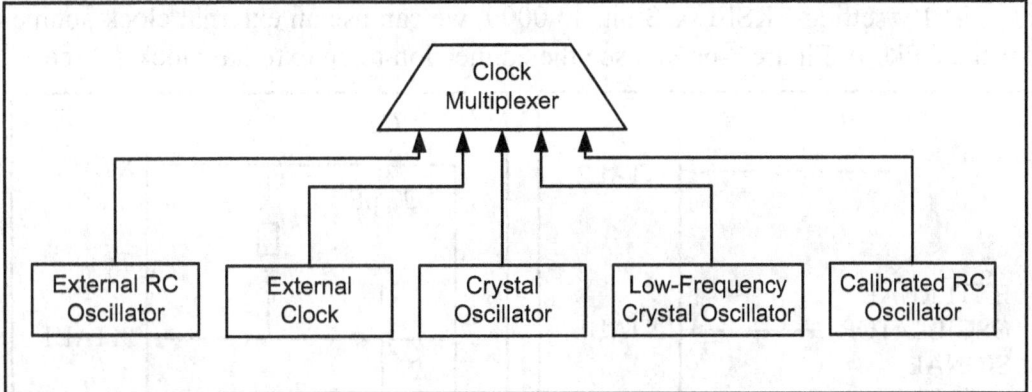

Figure 8-4. ATmega32 Clock Sources

Fuse bits and oscillator clock source

As you see in Figure 8-4, there are different clock sources in AVR. You can choose one by setting or clearing any of the bits CKSEL0 to CKSEL3.

CKSEL0–CKSEL3

The four bits of CKSEL3, CKSEL2, CKSEL1, and CKSEL0 are used to select the clock source to the CPU. The default choice is internal RC (0001), which uses the on-chip RC oscillator. In this option there is no need to connect an external crystal and capacitors to the chip. As you see in Table 8-8, by changing the values of CKSEL0–CKSEL3 we can choose among 1, 2, 4, or 8 MHz internal RC frequencies; but it must be noted that using an internal RC oscillator can cause about 3% inaccuracy and is not recommended in applications that need precise timing.

The external RC oscillator is another source to the CPU. As you see in Figure 8-5, to use the external RC oscillator, you have to connect an external resistor and capacitors to the XTAL1 pin. The values of R and C determine the clock speed. The frequency of the RC oscillator circuit is estimated by the equation $f = 1/(3RC)$. When you need a variable clock source you can use the external RC and replace the resistor with a potentiometer. By turning the potentiometer you will be able to change the frequency. Notice that the capacitor value should be at least 22 pF. Also, notice that by programming the CKOPT fuse, you can enable an internal 36 pF capacitor between XTAL1 and GND, and remove the external capacitor. As you see in Table 8-9, by changing the values of CKSEL0–CKSEL3, we can choose different frequency ranges.

Table 8-8: Internal RC Oscillator Operation Modes

CKSEL3...0	Frequency
0001	1 MHz
0010	2 MHz
0011	4 MHz
0100	8 MHz

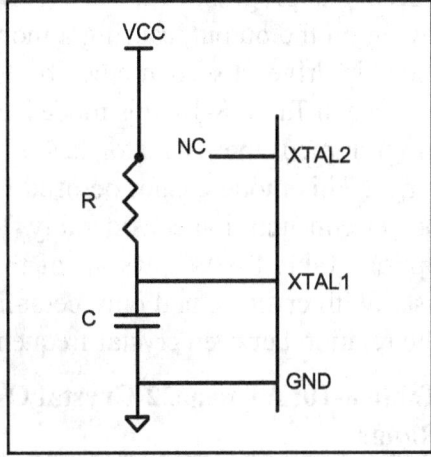

Figure 8-5 External RC

Table 8-9: External RC Oscillator Operation Modes

CKSEL3...0	Frequency (MHz)
0101	<0.9
0110	0.9–3.0
0111	3.0–8.0
1000	8.0–12.0

By setting CKSEL0...3 bits to 0000, we can use an external clock source for the CPU. In Figure 8-6a you see the connection to an external clock source.

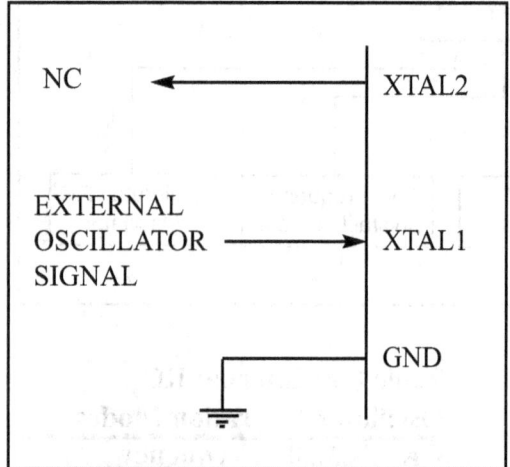

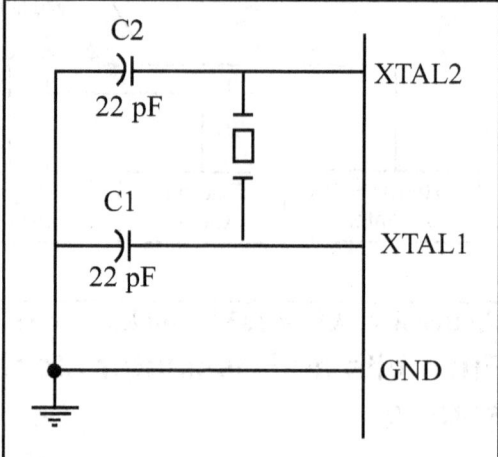

Figure 8-6a. XTAL1 Connection to an External Clock Source

Figure 8-6b. XTAL1–XTAL2 Connection to Crystal Oscillator

The most widely used option is to connect the XTAL1 and XTAL2 pins to a crystal (or ceramic) oscillator, as shown in Figure 8-6b. In this mode, when CKOPT is programmed, the oscillator output will oscillate with a full rail-to-rail swing on the output, causing a more powerful clock signal. This is suitable when the chip drives a second clock buffer or operates in a very noisy environment. As you see in Table 8-10, this mode has a wide frequency range. When CKOPT is not programmed, the oscillator has a smaller output swing and a limited frequency range. This mode cannot be used to drive other clock buffers, but it does reduce power consumption considerably. There are four choices for the crystal oscillator option. Table 8-10 shows all of these choices. Notice that mode 101 cannot be used with crystals, and only ceramic resonators can be used. Example 8-1 shows the relation between crystal frequency and instruction cycle time.

Table 8-10: ATmega32 Crystal Oscillator Frequency Choices and Capacitor Range

CKOPT	CKSEL3...1	Frequency (MHz)	C1 and C2 (pF)
1	101	0.4–0.9	Not for crystals
1	110	0.9–3.0	12–22
1	111	3.0–8.0	12–22
0	101, 110, 111	More than 1.0	12–22

Example 8-1

Find the instruction cycle time for the ATmega32 chip with the following crystal oscillators connected to the XTAL1 and XTAL2 pins.
(a) 4 MHz   (b) 8 MHz   (c) 10 MHz

Solution:
(a) Instruction cycle time is 1/(4 MHz) = 250 ns
(b) Instruction cycle time is 1/(8 MHz) = 125 ns
(c) Instruction cycle time is 1/(10 MHz) = 100 ns

Fuse bits and reset delay

The most difficult time for a system is during power-up. The CPU needs both a stable clock source and a stable voltage level to function properly. In AVRs, after all reset sources have gone inactive, a delay counter is activated to make the reset longer. This short delay allows the power to become stable before normal operation starts. You can choose the delay time through the SUT1, SUT0, and CKSEL0 fuses. Table 8-11 shows start-up times for the different values of SUT1, SUT0, and CKSEL fuse bits and also the recommended usage of each combination. Notice that the third column of Table 8-11 shows start-up time from power-down mode. Power-down mode is not discussed in this book.

Brown-out detector

Occasionally, the power source provided to the V_{CC} pin fluctuates, causing the CPU to malfunction. The ATmega family has a provision for this, called *brown-out detection*. The BOD circuit compares VCC with BOD-Level and resets the chip if VCC falls below the BOD-Level. The BOD-Level can be either 2.7 V when the BODLEVEL fuse bit is one (not programmed) or 4.0 V when the BODLEVEL fuse is zero (programmed). You can enable the BOD circuit by programming the BODEN fuse bit. When VCC increases above the trigger level, the BOD circuit releases the reset, and the MCU starts working after the time-out period has expired.

A good rule of thumb

There is a good rule of thumb for selecting the values of fuse bits. If you are using an external crystal with a frequency of more than 1 MHz you can set the CKSEL3, CKSEL2, CKSEL1, SUT1, and SUT0 bits to 1 (not programmed) and clear CKOPT to 0 (programmed).

Table 8-11: Startup Time for Crystal Oscillator and Recommended Usage

CKSEL0	SUT1...0	Start-Up Time from Power-Down	Delay from Reset (VCC = 5)	Recommended Usage
0	00	258 CK	4.1	Ceramic resonator, fast rising power
0	01	258 CK	65	Ceramic resonator, slowly rising power
0	10	1K CK	-	Ceramic resonator, BOD enabled
0	11	1K CK	4.1	Ceramic resonator, fast rising power
1	00	1K CK	65	Ceramic resonator, slowly rising power
1	01	16K CK	-	Crystal oscillator, BOD enabled
1	10	16K CK	4.1	Crystal oscillator, fast rising power
1	11	16K CK	65	Crystal oscillator, slowly rising power

Putting it all together

Many of the programs we showed in the first seven chapters were intended to be simulated. Now that we know what we should write in the fuse bits and how we should connect the ATmega32 pins, we can download the hex output file provided by the AVR Studio assembler into the Flash memory of the AVR chip using an AVR programmer.

We can use the following skeleton source code for the programs that we intend to download into a chip. Notice that you have to modify the first line if you use a chip other than ATmega32. As you can see in the comments, if you want to enable interrupts you have to modify ".ORG 0", and if you do not use call the instruction in your code, you can omit the codes that set the stack pointer.

```
.INCLUDE "M32DEF.INC"      ;change it according to your chip
.ORG 0                     ;change it if you use interrupt
LDI    R16,HIGH(RAMEND)    ;set the high byte of stack pointer to
OUT    SPH,R16             ;the high address of RAMEND
LDI    R16,LOW(RAMEND)     ;set the low byte of stack pointer to
OUT    SPL,R16             ;low address of RAMEND

...                        ;place your code here
```

As an example, examine Program 8-1. It will toggle all the bits of Port B with some delay between the "on" and "off" states.

```
;Test Program 8-1: Toggling PORTB for the Atmega32
.INCLUDE "M32DEF.INC"            ;using Atmega32
.ORG 0
        LDI    R16,HIGH(RAMEND)  ;set up stack
        OUT    SPH,R16
        LDI    R16,LOW(RAMEND)
        OUT    SPL,R16
        LDI    R16,0xFF          ;load R16 with 0xFF
        OUT    DDRB,R16          ;Port B is output
BACK:
        COM    R16               ;complement R16
        OUT    PORTB,R16         ;send it to Port B
        CALL   DELAY             ;time delay
        RJMP   BACK              ;keep doing this indefinitely

DELAY:
        LDI    R20,16
L1:     LDI    R21,200
L2:     LDI    R22,250
L3:
        NOP
        NOP
        DEC    R22
        BRNE   L3
        DEC    R21
        BRNE   L2

        DEC    R20
        BRNE   L1
        RET
```

Program 8-1: Toggling Port B in Assembly

In Chapter 7 we covered C programming of the AVR using the AVR GCC compiler. Program 8-2 shows the toggle program written in C. It will toggle all the bits of Port B with some delay between the "on" and "off" states.

```c
#include <avr/io.h>              //standard AVR header
#include <util/delay.h>

void delay_ms(int d);

int main(void)
{
    DDRB = 0xFF;                 //Port B is output
    while (1)
    {                            //do forever
        PORTB = 0x55;
        delay_ms(1000);          //delay 1 second
        PORTB = 0xAA;
        delay_ms(1000);          //delay 1 second
    }
    return 0;
}

void delay_ms(int d)
{
    _delay_ms(d);                //delay 1000 us
}
```

Program 8-2: Toggling Port B in C

Review Questions

1. A given ATmega32-based system has a crystal frequency of 16 MHz. What is the instruction cycle time for the CPU?
2. How many fuse bytes are available in ATmega32?
3. True or false. Upon power-up, both voltage and frequency are stable instantly.
4. The internal RC osciltor works for the frequency range of _____ to _____ MHz.
5. Which fuse bit is used to disable the BOD?
6. True or false. Upon power-up, the CPU starts working immediately.
7. What is the rule of thumb for ATmega32 fuse bits?
8. The brown-out detection voltage can be set at _____ or _____ by _____ fuse bit.
9. True or false. The higher the clock frequency for the system, the lower the power dissipation.

SECTION 8.3: EXPLAINING THE HEX FILE FOR AVR

Intel Hex is a widely used file format designed to standardize the loading (transferring) of executable machine code into a chip. Therefore, the loaders that come with every ROM burner (programmer) support the Intel Hex file format. In many Windows-based assemblers such as AVR Studio, the Intel Hex file is produced according to the settings you set. In the AVR Studio environment, the object file is fed into the linker program to produce the Intel hex file. The hex file is used by a programmer such as the AVRISP to transfer (load) the file into the Flash memory. The AVR Studio assembler can produce three types of hex files. They are (a) Intel Intellec 8/MDS (Intel Hex), (b) Motorola S-record, and (c) Generic. See Table 8-12. In this section we will explain Intel Hex with some examples. We recommend that you do not use AVR GCC if you want to test the programs in this section on your computer. It is better to use a simple .asm file like toggle.asm to understand this concept better.

Table 8-12: Intel Hex File Formats Produced by AVR Studio

Format Name	File Extension	Max. ROM Address
Extended Intel Hex file	.hex	20-bit address
Motorola S-record	.mot	32-bit address
Generic	.gen	24-bit address

Analyzing the Intel Hex file

We choose the hex type of Intel Hex, Motorola S-record, or Generic by using the command-line invocation options or setting the options in the AVR Studio assembler itself. If we do not choose one, the AVR Studio assembler selects Intel Hex by default. Intel Hex supports up to 16-bit addressing and is not applicable for programs more than 64K bytes in size. To overcome this limitation AVR Studio uses extended Intel Hex files, which support type 02 records to extend address space to 1M. We will explain extended Intel Hex file format in this section. Figure 8-10 shows the Intel Hex file of the test program whose list file is given in Figure 8-8. Since the programmer (loader) uses the Hex file to download the opcode into Flash, the hex file must provide the following: (1) the number of bytes of information to be loaded, (2) the information itself, and (3) the starting address where the information must be placed. Each record (line) of the Hex file consists of six parts as follows:

    :BBAAAATTHHHHH.......HHHHCC

The following describes each part:
1. ":" Each line starts with a colon.
2. BB, the count byte. This tells the loader how many bytes are in the line.
3. AAAA is for the record address. This is a 16-bit address. The loader places the first byte of record data into this Flash location. This is the case in files that are less than 64 KB. For files that are more than 64 KB the address field shows the record address in the current segment.

---

4. TT is for type. This field is 00, 01, or 02. If it is 00, it means that there are more lines to come after this line. If it is 01, it means that this is the last line and the loading should stop after this line. If it is 02, it indicates the current segment address. To calculate the absolute address of each record (line), we have to shift the current segment address 4 bits to left and then add it to the record address. Examples 8-2 and 8-3 show how to calculate the absolute address of a record in extended Intel hex file.

5. HH......H is the real information (data or code). The loader places this information into successive memory locations of Flash. The information in this field is presented as low byte followed by the high byte.

6. CC is a single byte. This last byte is the checksum byte for everything in that line. The checksum byte is used for error checking. Checksum bytes are discussed in detail in Chapters 6 and 7. Notice that the checksum byte at the end of each line represents the checksum byte for everything in that line, and not just for the data portion.

Example 8-2

What is the absolute address of the first byte of a record that has 0025 in the address field if the last type 02 record before it has the segment address 0030?

Solution:

To calculate the absolute address of each record (line), we have to shift the segment address (0030) four bits to the left and then add it to the record address (0025):

```
0030 (2 bytes segment address) shifted 4 bits to the left  -->      00300
0025 (record address)                                             +    25
                                                                  ---------
=>   (absolute address)                                             00325
```

Example 8-3

What is the absolute address of the first byte of the second record below?

```
:020000020000FC
:10000000008E00EBF0FE50DBF0FEF07BB05E500953C
```

Solution:

To calculate the absolute address of the first byte of the second record, we have to shift left the segment address (0000, as you see in the first record) four bits and then add it to the second record address (0000, as you see in the second record).

```
0000 (segment address) shift 4 bits to the left  -->   00000
                                                     + 0000    (record address)
                                                     ---------
                                                       000000    (absolute address)
```

Analyzing the bytes in the Flash memory vs. list file

The data in the Flash memory of the AVR is recorded in a way that is called *Little-endian*. This means that the high byte of the code is located in the higher address location of Flash memory, and the low byte of the code is located in the lower address location of Flash memory. Compare the first word of code (e008) in Figure 8-8 with the first two bytes of Flash memory (08e0) in Figure 8-7. As you see, 08, which is the low byte of the first instruction (LDI R16, HIGH(RAMEND)) in the code, is placed in the lower location of Flash memory, and e0, which is the high byte of the instruction in the code, is placed in the next location of program space just after 08.

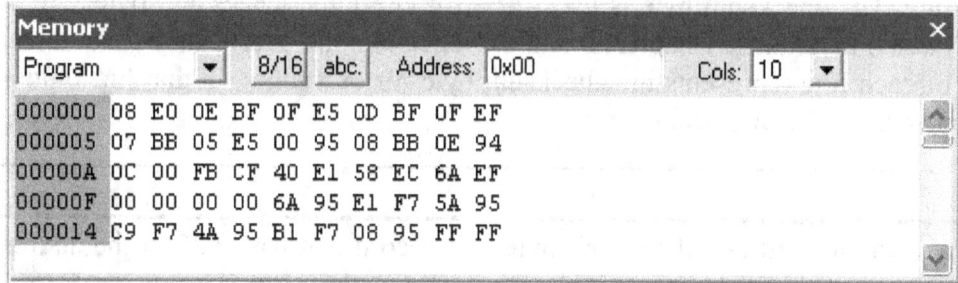

Figure 8-7. AVR Flash Memory Contents

```
LOC      OBJ           LINE
                       .ORG 0x000
000000  e008                LDI    R16,HIGH(RAMEND)
000001  bf0e                OUT    SPH,R16
000002  e50f                LDI    R16,LOW(RAMEND)
000003  bf0d                OUT    SPL,R16

000004  ef0f                LDI    R16,0xFF
000005  bb07                OUT DDRB,R16
000006  e505                LDI    R16,0x55
                       BACK:
000007  9500                COM         R16
000008  bb08                OUT    PORTB,R16
000009  940e  000c         CALL   DELAY_1S
00000b  cffb                RJMP   BACK
                       DELAY_1S:
00000c  e140                LDI    R20,16
00000d  ec58         L1:    LDI    R21,200
00000e  ef6a         L2:    LDI    R22,250
                       L3:
00000f  0000                NOP
000010  0000                NOP
000011  956a                DEC    R22
000012  f7e1                BRNE   L3
000013  955a                DEC    R21
000014  f7c9                BRNE   L2
000015  954a                DEC    R20
000016  f7b1                BRNE   L1
000017  9508                RET
```

Figure 8-8. List File for Test Program
(Comments and other lines are deleted, and some spaces are added for simplicity.)

As we mentioned in Chapter 2, each Flash location in the AVR is 2 bytes long. So, for example, the first byte of Flash location #2 is Byte #4 of the code. See Figure 8-9.

Flash Memory

Location #0	Byte #0	Byte #1
Location #1	Byte #2	Byte #3
Location #2	Byte #4	Byte #5
Location #3	Byte #6	Byte #7

Figure 8-9. AVR Flash Memory Locations

In Figure 8-10 you see the hex file of the toggle code. The first record (line) is a type 02 record and indicates the current segment address, which is 0000. The next record (line) is a type 00 record and contains the data (the code to be loaded into the chip). After ':' the record starts with 10, which means that the data field contains 10 (16 decimal) bytes of data. The next field is the address field (0000), and it indicates that the first byte of the data field will be placed in address location 0 in the current segment. So the first byte of code will be loaded into location 0 of Flash memory. (Reexamine Example 8-3 if needed.) Also, notice the use of .ORG 0x000 in the code. The next field is the data field, which contains the code to be loaded into the chip. The first byte of the data field is 08, which is the low byte of the first instruction (LDI R16,HIGH(RAMEND)). See Figure 8-8. The last field of the record is the checksum byte of the record. Notice that the checksum byte at the end of each line represents the checksum byte for everything in that line, and not just for the data portion.

Pay attention to the address field of the next record (0010) in Figure 8-10 and compare it with the address of the bb08 instruction in the list file in Figure 8-8. As you can see, the address in the list file is 000008, which is exactly half of the address of the bb08 instruction in the hex file, which is 0010. That is because each Flash location (word) contains 2 bytes.

```
:020000020000FC
:1000000008E00EBF0FE50DBF0FEF07BB05E500953C
:1000100008BB0E940C00FBCF40E158EC6AEF0000E7
:1000200000006A95E1F75A95C9F74A95B1F7089526
:00000001FF


Separating the fields, we get the following:


:BB AAAA TT HHHHHHHHHHHHHHHHHHHHHHHHHHHHHHHH          CC
:02 0000 02 0000                                      FC
:10 0000 00 08E00EBF0FE50DBF0FEF07BB05E50095          3C
:10 0010 00 08BB0E940C00FBCF40E158EC6AEF0000          E7
:10 0020 00 00006A95E1F75A95C9F74A95B1F70895          26
:00 0000 01                                           FF
```

Figure 8-10. Intel Hex File Test Program with the Intel Hex Option

Examine Examples 8-4 through 8-6 to gain insight into the Intel Hex file format.

Example 8-4

From Figure 8-10, analyze the six parts of line 3.

Solution:

After the colon (:), we have 10, which means that 16 bytes of data are in this line. 0010H is the record address, and means that 08, which is the first byte of the record, is placed in address location 10H (16 decimal). Next, 00 means that this is not the last line of the record. Then the data, which is 16 bytes, is as follows:
`08BB0E940C00FBCF40E158EC6AEF0000`. Finally, the last byte, E7, is the checksum byte.

Example 8-5

Compare the data portion of the Intel Hex file of Figure 8-10 with the opcodes in the list file of the test program given in Figure 8-8. Do they match?

Solution:

In the second line of Figure 8-10, the data portion starts with 08E0H, where the low byte is followed by the high byte. That means it is E008, the opcode for the instruction "`LDI   R16,HIGH(RAMEND)`", as shown in the list file of Figure 8-8. The last byte of the data in line 5 is 0895, which is the opcode for the "`RET`" instruction in the list file.

Example 8-6

(a) Verify the checksum byte for line 3 of Figure 8-10. (b) Verify also that the information is not corrupted.
Solution:

(a) `10 + 00 + 00 + 00 + 08 + E0 + 0E + BF + 0F + E5 + 0D + BF + 0F + EF + 07 + BB + 05 + E5 + 00 + 95 = 6C4` in hex. Dropping the carries (6) gives C4H, and its 2's complement is 3CH, which is the last byte of line 3.
(b) If we add all the information in line 2, including the checksum byte, and drop the carries we should get `10 + 00 + 00 + 00 + 08 + E0 + 0E + BF + 0F + E5 + 0D + BF + 0F + EF + 07 + BB + 05 + E5 + 00 + 95 + 3C = 700`. Dropping the carries (7) gives 00H, which means OK.

Review Questions

1. True or false. The Intel Hex file format does not use the checksum byte method to ensure data integrity.
2. The first byte of a line in an Intel Hex file represents ____.
3. The last byte of a line in an Intel Hex file represents ____.
4. In the TT field of an Intel Hex file, we have 00. What does it indicate?
5. Find the checksum byte for the following values: 22H, 76H, 5FH, 8CH, 99H.
6. In Question 5, add all the values and the checksum byte. What do you get?

SECTION 8.4: AVR PROGRAMMING AND TRAINER BOARD

In this section, we show various ways of loading a hex file into the AVR microcontroller. We also discuss the connection for a simple AVR trainer.

Atmel has skillfully designed AVR microcontrollers for maximum flexibility of loading programs. The three primary ways to load a program are:

1. Parallel programming. In this way a device burner loads the program into the microcontroller separate from the system. This is useful on a manufacturing floor where a gang programmer is used to program many chips at one time. Most mainstream device burners support the AVR families: EETools is a popular one. The device programming method is straightforward: The chip is programmed before it is inserted into the circuit. Or, the chip can be removed and reprogrammed if it is in a socket. A ZIF (zero insertion force) socket is even quicker and less damaging than a standard socket. When removing and reinserting, we must observe ESD (electrostatic discharge) procedures. Although AVR devices are rugged, there is always a risk when handling them. Using this method allows all of the device's resources to be utilized in the design. No pins are shared, nor are internal resources of the chip used as is the case in the other two methods. This allows the embedded designer to use the minimum board space for the design.

2. An in-circuit serial programmer (ISP) allows the developer to program and debug their microcontroller while it is in the system. This is done by a few wires with a system setup to accept this configuration. In-circuit serial programming is excellent for designs that change or require periodic updating. AVR has two methods of ISP. They are SPI and JTAG. Most of the ATmega family supports both methods. The SPI uses 3 pins, one for send, one for receive, and one for clock. These pins can be used as I/O after the device is programmed. The designer must make sure that these pins do not conflict with the programmer. Notice that SPI stands for "serial peripheral interface" and is a protocol. But ISP stands for "in-circuit serial programming" and is a method of code loading. AVRISP and many other devices support ISP. To connect AVRISP to your device you also need to connect VCC, GND, and RESET pins. You must bring the pins to a header on the board so that the programmer can connect to it. Figure 8-11 shows the pin connections.

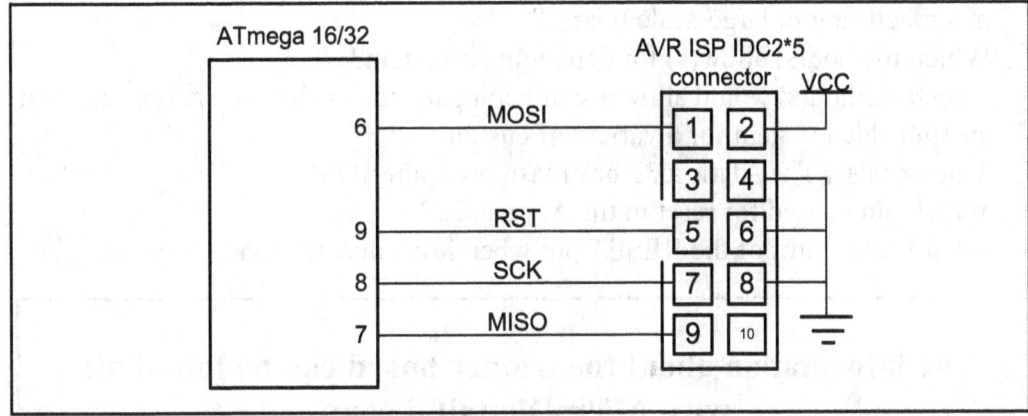

Figure 8-11. ISP 10-pin Connections (See www.Atmel.com for 6-pin version)

CHAPTER 8: AVR HARDWARE CONNECTION 305

Another method of ISP is JTAG. JTAG is another protocol that supports in-circuit programming and debugging. It means that in addition to programming you can trace your program on the chip line by line and watch or change the values of memory locations, ports, or registers while your program is running on the chip.

3. A boot loader is a piece of code burned into the microcontroller's program Flash. Its purpose is to communicate with the user's board to load the program. A boot loader can be written to communicate via a serial port, a CAN port, a USB port, or even a network connection. A boot loader can also be designed to debug a system, similar to the JTAG. This method of programming is excellent for the developer who does not always have a device programmer or a JTAG available. There are several application notes on writing boot loaders on the Web. The main drawback of the boot loader is that it does require a communication port and program code space on the microcontroller. Also, the boot loader has to be programmed into the device before it can be used, usually by one of the two previous ways.

The boot loader method is ideal for the developer who needs to quickly program and test code. This method also allows the update of devices in the field without the need of a programmer. All one needs is a computer with a port that is compatible with the board. (The serial port is one of the most commonly used and discussed, but a CAN or USB boot loader can also be written.) This method also consumes the largest amount of resources. Code space must be reserved and protected, and external devices are needed to connect and communicate with the PC. Developing projects using this method really helps programmers test their code. For mature designs that do not change, the other two methods are better suited.

AVR trainers

There are many popular trainers for the AVR chip. The vast majority of them have a built-in ISP programmer. See the following website for more information and support about the AVR trainers. For more information about how to use an AVR trainer you can visit the www.MicroDigitalEd.com website.

Review Questions

1. Which method(s) to program the AVR microcontroller is/are the best for the manufacturing of large-scale boards?
2. Which method(s) allow(s) for debugging a system?
3. Which method(s) would allow a small company to develop a prototype and test an embedded system for a variety of customers?
4. True or false. The ATmega32 has Flash program ROM.
5. Which pin is used for reset in the ATmega32?
6. What is the status of the RESET pin when it is not activated?

The information about the trainer board can be found at:
www.MicroDigitalEd.com

SUMMARY

This chapter began by describing the function of each pin of the ATmega32. A simple connection for ATmega32 was shown. Then, the fuse bytes were discussed. We use fuse bytes to enable features such as BOD and clock source and frequency. We also explained the Intel Hex file format and discussed each part of a record in a hex file using an example. Then, we explained list files in detail. The various ways of loading a hex file into a chip were discussed in the last section. The connections to a ISP device were shown.

PROBLEMS

SECTION 8.1: ATMEGA32 PIN CONNECTION

1. The ATmega32 DIP package is a(n) _____-pin package.
2. Which pins are assigned to VCC and GND?
3. In the ATmega32, how many pins are designated as I/O port pins?
4. The crystal oscillator is connected to pins _____ and _____ .
5. True or false. AVR chips comes only in DIP packages.
6. Indicate the pin number assigned to RESET in the DIP package.
7. The RESET pin is normally _____ (LOW, HIGH) and needs a _____ (LOW, HIGH) signal to be activated.
8. In the ATmega32, how many pins are set aside for the VCC?
9. In the ATmega32, how many pins are set aside for the GND?
10. True or false. In connecting VCC pins to power both must be connected.
11. RESET is an _____ (input, output) pin.
12. How many pins are designated as Port A and what are those in the 40-pin DIP package?
13. How many pins are designated as Port B and what are those in the 40-pin DIP package?
14. How many pins are designated as Port C and what are those in the 40-pin DIP package?
15. How many pins are designated as Port D and what are those in the 40-pin DIP package?
16. Upon reset, all the bits of ports are configured as _____ (input, output).

SECTION 8.2: AVR FUSE BITS

17. How many clock sources does the AVR have?
18. What fuse bits are used to select clock source?
19. Which clock source do you suggest if you need a variable clock source?
20. Which clock source do you suggest if you need to build a system with minimum external hardware?
21. Which clock source do you suggest if you need a precise clock source?
22. How many fuse bytes are there in the AVR?

23. Which fuse bit is used to set the brown-out detection voltage for the ATmega32?

24. Which fuse bit is used to enable and disable the brown-out detection voltage for the ATmega32?

25. If the brown-out detection voltage is set to 4.0 V, what does it mean to the system?

SECTION 8.3: EXPLAINING THE INTEL HEX FILE FOR AVR

26. True or false. The Hex option can be set in AVR Studio.

27. True or false. The extended Intel Hex file can be used for ROM sizes of less than 64 kilobytes.

28. True or false. The extended Intel Hex file can be used for ROM sizes of more than 64 kilobytes.

29. Analyze the six parts of line 3 of Figure 8-10.

30. Verify the checksum byte for line 3 of Figure 8-10. Verify also that the information is not corrupted.

31. What is the difference between Intel Hex files and extended Intel Hex files?

SECTION 8.4: AVR PROGRAMMING AND TRAINER BOARD

32. True or false. To use a parallel programmer, we must remove the AVR chip from the system and place it into the programmer.

33. True or false. ISP can work only with Flash chips.

34. What are the different ways of loading a code into an AVR chip?

35. True or false. A boot loader is a kind of parallel programmer.

ANSWERS TO REVIEW QUESTIONS

SECTION 8.1: ATMEGA32 PIN CONNECTION

1. 9
2. 0
3. False
4. LOW
5. 2.7 V–5.5 V

SECTION 8.2: AVR FUSE BITS

1. 1/16 MHz = 62.5 ns
2. 16 bits = 2 bytes
3. False
4. 1, 8
5. BODEN
6. False
7. If you are using an external crystal with a frequency of more than 1 MHz you can set the CKSEL3, CKSEL2, CKSEL1, SUT1, and SUT0 bits to 1 (not programmed) and clear CKOPT to 0 (programmed).
8. 2.7 V, 4 V, BODLEVEL
9. False

SECTION 8.3: EXPLAINING THE INTEL HEX FILE FOR AVR

1. False
2. The number of bytes of data in the line
3. The checksum byte of all the bytes in that line
4. 00 means this is not the last line and that more lines of data follow.
5. 22H + 76H + 5FH + 8CH + 99H = 21CH. Dropping the carries we have 1CH and its 2's complement, which is E4H.
6. 22H + 76H + 5FH + 8CH + 99H + E4H = 300H. Dropping the carries, we have 00, which means that the data is not corrupted.

SECTION 8.4: AVR PROGRAMMING AND TRAINER BOARD

1 Device burner
2. JTAG and boot loader
3. ISP
4. True
5. Pin 9
6. HIGH

CHAPTER 9

AVR TIMER PROGRAMMING IN ASSEMBLY AND C

OBJECTIVES

Upon completion of this chapter, you will be able to:

>> List the timers of the ATmega32 and their associated registers
>> Describe the Normal and CTC modes of the AVR timers
>> Program the AVR timers in Assembly and C to generate time delays
>> Program the AVR counters in Assembly and C as event counters

Many applications need to count an event or generate time delays. So, there are counter registers in microcontrollers for this purpose. See Figure 9-1. When we want to count an event, we connect the external event source to the clock pin of the counter register. Then, when an event occurs externally, the content of the counter is incremented; in this way, the content of the counter represents how many times an event has occurred. When we want to generate time delays, we connect the oscillator to the clock pin of the counter. So, when the oscillator ticks, the content of the counter is incremented. As a result, the content of the counter register represents how many ticks have occurred from the time we have cleared the counter. Since the speed of the oscillator in a microcontroller is known, we can calculate the tick period, and from the content of the counter register we will know how much time has elapsed.

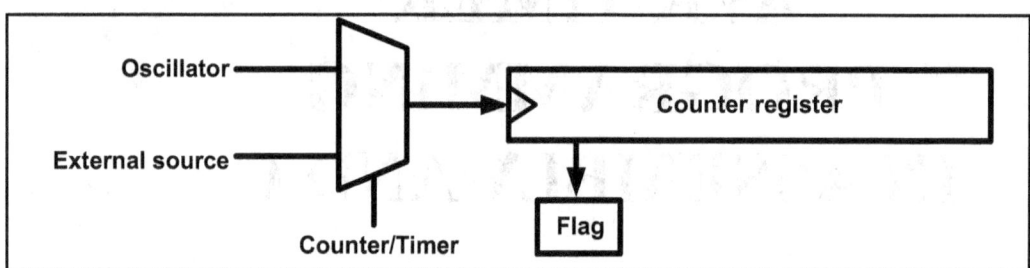

Figure 9-1. A General View of Counters and Timers in Microcontrollers

So, one way to generate a time delay is to clear the counter at the start time and wait until the counter reaches a certain number. For example, consider a microcontroller with an oscillator with frequency of 1 MHz; in the microcontroller, the content of the counter register increments once per microsecond. So, if we want a time delay of 100 microseconds, we should clear the counter and wait until it becomes equal to 100.

In the microcontrollers, there is a flag for each of the counters. The flag is set when the counter overflows, and it is cleared by software. The second method to generate a time delay is to load the counter register and wait until the counter overflows and the flag is set. For example, in a microcontroller with a frequency of 1 MHz, with an 8-bit counter register, if we want a time delay of 3 microseconds, we can load the counter register with $FD and wait until the flag is set after 3 ticks. After the first tick, the content of the register increments to $FE; after the second tick, it becomes $FF; and after the third tick, it overflows (the content of the register becomes $00) and the flag is set.

The AVR has one to six timers depending on the family member. They are referred to as Timers 0, 1, 2, 3, 4, and 5. They can be used as timers to generate a time delay or as counters to count events happening outside the microcontroller.

In the AVR some of the timers/counters are 8-bit and some are 16-bit. In ATmega32, there are three timers: Timer0, Timer1, and Timer2. Timer0 and Timer2 are 8-bit, while Timer1 is 16-bit. In this chapter we cover Timer0 and Timer2 as 8-bit timers, and Timer1 as a 16-bit timer.

If you learn to use the timers of ATmega32, you can easily use the timers of other AVRs. You can use the 8-bit timers like the Timer0 of ATmega32 and the 16-bit timers like the Timer1 of ATmega32.

SECTION 9.1: PROGRAMMING TIMERS 0, 1, AND 2

Every timer needs a clock pulse to tick. The clock source can be internal or external. If we use the internal clock source, then the frequency of the crystal oscillator is fed into the timer. Therefore, it is used for time delay generation and consequently is called a *timer*. By choosing the external clock option, we feed pulses through one of the AVR's pins. This is called a *counter*. In this section we discuss the AVR timer, and in the next section we program the timer as a counter.

Basic registers of timers

Examine Figure 9-2. In AVR, for each of the timers, there is a TCNTn (timer/counter) register. That means in ATmega32 we have TCNT0, TCNT1, and TCNT2. The TCNTn register is a counter. Upon reset, the TCNTn contains zero. It counts up with each pulse. The contents of the timers/counters can be accessed using the TCNTn. You can load a value into the TCNTn register or read its value.

Each timer has a TOVn (Timer Overflow) flag, as well. When a timer overflows, its TOVn flag will be set.

Each timer also has the TCCRn (timer/counter control register) register for setting modes of operation. For example, you can specify Timer0 to work as a timer or a counter by loading proper values into the TCCR0.

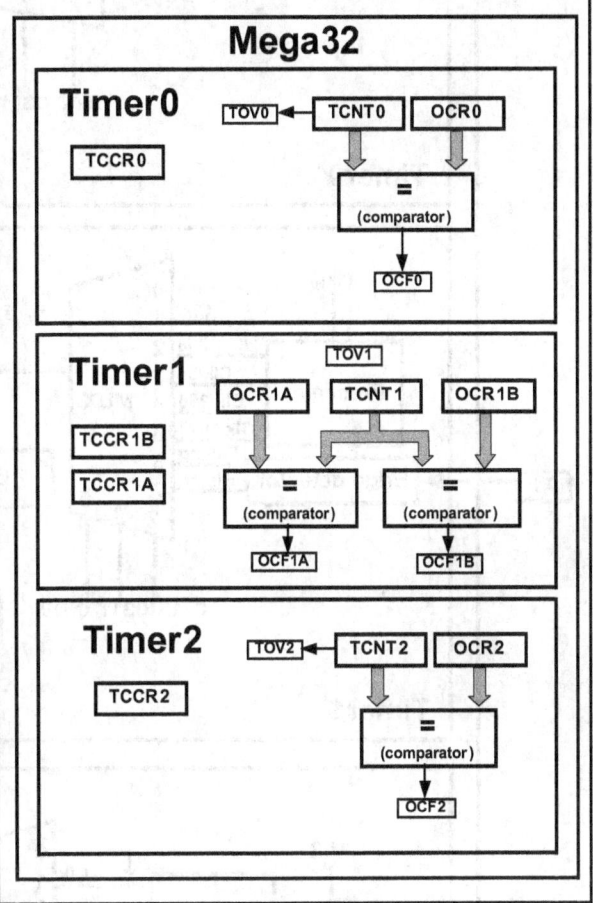

Figure 9-2. Timers in ATmega32

Each timer also has an OCRn (Output Compare Register) register. The content of the OCRn is compared with the content of the TCNTn. When they are equal the OCFn (Output Compare Flag) flag will be set.

The timer registers are located in the I/O register memory. Therefore, you can read or write from timer registers using IN and OUT instructions, like the other I/O registers. For example, the following instructions load TCNT0 with 25:

```
LDI R20,25      ;R20 = 25
OUT TCNT0,R20   ;TCNT0 = R20
```

or "IN R19,TCNT2" copies TCNT2 to R19.

The internal structure of the ATmega32 timers is shown in Figure 9-3. Next, we discuss each timer separately in more detail.

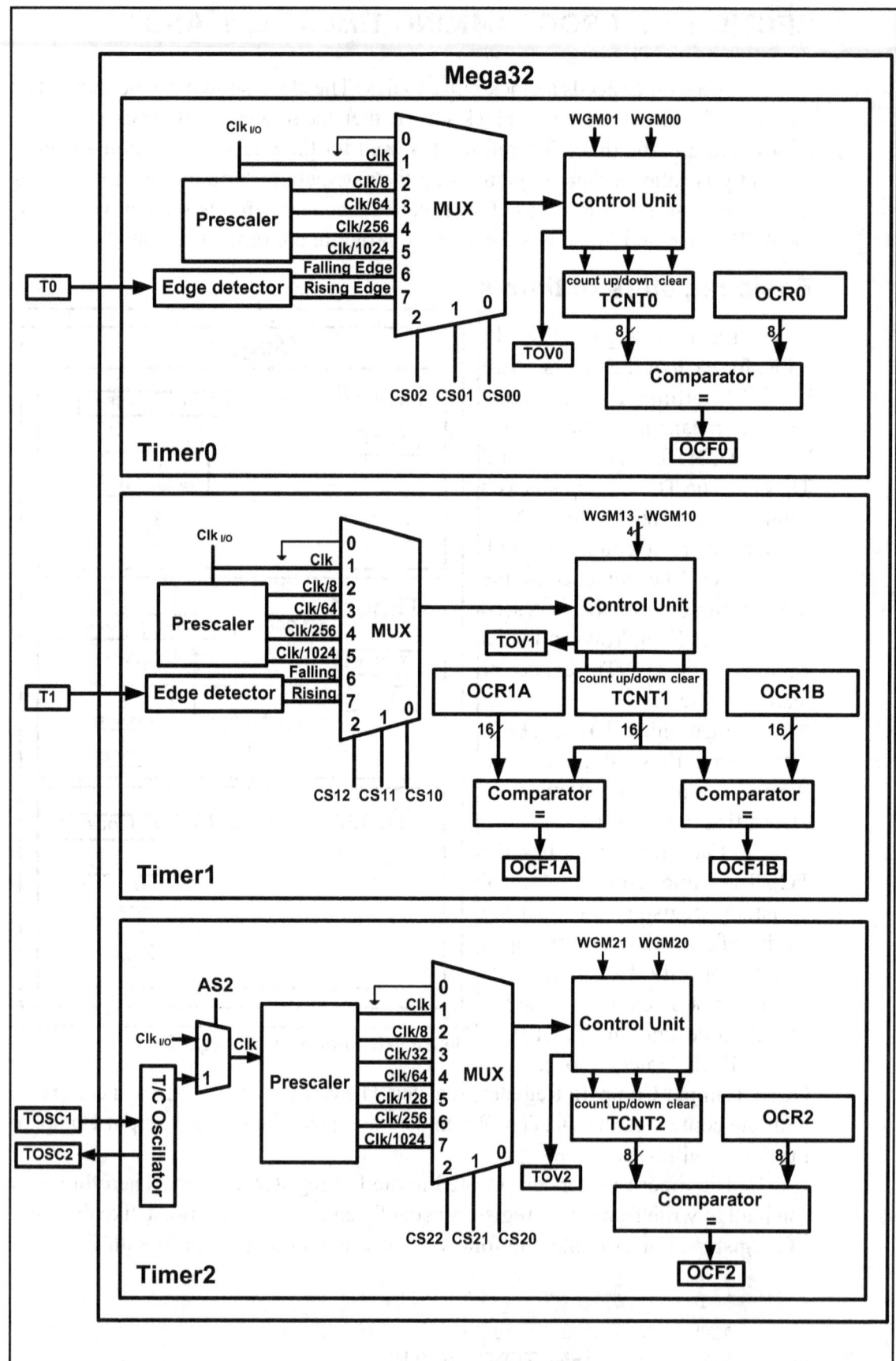

Figure 9-3. Timers in ATmega32

Timer0 programming

Timer0 is 8-bit in ATmega32; thus, TCNT0 is 8-bit as shown in Figure 9-4.

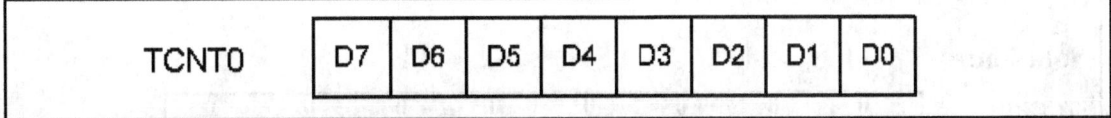

TCNT0	D7	D6	D5	D4	D3	D2	D1	D0

Figure 9-4. Timer/Counter 0 Register

TCCR0 (Timer/Counter Control Register) register

TCCR0 is an 8-bit register used for control of Timer0. The bits for TCCR0 are shown in Figure 9-5.

CS02:CS00 (Timer0 clock source)

These bits in the TCCR0 register are used to choose the clock source. If CS02:CS00 = 000, then the counter is stopped. If CS02–CS00 have values between 001 and 101, the oscillator is used as clock source and the timer/counter acts as a timer. In this case, the timers are often used for time delay generation. See Figure 9-3 and then see Examples 9-1 and 9-2.

Bit	7	6	5	4	3	2	1	0
	FOC0	WGM00	COM01	COM00	WGM01	CS02	CS01	CS00
Read/Write	W	RW	RW	RW	RW	RW	RW	RW
Initial Value	0	0	0	0	0	0	0	0

FOC0   D7   Force compare match: This is a write-only bit, which can be used while generating a wave. Writing 1 to it causes the wave generator to act as if a compare match had occurred.

WGM00, WGM01

D6	D3	Timer0 mode selector bits
0	0	Normal
0	1	CTC (Clear Timer on Compare Match)
1	0	PWM, phase correct
1	1	Fast PWM

COM01:00   D5 D4   Compare Output Mode:
These bits control the waveform generator (see Chapter 15).

CS02:00   D2 D1 D0   Timer0 clock selector

D2	D1	D0	
0	0	0	No clock source (Timer/Counter stopped)
0	0	1	clk (No Prescaling)
0	1	0	clk / 8
0	1	1	clk / 64
1	0	0	clk / 256
1	0	1	clk / 1024
1	1	0	External clock source on T0 pin. Clock on falling edge.
1	1	1	External clock source on T0 pin. Clock on rising edge.

Figure 9-5. TCCR0 (Timer/Counter Control Register) Register

Example 9-1

Find the value for TCCR0 if we want to program Timer0 in Normal mode, no prescaler. Use AVR's crystal oscillator for the clock source.

Solution:

TCCR0 =

0	0	0	0	0	0	0	1
FOC0	WGM00	COM01	COM00	WGM01	CS02	CS01	CS00

Example 9-2

Find the timer's clock frequency and its period for various AVR-based systems, with the following crystal frequencies. Assume that no prescaler is used.
(a) 10 MHz   (b) 8 MHz   (c) 1 MHz

Solution:
(a) F = 10 MHz and T = 1/10 MHz = 0.1 μs
(b) F = 8 MHz and T = 1/8 MHz = 0.125 μs
(c) F = 1 MHz and T = 1/1 MHz = 1 μs

If CS02–CS00 are 110 or 111, the external clock source is used and it acts as a counter. We will discuss Counter in the next section.

WGM01:00

Timer0 can work in four different modes: Normal, phase correct PWM, CTC, and Fast PWM. The WGM01 and WGM00 bits are used to choose one of them. We will discuss the PWM options in Chapter 16.

TIFR (Timer/counter Interrupt Flag Register) register

The TIFR register contains the flags of different timers, as shown in Figure 9-6. Next, we discuss the TOV0 flag, which is related to Timer0.

Bit	7	6	5	4	3	2	1	0
	OCF2	TOV2	ICF1	OCF1A	OCF1B	TOV1	OCF0	TOV0
Read/Write	R/W	R/W	R/W	R/W	R/W	R/W	R/W	R/W
Initial Value	0	0	0	0	0	0	0	0

TOV0	D0	Timer0 overflow flag bit
		0 = Timer0 did not overflow.
		1 = Timer0 has overflowed (going from $FF to $00).
OCF0	D1	Timer0 output compare flag bit
		0 = compare match did not occur.
		1 = compare match occurred.
TOV1	D2	Timer1 overflow flag bit
OCF1B	D3	Timer1 output compare B match flag
OCF1A	D4	Timer1 output compare A match flag
ICF1	D5	Input Capture flag
TOV2	D6	Timer2 overflow flag
OCF2	D7	Timer2 output compare match flag

Figure 9-6. TIFR (Timer/Counter Interrupt Flag Register)

The flag is set when the counter overflows, going from $FF to $00. As we will see soon, when the timer rolls over from $FF to 00, the TOV0 flag is set to 1 and it remains set until the software clears it. See Figure 9-6. The strange thing about this flag is that in order to clear it we need to write 1 to it. Indeed this rule applies to all flags of the AVR chip. In AVR, when we want to clear a given flag of a register we write 1 to it and 0 to the other bits. For example, the following program clears TOV0:

```
LDI   R20,0x01
OUT   TIFR,R20   ;TIFR = 0b00000001
```

Normal mode

In this mode, the content of the timer/counter increments with each clock. It counts up until it reaches its max of 0xFF. When it rolls over from 0xFF to 0x00, it sets high a flag bit called TOV0 (Timer Overflow). This timer flag can be monitored. See Figure 9-7.

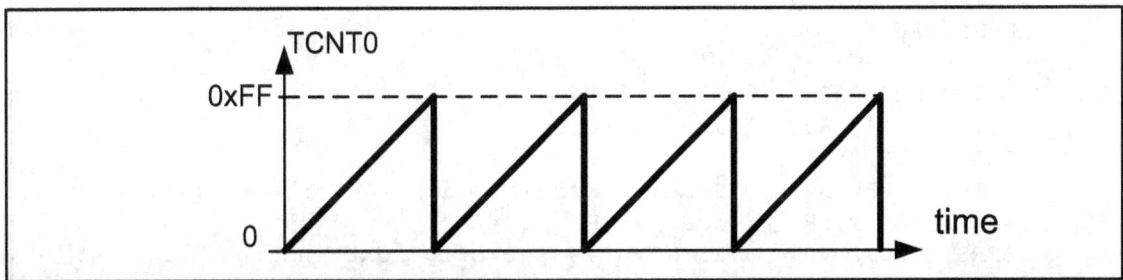

Figure 9-7. Timer/Counter 0 Normal Mode

Steps to program Timer0 in Normal mode

To generate a time delay using Timer0 in Normal mode, the following steps are taken:

1. Load the TCNT0 register with the initial count value.
2. Load the value into the TCCR0 register, indicating which mode (8-bit or 16-bit) is to be used and the prescaler option. When you select the clock source, the timer/counter starts to count, and each tick causes the content of the timer/counter to increment by 1.
3. Keep monitoring the timer overflow flag (TOV0) to see if it is raised. Get out of the loop when TOV0 becomes high.
4. Stop the timer by disconnecting the clock source, using the following instructions:
```
LDI   R20,0x00
OUT   TCCR0,R20   ;timer stopped, mode=Normal
```
5. Clear the TOV0 flag for the next round.
6. Go back to Step 1 to load TCNT0 again.

To clarify the above steps, see Example 9-3.

Example 9-3

In the following program, we are creating a square wave of 50% duty cycle (with equal portions high and low) on the PORTB.5 bit. Timer0 is used to generate the time delay. Analyze the program.

```
.INCLUDE "M32DEF.INC"
.MACRO      INITSTACK          ;set up stack
      LDI   R20,HIGH(RAMEND)
      OUT   SPH,R20
      LDI   R20,LOW(RAMEND)
      OUT   SPL,R20
.ENDMACRO
      INITSTACK
      LDI   R16,1<<5     ;R16 = 0x20 (0010 0000 for PB5)
      SBI   DDRB,5       ;PB5 as an output
      LDI   R17,0
      OUT   PORTB,R17    ;clear PORTB
BEGIN:RCALL DELAY        ;call timer delay
      EOR   R17,R16      ;toggle D5 of R17 by Ex-Oring with 1
      OUT   PORTB,R17    ;toggle PB5
      RJMP  BEGIN
;--------------------Time0 delay
DELAY:LDI   R20,0xF2     ;R20 = 0xF2
      OUT   TCNT0,R20    ;load timer0
      LDI   R20,0x01
      OUT   TCCR0,R20    ;Timer0, Normal mode, int clk, no prescaler
AGAIN:IN    R20,TIFR     ;read TIFR
      SBRS  R20,TOV0     ;if TOV0 is set skip next instruction
      RJMP  AGAIN
      LDI   R20,0x0
      OUT   TCCR0,R20    ;stop Timer0
      LDI   R20,(1<<TOV0)
      OUT   TIFR,R20     ;clear TOV0 flag by writing a 1 to TIFR
      RET
```

Solution:

In the above program notice the following steps:
1. 0xF2 is loaded into TCNT0.
2. TCCR0 is loaded and Timer0 is started.
3. Timer0 counts up with the passing of each clock, which is provided by the crystal oscillator. As the timer counts up, it goes through the states of F3, F4, F5, F6, F7, F8, F9, FA, FB, and so on until it reaches 0xFF. One more clock rolls it to 0, raising the Timer0 flag (TOV0 = 1). At that point, the "SBRS R20, TOV0" instruction bypasses the "RJMP AGAIN" instruction.
4. Timer0 is stopped.
5. The TOV0 flag is cleared.

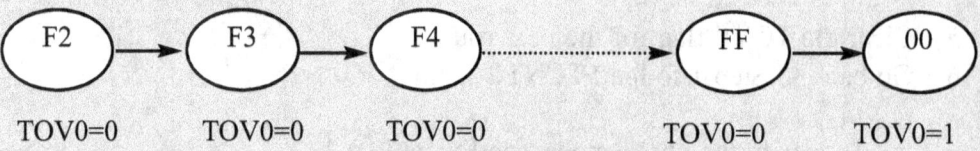

To calculate the exact time delay and the square wave frequency generated on pin PB5, we need to know the XTAL frequency. See Examples 9-4 and 9-5.

Example 9-4

In Example 9-3, calculate the amount of time delay generated by the timer. Assume that XTAL = 8 MHz.

Solution:

We have 8 MHz as the timer frequency. As a result, each clock has a period of T = 1 /8 MHz = 0.125 μs. In other words, Timer0 counts up each 0.125 μs resulting in delay = number of counts × 0.125 μs.
The number of counts for the rollover is 0xFF − 0xF2 = 0x0D (13 decimal). However, we add one to 13 because of the extra clock needed when it rolls over from FF to 0 and raises the TOV0 flag. This gives 14 × 0.125 μs = 1.75 μs for half the pulse.

Example 9-5

In Example 9-3, calculate the frequency of the square wave generated on pin PORTB.5. Assume that XTAL = 8 MHz.

Solution:

To get a more accurate timing, we need to add clock cycles due to the instructions.

```
                                              Cycles
       LDI    R16,0x20
       SBI    DDRB,5
       LDI    R17,0
       OUT    PORTB,R17
BEGIN:RCALL   DELAY                              3
       EOR    R17,R16                            1
       OUT    PORTB,R17                          1
       RJMP   BEGIN                              2
DELAY:LDI     R20,0xF2                           1
       OUT    TCNT0,R20                          1
       LDI    R20,0x01                           1
       OUT    TCCR0,R20                          1
AGAIN:IN      R20,TIFR                           1
       SBRS   R20,0                          1 / 2
       RJMP   AGAIN                              2
       LDI    R20,0x0                            1
       OUT    TCCR0,R20                          1
       LDI    R20,0x01                           1
       OUT    TIFR,R20                           1
       RET                                       4
                                                24
```

T = 2 × (14 + 24) × 0.125 μs = 9.5 μs and F = 1 / T = 105.263 kHz.

(a) in hex	(b) in decimal
(FF - XX + 1) × 0.125 μs where XX is the TCNT0, initial value. Notice that XX value is in hex.	Convert XX value of the TCNT0 register to decimal to get a NNN decimal number, then (256 - NNN) × 0.125 μs

Figure 9-8. Timer Delay Calculation for XTAL = 8 MHz with No Prescaler

We can develop a formula for delay calculations using the Normal mode of the timer for a crystal frequency of XTAL = 8 MHz. This is given in Figure 9-8. The scientific calculator in the Accessories menu directory of Microsoft Windows can help you find the TCNT0 value. This calculator supports decimal, hex, and binary calculations. See Example 9-6.

Example 9-6

Find the delay generated by Timer0 in the following code, using both of the methods of Figure 9-8. Do not include the overhead due to instructions. (XTAL = 8 MHz)

```
.INCLUDE "M32DEF.INC"
        INITSTACK              ;add its definition from Example 9-3
        LDI     R16,0x20
        SBI     DDRB,5         ;PB5 as an output
        LDI     R17,0
        OUT     PORTB,R17
BEGIN:RCALL     DELAY
        EOR     R17,R16        ;toggle D5 of R17
        OUT     PORTB,R17      ;toggle PB5
        RJMP    BEGIN
DELAY:LDI       R20,0x3E
        OUT     TCNT0,R20      ;load timer0
        LDI     R20,0x01
        OUT     TCCR0,R20      ;Timer0, Normal mode, int clk, no prescaler
AGAIN:IN        R20,TIFR       ;read TIFR
        SBRS    R20,TOV0       ;if TOV0 is set skip next instruction
        RJMP    AGAIN
        LDI     R20,0x00
        OUT     TCCR0,R20      ;stop Timer0
        LDI     R20,(1<<TOV0)  ;R20 = 0x01
        OUT     TIFR,R20       ;clear TOV0 flag
        RET
```

Solution:

(a) (FF − 3E + 1) = 0xC2 = 194 in decimal and 194 × 0.125 μs = 24.25 μs.

(b) Because TCNT0 = 0x3E = 62 (in decimal) we have 256 − 62 = 194. This means that the timer counts from 0x3E to 0xFF. This plus rolling over to 0 goes through a total of 194 clock cycles, where each clock is 0.125 μs in duration. Therefore, we have 194 × 0.125 μs = 24.25 μs as the width of the pulse.

Finding values to be loaded into the timer

Assuming that we know the amount of timer delay we need, the question is how to find the values needed for the TCNT0 register. To calculate the values to be loaded into the TCNT0 registers, we can use the following steps:

1. Calculate the period of the timer clock using the following formula:

$$T_{clock} = 1/F_{Timer}$$

where F_{Timer} is the frequency of the clock used for the timer. For example, in no prescaler mode, $F_{Timer} = F_{oscillator}$. T_{clock} gives the period at which the timer increments.

2. Divide the desired time delay by T_{clock}. This says how many clocks we need.

3. Perform 256 – n, where n is the decimal value we got in Step 2.

4. Convert the result of Step 3 to hex, where xx is the initial hex value to be loaded into the timer's register.

5. Set TCNT0 = xx.

Look at Examples 9-7 and 9-8, where we use a crystal frequency of 8 MHz for the AVR system.

Example 9-7

Assuming that XTAL = 8 MHz, write a program to generate a square wave with a period of 12.5 μs on pin PORTB.3.

Solution:

For a square wave with T = 12.5 μs we must have a time delay of 6.25 μs. Because XTAL = 8 MHz, the counter counts up every 0.125 μs. This means that we need 6.25 μs / 0.125 μs = 50 clocks. 256 – 50 = 206 = 0xCE. Therefore, we have TCNT0 = 0xCE.

```
.INCLUDE "M32DEF.INC"
      INITSTACK           ;add its definition from Example 9-3
      LDI    R16,0x08
      SBI    DDRB,3        ;PB3 as an output
      LDI    R17,0
      OUT    PORTB,R17
BEGIN:RCALL DELAY
      EOR    R17,R16       ;toggle D3 of R17
      OUT    PORTB,R17     ;toggle PB3
      RJMP   BEGIN
;---------------- Timer0 Delay
DELAY:LDI    R20,0xCE
      OUT    TCNT0,R20     ;load Timer0
      LDI    R20,0x01
      OUT    TCCR0,R20     ;Timer0, Normal mode, int clk, no prescaler
AGAIN:IN     R20,TIFR      ;read TIFR
      SBRS   R20,TOV0      ;if TOV0 is set skip next instruction
      RJMP   AGAIN
      LDI    R20,0x00
      OUT    TCCR0,R20     ;stop Timer0
      LDI    R20,(1<<TOV0)
      OUT    TIFR,R20      ;clear TOV0 flag
      RET
```

Example 9-8

Assuming that XTAL = 8 MHz, modify the program in Example 9-7 to generate a square wave of 16 kHz frequency on pin PORTB.3.

Solution:

Look at the following steps.
(a) T = 1 / F = 1 / 16 kHz = 62.5 µs the period of the square wave.
(b) 1/2 of it for the high and low portions of the pulse is 31.25 µs.
(c) 31.25 µs / 0.125 µs = 250 and 256 – 250 = 6, which in hex is 0x06.
(d) TCNT0 = 0x06.

Using the Windows calculator to find TCNT0

The scientific calculator in Microsoft Windows is a handy and easy-to-use tool to find the TCNT0 value. Assume that we would like to find the TCNT0 value for a time delay that uses 135 clocks of 0.125 µs. The following steps show the calculation:

1. Bring up the scientific calculator in MS Windows and select decimal.
2. Enter 135.
3. Select hex. This converts 135 to hex, which is 0x87.
4. Select +/– to give –135 decimal (0x79).
5. The lowest two digits (79) of this hex value are for TCNT0. We ignore all the Fs on the left because our number is 8-bit data.

Prescaler and generating a large time delay

As we have seen in the examples so far, the size of the time delay depends on two factors, (a) the crystal frequency, and (b) the timer's 8-bit register. Both of

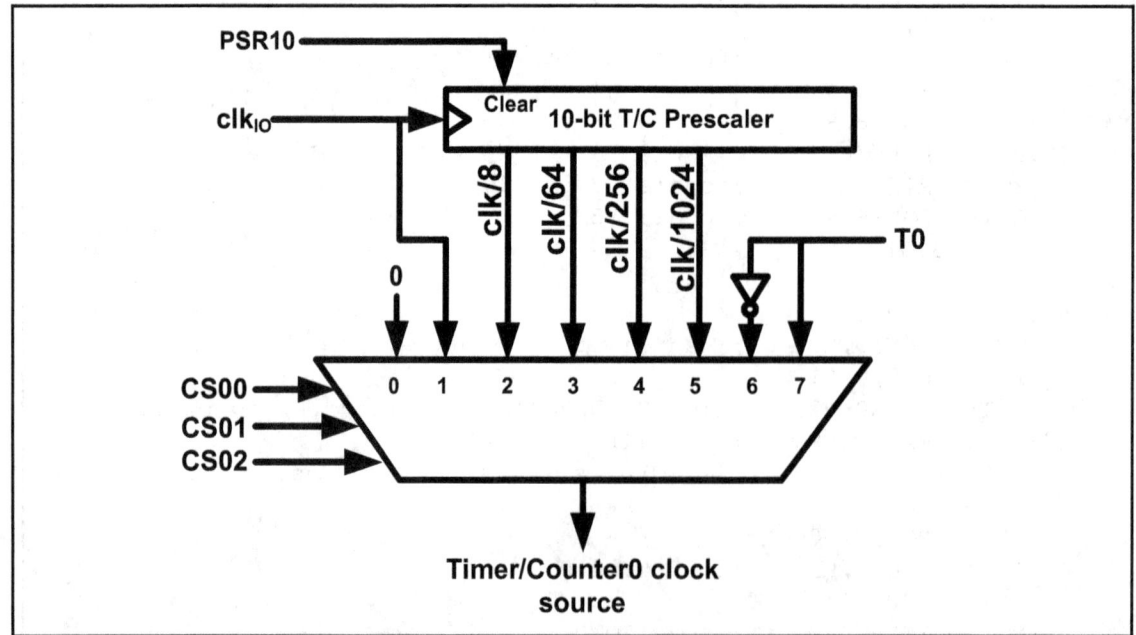

Figure 9-9. Timer/Counter 0 Prescaler

Example 9-9

Modify TCNT0 in Example 9-7 to get the largest time delay possible. Find the delay in ms. In your calculation, exclude the overhead due to the instructions in the loop.

Solution:

To get the largest delay we make TCNT0 zero. This will count up from 00 to 0xFF and then roll over to zero.

```
.INCLUDE "M32DEF.INC"
        INITSTACK           ;add its definition from Example 9-3
        LDI     R16,0x08
        SBI     DDRB,3      ;PB3 as an output
        LDI     R17,0
        OUT     PORTB,R17
BEGIN:RCALL     DELAY
        EOR     R17,R16     ;toggle D3 of R17
        OUT     PORTB,R17   ;toggle PB3
        RJMP    BEGIN
;---------------- Timer0 Delay
DELAY:LDI       R20,0x00
        OUT     TCNT0,R20   ;load Timer0 with zero
        LDI     R20,0x01
        OUT     TCCR0,R20   ;Timer0, Normal mode, int clk, no prescaler
AGAIN:IN        R20,TIFR    ;read TIFR
        SBRS    R20,TOV0    ;if TOV0 is set skip next instruction
        RJMP    AGAIN
        LDI     R20,0x00
        OUT     TCCR0,R20   ;stop Timer0
        LDI     R20,(1<<TOV0)
        OUT     TIFR,R20    ;clear TOV0 flag
        RET
```

Making TCNT0 zero means that the timer will count from 00 to 0xFF, and then will roll over to raise the TCNT0 flag. As a result, it goes through a total of 256 states. Therefore, we have delay = $(256 - 0) \times 0.125$ μs = 32 μs. That gives us the smallest frequency of $1 / (2 \times 32$ μs$) = 1 / (64$ μs$) = 15.625$ kHz.

these factors are beyond the control of the AVR programmer. We saw in Example 9-9 that the largest time delay is achieved by making TCNT0 zero. What if that is not enough? We can use the prescaler option in the TCCR0 register to increase the delay by reducing the period. The prescaler option of TCCR0 allows us to divide the instruction clock by a factor of 8 to 1024 as was shown in Figure 9-5. The prescaler of Timer/Counter 0 is shown in Figure 9-9.

As we have seen so far, with no prescaler enabled, the crystal oscillator frequency is fed directly into Timer0. If we enable the prescaler bit in the TCCR0 register, however, then we can divide the clock before it is fed into Timer0. The lower 3 bits of the TCCR0 register give the options of the number we can divide by. As shown in Figure 9-9, this number can be 8, 64, 256, and 1024. Notice that the lowest number is 8 and the highest number is 1024. Examine Examples 9-10 through 9-14 to see how the prescaler options are programmed.

Example 9-10

Find the timer's clock frequency and its period for various AVR-based systems, with the following crystal frequencies. Assume that a prescaler of 1:64 is used.

(a) 8 MHz   (b) 16 MHz   (c) 10 MHz

Solution:

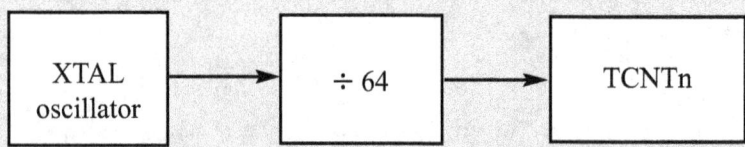

(a) $1/64 \times$ 8 MHz = 125 kHz due to 1:64 prescaler and T = 1/125 kHz = 8 μs
(b) $1/64 \times$ 16 MHz = 250 kHz due to prescaler and T = 1/250 kHz = 4 μs
(c) $1/64 \times$ 10 MHz = 156.2 kHz due to prescaler and T = 1/156 kHz = 6.4 μs

Example 9-11

Find the value for TCCR0 if we want to program Timer0 in Normal mode with a prescaler of 64 using internal clock for the clock source.

Solution:

From Figure 9-5 we have TCCR0 = 0000 0011; XTAL clock source, prescaler of 64.

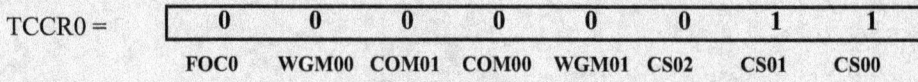

Example 9-12

Examine the following program and find the time delay in seconds. Exclude the overhead due to the instructions in the loop. Assume XTAL = 8 MHz.

```
.INCLUDE "M32DEF.INC"
      INITSTACK           ;add its definition from Example 9-3
      LDI    R16,0x08
      SBI    DDRB,3        ;PB3 as an output
      LDI    R17,0
      OUT    PORTB,R17
BEGIN:RCALL  DELAY
      EOR    R17,R16       ;toggle D3 of R17
      OUT    PORTB,R17     ;toggle PB3
      RJMP   BEGIN
;---------------- Timer0 Delay
DELAY:LDI    R20,0x10
      OUT    TCNT0,R20     ;load Timer0
      LDI    R20,0x03
      OUT    TCCR0,R20     ;Timer0, Normal mode, int clk, prescaler 64
AGAIN:IN     R20,TIFR      ;read TIFR
      SBRS   R20,TOV0      ;if TOV0 is set skip next instruction
      RJMP   AGAIN
      LDI    R20,0x0
```

Example 9-12 (Cont.)

```
        OUT   TCCR0,R20    ;stop Timer0
        LDI   R20,1<<TOV0
        OUT   TIFR,R20     ;clear TOV0 flag
        RET
```

Solution:

TCNT0 = 0x10 = 16 in decimal and 256 – 16 = 240. Now $240 \times 64 \times 0.125\ \mu s = 1920$ μs, or from Example 9-10, we have $240 \times 8\ \mu s = 1920\ \mu s$.

Example 9-13

Assume XTAL = 8 MHz. (a) Find the clock period fed into Timer0 if a prescaler option of 1024 is chosen. (b) Show what is the largest time delay we can get using this prescaler option and Timer0.

Solution:

(a) 8 MHz × 1/1024 = 7812.5 Hz due to 1:1024 prescaler and T = 1/7812.5 Hz = 128 ms = 0.128 ms

(b) To get the largest delay, we make TCNT0 zero. Making TCNT0 zero means that the timer will count from 00 to 0xFF, and then roll over to raise the TOV0 flag. As a result, it goes through a total of 256 states. Therefore, we have delay = (256 – 0) × 128 μs = 32,768 μs = 0.032768 seconds.

Example 9-14

Assuming XTAL = 8 MHz, write a program to generate a square wave of 125 Hz frequency on pin PORTB.3. Use Timer0, Normal mode, with prescaler = 256.

Solution:

Look at the following steps:
(a) T = 1 / 125 Hz = 8 ms, the period of the square wave.
(b) 1/2 of it for the high and low portions of the pulse = 4 ms
(c) (4 ms / 0.125 μs) / 256 = 125 and 256 – 125 = 131 in decimal, and in hex it is 0x83.
(d) TCNT0 = 83 (hex)

```
.INCLUDE "M32DEF.INC"
     .MACRO INITSTACK          ;set up stack
     LDI   R20,HIGH(RAMEND)
     OUT   SPH,R20
     LDI   R20,LOW(RAMEND)
     OUT   SPL,R20
.ENDMACRO
```

Example 9-14 (Cont.)

```
        INITSTACK
        LDI    R16,0x08
        SBI    DDRB,3        ;PB3 as an output
        LDI    R17,0
BEGIN:OUT      PORTB,R17     ;PORTB = R17
        CALL   DELAY
        EOR    R17,R16       ;toggle D3 of R17
        RJMP   BEGIN

;---------------- Timer0 Delay
DELAY:LDI      R20,0x83
        OUT    TCNT0,R20     ;load Timer0
        LDI    R20,0x04
        OUT    TCCR0,R20     ;Timer0, Normal mode, int clk, prescaler 256

AGAIN:IN       R20,TIFR      ;read TIFR
        SBRS   R20,TOV0      ;if TOV0 is set skip next instruction
        RJMP   AGAIN

        LDI    R20,0x0
        OUT    TCCR0,R20     ;stop Timer0
        LDI    R20,1<<TOV0
        OUT    TIFR,R20      ;clear TOV0 flag
        RET
```

Assemblers and negative values

Because the timer is in 8-bit mode, we can let the assembler calculate the value for TCNT0. For example, in the "LDI R20, -100" instruction, the assembler will calculate the –100 = 9C and make R20 = 9C in hex. This makes our job easier. See Examples 9-15 and 9-16.

Example 9-15

Find the value (in hex) loaded into TCNT0 for each of the following cases.

(a)   LDI R20, -200   (b)   LDI R17,-60   (c)   LDI R25,-12
       OUT TCNT0,R20   OUT TCNT0,R17   OUT TCNT0,R25

Solution:

You can use the Windows scientific calculator to verify the results provided by the assembler. In the Windows calculator, select decimal and enter 200. Then select hex, then +/– to get the negative value. The following is what we get.

Decimal	*2's complement (TCNT0 value)*
–200	0x38
–60	0xC4
–12	0xF4

Example 9-16

Find (a) the frequency of the square wave generated in the following code, and (b) the duty cycle of this wave. Assume XTAL = 8 MHz.

```
.INCLUDE "M32DEF.INC"
      LDI    R16,HIGH(RAMEND)
      OUT    SPH,R16
      LDI    R16,LOW(RAMEND)
      OUT    SPL,R16           ;initialize stack pointer
      LDI    R16,0x20
      SBI    DDRB,5            ;PB5 as an output
      LDI    R18,-150
BEGIN:SBI    PORTB,5           ;PB5 = 1
      OUT    TCNT0,R18         ;load Timer0 byte
      CALL   DELAY
      OUT    TCNT0,R18         ;reload Timer0 byte
      CALL   DELAY
      CBI    PORTB,5           ;PB5 = 0
      OUT    TCNT0,R18         ;reload Timer0 byte
      CALL   DELAY
      RJMP   BEGIN

;----- Delay using Timer0
DELAY:LDI    R20,0x01
      OUT    TCCR0,R20   ;start Timer0, Normal mode, int clk, no prescaler
AGAIN:IN     R20,TIFR    ;read TIFR
      SBRS   R20,TOV0    ;monitor TOV0 flag and skip if high
      RJMP   AGAIN
      LDI    R20,0x0
      OUT    TCCR0,R20   ;stop Timer0
      LDI    R20,1<<TOV0
      OUT    TIFR,R20    ;clear TOV0 flag bit
      RET
```

Solution:

For the TCNT0 value in 8-bit mode, the conversion is done by the assembler as long as we enter a negative number. This also makes the calculation easy. Because we are using 150 clocks, we have time for the DELAY subroutine = 150 × 0.125 μs = 18.75 μs. The high portion of the pulse is twice the size of the low portion (66% duty cycle). Therefore, we have: T = high portion + low portion = 2 × 18.75 μs + 18.75 μs = 56.25 μs and frequency = 1 / 56.25 μs = 17.777 kHz.

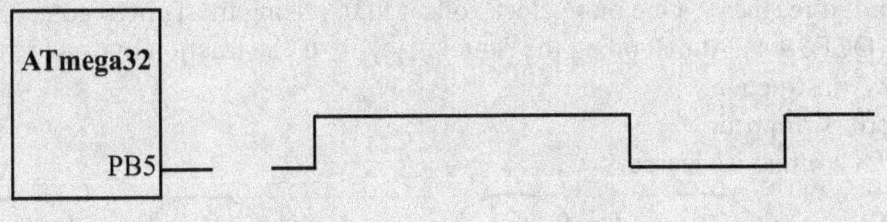

Clear Timer0 on compare match (CTC) mode programming

Examining Figure 9-2 once more, we see the OCR0 register. The OCR0 register is used with CTC mode. As with the Normal mode, in the CTC mode, the timer is incremented with a clock. But it counts up until the content of the TCNT0 register becomes equal to the content of OCR0 (compare match occurs); then, the timer will be cleared and the OCF0 flag will be set when the next clock occurs. The OCF0 flag is located in the TIFR register. See Figure 9-10 and Examples 9-17 through 9-21.

Example 9-17

In the following program, we are creating a square wave of 50% duty cycle (with equal portions high and low) on the PORTB.5 bit. Timer0 is used to generate the time delay. Analyze the program.

```
.INCLUDE "M32DEF.INC"
        INITSTACK               ;add its definition from Example 9-3
        LDI    R16,0x08
        SBI    DDRB,3           ;PB3 as an output
        LDI    R17,0
BEGIN:OUT      PORTB,R17        ;PORTB = R17
        RCALL  DELAY
        EOR    R17,R16          ;toggle D3 of R17
        RJMP   BEGIN
;--------------- Timer0 Delay
DELAY:LDI      R20,0
        OUT    TCNT0,R20
        LDI    R20,9
        OUT    OCR0,R20         ;load OCR0
        LDI    R20,0x09
        OUT    TCCR0,R20        ;Timer0, CTC mode, int clk
AGAIN:IN       R20,TIFR         ;read TIFR
        SBRS   R20,OCF0         ;if OCF0 is set skip next inst.
        RJMP   AGAIN
        LDI    R20,0x0
        OUT    TCCR0,R20        ;stop Timer0
        LDI    R20,1<<OCF0
        OUT    TIFR,R20         ;clear OCF0 flag
        RET
```

Solution:

In the above program notice the following steps:

1. 9 is loaded into OCR0.
2. TCCR0 is loaded and Timer0 is started.
3. Timer0 counts up with the passing of each clock, which is provided by the crystal oscillator. As the timer counts up, it goes through the states of 00, 01, 02, 03, and so on until it reaches 9. One more clock rolls it to 0, raising the Timer0 compare match flag (OCF0 = 1). At that point, the "SBRS R20,OCF0" instruction bypasses the "RJMP AGAIN" instruction.
4. Timer0 is stopped.
5. The OCF0 flag is cleared.

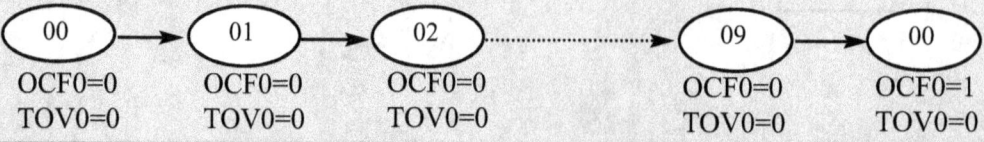

328

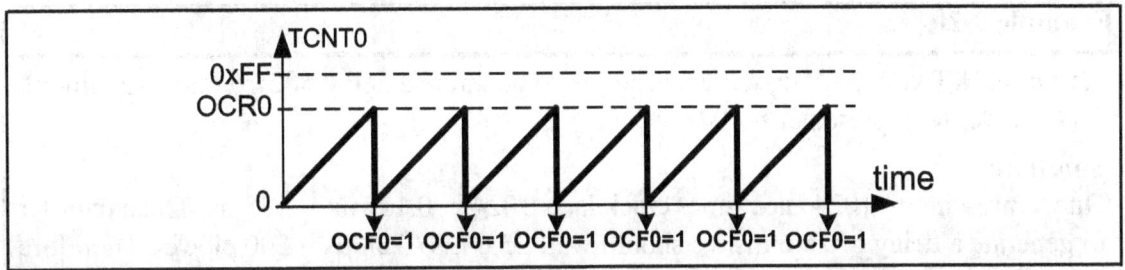

Figure 9-10. Timer/Counter 0 CTC Mode

Example 9-18

Find the delay generated by Timer0 in Example 9-17. Do not include the overhead due to instructions. (XTAL = 8 MHz)

Solution:

OCR0 is loaded with 9 and TCNT0 is cleared; Thus, after 9 clocks TCNT0 becomes equal to OCR0. On the next clock, the OCF0 flag is set and the reset occurs. That means the TCNT0 is cleared after $9 + 1 = 10$ clocks. Because XTAL = 8 MHz, the counter counts up every 0.125 µs. Therefore, we have 10×0.125 µs = 1.25 µs.

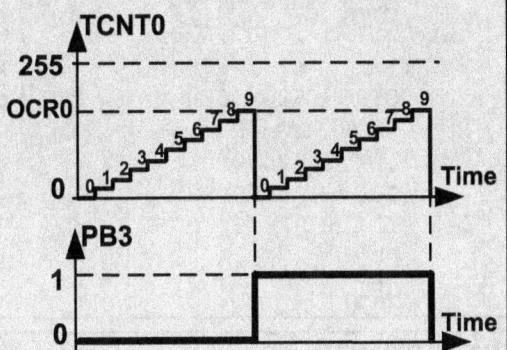

Example 9-19

Find the delay generated by Timer0 in the following program. Do not include the overhead due to instructions. (XTAL = 8 MHz)

```
.INCLUDE "M32DEF.INC"
      LDI    R16,0x08
      SBI    DDRB,3        ;PB3 as an output
      LDI    R17,0
      OUT    PORTB,R17
      LDI    R20,89
      OUT    OCR0,R20       ;load Timer0
BEGIN:LDI    R20,0x0B
      OUT    TCCR0,R20      ;Timer0, CTC mode, prescaler = 64
AGAIN:IN     R20,TIFR       ;read TIFR
      SBRS   R20,OCF0       ;if OCF0 flag is set skip next instruction
      RJMP   AGAIN
      LDI    R20,0x0
      OUT    TCCR0,R20      ;stop Timer0 (This line can be omitted)
      LDI    R20,1<<OCF0
      OUT    TIFR,R20       ;clear OCF0 flag
      EOR    R17,R16        ;toggle D3 of R17
      OUT    PORTB,R17      ;toggle PB3
      RJMP   BEGIN
```

Solution:

Due to prescaler = 64 each timer clock lasts 64×0.125 µs = 8 µs. OCR0 is loaded with 89; thus, after 90 clocks OCF0 is set. Therefore we have 90×8 µs = 720 µs.

Example 9-20

Assuming XTAL = 8 MHz, write a program to generate a delay of 25.6 ms. Use Timer0, CTC mode, with prescaler = 1024.

Solution:

Due to prescaler = 1024 each timer clock lasts $1024 \times 0.125\ \mu s = 128\ \mu s$. Thus, in order to generate a delay of 25.6 ms we should wait 25.6 ms / 128 μs = 200 clocks. Therefore the OCR0 register should be loaded with 200 − 1 = 199.

```
DELAY:LDI   R20,0
      OUT   TCNT0,R20
      LDI   R20,199
      OUT   OCR0,R20         ;load OCR0
      LDI   R20,0x0D
      OUT   TCCR0,R20        ;Timer0, CTC mode, prescaler = 1024
AGAIN:IN    R20,TIFR         ;read TIFR
      SBRS  R20,OCF0         ;if OCF0 is set skip next inst.
      RJMP  AGAIN
      LDI   R20,0x0
      OUT   TCCR0,R20        ;stop Timer0
      LDI   R20,1<<OCF0
      OUT   TIFR,R20         ;clear OCF0 flag
      RET
```

Example 9-21

Assuming XTAL = 8 MHz, write a program to generate a delay of 1 ms.

Solution:

As XTAL = 8 MHz, the different outputs of the prescaler are as follows:

Prescaler	Timer Clock	Timer Period	Timer Value
None	8 MHz	1/8 MHz = 0.125 μs	1 ms/0.125 μs = 8000
8	8 MHz/8 = 1 MHz	1/1 MHz = 1 μs	1 ms/1 μs = 1000
64	8 MHz/64 = 125 kHz	1/125 kHz = 8 μs	1 ms/8 μs = **125**
256	8 MHz/256 = 31.25 kHz	1/31.25 kHz = 32 μs	1 ms/32 μs = **31.25**
1024	8 MHz/1024 = 7.8125 kHz	1/7.8125 kHz= 128 μs	1 ms/128 μs = **7.8125**

From the above calculation we can only use the options Prescaler = 64, Prescaler = 256, or Prescaler = 1024. We should use the option Prescaler = 64 since we cannot use a decimal point. To wait 125 clocks we should load OCR0 with 125 − 1 = 124.

```
DELAY:LDI   R20,0
      OUT   TCNT0,R20        ;TCNT0 = 0
      LDI   R20,124
      OUT   OCR0,R20         ;OCR0 = 124
      LDI   R20,0x0B
      OUT   TCCR0,R20        ;Timer0, CTC mode, prescaler = 64
AGAIN:IN    R20,TIFR         ;read TIFR
      SBRS  R20,OCF0         ;if OCF0 is set skip next instruction
      RJMP  AGAIN
      LDI   R20,0x0
      OUT   TCCR0,R20        ;stop Timer0
      LDI   R20,1<<OCF0
      OUT   TIFR,R20         ;clear OCF0 flag
      RET
```

Notice that the comparator checks for equality; thus, if we load the OCR0 register with a value that is smaller than TCNT0's value, the counter will miss the compare match and will count up until it reaches the maximum value of $FF and rolls over. This causes a big delay and is not desirable in many cases. See Example 9-22.

Example 9-22

In the following program, how long does it take for the PB3 to become one? Do not include the overhead due to instructions. (XTAL = 8 MHz)

```
.INCLUDE "M32DEF.INC"
        SBI    DDRB,3          ;PB3 as an output
        CBI    PORTB,3         ;PB3 = 0
        LDI    R20,89
        OUT    OCR0,R20        ;OCR0 = 89
        LDI    R20,95
        OUT    TCNT0,R20       ;TCNT0 = 95
BEGIN:LDI    R20,0x09
        OUT    TCCR0,R20       ;Timer0, CTC mode, prescaler = 1
AGAIN:IN     R20,TIFR         ;read TIFR
        SBRS   R20,OCF0        ;if OCF0 flag is set skip next inst.
        RJMP   AGAIN
        LDI    R20,0x0
        OUT    TCCR0,R20       ;stop Timer0 (This line can be omitted)
        LDI    R20,1<<OCF0
        OUT    TIFR,R20    ;clear OCF0 flag
        EOR    R17,R16     ;toggle D3 of R17
        OUT    PORTB,R17   ;toggle PB3
        RJMP   BEGIN
```

Solution:

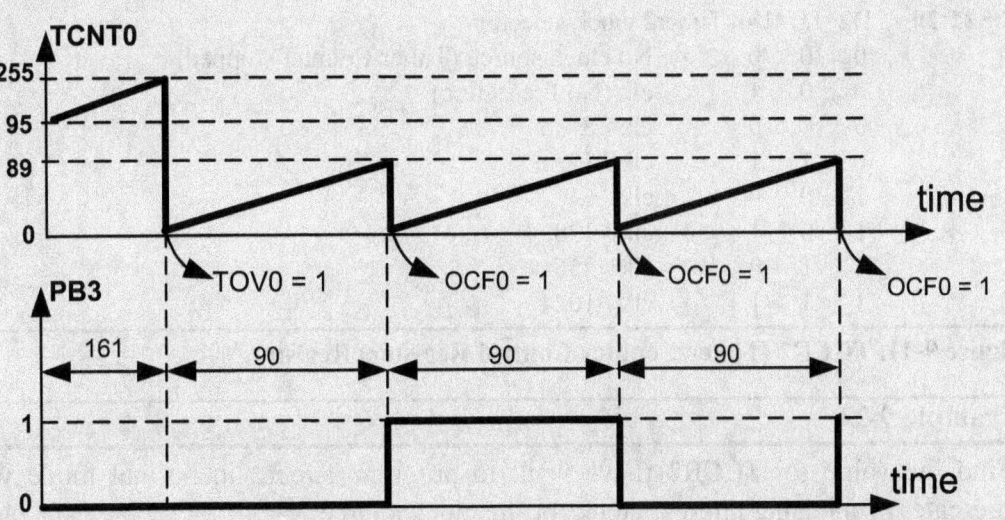

Since the value of TCNT0 (95) is bigger than the content of OCR0 (89), the timer counts up until it gets to $FF and rolls over to zero. The TOV0 flag will be set as a result of the overflow. Then, the timer counts up until it becomes equal to 89 and compare match occurs. Thus, the first compare match occurs after 161 + 90 = 251 clocks, which means after 251 × 0.125 μs = 31.375 μs. The next compare matches occur after 90 clocks, which means after 90 × 0.125 μs = 11.25 μs.

Timer2 programming

See Figure 9-12. Timer2 is an 8-bit timer. Therefore it works the same way as Timer0. But there are two differences between Timer0 and Timer2:

1. Timer2 can be used as a real time counter. To do so, we should connect a crystal of 32.768 kHz to the TOSC1 and TOSC2 pins of AVR and set the AS2 bit. See Figure 9-12. For more information about this feature, see the AVR datasheet.

2. In Timer0, when CS02–CS00 have values 110 or 111, Timer0 counts the external events. But in Timer2, the multiplexer selects between the different scales of the clock. In other words, the same values of the CS bits can have different meanings for Timer0 and Timer2. Compare Figure 9-11 with Figure 9-5 and examine Examples 9-23 through 9-25.

Bit	7	6	5	4	3	2	1	0
	FOC2	WGM20	COM21	COM20	WGM21	CS22	CS21	CS20
Read/Write	W	RW	RW	RW	RW	RW	RW	RW
Initial Value	0	0	0	0	0	0	0	0

FOC2　　　D7　　　Force compare match: a write-only bit, which can be used while generating a wave. Writing 1 to it causes the wave generator to act as if a compare match had occurred.

WGM20, WGM21

D6	D3	
		Timer2 mode selector bits
0	0	Normal
0	1	CTC (Clear Timer on Compare Match)
1	0	PWM, phase correct
1	1	Fast PWM

COM21:20　　D5 D4　　Compare Output Mode:
These bits control the waveform generator (see Chapter 15).

CS22:20　　D2 D1 D0 Timer2 clock selector

D2	D1	D0	
0	0	0	No clock source (Timer/Counter stopped)
0	0	1	clk (No Prescaling)
0	1	0	clk / 8
0	1	1	clk / 32
1	0	0	clk / 64
1	0	1	clk / 128
1	1	0	clk / 256
1	1	1	clk / 1024

Figure 9-11. TCCR2 (Timer/Counter Control Register) Register

Example 9-23

Find the value for TCCR2 if we want to program Timer2 in normal mode with a prescaler of 64 using internal clock for the clock source.

Solution:

From Figure 9-11 we have TCCR2 = 0000 0100; XTAL clock source, prescaler of 64.

TCCR2 =	0	0	0	0	0	1	0	0
	FOC2	WGM20	COM21	COM20	WGM21	CS22	CS21	CS20

Compare the answer with Example 9-11.

Bit	7	6	5	4	3	2	1	0
	-	-	-	-	AS2	TCN2UB	OCR2UB	TCR2UB

AS2 When it is zero, Timer2 is clocked from clk$_{I/O}$. When it is set, Timer2 works as RTC.

Figure 9-12. ASSR (Asynchronous Status Register)

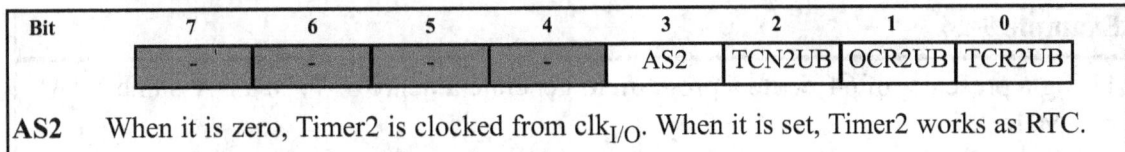

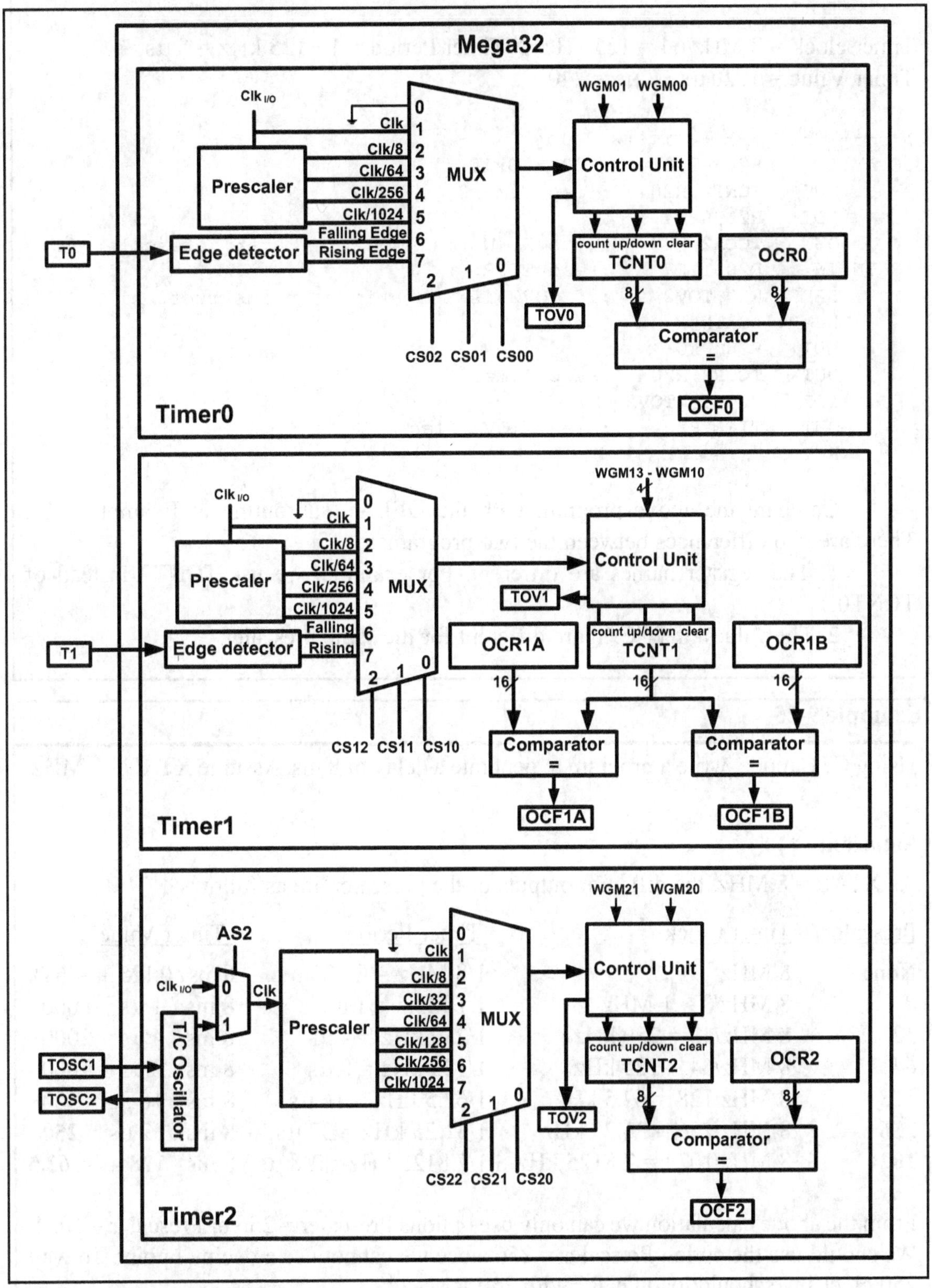

Figure 9-13. Timers in ATmega32

CHAPTER 9: AVR TIMER PROGRAMMING IN ASSEMBLY AND C 333

Example 9-24

Using a prescaler of 64, write a program to generate a delay of 1920 μs. Assume XTAL = 8 MHz.

Solution:

Timer clock = 8 MHz/64 = 125 kHz ➔ Timer Period = 1 / 125 kHz = 8 μs ➔ Timer Value = 1920 μs / 8 μs = 240

```
;---------------- Timer2 Delay
DELAY:LDI    R20,-240    ;R20 = 0x10
      OUT    TCNT2,R20   ;load Timer2
      LDI    R20,0x04
      OUT    TCCR2,R20   ;Timer2, Normal mode, int clk, prescaler 64
AGAIN:IN     R20,TIFR    ;read TIFR
      SBRS   R20,TOV2    ;if TOV2 is set skip next instruction
      RJMP   AGAIN
      LDI    R20,0x0
      OUT    TCCR2,R20   ;stop Timer2
      LDI    R20,1<<TOV2
      OUT    TIFR,R20    ;clear TOV2 flag
      RET
```

Compare the above program with the DELAY subroutine in Example 9-12. There are two differences between the two programs:

1. The register names are different. For example, we use TCNT2 instead of TCNT0.

2. The values of TCCRn are different for the same prescaler.

Example 9-25

Using CTC mode, write a program to generate a delay of 8 ms. Assume XTAL = 8 MHz.

Solution:

As XTAL = 8 MHz, the different outputs of the prescaler are as follows:

Prescaler	Timer Clock	Timer Period	Timer Value
None	8 MHz	1/8 MHz = 0.125 μs	8 ms / 0.125 μs = 64 k
8	8 MHz/8 = 1 MHz	1/1 MHz = 1 μs	8 ms / 1 μs = 8000
32	8 MHz/32 = 250 kHz	1/250 kHz = 4 μs	8 ms / 4 μs = 2000
64	8 MHz/64 = 125 kHz	1/125 kHz = 8 μs	8 ms / 8 μs = 1000
128	8 MHz/128 = 62.5 kHz	1/62.5 kHz = 16 μs	8 ms / 16 μs = 500
256	8 MHz/256 = 31.25 kHz	1/31.25 kHz = 32 μs	8 ms / 32 μs = **250**
1024	8 MHz/1024 = 7.8125 kHz	1/7.8125 kHz = 128 μs	8 ms / 128 μs = **62.5**

From the above calculation we can only use options Prescaler = 256 or Prescaler = 1024. We should use the option Prescaler = 256 since we cannot use a decimal point. To wait 250 clocks we should load OCR2 with 250 − 1 = 249.

Example 9-25 (Cont.)

TCCR2 =	0	0	0	0	1	1	1	0
	FOC2	WGM20	COM21	COM20	WGM21	CS22	CS21	CS20

```
;---------------- Timer2 Delay
DELAY:LDI    R20,0
      OUT    TCNT2,R20          ;TCNT2 = 0
      LDI    R20,249
      OUT    OCR0,R20           ;OCR0 = 249
      LDI    R20,0x0E
      OUT    TCCR0,R20          ;Timer0,CTC mode,prescaler = 256
AGAIN:IN     R20,TIFR           ;read TIFR
      SBRS   R20,OCF2           ;if OCF2 is set skip next inst.
      RJMP   AGAIN
      LDI    R20,0x0
      OUT    TCCR2,R20          ;stop Timer2
      LDI    R20,1<<OCF2
      OUT    TIFR,R20           ;clear OCF2 flag
      RET
```

Timer1 programming

Timer1 is a 16-bit timer and has lots of capabilities. Next, we discuss Timer1 and its capabilities.

Since Timer1 is a 16-bit timer its 16-bit register is split into two bytes. These are referred to as TCNT1L (Timer1 low byte) and TCNT1H (Timer1 high byte). See Figure 9-15. Timer1 also has two control registers named TCCR1A (Timer/counter 1 control register) and TCCR1B. The TOV1 (timer overflow) flag bit goes HIGH when overflow occurs. Timer1 also has the prescaler options of 1:1, 1:8, 1:64, 1:256, and 1:1024. See Figure 9-14 for the Timer1 block diagram and Figures 9-15 and 9-16 for TCCR1 register options. There are two OCR registers in Timer1: OCR1A and OCR1B. There are two separate flags for each of the OCR registers, which act independently of each other. Whenever TCNT1 equals OCR1A, the OCF1A flag will be set on

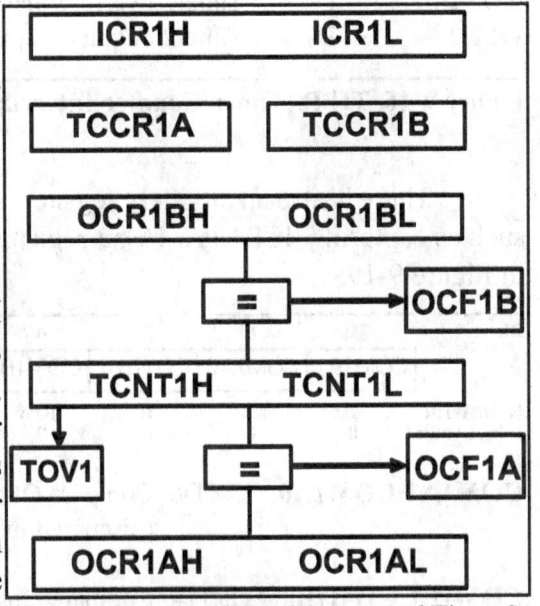

Figure 9-14. Simplified Diagram of Timer1

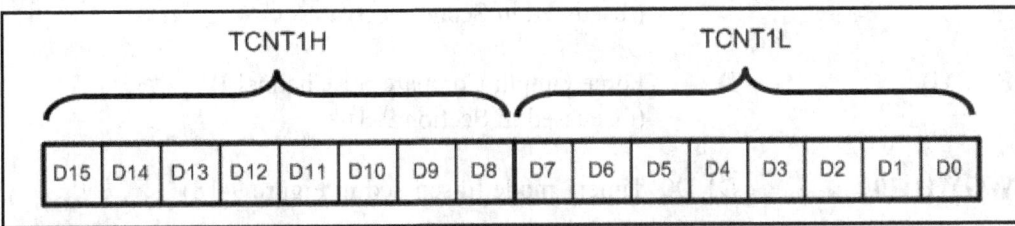

Figure 9-15. Timer1 High and Low Registers

---

the next timer clock. When TCNT equals OCR1B, the OCF1B flag will be set on the next clock. As Timer1 is a 16-bit timer, the OCR registers are 16-bit registers as well and they are made of two 8-bit registers. For example, OCR1A is made of OCR1AH (OCR1A high byte) and OCR1AL (OCR1A low byte). For a detailed view of Timer1 see Figure 9-13.

The TIFR register contains the TOV1, OCF1A, and OCF1B flags. See Figure 9-16.

Bit	7	6	5	4	3	2	1	0
	OCF2	TOV2	ICF1	OCF1A	OCF1B	TOV1	OCF0	TOV0
Read/Write	R/W	R/W	R/W	R/W	R/W	R/W	R/W	R/W
Initial Value	0	0	0	0	0	0	0	0

TOV0   D0   Timer0 overflow flag bit
        0 = Timer0 did not overflow.
        1 = Timer0 has overflowed (going from $FF to $00).

OCF0   D1   Timer0 output compare flag bit
        0 = compare match did not occur.
        1 = compare match occurred.

TOV1   D2   Timer1 overflow flag bit

OCF1B   D3   Timer1 output compare B match flag

OCF1A   D4   Timer1 output compare A match flag

ICF1   D5   Input Capture flag

TOV2   D6   Timer2 overflow flag

OCF2   D7   Timer2 output compare match flag

Figure 9-16. TIFR (Timer/Counter Interrupt Flag Register)

There is also an auxiliary register named ICR1, which is used in operations such as capturing. ICR1 is a 16-bit register made of ICR1H and ICR1L, as shown in Figure 9-19.

Bit	7	6	5	4	3	2	1	0
	COM1A1	COM1A0	COM1B1	COM1B0	FOC1A	FOC1B	WGM11	WGM10
Read/Write	R/W	R/W	R	R/W	R/W	R/W	R/W	R/W
Initial Value	0	0	0	0	0	0	0	0

COM1A1:COM1A0   D7 D6   Compare Output Mode for Channel A
                                (discussed in Section 9-3)

COM1B1:COM1B0   D5 D4   Compare Output Mode for Channel B
                                (discussed in Section 9-3)

FOC1A   D3   Force Output Compare for Channel A
                                (discussed in Section 9-3)

FOC1B   D2   Force Output Compare for Channel B
                                (discussed in Section 9-3)

WGM11:10   D1 D0   Timer1 mode (discussed in Figure 9-18)

Figure 9-17. TCCR1A (Timer 1 Control) Register

Bit	7	6	5	4	3	2	1	0	
	ICNC1	ICES1	-	WGM13	WGM12	CS12	CS11	CS10	TCCR1B
Read/Write	R/W	R/W	R	R/W	R/W	R/W	R/W	R/W	
Initial Value	0	0	0	0	0	0	0	0	

ICNC1 D7 Input Capture Noise Canceler
  0 = Input Capture is disabled.
  1 = Input Capture is enabled.

ICES1 D6 Input Capture Edge Select
  0 = Capture on the falling (negative) edge
  1 = Capture on the rising (positive) edge

  D5 Not used

WGM13:WGM12 D4 D3 Timer1 mode

Mode	WGM13	WGM12	WGM11	WGM10	Timer/Counter Mode of Operation	Top	Update of OCR1x	TOV1 Flag Set on
0	0	0	0	0	Normal	0xFFFF	Immediate	MAX
1	0	0	0	1	PWM, Phase Correct, 8-bit	0x00FF	TOP	BOTTOM
2	0	0	1	0	PWM, Phase Correct, 9-bit	0x01FF	TOP	BOTTOM
3	0	0	1	1	PWM, Phase Correct, 10-bit	0x03FF	TOP	BOTTOM
4	0	1	0	0	CTC	OCR1A	Immediate	MAX
5	0	1	0	1	Fast PWM, 8-bit	0x00FF	TOP	TOP
6	0	1	1	0	Fast PWM, 9-bit	0x01FF	TOP	TOP
7	0	1	1	1	Fast PWM, 10-bit	0x03FF	TOP	TOP
8	1	0	0	0	PWM, Phase and Frequency Correct	ICR1	BOTTOM	BOTTOM
9	1	0	0	1	PWM, Phase and Frequency Correct	OCR1A	BOTTOM	BOTTOM
10	1	0	1	0	PWM, Phase Correct	ICR1	TOP	BOTTOM
11	1	0	1	1	PWM, Phase Correct	OCR1A	TOP	BOTTOM
12	1	1	0	0	CTC	ICR1	Immediate	MAX
13	1	1	0	1	Reserved	-	-	-
14	1	1	1	0	Fast PWM	ICR1	TOP	TOP
15	1	1	1	1	Fast PWM	OCR1A	TOP	TOP

CS12:CS10 D2D1D0 Timer1 clock selector
  0 0 0 No clock source (Timer/Counter stopped)
  0 0 1 clk (no prescaling)
  0 1 0 clk / 8
  0 1 1 clk / 64
  1 0 0 clk / 256
  1 0 1 clk / 1024
  1 1 0 External clock source on T1 pin. Clock on falling edge.
  1 1 1 External clock source on T1 pin. Clock on rising edge.

Figure 9-18. TCCR1B (Timer 1 Control) Register

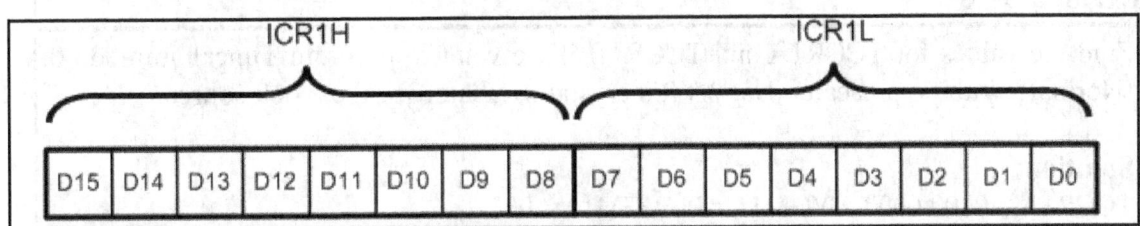

Figure 9-19. Input Capture Register (ICR) for Timer1

WGM13:10

The WGM13, WGM12, WGM11, and WGM10 bits define the mode of Timer1, as shown in Figure 9-18. Timer1 has 16 different modes. One of them (mode 13) is reserved (not implemented). In this chapter, we cover mode 0 (Normal mode) and mode 4 (CTC mode). The other modes will be covered in Chapters 15 and 16.

Timer1 operation modes

Normal mode (WGM13:10 = 0000)

In this mode, the timer counts up until it reaches $FFFF (which is the maximum value) and then it rolls over from $FFFF to 0000. When the timer rolls over from $FFFF to 0000, the TOV1 flag will be set. See Figure 9-20 and Examples 9-26 and 9-27. In Example 9-27, a delay is generated using Normal mode.

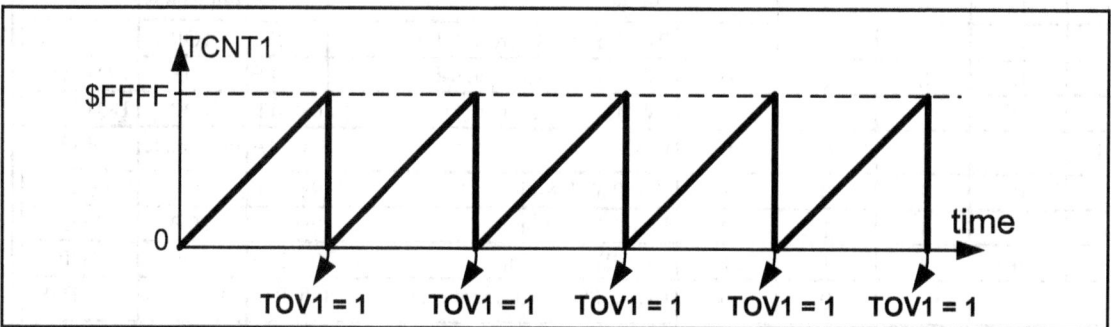

Figure 9-20. TOV in Normal and Fast PWM

CTC mode (WGM13:10 = 0100)

In mode 4, the timer counts up until the content of the TCNT1 register becomes equal to the content of OCR1A (compare match occurs); then, the timer will be cleared when the next clock occurs. The OCF1A flag will be set as a result of the compare match as well. See Figure 9-21 and Examples 9-28 and 9-29.

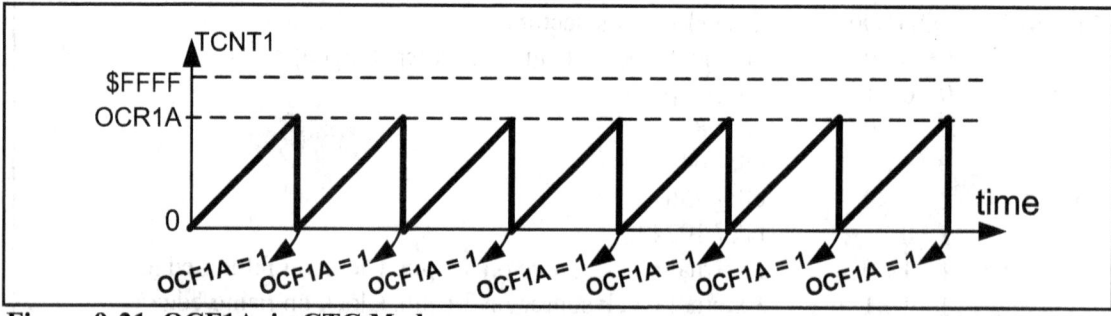

Figure 9-21. OCF1A in CTC Mode

Example 9-26

Find the values for TCCR1A and TCCR1B if we want to program Timer1 in mode 0 (Normal), with no prescaler. Use AVR's crystal oscillator for the clock source.

Solution:
TCCR1A = 0000 0000 WGM11 = 0, WGM10 = 0
TCCR1B = 0000 0001 WGM13 = 0, WGM12 = 0, oscillator clock source, no prescaler

Example 9-27

Find the frequency of the square wave generated by the following program if XTAL = 8 MHz. In your calculation do not include the overhead due to instructions in the loop.

```
.INCLUDE "M32DEF.INC"
      INITSTACK            ;add its definition from Example 9-3
      LDI    R16,0x20
      SBI    DDRB,5        ;PB5 as an output
      LDI    R17,0
      OUT    PORTB,R17     ;PB5 = 0
BEGIN:RCALL DELAY
      EOR    R17,R16       ;toggle D5 of R17
      OUT    PORTB,R17     ;toggle PB5
      RJMP   BEGIN
;-------------- Timer1 delay
DELAY:LDI    R20,0xD8
      OUT    TCNT1H,R20    ;TCNT1H = 0xD8
      LDI    R20,0xF0
      OUT    TCNT1L,R20    ;TCNT1L = 0xF0
      LDI    R20,0x00
      OUT    TCCR1A,R20    ;WGM11:10 = 00
      LDI    R20,0x01
      OUT    TCCR1B,R20    ;WGM13:12 = 00, Normal mode, prescaler = 1
AGAIN:IN     R20,TIFR      ;read TIFR
      SBRS   R20,TOV1      ;if TOV1 is set skip next instruction
      RJMP   AGAIN
      LDI    R20,0x00
      OUT    TCCR1B,R20    ;stop Timer1
      LDI    R20,0x04
      OUT    TIFR,R20      ;clear TOV1 flag
      RET
```

Solution:

WGM13:10 = 0000 = 0x00, so Timer1 is working in mode 0, which is Normal mode, and the top is 0xFFFF.

FFFF + 1 – D8F0 = 0x2710 = 10,000 clocks, which means that it takes 10,000 clocks. As XTAL = 8 MHz each clock lasts $1/(8M) = 0.125\ \mu s$ and delay = $10,000 \times 0.125\ \mu s$ = $1250\ \mu s = 1.25$ ms and frequency = $1 / (1.25\ ms \times 2) = 400$ Hz.

In this calculation, the overhead due to all the instructions in the loop is not included.

Notice that instead of using hex numbers we can use HIGH and LOW directives, as shown below:

```
      LDI    R20,HIGH (65536-10000)      ;load Timer1 high byte
      OUT    TCNT1H,R20  ;TCNT1H = 0xD8
      LDI    R20,LOW (65536-10000)       ;load Timer1 low byte
      OUT    TCNT1L,R20  ;TCNT1L = 0xF0
```

or we can simply write it as follows:

```
      LDI    R20,HIGH (-10000)           ;load Timer1 high byte
      OUT    TCNT1H,R20  ;TCNT1H = 0xD8
      LDI    R20,LOW (-10000)            ;load Timer1 low byte
      OUT    TCNT1L,R20  ;TCNT1L = 0xF0
```

Example 9-28

Find the values for TCCR1A and TCCR1B if we want to program Timer1 in mode 4 (CTC, Top = OCR1A), no prescaler. Use AVR's crystal oscillator for the clock source.

Solution:

TCCR1A = 0000 0000 WGM11 = 0, WGM10 = 0
TCCR1B = 0000 1001 WGM13 = 0, WGM12 = 1, oscillator clock source, no prescaler

Example 9-29

Find the frequency of the square wave generated by the following program if XTAL = 8 MHz. In your calculation do not include the overhead due to instructions in the loop.

```
.INCLUDE "M32DEF.INC"
      SBI    DDRB,5           ;PB5 as an output
BEGIN:SBI    PORTB,5          ;PB5 = 1
      RCALL  DELAY
      CBI    PORTB,5          ;PB5 = 0
      RCALL  DELAY
      RJMP   BEGIN
;--------------- Timer1 delay
DELAY:LDI    R20,0x00
      OUT    TCNT1H,R20
      OUT    TCNT1L,R20       ;TCNT1 = 0
      LDI    R20,0
      OUT    OCR1AH,R20
      LDI    R20,159
      OUT    OCR1AL,R20       ;OCR1A = 159 = 0x9F
      LDI    R20,0x0
      OUT    TCCR1A,R20       ;WGM11:10 = 00
      LDI    R20,0x09
      OUT    TCCR1B,R20       ;WGM13:12 = 01,CTC mode, prescaler = 1
AGAIN:IN     R20,TIFR         ;read TIFR
      SBRS   R20,OCF1A        ;if OCF1A is set skip next instruction
      RJMP   AGAIN
      LDI    R20,1<<OCF1A
      OUT    TIFR,R20         ;clear OCF1A flag
      LDI    R19,0
      OUT    TCCR1B,R19       ;stop timer
      OUT    TCCR1A,R19
      RET
```

Solution:

WGM13:10 = 0100 = 0x04 therefore, Timer1 is working in mode 4, which is a CTC mode, and max is defined by OCR1A.
159 + 1 = 160 clocks
XTAL = 8 MHz, so each clock lasts $1/(8M) = 0.125$ μs.
Delay = 160×0.125 μs = 20 μs and frequency = $1 / (20$ μs $\times 2) = 25$ kHz.
In this calculation, the overhead due to all the instructions in the loop is not included.

Accessing 16-bit registers

The AVR is an 8-bit microcontroller, which means it can manipulate data 8 bits at a time, only. But some Timer1 registers, such as TCNT1, OCR1A, ICR1, and so on, are 16-bit; in this case, the registers are split into two 8-bit registers, and each one is accessed individually. This is fine for most cases. For example, when we want to load the content of SP (stack pointer), we first load one half and then the other half, as shown below:

```
LDI   R16, 0x12
OUT   SPL, R16
LDI   R16, 0x34
OUT   SPH, R16    ;SP = 0x3412
```

In 16-bit timers, however, we should read/write the entire content of a register at once, otherwise we might have problems. For example, imagine the following scenario:

The TCNT1 register contains 0x15FF. We read the low byte of TCNT1, which is 0xFF, and store it in R20. At the same time a timer clock occurs, and the content of TCNT1 becomes 0x1600; now we read the high byte of TCNT1, which is now 0x16, and store it in R21. If we look at the value we have read, R21:R20 = 0x16FF. So, we believe that TCNT1 contains 0x16FF, although it actually contains 0x15FF.

This problem exists in many 8-bit microcontrollers. But the AVR designers have resolved this issue with an 8-bit register called TEMP, which is used as a buffer. See Figure 9-22. When we write or read the high byte of a 16-bit register, the value will be written into the TEMP register. When we write into the low byte of a 16-bit register, the content of TEMP will be written into the high byte of the 16-bit register as well. For example, consider the following program:

```
LDI   R16, 0x15
OUT   TCNT1H, R16    ;store 0x15 in TEMP of Timer1
LDI   R16, 0xFF
OUT   TCNT1L, R16    ;TCNT1L = R16, TCNT1H = TEMP
```

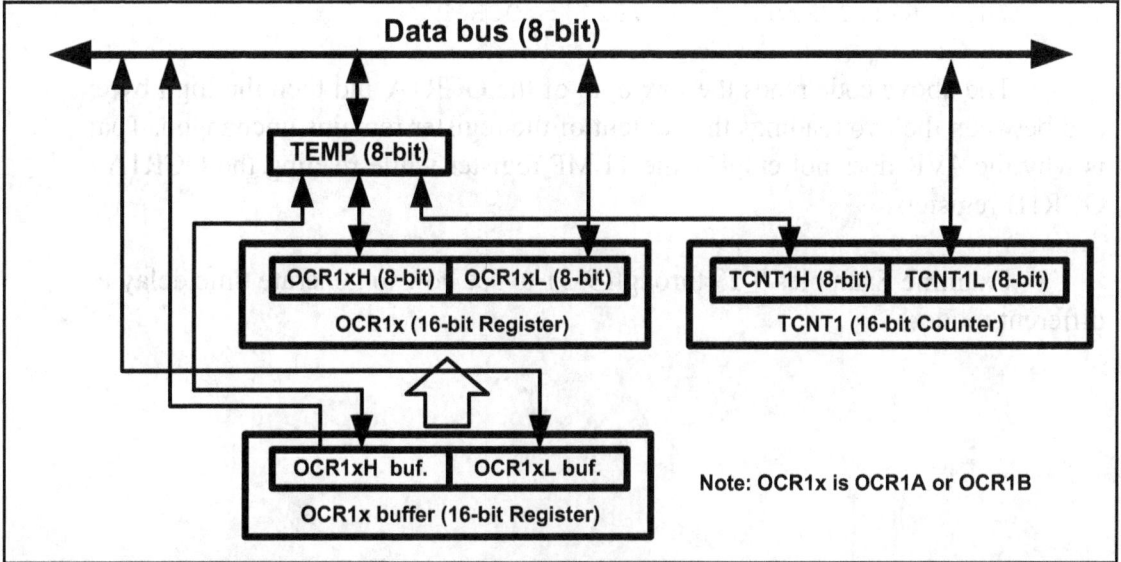

Figure 9-22. Accessing 16-bit Registers through TEMP

After the execution of "OUT TCNT1H, R16", the content of R16, 0x15, will be stored in the TEMP register. When the instruction "OUT TCNT1L, R16" is executed, the content of R16, 0xFF, is loaded into TCNT1L, and the content of the TEMP register, 0x15, is loaded into TCNT1H. So, 0x15FF will be loaded into the TCNT1 register at once.

Notice that according to the internal circuitry of the AVR, we should first write into the high byte of the 16-bit registers and then write into the lower byte. Otherwise, the program does not work properly. For example, the following code:

```
LDI   R16, 0xFF
OUT   TCNT1L, R16      ;TCNT1L = R16, TCNT1H = TEMP
LDI   R16, 0x15
OUT   TCNT1H, R16      ;store 0x15 in TEMP of Timer1
```

does not work properly. This is because, when the TCNT1L is loaded, the content of TEMP will be loaded into TCNT1H. But when the TCNT1L register is loaded, TEMP contains garbage (improper data), and this is not what we want.

When we read the low byte of 16-bit registers, the content of the high byte will be copied to the TEMP register. So, the following program reads the content of TCNT1:

```
IN    R20,TCNT1L      ;R20 = TCNT1L, TEMP = TCNT1H
IN    R21,TCNT1H      ;R21 = TEMP of Timer1
```

We must pay attention to the order of reading the high and low bytes of the 16-bit registers. Otherwise, the result is erroneous.

Notice that reading the OCR1A and OCR1B registers does not involve using the temporary register. You might be wondering why. It is because the AVR microcontroller does not update the content of OCR1A nor OCR1B unless we update them. For example, consider the following program:

```
IN    R20,OCR1AL      ;R20 = OCR1L
IN    R21,OCR1AH      ;R21 = OCR1H
```

The above code reads the low byte of the OCR1A and then the high byte, and between the two readings the content of the register remains unchanged. That is why the AVR does not employ the TEMP register while reading the OCR1A / OCR1B registers.

Examine Examples 9-29 through 9-31 to see how to generate time delay in different modes.

Example 9-30

Assuming XTAL = 8 MHz, write a program that toggles PB5 once per millisecond.

Solution:

XTAL = 8 MHz means that each clock takes 0.125 µs. Now for 1 ms delay, we need 1 ms/0.125 µs = 8000 clocks = 0x1F40 clocks. We initialize the timer so that after 8000 clocks the OCF1A flag is raised, and then we will toggle the PB5.

```
.INCLUDE "M32DEF.INC"
      LDI    R16,HIGH(RAMEND)
      OUT    SPH,R16
      LDI    R16,LOW(RAMEND)
      OUT    SPL,R16             ;initialize the stack
      SBI    DDRB,5              ;PB5 as an output
BEGIN:SBI    PORTB,5             ;PB5 = 1
      RCALL  DELAY_1ms
      CBI    PORTB,5             ;PB5 = 0
      RCALL  DELAY_1ms
      RJMP   BEGIN
;---------------------Timer1 delay
DELAY_1ms:
      LDI    R20,0x00
      OUT    TCNT1H,R20          ;TEMP = 0
      OUT    TCNT1L,R20          ;TCNT1L = 0, TCNT1H = TEMP

      LDI    R20,HIGH(8000-1)
      OUT    OCR1AH,R20          ;TEMP = 0x1F
      LDI    R20,LOW(8000-1)
      OUT    OCR1AL,R20          ;OCR1AL = 0x3F, OCR1AH = TEMP

      LDI    R20,0x0
      OUT    TCCR1A,R20          ;WGM11:10 = 00
      LDI    R20,0x09
      OUT    TCCR1B,R20          ;WGM13:12 = 01, CTC mode, CS = 1
AGAIN:
      IN     R20,TIFR            ;read TIFR
      SBRS   R20,OCF1A           ;if OCF1A is set skip next instruction
      RJMP   AGAIN
      LDI    R20,1<<OCF1A
      OUT    TIFR,R20            ;clear OCF1A flag
      LDI    R19,0
      OUT    TCCR1B,R19          ;stop timer
      OUT    TCCR1A,R19
      RET
```

Example 9-31

Rewrite Example 9-30 using the TOV1 flag.

Solution:

To wait 1 ms we should load the TCNT1 register so that it rolls over after 8000 = 0x1F40 clocks. In Normal mode the top value is 0xFFFF = 65535.
65535 + 1 – 8000 = 57536 = 0xE0C0. Thus, we should load TCNT1 with 57536, or 0xE0C0 in hex, or we can simply use 65536 – 8000, as shown below:

```
.INCLUDE "M32DEF.INC"
      LDI   R16,HIGH(RAMEND)   ;initialize stack pointer
      OUT   SPH,R16
      LDI   R16,LOW(RAMEND)
      OUT   SPL,R16
      SBI   DDRB,5             ;PB5 as an output
BEGIN:SBI   PORTB,5            ;PB5 = 1
      RCALL DELAY_1ms
      CBI   PORTB,5            ;PB5 = 0
      RCALL DELAY_1ms
      RJMP  BEGIN
;----------------------Timer1 delay
DELAY_1ms:
      LDI   R20,HIGH(65536-8000)    ;R20 = high byte of 57536
      OUT   TCNT1H,R20              ;TEMP = 0xE0
      LDI   R20,LOW(65536-8000)     ;R20 = low byte of 57536
      OUT   TCNT1L,R20         ;TCNT1L = 0xC1, TCNT1H = TEMP
      LDI   R20,0x0
      OUT   TCCR1A,R20         ;WGM11:10 = 00
      LDI   R20,0x1
      OUT   TCCR1B,R20         ;WGM13:12 = 00, Normal mode, CS = 1
AGAIN:
      IN    R20,TIFR           ;read TIFR
      SBRS  R20,TOV1           ;if OCF1A is set skip next instruction
      RJMP  AGAIN
      LDI   R20,1<<TOV1
      OUT   TIFR,R20           ;clear TOV1 flag
      LDI   R19,0
      OUT   TCCR1B,R19         ;stop timer
      OUT   TCCR1A,R19
      RET
```

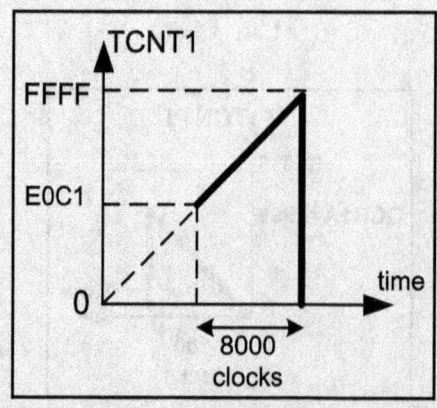

344

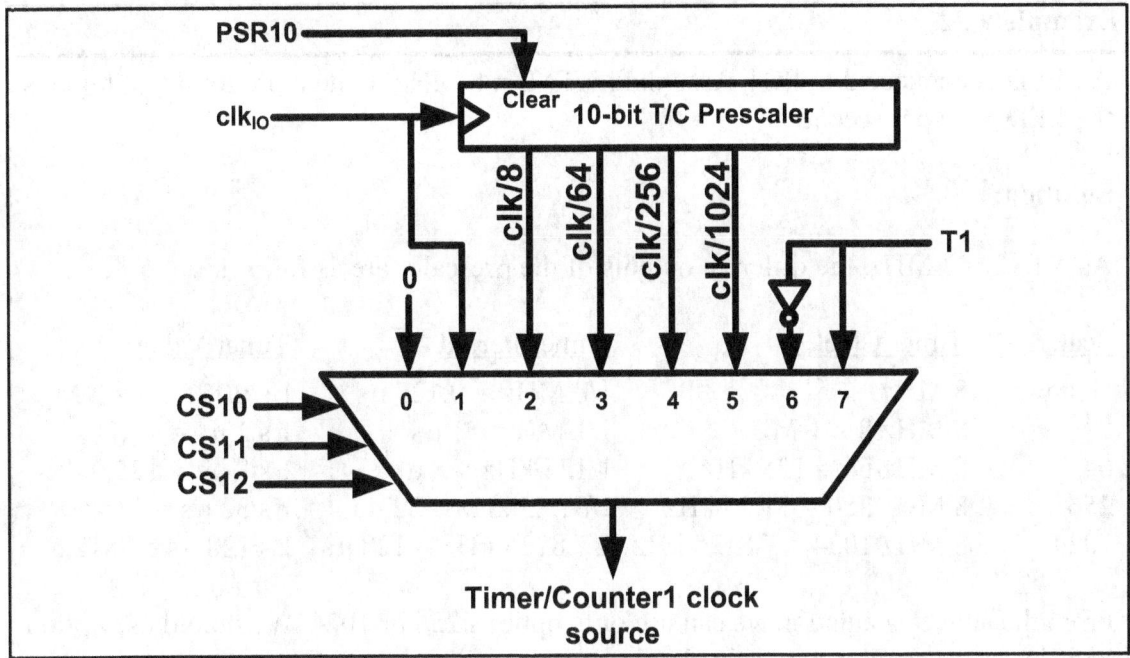

Figure 9-23. Timer/Counter 1 Prescaler

Generating a large time delay using prescaler

As we have seen in the examples so far, the size of the time delay depends on two factors: (a) the crystal frequency, and (b) the timer's 16-bit register. Both of these factors are beyond the control of the AVR programmer. We can use the prescaler option in the TCCR1B register to increase the delay by reducing the period. The prescaler option of TCCR1B allows us to divide the instruction clock by a factor of 8 to 1024, as was shown in Figure 9-16. The prescaler of Timer/Counter 1 is shown in Figure 9-23.

As we have seen so far, with no prescaler enabled, the crystal oscillator frequency is fed directly into Timer1. If we enable the prescaler bit in the TCCR1B register, then we can divide the instruction clock before it is fed into Timer1. The lower 3 bits of the TCCR1B register give the options of the number we can divide the clock by before it is fed to timer. As shown in Figure 9-23, this number can be 8, 64, 256, or 1024. Notice that the lowest number is 8, and the highest number is 1024. Examine Examples 9-32 and 9-33 to see how the prescaler options are programmed.

Review Questions

1. How many timers do we have in the ATmega32?
2. True or false. Timer0 is a 16-bit timer.
3. True or false. Timer1 is a 16-bit timer.
4. True or false. The TCCR0 register is a bit-addressable register.
5. In Normal mode, when the counter rolls over it goes from ____ to ____.
6. In CTC mode, the counter rolls over when the counter reaches____.
7. To get a 5-ms delay, what numbers should be loaded into TCNT1H and TCNT1L using Normal mode and the TOV1 flag? Assume that XTAL = 8 MHz.
8. To get a 20-μs delay, what number should be loaded into the TCNT0 register using Normal mode and the TOV0 flag? Assume that XTAL = 1 MHz.

Example 9-32

An LED is connected to PC4. Assuming XTAL = 8 MHz, write a program that toggles the LED once per second.

Solution:

As XTAL = 8 MHz, the different outputs of the prescaler are as follows:

Scaler	Timer Clock	Timer Period	Timer Value
None	8 MHz	1/8 MHz = 0.125 μs	1 s/0.125 μs = 8 M
8	8 MHz/8 = 1 MHz	1/1 MHz = 1 μs	1 s/1 μs = 1 M
64	8 MHz/64 = 125 kHz	1/125 kHz = 8 μs	1 s/8 μs = 125,000
256	8 MHz/256 = 31.25 kHz	1/31.25 kHz = 32 μs	1 s/32 μs = 31,250
1024	8 MHz/1024 = 7.8125 kHz	1/7.8125 kHz = 128 μs	1 s/128 μs = 7812.5

From the above calculation we can use only options 256 or 1024. We should use option 256 since we cannot use a decimal point.

```
.INCLUDE "M32DEF.INC"
      LDI    R16,HIGH(RAMEND)    ;initialize stack pointer
      OUT    SPH,R16
      LDI    R16,LOW(RAMEND)
      OUT    SPL,R16
      SBI    DDRC,4              ;PC4 as an output
BEGIN:SBI    PORTC,4             ;PC4 = 1
      RCALL  DELAY_1s
      CBI    PORTC,4             ;PC4 = 0
      RCALL  DELAY_1s
      RJMP   BEGIN
;------------------- Timer1 delay
DELAY_1s:
      LDI    R20,HIGH (31250-1)
      OUT    OCR1AH,R20          ;TEMP = $7A (since 31249 = $7A11)
      LDI    R20,LOW (31250-1)
      OUT    OCR1AL,R20          ;OCR1AL = $11 (since 31249 = $7A11)
      LDI    R20,0
      OUT    TCNT1H,R20          ;TEMP = 0x00
      OUT    TCNT1L,R20          ;TCNT1L = 0x00, TCNT1H = TEMP
      LDI    R20,0x00
      OUT    TCCR1A,R20          ;WGM11:10 = 00
      LDI    R20,0x4
      OUT    TCCR1B,R20          ;WGM13:12 = 00, Normal mode,CS = CLK/256
AGAIN:IN     R20,TIFR            ;read TIFR
      SBRS   R20,OCF1A           ;if OCF1A is set skip next instruction
      RJMP   AGAIN
      LDI    R20,1<<OCF1A
      OUT    TIFR,R20            ;clear OCF1A flag
      LDI    R19,0
      OUT    TCCR1B,R19          ;stop timer
      OUT    TCCR1A,R19
      RET
```

Example 9-33

Assuming XTAL = 8 MHz, write a program to generate 1 Hz frequency on PC4.

Solution:

With 1 Hz we have T = 1 / F = 1 / 1 Hz = 1 second, half of which is high and half low. Thus we need a delay of 0.5 second duration.

Since XTAL = 8 MHz, the different outputs of the prescaler are as follows:

Scaler	Timer Clock	Timer Period	Timer Value
None	8 MHz	1/8 MHz = 0.125 µs	0.5 s/0.125 µs = 4 M
8	8 MHz/8 = 1 MHz	1/1 MHz = 1 µs	0.5 s/1 µs = 500 k
64	8 MHz/64 = 125 kHz	1/125 kHz = 8 µs	0.5 s/8 µs = 62,500
256	8 MHz/256 = 31.25 kHz	1/31.25 kHz = 32 µs	0.5 s/32 µs = 15,625
1024	8 MHz/1024 = 7.8125 kHz	1/7.8125 kHz = 128 µs	0.5 s/128 µs = 3906.25

From the above calculation we can use options 64 or 256. We choose 64 in this Example.

```
.INCLUDE "M32DEF.INC"
     LDI   R16,HIGH(RAMEND)   ;initialize stack pointer
     OUT   SPH,R16
     LDI   R16,LOW(RAMEND)
     OUT   SPL,R16
     SBI   DDRC,4             ;PC4 as an output
BEGIN:SBI  PORTC,4            ;PC4 = 1
     RCALL DELAY_1s
     CBI   PORTC,4            ;PC4 = 0
     RCALL DELAY_1s
     RJMP  BEGIN
;-------------------- Timer1 delay
DELAY_1s:
     LDI   R20,HIGH (62500-1)
     OUT   OCR1AH,R20         ;TEMP = $F4 (since 62499 = $F423)
     LDI   R20,LOW (62500-1)
     OUT   OCR1AL,R20         ;OCR1AL = $23 (since 62499 = $F423)
     LDI   R20,0x00
     OUT   TCNT1H,R20         ;TEMP = 0x00
     OUT   TCNT1L,R20         ;TCNT1L = 0x00, TCNT1H = TEMP
     LDI   R20,0x00
     OUT   TCCR1A,R20         ;WGM11:10 = 00
     LDI   R20,0x3
     OUT   TCCR1B,R20         ;WGM13:12 = 00, Normal mode, CS = CLK/64
AGAIN:IN   R20,TIFR           ;read TIFR
     SBRS  R20,OCF1A          ;if OCF1A is set skip next instruction
     RJMP  AGAIN
     LDI   R20,1<<OCF1A
     OUT   TIFR,R20           ;clear OCF1A flag
     LDI   R19,0
     OUT   TCCR1B,R19         ;stop timer
     OUT   TCCR1A,R19
     RET
```

SECTION 9.2: COUNTER PROGRAMMING

In the previous section, we used the timers of the AVR to generate time delays. The AVR timer can also be used to count, detect, and measure the time of events happening outside the AVR. The use of the timer as an event counter is covered in this section. When the timer is used as a timer, the AVR's crystal is used as the source of the frequency. When it is used as a counter, however, it is a pulse outside the AVR that increments the TCNTx register. Notice that, in counter mode, registers such as TCCR, OCR0, and TCNT are the same as for the timer discussed in the previous section; they even have the same names.

CS00, CS01, and CS02 bits in the TCCR0 register

Recall from the previous section that the CS bits (clock selector) in the TCCR0 register decide the source of the clock for the timer. If CS02:00 is between 1 and 5, the timer gets pulses from the crystal oscillator. In contrast, when CS02:00 is 6 or 7, the timer is used as a counter and gets its pulses from a source outside the AVR chip. See Figure 9-24. Therefore, when CS02:00 is 6 or 7, the TCNT0 counter counts up as pulses are fed from pin T0 (Timer/Counter 0 External Clock input). In ATmega32/ATmega16, T0 is the alternative function of PORTB.0. In the

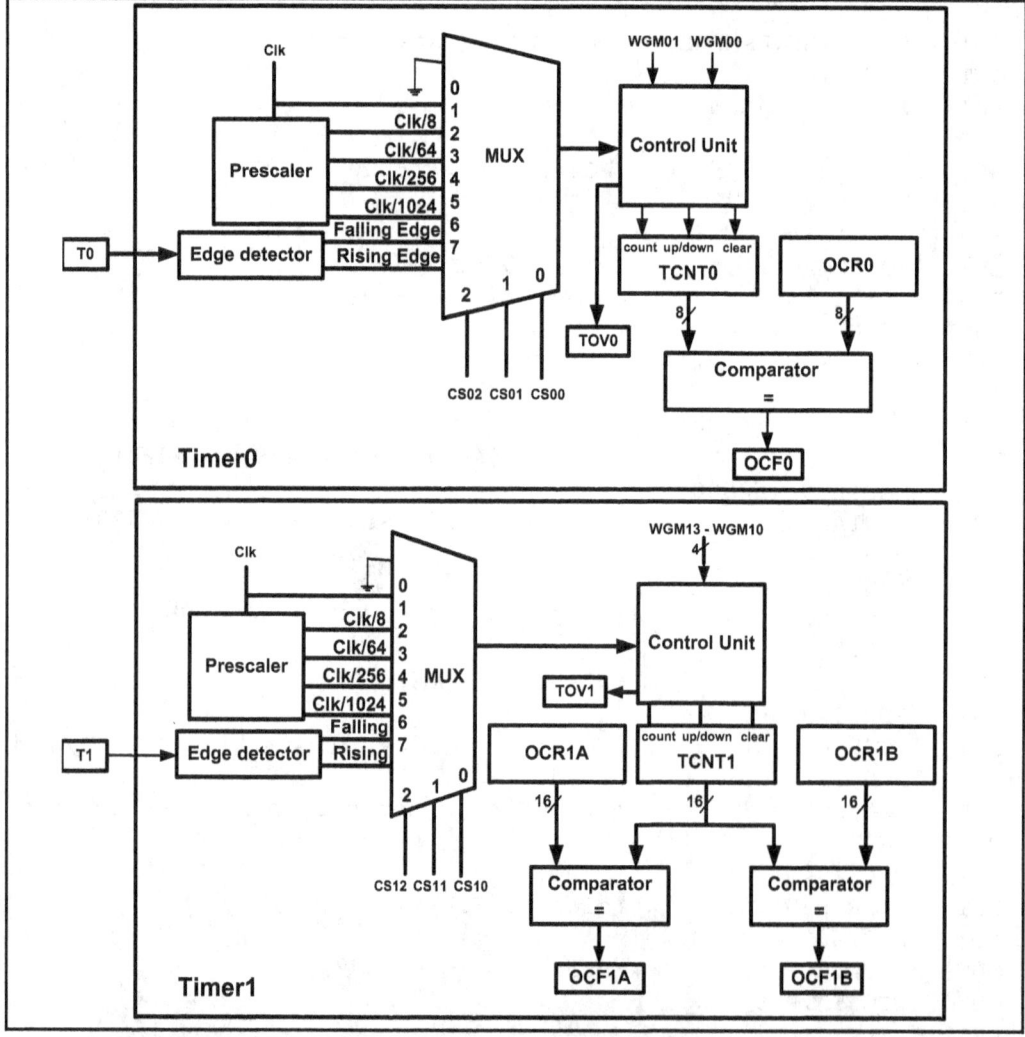

Figure 9-24. Timer/Counters 0 and 1 Prescalers

Example 9-34

Find the value for TCCR0 if we want to program Timer0 as a Normal mode counter. Use an external clock for the clock source and increment on the positive edge.

Solution:

TCCR0 = 0000 0111 Normal, external clock source, no prescaler

case of Timer0, when CS02:00 is 6 or 7, pin T0 provides the clock pulse and the counter counts up after each clock pulse coming from that pin. Similarly, for Timer1, when CS12:10 is 6 or 7, the clock pulse coming in from pin T1 (Timer/Counter 1 External Clock input) makes the TCNT1 counter count up. When CS12:10 is 6, the counter counts up on the negative (falling) edge. When CS12:10 is 7, the counter counts up on the positive (rising) edge. In ATmega32/ATmega16, T1 is the alternative function of PORTB.1. See Example 9-34.

In Example 9-35, we are using Timer0 as an event counter that counts up as clock pulses are fed into PB0. These clock pulses could represent the number of people passing through an entrance, or of wheel rotations, or any other event that can be converted to pulses.

Example 9-35

Assuming that a 1 Hz clock pulse is fed into pin T0 (PB0), write a program for Counter0 in normal mode to count the pulses on falling edge and display the state of the TCNT0 count on PORTC.

Solution:

```
.INCLUDE  "M32DEF.INC"
      CBI    DDRB,0            ;make T0 (PB0) input
      LDI    R20,0xFF
      OUT    DDRC,R20          ;make PORTC output
      LDI    R20,0x06
      OUT    TCCR0,R20         ;counter, falling edge
AGAIN:
      IN     R20,TCNT0
      OUT    PORTC,R20         ;PORTC = TCNT0
      IN     R16,TIFR
      SBRS   R16,TOV0          ;monitor TOV0 flag
      RJMP   AGAIN             ;keep doing if Timer0 flag is low
      LDI    R16,1<<TOV0
      OUT    TIFR, R16         ;clear TOV0 flag
      RJMP   AGAIN             ;keep doing it
```

PORTC is connected to 8 LEDs and input T0 (PB0) to 1 Hz pulse.

In Example 9-35, the TCNT0 data was displayed in binary. In Example 9-36, the TCNT0 register is extended to a 16-bit counter using the TOV0 flag. See Examples 9-37 and 9-38.

As another example of the application of the counter, we can feed an external square wave of 60 Hz frequency into the timer. The program will generate the second, the minute, and the hour out of this input frequency and display the result on an LCD. This will be a nice looking digital clock, although not a very accurate one.

Before we finish this section, we need to state an important point. You might think monitoring the TOV and OCR flags is a waste of the microcontroller's time. You are right. There is a solution to this: the use of interrupts. Using interrupts enables us to do other things with the microcontroller. When a timer Interrupt flag such as TOV0 is raised it will inform us. This important and powerful feature of the AVR is discussed in Chapter 10.

Example 9-36

Assuming that a 1 Hz clock pulse is fed into pin T0, use the TOV0 flag to extend Timer0 to a 16-bit counter and display the counter on PORTC and PORTD.

Solution:

```
.INCLUDE "M32DEF.INC"
        LDI    R19,0            ;R19 = 0
        CBI    DDRB,0           ;make T0 (PB0) input
        LDI    R20,0xFF
        OUT    DDRC,R20         ;make PORTC output
        OUT    DDRD,R20         ;make PORTD output
        LDI    R20,0x06
        OUT    TCCR0,R20        ;counter, falling edge
AGAIN:
        IN     R20,TCNT0
        OUT    PORTC,R20        ;PORTC = TCNT0
        IN     R16,TIFR
        SBRS   R16,TOV0
        RJMP   AGAIN            ;keep doing it
        LDI    R16,1<<TOV0      ;clear TOV0 flag
        OUT    TIFR, R16
        INC    R19              ;R19 = R19 + 1
        OUT    PORTD,R19        ;PORTD = R19
        RJMP   AGAIN            ;keep doing it
```

PORTC and PORTD are connected to 16 LEDs and input T0 (PB0) to 1 Hz pulse.

Example 9-37

Assuming that clock pulses are fed into pin T1 (PB1), write a program for Counter1 in Normal mode to count the pulses on falling edge and display the state of the TCNT1 count on PORTC and PORTD.

Solution:

```
.INCLUDE "M32DEF.INC"

        CBI    DDRB,1              ;make T1 (PB1) input
        LDI    R20,0xFF
        OUT    DDRC,R20            ;make PORTC output
        OUT    DDRD,R20            ;make PORTD output
        LDI    R20,0x0
        OUT    TCCR1A,R20
        LDI    R20,0x06
        OUT    TCCR1B,R20          ;counter, falling edge

AGAIN:
        IN     R20,TCNT1L          ;R20 = TCNT1L, TEMP = TCNT1H
        OUT    PORTC,R20           ;PORTC = TCNT0
        IN     R20,TCNT1H          ;R20 = TEMP
        OUT    PORTD,R20           ;PORTD = TCNT0
        IN     R16,TIFR
        SBRS   R16,TOV1
        RJMP   AGAIN               ;keep doing it
        LDI    R16,1<<TOV1         ;clear TOV1 flag
        OUT    TIFR, R16
        RJMP   AGAIN               ;keep doing it
```

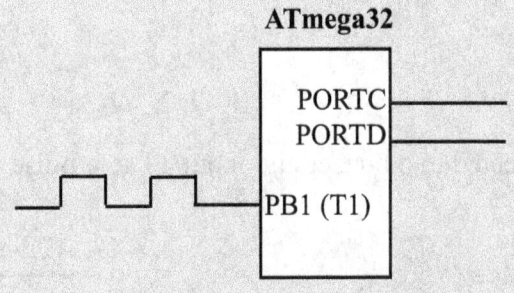

CHAPTER 9: AVR TIMER PROGRAMMING IN ASSEMBLY AND C

351

Example 9-38

Assuming that clock pulses are fed into pin T1 (PB1) and a buzzer is connected to pin PORTC.0, write a program for Counter 1 in CTC mode to sound the buzzer every 100 pulses.

Solution:

To sound the buzzer every 100 pulses, we set the OCR1A value to 99 (63 in hex), and then the counter counts up until it reaches OCR1A. Upon compare match, we can sound the buzzer by toggling the PORTC.0 pin.

```
.INCLUDE "M32DEF.INC"

        CBI    DDRB,1              ;make T1 (PB1) input
        SBI    DDRC,0              ;PC0 as an output
        LDI    R16,0x1
        LDI    R17,0

        LDI    R20,0x0
        OUT    TCCR1A,R20
        LDI    R20,0x0E
        OUT    TCCR1B,R20          ;CTC, counter, falling edge
AGAIN:
        LDI    R20,0
        OUT    OCR1AH,R20          ;TEMP = 0
        LDI    R20,99
        OUT    OCR1AL,R20          ;ORC1L = R20, OCR1H = TEMP
L1:     IN     R20,TIFR
        SBRS   R20,OCF1A
        RJMP   L1                  ;keep doing it
        LDI    R20,1<<OCF1A        ;clear OCF1A flag
        OUT    TIFR, R20

        EOR    R17,R16             ;toggle D0 of R17
        OUT    PORTC,R17           ;toggle PC0
        RJMP   AGAIN               ;keep doing it
```

PC0 is connected to a buzzer and input T1 to a pulse.

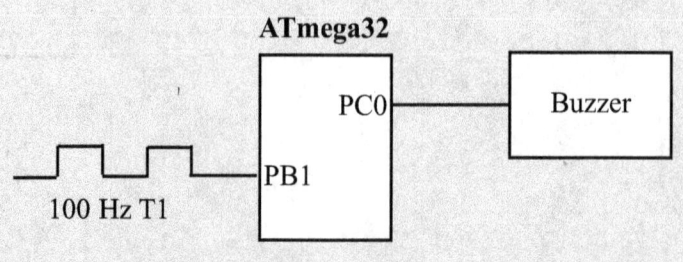

352

Review Questions

1. Which resource provides the clock pulses to AVR timers if CS02:00 = 6?
2. For Counter 0, which pin is used for the input clock?
3. To allow PB1 to be used as an input for the Timer1 clock, what must be done, and why?
4. Do we have a choice of counting up on the positive or negative edge of the clock?

SECTION 9.3: PROGRAMMING TIMERS IN C

In Chapter 7 we showed some examples of C programming for the AVR. In this section we show C programming for the AVR timers. As we saw in the examples in Chapter 7, the general-purpose registers of the AVR are under the control of the C compiler and are not accessed directly by C statements. All of the SFRs (Special Function Registers), however, are accessible directly using C statements. As an example of accessing the SFRs directly, we saw how to access ports PORTB–PORTD in Chapter 7.

In C we can access timer registers such as TCNT0, OCR0, and TCCR0 directly using their names. See Example 9-39.

Example 9-39

Write a C program to toggle all the bits of PORTB continuously with some delay. Use Timer0, Normal mode, and no prescaler options to generate the delay.

Solution:

```
#include "avr/io.h"
void T0Delay ( );
int main ( )
{
    DDRB = 0xFF;        //PORTB output port

    while (1)
    {
        PORTB = 0x55;      //repeat forever
        T0Delay ( );       //delay size unknown
        PORTB = 0xAA;      //repeat forever
        T0Delay ( );
    }
}

void T0Delay ( )
{
    TCNT0 = 0x20;             //load TCNT0
    TCCR0 = 0x01;             //Timer0, Normal mode, no prescaler
    while ((TIFR&0x1)==0);    //wait for TF0 to roll over
    TCCR0 = 0;
    TIFR = 0x1;               //clear TF0
}
```

Calculating delay length using timers

As we saw in the last two sections, the delay length depends primorily on two factors: (a) the crystal frequency, and (b) the prescaler factor. A third factor in the delay size is the C compiler because various C compilers generate different hex code sizes, and the amount of overhead due to the instructions varies by compiler. Study Examples 9-40 through 9-42 and verify them using an oscilloscope.

Example 9-40

Write a C program to toggle only the PORTB.4 bit continuously every 70 μs. Use Timer0, Normal mode, and 1:8 prescaler to create the delay. Assume XTAL = 8 MHz.

Solution:

XTAL = 8MHz ➔ $T_{machine\ cycle}$ = 1/8 MHz

Prescaler = 1:8 ➔ T_{clock} = 8 × 1/8 MHz = 1 μs

70 μs/1 μs = 70 clocks ➔ 1 + 0xFF − 70 = 0x100 − 0x46 = 0xBA = **186**

```c
#include "avr/io.h"

void T0Delay ( );

int main ( )
{
    DDRB = 0xFF;        //PORTB output port

    while (1)
    {
        T0Delay ( );            //Timer0, Normal mode
        PORTB = PORTB ^ 0x10;   //toggle PORTB.4
    }
}

void T0Delay ( )
{
    TCNT0 = 186;        //load TCNT0
    TCCR0 = 0x02;       //Timer0, Normal mode, 1:8 prescaler
    while ((TIFR&(1<<TOV0))==0); //wait for TOV0 to roll over

    TCCR0 = 0;          //turn off Timer0
    TIFR = 0x1;         //clear TOV0
}
```

Example 9-41

Write a C program to toggle only the PORTB.4 bit continuously every 2 ms. Use Timer1, Normal mode, and no prescaler to create the delay. Assume XTAL = 8 MHz.

Solution:

XTAL = 8 MHz $\rightarrow$ $T_{machine\ cycle}$ = 1/8 MHz = 0.125 µs
Prescaler = 1:1 $\rightarrow$ T_{clock} = 0.125 µs
2 ms/0.125 µs = 16,000 clocks = 0x3E80 clocks

1 + 0xFFFF − 0x3E80 = 0xC180

```c
#include "avr/io.h"

void T1Delay ( );

int main ( )
{
    DDRB = 0xFF;                //PORTB output port

    while (1)
    {
        PORTB = PORTB ^ (1<<PB4); //toggle PB4
        T1Delay ( );           //delay size unknown
    }
}


void T1Delay ( )
{
    TCNT1H = 0xC1;      //TEMP = 0xC1
    TCNT1L = 0x80;

    TCCR1A = 0x00;     //Normal mode
    TCCR1B = 0x01;     //Normal mode, no prescaler

    while ((TIFR&(0x1<<TOV1))==0);      //wait for TOV1 to roll over

    TCCR1B = 0;
    TIFR = 0x1<<TOV1;        //clear TOV1
}
```

Example 9-42 (C version of Example 9-32)

Write a C program to toggle only the PORTB.4 bit continuously every second. Use Timer1, Normal mode, and 1:256 prescaler to create the delay. Assume XTAL = 8 MHz.

Solution:

XTAL = 8 MHz ➜ $T_{machine\ cycle}$ = 1/8 MHz = 0.125 μs = T_{clock}

Prescaler = 1:256 ➜ T_{clock} = 256 × 0.125 μs = 32 μs

1 s/32 μs = 31,250 clocks = 0x7A12 clocks ➜ 1 + 0xFFFF − 0x7A12 = **0x85EE**

```
#include "avr/io.h"

void T1Delay ( );

int main ( )
{
    DDRB = 0xFF;        //PORTB output port

    while (1)
    {
        PORTB = PORTB ^ (1<<PB4); //toggle PB4
        T1Delay ( );        //delay size unknown
    }
}

void T1Delay ( )
{
    TCNT1H = 0x85;     //TEMP = 0x85
    TCNT1L = 0xEE;

    TCCR1A = 0x00;     //Normal mode
    TCCR1B = 0x04;     //Normal mode, 1:256 prescaler

    while ((TIFR&(0x1<<TOV1))==0);       //wait for TF0 to roll over

    TCCR1B = 0;
    TIFR = 0x1<<TOV1;       //clear TOV1
}
```

C programming of Timers 0 and 1 as counters

In Section 9.2 we showed how to use Timers 0 and 1 as event counters. Timers can be used as counters if we provide pulses from outside the chip instead of using the frequency of the crystal oscillator as the clock source. By feeding pulses to the T0 (PB0) and T1 (PB1) pins, we use Timer0 and Timer1 as Counter 0 and Counter 1, respectively. Study Examples 9-43 and 9-44 to see how Timers 0 and 1 are programmed as counters using C language.

Example 9-43 (C version of Example 9-36)

Assuming that a 1 Hz clock pulse is fed into pin T0, use the TOV0 flag to extend Timer0 to a 16-bit counter and display the counter on PORTC and PORTD.

Solution:

```c
#include "avr/io.h"

int main ( )
{
    PORTB = 0x01;           //activate pull-up of PB0
    DDRC = 0xFF;            //PORTC as output
    DDRD = 0xFF;            //PORTD as output

    TCCR0 = 0x06;          //output clock source
    TCNT0 = 0x00;

    while (1)
    {
        do
        {
            PORTC = TCNT0;
        } while((TIFR&(0x1<<TOV0))==0);//wait for TOV0 to roll over

        TIFR = 0x1<<TOV0;       //clear TOV0
        PORTD ++;               //increment PORTD
    }
}
```

ATmega32

PORTC and PORTD are connected to 16 LEDs.
T0 (PB0) is connected to a
1-Hz external clock.

1 Hz   T0

PD

PC

PB0

to
LEDs

Example 9-44 (C version of Example 9-37)

Assume that a 1-Hz external clock is being fed into pin T1 (PB1). Write a C program for Counter1 in rising edge mode to count the pulses and display the TCNT1H and TCNT1L registers on PORTD and PORTC, respectively.

Solution:

```c
#include "avr/io.h"

int main ( )
{
    PORTB = 0x01;           //activate pull-up of PB0
    DDRC = 0xFF;            //PORTC as output
    DDRD = 0xFF;            //PORTD as output

    TCCR1A = 0x00;          //output clock source
    TCCR1B = 0x06;          //output clock source

    TCNT1H = 0x00;          //set count to 0
    TCNT1L = 0x00;          //set count to 0

    while (1)               //repeat forever
    {
        do
        {
            PORTC = TCNT1L;
            PORTD = TCNT1H;         //place value on pins
        } while((TIFR&(0x1<<TOV1))==0);//wait for TOV1

        TIFR = 0x1<<TOV1;           //clear TOV1
    }
}
```

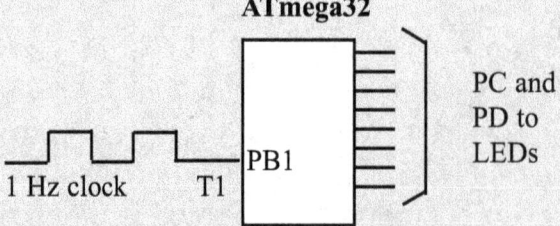

ATmega32

1 Hz clock   T1   PB1

PC and PD to LEDs

SUMMARY

The AVR has one to six timers/counters depending on the family member. When used as timers, they can generate time delays. When used as counters, they can serve as event counters.

Some of the AVR timers are 8-bit and some are 16-bit. The 8-bit timers are accessed as TCNTn (like TCNT0 for Timer0), whereas 16-bit timers are accessed as two 8-bit registers (TCNTnH, TCNTnL).

Each timer has its own TCCR (Timer/Counter Control Register) register, allowing us to choose various operational modes. Among the modes are the prescaler and timer/counter options. When the timer is used as a timer, the AVR crystal is used as the source of the frequency; however, when it is used as a counter, it is a pulse outside of the AVR that increments the TCNT register.

This chapter showed how to program the timers/counters to generate delays and count events using Normal and CTC modes.

PROBLEMS

SECTION 9.1: PROGRAMMING TIMERS 0, 1, AND 2

1. How many timers are in the ATmega32?
2. Timer0 of the ATmega32 is _____-bit, accessed as _____.
3. Timer1 of the ATmega32 is _____-bit, accessed as _____ and _____.
4. Timer0 supports the highest prescaler value of _____.
5. Timer1 supports the highest prescaler value of _____.
6. The TCCR0 register is a(n) ____-bit register.
7. What is the job of the TCCR0 register?
8. True or false. TCCR0 is a bit-addressable register.
9. True or false. TIFR is a bit-addressable register.
10. Find the TCCR0 value for Normal mode, no prescaler, with the clock coming from the AVR's crystal.
11. Find the frequency and period used by the timer if the crystal attached to the AVR has the following values:
    (a) XTAL = 8 MHz   (b) XTAL = 16 MHz
    (c) XTAL = 1 MHz   (d) XTAL = 10 MHz
12. Which register holds the TOV0 (timer overflow flag) and TOV1 bits?
13. Indicate the rollover value (in hex and decimal) of the timer for each of the following cases:
    (a) Timer0 and Normal mode   (b) Timer1 and Normal mode
14. Indicate when the TOVx flag is raised for each of the following cases:
    (a) Timer0 and Normal mode   (b) Timer1 and Normal mode
15. True or false. Both Timer0 and Timer1 have their own timer overflow flags.
16. True or false. Both Timer0 and Timer1 have their own timer compare match flags.

---

17. Assume that XTAL = 8 MHz. Find the TCNT0 value needed to generate a time delay of 20 μs. Use Normal mode, no prescaler mode.

18. Assume that XTAL = 8 MHz. Find the TCNT0 value needed to generate a time delay of 5 ms. Use Normal mode, and the largest prescaler possible.

19. Assume that XTAL = 1 MHz. Find the TCNT1H,TCNT1L value needed to generate a time delay of 2.5 ms. Use Normal mode, no prescaler mode.

20. Assume that XTAL = 1 MHz. Find the OCR0 value needed to generate a time delay of 0.2 ms. Use CTC mode, no prescaler mode.

21. Assume that XTAL = 1 MHz. Find the OCR1H,OCR1L value needed to generate a time delay of 2 ms. Use CTC mode, and no prescaler mode.

22. Assuming that XTAL = 8 MHz, and we are generating a square wave on pin PB7, find the lowest square wave frequency that we can generate using Timer1 in Normal mode.

23. Assuming that XTAL = 8 MHz, and we are generating a square wave on pin PB2, find the highest square wave frequency that we can generate using Timer1 in Normal mode.

24. Repeat Problems 22 and 23 for Timer0.

25. Assuming that TCNT0 = $F1, indicate which states Timer0 goes through until TOV0 is raised. How many states is that?

26. Program Timer0 to generate a square wave of 1 kHz. Assume that XTAL = 8 MHz.

27. Program Timer1 to generate a square wave of 3 kHz. Assume that XTAL = 8 MHz.

28. State the differences between Timer0 and Timer1.

29. Find the value (in hex) loaded into R16 in each of the following:

(a)	LDI R16,-12	(b)	LDI R16,-22
(c)	LDI R16,-34	(d)	LDI R16,-92
(e)	LDI R16,-120	(f)	LDI R16,-104

SECTION 9.2: COUNTER PROGRAMMING

30. To use a timer as an event counter we must set the _____ bits in the TCCR register to _____.

31. Can we use both Timer0 and Timer1 as event counters?

32. For Counter 0, which pin is used for the input clock?

33. For Counter 1, which pin is used for the input clock?

34. Program Timer1 to be an event counter. Use Normal mode, and display the binary count on PORTC and PORTD continuously. Set the initial count to 20,000.

35. Program Timer0 to be an event counter. Use Normal mode and display the binary count on PORTC continuously. Set the initial count to 20.

SECTION 9.3: PROGRAMMING TIMERS IN C

36. Program Timer0 in C to generate a square wave of 1 kHz. Assume that XTAL = 1 MHz.

37. Program Timer1 in C to generate a square wave of 1 kHz. Assume that XTAL = 8 MHz.

38. Program Timer0 in C to generate a square wave of 3 kHz. Assume that XTAL = 16 MHz.

39. Program Timer1 in C to generate a square wave of 3 kHz. Assume that XTAL = 10 MHz.

40. Program Timer1 in C to be an event counter. Use Normal mode and display the binary count on PORTB and PORTD continuously. Set the initial count to 20,000.

41. Program Timer0 in C to be an event counter. Use Normal mode and display the binary count on PORTD continuously. Set the initial count to 20.

ANSWERS TO REVIEW QUESTIONS

SECTION 9.1: PROGRAMMING TIMERS 0, 1, AND 2

1. 3
2. False
3. True
4. False
5. Max ($FFFF for 16-bit timers and $FF for 8-bit timers), 0000
6. OCR1A
7. $10000 - (5000 \times 8) = 25536 = 63C0, TCNT1H = 0x64 and TCNT1L = 0xC0
8. XTAL = 1 MHz → $T_{machine\ cycle}$ = 1/1 M = 1 μs → 20 μs / 1 μs = 20
   −20 = $100 − 20 = 256 − 20 = 236 = 0xEC

SECTION 9.2: COUNTER PROGRAMMING

1. External clock (falling edge)
2. PORTB.0 (T0)
3. DDRB.0 must be cleared to turn the output circuit off and use the pin as input.
4. Yes

CHAPTER 10

AVR INTERRUPT PROGRAMMING IN ASSEMBLY AND C

OBJECTIVES

Upon completion of this chapter, you will be able to:

>> Contrast and compare interrupts versus polling
>> Explain the purpose of the ISR (interrupt service routine)
>> List all the major interrupts of the AVR
>> Explain the purpose of the interrupt vector table
>> Enable or disable AVR interrupts
>> Program the AVR timers using interrupts
>> Describe the external hardware interrupts of the AVR
>> Define the interrupt priority of the AVR
>> Program AVR interrupts in C

In this chapter we explore the concept of the interrupt and interrupt programming. In Section 10.1, the basics of AVR interrupts are discussed. In Section 10.2, interrupts belonging to timers are discussed. External hardware interrupts are discussed in Section 10.3. In Section 10.4, we cover interrupt priority. In Section 10.5, we provide AVR interrupt programming examples in C.

SECTION 10.1: AVR INTERRUPTS

In this section, we first examine the difference between polling and interrupt and then describe the various interrupts of the AVR.

Interrupts vs. polling

A single microcontroller can serve several devices. There are two methods by which devices receive service from the microcontroller: interrupts or polling. In the *interrupt* method, whenever any device needs the microcontroller's service, the device notifies it by sending an interrupt signal. Upon receiving an interrupt signal, the microcontroller stops whatever it is doing and serves the device. The program associated with the interrupt is called the *interrupt service routine* (ISR) or *interrupt handler*. In *polling*, the microcontroller continuously monitors the status of a given device; when the status condition is met, it performs the service. After that, it moves on to monitor the next device until each one is serviced. Although polling can monitor the status of several devices and serve each of them as certain conditions are met, it is not an efficient use of the microcontroller. The advantage of interrupts is that the microcontroller can serve many devices (not all at the same time, of course); each device can get the attention of the microcontroller based on the priority assigned to it. The polling method cannot assign priority because it checks all devices in a round-robin fashion. More importantly, in the interrupt method the microcontroller can also ignore (mask) a device request for service. This also is not possible with the polling method. The most important reason that the interrupt method is preferable is that the polling method wastes much of the microcontroller's time by polling devices that do not need service. So interrupts are used to avoid tying down the microcontroller. For example, in discussing timers in Chapter 9 we used the bit test instruction "SBRS R20, TOV0" and waited until the timer rolled over, and while we were waiting we could not do anything else. That is a waste of microcontroller time that could have been used to perform some useful tasks. In the case of the timer, if we use the interrupt method, the microcontroller can go about doing other tasks, and when the TOV0 flag is raised, the timer will interrupt the microcontroller in whatever it is doing.

Interrupt service routine

For every interrupt, there must be an interrupt service routine (ISR), or interrupt handler. When an interrupt is invoked, the microcontroller runs the interrupt service routine. Generally, in most microprocessors, for every interrupt there is a fixed location in memory that holds the address of its ISR. The group of memory locations set aside to hold the addresses of ISRs is called the *interrupt vector table*, as shown in Table 10-1.

Steps in executing an interrupt

Upon activation of an interrupt, the microcontroller goes through the following steps:

1. It finishes the instruction it is currently executing and saves the address of the next instruction (program counter) on the stack.
2. It jumps to a fixed location in memory called the *interrupt vector table*. The interrupt vector table directs the microcontroller to the address of the interrupt service routine (ISR).
3. The microcontroller starts to execute the interrupt service subroutine until it reaches the last instruction of the subroutine, which is RETI (return from interrupt).
4. Upon executing the RETI instruction, the microcontroller returns to the place where it was interrupted. First, it gets the program counter (PC) address from the stack by popping the top bytes of the stack into the PC. Then it starts to execute from that address.

Notice from Step 4 the critical role of the stack. For this reason, we must be careful in manipulating the stack contents in the ISR. Specifically, in the ISR, just as in any CALL subroutine, the number of pushes and pops must be equal.

Table 10-1: Interrupt Vector Table for the ATmega32 AVR

Interrupt	ROM Location (Hex)
Reset	0000
External Interrupt request 0	0002
External Interrupt request 1	0004
External Interrupt request 2	0006
Time/Counter2 Compare Match	0008
Time/Counter2 Overflow	000A
Time/Counter1 Capture Event	000C
Time/Counter1 Compare Match A	000E
Time/Counter1 Compare Match B	0010
Time/Counter1 Overflow	0012
Time/Counter0 Compare Match	0014
Time/Counter0 Overflow	0016
SPI Transfer complete	0018
USART, Receive complete	001A
USART, Data Register Empty	001C
USART, Transmit Complete	001E
ADC Conversion complete	0020
EEPROM ready	0022
Analog Comparator	0024
Two-wire Serial Interface (I2C)	0026
Store Program Memory Ready	0028

Sources of interrupts in the AVR

There are many sources of interrupts in the AVR, depending on which peripheral is incorporated into the chip. The following are some of the most widely used sources of interrupts in the AVR:

1. There are at least two interrupts set aside for each of the timers, one for overflow and another for compare match. See Section 10.2.
2. Three interrupts are set aside for external hardware interrupts. Pins PD2 (PORTD.2), PD3 (PORTD.3), and PB2 (PORTB.2) are for the external hardware interrupts INT0, INT1, and INT2, respectively. See Section 10.3.
3. Serial communication's USART has three interrupts, one for receive and two interrupts for transmit. See Chapter 11.
4. The SPI interrupts. See Chapter 17.
5. The ADC (analog-to-digital converter). See Chapter 13.

The AVR has many more interrupts than the list shows. We will cover them throughout the book as we study the peripherals of the AVR. Notice in Table 10-1 that a limited number of bytes is set aside for interrupts. For example, a total of 2 words (4 bytes), from locations 0016 to 0018, are set aside for Timer0 overflow interrupt. Normally, the service routine for an interrupt is too long to fit into the memory space allocated. For that reason, a JMP instruction is placed in the vector table to point to the address of the ISR. In upcoming sections of this chapter, we will see many examples of interrupt programming that clarify these concepts.

From Table 10-1, also notice that only 2 words (4 bytes) of ROM space are assigned to the reset pin. They are ROM address locations 0–1. For this reason, in our program we put the JMP as the first instruction and redirect the processor away from the interrupt vector table, as shown in Figure 10-1. In the next section we will see how this works in the context of some examples.

```
            .ORG   0      ;wake-up ROM reset location
            JMP    MAIN   ;bypass interrupt vector table

;---- the wake-up program
            .ORG   $100
MAIN:       ....          ;enable interrupt flags
            ....
```

Figure 10-1. Redirecting the AVR from the Interrupt Vector Table at Power-up

Enabling and disabling an interrupt

Upon reset, all interrupts are disabled (masked), meaning that none will be responded to by the microcontroller if they are activated. The interrupts must be enabled (unmasked) by software in order for the microcontroller to respond to them. The D7 bit of the SREG (Status Register) register is responsible for enabling and disabling the interrupts globally. Figure 10-2 shows the SREG register. The I bit makes the job of disabling all the interrupts easy. With a single instruction "CLI" (Clear Interrupt), we can make I = 0 during the operation of a critical task.

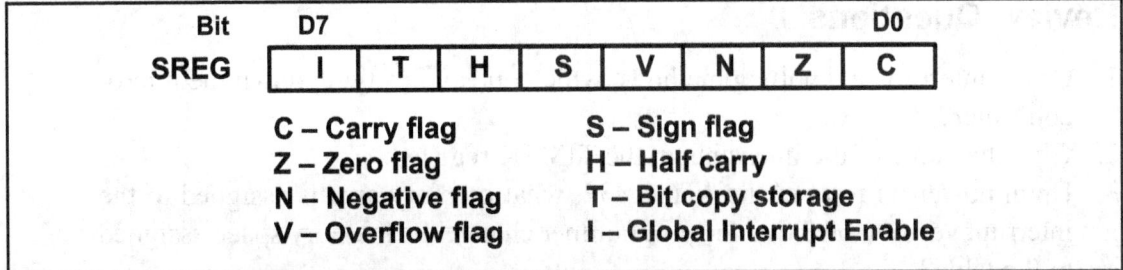

Bit	D7							D0
SREG	I	T	H	S	V	N	Z	C

C – Carry flag   S – Sign flag
Z – Zero flag   H – Half carry
N – Negative flag   T – Bit copy storage
V – Overflow flag   I – Global Interrupt Enable

Figure 10-2. Bits of Status Register (SREG)

Steps in enabling an interrupt

To enable any one of the interrupts, we take the following steps:

1. Bit D7 (I) of the SREG register must be set to HIGH to allow the interrupts to happen. This is done with the "SEI" (Set Interrupt) instruction.
2. If I = 1, each interrupt is enabled by setting to HIGH the interrupt enable (IE) flag bit for that interrupt. There are some I/O registers holding the interrupt enable bits. Figure 10-3 shows that the TIMSK register has interrupt enable bits for Timer0, Timer1, and Timer2. As we study each of peripherals throughout the book we will examine the registers holding the interrupt enable bits. It must be noted that if I = 0, no interrupt will be responded to, even if the corresponding interrupt enable bit is high. To understand this important point look at Example 10-1.

Example 10-1

Show the instructions to (a) enable (unmask) the Timer0 overflow interrupt and Timer2 compare match interrupt, and (b) disable (mask) the Timer0 overflow interrupt, then (c) show how to disable (mask) all the interrupts with a single instruction.

Solution:

```
(a)   LDI R20,(1<<TOIE0)|(1<<OCIE2) ;TOIE0 = 1, OCIE2 = 1
      OUT TIMSK,R20  ;enable Timer0 overflow and Timer2 compare match
      SEI                            ;allow interrupts to come in

(b)   IN    R20,TIMSK               ;R20 = TIMSK
      ANDI  R20,0xFF^(1<<TOIE0)      ;TOIE0 = 0
      OUT   TIMSK,R20                ;mask (disable) Timer0 interrupt
```

We can perform the above actions with the following instructions, as well:

```
      IN    R20,TIMSK        ;R20 = TIMSK
      CBR   R20,1<<TOIE0      ;TOIE0 = 0
      OUT   TIMSK,R20         ;mask (disable) Timer0 interrupt

(c)   CLI                    ;mask all interrupts globally
```

Notice that in part (a) we can use "LDI,0x81" in place of the following instruction:
"LDI R20,(1<<TOIE0)|(1<<OCIE2)"

Review Questions

1. Of the interrupt and polling methods, which one avoids tying down the micro-controller?
2. Give the name of the interrupts in the TIMSK register.
3. Upon power-on reset of the ATmega32, what memory area is assigned to the interrupt vector table? Can the programmer change the memory space assigned to the table?
4. What is the content of D7 (I) of the SREG register upon reset, and what does it mean?
5. Show the instructions needed to enable the Timer1 compare A match interrupt.
6. What address in the interrupt vector table is assigned to the Timer1 overflow and INT0 interrupts?

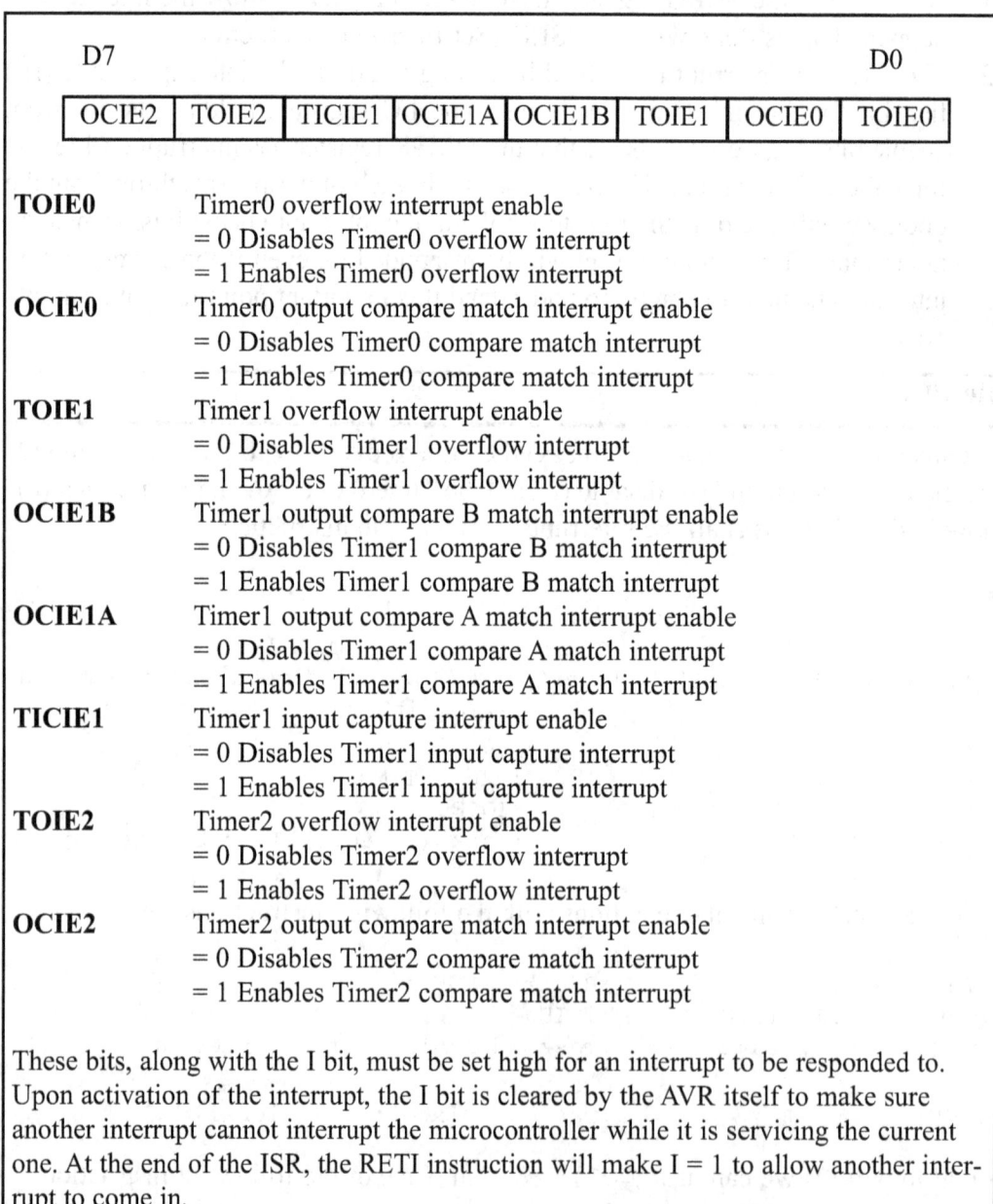

D7							D0
OCIE2	TOIE2	TICIE1	OCIE1A	OCIE1B	TOIE1	OCIE0	TOIE0

TOIE0   Timer0 overflow interrupt enable
                    = 0 Disables Timer0 overflow interrupt
                    = 1 Enables Timer0 overflow interrupt

OCIE0   Timer0 output compare match interrupt enable
                    = 0 Disables Timer0 compare match interrupt
                    = 1 Enables Timer0 compare match interrupt

TOIE1   Timer1 overflow interrupt enable
                    = 0 Disables Timer1 overflow interrupt
                    = 1 Enables Timer1 overflow interrupt

OCIE1B   Timer1 output compare B match interrupt enable
                    = 0 Disables Timer1 compare B match interrupt
                    = 1 Enables Timer1 compare B match interrupt

OCIE1A   Timer1 output compare A match interrupt enable
                    = 0 Disables Timer1 compare A match interrupt
                    = 1 Enables Timer1 compare A match interrupt

TICIE1   Timer1 input capture interrupt enable
                    = 0 Disables Timer1 input capture interrupt
                    = 1 Enables Timer1 input capture interrupt

TOIE2   Timer2 overflow interrupt enable
                    = 0 Disables Timer2 overflow interrupt
                    = 1 Enables Timer2 overflow interrupt

OCIE2   Timer2 output compare match interrupt enable
                    = 0 Disables Timer2 compare match interrupt
                    = 1 Enables Timer2 compare match interrupt

These bits, along with the I bit, must be set high for an interrupt to be responded to. Upon activation of the interrupt, the I bit is cleared by the AVR itself to make sure another interrupt cannot interrupt the microcontroller while it is servicing the current one. At the end of the ISR, the RETI instruction will make I = 1 to allow another interrupt to come in.

Figure 10-3. TIMSK (Timer Interrupt Mask) Register

SECTION 10.2: PROGRAMMING TIMER INTERRUPTS

In Chapter 9 we discussed how to use Timers 0, 1, and 2 with the polling method. In this section we use interrupts to program the AVR timers. Please review Chapter 9 before you study this section.

Rollover timer flag and interrupt

In Chapter 9 we stated that the timer overflow flag is raised when the timer rolls over. In that chapter, we also showed how to monitor the timer flag with the instruction "SBRS R20,TOV0". In polling TOV0, we have to wait until TOV0 is raised. The problem with this method is that the microcontroller is tied down waiting for TOV0 to be raised, and cannot do anything else. Using interrupts avoids tying down the controller. If the timer interrupt in the interrupt register is enabled, TOV0 is raised whenever the timer rolls over and the microcontroller jumps to the interrupt vector table to service the ISR. In this way, the microcontroller can do other things until it is notified that the timer has rolled over. To use an interrupt in place of polling, first we must enable the interrupt because all the interrupts are masked upon reset. The TOIEx bit enables the interrupt for a given timer. TOIEx bits are held by the TIMSK register as shown in Table 10-2. See Figure 10-4 and Program 10-1.

Table 10-2: Timer Interrupt Flag Bits and Associated Registers

Interrupt	Overflow Flag Bit	Register	Enable Bit	Register
Timer0	TOV0	TIFR	TOIE0	TIMSK
Timer1	TOV1	TIFR	TOIE1	TIMSK
Timer2	TOV2	TIFR	TOIE2	TIMSK

Notice the following points about Program 10-1:
1. We must avoid using the memory space allocated to the interrupt vector table. Therefore, we place all the initialization codes in memory starting at an address such as $100. The JMP instruction is the first instruction that the AVR executes when it is awakened at address 0000 upon reset. The JMP instruction at address 0000 redirects the controller away from the interrupt vector table.
2. In the MAIN program, we enable (unmask) the Timer0 interrupt with the following instructions:

```
LDI     R16,1<<TOV0
OUT     TIMSK,R16    ;enable Timer0 overflow interrupt
SEI                  ;set I (enable interrupts globally)
```

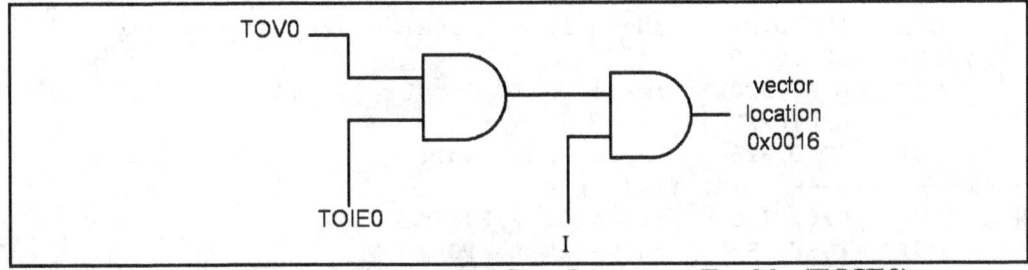

Figure 10-4. The Role of Timer Overflow Interrupt Enable (TOIE0)

---

3. In the MAIN program, we initialize the Timer0 register and then enter into an infinite loop to keep the CPU busy. The loop could be replaced with a real-world application being executed by the CPU. In this case, the loop gets data from PORTC and sends it to PORTD. While the PORTC data is brought in and issued to PORTD continuously, the TOIE0 flag is raised as soon as Timer0 rolls over, and the microcontroller gets out of the loop and goes to $0016 to execute the ISR associated with Timer0. At this point, the AVR clears the I bit (D7 of SREG) to indicate that it is currently serving an interrupt and cannot be interrupted again; in other words, no interrupt inside the interrupt. In Section 10.6, we show how to allow an interrupt inside an interrupt.

4. The ISR for Timer0 is located starting at memory location $200 because it is too large to fit into address space $16–$18, the address allocated to the Timer0 overflow interrupt in the interrupt vector table.

5. RETI must be the last instruction of the ISR. Upon execution of the RETI instruction, the AVR automatically enables the I bit (D7 of the SREG register) to indicate that it can accept new interrupts.

6. In the ISR for Timer0, notice that there is no need for clearing the TOV0 flag since the AVR clears the TOV0 flag internally upon jumping to the interrupt vector table.

Program 10-1: For this program, we assume that PORTC is connected to 8 switches and PORTD to 8 LEDs. This program uses Timer0 to generate a square wave on pin PORTB.5, while at the same time data is being transferred from PORTC to PORTD.

```
;Program 10-1
.INCLUDE "M32DEF.INC"
.ORG  0x0           ;location for reset
      JMP   MAIN
.ORG  0x16          ;location for Timer0 overflow (see Table 10.1)
      JMP   T0_OV_ISR         ;jump to ISR for Timer0
;-main program for initialization and keeping CPU busy
.ORG  0x100
MAIN: LDI   R20,HIGH(RAMEND)
      OUT   SPH,R20
      LDI   R20,LOW(RAMEND)
      OUT   SPL,R20           ;initialize stack
      SBI   DDRB,5            ;PB5 as an output
      LDI   R20,(1<<TOIE0)
      OUT   TIMSK,R20   ;enable Timer0 overflow interrupt
      SEI               ;set I (enable interrupts globally)
      LDI   R20,-32     ;timer value for 4 µs
      OUT   TCNT0,R20   ;load Timer0 with -32
      LDI   R20,0x01
      OUT   TCCR0,R20   ;Normal, internal clock, no prescaler
      LDI   R20,0x00
      OUT   DDRC,R20    ;make PORTC input
      LDI   R20,0xFF
      OUT   DDRD,R20    ;make PORTD output
;--------------- Infinite loop
HERE: IN    R20,PINC    ;read from PORTC
      OUT   PORTD,R20   ;give it to PORTD
      JMP   HERE        ;keeping CPU busy waiting for interrupt
```

```
;---------------ISR for Timer0 (it is executed every 4 µs)
.ORG   0x200
T0_OV_ISR:
       IN    R16,PORTB    ;read PORTB
       LDI   R17,0x20     ;00100000 for toggling PB5
       EOR   R16,R17
       OUT   PORTB,R16    ;toggle PB5
       LDI   R16,-32      ;timer value for 4 µs
       OUT   TCNT0,R16    ;load Timer0 with -32 (for next round)
       RETI               ;return from interrupt
```

See Example 10-2 to understand the difference between RET and RETI.

Example 10-2

What is the difference between the RET and RETI instructions? Explain why we cannot use RET instead of RETI as the last instruction of an ISR.

Solution:

Both perform the same actions of popping off the top bytes of the stack into the program counter, and making the AVR return to where it left off. However, RETI also performs the additional task of setting the I flag, indicating that the servicing of the interrupt is over and the AVR now can accept a new interrupt. If you use RET instead of RETI as the last instruction of the interrupt service routine, you simply block any new interrupt after the first interrupt, because the I would indicate that the interrupt is still being serviced.

See Program 10-2. Program 10-2 uses Timer0 and Timer1 interrupts simultaneously, to generate square waves on pins PB1 and PB7 respectively, while data is being transferred from PORTC to PORTD.

```
;Program 10-2
.INCLUDE "M32DEF.INC"
.ORG   0x0                ;location for reset
       JMP   MAIN         ;bypass interrupt vector table
.ORG   0x12               ;ISR location for Timer1 overflow
       JMP   T1_OV_ISR    ;go to an address with more space
.ORG   0x16               ;ISR location for Timer0 overflow
       JMP   T0_OV_ISR    ;go to an address with more space
;----main program for initialization and keeping CPU busy
.ORG   0x100
MAIN:  LDI   R20,HIGH(RAMEND)
       OUT   SPH,R20
       LDI   R20,LOW(RAMEND)
       OUT   SPL,R20       ;initialize stack point
       SBI   DDRB,1        ;PB1 as an output
       SBI   DDRB,7        ;PB7 as an output
       LDI   R20,(1<<TOIE0)|(1<<TOIE1)
       OUT   TIMSK,R20     ;enable Timer0 overflow interrupt
       SEI                 ;set I (enable interrupts globally)
       LDI   R20,-160      ;value for 20 µs
       OUT   TCNT0,R20     ;load Timer0 with -160
       LDI   R20,0x01
       OUT   TCCR0,R20     ;Normal mode, int clk, no prescaler
       LDI   R20,HIGH(-640) ;the high byte
       OUT   TCNT1H,R20    ;load Timer1 high byte
```

```
        LDI     R20,LOW(-640)   ;the low byte
        OUT     TCNT1L,R20  ;load Timer1 low byte
        LDI     R20,0x00
        OUT     TCCR1A,R20  ;Normal mode
        LDI     R20,0x01
        OUT     TCCR1B,R20  ;internal clk, no prescaler
        LDI     R20,0x00
        OUT     DDRC,R20    ;make PORTC input
        LDI     R20,0xFF
        OUT     DDRD,R20    ;make PORTD output
;--------------- Infinite loop
HERE: IN        R20,PINC    ;read from PORTC
        OUT     PORTD,R20   ;and give it to PORTD
        JMP     HERE        ;keeping CPU busy waiting for interrupt
;------ISR for Timer0 (It comes here after elapse of 20 μs time)
.ORG  0x200
T0_OV_ISR:
        LDI     R16,-160    ;value for 20 μs
        OUT     TCNT0,R16   ;load Timer0 with -160 (for next round)
        IN      R16,PORTB   ;read PORTB
        LDI     R17,0x02    ;00000010 for toggling PB1
        EOR     R16,R17
        OUT     PORTB,R16   ;toggle PB1
        RETI                ;return from interrupt
;------ISR for Timer1 (It comes here after elapse of 80 μs time)
.ORG  0x300
T1_OV_ISR:
        LDI     R18,HIGH(-640)
        OUT     TCNT1H,R18  ;load Timer1 high byte
        LDI     R18,LOW(-640)
        OUT     TCNT1L,R18  ;load Timer1 low byte (for next round)
        IN      R18,PORTB   ;read PORTB
        LDI     R19,0x80    ;10000000 for toggling PB7
        EOR     R18,R19
        OUT     PORTB,R18   ;toggle PB7
        RETI                ;return from interrupt
```

Notice that the addresses $0100, $0200, and $0300 that we used in Program 10-2 are all arbitrary and can be changed to any addresses we want. The only addresses that we cannot change are the reset location of 0000, the Timer0 overflow address of $0016, and the Timer1 overflow address of $0012 in the interrupt vector table because they were fixed at the time of the ATmega32 design.

Program 10-3 has two interrupts: (1) PORTA counts up every time Timer1 overflows. It overflows once per second. (2) A pulse is fed into Timer0, where Timer0 is used as counter and counts up. Whenever the counter reaches 200, it will toggle the pin PORTB.6.

```
;Program 10-3
.INCLUDE "M32DEF.INC"
.ORG  0x0                   ;location for reset
        JMP     MAIN        ;bypass interrupt vector table
.ORG  0x12                  ;ISR location for Timer1 overflow
        JMP     T1_OV_ISR   ;go to an address with more space
.ORG  0x16                  ;ISR location for Timer0 overflow
        JMP     T0_OV_ISR   ;go to an address with more space
```

```
;---main program for initialization and keeping CPU busy
.ORG  0x40
MAIN: LDI   R20,HIGH(RAMEND)
      OUT   SPH,R20
      LDI   R20,LOW(RAMEND)
      OUT   SPL,R20      ;initialize SP

      LDI   R18,0        ;R18 = 0
      OUT   PORTA,R18    ;PORTA = 0
      LDI   R20,0
      OUT   DDRC,R20     ;PORTC as input
      LDI   R20,0xFF
      OUT   DDRA,R20     ;PORTA as output
      OUT   DDRD,R20     ;PORTD as output
      SBI   DDRB,6       ;PB6 as an output
      SBI   PORTB,0      ;activate pull-up of PB0

      LDI   R20,0x06
      OUT   TCCR0,R20    ;Normal, T0 pin falling edge, no scale
      LDI   R16,-200
      OUT   TCNT0,R16    ;load Timer0 with -200
      LDI   R19,HIGH(-31250)  ;timer value for 1 second
      OUT   TCNT1H,R19   ;load Timer1 high byte
      LDI   R19,LOW(-31250)
      OUT   TCNT1L,R19   ;load Timer1 low byte
      LDI   R20,0
      OUT   TCCR1A,R20   ;Timer1 Normal mode
      LDI   R20,0x04
      OUT   TCCR1B,R20   ;int clk, prescale 1:256
      LDI   R20,(1<<TOIE0)|(1<<TOIE1)
      OUT   TIMSK,R20    ;enable Timer0 & Timer1 overflow ints
      SEI                ;set I (enable interrupts globally)
;--------------- Infinite loop
HERE: IN    R20,PINC     ;read from PORTC
      OUT   PORTD,R20    ;and send it to PORTD
      JMP   HERE         ;waiting for interrupt
;-------ISR for Timer0 to toggle after 200 clocks
.ORG 0x200
T0_OV_ISR:
      IN    R16,PORTB    ;read PORTB
      LDI   R17,0x40     ;0100 0000 for toggling PB7
      EOR   R16,R17
      OUT   PORTB,R16    ;toggle PB6
      LDI   R16,-200     ;setup for next round
      OUT   TCNT0,R16    ;load Timer0 with -200 for next round
      RETI               ;return from interrupt
;---------ISR for Timer1 (It comes here after elapse of 1s time)
.ORG 0x300
T1_OV_ISR:
      INC   R18          ;increment upon overflow
      OUT   PORTA,R18    ;display it on PORTA
      LDI   R19,HIGH(-31250)
      OUT   TCNT1H,R19   ;load Timer1 high byte
      LDI   R19,LOW(-31250)
      OUT   TCNT1L,R19   ;load Timer1 low byte (for next round)
      RETI               ;return from interrupt
```

Compare match timer flag and interrupt

Sometimes a task should be done periodically, as in the previous examples. The programs can be written using the CTC mode and compare match (OCF) flag. To do so, we load the OCR register with the proper value and initialize the timer to the CTC mode. When the content of TCNT matches with OCR, the OCF flag is set, which causes the compare match interrupt to occur.

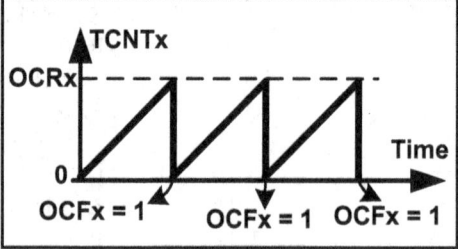

Figure 10-5. CTC Mode

Example 10-3

Using Timer0, write a program that toggles pin PORTB.5 every 40 µs, while at the same time transferring data from PORTC to PORTD. Assume XTAL = 1 MHz.

Solution:

1/1 MHz = 1 µs and 40 µs/1 µs = 40. That means we must have OCR0 = 40 − 1 = 39

```
.INCLUDE "M32DEF.INC"
.ORG  0x0    ;location for reset
      JMP   MAIN
.ORG  0x14   ;ISR location for Timer0 compare match
      JMP   T0_CM_ISR
;main program for initialization and keeping CPU busy
.ORG  0x100
MAIN: LDI   R20,HIGH(RAMEND)
      OUT   SPH,R20
      LDI   R20,LOW(RAMEND)
      OUT   SPL,R20      ;set up stack
      SBI   DDRB,5       ;PB5 as an output
      LDI   R20,(1<<OCIE0)
      OUT   TIMSK,R20    ;enable Timer0 compare match interrupt
      SEI                ;set I (enable interrupts globally)
      LDI   R20,39
      OUT   OCR0,R20     ;load Timer0 with 39
      LDI   R20,0x09
      OUT   TCCR0,R20    ;start Timer0, CTC mode, int clk, no prescaler
      LDI   R20,0x00
      OUT   DDRC,R20     ;make PORTC input
      LDI   R20,0xFF
      OUT   DDRD,R20     ;make PORTD output
;--------------- Infinite loop
HERE: IN    R20,PINC     ;read from PORTC
      OUT   PORTD,R20    ;and send it to PORTD
      JMP   HERE         ;keeping CPU busy waiting for interrupt
;----------------ISR for Timer0 (it is executed every 40 µs)
T0_CM_ISR:
      IN    R16,PORTB    ;read PORTB
      LDI   R17,0x20     ;00100000 for toggling PB5
      EOR   R16,R17
      OUT   PORTB,R16    ;toggle PB5
      RETI               ;return from interrupt
```

Because the timer is in the CTC mode, the timer will be loaded with zero as well. So, the compare match interrupt occurs periodically. See Figure 10-5 and Examples 10-3 and 10-4. Notice that the AVR chip clears the OCF flag upon jumping to the interrupt vector table.

Example 10-4

Using Timer1, write a program that toggles pin PORTB.5 every second, while at the same time transferring data from PORTC to PORTD. Assume XTAL = 8 MHz.

Solution:

For prescaler = 1024 we have $T_{Clock}= (1 / 8\ MHz) \times 1024 = 128\ \mu s$ and $1\ s/128\ \mu s = 7812$. That means we must have OCR1A = 7811 = 0x1E83

```
.INCLUDE "M32DEF.INC"
.ORG  0x0    ;location for reset
      JMP    MAIN
.ORG  0x14   ;location for Timer1 compare match
      JMP    T1_CM_ISR
;------main program for initialization and keeping CPU busy
MAIN: LDI    R20,HIGH(RAMEND)
      OUT    SPH,R20
      LDI    R20,LOW(RAMEND)
      OUT    SPL,R20      ;set up stack
      SBI    DDRB,5       ;PB5 as an output
      LDI    R20,(1<<OCIE1A)
      OUT    TIMSK,R20    ;enable Timer1 compare match interrupt
      SEI                 ;set I (enable interrupts globally)
      LDI    R20,0x00
      OUT    TCCR1A,R20
      LDI    R20,0xD
      OUT    TCCR1B,R20          ;prescaler 1:1024, CTC mode
      LDI    R20,HIGH(7811)      ;the high byte
      OUT    OCR1AH,R20          ;Temp = 0x1E (high byte of 7811)
      LDI    R20,LOW(7811)       ;the low byte
      OUT    OCR1AL,R20          ;OCR1A = 7811
      LDI    R20,0x00
      OUT    DDRC,R20            ;make PORTC input
      LDI    R20,0xFF
      OUT    DDRD,R20            ;make PORTD output
;--------------- Infinite loop
HERE: IN     R20,PINC     ;read from PORTC
      OUT    PORTD,R20    ;PORTD = R20
      JMP    HERE         ;keeping CPU busy waiting for interrupt
;---ISR for Timer1 (It comes here after elapse of 1 second time)
T1_CM_ISR:
      IN     R16,PORTB
      LDI    R17,0x20     ;00100000 for toggling PB5
      EOR    R16,R17
      OUT    PORTB,R16    ;toggle PB5
      RETI                ;return from interrupt
```

Review Questions

1. True or false. There is a single interrupt in the interrupt vector table assigned to both Timer0 and Timer1.
2. What address in the interrupt vector table is assigned to Timer0 overflow?
3. Which register does TOIE1 belong to? Show how it is enabled.
4. Assume that Timer0 is programmed in Normal mode, TCNT0 = 0xF1, and the TOIE0 bit is enabled. Explain how the interrupt for the timer works.
5. True or false. The last two instructions of the ISR for Timer0 are:
   ```
   OUT    TIFR, 1<<TOV0      ;clear TOV0 flag
   RETI
   ```
6. Assume that Timer0 is programmed in CTC mode, OCR0 = 0x21, and the compare match interrupt is enabled. Explain how the interrupt for the timer works.
7. In the previous problem, assume XTAL = 8 MHz, and the timer is in no prescaler mode. How often is the ISR executed?

SECTION 10.3: PROGRAMMING EXTERNAL HARDWARE INTERRUPTS

The number of external hardware interrupt interrupts varies in different AVRs. The ATmega32 has three external hardware interrupts: pins PD2 (PORTD.2), PD3 (PORTD.3), and PB2 (PORTB.2), designated as INT0, INT1, and INT2, respectively. Upon activation of these pins, the AVR is interrupted in whatever it is doing and jumps to the vector table to perform the interrupt service routine. In this section we study these three external hardware interrupts of the AVR with some examples in Assembly language.

External interrupts INT0, INT1, and INT2

There are three external hardware interrupts in the ATmega32: INT0, INT1, and INT2. They are located on pins PD2, PD3, and PB2, respectively. As we saw in Table 10-1, the interrupt vector table locations $2, $4, and $6 are set aside for INT0, INT1, and INT2, respectively. The hardware interrupts must be enabled before they can take effect. This is done using the INTx bit located in the GICR register. See Figure 10-6. For example, the following instructions enable INT0:

```
LDI    R20,0x40
OUT    GICR,R20
```

The INT0 is a low-level-triggered interrupt by default, which means, when a low signal is applied to pin PD2 (PORTD.2), the controller will be interrupted and jump to location $0002 in the vector table to service the ISR.

Study Example 10-5 to gain insight into external hardware interrupts. In this program, the microcontroller is looping continuously in the HERE loop. Whenever the switch on INT0 (pin PD2) is activated, the microcontroller gets out of the loop and jumps to vector location $0002. The ISR for INT0 toggles the PC0. If, by the time it executes the RETI instruction, the INT0 pin is still low, the microcontroller initiates the interrupt again. Therefore, if we want the ISR to be executed once, the

INT0 pin must be brought back to high before RETI is executed, or we should make the interrupt edge-triggered, as discussed next.

D7							D0
INT1	INT0	INT2	-	-	-	IVSEL	IVCE

INT0   External Interrupt Request 0 Enable
                = 0 Disables external interrupt 0
                = 1 Enables external interrupt 0
INT1   External Interrupt Request 1 Enable
                = 0 Disables external interrupt 1
                = 1 Enables external interrupt 1
INT2   External Interrupt Request 2 Enable
                = 0 Disables external interrupt 2
                = 1 Enables external interrupt 2

These bits, along with the I bit, must be set high for an interrupt to be responded to.

Figure 10-6. GICR (General Interrupt Control Register) Register

---

Example 10-5

Assume that the INT0 pin is connected to a switch that is normally high. Write a program that toggles PORTC.3 whenever the INT0 pin goes low.

Solution:

```
.INCLUDE "M32DEF.INC"
.ORG 0                  ;location for reset
     JMP    MAIN
.ORG 0x02               ;vector location for external interrupt 0
     JMP    EX0_ISR
MAIN: LDI   R20,HIGH(RAMEND)
     OUT    SPH,R20
     LDI    R20,LOW(RAMEND)
     OUT    SPL,R20           ;initialize stack
     SBI    DDRC,3            ;PORTC.3 = output
     SBI    PORTD,2           ;pull-up activated
     LDI    R20,1<<INT0       ;enable INT0
     OUT    GICR,R20
     SEI                      ;enable interrupts
HERE:JMP    HERE              ;stay here forever
EX0_ISR:
     IN     R21,PINC   ;read PINC
     LDI    R22,0x08   ;00001000
     EOR    R21,R22
     OUT    PORTC,R21
     RETI
```

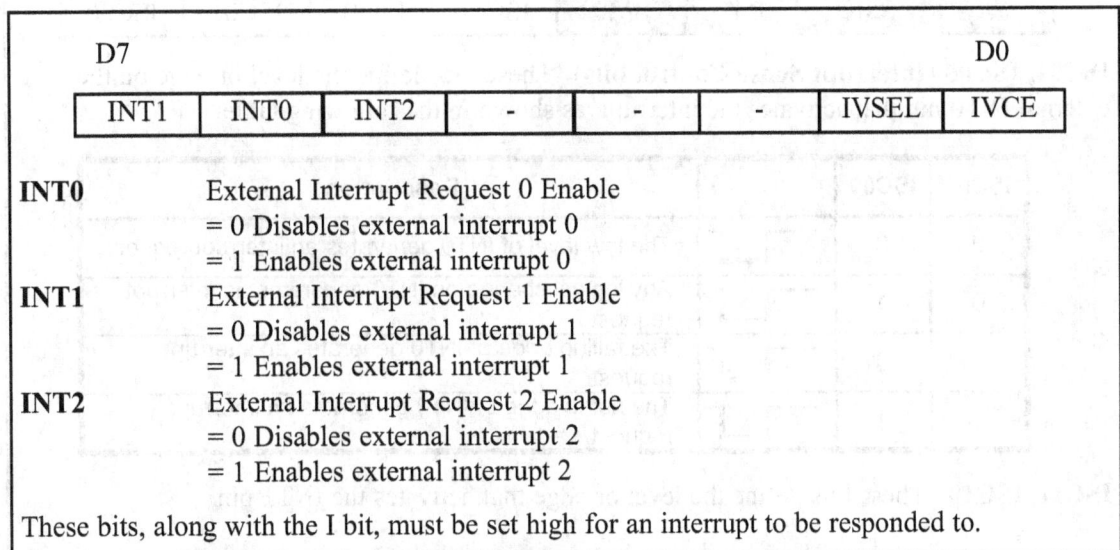

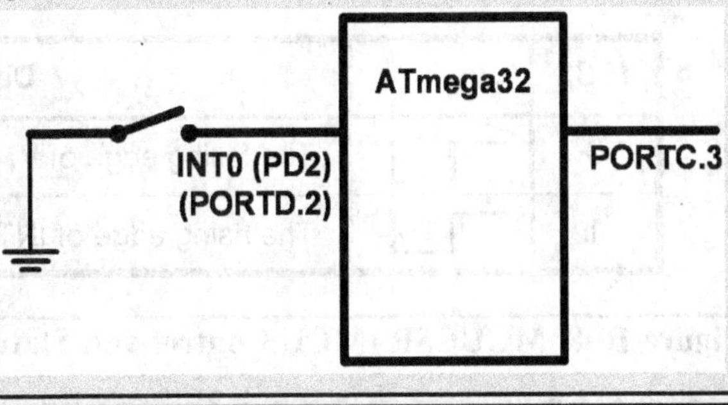

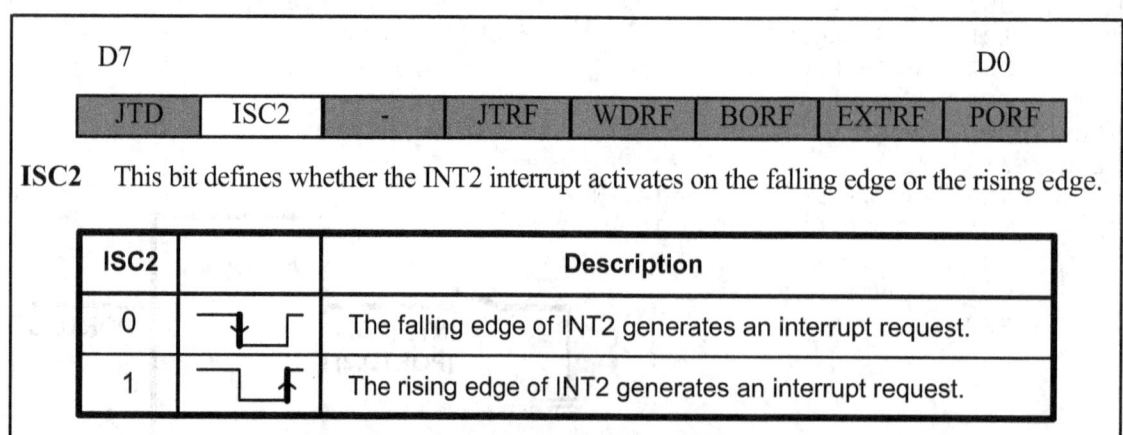

D7							D0
SE	SM2	SM1	SM0	ISC11	ISC10	ISC01	ISC00

ISC01, ISC00 (Interrupt Sense Control bits) These bits define the level or edge on the external INT0 pin that activates the interrupt, as shown in the following table:

ISC01	ISC00		Description
0	0		The low level of INT0 generates an interrupt request.
0	1		Any logical change on INT0 generates an interrupt request.
1	0		The falling edge of INT0 generates an interrupt request.
1	1		The rising edge of INT0 generates an interrupt request.

ISC11, ISC10 These bits define the level or edge that activates the INT1 pin.

ISC11	ISC10		Description
0	0		The low level of INT1 generates an interrupt request.
0	1		Any logical change on INT1 generates an interrupt request.
1	0		The falling edge of INT1 generates an interrupt request.
1	1		The rising edge of INT1 generates an interrupt request.

Figure 10-7. MCUCR (MCU Control Register) Register

Edge-triggered vs. level-triggered interrupts

There are two types of activation for the external hardware interrupts: (1) level triggered, and (2) edge triggered. INT2 is only edge triggered, while INT0 and INT1 can be level or edge triggered.

As stated before, upon reset INT0 and INT1 are low-level-triggered interrupts. The bits of the MCUCR register indicate the trigger options of INT0 and INT1, as shown in Figure 10-7.

D7							D0
JTD	ISC2	-	JTRF	WDRF	BORF	EXTRF	PORF

ISC2 This bit defines whether the INT2 interrupt activates on the falling edge or the rising edge.

ISC2		Description
0		The falling edge of INT2 generates an interrupt request.
1		The rising edge of INT2 generates an interrupt request.

Figure 10-8. MCUCSR (MCU Control and Status Register) Register

The ISC2 bit of the MCUCSR register defines whether INT2 activates in the falling edge or the rising edge (see Figure 10-8). Upon reset ISC2 is 0, meaning that the external hardware interrupt of INT2 is falling edge triggered. See Examples 10-6 and 10-7.

Example 10-6

Show the instructions to (a) make INT0 falling edge triggered, (b) make INT1 triggered on any change, and (c) make INT2 rising edge triggered.

Solution:

```
(a)    LDI    R20,0x02
       OUT    MCUCR,R20


(b)    LDI    R20,1<<ISC10        ;R20 = 0x04
       OUT    MCUCR,R20


(c)    LDI    R20,1<<ISC2         ;R20 = 0x40
       OUT    MCUCSR,R20
```

Example 10-7

Rewrite Example 10-5, so that whenever INT0 goes low, it toggles PORTC.3 only once.

Solution:

```
.INCLUDE "M32DEF.INC"
.ORG 0                          ;location for reset
     JMP    MAIN
.ORG 0x02                       ;location for external interrupt 0
     JMP    EX0_ISR
MAIN: LDI   R20,HIGH(RAMEND)
      OUT   SPH,R20
      LDI   R20,LOW(RAMEND)
      OUT   SPL,R20             ;initialize stack
      LDI   R20,0x2             ;make INT0 falling edge triggered
      OUT   MCUCR,R20
      SBI   DDRC,3              ;PORTC.3 = output
      SBI   PORTD,2             ;pull-up activated
      LDI   R20,1<<INT0         ;enable INT0
      OUT   GICR,R20
      SEI                       ;enable interrupts
HERE: JMP   HERE
EX0_ISR:
      IN    R21,PORTC
      LDI   R22,0x08            ;00001000 for toggling PC3
      EOR   R21,R22
      OUT   PORTC,R21
      RETI
```

In Example 10-7, notice that the only difference between it and the program in Example 10-5 is in the following instructions:

```
LDI   R20,0x2          ;make INT0 falling edge triggered
OUT   MCUCR,R20
```

which makes INT0 an edge-triggered interrupt. When the falling edge of the signal is applied to pin INT0, PORTC.3 will toggle. To toggle the LED again, another high-to-low pulse must be applied to INT0. This is the opposite of Example 10-5. In Example 10-5, due to the level-triggered nature of the interrupt, as long as INT0 is kept at a low level, PORTC.3 toggles. But in this example, to turn on PORTC.3 again, the INT0 pulse must be brought back high and then low to create a falling edge to activate the interrupt.

Sampling the edge-triggered and level-triggered interrupts

Examine Figure 10-9. The edge interrupt (the falling edge, the rising edge, or the change level) is latched by the AVR and is held by the INTFx bits of the GIFR register. This means that when an external interrupt is in an edge-triggered mode (falling edge, rising edge, or change level), upon triggering an interrupt request, the related INTFx flag becomes set. If the interrupt is active (the INTx bit is set and the I-bit in SREG is one), the AVR will jump to the corresponding interrupt vector location and the INTFx flag will be cleared automatically, otherwise,

Bit	D7							D0
	INTF1	INTF0	INTF2	-	-	-	-	-

Figure 10-9. GIFR (General Interrupt Flag Register) Register

the flag remains set. The flag can be cleared by writing a one to it. For example, the INTF1 flag can be cleared using the following instructions:

```
LDI   R20,(1<<INTF1)    ;R20 = 0x80
OUT   GIFR,R20          ;clear the INTF1 flag
```

Notice that in edge-triggered interrupts (falling edge, rising edge, and change level interrupts), the pulse must last at least 1 instruction cycle to ensure that the transition is seen by the microcontroller. This means that pulses shorter than 1 machine cycle are not guaranteed to generate an interrupt.

When an external interrupt is in level-triggered mode, the interrupt is not latched, meaning that the INTFx flag remains unchanged when an interrupt occurs, and the state of the pin is read directly. As a result, when an interrupt is in level-triggered mode, the pin must be held low for a minimum time of 5 machine cycles to be recognized.

Review Questions

1. True or false. Upon reset, the external hardware interrupts INT0–INT2 are edge triggered.

2. For ATmega32, what pins are assigned to INT0–INT2?
3. Show how to enable the INT1 interrupt.
4. Assume that the external hardware interrupt INT0 is enabled, and is set to the low-edge trigger. Explain how this interrupt works when it is activated.
5. True or false. Upon reset, the INT2 interrupt is falling edge triggered.
6. Assume that INT0 is falling edge triggered. How do we make sure that a single interrupt is not recognized as multiple interrupts?
7. Using polling and INT0, write a program that upon falling edges toggles PORTC.3. Compare it with Example 10-7; which program is better?

SECTION 10.4: INTERRUPT PRIORITY IN THE AVR

The next topic that we must deal with is what happens when two interrupts are activated at the same time. Which of these two interrupts is responded to first?

Interrupt priority

If two interrupts are activated at the same time, the interrupt with the higher priority is served first. The priority of each interrupt is related to the address of that interrupt in the interrupt vector. The interrupt that has a lower address, has a higher priority. See Table 10-1. For example, the address of external interrupt 0 is 2, while the address of external interrupt 2 is 6; thus, external interrupt 0 has a higher priority, and if both of these interrupts are activated at the same time, external interrupt 0 is served first.

Interrupt inside an interrupt

What happens if the AVR is executing an ISR belonging to an interrupt and another interrupt is activated? When the AVR begins to execute an ISR, it disables the I bit of the SREG register, causing all the interrupts to be disabled, and no other interrupt occurs while serving the interrupt. When the RETI instruction is executed, the AVR enables the I bit, causing the other interrupts to be served. If you want another interrupt (with any priority) to be served while the current interrupt is being served you can set the I bit using the SEI instruction. But do it with care. For example, in a low-level-triggered external interrupt, enabling the I bit while the pin is still active will cause the ISR to be reentered infinitely, causing the stack to overflow with unpredictable consequences.

Context saving in task switching

In multitasking systems, such as multitasking real-time operating systems (RTOS), the CPU serves one task (job or process) at a time and then moves to the next one. In simple systems, the tasks can be organized as the interrupt service routine. For example, in Example 10-3, the program does two different tasks:
(1) copying the contents of PORTC to PORTD,
(2) toggling PORTC.2 every 5 µs
While writing a program for a multitasking system, we should manage the resources carefully so that the tasks do not conflict with each other. For example, consider a system that should perform the following tasks: (1) increasing the con-

tents of PORTC continuously, and (2) increasing the content of PORTD once every 5 μs. Read the following program. Does it work?

```
;Program 10-4
.INCLUDE "M32DEF.INC"
.ORG  0x0    ;location for reset
      JMP  MAIN
.ORG  0x14   ;location for Timer0 compare match
      JMP  T0_CM_ISR
;-main program for initialization and keeping CPU busy
.ORG  0x100
MAIN: LDI  R20,HIGH(RAMEND)
      OUT  SPH,R20
      LDI  R20,LOW(RAMEND)
      OUT  SPL,R20    ;set up stack
      SBI  DDRB,5      ;PB5 as an output
      LDI  R20,(1<<OCIE0)
      OUT  TIMSK,R20   ;enable Timer0 compare match interrupt
      SEI              ;set I (enable interrupts globally)
      LDI  R20,160
      OUT  OCR0,R20    ;load Timer0 with 160
      LDI  R20,0x09
      OUT  TCCR0,R20   ;CTC mode, int clk, no prescaler
      LDI  R20,0xFF
      OUT  DDRC,R20    ;make PORTC output
      OUT  DDRD,R20    ;make PORTD output
      LDI  R20, 0
HERE: OUT  PORTC,R20   ;PORTC = R20
      INC  R20
      JMP  HERE        ;keeping CPU busy waiting for interrupt
;-------------------------ISR for Timer0
T0_CM_ISR:
      IN   R20,PIND
      INC  R20
      OUT  PORTD,R20   ;PORTD = R20
      RETI             ;return from interrupt
```

The tasks do not work properly, since they have resource conflict and they interfere with each other. R20 is used and changed by both tasks, which causes the program not to work properly. For example, consider the following scenario: The content of R20 increases in the main program, at first becoming 0, then 1, and so on. When the timer interrupt occurs, R20 is 95, and PORTC is 95 as well. In the ISR, the R20 is loaded with the content of PORTD, which is 0. So, when it goes back to the main program, the content of R20 is 1 and PORTC will be loaded by 2. But if the program worked properly, PORTC would be loaded with 96.

We can solve such problems in the following two ways:

(1) Using different registers for different tasks. In the program discussed above, if we use different registers in the main program and in the ISR, the program will work properly.

```
;Program 10-5
.INCLUDE "M32DEF.INC"
.ORG  0x0                ;location for reset
      JMP  MAIN
```

```
      .ORG   0x14                 ;location for Timer0 compare match
             JMP    T0_CM_ISR
;------main program for initialization and keeping CPU busy
      .ORG   0x100
MAIN: LDI    R20,HIGH(RAMEND)
      OUT    SPH,R20
      LDI    R20,LOW(RAMEND)
      OUT    SPL,R20       ;set up stack
      SBI    DDRB,5        ;PB5 as an output
      LDI    R20,(1<<OCIE0)
      OUT    TIMSK,R20     ;enable Timer0 compare match interrupt
      SEI                  ;set I (enable interrupts globally)
      LDI    R20,160
      OUT    OCR0,R20      ;load Timer0 with 160
      LDI    R20,0x09
      OUT    TCCR0,R20     ;start timer,CTC mode,int clk,no prescaler
      LDI    R20,0xFF
      OUT    DDRC,R20      ;make PORTC output
      OUT    DDRD,R20      ;make PORTD output
      LDI    R20, 0
HERE: OUT    PORTC,R20     ;PORTC = R20
      INC    R20
      JMP    HERE          ;keeping CPU busy waiting for int.
;-------------------------ISR for Timer0
T0_CM_ISR:
      IN     R21,PIND
      INC    R21
      OUT    PORTD,R21     ;toggle PB5
      RETI                 ;return from interrupt
```

(2) Context saving. In big programs we might not have enough registers to use separate registers for different tasks. In these cases, we can save the contents of registers on the stack before execution of each task, and reload the registers at the end of the task. This saving of the CPU contents before switching to a new task is called *context saving* (or *context switching*). See the following program:

```
;Program 10-6
.INCLUDE "M32DEF.INC"
.ORG  0x0    ;location for reset
      JMP    MAIN
.ORG  0x14   ;location for Timer0 compare match
      JMP    T0_CM_ISR
;main program for initialization and keeping CPU busy
.ORG  0x100
MAIN: LDI    R20,HIGH(RAMEND)
      OUT    SPH,R20
      LDI    R20,LOW(RAMEND)
      OUT    SPL,R20       ;set up stack
      SBI    DDRB,5        ;PB5 as an output
      LDI    R20,(1<<OCIE0)
      OUT    TIMSK,R20     ;enable Timer0 compare match interrupt
      SEI                  ;set I (enable interrupts globally)
      LDI    R20,160
      OUT    OCR0,R20      ;load Timer0 with 160
      LDI    R20,0x09
```

```
        OUT    TCCR0,R20     ;CTC mode, int clk, no prescaler
        LDI    R20,0xFF
        OUT    DDRC,R20      ;make PORTC output
        OUT    DDRD,R20      ;make PORTD output
        LDI    R20, 0
HERE:   OUT  PORTC,R20       ;PORTC = R20
        INC    R20
        JMP    HERE          ;keeping CPU busy waiting for interrupt
;-------------------------ISR for Timer0
T0_CM_ISR:
        PUSH   R20           ;save R20 on stack
        IN     R20,PIND
        INC    R20
        OUT    PORTD,R20     ;toggle PB5
        POP    R20           ;restore value for R20
        RETI                 ;return from interrupt
```

Notice that using the stack as a place to save the CPU's contents is tedious, time consuming, and slow. So, we might want to use the first solution, whenever we have enough registers.

Saving flags of the SREG register

The flags of SREG are important especially when there are conditional jumps in our program. We should save the SREG register if the flags are changed in a task. See Figure 10-10.

Interrupt latency

The time from the moment an interrupt is activated to the moment the CPU starts to execute the task is called the *interrupt latency*. This latency is 4 machine cycle times. During this time the PC register is pushed on the stack and the I bit of the SREG register clears, causing all the interrupts to be disabled. The duration of an interrupt latency can be affected by the type of instruction that the CPU is executing when the interrupt comes in, since the CPU finishes the execution of the current instruction before it serves the interrupt. It takes slightly longer in cases where the instruction being executed lasts for two (or more) machine cycles (e.g., MUL) compared to the instructions that last for only one instruction cycle (e.g., ADD). See the AVR datasheet for the timing.

```
Sample_ISR:
      PUSH   R20
      IN     R20,SREG
      PUSH   R20
      ...
      POP    R20
      OUT    SREG,R20
      POP    R20
      RETI
```

Figure 10-10. Saving the SREG Register

Review Questions

1. True or false. In ATmega32, if the Timer1 and Timer0 interrupts are activated at the same time, the Timer0 interrupt is served first.
2. What happens if two interrupts are activated at the same time?
3. What happens if an interrupt is activated while the CPU is serving another interrupt?
4. What is context saving?

SECTION 10.5: INTERRUPT PROGRAMMING IN C

So far all the programs in this chapter have been written in Assembly. In this section we show how to program the AVR's interrupts in WinAVR C language.

In C language there is no instruction to manage the interrupts. So, in WinAVR the following have been added to manage the interrupts:

1. **Interrupt include file**: We should include the interrupt header file if we want to use interrupts in our program. Use the following instruction:

   ```
   #include <avr\interrupt.h>
   ```

2. **cli( ) and sei( )**: In Assembly, the CLI and SEI instructions clear and set the I bit of the SREG register, respectively. In WinAVR, the cli() and sei() macros do the same tasks.

Table 10-3: Interrupt Vector Name for the ATmega32/ATmega16 in WinAVR

Interrupt	Vector Name in WinAVR
External Interrupt request 0	INT0_vect
External Interrupt request 1	INT1_vect
External Interrupt request 2	INT2_vect
Time/Counter2 Compare Match	TIMER2_COMP_vect
Time/Counter2 Overflow	TIMER2_OVF_vect
Time/Counter1 Capture Event	TIMER1_CAPT_vect
Time/Counter1 Compare Match A	TIMER1_COMPA_vect
Time/Counter1 Compare Match B	TIMER1_COMPB_vect
Time/Counter1 Overflow	TIMER1_OVF_vect
Time/Counter0 Compare Match	TIMER0_COMP_vect
Time/Counter0 Overflow	TIMER0_OVF_vect
SPI Transfer complete	SPI_STC_vect
USART, Receive complete	USART0_RX_vect
USART, Data Register Empty	USART0_UDRE_vect
USART, Transmit Complete	USART0_TX_vect
ADC Conversion complete	ADC_vect
EEPROM ready	EE_RDY_vect
Analog Comparator	ANALOG_COMP_vect
Two-wire Serial Interface	TWI_vect
Store Program Memory Ready	SPM_RDY_vect

3. **Defining ISR**: To write an ISR (interrupt service routine) for an interrupt we use the following structure:

```
ISR(interrupt vector name)
{
        //our program
}
```

For the *interrupt vector name* we must use the ISR names in Table 10-3. For example, the following ISR serves the Timer0 compare match interrupt:

```
ISR (TIMER0_COMP_vect)
{
}
```

See Example 10-8.

Example 10-8 (C version of Program 10-1)

Using Timer0 generate a square wave on pin PORTB.5, while at the same time transferring data from PORTC to PORTD.

Solution:

```
#include "avr/io.h"
#include "avr/interrupt.h"

int main ()
{
     DDRB |= 0x20;            //DDRB.5 = output

     TCNT0 = -32;             //timer value for 4 µs
     TCCR0 = 0x01;            //Normal mode, int clk, no prescaler

     TIMSK = (1<<TOIE0);      //enable Timer0 overflow interrupt
     sei ();                  //enable interrupts

     DDRC = 0x00;             //make PORTC input
     DDRD = 0xFF;             //make PORTD output

     while (1)                //wait here
          PORTD = PINC;
}

ISR (TIMER0_OVF_vect)        //ISR for Timer0 overflow
{
     TCNT0 = -32;
     PORTB ^= 0x20;           //toggle PORTB.5
}
```

Context saving

The C compiler automatically adds instructions to the beginning of the ISRs, which save the contents of all of the general purpose registers and the SREG register on the stack. Some instructions are also added to the end of the ISRs to reload the registers. See Examples 10-9 through 10-13.

Example 10-9 (C version of Program 10-2)

Using Timer0 and Timer1 interrupts, generate square waves on pins PB1 and PB7 respectively, while transferring data from PORTC to PORTD.

Solution:

```c
#include "avr/io.h"
#include "avr/interrupt.h"

int main ( )
{
    DDRB |= 0x82;            //make DDRB.1 and DDRB.7 output
    DDRC = 0x00;             //make PORTC input
    DDRD = 0xFF;             //make PORTD output

    TCNT0 = -160;
    TCCR0 = 0x01;            //Normal mode, int clk, no prescaler

    TCNT1H = (-640)>>8;      //the high byte
    TCNT1L = (-640);         //the low byte
    TCCR1A = 0x00;
    TCCR1B = 0x01;
    TIMSK = (1<<TOIE0)|(1<<TOIE1); //enable Timers 0 and 1 int.
    sei ();                  //enable interrupts

    while (1)                //wait here
        PORTD = PINC;
}

ISR (TIMER0_OVF_vect)        //ISR for Timer0 overflow
{
    TCNT0 = -160;            //TCNT0 = -160 (reload for next round)
    PORTB ^= 0x02;           //toggle PORTB.1
}

ISR (TIMER1_OVF_vect)        //ISR for Timer0 overflow
{
    TCNT1H = (-640)>>8;
    TCNT1L = (-640);         //TCNT1 = -640 (reload for next round)

    PORTB ^= 0x80;           //toggle PORTB.7
}
```

Note: We can use "TCNT1 = -640;" in place of the following instructions:
```c
        TCNT1H = (-640)>>8;
        TCNT1L = (-640);
```

Example 10-10 (C version of Program 10-3)

Using Timer0 and Timer1 interrupts, write a program in which:
(a) PORTA counts up everytime Timer1 overflows. It overflows once per second.
(b) A pulse is fed into Timer0 where Timer0 is used as counter and counts up. Whenever the counter reaches 200, it will toggle the pin PORTB.6.

Solution:

```c
#include "avr/io.h"
#include "avr/interrupt.h"

int main ()
{
    DDRA = 0xFF;              //make PORTA output
    DDRD = 0xFF;              //make PORTD output
    DDRB |= 0x40;            //PORTB.6 as an output
    PORTB |= 0x01;          //activate pull-up

    TCNT0 = -200;            //load Timer0 with -200
    TCCR0 = 0x06;            //Normal mode, falling edge, no prescaler

    TCNT1H = (-31250)>>8;   //the high byte
    TCNT1L = (-31250)&0xFF; //overflow after 31250 clocks
    TCCR1A = 0x00;          //Normal mode
    TCCR1B = 0x04;          //internal clock, prescaler 1:256

    TIMSK = (1<<TOIE0)|(1<<TOIE1); //enable Timers 0 & 1 int.
    sei ();                        //enable interrupts

    DDRC = 0x00;            //make PORTC input
    DDRD = 0xFF;            //make PORTD output

    while (1)               //wait here
        PORTD = PINC;
}

ISR (TIMER0_OVF_vect)       //ISR for Timer0 overflow
{
    TCNT0 = -200;           //TCNT0 = -200
    PORTB ^= 0x40;          //toggle PORTB.6
}

ISR (TIMER1_OVF_vect)       //ISR for Timer1 overflow
{
    TCNT1H = (-31250)>>8;   //the high byte
    TCNT1L = (-31250)&0xFF; //overflow after 31250 clocks
    PORTA ++;               //increment PORTA
}
```

Example 10-11 (C version of Example 10-4)

Using Timer1, write a program that toggles pin PORTB.5 every second, while at the same time transferring data from PORTC to PORTD. Assume XTAL = 8 MHz.

Solution:

```c
#include "avr/io.h"
#include "avr/interrupt.h"

int main ()
{
    DDRB |= 0x20;                //make DDRB.5 output

    OCR0 = 40;
    TCCR0 = 0x09;                //CTC mode, internal clk, no prescaler

    TIMSK = (1<<OCIE0);          //enable Timer0 compare match int.
    sei ();                      //enable interrupts

    DDRC = 0x00;                 //make PORTC input
    DDRD = 0xFF;                 //make PORTD output

    while (1)                    //wait here
        PORTD = PINC;
}

ISR (TIMER0_COMP_vect)           //ISR for Timer0 compare match
{
    PORTB ^= 0x20;               //toggle PORTB.5
}
```

Example 10-12 (C version of Example 10-5)

Assume that the INT0 pin is connected to a switch that is normally high. Write a program that toggles PORTC.3, whenever INT0 pin goes low. Use the external interrupt in level-triggered mode.

Solution:

```c
#include "avr/io.h"
#include "avr/interrupt.h"

int main ()
{
    DDRC = 1<<3;              //PC3 as an output
    PORTD = 1<<2;             //pull-up activated
    GICR = (1<<INT0);         //enable external interrupt 0
    sei ();                   //enable interrupts

    while (1);                //wait here
}

ISR (INT0_vect)              //ISR for external interrupt 0
{
    PORTC ^= (1<<3);         //toggle PORTC.3
}
```

Example 10-13 (C version of Example Example 10-7)

Rewrite Example 10-12 so that whenever INT0 goes low, it toggles PORTC.3 only once.

Solution:

```c
#include "avr/io.h"
#include "avr/interrupt.h"

int main ()
{
    DDRC = 1<<3;              //PC3 as an output
    PORTD = 1<<2;             //pull-up activated
    MCUCR = 0x02;             //make INT0 falling edge triggered
    GICR = (1<<INT0);         //enable external interrupt 0
    sei ();                   //enable interrupts

    while (1);                //wait here
}

ISR (INT0_vect)              //ISR for external interrupt 0
{
    PORTC ^= (1<<3);         //toggle PORTC.3
}
```

SUMMARY

An interrupt is an external or internal event that interrupts the microcontroller to inform it that a device needs its service. Every interrupt has a program associated with it called the ISR, or interrupt service routine. The AVR has many sources of interrupts, depending on the family member. Some of the most widely used interrupts are for the timers, external hardware interrupts, and serial communication. When an interrupt is activated, the IF (interrupt flag) bit is raised.

The AVR can be programmed to enable (unmask) or disable (mask) an interrupt, which is done with the help of the I (global interrupt enable) and IE (interrupt enable) bits. This chapter also showed how to program AVR interrupts in both Assembly and C languages.

PROBLEMS

SECTION 10.1: AVR INTERRUPTS

1. Which technique, interrupt or polling, avoids tying down the microcontroller?
2. List some of the interrupt sources in the AVR.
3. In the ATmega32 what memory area is assigned to the interrupt vector table?
4. True or false. The AVR programmer cannot change the memory address location assigned to the interrupt vector table.
5. What memory address is assigned to the Timer0 overflow interrupt in the interrupt vector table?
6. What memory address is assigned to the Timer1 overflow interrupt in the interrupt vector table?
7. Do we have a memory address assigned to the Time0 compare match interrupt in the interrupt vector table?
8. Do we have a memory address assigned to the external INT0 interrupt in the interrupt vector table?
9. To which register does the I bit belong?
10. Why do we put a JMP instruction at address 0?
11. What is the state of the I bit upon power-on reset, and what does it mean?
12. Show the instruction to enable the Timer0 compare match interrupt.
13. Show the instruction to enable the Timer1 overflow interrupt.
14. The TOIE0 bit belongs to register_____.
15. True or false. The TIMSK register is not a bit-addressable register.
16. With a single instruction, show how to disable all the interrupts.
17. Show how to disable the INT0 interrupt.
18. True or false. Upon reset, all interrupts are enabled by the AVR.
19. In the AVR, how many bytes of program memory are assigned to the reset?

SECTION 10.2: PROGRAMMING TIMER INTERRUPTS

20. True or false. For each of Timer0 and Timer1, there is a unique address in the interrupt vector table.

21. What address in the interrupt vector table is assigned to Timer2 overflow?
22. Show how to enable the Timer2 overflow interrupt.
23. Which bit of TIMSK belongs to the Timer0 overflow interrupt? Show how it is enabled.
24. Assume that Timer0 is programmed in Normal mode, TCNT0 = $E0, and the TOIE0 bit is enabled. Explain how the interrupt for the timer works.
25. True or false. The last three instructions of the ISR for Timer0 are:

```
LDI    R20,0x01
OUT    TIFR,R20    ;clear TOV0 flag
RETI
```

26. Assume that Timer1 is programmed for CTC mode, TCNT1H = $01, TCNT1L = $00, OCR1AH = $01, OCR1AL = $F5, and the OCIE1A bit is enabled. Explain how the interrupt is activated.
27. Assume that Timer1 is programmed for Normal mode, TCNT1H = $FF, TCNT1L = $E8, and the TOIE1 bit is enabled. Explain how the interrupt is activated.
28. Write a program using the Timer1 interrupt to create a square wave of 1 Hz on pin PB7 while sending data from PORTC to PORTD. Assume XTAL = 8 MHz.
29. Write a program using the Timer0 interrupt to create a square wave of 3 kHz on pin PB7 while sending data from PORTC to PORTD. Assume XTAL = 1 MHz.

SECTION 10.3: PROGRAMMING EXTERNAL HARDWARE INTERRUPTS

30. True or false. An address location is assigned to each of the external hardware interrupts INT0, INT1, and INT2.
31. What address in the interrupt vector table is assigned to INT0, INT1 and INT2? How about the pins?
32. To which register does the INT0 bit belong? Show how it is enabled.
33. To which register does the INT1 bit belong? Show how it is enabled.
34. Show how to enable all three external hardware interrupts.
35. Assume that the INT0 bit for external hardware interrupt is enabled and is negative edge-triggered. When is the interrupt activated? How does this interrupt work when it is activated.
36. True or false. Upon reset, all the external hardware interrupts are negative edge triggered.
37. The INTF0 bit belongs to the _____ register.
38. The INTF1 bit belongs to the _____ register.
39. Explain the role of INTF0 and INT0 in the execution of external interrupt 0.
40. Explain the role of I in the execution of external interrupts.
41. True or false. Upon power-on reset, all of INT0–INT2 are positive edge triggered.
42. Explain the difference between low-level and falling edge-triggered interrupts.
43. Show how to make the external INT0 negative edge triggered.
44. True or false. INT0–INT2 must be configured as an input pin for a hardware interrupt to come in.
45. Assume that the INT0 pin is connected to a switch. Write a program in which, whenever it goes low, the content of PORTC increases by one.
46. Assume that the INT0 and INT1 are connected to two switches named S1 and

S2. Write a program in which, whenever S1 goes low, the content of PORTC increases by one; and when S2 goes low, the content of PORTC decreases by one. When the value of PORTC is bigger than 100, PD7 is high; otherwise, it is low.

SECTION 10.4: INTERRUPT PRIORITY IN THE AVR

47. Explain what happens if both INT1F and INT2F are activated at the same time.
48. Assume that the Timer1 and Timer0 overflow interrupts are both enabled. Explain what happens if both TOV1 and TOV0 are activated at the same time.
49. Explain what happens if an interrupt is activated while the AVR is serving an interrupt.
50. True or false. In the AVR, an interrupt inside an interrupt is not allowed.

ANSWERS TO REVIEW QUESTIONS

SECTION 10.1: AVR INTERRUPTS

1. Interrupt
2. Timer0 overflow, Timer0 compare match, Timer1 overflow, Timer1 compare B match, Timer1 compare A match, Timer1 input capture, Timer2 overflow, Timer2 output compare match
3. Address locations 0x00 to 0x28. No. It is set when the processor is designed.
4. I = 0 means that all interrupts are masked, and as a result no interrupts will be responded to by the AVR.
5. Assuming I = 1, we need:
   ```
   LDI  R16,(1<<OCIE1A)
   OUT  TIMSK,R16
   ```
6. $12 for Timer1 overflow interrupt and 0x02 for INT0.

SECTION 10.2: PROGRAMMING TIMER INTERRUPTS

1. False. For each of the interrupts there is a separate address.
2. 0x16
3. TIMSK
   ```
   LDI  R16,(1<<TOIE0)
   OUT  TIMSK,R16
   ```
4. After Timer0 is started, the timer will count up from $F1 to $FF on its own while the AVR is executing other tasks. Upon rolling over from $FF to 00, the TOV0 flag is raised, which will interrupt the AVR in whatever it is doing and force it to jump to memory location $0016 to execute the ISR belonging to this interrupt.
5. False. There is no need to clear the TOV0 flag since the AVR clears the TOV0 flag internally upon jumping to the interrupt vector table.
6. The timer counts from 0 to 21. Then TCNT0 is loaded with 0 and the OCF0 flag is set. If Timer0 compare match interrupt is enabled, the ISR of the compare match interrupt is executed on each compare match.
7. $1/8$ MHz = 125 ns ➔ 125 ns × (21 + 1) = 2.75 µs

SECTION 10.3: PROGRAMMING EXTERNAL HARDWARE INTERRUPTS

1. False. Only INT2 is in edge-triggered mode.
2. Bits PD2 (PORTD.2), PD3 (PORTD.3), and PB2 (PORTB.2) are assigned to INT0, INT1, and INT2, respectively.

3.
```
LDI R20,(1<<INT1)
OUT GICR,R20
```
4. Upon application of a high-to-low pulse to pin PD2, the INTF0 flag will be set; as a result, the AVR is interrupted in whatever it is doing, clears the INTF0 flag, and jumps to ROM location 0x02 to execute the ISR.
5. True
6. When the CPU jumps to the interrupt vector to execute the ISR, it clears the flag that has caused the interrupt (the INTF0 flag in this case). The INTF0 flag will be set only if a new high-to-low pulse is applied to the pin.
7.
```
        .INCLUDE "M32DEF.INC"
        LDI    R16,0x2
        OUT    MCUCR,R16       ;make INT0 falling edge triggered
    L1: IN     R20,GIFR
        SBRS   R20,INTF0 ;skip next instruct. if the INTF0 bit of GIFR is set
        RJMP   L1              ;go to L1
        IN     R21,PORTC       ;R21 = PORTC
        LDI    R22,0x08
        EOR    R21,R22         ;R21 = R21 xor 0x08  (toggle bit 3)
        OUT    PORTC,R21       ;PORTC = R21
        LDI    R20,1<<INTF0
        OUT    GIFR,R20        ;clear INTF0 flag
        RJMP   L1
```

SECTION 10.4: INTERRUPT PRIORITY IN THE AVR

1. False. As shown in Table 10-1, the address of the Timer0 overflow interrupt is $16, while the address of Timer1 overflow is $12. Thus, the Timer1 overflow has a higher priority.
2. The interrupt whose vector is first in the interrupt vector is served first.
3. The flag of the interrupt will be set, but since I is 0, the new interrupt will not be served. The last instruction of the old interrupt is RETI, which causes the I flag to be set and the new interrupt to be served.
4. Context saving is the saving of the CPU contents before switching to a new task.

CHAPTER 11

AVR SERIAL PORT PROGRAMMING IN ASSEMBLY AND C

OBJECTIVES

Upon completion of this chapter, you will be able to:

>> Contrast and compare serial versus parallel data transfer
>> List the advantages of serial communication over parallel
>> Explain serial communication protocol
>> Contrast synchronous versus asynchronous communication
>> Contrast half- versus full-duplex transmission
>> Explain the process of data framing
>> Describe data transfer rate and bps rate
>> Define the RS232 standard
>> Explain the use of the MAX232 and MAX233 chips
>> Interface the AVR with an RS232 connector
>> Discuss the baud rate of the AVR
>> Describe serial communication features of the AVR
>> Describe the main registers used by serial communication of the AVR
>> Program the ATmega32 serial port in Assembly and C

Computers transfer data in two ways: parallel and serial. In parallel data transfers, often eight or more lines (wire conductors) are used to transfer data to a device that is only a few feet away. Devices that use parallel transfers include printers and IDE hard disks; each uses cables with many wires. Although a lot of data can be transferred in a short amount of time by using many wires in parallel, the distance cannot be great. To transfer to a device located many meters away, the serial method is used. In serial communication, the data is sent one bit at a time, in contrast to parallel communication, in which the data is sent a byte or more at a time. Serial communication of the AVR is the topic of this chapter. The AVR has serial communication capability built into it, thereby making possible fast data transfer using only a few wires.

In this chapter we first discuss the basics of serial communication. In Section 11.2, AVR interfacing to RS232 connectors via MAX232 line drivers is discussed. Serial port programming of the AVR is discussed in Section 11.3. Section 11.4 covers AVR C programming for the serial port using the Win AVR compiler. In Section 11.5 interrupt-based serial port programming is discussed.

SECTION 11.1: BASICS OF SERIAL COMMUNICATION

When a microprocessor communicates with the outside world, it provides the data in byte-sized chunks. For some devices, such as printers, the information is simply grabbed from the 8-bit data bus and presented to the 8-bit data bus of the device. This can work only if the cable is not too long, because long cables diminish and even distort signals. Furthermore, an 8-bit data path is expensive. For these reasons, serial communication is used for transferring data between two systems located at distances of hundreds of feet to millions of miles apart. Figure 11-1 diagrams serial versus parallel data transfers.

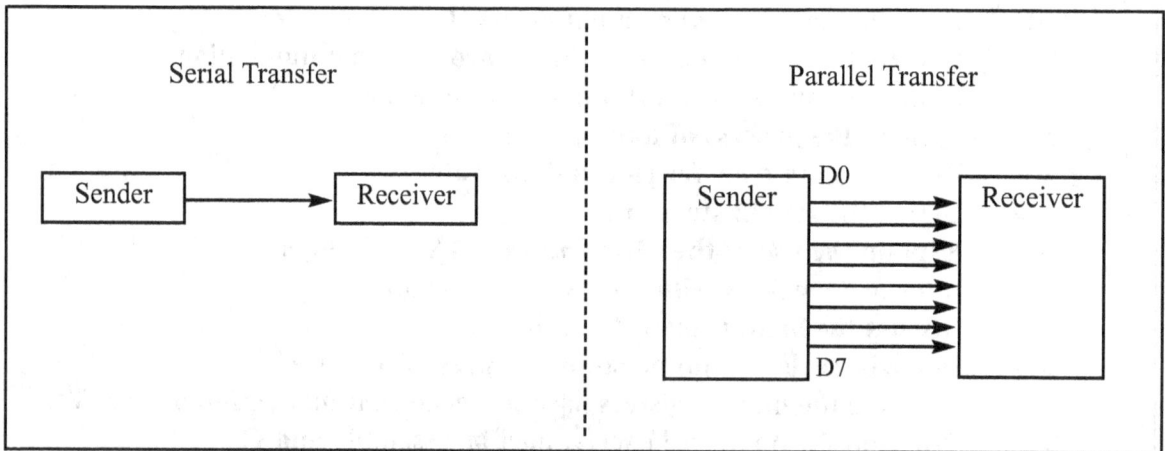

Figure 11-1. Serial versus Parallel Data Transfer

The fact that a single data line is used in serial communication instead of the 8-bit data line of parallel communication makes serial transfer not only much cheaper but also enables two computers located in two different cities to communicate over the telephone.

For serial data communication to work, the byte of data must be converted to serial bits using a parallel-in-serial-out shift register; then it can be transmitted over a single data line. This also means that at the receiving end there must be a serial-in-parallel-out shift register to receive the serial data and pack them into a byte. Of course, if data is to be transmitted on the telephone line, it must be converted from 0s and 1s to audio tones, which are sinusoidal signals. This conversion is performed by a peripheral device called a *modem*, which stands for "modulator/demodulator."

When the distance is short, the digital signal can be transmitted as it is on a simple wire and requires no modulation. This is how x86 PC keyboards transfer data to the motherboard. For long-distance data transfers using communication lines such as a telephone, however, serial data communication requires a modem to *modulate* (convert from 0s and 1s to audio tones) and *demodulate* (convert from audio tones to 0s and 1s).

Serial data communication uses two methods, asynchronous and synchronous. The *synchronous* method transfers a block of data (characters) at a time, whereas the *asynchronous* method transfers a single byte at a time. It is possible to write software to use either of these methods, but the programs can be tedious and long. For this reason, special IC chips are made by many manufacturers for serial data communications. These chips are commonly referred to as UART (universal asynchronous receiver-transmitter) and USART (universal synchronous-asynchronous receiver-transmitter). The AVR chip has a built-in USART, which is discussed in detail in Section 11.3.

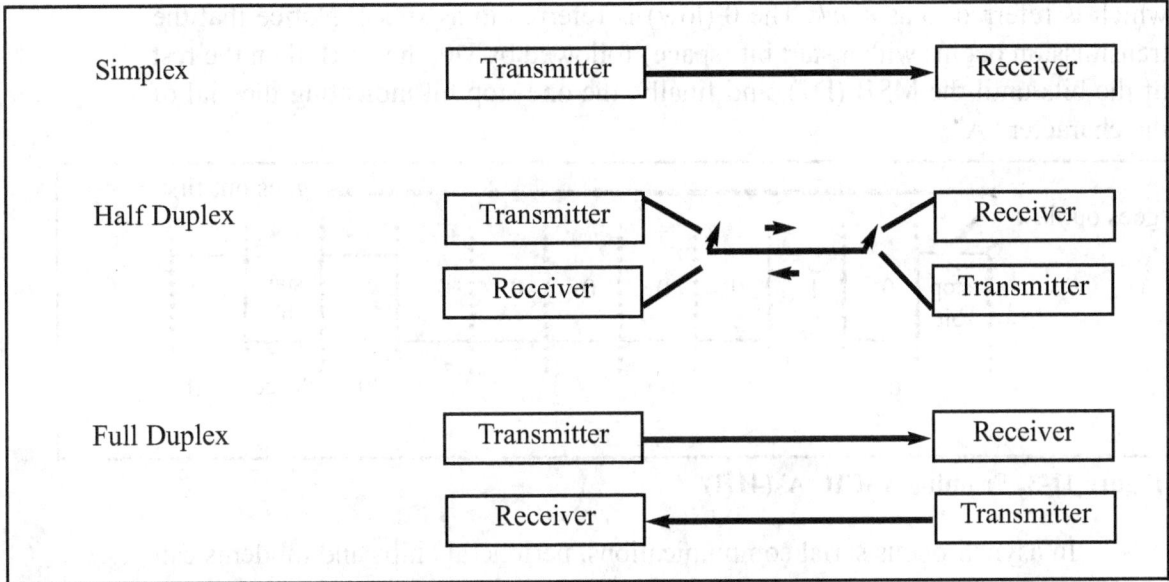

Figure 11-2. Simplex, Half-, and Full-Duplex Transfers

Half- and full-duplex transmission

In data transmission, if the data can be both transmitted and received, it is a *duplex* transmission. This is in contrast to *simplex* transmissions such as with printers, in which the computer only sends data. Duplex transmissions can be half or full duplex, depending on whether or not the data transfer can be simultaneous. If data is transmitted one way at a time, it is referred to as *half duplex*. If the data

---

can go both ways at the same time, it is *full duplex*. Of course, full duplex requires two wire conductors for the data lines (in addition to the signal ground), one for transmission and one for reception, in order to transfer and receive data simultaneously. See Figure 11-2.

Asynchronous serial communication and data framing

The data coming in at the receiving end of the data line in a serial data transfer is all 0s and 1s; it is difficult to make sense of the data unless the sender and receiver agree on a set of rules, a *protocol*, on how the data is packed, how many bits constitute a character, and when the data begins and ends.

Start and stop bits

Asynchronous serial data communication is widely used for character-oriented transmissions, while block-oriented data transfers use the synchronous method. In the asynchronous method, each character is placed between start and stop bits. This is called *framing*. In data framing for asynchronous communications, the data, such as ASCII characters, are packed between a start bit and a stop bit. The start bit is always one bit, but the stop bit can be one or two bits. The start bit is always a 0 (low), and the stop bit(s) is 1 (high). For example, look at Figure 11-3 in which the ASCII character "A" (8-bit binary 0100 0001) is framed between the start bit and a single stop bit. Notice that the LSB is sent out first.

Notice in Figure 11-3 that when there is no transfer, the signal is 1 (high), which is referred to as *mark*. The 0 (low) is referred to as *space*. Notice that the transmission begins with a start bit (space) followed by D0, the LSB, then the rest of the bits until the MSB (D7), and finally, the one stop bit indicating the end of the character "A".

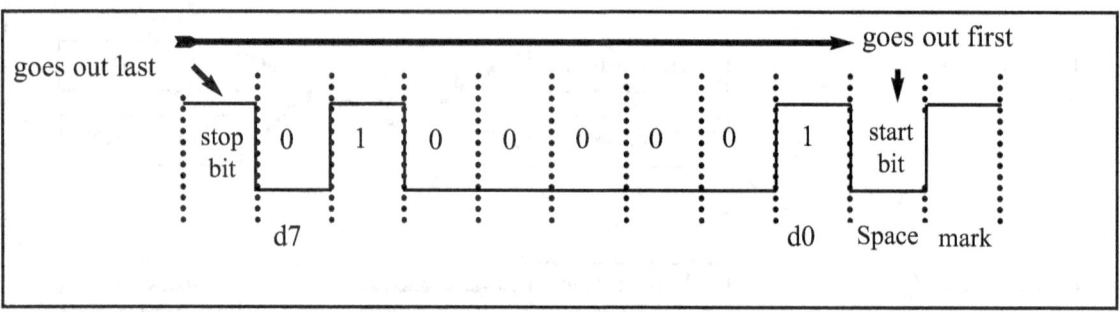

Figure 11-3. Framing ASCII 'A' (41H)

In asynchronous serial communications, peripheral chips and modems can be programmed for data that is 7 or 8 bits wide. This is in addition to the number of stop bits, 1 or 2. While in older systems ASCII characters were 7-bit, in recent years, 8-bit data has become common due to the extended ASCII characters. In some older systems, due to the slowness of the receiving mechanical device, two stop bits were used to give the device sufficient time to organize itself before transmission of the next byte. In modern PCs, however, the use of one stop bit is standard. Assuming that we are transferring a text file of ASCII characters using 1 stop bit, we have a total of 10 bits for each character: 8 bits for the ASCII code, and 1 bit each for the start and stop bits. Therefore, each 8-bit character has an extra 2 bits, which gives 25% overhead.

In some systems, the parity bit of the character byte is included in the data frame in order to maintain data integrity. This means that for each character (7- or 8-bit, depending on the system) we have a single parity bit in addition to start and stop bits. The parity bit is odd or even. In the case of an odd parity bit the number of 1s in the data bits, including the parity bit, is odd. Similarly, in an even parity bit system the total number of bits, including the parity bit, is even. For example, the ASCII character "A", binary 0100 0001, has 0 for the even parity bit. UART chips allow programming of the parity bit for odd-, even-, and no-parity options.

Data transfer rate

The rate of data transfer in serial data communication is stated in *bps* (bits per second). Another widely used terminology for bps is *baud rate*. However, the baud and bps rates are not necessarily equal. This is because baud rate is the modem terminology and is defined as the number of signal changes per second. In modems, sometimes a single change of signal transfers several bits of data. As far as the conductor wire is concerned, the baud rate and bps are the same, and for this reason in this book we use the terms bps and baud interchangeably.

The data transfer rate of a given computer system depends on communication ports incorporated into that system. For example, the early IBM PC/XT could transfer data at the rate of 100 to 9600 bps. In recent years, however, Pentium-based PCs transfer data at rates as high as 56K. Notice that in asynchronous serial data communication, the baud rate is generally limited to 100,000 bps.

RS232 standards

To allow compatibility among data communication equipment made by various manufacturers, an interfacing standard called RS232 was set by the Electronics Industries Association (EIA) in 1960. In 1963 it was modified and called RS232A. RS232B and RS232C were issued in 1965 and 1969, respectively. In this book we refer to it simply as RS232. Today, RS232 is one of the most widely used serial I/O interfacing standards. This standard is used in PCs and numerous types of equipment. Because the standard was set long before the advent of the TTL logic family, however, its input and output voltage levels are not TTL compatible. In RS232, a 1 is represented by –3 to –25 V, while a 0 bit is +3 to +25 volts, making –3 to +3 undefined. For this reason, to connect any RS232 to a microcontroller system we must use voltage converters such as MAX232 to convert the TTL logic levels to the RS232 voltage levels, and vice versa. MAX232 IC chips are commonly referred to as line drivers. Original RS232 connection to MAX232 is discussed in Section 11.2.

RS232 pins

Table 11-1 shows the pins for the original RS232 cable and their labels, commonly referred to as the DB-25 connector. In labeling, DB-25P refers to the plug connector (male), and DB-25S is for the socket connector (female). See Figure 11-4.

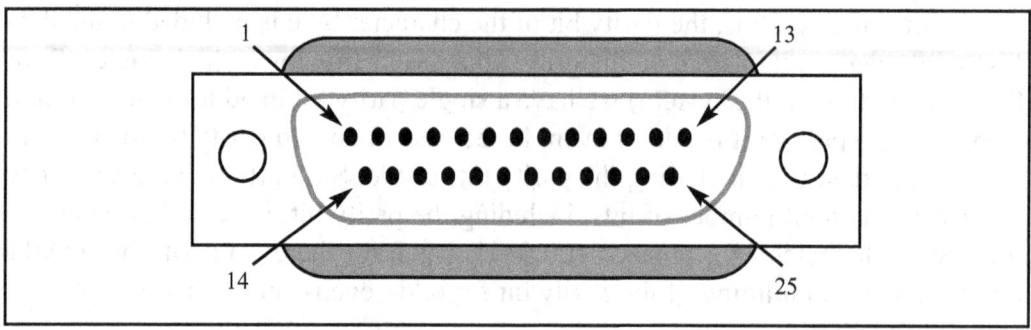

Figure 11-4. The Original RS232 Connector DB-25 (No longer in use)

Because not all the pins were used in PC cables, IBM introduced the DB-9 version of the serial I/O standard, which uses only 9 pins, as shown in Table 11-2. The DB-9 pins are shown in Figure 11-5.

Data communication classification

Current terminology classifies data communication equipment as DTE (data terminal equipment) or DCE (data communication equipment). DTE refers to terminals and computers that send and receive data, while DCE refers to communication equipment, such as modems, that are responsible for transferring the data. Notice that all the RS232 pin function definitions of Tables 11-1 and 11-2 are from the DTE point of view.

The simplest connection between a PC and a microcontroller requires a minimum of three pins, TX, RX, and ground, as shown in Figure 11-6. Notice in that figure that the RX and TX pins are interchanged.

Examining RS232 handshaking signals

To ensure fast and reliable data transmission between two devices, the data transfer must be coordinated. Just as in the case of the printer, because the receiving device may have no room for the data in serial data communication, there must be a way to inform the sender to stop sending data. Many of the pins of the RS-232 connector are used for handshaking signals. Their description is provided below only as a reference, and they can be bypassed

Table 11-1: RS232 Pins (DB-25)

Pin	Description
1	Protective ground
2	Transmitted data (TxD)
3	Received data (RxD)
4	Request to send ($\overline{RTS}$)
5	Clear to send ($\overline{CTS}$)
6	Data set ready ($\overline{DSR}$)
7	Signal ground (GND)
8	Data carrier detect ($\overline{DCD}$)
9/10	Reserved for data testing
11	Unassigned
12	Secondary data carrier detect
13	Secondary clear to send
14	Secondary transmitted data
15	Transmit signal element timing
16	Secondary received data
17	Receive signal element timing
18	Unassigned
19	Secondary request to send
20	Data terminal ready ($\overline{DTR}$)
21	Signal quality detector
22	Ring indicator
23	Data signal rate select
24	Transmit signal element timing
25	Unassigned

because they are not supported by the AVR UART chip.

1. DTR (data terminal ready). When the terminal (or a PC COM port) is turned on, after going through a self-test, it sends out signal DTR to indicate that it is ready for communication. If there is something wrong with the COM port, this signal will not be activated. This is an active-LOW signal and can be used to inform the modem that the computer is alive and kicking. This is an output pin from DTE (PC COM port) and an input to the modem.

2. DSR (data set ready). When the DCE (modem) is turned on and has gone through the self-test, it asserts DSR to indicate that it is ready to communicate. Thus, it is an output from the modem (DCE) and an input to the PC (DTE). This is an active-LOW signal. If for any reason the modem cannot make a connection to the telephone, this signal remains inactive, indicating to the PC (or terminal) that it cannot accept or send data.

3. RTS (request to send). When the DTE device (such as a PC) has a byte to transmit, it asserts RTS to signal the modem that it has a byte of data to transmit. RTS is an active-LOW output from the DTE and an input to the modem.

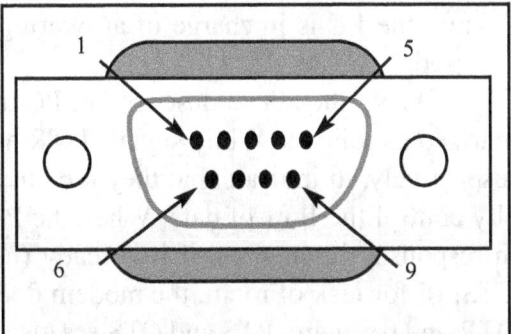

Figure 11-5. 9-Pin Connector for DB-9

Table 11-2: IBM PC DB-9 Signals

Pin	Description
1	Data carrier detect (DCD)
2	Received data (RxD)
3	Transmitted data (TxD)
4	Data terminal ready (DTR)
5	Signal ground (GND)
6	Data set ready (DSR)
7	Request to send (RTS)
8	Clear to send (CTS)
9	Ring indicator (RI)

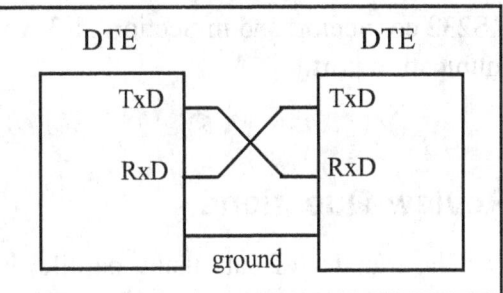

Figure 11-6. Null Modem Connection

4. CTS (clear to send). In response to RTS, when the modem has room to store the data it is to receive, it sends out signal CTS to the DTE (PC) to indicate that it can receive the data now. This input signal to the DTE is used by the DTE to start transmission.

5. DCD (data carrier detect). The modem asserts signal DCD to inform the DTE (PC) that a valid carrier has been detected and that contact between it and the other modem is established. Therefore, DCD is an output from the modem and an input to the PC (DTE).

6. RI (ring indicator). An output from the modem (DCE) and an input to a PC (DTE) indicates that the telephone is ringing. RI goes on and off in synchronization with the ringing sound. Of the six handshake signals, this is the least often used because modems take care of answering the phone. If in a given sys-

tem the PC is in charge of answering the phone, however, this signal can be used.

From the above description, PC and modem communication can be summarized as follows: While signals DTR and DSR are used by the PC and modem, respectively, to indicate that they are alive and well, it is RTS and CTS that actually control the flow of data. When the PC wants to send data it asserts RTS, and in response, the modem, if it is ready (has room) to accept the data, sends back CTS. If, for lack of room, the modem does not activate CTS, the PC will deassert DTR and try again. RTS and CTS are also referred to as hardware control flow signals.

This concludes the description of the most important pins of the RS232 handshake signals plus TX, RX, and ground. Ground is also referred to as SG (signal ground).

x86 PC COM ports

The x86 PCs (based on 8086, 286, 386, 486, and all Pentium microprocessors) used to have two COM ports. Both COM ports were RS232-type connectors. The COM ports were designated as COM 1 and COM 2. In recent years, one of these has been replaced with the USB port, and COM 1 is the only serial port available, if any. We can connect the AVR serial port to the COM 1 port of a PC for serial communication experiments. In the absence of a COM port, we can use a COM-to-USB converter module.

With this background in serial communication, we are ready to look at the AVR. In the next section we discuss the physical connection of the AVR and RS232 connector, and in Section 11.3 we see how to program the AVR serial communication port.

Review Questions

1. The transfer of data using parallel lines is _____ (faster, slower) but _____ (more expensive, less expensive).
2. True or false. Sending data from a radio station is duplex.
3. True or false. In full duplex we must have two data lines, one for transfer and one for receive.
4. The start and stop bits are used in the _____ (synchronous, asynchronous) method.
5. Assuming that we are transmitting the ASCII letter "E" (0100 0101 in binary) with no parity bit and one stop bit, show the sequence of bits transferred serially.
6. In Question 5, find the overhead due to framing.
7. Calculate the time it takes to transfer 10,000 characters as in Question 5 if we use 9600 bps. What percentage of time is wasted due to overhead?
8. True or false. RS232 is not TTL compatible.
9. What voltage levels are used for binary 0 in RS232?
10. True or false. The AVR has a built-in UART.

SECTION 11.2: ATMEGA32 CONNECTION TO RS232

In this section, the details of the physical connections of the ATmega32 to RS232 connectors are given. As stated in Section 11.1, the RS232 standard is not TTL compatible; therefore, a line driver such as the MAX232 chip is required to convert RS232 voltage levels to TTL levels, and vice versa. The interfacing of ATmega32 with RS232 connectors via the MAX232 chip is the main topic of this section.

RX and TX pins in the ATmega32

The ATmega32 has two pins that are used specifically for transferring and receiving data serially. These two pins are called TX and RX and are part of the Port D group (PD0 and PD1) of the 40-pin package. Pin 15 of the ATmega32 is assigned to TX and pin 14 is designated as RX. These pins are TTL compatible; therefore, they require a line driver to make them RS232 compatible. One such line driver is the MAX232 chip. This is discussed next.

MAX232

Because the RS232 is not compatible with today's microprocessors and microcontrollers, we need a line driver (voltage converter) to convert the RS232's signals to TTL voltage levels that will be acceptable to the AVR's TX and RX pins. One example of such a converter is MAX232 from Maxim Corp. (www.maxim-ic.com). The MAX232 converts from RS232 voltage levels to TTL voltage levels, and vice versa. One advantage of the MAX232 chip is that it uses a +5 V power source, which is the same as the source voltage for the AVR. In other words, with a single +5 V power supply we can power both the AVR and MAX232, with no need for the dual power supplies that are common in many older systems.

The MAX232 has two sets of line drivers for transferring and receiving data, as shown in Figure 11-7. The line drivers used for TX are called T1 and T2,

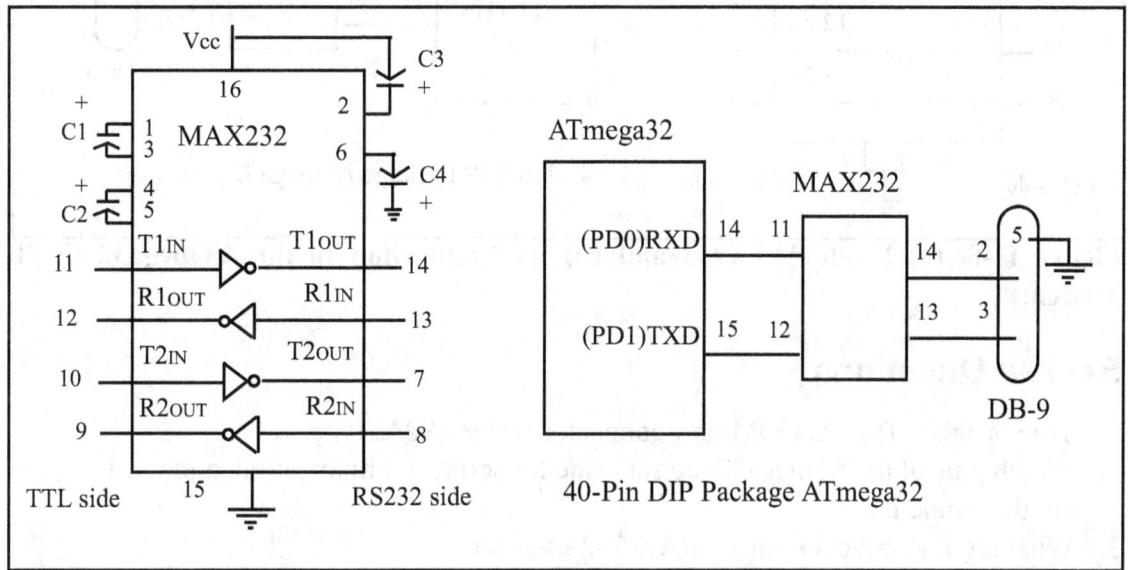

Figure 11-7. (a) Inside MAX232 and (b) Its Connection to the ATmega32 (Null Modem)

while the line drivers for RX are designated as R1 and R2. In many applications only one of each is used. For example, T1 and R1 are used together for TX and RX of the AVR, and the second set is left unused. Notice in MAX232 that the T1 line driver has a designation of T1in and T1out on pin numbers 11 and 14, respectively. The T1in pin is the TTL side and is connected to TX of the microcontroller, while T1out is the RS232 side that is connected to the RX pin of the RS232 DB connector. The R1 line driver has a designation of R1in and R1out on pin numbers 13 and 12, respectively. The R1in (pin 13) is the RS232 side that is connected to the TX pin of the RS232 DB connector, and R1out (pin 12) is the TTL side that is connected to the RX pin of the microcontroller. See Figure 11-7. Notice the null modem connection where RX for one is TX for the other.

MAX232 requires four capacitors ranging from 0.1 to 22 μF. The most widely used value for these capacitors is 22 μF.

MAX233

To save board space, some designers use the MAX233 chip from Maxim. The MAX233 performs the same job as the MAX232 but eliminates the need for capacitors. However, the MAX233 chip is much more expensive than the MAX232. Notice that MAX233 and MAX232 are not pin compatible. You cannot take a MAX232 out of a board and replace it with a MAX233. See Figure 11-8 for MAX233 with no capacitor used.

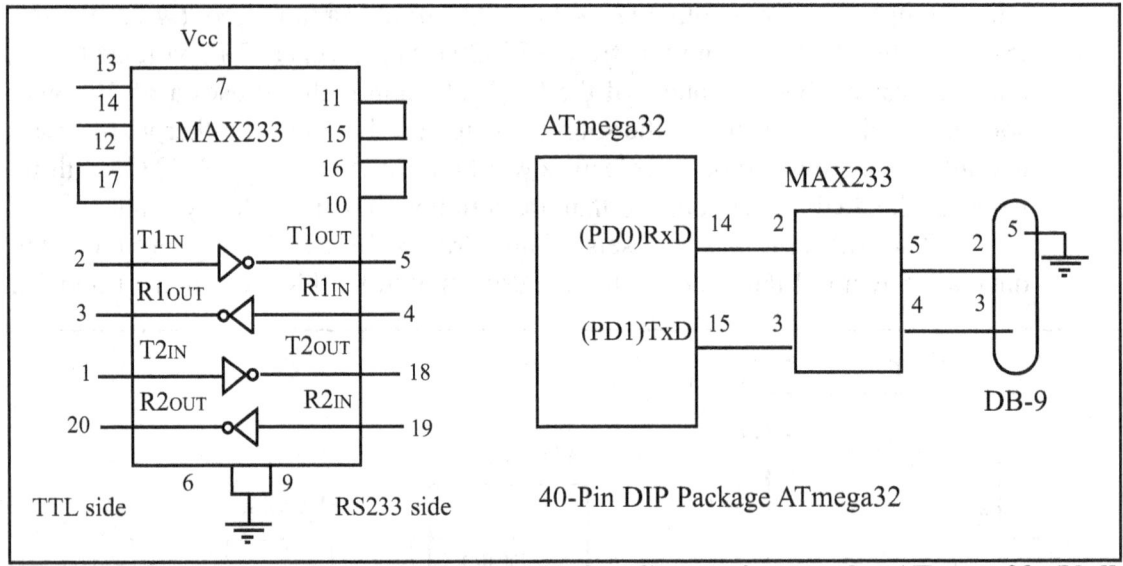

Figure 11-8. (a) Inside MAX233 and (b) Its Connection to the ATmega32 (Null Modem)

Review Questions

1. True or false. The PC COM port connector is the RS232 type.
2. Which pins of the ATmega32 are set aside for serial communication, and what are their functions?
3. What are line drivers such as MAX 232 used for?
4. MAX232 can support ____ lines for TX and ____ lines for RX.
5. What is the advantage of the MAX233 over the MAX232 chip?

SECTION 11.3: AVR SERIAL PORT PROGRAMMING IN ASSEMBLY

In this section we discuss the serial communication registers of the ATmega32 and show how to program them to transfer and receive data using asynchronous mode. The USART (universal synchronous asynchronous receiver/transmitter) in the AVR has normal asynchronous, double-speed asynchronous, master synchronous, and slave synchronous mode features. The synchronous mode can be used to transfer data between the AVR and external peripherals such as ADC and EEPROMs. The asynchronous mode is the one we will use to connect the AVR-based system to the x86 PC serial port for the purpose of full-duplex serial data transfer. In this section we examine the asynchronous mode only.

In the AVR microcontroller five registers are associated with the USART that we deal with in this chapter. They are UDR (USART Data Register), UCSRA, UCSRB, UCSRC (USART Control Status Register), and UBRR (USART Baud Rate Register). We examine each of them and show how they are used in full-duplex serial data communication.

UBRR register and baud rate in the AVR

Because the x86 PCs are so widely used to communicate with AVR-based systems, we will emphasize serial communications of the AVR with the COM port of the x86 PC. Some of the baud rates supported by PC HyperTerminal are listed in Table 11-3. You can examine these baud rates by going to the Microsoft Windows HyperTerminal program and clicking on the Communication Settings option. The AVR transfers and receives data serially at many different baud rates. The baud rate in the AVR is programmable. This is done with the help of the 8-bit register called UBRR. For a given crystal frequency, the value loaded into the UBRR decides the baud rate. The relation between the value loaded into UBBR and the Fosc (frequency of oscillator connected to the XTAL1 and XTAL2 pins) is dictated by the following formula:

Table 11-3: Some PC Baud Rates in HyperTerminal

1,200
2,400
4,800
9,600
19,200
38,400
57,600
115,200

$$\text{Desired Baud Rate} = \text{Fosc}/ (16(X + 1))$$

where X is the value we load into the UBRR register. To get the X value for different baud rates we can solve the equation as follows:

$$X = (\text{Fosc}/ (16(\text{Desired Baud Rate}))) - 1$$

Assuming that Fosc = 8 MHz, we have the following:

$$\text{Desired Baud Rate} = \text{Fosc}/ (16(X + 1)) = 8 \text{ MHz}/16(X + 1) = 500 \text{ kHz}/(X + 1)$$

$$X = (500 \text{ kHz}/ \text{Desired Baud Rate}) - 1$$

Table 11-4: UBRR Values for Various Baud Rates (Fosc = 8 MHz, U2X = 0)

Baud Rate	UBRR (Decimal Value)	UBRR (Hex Value)
38400	12	C
19200	25	19
9600	51	33
4800	103	67
2400	207	CF
1200	415	19F

Note: For Fosc = 8 MHz we have UBRR = (500000/BaudRate) – 1

Table 11-4 shows the X values for the different baud rates if Fosc = 8 MHz. Another way to understand the UBRR values listed in Table 11-4 is to look at Figure 11-9. The UBRR is connected to a down-counter, which functions as a programmable prescaler to generate baud rate. The system clock (Fosc) is the clock input to the down-counter. The down-counter is loaded with the UBRR value each time it counts down to zero. When the counter reaches zero, a clock is generated. This makes a frequency divider that divides the OSC frequency by UBRR + 1. Then the frequency is divided by 2, 4, and 2. See Example 11-1. As you can see

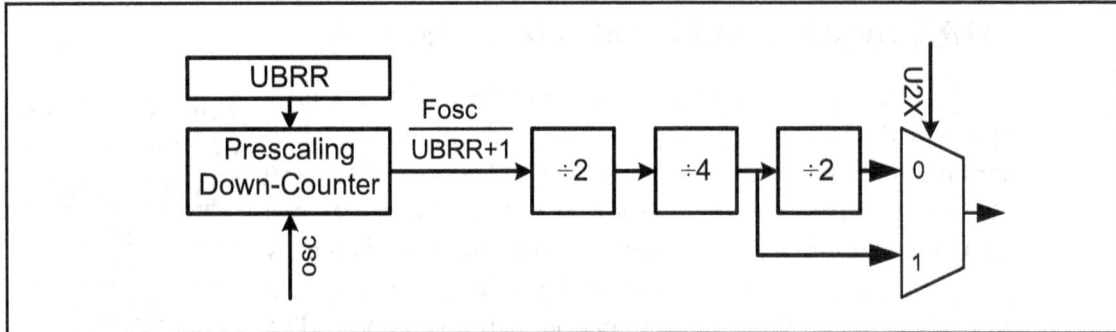

Figure 11-9. Baud Rate Generation Block Diagram

Example 11-1

With Fosc = 8 MHz, find the UBRR value needed to have the following baud rates:
(a) 9600   (b) 4800   (c) 2400   (d) 1200

Solution:

Fosc = 8 MHz => X = (8 MHz/16(Desired Baud Rate)) – 1
                               => X = (500 kHz/(Desired Baud Rate)) – 1

(a) (500 kHz/ 9600) – 1 = 52.08 – 1 = 51.08 = 51 = 33 (hex) is loaded into UBRR
(b) (500 kHz/ 4800) – 1 = 104.16 – 1 = 103.16 = 103 = 67 (hex) is loaded into UBRR
(c) (500 kHz/ 2400) – 1 = 208.33 – 1 = 207.33 = 207 = CF (hex) is loaded into UBRR
(d) (500 kHz/ 1200) – 1 = 416.66 – 1 = 415.66 = 415 = 19F (hex) is loaded into UBRR

Notice that dividing the output of the prescaling down-counter by 16 is the default setting upon Reset. We can get a higher baud rate with the same crystal by changing this default setting. This is explained at the end of this section.

406

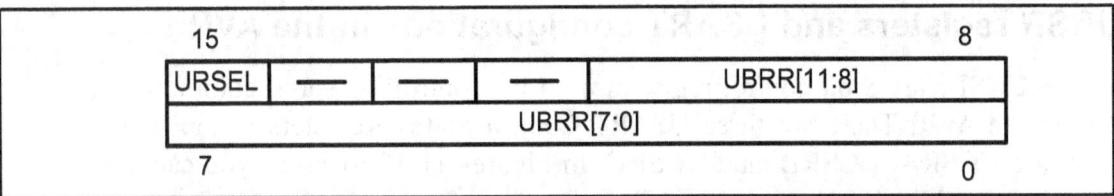

Figure 11-10. UBRR Register

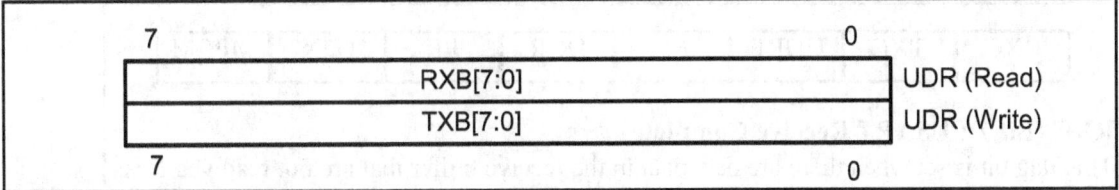

Figure 11-11. UDR Register

in Figure 11-9, we can choose to bypass the last divider and double the baud rate. In the next section we learn more about it.

As you see in Figure 11-10, UBRR is a 16-bit register but only 12 bits of it are used to set the USART baud rate. Bit 15 is URSEL and, as we will see in the next section, selects between accessing the UBRRH or the UCSRC register. The other bits are reserved.

UDR registers and USART data I/O in the AVR

In the AVR, to provide a full-duplex serial communication, there are two shift registers referred to as *Transmit Shift Register* and *Receive Shift Register*. Each shift register has a buffer that is connected to it directly. These buffers are called *Transmit Data Buffer Register* and *Receive Data Buffer Register*. The USART Transmit Data Buffer Register and USART Receive Data Buffer Register share the same I/O address, which is called *USART Data Register* or *UDR*. When you write data to UDR, it will be transferred to the Transmit Data Buffer Register (TXB), and when you read data from UDR, it will return the contents of the Receive Data Buffer Register (RXB). See Figure 11-12.

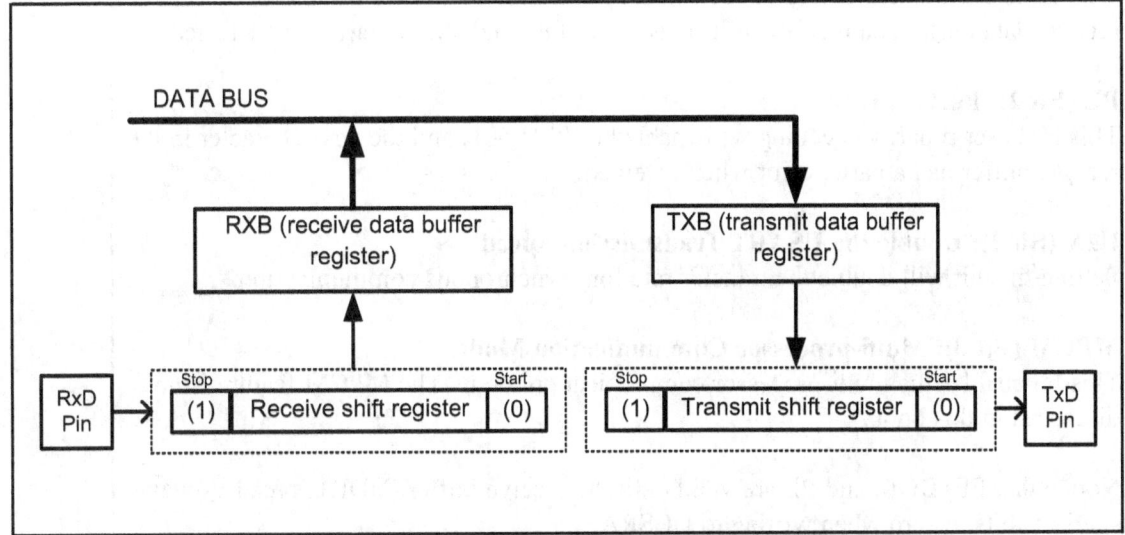

Figure 11-12. Simplified USART Transmit Block Diagram

UCSR registers and USART configurations in the AVR

UCSRs are 8-bit control registers used for controlling serial communication in the AVR. There are three USART Control Status Registers in the AVR. They are UCSRA, UCSRB, and UCSRC. In Figures 11-13 to 11-15 you can see the role of each bit in these registers. Examine these figures carefully before continuing to read this chapter.

RXC	TXC	UDRE	FE	DOR	PE	U2X	MPCM

RXC (Bit 7): USART Receive Complete
This flag bit is set when there are new data in the receive buffer that are not read yet. It is cleared when the receive buffer is empty. It also can be used to generate a receive complete interrupt.

TXC (Bit 6): USART Transmit Complete
This flag bit is set when the entire frame in the transmit shift register has been transmitted and there are no new data available in the transmit data buffer register (TXB). It can be cleared by writing a one to its bit location. Also it is automatically cleared when a transmit complete interrupt is executed. It can be used to generate a transmit complete interrupt.

UDRE (Bit 5): USART Data Register Empty
This flag is set when the transmit data buffer is empty and it is ready to receive new data. If this bit is cleared you should not write to UDR because it overrides your last data. The UDRE flag can generate a data register empty interrupt.

FE (Bit 4): Frame Error
This bit is set if a frame error has occurred in receiving the next character in the receive buffer. A frame error is detected when the first stop bit of the next character in the receive buffer is zero.

DOR (Bit 3): Data OverRun
This bit is set if a data overrun is detected. A data overrun occurs when the receive data buffer and receive shift register are full, and a new start bit is detected.

PE (Bit 2): Parity Error
This bit is set if parity checking was enabled (UPM1 = 1) and the next character in the receive buffer had a parity error when received.

U2X (Bit 1): Double the USART Transmission Speed
Setting this bit will double the transfer rate for asynchronous communication.

MPCM (Bit 0): Multi-processor Communication Mode
This bit enables the multi-processor communication mode. The MPCM feature is not discussed in this book.

Notice that FE, DOR, and PE are valid until the receive buffer (UDR) is read. Always set these bits to zero when writing to UCSRA.

Figure 11-13. UCSRA: USART Control and Status Register A

RXCIE	TXCIE	UDRIE	RXEN	TXEN	UCSZ2	RXB8	TXB8

RXCIE (Bit 7): Receive Complete Interrupt Enable
To enable the interrupt on the RXC flag in UCSRA you should set this bit to one.

TXCIE (Bit 6): Transmit Complete Interrupt Enable
To enable the interrupt on the TXC flag in UCSRA you should set this bit to one.

UDRIE (Bit 5): USART Data Register Empty Interrupt Enable
To enable the interrupt on the UDRE flag in UCSRA you should set this bit to one.

RXEN (Bit 4): Receive Enable
To enable the USART receiver you should set this bit to one.

TXEN (Bit 3): Transmit Enable
To enable the USART transmitter you should set this bit to one.

UCSZ2 (Bit 2): Character Size
This bit combined with the UCSZ1:0 bits in UCSRC sets the number of data bits (character size) in a frame.

RXB8 (Bit 1): Receive data bit 8
This is the ninth data bit of the received character when using serial frames with nine data bits. This bit is not used in this book.

TXB8 (Bit 0): Transmit data bit 8
This is the ninth data bit of the transmitted character when using serial frames with nine data bits. This bit is not used in this book.

Figure 11-14. UCSRB: USART Control and Status Register B

Three of the UCSRB register bits are related to interrupt. They are RXCIE, TXCIE, and UDRIE. See Figure 11-14. In Section 11-5 we will see how these flags are used with interrupts. In this section we monitor (poll) the UDRE flag bit to make sure that the transmit data buffer is empty and it is ready to receive new data. By the same logic, we monitor the RXC flag to see if a byte of data has come in yet.

Before you start serial communication you have to enable the USART receiver or USART transmitter by writing one to the RXEN or TXEN bit of UCSRB. As we mentioned before, in the AVR you can use either synchronous or asynchronous operating mode. The UMSEL bit of the UCSRC register selects the USART operating mode. Since we want to use synchronous USART operating mode, we have to set the UMSEL bit to one. Also you have to set an identical character size for both transmitter and the receiver. If the character size of the receiver does not match the character size of the transmitter, data transfer would fail. Parity mode and number of stop bits are other factors that the receiver and transmitter must agree on before starting USART communication.

URSEL	UMSEL	UPM1	UPM0	USBS	UCSZ1	UCSZ0	UCPOL

URSEL (Bit 7): Register Select
This bit selects to access either the UCSRC or the UBRRH register and will be discussed more in this section.

UMSEL (Bit 6): USART Mode Select
This bit selects to operate in either the asynchronous or synchronous mode of operation.
> 0 = Asynchronous operation
> 1 = Synchronous operation

UPM1:0 (Bit 5:4): Parity Mode
These bits disable or enable and set the type of parity generation and check.
> 00 = Disabled
> 01 = Reserved
> 10 = Even Parity
> 11 = Odd Parity

USBS (Bit 3): Stop Bit Select
This bit selects the number of stop bits to be transmitted.
> 0 = 1 bit
> 1 = 2 bits

UCSZ1:0 (Bit 2:1): Character Size
These bits combined with the UCSZ2 bit in UCSRB set the character size in a frame and will be discussed more in this section.

UCPOL (Bit 2): Clock Polarity
This bit is used for synchronous mode only and will not be covered in this section.

Figure 11-15. UCSRC: USART Control and Status Register C

To set the number of data bits (character size) in a frame you must set the values of the UCSZ1 and USCZ0 bits in the UCSRB and UCSZ2 bits in UCSRC. Table 11-5 shows the values of UCSZ2, UCSZ1, and UCSZ0 for different character sizes. In this book we use the 8-bit character size because it is the most common in x86 serial communications. If you want to use 9-bit data, you have to use the RXB8 and TXB8 bits in UCSRB as the 9th bit of UDR (USART Data

Table 11-5: Values of UCSZ2:0 for Different Character Sizes

UCSZ2	UCSZ1	UCSZ0	Character Size
0	0	0	5
0	0	1	6
0	1	0	7
0	1	1	8
1	1	1	9

Note: Other values are reserved. Also notice that UCSZ0 and UCSZ1 belong to UCSRC and UCSZ2 belongs to UCSRB

Register).

Because of some technical considerations, the UCSRC register shares the same I/O location as the UBRRH, and therefore some care must be taken when accessing these I/O locations. When you write to UCSRC or UBRRH, the high bit of the written value (URSEL) controls which of the two registers will be the target of the write operation. If URSEL is zero during a write operation, the UBRRH value will be updated; otherwise, UCSRC will be updated. See Examples 11-2 and 11-3 to learn how we access the bits of UCSRC and UBRR.

Example 11-2

(a) What are the values of UCSRB and UCSRC needed to configure USART for asynchronous operating mode, 8 data bits (character size), no parity, and 1 stop bit? Enable both receive and transmit.
(b) Write a program for the AVR to set the values of UCSRB and UCSRC for this configuration.

Solution:

(a) RXEN and TXEN have to be 1 to enable receive and transmit. UCSZ2:0 should be 011 for 8-bit data, UMSEL should be 0 for asynchronous operating mode, UPM1:0 have to be 00 for no parity, and USBS should be 0 for one stop bit.
(b)
```
      .INCLUDE "M32DEF.INC"

      LDI   R16,(1<<RXEN)|(1<<TXEN)
      OUT   UCSRB, R16
;In the next line URSEL = 1 to access UCSRC. Note that instead
;of using shift operator, you can write "LDI R16, 0b10000110"
      LDI   R16,(1<<UCSZ1)|(1<<UCSZ0)|(1<<URSEL)
      OUT   UCSRC, R16
```

Example 11-3

In Example 11-2, set the baud rate to 1200 and write a program for the AVR to set up the values of UCSRB, UCSRC, and UBRR. (Focs = 8 MHz)

Solution:

```
      .INCLUDE "M32DEF.INC"

      LDI   R16,(1<<RXEN)|(1<<TXEN)
      OUT   UCSRB, R16
;In the next line URSEL = 1 to access UCSRC. Note that instead
;of using shift operator, you can write "LDI R16, 0b10000110"
      LDI   R16,(1<<UCSZ1)|(1<<UCSZ0)|(1<<URSEL)
      OUT   UCSRC, R16                    ;move R16 to UCSRC
      LDI   R16,0x9F                      ;see Table 11-4
      OUT   UBRRL,R16                     ;1200 baud rate
      LDI   R16,0x1                       ;URSEL= 0 to
      OUT   UBRRH,R16                     ;access UBRRH
```

FE and PE flag bits

When the AVR USART receives a byte, we can check the parity bit and stop bit. If the parity bit is not correct, the AVR will set PE to one, indicating that an parity error has occurred. We can also check the stop bit. As we mentioned before, the stop bit must be one, otherwise the AVR would generate a stop bit error and set the FE flag bit to one, indicating that a stop bit error has occurred. We can check these flags to see if the received data is valid and correct. Notice that FE and PE are valid until the receive buffer (UDR) is read. So we have to read FE and PE bits before reading UDR. You can explore this on your own.

Programming the AVR to transfer data serially

In programming the AVR to transfer character bytes serially, the following steps must be taken:

1. The UCSRB register is loaded with the value 08H, enabling the USART transmitter. The transmitter will override normal port operation for the TxD pin when enabled.
2. The UCSRC register is loaded with the value 06H, indicating asynchronous mode with 8-bit data frame, no parity, and one stop bit.
3. The UBRR is loaded with one of the values in Table 11-4 (if Fosc = 8 MHz) to set the baud rate for serial data transfer.
4. The character byte to be transmitted serially is written into the UDR register.
5. Monitor the UDRE bit of the UCSRA register to make sure UDR is ready for the next byte.
6. To transmit the next character, go to Step 4.

Importance of monitoring the UDRE flag

By monitoring the UDRE flag, we make sure that we are not overloading

---

Example 11-4

Write a program for the AVR to transfer the letter 'G' serially at 9600 baud, continuously. Assume XTAL = 8 MHz.

Solution:

```
.INCLUDE "M32DEF.INC"
    LDI   R16,(1<<TXEN)                    ;enable transmitter
    OUT   UCSRB, R16
    LDI   R16,(1<<UCSZ1)|(1<<UCSZ0)|(1<<URSEL);8-bit data
    OUT   UCSRC, R16                        ;no parity, 1 stop bit
    LDI   R16,0x33                          ;9600 baud rate
    OUT   UBRRL,R16                         ;for XTAL = 8 MHz
AGAIN:
    SBIS  UCSRA,UDRE                        ;is UDR empty
    RJMP  AGAIN                             ;wait more
    LDI   R16,'G'                           ;send 'G'
    OUT   UDR,R16                           ;to UDR
    RJMP  AGAIN                             ;do it again
```

412

the UDR register. If we write another byte into the UDR register before it is empty, the old byte could be lost before it is transmitted.

We conclude that by checking the UDRE flag bit, we know whether or not the AVR is ready to accept another byte to transmit. The UDRE flag bit can be checked by the instruction "SBIS UCSRA, UDRE" or we can use an interrupt, as we will see in Section 11.5. In Section 11.5 we will show how to use interrupts to transfer data serially, and avoid tying down the microcontroller with instructions such as "SBIS UCSRA, UDRE". Example 11-4 shows the program to transfer 'G' serially at 9600 baud.

Example 11-5 shows how to transfer "YES " continuously.

Example 11-5

Write a program to transmit the message "YES " serially at 9600 baud, 8-bit data, and 1 stop bit. Do this forever.

Solution:

```
        .INCLUDE "M32DEF.INC"

        LDI     R21,HIGH(RAMEND)              ;initialize high
        OUT     SPH,R21                       ;byte of SP
        LDI     R21,LOW(RAMEND)               ;initialize low
        OUT     SPL,R21                       ;byte of SP

        LDI     R16,(1<<TXEN)                 ;enable transmitter
        OUT     UCSRB, R16
        LDI     R16,(1<<UCSZ1)|(1<<UCSZ0)|(1<<URSEL); 8-bit data
        OUT     UCSRC, R16                    ;no parity, 1 stop bit
        LDI     R16,0x33                      ;9600 baud rate
        OUT     UBRRL,R16
AGAIN:
        LDI     R17,'Y'                       ;move 'Y' to R17
        CALL    TRNSMT                        ;transmit r17 to TxD
        LDI     R17,'E'                       ;move 'E' to R17
        CALL    TRNSMT                        ;transmit r17 to TxD
        LDI     R17,'S'                       ;move 'S' to R17
        CALL    TRNSMT                        ;transmit r17 to TxD
        LDI     R17,' '                       ;move ' ' to R17
        CALL    TRNSMT                        ;transmit space to TxD
        RJMP    AGAIN                         ;do it again
TRNSMT:
        SBIS    UCSRA,UDRE                    ;is UDR empty?
        RJMP    TRNSMT                        ;wait more
        OUT     UDR,R17                       ;send R17 to UDR
        RET
```

Programming the AVR to receive data serially

In programming the AVR to receive character bytes serially, the following steps must be taken:

1. The UCSRB register is loaded with the value 10H, enabling the USART receiver. The receiver will override normal port operation for the RxD pin when enabled.
2. The UCSRC register is loaded with the value 06H, indicating asynchronous mode with 8-bit data frame, no parity, and one stop bit.
3. The UBRR is loaded with one of the values in Table 11-4 (if Fosc = 8 MHz) to set the baud rate for serial data transfer.
4. The RXC flag bit of the UCSRA register is monitored for a HIGH to see if an entire character has been received yet.
5. When RXC is raised, the UDR register has the byte. Its contents are moved into a safe place.
6. To receive the next character, go to Step 5.
   Example 11-6 shows the coding of the above steps.

Example 11-6

Program the ATmega32 to receive bytes of data serially and put them on Port B. Set the baud rate at 9600, 8-bit data, and 1 stop bit.

Solution:

```
.INCLUDE "M32DEF.INC"
        LDI   R16,(1<<RXEN)                    ;enable receiver
        OUT   UCSRB, R16
        LDI   R16,(1<<UCSZ1)|(1<<UCSZ0)|(1<<URSEL);8-bit data
        OUT   UCSRC, R16                        ;no parity, 1 stop bit
        LDI   R16,0x33                          ;9600 baud rate
        OUT   UBRRL,R16
        LDI   R16,0xFF                          ;Port B is output
        OUT   DDRB,R16
RCVE:
        SBIS  UCSRA,RXC                         ;is any byte in UDR?
        RJMP  RCVE                              ;wait more
        IN    R17,UDR                           ;send UDR to R17
        OUT   PORTB,R17                         ;send R17 to PORTB
        RJMP  RCVE                              ;do it again
```

Transmit and receive

In previous examples we showed how to transmit or receive data serially. Next we show how do both send and receive at the same time in a program. Assume that the AVR serial port is connected to the COM port of the x86 PC, and we are using the HyperTerminal program on the PC to send and receive data serially. Ports A and B of the AVR are connected to LEDs and switches, respectively. Example 11-7 shows an AVR program with the following parts: (a) sends the message "YES" once to the PC screen, (b) gets data on switches and transmits it via the serial port to the PC's screen, and (c) receives any key press sent by HyperTerminal and puts it on LEDs.

Example 11-7

Write an AVR program with the following parts: (a) send the message "YES" once to the PC screen, (b) get data from switches on Port A and transmit it via the serial port to the PC's screen, and (c) receive any key press sent by HyperTerminal and put it on LEDs. The programs must do parts (b) and (c) repeatedly.

Solution:

```
.INCLUDE "M32DEF.INC"

        LDI     R21,0x00
        OUT     DDRA,R21                    ;Port A is input
        LDI     R21,0xFF
        OUT     DDRB,R21                    ;Port B is output

        LDI     R21,HIGH(RAMEND)            ;initialize high
        OUT     SPH,R21                     ;byte of SP
        LDI     R21,LOW(RAMEND)             ;initialize low
        OUT     SPL,R21                     ;byte of SP

        LDI     R16,(1<<TXEN)|(1<<RXEN)     ;enable transmitter
        OUT     UCSRB, R16                  ;and receiver
        LDI     R16,(1<<UCSZ1)|(1<<UCSZ0)|(1<<URSEL)  ;8-bit data
        OUT     UCSRC, R16                  ;no parity, 1 stop bit
        LDI     R16,0x33                    ;9600 baud rate
        OUT     UBRRL,R16

        LDI     R17,'Y'                     ;move 'Y' to R17
        CALL    TRNSMT                      ;transmit r17 to TxD
        LDI     R17,'E'                     ;move 'E' to R17
        CALL    TRNSMT                      ;transmit r17 to TxD
        LDI     R17,'S'                     ;move 'S' to R17
        CALL    TRNSMT                      ;transmit r17 to TxD
AGAIN:
        SBIS    UCSRA,RXC                   ;is there new data?
        RJMP    SKIP_RX                     ;skip receive cmnds
        IN      R17,UDR                     ;move UDR to R17
        OUT     PORTB,R17                   ;move R17 TO PORTB

SKIP_RX:
        SBIS    UCSRA,UDRE                  ;is UDR empty?
        RJMP    SKIP_TX                     ;skip transmit cmnds
        IN      R17,PINA                    ;move Port A to R17
        OUT     UDR,R17                     ;send R17 to UDR
SKIP_TX:
        RJMP    AGAIN                       ;do it again

TRNSMT:
        SBIS    UCSRA,UDRE                  ;is UDR empty?
        RJMP    TRNSMT                      ;wait more
        OUT     UDR,R17                     ;send R17 to UDR
        RET
```

Doubling the baud rate in the AVR

There are two ways to increase the baud rate of data transfer in the AVR:

1. Use a higher-frequency crystal.
2. Change a bit in the UCSRA register, as shown below.

Option 1 is not feasible in many situations because the system crystal is fixed. Therefore, we will explore option 2. There is a software way to double the baud rate of the AVR while the crystal frequency stays the same. This is done with the U2X bit of the UCSRA register. When the AVR is powered up, the U2X bit of the UCSRA register is zero. We can set it to high by software and thereby double the baud rate.

To see how the baud rate is doubled with this method, look again at Figure 11-9. If we set the U2X bit to HIGH, the third frequency divider will be bypassed. In the case of XTAL = 8 MHz and U2X bit set to HIGH, we would have:

Desired Baud Rate = Fosc / (8 (X + 1)) = 8 MHz / 8 (X + 1) = 1 MHz / (X + 1)

To get the X value for different baud rates we can solve the equation as follows:

X = (1 kHz / Desired Baud Rate) – 1

In Table 11-6 you can see values of UBRR in hex and decimal for different baud rates for U2X = 0 and U2X = 1. (XTAL = 8 MHz).

Table 11-6: UBRR Values for Various Baud Rates (XTAL = 8 MHz)

	U2X = 0		U2X = 1	
Baud Rate	**UBRR**	**UBR (HEX)**	**UBRR**	**UBR (HEX)**
38400	12	C	25	19
19200	25	19	51	33
9,600	51	33	103	67
4,800	103	67	207	CF

UBRR = (500 kHz / Baud rate) – 1   *UBRR = (1 kHz / Baud rate) – 1*

Baud rate error calculation

In calculating the baud rate we have used the integer number for the UBRR register values because AVR microcontrollers can only use integer values. By dropping the decimal portion of the calculated values we run the risk of introducing error into the baud rate. There are several ways to calculate this error. One way would be to use the following formula.

Error = (Calculated value for the UBRR – Integer part) / Integer part

For example, with XTAL = 8 MHz and U2X = 0 we have the following for the 9600 baud rate:

$$\text{UBRR value} = (500,000/ 9600) - 1 = 52.08 - 1 = 51.08 = 51$$
$$\Rightarrow \qquad \text{Error} = (51.08 - 51)/ 51 = 0.16\%$$

Another way to calculate the error rate is as follows:

Error = (Calculated baud rate – desired baud rate) / desired baud rate

Where the desired baud rate is calculated using X = (Fosc / (16(Desired Baud rate))) – 1, and then the integer X (value loaded into UBRR reg) is used for the calculated baud rate as follows:

Calculated baud rate = Fosc / (16(X + 1))   (for U2X = 0)

For XTAL = 8 MHz and 9600 baud rate, we got X = 51. Therefore, we get the calculated baud rate of 8 MHz/(16(51 + 1)) = 9765. Now the error is calculated as follows:

Error = (9615 – 9600) / 9600 = 0.16%

which is the same as what we got earlier using the other method. Table 11-7 provides the error rates for UBRR values of 8 MHz crystal frequencies.

Table 11-7: UBRR Values for Various Baud Rates (XTAL = 8 MHz)

	U2X = 0		U2X = 1	
Baud Rate	**UBRR**	**Error**	**UBRR**	**Error**
38400	12	0.2%	25	0.2%
19200	25	0.2%	51	0.2%
9,600	51	0.2%	103	0.2%
4,800	103	0.2%	207	0.2%
UBRR = (500,000 / Baud rate) – 1			*UBRR = (1,000,000 / Baud rate) – 1*	

In some applications we need very accurate baud rate generation. In these cases we can use a 7.3728 MHz or 11.0592 MHz crystal. As you can see in Table 11-8, the error is 0% if we use a 7.3728 MHz crystal. In the table there are values of UBRR for different baud rates for U2X = 0 and U2X = 1.

Table 11-8: UBRR Values for Various Baud Rates (XTAL = 7.3728 MHz)

	U2X = 0		U2X = 1	
Baud Rate	**UBRR**	**Error**	**UBRR**	**Error**
38400	11	0%	23	0%
19200	23	0%	47	0%
9,600	47	0%	95	0%
4,800	95	0%	191	0%
UBRR = (460,800 / Baud rate) – 1			*UBRR = (921,600 / Baud rate) – 1*	

See Example 11-8 to see how to calculate the error for different baud rates.

Example 11-8

Assuming XTAL = 10 MHz, calculate the baud rate error for each of the following:
(a) 2400   (b) 9600   (c) 19,200   (d) 57,600
Use the U2X = 0 mode.

Solution:

UBRR Value = (Fosc / 16(baud rate)) – 1 , Fosc = 10 MHz =>

(a) UBRR Value = (625,000 / 2400) – 1 = 260.41 – 1 = 259.41 = 259

        Error = (259.41c259) / 259 = 0.158%

(b) UBRR Value (625,000 / 9600) – 1 = 65.104 – 1 = 64.104 = 64

        Error = (64.104 – 64)/64 = 0.162%

(c) UBRR Value (625,000 / 19,200) – 1 = 32.55 – 1 = 31.55 = 31

        Error = (31.55 – 31) / 31 = 1.77%

(d) UBRR Value (625,000 / 57,600) – 1 = 10.85 – 1 = 9.85 = 9

        Error = (9.85 – 9) / 9 = 9.4%

Review Questions

1. Which registers of the AVR are used to set the baud rate?
2. If XTAL = 10 MHz, what value should be loaded into UBRR to have a 14,400 baud rate? Give the answer in both decimal and hex.
3. What is the baud rate error in the last question?
4. With XTAL = 7.3728 MHz, what value should be loaded into UBRR to have a 9600 baud rate? Give the answer in both decimal and hex.
5. To transmit a byte of data serially, it must be placed in register _____.
6. UCSRA stands for _____ .
7. Which bits are used to set the data frame size?
8. True or false. UCSRA and UCSRB share the same I/O address.
9. What parameters should the transmitter and receiver agree on before starting a serial transmission?
10. Which register has the U2X bit, and why do we use the U2X bit?

SECTION 11.4: AVR SERIAL PORT PROGRAMMING IN C

As we have seen in previous chapters, all the special function registers of the AVR are accessible directly in C compilers by using the appropriate header file. Examples 11-9 through 11-14 show how to program the serial port in C. Connect your AVR trainer to the PC's COM port and use HyperTerminal to test the operation of these examples.

Example 11-9

Write a C function to initialize the USART to work at 9600 baud, 8-bit data, and 1 stop bit. Assume XTAL = 8 MHz.

Solution:

```c
void usart_init (void)
{
  UCSRB = (1<<TXEN);
  UCSRC = (1<< UCSZ1)|(1<<UCSZ0)|(1<<URSEL);
  UBRRL = 0x33;
}
```

Example 11-10 (C Version of Example **11-4**)

Write a C program for the AVR to transfer the letter 'G' serially at 9600 baud, continuously. Use 8-bit data and 1 stop bit. Assume XTAL = 8 MHz.

Solution:

```c
#include <avr/io.h>                     //standard AVR header
void usart_init (void)
{
  UCSRB = (1<<TXEN);
  UCSRC = (1<< UCSZ1)|(1<<UCSZ0)|(1<<URSEL);
  UBRRL = 0x33;
}
void usart_send (unsigned char ch)
{                                        //wait until UDR
  while (! (UCSRA & (1<<UDRE)));         //is empty
  UDR = ch;                              //transmit 'G'
}

int main (void)
{
  usart_init();                          //initialize the USART
  while(1)                               //do forever
    usart_send ('G' );                   //transmit 'G' letter
  return 0;
}
```

Example 11-11

Write a program to send the message "The Earth is but One Country. " to the serial port continuously. Using the settings in the last example.

Solution:

```c
#include <avr/io.h>                         //standard AVR header

void usart_init (void)
{  UCSRB = (1<<TXEN);
   UCSRC = (1<< UCSZ1)|(1<<UCSZ0)|(1<<URSEL);
   UBRRL = 0x33;
}
void usart_send (unsigned char ch)
{  while (! (UCSRA & (1<<UDRE)));
   UDR = ch;
}
int main (void)
{  unsigned char str[ 30] = "The Earth is but One Country. ";
   unsigned char strLenght = 30;
   unsigned char i = 0;
   usart_init();
   while(1)
   {
     usart_send(str[ i++] );
     if (i >= strLenght)
       i = 0;
   }
   return 0;
}
```

Example 11-12

Program the AVR in C to receive bytes of data serially and put them on Port A. Set the baud rate at 9600, 8-bit data, and 1 stop bit.

Solution:

```c
#include <avr/io.h>                         //standard AVR header
int main (void)
{
  DDRA = 0xFF;                              //Port A is input
  UCSRB = (1<<RXEN);                        //initialize USART
  UCSRC = (1<< UCSZ1)|(1<<UCSZ0)|(1<<URSEL);
  UBRRL = 0x33;
  while(1)
  {
    while (! (UCSRA & (1<<RXC)));           //wait until new data
      PORTA = UDR;
  }
  return 0;
}
```

Example 11-13

Write an AVR C program to receive a character from the serial port. If it is 'a' – 'z' change it to capital letters and transmit it back. Use the settings in the last example.

Solution:

```c
#include <avr/io.h>                              //standard AVR header

void transmit (unsigned char data);

int main (void)
{
                   // initialize USART transmitter and receiver
  UCSRB = (1<<TXEN)|(1<<RXEN);

  UCSRC = (1<< UCSZ1)|(1<<UCSZ0)|(1<<URSEL);
  UBRRL = 0x33;

  unsigned char ch;

  while(1)
  {
    while(!(UCSRA&(1<<RXC)));                    //while new data received
      ch = UDR;
    if (ch >= 'a' && ch<='z')
    {
      ch+=('A'-'a');
      while (! (UCSRA & (1<<UDRE)));
        UDR = ch;
    }
  }
  return 0;
}
```

Example 11-14

In the last five examples, what is the baud rate error?

Solution:

According to Table 11-8, for 9600 baud rate and XTAL = 8 MHz, the baud rate error is about 2%.

Review Questions

1. True or false. All the SFR registers of AVR are accessible in the C compiler.
2. True or false. The FE flag is cleared the moment we read from the UDR register.

SECTION 11.5: AVR SERIAL PORT PROGRAMMING IN ASSEMBLY AND C USING INTERRUPTS

By now you might have noticed that it is a waste of the microcontroller's time to poll the TXIF and RXIF flags. In order to avoid wasting the microcontroller's time we use interrupts instead of polling. In this section, we will show how to use interrupts to program the AVR's serial communication port.

Interrupt-based data receive

To program the serial port to receive data using the interrupt method, we need to set HIGH the Receive Complete Interrupt Enable (RXCIE) bit in UCSRB. Setting this bit enables the interrupt on the RXC flag in UCSRA. Upon completion of the receive, the RXC (USART receive complete flag) becomes HIGH. If RXCIE = 1, changing RXC to one will force the CPU to jump to the interrupt vector. Program 11-15 shows how to receive data using interrupts.

Example 11-15

Program the ATmega32 to receive bytes of data serially and put them on Port B. Set the baud rate at 9600, 8-bit data, and 1 stop bit. Use Receive Complete Interrupt instead of the polling method.

Solution:

```
.INCLUDE "M32DEF.INC"
.CSEG                                   ;put in code segment
     RJMP  MAIN                         ;jump main after reset
.ORG URXCaddr                           ;int-vector of URXC int.
     RJMP  URXC_INT_HANDLER             ;jump to URXC_INT_HANDLER
.ORG 40                                 ;start main after
                                        ;interrupt vector
MAIN: LDI  R16, HIGH(RAMEND)            ;initialize high byte of
      OUT  SPH, R16                     ;stack pointer
      LDI  R16, LOW(RAMEND)             ;initialize low byte of
      OUT  SPL,R16                      ;stack pointer
      LDI  R16,(1<<RXEN)|(1<<RXCIE)     ;enable receiver
      OUT  UCSRB, R16                   ;and RXC interrupt
      LDI  R16,(1<<UCSZ1)|(1<<UCSZ0)|(1<<URSEL);sync,8-bit data
      OUT  UCSRC, R16                   ;no parity, 1 stop bit
      LDI  R16,0x33                     ;9600 baud rate
      OUT  UBRRL,R16
      LDI  R16,0xFF                     ;set Port B as an
      OUT  DDRB,R16                     ;input
      SEI                              ;enable interrupts
WAIT_HERE:
      RJMP WAIT_HERE                    ;stay here
URXC_INT_HANDLER:
      IN   R17,UDR                      ;send UDR to R17
      OUT  PORTB,R17                    ;send R17 to PORTB
      RETI
```

Interrupt-based data transmit

To program the serial port to transmit data using the interrupt method, we need to set HIGH the USART Data Register Empty Interrupt Enable (UDRIE) bit in UCSRB. Setting this bit enables the interrupt on the UDRE flag in UCSRA. When the UDR register is ready to accept new data, the UDRE (USART Data Register Empty flag) becomes HIGH. If UDRIE = 1, changing UDRE to one will force the CPU to jump to the interrupt vector. Example 11-16 shows how to transmit data using interrupts. To transmit data using the interrupt method, there is another source of interrupt; it is Transmit Complete Interrupt. Try to clarify the difference between these two interrupts for yourself. Can you provide some example that the two interrupts can be used interchangeably?

Example 11-16

Write a program for the AVR to transmit the letter 'G' serially at 9600 baud, continuously. Assume XTAL = 8 MHz. Use interrupts instead of the polling method.

Solution:

```
.INCLUDE "M32DEF.INC"

.CSEG                                    ;put in code segment
    RJMP  MAIN                           ;jump main after reset
.ORG UDREaddr                            ;int. vector of UDRE int.
    RJMP  UDRE_INT_HANDLER               ;jump to UDRE_INT_HANDLER
.ORG 40                                  ;start main after
                                         ;interrupt vector
;*****************************
MAIN:
    LDI   R16, HIGH(RAMEND)              ;initialize high byte of
    OUT   SPH, R16                       ;stack pointer
    LDI   R16, LOW(RAMEND)               ;initialize low byte of
    OUT   SPL,R16                        ;stack pointer
    LDI   R16,(1<<TXEN)|(1<<UDRIE)       ;enable transmitter
    OUT   UCSRB, R16                     ;and UDRE interrupt
    LDI   R16,(1<<UCSZ1)|(1<<UCSZ0)|(1<<URSEL); sync., 8-bit
    OUT   UCSRC, R16                     ;data no parity, 1 stop bit
    LDI   R16,0x33                       ;9600 baud rate
    OUT   UBRRL,R16
    SEI                                  ;enable interrupts
WAIT_HERE:
    RJMP  WAIT_HERE                      ;stay here
;*****************************
UDRE_INT_HANDLER:
    LDI   R26,'G'                        ;send 'G'
    OUT   UDR,R26                        ;to UDR
    RETI
```

Examples 11-17 and 11-18 are the C versions of Examples 11-15 and 11-16, respectively.

Example 11-17

Write a C program to receive bytes of data serially and put them on Port B. Use Receive Complete Interrupt instead of the polling method.

Solution:

```c
#include <avr\io.h>
#include <avr\interrupt.h>

ISR(USART_RXC_vect)
{
  PORTB = UDR;
}

int main (void)
{
  DDRB = 0xFF;                      //make Port B an output
  UCSRB = (1<<RXEN)|(1<<RXCIE); //enable receive and RXC int.
  UCSRC = (1<< UCSZ1)|(1<<UCSZ0)|(1<<URSEL);
  UBRRL = 0x33;
  sei();                            //enable interrupts
  while (1);                        //wait forever
  return 0;
}
```

Example 11-18

Write a C program to transmit the letter 'G' serially at 9600 baud, continuously. Assume XTAL = 8 MHz. Use interrupts instead of the polling method.

Solution:

```c
#include <avr\io.h>
#include <avr\interrupt.h>
ISR(USART_UDRE_vect)
{
   UDR = 'G';
}

int main (void)
{
  UCSRB = (1<<TXEN)|(1<<UDRIE);
  UCSRC = (1<< UCSZ1)|(1<<UCSZ0)|(1<<URSEL);
  UBRRL = 0x33;

  sei();                    //enable interrupts
  while (1);                //wait forever
  return 0;
}
```

Review Questions

1. What is the advantage of interrupt-based programming over polling?
2. How do you enable transmit or receive interrupts in AVR?

SUMMARY

This chapter began with an introduction to the fundamentals of serial communication. Serial communication, in which data is sent one bit a time, is used in situations where data is sent over significant distances. (In parallel communication, where data is sent a byte or more at a time, great distances can cause distortion of the data.) Serial communication has the additional advantage of allowing transmission over phone lines. Serial communication uses two methods: synchronous and asynchronous. In synchronous communication, data is sent in blocks of bytes; in asynchronous, data is sent one byte at a time. Data communication can be simplex (can send but cannot receive), half duplex (can send and receive, but not at the same time), or full duplex (can send and receive at the same time). RS232 is a standard for serial communication connectors.

The AVR's UART was discussed. We showed how to interface the ATmega32 with an RS232 connector and change the baud rate of the ATmega32. In addition, we described the serial communication features of the AVR, and programmed the ATmega32 for serial data communication. We also showed how to program the serial port of the ATmega32 chip in Assembly and C.

PROBLEMS

SECTION 11.1: BASICS OF SERIAL COMMUNICATION

1. Which is more expensive, parallel or serial data transfer?
2. True or false. 0- and 5-V digital pulses can be transferred on the telephone without being converted (modulated).
3. Show the framing of the letter ASCII 'Z' (0101 1010), no parity, 1 stop bit.
4. If there is no data transfer and the line is high, it is called _____ (mark, space).
5. True or false. The stop bit can be 1, 2, or none at all.
6. Calculate the overhead percentage if the data size is 7, 1 stop bit, and no parity bit.
7. True or false. The RS232 voltage specification is TTL compatible.
8. What is the function of the MAX 232 chip?
9. True or false. DB-25 and DB-9 are pin compatible for the first 9 pins.
10. How many pins of the RS232 are used by the IBM serial cable, and why?
11. True or false. The longer the cable, the higher the data transfer baud rate.
12. State the absolute minimum number of signals needed to transfer data between two PCs connected serially. What are those signals?
13. If two PCs are connected through the RS232 without a modem, both are con-

figured as a _____ (DTE, DCE) -to- _____ (DTE, DCE) connection.

14. State the nine most important signals of the RS232.
15. Calculate the total number of bits transferred if 200 pages of ASCII data are sent using asynchronous serial data transfer. Assume a data size of 8 bits, 1 stop bit, and no parity. Assume each page has 80×25 of text characters.
16. In Problem 15, how long will the data transfer take if the baud rate is 9600?

SECTION 11.2: ATMEGA32 CONNECTION TO RS232

17. The MAX232 DIP package has _____ pins.
18. For the MAX232, indicate the V_{CC} and GND pins.
19. The MAX233 DIP package has _____ pins.
20. For the MAX233, indicate the V_{CC} and GND pins.
21. Is the MAX232 pin compatible with the MAX233?
22. State the advantages and disadvantages of the MAX232 and MAX233.
23. MAX232/233 has _____ line driver(s) for the RX wire.
24. MAX232/233 has _____ line driver(s) for the TX wire.
25. Show the connection of pins TX and RX of the ATmega32 to a DB-9 RS232 connector via the second set of line drivers of MAX232.
26. Show the connection of the TX and RX pins of the ATmega32 to a DB-9 RS232 connector via the second set of line drivers of MAX233.
27. What is the advantage of the MAX233 over the MAX232 chip?
28. Which pins of the ATmega32 are set aside for serial communication, and what are their functions?

SECTION 11.3: AVR SERIAL PORT PROGRAMMING IN ASSEMBLY

29. Which of the following baud rates are supported by the HyperTerminal program in PC?
    (a) 4800   (b) 3600   (c) 9600
    (d) 1800   (e) 1200   (f) 19,200
30. Which register of ATmega32 is used for baud rate programming?
31. Which bit of the UCSRA is used for baud rate speed?
32. What is the role of the UDR register in serial data transfer?
33. UDR is a(n) _____-bit register.
34. For XTAL = 10 MHz, find the UBRR value (in both decimal and hex) for each of the following baud rates.
    (a) 9600   (b) 4800   (c) 1200
35. What is the baud rate if we use UBRR = 15 to program the baud rate? Assume XTAL = 10 MHz.
36. Write an AVR program to transfer serially the letter 'Z' continuously at 9600 baud rate. Assume XTAL = 10 MHz.
37. When is the PE flag bit raised?
38. When is the RXC flag bit raised or cleared?
39. When is the UDRE flag bit raised or cleared?
40. To which register do RXC and UDRE belong?
41. Find the UBRR for the following baud rates if XTAL = 16 MHz and U2X = 0.

       (a) 9600   (b) 19200

       (c) 38400   (d) 57600

42. Find the UBRR for the following baud rates if XTAL = 16 MHz and U2X = 1.

       (a) 9600   (b) 19200

       (c) 38400   (d) 57600

43. Find the UBRR for the following baud rates if XTAL = 11.0592 MHz and U2X = 0.

       (a) 9600   (b) 19200

       (c) 38400   (d) 57600

44. Find the UBRR for the following baud rates if XTAL = 11.0592 MHz and U2X = 1.

       (a) 9600   (b) 19200

       (c) 38400   (d) 57600

45. Find the baud rate error for Problem 41.

46. Find the baud rate error for Problem 42.

47. Find the baud rate error for Problem 43.

48. Find the baud rate error for Problem 44.

SECTION 11.4: AVR SERIAL PORT PROGRAMMING IN C

49. Write an AVR C program to transmit serially the letter 'Z' continuously at 9600 baud rate.

50. Write an AVR C program to transmit serially the message "The earth is but one country and mankind its citizens" continuously at 57,600 baud rate.

ANSWERS TO REVIEW QUESTIONS

SECTION 11.1: BASICS OF SERIAL COMMUNICATION

1. Faster, more expensive
2. False; it is simplex.
3. True
4. Asynchronous
5. With 0100 0101 binary the bits are transmitted in the sequence:
   (a) 0 (start bit) (b) 1 (c) 0 (d) 1 (e) 0 (f) 0 (g) 0 (h) 1 (i) 0 (j) 1 (stop bit)
6. 2 bits (one for the start bit and one for the stop bit). Therefore, for each 8-bit character, a total of 10 bits is transferred.
7. $10,000 \times 10 = 100,000$ total bits transmitted. $100,000 / 9600 = 10.4$ seconds; $2 / 10 = 20\%$.
8. True
9. +3 to +25 V
10. True

SECTION 11.2: ATMEGA32 CONNECTION TO RS232

1. True
2. Pin 14, which is RxD, and pin15, which is TXD .
3. They convert different voltage levels to each other to make two different standards compatible.
4. 2, 2
5. It has a built-in capacitor.

SECTION 11.3: AVR SERIAL PORT PROGRAMMING IN ASSEMBLY

1. UBRRL and UBRRH.
2. (Fosc / 16 (baud rate)) − 1 = (10M / 16 (14400)) − 1 = 42.4 = 42 or 2AH
3. (42.4 − 42) / 42 = 0.95%
4. (Fosc / 16 (baud rate)) − 1 = (7372800 / 16 (9600)) − 1 = 47 or 2FH
5. UDR
6. USART Control Status Register A
7. UCSZ0 and UCSZ1 bits in UCSRB and UCSZ2 in UCSRC
8. False
9. Baud rate, frame size, stop bit, parity
10. U2X is bit1 of UCSRA and doubles the transfer rate for asynchronous communication.

SECTION 11.4 : AVR SERIAL PORT PROGRAMMING IN C

1. True
2. True

SECTION 11.5 : AVR SERIAL PORT PROGRAMMING IN ASSEMBLY AND C USING INTERRUPTS

1. In interrupt-based programming, CPU time is not wasted.
2. By writing the value of 1 to the UDRIE and RXCIE bits

CHAPTER 12

LCD AND KEYBOARD
INTERFACING

OBJECTIVES

Upon completion of this chapter, you will be able to:

>> List reasons that LCDs are gaining widespread use, replacing LEDs
>> Describe the functions of the pins of a typical LCD
>> List instruction command codes for programming an LCD
>> Interface an LCD to the AVR
>> Program an LCD in Assembly and C
>> Explain the basic operation of a keyboard
>> Describe the key press and detection mechanisms
>> Interface a 4 × 4 keypad to the AVR using C and Assembly

This chapter explores some real-world applications of the AVR. We explain how to interface the AVR to devices such as an LCD and a keyboard. In Section 12.1, we show LCD interfacing with the AVR. In Section 12.2, keyboard interfacing with the AVR is shown. We use C and Assembly for both sections.

SECTION 12.1: LCD INTERFACING

This section describes the operation modes of LCDs and then describes how to program and interface an LCD to an AVR using Assembly and C.

LCD operation

In recent years the LCD is finding widespread use replacing LEDs (seven-segment LEDs or other multisegment LEDs). This is due to the following reasons:
1. The declining prices of LCDs.
2. The ability to display numbers, characters, and graphics. This is in contrast to LEDs, which are limited to numbers and a few characters.
3. Incorporation of a refreshing controller into the LCD, thereby relieving the CPU of the task of refreshing the LCD. In contrast, the LED must be refreshed by the CPU (or in some other way) to keep displaying the data.
4. Ease of programming for characters and graphics.

LCD pin descriptions

The LCD discussed in this section has 14 pins. The function of each pin is given in Table 12-1. Figure 12-1 shows the pin positions for various LCDs.

V_{CC}, V_{SS}, and V_{EE}

While V_{CC} and V_{SS} provide +5 V and ground, respectively, V_{EE} is used for controlling LCD contrast.

RS, register select

There are two very important registers inside the LCD. The RS pin is used for their selection as follows. If RS = 0, the instruction command code register is selected, allowing the user to send commands such as clear display, cursor at home, and so on. If RS = 1 the data register is selected, allowing the user to send data to be displayed on the LCD.

R/W, read/write

R/W input allows the user to write information to the LCD or read information from it. R/W = 1 when reading; R/W = 0 when writing.

E, enable

The enable pin is used by the LCD to latch information presented to its data pins.

Table 12-1: Pin Descriptions for LCD

Pin	Symbol	I/O	Description
1	V_{SS}	--	Ground
2	V_{CC}	--	+5 V power supply
3	V_{EE}	--	Power supply to control contrast
4	RS	I	RS = 0 to select command register, RS = 1 to select data register
5	R/W	I	R/W = 0 for write, R/W = 1 for read
6	E	I/O	Enable
7	DB0	I/O	The 8-bit data bus
8	DB1	I/O	The 8-bit data bus
9	DB2	I/O	The 8-bit data bus
10	DB3	I/O	The 8-bit data bus
11	DB4	I/O	The 8-bit data bus
12	DB5	I/O	The 8-bit data bus
13	DB6	I/O	The 8-bit data bus
14	DB7	I/O	The 8-bit data bus

When data is supplied to data pins, a high-to-low pulse must be applied to this pin in order for the LCD to latch in the data present at the data pins. This pulse must be a minimum of 450 ns wide.

D0–D7

The 8-bit data pins, D0–D7, are used to send information to the LCD or read the contents of the LCD's internal registers.

To display letters and numbers, we send ASCII codes for the letters A–Z, a–z, and numbers 0–9 to these pins while making RS = 1.

There are also instruction command codes that can be sent to the LCD to clear the display or force the cursor to the home position or blink the cursor. Table 12-2 lists the instruction command codes.

In this section you will see how to interface an LCD to the AVR in two different ways. We can use 8-bit data or 4-bit data options. The 8-bit data interfacing is easier to program but uses 4 more pins.

Table 12-2: LCD Command Codes

Code (Hex)	Command to LCD Instruction Register
1	Clear display screen
2	Return home
4	Decrement cursor (shift cursor to left)
6	Increment cursor (shift cursor to right)
5	Shift display right
7	Shift display left
8	Display off, cursor off
A	Display off, cursor on
C	Display on, cursor off
E	Display on, cursor blinking
F	Display on, cursor blinking
10	Shift cursor position to left
14	Shift cursor position to right
18	Shift the entire display to the left
1C	Shift the entire display to the right
80	Force cursor to beginning of 1st line
C0	Force cursor to beginning of 2nd line
28	2 lines and 5 × 7 matrix (D4–D7, 4-bit)
38	2 lines and 5 × 7 matrix (D0–D7, 8-bit)

Note: This table is extracted from Table 12-4.

Dot matrix character LCDs are available in different packages. Figure 12-1 shows the position of each pin in different packages.

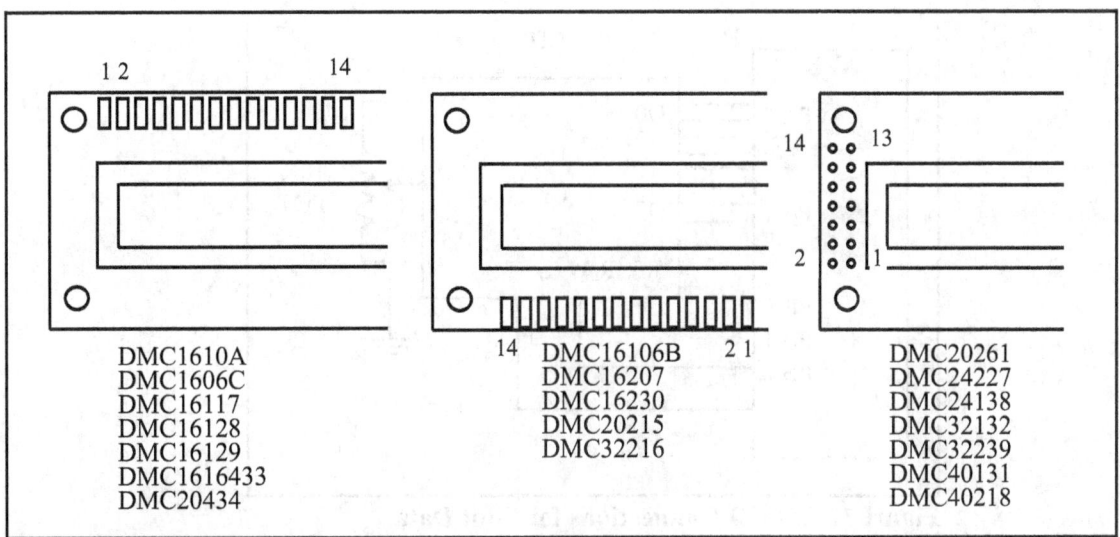

Figure 12-1. Pin Positions for Various LCDs from Optrex

CHAPTER 12: LCD AND KEYBOARD INTERFACING   **431**

Sending commands and data to LCDs

To send data and commands to LCDs you should do the following steps. Notice that steps 2 and 3 can be repeated many times:

1. Initialize the LCD.
2. Send any of the commands from Table 12-2 to the LCD.
3. Send the character to be shown on the LCD.

Initializing the LCD

To initialize the LCD for 5 × 7 matrix and 8-bit operation, the following sequence of commands should be sent to the LCD: 0x38, 0x0E, and 0x01. Next we will show how to send a command to the LCD. After power-up you should wait about 15 ms before sending initializing commands to the LCD. If the LCD initializer function is not the first function in your code you can omit this delay.

Sending commands to the LCD

To send any of the commands from Table 12-2 to the LCD, make pins RS and R/W = 0 and put the command number on the data pins (D0–D7). Then send a high-to-low pulse to the E pin to enable the internal latch of the LCD. Notice that after each command you should wait about 100 μs to let the LCD module run the command. Clear LCD and Return Home commands are exceptions to this rule. After the 0x01 and 0x02 commands you should wait for about 2 ms. Table 12-3 shows the details of commands and their execution times.

Sending data to the LCD

To send data to the LCD, make pins RS = 1 and R/W = 0. Then put the data on the data pins (D0–D7) and send a high-to-low pulse to the E pin to enable the internal latch of the LCD. Notice that after sending data you should wait about 100 μs to let the LCD module write the data on the screen.

Program 12-1 shows how to write "Hi" on the LCD using 8-bit data. The AVR connection to the LCD for 8-bit data is shown in Figure 12-2.

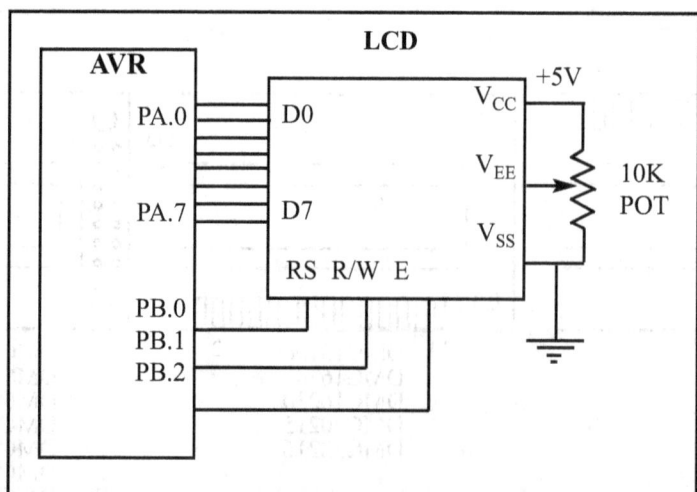

Figure 12-2. LCD Connections for 8-bit Data

```
.INCLUDE "M32DEF.INC"
.EQU       LCD_DPRT = PORTA       ;LCD DATA PORT
.EQU       LCD_DDDR = DDRA        ;LCD DATA DDR
.EQU       LCD_DPIN = PINA        ;LCD DATA PIN
.EQU       LCD_CPRT = PORTB       ;LCD COMMANDS PORT
.EQU       LCD_CDDR = DDRB        ;LCD COMMANDS DDR
.EQU       LCD_CPIN = PINB        ;LCD COMMANDS PIN
.EQU       LCD_RS = 0             ;LCD RS
.EQU       LCD_RW = 1             ;LCD RW
.EQU       LCD_EN = 2             ;LCD EN


           LDI   R21,HIGH(RAMEND)
           OUT   SPH,R21          ;set up stack
           LDI   R21,LOW(RAMEND)
           OUT   SPL,R21


           LDI   R21,0xFF;
           OUT   LCD_DDDR, R21    ;LCD data port is output
           OUT   LCD_CDDR, R21    ;LCD command port is output
           CBI   LCD_CPRT,LCD_EN  ;LCD_EN = 0
           CALL  DELAY_2ms        ;wait for power on
           LDI   R16,0x38         ;init LCD 2 lines,5×7 matrix
           CALL  CMNDWRT          ;call command function
           CALL  DELAY_2ms        ;wait 2 ms
           LDI   R16,0x0E         ;display on, cursor on
           CALL  CMNDWRT          ;call command function
           LDI   R16,0x01         ;clear LCD
           CALL  CMNDWRT          ;call command function
           CALL  DELAY_2ms        ;wait 2 ms
           LDI   R16,0x06         ;shift cursor right
           CALL  CMNDWRT          ;call command function
           LDI   R16,'H'          ;display letter 'H'
           CALL  DATAWRT          ;call data write function
           LDI   R16,'i'          ;display letter 'i'
           CALL  DATAWRT          ;call data write function
HERE:      JMP HERE               ;stay here
;-------------------------------------------------------------
CMNDWRT:
           OUT   LCD_DPRT,R16          ;LCD data port = R16
           CBI   LCD_CPRT,LCD_RS       ;RS = 0 for command
           CBI   LCD_CPRT,LCD_RW       ;RW = 0 for write
           SBI   LCD_CPRT,LCD_EN       ;EN = 1
           CALL  SDELAY               ;make a wide EN pulse
           CBI   LCD_CPRT,LCD_EN       ;EN=0 for H-to-L pulse
           CALL  DELAY_100us          ;wait 100 us
           RET
```

Program 12-1: Communicating with LCD *(continued on next page)*

```
DATAWRT:
        OUT     LCD_DPRT,R16        ;LCD data port = R16
        SBI     LCD_CPRT,LCD_RS     ;RS = 1 for data
        CBI     LCD_CPRT,LCD_RW     ;RW = 0 for write
        SBI     LCD_CPRT,LCD_EN     ;EN = 1
        CALL    SDELAY              ;make a wide EN pulse
        CBI     LCD_CPRT,LCD_EN     ;EN=0 for H-to-L pulse
        CALL    DELAY_100us         ;wait 100 us
        RET
;----------------------------------------------------------
SDELAY: NOP
        NOP
        RET
;----------------------------------------------------------
DELAY_100us:
        PUSH    R17
        LDI     R17,60
DR0:    CALL    SDELAY
        DEC     R17
        BRNE    DR0
        POP     R17
        RET
;----------------------------------------------------------
DELAY_2ms:
        PUSH    R17
        LDI     R17,20
LDR0:   CALL    DELAY_100US
        DEC     R17
        BRNE    LDR0
        POP     R17
        RET
```

Program 12-1: Communicating with LCD *(continued from previous page)*

Sending code or data to the LCD 4 bits at a time

The above code showed how to send commands to the LCD with 8 bits for the data pin. In most cases it is preferred to use 4-bit data to save pins. The LCD may be forced into the 4-bit mode as shown in Program 12-2. Notice that its initialization differs from that of the 8-bit mode and that data is sent out on the high nibble of Port A, high nibble first.

In 4-bit mode, we initialize the LCD with the series 33, 32, and 28 in

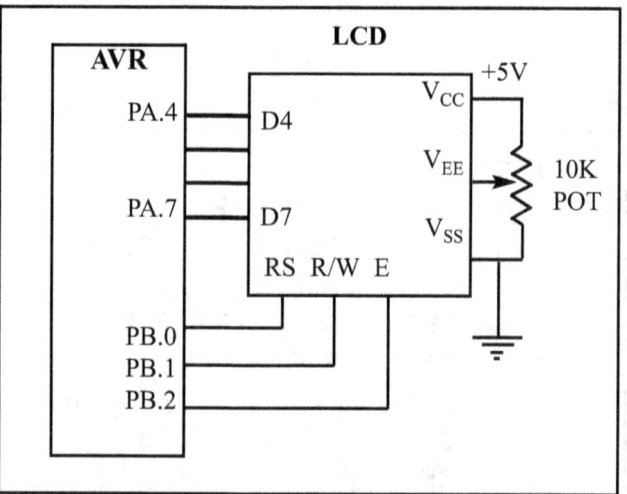

Figure 12-3. LCD Connections Using 4-bit Data

hex. This represents nibbles 3, 3, 3, and 2, which tells the LCD to go into 4-bit mode. The value $28 initializes the display for 5 × 7 matrix and 4-bit operation as required by the LCD datasheet. The write routines (CMNDWRT and DATAWRT) send the high nibble first, then swap the low nibble with the high nibble before it is sent to data pins D4–D7. The delay function of the program is the same as in Program 12-1.

```
.INCLUDE "M32DEF.INC"

.EQU    LCD_DPRT = PORTA      ;LCD DATA PORT
.EQU    LCD_DDDR = DDRA       ;LCD DATA DDR
.EQU    LCD_DPIN = PINA       ;LCD DATA PIN
.EQU    LCD_CPRT = PORTB      ;LCD COMMANDS PORT
.EQU    LCD_CDDR = DDRB       ;LCD COMMANDS DDR
.EQU    LCD_CPIN = PINB       ;LCD COMMANDS PIN
.EQU    LCD_RS = 0            ;LCD RS
.EQU    LCD_RW = 1            ;LCD RW
.EQU    LCD_EN = 2            ;LCD EN

        LDI    R21,HIGH(RAMEND)
        OUT    SPH,R21           ;set up stack
        LDI    R21,LOW(RAMEND)
        OUT    SPL,R21
        LDI    R21,0xFF;
        OUT    LCD_DDDR, R21     ;LCD data port is output
        OUT    LCD_CDDR, R21     ;LCD command port is output
        LDI    R16,0x33          ;init. LCD for 4-bit data
        CALL   CMNDWRT           ;call command function
        CALL   DELAY_2ms         ;init. hold
        LDI    R16,0x32          ;init. LCD for 4-bit data
        CALL   CMNDWRT           ;call command function
        CALL   DELAY_2ms         ;init. hold
        LDI    R16,0x28          ;init. LCD 2 lines,5×7 matrix
        CALL   CMNDWRT           ;call command function
        CALL   DELAY_2ms         ;init. hold
        LDI    R16,0x0E          ;display on, cursor on
        CALL   CMNDWRT           ;call command function
        LDI    R16,0x01          ;clear LCD
        CALL   CMNDWRT           ;call command function
        CALL   DELAY_2ms         ;delay 2 ms for clear LCD
        LDI    R16,0x06          ;shift cursor right
        CALL   CMNDWRT           ;call command function
        LDI    R16,'H'           ;display letter 'H'
        CALL   DATAWRT           ;call data write function
        LDI    R16,'i'           ;display letter 'i'
        CALL   DATAWRT           ;call data write function
HERE:   JMP HERE                 ;stay here
```

Program 12-2: Communicating with LCD Using 4-bit Mode *(continued on next page)*

```
;-------------------------------------------------------
CMNDWRT:
        MOV     R27,R16
        ANDI    R27,0xF0
        OUT     LCD_DPRT,R27            ;send the high nibble
        CBI     LCD_CPRT,LCD_RS         ;RS = 0 for command
        CBI     LCD_CPRT,LCD_RW         ;RW = 0 for write
        SBI     LCD_CPRT,LCD_EN         ;EN = 1 for high pulse
        CALL    SDELAY                  ;make a wide EN pulse
        CBI     LCD_CPRT,LCD_EN         ;EN=0 for H-to-L pulse
        CALL    DELAY_100us             ;make a wide EN pulse

        MOV     R27,R16
        SWAP    R27                     ;swap the nibbles
        ANDI    R27,0xF0                ;mask D0-D3
        OUT     LCD_DPRT,R27            ;send the low nibble
        SBI     LCD_CPRT,LCD_EN         ;EN = 1 for high pulse
        CALL    SDELAY                  ;make a wide EN pulse
        CBI     LCD_CPRT,LCD_EN         ;EN=0 for H-to-L pulse
        CALL    DELAY_100us             ;wait 100 us
        RET
;-------------------------------------------------------
DATAWRT:
        MOV     R27,R16
        ANDI    R27,0xF0
        OUT     LCD_DPRT,R27            ;;send the high nibble
        SBI     LCD_CPRT,LCD_RS         ;RS = 1 for data
        CBI     LCD_CPRT,LCD_RW         ;RW = 0 for write
        SBI     LCD_CPRT,LCD_EN         ;EN = 1 for high pulse
        CALL    SDELAY                  ;make a wide EN pulse
        CBI     LCD_CPRT,LCD_EN         ;EN=0 for H-to-L pulse

        MOV     R27,R16
        SWAP    R27                     ;swap the nibbles
        ANDI    R27,0xF0                ;mask D0-D3
        OUT     LCD_DPRT,R27            ;send the low nibble
        SBI     LCD_CPRT,LCD_EN         ;EN = 1 for high pulse
        CALL    SDELAY                  ;make a wide EN pulse
        CBI     LCD_CPRT,LCD_EN         ;EN=0 for H-to-L pulse

        CALL    DELAY_100us             ;wait 100 us
        RET
;-------------------------------------------------------
;delay functions are the same as last program and should
;be placed here.
;-------------------------------------------------------
```

Program 12-2: Communicating with LCD Using 4-bit Mode (continued from previous page)

Sending code or data to the LCD using a single port

The above code showed how to send commands to the LCD with 4-bit data but we used two different ports for data and commands. In most cases it is preferred to use a single port. Program 12-3 shows Program 12-2 modified to use a single port for LCD interfacing.

Figure 12-4 shows the hardware connection.

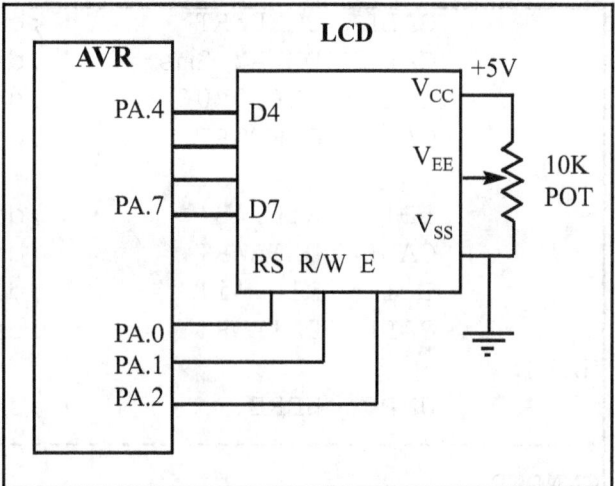

Figure 12-4. LCD Connections Using a Single Port

```
.INCLUDE  "M32DEF.INC"

.EQU     LCD_PRT = PORTA        ;LCD DATA PORT
.EQU     LCD_DDR = DDRA         ;LCD DATA DDR
.EQU     LCD_PIN = PINA         ;LCD DATA PIN
.EQU     LCD_RS = 0             ;LCD RS
.EQU     LCD_RW = 1             ;LCD RW
.EQU     LCD_EN = 2             ;LCD EN


         LDI   R21,HIGH(RAMEND)
         OUT   SPH,R21          ;set up stack
         LDI   R21,LOW(RAMEND)
         OUT   SPL,R21


         LDI   R21,0xFF;
         OUT   LCD_DDR, R21     ;LCD data port is output
         OUT   LCD_DDR, R21     ;LCD command port is output

         LDI   R16,0x33         ;init. LCD for 4-bit data
         CALL  CMNDWRT          ;call command function
         CALL  DELAY_2ms        ;init. hold
         LDI   R16,0x32         ;init. LCD for 4-bit data
         CALL  CMNDWRT          ;call command function
         CALL  DELAY_2ms        ;init. hold
         LDI   R16,0x28         ;init. LCD 2 lines,5×7 matrix
         CALL  CMNDWRT          ;call command function
         CALL  DELAY_2ms        ;init. hold
         LDI   R16,0x0E         ;display on, cursor on
         CALL  CMNDWRT          ;call command function
         LDI   R16,0x01         ;clear LCD
```

Program 12-3: Communicating with LCD Using a Single Port *(continued on next page)*

```
              CALL    CMNDWRT              ;call command function
              CALL    DELAY_2ms            ;delay 2 ms for clear LCD
              LDI     R16,0x06             ;shift cursor right
              CALL    CMNDWRT              ;call command function

              LDI     R16,'H'              ;display letter 'H'
              CALL    DATAWRT              ;call data write function
              LDI     R16,'i'              ;display letter 'i'
              CALL    DATAWRT              ;call data write function
HERE:
              JMP     HERE                 ;stay here
;-------------------------------------------------------------
CMNDWRT:
              MOV     R27,R16
              ANDI    R27,0xF0
              IN      R26,LCD_PRT
              ANDI    R26,0x0F
              OR      R26,R27
              OUT     LCD_PRT,R26          ;LCD data port = R16
              CBI     LCD_PRT,LCD_RS       ;RS = 0 for command
              CBI     LCD_PRT,LCD_RW       ;RW = 0 for write
              SBI     LCD_PRT,LCD_EN       ;EN = 1 for high pulse
              CALL    SDELAY               ;make a wide EN pulse
              CBI     LCD_PRT,LCD_EN       ;EN=0 for H-to-L pulse

              CALL    DELAY_100us          ;make a wide EN pulse

              MOV     R27,R16
              SWAP    R27
              ANDI    R27,0xF0
              IN      R26,LCD_PRT
              ANDI    R26,0x0F
              OR      R26,R27
              OUT     LCD_PRT,R26          ;LCD data port = R16
              SBI     LCD_PRT,LCD_EN       ;EN = 1 for high pulse
              CALL    SDELAY               ;make a wide EN pulse
              CBI     LCD_PRT,LCD_EN       ;EN=0 for H-to-L pulse

              CALL    DELAY_100us          ;wait 100 us
              RET
;-------------------------------------------------------------
DATAWRT:
              MOV     R27,R16
              ANDI    R27,0xF0
              IN      R26,LCD_PRT
              ANDI    R26,0x0F
```

Program 12-3: Communicating with LCD Using a Single Port *(continued from previous page)*

```
            OR      R26,R27
            OUT     LCD_PRT,R26             ;LCD data port = R16
            SBI     LCD_PRT,LCD_RS          ;RS = 1 for data
            CBI     LCD_PRT,LCD_RW          ;RW = 0 for write
            SBI     LCD_PRT,LCD_EN          ;EN = 1 for high pulse
            CALL    SDELAY                  ;make a wide EN pulse
            CBI     LCD_PRT,LCD_EN          ;EN=0 for H-to-L pulse

            MOV     R27,R16
            SWAP    R27
            ANDI    R27,0xF0
            IN      R26,LCD_PRT
            ANDI    R26,0x0F
            OR      R26,R27
            OUT     LCD_PRT,R26             ;LCD data port = R16
            SBI     LCD_PRT,LCD_EN          ;EN = 1 for high pulse
            CALL    SDELAY                  ;make a wide EN pulse
            CBI     LCD_PRT,LCD_EN          ;EN=0 for H-to-L pulse

            CALL    DELAY_100us             ;wait 100 us
            RET
;--------------------------------------------------------------
SDELAY:
            NOP
            NOP
            RET
;--------------------------------------------------------------
DELAY_100us:
            PUSH    R17
            LDI             R17,60
DR0:        CALL    SDELAY
            DEC             R17
            BRNE    DR0
            POP             R17
            RET
;--------------------------------------------------------------
DELAY_2ms:
            PUSH    R17
            LDI             R17,20
LDR0:       CALL    DELAY_100us
            DEC             R17
            BRNE    LDR0
            POP             R17
            RET
```

Program 12-3: Communicating with LCD Using a Single Port *(continued from previous page)*

Sending information to LCD using the LPM instruction

Program 12-4 shows how to use the LPM instruction to send a long string of characters to an LCD. Program 12-4 shows only the main part of the code. The other functions do not change. If you want to use a single port you have to change the port definition in the beginning of the code according to Program 12-2.

```
.INCLUDE "M32DEF.INC"
.EQU      LCD_DPRT = PORTA      ;LCD DATA PORT
.EQU      LCD_DDDR = DDRA       ;LCD DATA DDR
.EQU      LCD_DPIN = PINA       ;LCD DATA PIN
.EQU      LCD_CPRT = PORTB      ;LCD COMMANDS PORT
.EQU      LCD_CDDR = DDRB       ;LCD COMMANDS DDR
.EQU      LCD_CPIN = PINB       ;LCD COMMANDS PIN
.EQU      LCD_RS = 0                    ;LCD RS
.EQU      LCD_RW = 1                    ;LCD RW
.EQU      LCD_EN = 2                    ;LCD EN

          LDI    R21,HIGH(RAMEND)
          OUT    SPH,R21          ;set up stack
          LDI    R21,LOW(RAMEND)
          OUT    SPL,R21
          LDI    R21, 0xFF;
          OUT    LCD_DDDR, R21   ;LCD data port is output
          OUT    LCD_CDDR, R21   ;LCD command port is output
          CBI    LCD_CPRT,LCD_EN;LCD_EN = 0
          CALL   LDELAY           ;wait for init.
          LDI    R16,0x38         ;init LCD 2 lines, 5×7 matrix
          CALL   CMNDWRT          ;call command function
          CALL   LDELAY           ;init. hold
          LDI    R16,0x0E         ;display on, cursor on
          CALL   CMNDWRT          ;call command function
          LDI    R16,0x01         ;clear LCD
          CALL   CMNDWRT          ;call command function
          LDI    R16,0x06         ;shift cursor right
          CALL   CMNDWRT          ;call command function
          LDI    R16,0x84         ;cursor at line 1 pos. 4
          CALL   CMNDWRT          ;call command function
          LDI    R31,HIGH(MSG<<1)
          LDI    R30,LOW(MSG<<1) ;Z points to MSG
LOOP:     LPM    R16,Z+
          CPI    R16,0            ;compare R16 with 0
          BREQ   HERE             ;if R16 equals 0 exit
          CALL   DATAWRT          ;call data write function
          RJMP   LOOP             ;jump to loop
HERE:     JMP HERE                ;stay here
MSG:      .DB "Hello World!",0
```

Program 12-4: Communicating with LCD Using the LPM Instruction

LCD data sheet

Here we deepen your understanding of LCDs by concentrating on two important concepts. First we will show you the timing diagram of the LCD; then we will discuss how to put data at any location.

LCD timing diagrams

In Figures 12-5 and 12-6 you can study and contrast the Write timing for the 8-bit and 4-bit modes. Notice that in the 4-bit operating mode, the high nibble is transmitted. Also notice that each nibble is followed by a high-to-low pulse to enable the internal latch of the LCD.

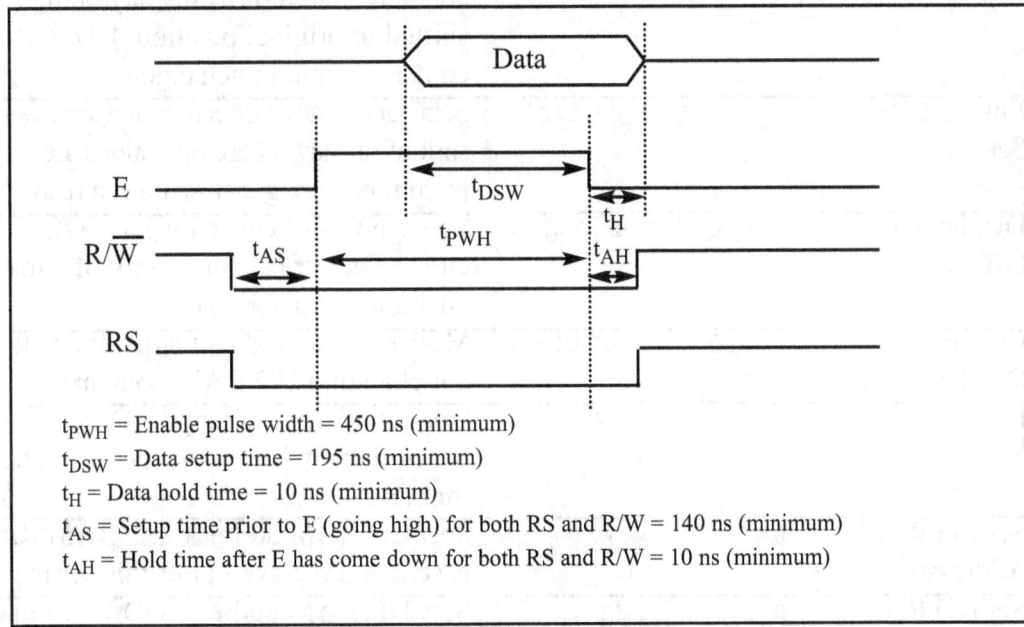

t_{PWH} = Enable pulse width = 450 ns (minimum)
t_{DSW} = Data setup time = 195 ns (minimum)
t_H = Data hold time = 10 ns (minimum)
t_{AS} = Setup time prior to E (going high) for both RS and R/W = 140 ns (minimum)
t_{AH} = Hold time after E has come down for both RS and R/W = 10 ns (minimum)

Figure 12-5. LCD Timing for Write (H-to-L for E line)

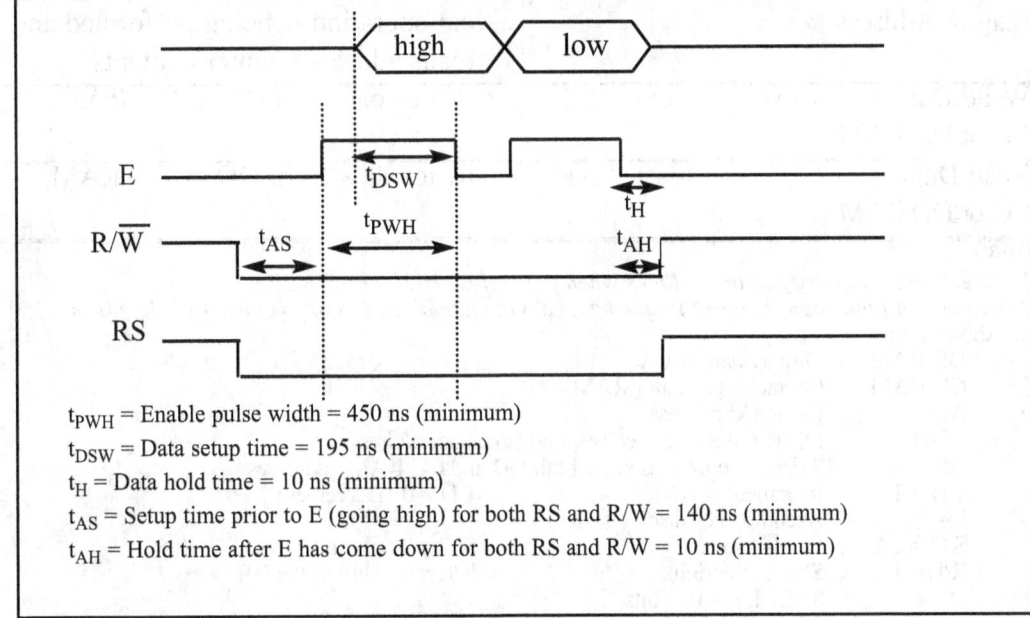

t_{PWH} = Enable pulse width = 450 ns (minimum)
t_{DSW} = Data setup time = 195 ns (minimum)
t_H = Data hold time = 10 ns (minimum)
t_{AS} = Setup time prior to E (going high) for both RS and R/W = 140 ns (minimum)
t_{AH} = Hold time after E has come down for both RS and R/W = 10 ns (minimum)

Figure 12-6. LCD Timing for 4-bit Write

Table 12-3 provides a detailed list of LCD commands and instructions.

Table 12-3: List of LCD Instructions

Instruction	RS	R/W	DB7	DB6	DB5	DB4	DB3	DB2	DB1	DB0	Description	Execution Time (Max)
Clear Display	0	0	0	0	0	0	0	0	0	1	Clears entire display and sets DD RAM address 0 in address counter.	1.64 ms
Return Home	0	0	0	0	0	0	0	0	1	-	Sets DD RAM address 0 as address counter. Also returns display being shifted to original position. DD RAM contents remain unchanged.	1.64 ms
Entry Mode Set	0	0	0	0	0	0	0	1	1/D	S	Sets cursor move direction and specifies shift of display. These operations are performed during data write and read.	40 μs
Display On/ Off Control	0	0	0	0	0	0	1	D	C	B	Sets On/Off of entire display (D), cursor On/Off (C), and blink of cursor position character (B).	40 μs
Cursor or Display Shift	0	0	0	0	0	1	S/C	R/L	-	-	Moves cursor and shifts display without changing DD RAM contents.	40 μs
Function Set	0	0	0	0	1	DL	N	F	-	-	Sets interface data length (DL), number of display lines (L), and character font (F).	40 μs
Set CG RAM Address	0	0	0	1		AGC					Sets CG RAM address. CG RAM data is sent and received after this setting.	40 μs
Set DD RAM Address	0	0	1		ADD						Sets DD RAM address. DD RAM data is sent and received after this setting.	40 μs
Read Busy Flag & Address	0	1	BF		AC						Reads Busy flag (BF) indicating internal operation is being performed and reads address counter contents.	40 μs
Write Data CG or DD RAM	1	0			Write Data						Writes data into DD or CG RAM.	40 μs
Read Data CG or DD RAM	1	1			Read Data						Reads data from DD or CG RAM.	40 μs

Notes:

1. *Execution times are maximum times when fcp or fosc is 250 kHz.*
2. *Execution time changes when frequency changes. Ex: When fcp or fosc is 270 kHz: 40 μs × 250 / 270 = 37 μs.*
3. Abbreviations:

DD RAM	Display data RAM
CG RAM	Character generator RAM
ACC	CG RAM address
ADD	DD RAM address, corresponds to cursor address
AC	Address counter used for both DD and CG RAM addresses

1/D = 1	Increment	1/D = 0	Decrement
S = 1	Accompanies display shift		
S/C = 1	Display shift;	S/C = 0	Cursor move
R/L = 1	Shift to the right;	R/L = 0	Shift to the left
DL = 1	8 bits, DL = 0: 4 bits		
N = 1	1 line, N = 0: 1 line		
F = 1	5 × 10 dots, F = 0: 5 × 7 dots		
BF = 1	Internal operation;	BF = 0	Can accept instruction

(Table 12-2 is extracted from this table.) As you see in the eighth row of Table 12-3, you can set the DD RAM address. It lets you put data at any location. The following shows how to set DD RAM address locations.

RS	R/W	DB7	DB6	DB5	DB4	DB3	DB2	DB1	DB0
0	0	1	A	A	A	A	A	A	A

Where AAAAAAA = 0000000 to 0100111 for line 1 and AAAAAAA = 1000000 to 1100111 for line 2.

The upper address range can go as high as 0100111 for the 40-character-wide LCD, while for the 20-character-wide LCD it goes up to 010011 (19 decimal = 10011 binary). Notice that the upper range 0100111 (binary) = 39 decimal, which corresponds to locations 0 to 39 for the LCDs of 40×2 size.

From the above discussion we can get the addresses of cursor positions for various sizes of LCDs. See Table 12-4 for the cursor addresses for common types of LCDs. Notice that all the addresses are in hex. See Example 12-1.

LCD Type	Line	Address Range					
16 × 2 LCD	Line 1:	80	81	82	83	through	8F
	Line 2:	C0	C1	C2	C3	through	CF
20 × 1 LCD	Line 1:	80	81	82	83	through	93
20 × 2 LCD	Line 1:	80	81	82	83	through	93
	Line 2:	C0	C1	C2	C3	through	D3
20 × 4 LCD	Line 1:	80	81	82	83	through	93
	Line 2:	C0	C1	C2	C3	through	D3
	Line 3:	94	95	96	97	through	A7
	Line 4:	D4	D5	D6	D7	through	E7
40 × 2 LCD	Line 1:	80	81	82	83	through	A7
	Line 2:	C0	C1	C2	C3	through	E7

Note: All data is in hex.

Table 12-4: Cursor Addresses for Some LCDs

Example 12-1

What is the cursor address for the following positions in a 20 × 4 LCD?
(a) Line 1, Column 1
(b) Line 2, Column 1
(c) Line 3, Column 2
(d) Line 4, Column 3

Solution:

(a) 80
(b) C0
(c) 95
(d) D6

LCD programming in C

Programs 12-5, 12-6, and 12-7 show how to interface an LCD to the AVR using C programming. The codes are modular to improve code clarity.

Program 12-5 shows how to use 8-bit data to interface an LCD to the AVR in C language.

```c
// YOU HAVE TO SET THE CPU FREQUENCY IN AVR STUDIO
// BECAUSE YOU ARE USING PREDEFINED DELAY FUNCTION

#include <avr/io.h>              //standard AVR header
#include <util/delay.h>          //delay header

#define   LCD_DPRT   PORTA       //LCD DATA PORT
#define   LCD_DDDR   DDRA        //LCD DATA DDR
#define   LCD_DPIN   PINA        //LCD DATA PIN
#define   LCD_CPRT   PORTB       //LCD COMMANDS PORT
#define   LCD_CDDR   DDRB        //LCD COMMANDS DDR
#define   LCD_CPIN   PINB        //LCD COMMANDS PIN
#define   LCD_RS   0             //LCD RS
#define   LCD_RW   1             //LCD RW
#define   LCD_EN   2             //LCD EN

//**********************************************************
void delay_us(unsigned int d)
{
   _delay_us(d);
}

//**********************************************************
void lcdCommand( unsigned char cmnd )
{
   LCD_DPRT = cmnd;               //send cmnd to data port
   LCD_CPRT &= ~ (1<<LCD_RS);     //RS = 0 for command
   LCD_CPRT &= ~ (1<<LCD_RW);     //RW = 0 for write
   LCD_CPRT |= (1<<LCD_EN);       //EN = 1 for H-to-L pulse
   delay_us(1);                   //wait to make enable wide
   LCD_CPRT &= ~ (1<<LCD_EN);     //EN = 0 for H-to-L pulse
   delay_us(100);                 //wait to make enable wide
}

//**********************************************************
void lcdData( unsigned char data )
{
   LCD_DPRT = data;               //send data to data port
   LCD_CPRT |= (1<<LCD_RS);       //RS = 1 for data
   LCD_CPRT &= ~ (1<<LCD_RW);     //RW = 0 for write
   LCD_CPRT |= (1<<LCD_EN);       //EN = 1 for H-to-L pulse
   delay_us(1);                   //wait to make enable wide
```

Program 12-5: Communicating with LCD Using 8-bit Data in C *(continued on next page)*

```
    LCD_CPRT &= ~ (1<<LCD_EN);           //EN = 0 for H-to-L pulse
    delay_us(100);                       //wait to make enable wide
}


//*********************************************************
void lcd_init()
{
  LCD_DDDR = 0xFF;
  LCD_CDDR = 0xFF;

  LCD_CPRT &=~(1<<LCD_EN);             //LCD_EN = 0
  delay_us(2000);                      //wait for init.
  lcdCommand(0x38);                    //init. LCD 2 line, 5 × 7 matrix
  lcdCommand(0x0E);                    //display on, cursor on
  lcdCommand(0x01);                    //clear LCD
  delay_us(2000);                      //wait
  lcdCommand(0x06);                    //shift cursor right
}


//*********************************************************
void lcd_gotoxy(unsigned char x, unsigned char y)
{
 unsigned char firstCharAdr[]={0x80,0xC0,0x94,0xD4};//Table 12-5
 lcdCommand(firstCharAdr[y-1] + x - 1);
 delay_us(100);
}


//*********************************************************
void lcd_print( char * str )
{
  unsigned char i = 0;
  while(str[i] !=0)
  {
    lcdData(str[i]);
    i++ ;
  }
}


//*********************************************************
int main(void)
{
        lcd_init();
        lcd_gotoxy(1,1);
        lcd_print("The world is but");
        lcd_gotoxy(1,2);
        lcd_print("one country");

        while(1);                        //stay here forever
        return 0;
}
```

Program 12-5: Communicating with LCD Using 8-bit Data in C

Program 12-6 shows how to use 4-bit data to interface an LCD to the AVR in C language.

```c
#include <avr/io.h>              //standard AVR header
#include <util/delay.h>          //delay header
#define  LCD_DPRT   PORTA        //LCD DATA PORT
#define  LCD_DDDR   DDRA         //LCD DATA DDR
#define  LCD_DPIN   PINA         //LCD DATA PIN
#define  LCD_CPRT   PORTB        //LCD COMMANDS PORT
#define  LCD_CDDR   DDRB         //LCD COMMANDS DDR
#define  LCD_CPIN   PINB         //LCD COMMANDS PIN
#define  LCD_RS   0              //LCD RS
#define  LCD_RW   1              //LCD RW
#define  LCD_EN   2              //LCD EN

void delay_us(int d)
{
   _delay_us(d);
}

void lcdCommand( unsigned char cmnd )
{
  LCD_DPRT = cmnd & 0xF0;        //send high nibble to D4-D7
  LCD_CPRT &= ~ (1<<LCD_RS);     //RS = 0 for command
  LCD_CPRT &= ~ (1<<LCD_RW);     //RW = 0 for write
  LCD_CPRT |= (1<<LCD_EN);       //EN = 1 for H-to-L pulse
  delay_us(1);                   //make EN pulse wider
  LCD_CPRT &= ~ (1<<LCD_EN);     //EN = 0 for H-to-L pulse
  delay_us(100);                 //wait
  LCD_DPRT = cmnd<<4;            //send low nibble to D4-D7
  LCD_CPRT |= (1<<LCD_EN);       //EN = 1 for H-to-L pulse
  delay_us(1);                   //make EN pulse wider
  LCD_CPRT &= ~ (1<<LCD_EN);     //EN = 0 for H-to-L pulse
  delay_us(100);                 //wait
}

void lcdData( unsigned char data )
{
  LCD_DPRT = data & 0xF0;        //send high nibble to D4-D7
  LCD_CPRT |= (1<<LCD_RS);       //RS = 1 for data
  LCD_CPRT &= ~ (1<<LCD_RW);     //RW = 0 for write
  LCD_CPRT |= (1<<LCD_EN);       //EN = 1 for H-to-L pulse
  delay_us(1);                   //make EN pulse wider
  LCD_CPRT &= ~ (1<<LCD_EN);     //EN = 0 for H-to-L pulse
  LCD_DPRT = data<<4;            //send low nibble to D4-D7
  LCD_CPRT |= (1<<LCD_EN);       //EN = 1 for H-to-L pulse
```

Program 12-6: Communicating with LCD Using 4-bit Data in C *(continued on next page)*

```c
      delay_us(1);                        //make EN pulse wider
      LCD_CPRT &= ~ (1<<LCD_EN);  //EN = 0 for H-to-L pulse
      delay_us(100);                      //wait
}


void lcd_init()
{
   LCD_DDDR = 0xFF;
   LCD_CDDR = 0xFF;
   LCD_CPRT &=~(1<<LCD_EN);        //LCD_EN = 0
   lcdCommand(0x33);               //send $33 for init.
   lcdCommand(0x32);               //send $32 for init.
   lcdCommand(0x28);               //init. LCD 2 line,5×7 matrix
   lcdCommand(0x0e);               //display on, cursor on
   lcdCommand(0x01);               //clear LCD
   delay_us(2000);
   lcdCommand(0x06);               //shift cursor right
}


void lcd_gotoxy(unsigned char x, unsigned char y)
{
   unsigned char firstCharAdr[]={ 0x80,0xC0,0x94,0xD4};
   lcdCommand(firstCharAdr[ y-1] + x - 1);
   delay_us(100);
}


void lcd_print(char * str )
{
   unsigned char i = 0;
   while(str[ i] !=0)
   {
      lcdData(str[ i] );
      i++ ;
   }
}


int main(void)
{
   lcd_init();
   lcd_gotoxy(1,1);
   lcd_print("The world is but");
   lcd_gotoxy(1,2);
   lcd_print("one country");
   while(1);                       //stay here forever
   return 0;
}
```

Program 12-6: Communicating with LCD Using 4-bit Data in C

Program 12-7 shows how to use 4-bit data to interface an LCD to the AVR in C language. It uses only a single port. Also there are some useful functions to print a string (array of chars) or to move the cursor to a specific location.

```c
#include <avr/io.h>                    //standard AVR header
#include <util/delay.h>                //delay header
#define  LCD_PRT   PORTA               //LCD DATA PORT
#define  LCD_DDR   DDRA                //LCD DATA DDR
#define  LCD_PIN   PINA                //LCD DATA PIN
#define  LCD_RS   0                    //LCD RS
#define  LCD_RW   1                    //LCD RW
#define  LCD_EN   2                    //LCD EN

void delay_us(int d)
{
   _delay_us(d);
}

void delay_ms(int d)
{
   _delay_ms(d);
}

void lcdCommand( unsigned char cmnd ){
   LCD_PRT = (LCD_PRT & 0x0F) | (cmnd & 0xF0);
   LCD_PRT &= ~ (1<<LCD_RS);           //RS = 0 for command
   LCD_PRT &= ~ (1<<LCD_RW);           //RW = 0 for write
   LCD_PRT |= (1<<LCD_EN);             //EN = 1 for H-to-L
   delay_us(1);                        //wait to make EN wider
   LCD_PRT &= ~ (1<<LCD_EN);           //EN = 0 for H-to-L

   delay_us(20);                       //wait

   LCD_PRT = (LCD_PRT & 0x0F) | (cmnd << 4);
   LCD_PRT |= (1<<LCD_EN);             //EN = 1 for H-to-L
   delay_us(1);                        //wait to make EN wider
   LCD_PRT &= ~ (1<<LCD_EN);           //EN = 0 for H-to-L
}

void lcdData( unsigned char data ){
   LCD_PRT = (LCD_PRT & 0x0F) | (data & 0xF0);
   LCD_PRT |= (1<<LCD_RS);             //RS = 1 for data
   LCD_PRT &= ~ (1<<LCD_RW);           //RW = 0 for write
   LCD_PRT |= (1<<LCD_EN);             //EN = 1 for H-to-L
```

Program 12-7: Communicating with LCD Using 4-bit Data in C *(continued on next page)*

```c
    delay_us(1);                        //wait to make EN wider
    LCD_PRT &= ~ (1<<LCD_EN);           //EN = 0 for H-to-L


    LCD_PRT = (LCD_PRT & 0x0F) | (data << 4);
    LCD_PRT |= (1<<LCD_EN);             //EN = 1 for H-to-L
    delay_us(1);                        //wait to make EN wider
    LCD_PRT &= ~ (1<<LCD_EN);           //EN = 0 for H-to-L
}


void lcd_init(){
    LCD_DDR = 0xFF;                     //LCD port is output


    LCD_PRT &=~(1<<LCD_EN);             //LCD_EN = 0
    delay_us(2000);                     //wait for stable power
    lcdCommand(0x33);                   //$33 for 4-bit mode
    delay_us(100);                      //wait
    lcdCommand(0x32);                   //$32 for 4-bit mode
    delay_us(100);                      //wait
    lcdCommand(0x28);                   //$28 for 4-bit mode
    delay_us(100);                      //wait
    lcdCommand(0x0e);                   //display on, cursor on
    delay_us(100);                      //wait
    lcdCommand(0x01);                   //clear LCD
    delay_us(2000);                     //wait
    lcdCommand(0x06);                   //shift cursor right
    delay_us(100);
}


void lcd_gotoxy(unsigned char x, unsigned char y)
{   //Table 12-5
    unsigned char firstCharAdr[] = { 0x80, 0xC0, 0x94, 0xD4};

    lcdCommand(firstCharAdr[ y-1] + x - 1);
    delay_us(100);
}


void lcd_print( char * str )
{
    unsigned char i = 0;

    while(str[ i] !=0)
    {
        lcdData(str[ i] );
        i++;
    }
}
```

Program 12-7: Communicating with LCD Using 4-bit Data in C

```
int main(void)
{
  lcd_init();
  while(1)
  {                              //stay here forever
        lcd_gotoxy(1,1);
        lcd_print("The world is but");
        lcd_gotoxy(1,2);
        lcd_print("one country     ");
        delay_ms(1000);
        lcd_gotoxy(1,1);
        lcd_print("and mankind its ");
        lcd_gotoxy(1,2);
        lcd_print("citizens        ");
        delay_ms(1000);
  }
  return 0;
}
```

Program 12-7: Communicating with LCD Using 4-bit Data in C *(cont. from previous page)*

You can purchase the LCD expansion board of the MDE AVR trainer from the following websites:

www.digilentinc.com
www.MicroDigitalEd.com

The LCDs can be purchased from the following websites:

www.digikey.com
www.jameco.com
www.elexp.com

Review Questions

1. The RS pin is an _____ (input, output) pin for the LCD.
2. The E pin is an _____ (input, output) pin for the LCD.
3. The E pin requires an _____ (H-to-L, L-to-H) pulse to latch in information at the data pins of the LCD.
4. For the LCD to recognize information at the data pins as data, RS must be set to _____ (high, low).
5. What is the 0x06 command ?
6. Which of the following commands takes more than 100 microseconds to run?
   (a) Shift cursor left
   (b) Shift cursor right
   (c) Set address location of DDRAM
   (d) Clear screen
7. Which of the following initialization commands initializes an LCD for 5 × 7 matrix characters in 8-bit operating mode?
   (a) 0x38, 0x0E, 0x0, 0x06
   (b) 0x0E, 0x0, 0x06
   (c) 0x33, 0x32, 0x28, 0x0E, 0x01, 0x06
   (d) 0x01, 0x06
8. Which of the following initialization commands initializes an LCD for 5 × 7 matrix characters in 4-bit operating mode?
   (a) 0x38, 0x0E, 0x0, 0x06
   (b) 0x0E, 0x0, 0x06
   (c) 0x33, 0x32, 0x28, 0x0E, 0x01, 0x06
   (d) 0x01, 0x06
9. Which of the following is the address of the second column of the second row in a 2 × 20 LCD?
   (a) 0x80
   (b) 0x81
   (c) 0xC0
   (d) 0xC1
10. Which of the following is the address of the second column of the second row in a 4 × 20 LCD?
    (a) 0x80
    (b) 0x81
    (c) 0xC0
    (d) 0xC1
11. Which of the following is the address of the first column of the second row in a 4 × 20 LCD?
    (a) 0x80
    (b) 0x81
    (c) 0xC0
    (d) 0xC1

SECTION 12.2: KEYBOARD INTERFACING

Keyboards and LCDs are the most widely used input/output devices in microcontrollers such as the AVR and a basic understanding of them is essential. In the previous section, we discussed how to interface an LCD with an AVR using some examples. In this section, we first discuss keyboard fundamentals, along with key press and key detection mechanisms. Then we show how a keyboard is interfaced to an AVR.

Interfacing the keyboard to the AVR

At the lowest level, keyboards are organized in a matrix of rows and columns. The CPU accesses both rows and columns through ports; therefore, with two 8-bit ports, an 8 × 8 matrix of keys can be connected to a microcontroller. When a key is pressed, a row and a column make a contact; otherwise, there is no connection between rows and columns. In x86 PC keyboards, a single microcontroller takes care of hardware and software interfacing of the keyboard. In such systems, it is the function of programs stored in the Flash of the microcontroller to scan the keys continuously, identify which one has been activated, and present it to the motherboard. In this section we look at the mechanism by which the AVR scans and identifies the key.

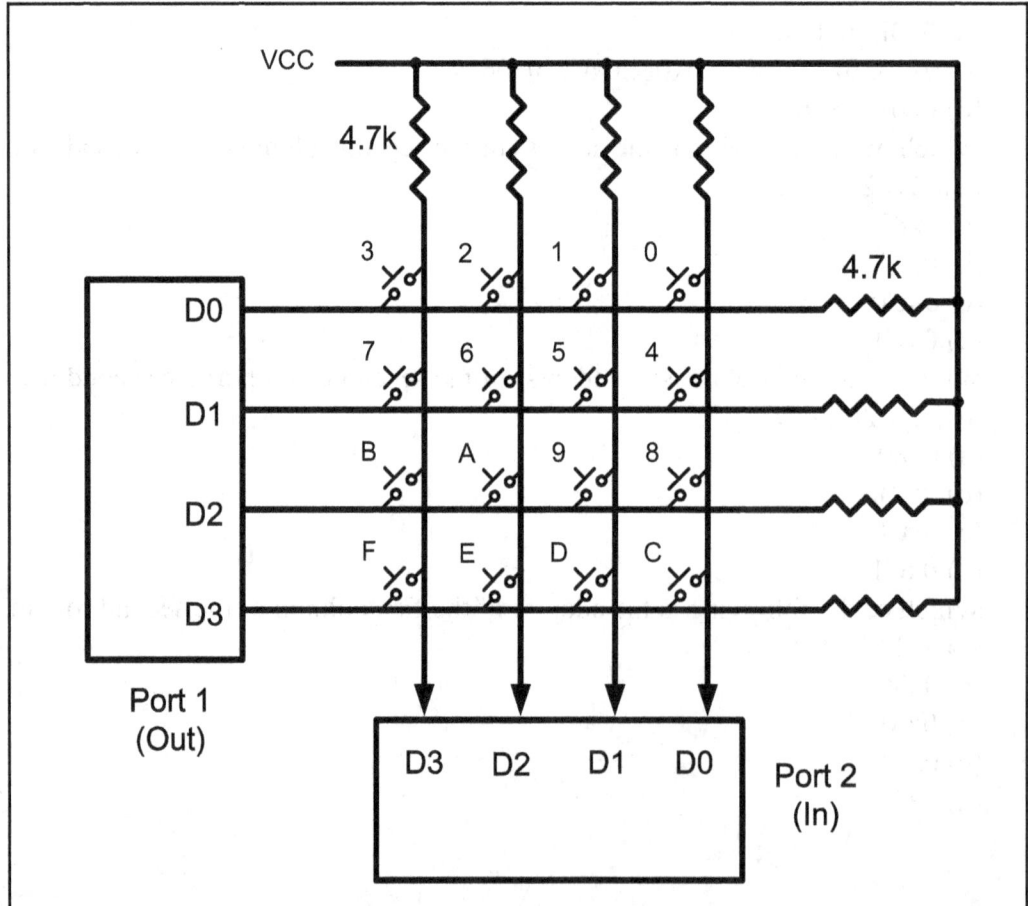

Figure 12-7. Matrix Keyboard Connection to Ports

Scanning and identifying the key

Figure 12-7 shows a 4 × 4 matrix connected to two ports. The rows are connected to an output port and the columns are connected to an input port. If no key has been pressed, reading the input port will yield 1s for all columns since they are all connected to high (VCC). If all the rows are grounded and a key is pressed, one of the columns will have 0 since the key pressed provides the path to ground. It is the function of the microcontroller to scan the keyboard continuously to detect and identify the key pressed. How this is done is explained next.

Grounding rows and reading the columns

To detect a pressed key, the microcontroller grounds all rows by providing 0 to the output latch, and then it reads the columns. If the data read from the columns is D3–D0 = 1111, no key has been pressed and the process continues until a key press is detected. However, if one of the column bits has a zero, this means that a key press has occurred. For example, if D3–D0 = 1101, this means that a key in the D1 column has been pressed. After a key press is detected, the microcontroller will go through the process of identifying the key. Starting with the top row, the microcontroller grounds it by providing a low to row D0 only; then it reads the columns. If the data read is all 1s, no key in that row is activated and the process is moved to the next row. It grounds the next row, reads the columns, and checks for any zero. This process continues until the row is identified. After identification of the row in which the key has been pressed, the next task is to find out which column the pressed key belongs to. This should be easy since the microcontroller knows at any time which row and column are being accessed. Look at Example 12-2.

Example 12-2

From Figure 12-7 identify the row and column of the pressed key for each of the following.
(a) D3–D0 = 1110 for the row, D3–D0 = 1011 for the column
(b) D3–D0 = 1101 for the row, D3–D0 = 0111 for the column

Solution:

From Figure 12-7 the row and column can be used to identify the key.
(a) The row belongs to D0 and the column belongs to D2; therefore, key number 2 was pressed.
(b) The row belongs to D1 and the column belongs to D3; therefore, key number 7 was pressed.

Program 12-8 is the AVR Assembly language program for detection and identification of key activation. In this program, it is assumed that PC0–PC3 are connected to the rows and PC4–PC7 are connected to the columns.

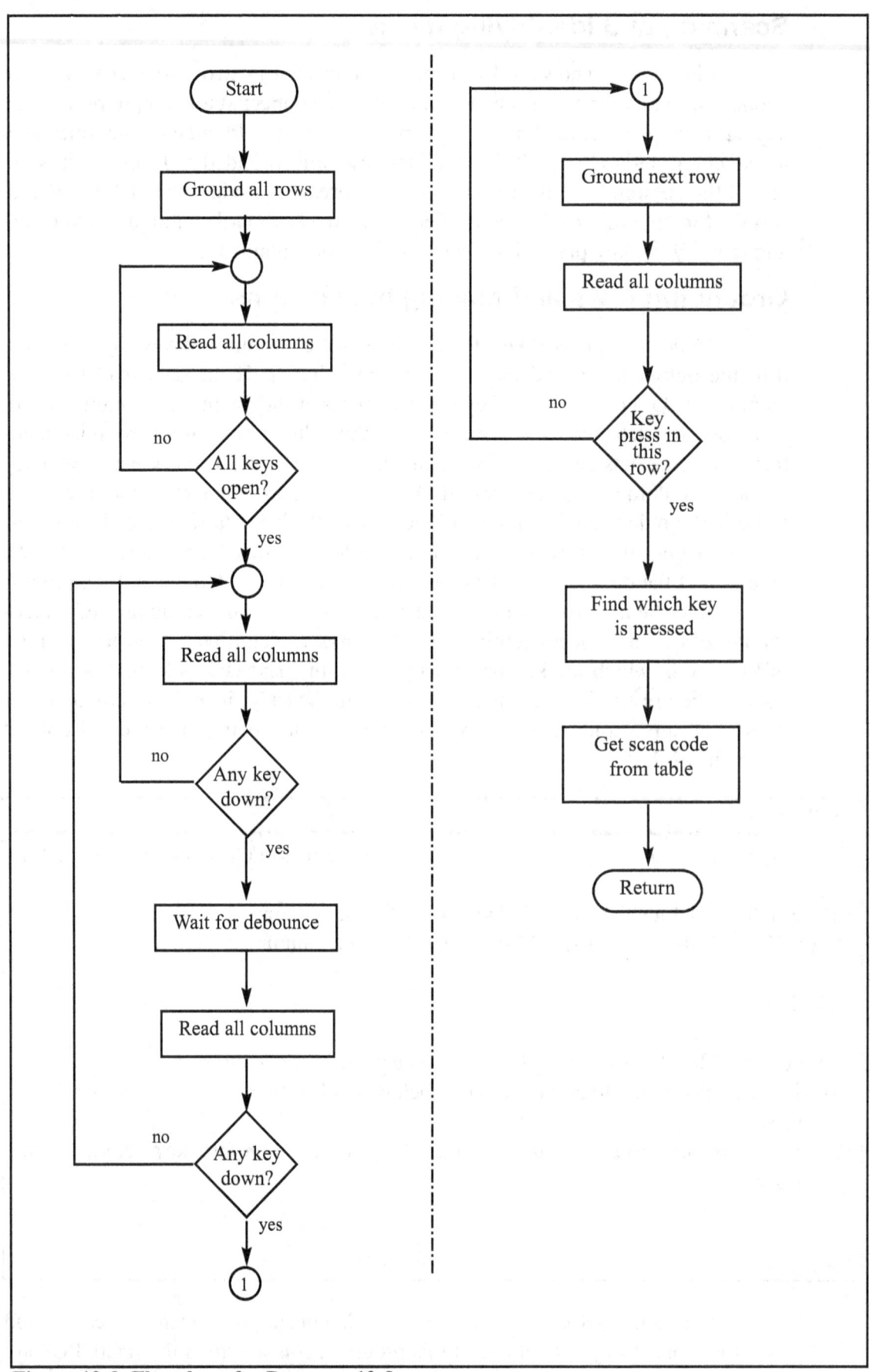

Figure 12-8. Flowchart for Program 12-8

Program 12-8 goes through the following four major stages (Figure 12-8 flowcharts this process):

1. To make sure that the preceding key has been released, 0s are output to all rows at once, and the columns are read and checked repeatedly until all the columns are high. When all columns are found to be high, the program waits for a short amount of time before it goes to the next stage of waiting for a key to be pressed.

2. To see if any key is pressed, the columns are scanned over and over in an infinite loop until one of them has a 0 on it. Remember that the output latches connected to rows still have their initial zeros (provided in stage 1), making them grounded. After the key press detection, the microcontroller waits 20 ms for the bounce and then scans the columns again. This serves two functions: (a) it ensures that the first key press detection was not an erroneous one due to a spike noise, and (b) the 20-ms delay prevents the same key press from being interpreted as a multiple key press. Look at Figure 12-9. If after the 20-ms delay the key is still pressed, it goes to the next stage to detect which row it belongs to; otherwise, it goes back into the loop to detect a real key press.

3. To detect which row the key press belongs to, the microcontroller grounds one row at a time, reading the columns each time. If it finds that all columns are high, this means that the key press cannot belong to that row; therefore, it grounds the next row and continues until it finds the row the key press belongs to. Upon finding the row that the key press belongs to, it sets up the starting address for the look-up table holding the scan codes (or the ASCII value) for that row and goes to the next stage to identify the key.

4. To identify the key press, the microcontroller rotates the column bits, one bit at a time, into the carry flag and checks to see if it is low. Upon finding the zero, it pulls out the ASCII code for that key from the look-up table; otherwise, it increments the pointer to point to the next element of the look-up table.

While the key press detection is standard for all keyboards, the process for determining which key is pressed varies. The look-up table method shown in Program 12-8 can be modified to work with any matrix up to 8×8. Example 12-3 shows keypad programming in C.

There are IC chips such as National Semiconductor's MM74C923 that incorporate keyboard scanning and decoding all in one chip. Such chips use combinations of counters and logic gates (no microcontroller) to implement the underlying concepts presented in Program 12-8.

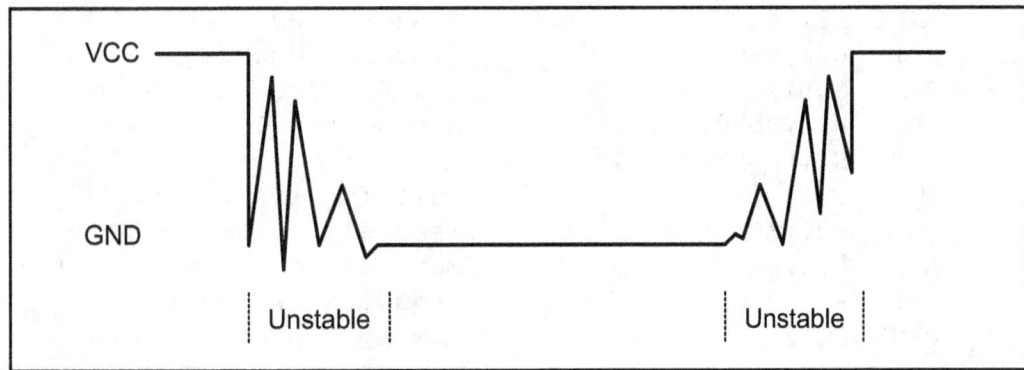

Figure 12-9. Keyboard Debounce

CHAPTER 12: LCD AND KEYBOARD INTERFACING 455

```
;Keyboard Program. This program sends the ASCII code
;for pressed key to Port D
;PC0-PC3 connected to rows PC4-PC7 connected to columns

.INCLUDE "M32DEF.INC"
.EQU  KEY_PORT = PORTC
.EQU  KEY_PIN = PINC
.EQU  KEY_DDR = DDRC
      LDI   R20,HIGH(RAMEND)
      OUT   SPH,R20
      LDI   R20,LOW(RAMEND)       ;init. stack pointer
      OUT   SPL,R20
      LDI   R21,0xFF
      OUT   DDRD,R21
      LDI   R20,0xF0
      OUT   KEY_DDR,R20
GROUND_ALL_ROWS:
      LDI   R20,0x0F
      OUT   KEY_PORT,R20
WAIT_FOR_RELEASE:
      NOP
      IN    R21,KEY_PIN           ;read key pins
      ANDI  R21,0x0F              ;mask unused bits
      CPI   R21,0x0F              ;(equal if no key)
      BRNE  WAIT_FOR_RELEASE      ;do again until keys released
WAIT_FOR_KEY:
      NOP                         ;wait for sync. circuit
      IN    R21,KEY_PIN           ;read key pins
      ANDI  R21,0x0F              ;mask unused bits
      CPI   R21,0x0F              ;(equal if no key)
      BREQ  WAIT_FOR_KEY          ;do again until a key pressed
      CALL  WAIT15MS              ;wait 15 ms
      IN    R21,KEY_PIN           ;read key pins
      ANDI  R21,0x0F              ;mask unused bits
      CPI   R21,0x0F              ;(equal if no key)
      BREQ  WAIT_FOR_KEY          ;do again until a key pressed
      LDI   R21,0b01111111        ;ground row 0
      OUT   KEY_PORT,R21
      NOP                         ;wait for sync. circuit
      IN    R21,KEY_PIN           ;read all columns
      ANDI  R21,0x0F              ;mask unused bits
      CPI   R21,0x0F              ;(equal if no key)
      BRNE  COL1                  ;row 0, find the colum
      LDI   R21,0b10111111        ;ground row 1
      OUT   KEY_PORT,R21
      NOP                         ;wait for sync. circuit
      IN    R21,KEY_PIN           ;read all columns
      ANDI  R21,0x0F              ;mask unused bits
      CPI   R21,0x0F              ;(equal if no key)
      BRNE  COL2                  ;row 1, find the colum
```

Program 12-8: Keyboard Interfacing Program *(continued on next page)*

```
            LDI    R21,0b11011111        ;ground row 2
            OUT    KEY_PORT,R21
            NOP                          ;wait for sync. circuit
            IN     R21,KEY_PIN           ;read all columns
            ANDI   R21,0x0F              ;mask unused bits
            CPI    R21,0x0F              ;(equal if no key)
            BRNE   COL3                  ;row 2, find the colum
            LDI    R21,0b11101111        ;ground row 3
            OUT    KEY_PORT,R21
            NOP                          ;wait for sync. circuit
            IN     R21,KEY_PIN           ;read all columns
            ANDI   R21,0x0F              ;mask unused bits
            CPI    R21,0x0F              ;(equal if no key)
            BRNE   COL4                  ;row 3, find the colum
COL1:
            LDI    R30,LOW(KCODE0<<1)
            LDI    R31,HIGH(KCODE0<<1)
            RJMP   FIND
COL2:
            LDI    R30,LOW(KCODE1<<1)
            LDI    R31,HIGH(KCODE1<<1)
            RJMP   FIND
COL3:
            LDI    R30,LOW(KCODE2<<1)
            LDI    R31,HIGH(KCODE2<<1)
            RJMP   FIND
COL4:
            LDI    R30,LOW(KCODE3<<1)
            LDI    R31,HIGH(KCODE3<<1)
            RJMP   FIND
FIND:
            LSR    R21
            BRCC   MATCH                 ;if Carry is low go to match
            LPM    R20,Z+                ;INC Z
            RJMP   FIND
MATCH:
            LPM    R20,Z
            OUT    PORTD,R20
            RJMP   GROUND_ALL_ROWS
WAIT15MS:                               ;place a code to wait 15 ms
                                        ;here
            RET

.ORG 0x300

KCODE0:       .DB  '0','1','2','3'      ;ROW 0
KCODE1:       .DB  '4','5','6','3'      ;ROW 1
KCODE2:       .DB  '8','9','A','B'      ;ROW 2
KCODE3:       .DB  'C','D','E','F'      ;ROW 3
```

Program 12-8. Keyboard Interfacing Program *(continued from previous page)*

Example 12-3

Write a C program to read the keypad and send the result to Port D.
PC0–PC3 connected to columns
PC4–PC7 connected to rows

Solution:

```c
#include <avr/io.h>                      //standard AVR header
#include <util/delay.h>                  //delay header


#define     KEY_PRT   PORTC              //keyboard PORT
#define     KEY_DDR   DDRC               //keyboard DDR
#define     KEY_PIN   PINC               //keyboard PIN

void delay_ms(unsigned int d)
{
  _delay_ms(d);
}

unsigned char keypad[4][4]  ={ '0','1','2','3',
                               '4','5','6','7',
                               '8','9','A','B',
                               'C','D','E','F'};

int main(void)
{
  unsigned char colloc, rowloc;

  //keyboard routine. This sends the ASCII
  //code for pressed key to  port c
  DDRD = 0xFF;
  KEY_DDR = 0xF0;                        //
  KEY_PRT = 0xFF;
  while(1)                               //repeat forever
  {
    do
    {
      KEY_PRT &= 0x0F;                   //ground all rows at once
      colloc = (KEY_PIN & 0x0F);         //read the columns
    } while(colloc != 0x0F);             //check until all keys released

    do
    {
      do
      {
        delay_ms(20);                    //call delay
        colloc =(KEY_PIN&0x0F);          //see if any key is pressed
      } while(colloc == 0x0F);           //keep checking for key press

      delay_ms(20);                      //call delay for debounce
      colloc = (KEY_PIN & 0x0F);         //read columns
    } while(colloc == 0x0F);             //wait for key press

    while(1)
    {
      KEY_PRT = 0xEF;                    //ground row 0
      colloc = (KEY_PIN & 0x0F);         //read the columns
```

458

Example 12-3 *(continued from previous page)*

```
        if(colloc != 0x0F)              //column detected
        {
          rowloc = 0;                   //save row location
          break;                        //exit while loop
        }

        KEY_PRT = 0xDF;                 //ground row 1
        colloc = (KEY_PIN & 0x0F);      //read the columns

        if(colloc != 0x0F)              //column detected
        {
          rowloc = 1;                   //save row location
          break;                        //exit while loop
        }

        KEY_PRT = 0xBF;                 //ground row 2
         colloc = (KEY_PIN & 0x0F);     //read the columns
        if(colloc != 0x0F)              //column detected
        {
          rowloc = 2;                   //save row location
          break;                        //exit while loop
        }

        KEY_PRT = 0x7F;                 //ground row 3
        colloc = (KEY_PIN & 0x0F);      //read the columns
        rowloc = 3;                     //save row location
        break;                          //exit while loop
      }

    //check column and send result to Port D
    if(colloc == 0x0E)
      PORTD = (keypad[ rowloc][ 0] );
    else if(colloc == 0x0D)
      PORTD = (keypad[ rowloc][ 1] );
    else if(colloc == 0x0B)
      PORTD = (keypad[ rowloc][ 2] );
    else
      PORTD = (keypad[ rowloc][ 3] );
  }
  return 0 ;

}
```

Review Questions

1. True or false. To see if any key is pressed, all rows are grounded.
2. If D3–D0 = 0111 is the data read from the columns, which column does the pressed key belong to?
3. True or false. Key press detection and key identification require two different processes.
4. In Figure 12-7, if the rows are D3–D0 = 1110 and the columns are D3–D0 = 1110, which key is pressed?
5. True or false. To identify the pressed key, one row at a time is grounded.

SUMMARY

This chapter showed how to interface real-world devices such as LCDs and keypads to the AVR. First, we described the operation modes of LCDs, and then described how to program the LCD by sending data or commands to it via its interface to the AVR.

Keyboards are one of the most widely used input devices for AVR projects. This chapter also described the operation of keyboards, including key press and detection mechanisms. Then the AVR was shown interfacing with a keyboard. AVR programs were written to return the ASCII code for the pressed key.

PROBLEMS

SECTION 12.1: LCD INTERFACING

1. The LCD discussed in this section has ____ pins.
2. Describe the function of pins E, R/W, and RS in the LCD.
3. What is the difference between the V_{CC} and V_{EE} pins on the LCD?
4. "Clear LCD" is a _____ (command code, data item) and its value is ___ hex.
5. What is the hex value of the command code for "display on, cursor on"?
6. Give the state of RS, E, and R/W when sending a command code to the LCD.
7. Give the state of RS, E, and R/W when sending data character 'Z' to the LCD.
8. Which of the following is needed on the E pin in order for a command code (or data) to be latched in by the LCD?
   (a) H-to-L pulse (b) L-to-H pulse
9. True or false. For the above to work, the value of the command code (data) must already be at the D0–D7 pins.
10. There are two methods of sending commands and data to the LCD: (1) 4-bit mode or (2) 8-bit mode. Explain the difference and the advantages and disadvantages of each method.
11. For a 16 × 2 LCD, the location of the last character of line 1 is 8FH (its command code). Show how this value was calculated.
12. For a 16 × 2 LCD, the location of the first character of line 2 is C0H (its command code). Show how this value was calculated.
13. For a 20 × 2 LCD, the location of the last character of line 2 is 93H (its command code). Show how this value was calculated.
14. For a 20 × 2 LCD, the location of the third character of line 2 is C2H (its command code). Show how this value was calculated.
15. For a 40 × 2 LCD, the location of the last character of line 1 is A7H (its command code). Show how this value was calculated.
16. For a 40 × 2 LCD, the location of the last character of line 2 is E7H (its command code). Show how this value was calculated.
17. Show the value (in hex) for the command code for the 10th location, line 1 on a 20 × 2 LCD. Show how you got your value.
18. Show the value (in hex) for the command code for the 20th location, line 2 on

a 40×2 LCD. Show how you got your value.

SECTION 12.2: KEYBOARD INTERFACING

19. In reading the columns of a keyboard matrix, if no key is pressed we should get all _____ (1s, 0s).
20. In the 4×4 keyboard interfacing, to detect the key press, which of the following is grounded?
    (a) all rows   (b) one row at time   (c) both (a) and (b)
21. In the 4×4 keyboard interfacing, to identify the key pressed, which of the following is grounded?
    (a) all rows   (b) one row at time   (c) both (a) and (b)
22. For the 4×4 keyboard interfacing (Figure 12-7), indicate the column and row for each of the following.
    (a) D3–D0 = 0111   (b) D3–D0 = 1110
23. Indicate the steps to detect the key press.
24. Indicate the steps to identify the key pressed.
25. Indicate an advantage and a disadvantage of using an IC chip for keyboard scanning and decoding instead of using a microcontroller.
26. What is the best compromise for the answer to Problem 25?

ANSWERS TO REVIEW QUESTIONS

SECTION 12.1: LCD INTERFACING

1. Input
2. Input
3. H-to-L
4. High
5. Shift cursor to right
6. d
7. a
8. c
9. d
10. d
11. c

SECTION 12.2: KEYBOARD INTERFACING

1. True
2. Column 3
3. True
4. 0
5. True

CHAPTER 13

ADC, DAC, AND SENSOR INTERFACING

OBJECTIVES

Upon completion of this chapter, you will be able to:

>> Discuss the ADC (analog-to-digital converter) section of the AVR chip
>> Interface temperature sensors to the AVR
>> Explain the process of data acquisition using ADC
>> Describe factors to consider in selecting an ADC chip
>> Program the AVR's ADC in C and Assembly
>> Describe the basic operation of a DAC (digital-to-analog converter) chip
>> Interface a DAC chip to the AVR
>> Program DAC chips in AVR C and Assembly
>> Explain the function of precision IC temperature sensors
>> Describe signal conditioning and its role in data acquisition

This chapter explores more real-world devices such as ADCs (analog-to-digital converters), DACs (digital-to-analog converters), and sensors. We will also explain how to interface the AVR to these devices. In Section 13.1, we describe analog-to-digital converter (ADC) chips. We will program the ADC portion of the AVR chip in Section 13.2. In Section 13.3, we show the interfacing of sensors and discuss the issue of signal conditioning. The characteristics of DAC chips are discussed in Section 13.4.

SECTION 13.1: ADC CHARACTERISTICS

This section will explore ADC generally. First, we describe some general aspects of the ADC itself, then focus on the functionality of some important pins in ADC.

ADC devices

Analog-to-digital converters are among the most widely used devices for data acquisition. Digital computers use binary (discrete) values, but in the physical world everything is analog (continuous). Temperature, pressure (wind or liquid), humidity, and velocity are a few examples of physical quantities that we deal with every day. A physical quantity is converted to electrical (voltage, current) signals using a device called a *transducer*. Transducers are also referred to as *sensors*. Sensors for temperature, velocity, pressure, light, and many other natural quantities produce an output that is voltage (or current). Therefore, we need an analog-to-digital converter to translate the analog signals to digital numbers so that the microcontroller can read and process them. See Figures 13-1 and 13-2.

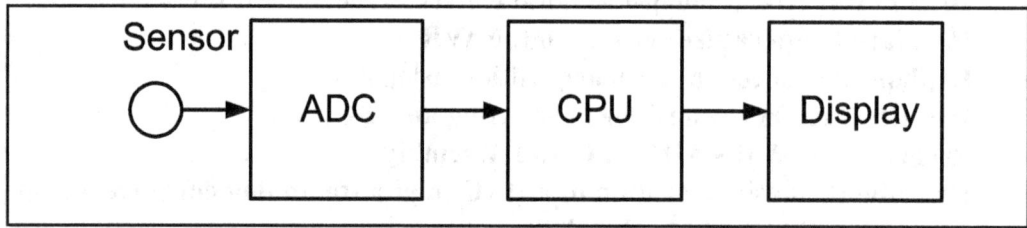

Figure 13-1. Microcontroller Connection to Sensor via ADC

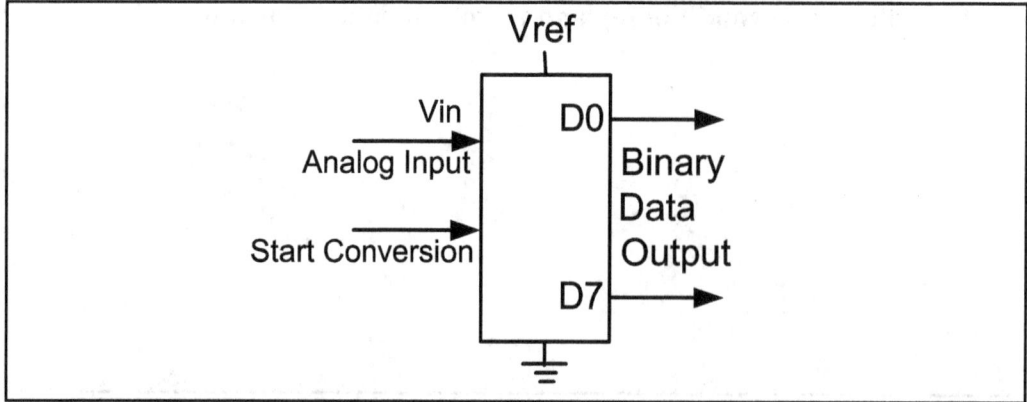

Figure 13-2. An 8-bit ADC Block Diagram

Table 13-1: Resolution versus Step Size for ADC (V_{ref} = 5 V)

n-bit	Number of steps	Step size (mV)
8	256	5/256 = 19.53
10	1024	5/1024 = 4.88
12	4096	5/4096 = 1.2
16	65,536	5/65,536 = 0.076

Notes: V_{CC} = 5 V

Step size (resolution) is the smallest change that can be discerned by an ADC.

Some of the major characteristics of the ADC

Resolution

The ADC has n-bit resolution, where n can be 8, 10, 12, 16, or even 24 bits. Higher-resolution ADCs provide a smaller step size, where *step size* is the smallest change that can be discerned by an ADC. Some widely used resolutions for ADCs are shown in Table 13-1. Although the resolution of an ADC chip is decided at the time of its design and cannot be changed, we can control the step size with the help of what is called V_{ref}. This is discussed below.

Conversion time

In addition to resolution, conversion time is another major factor in judging an ADC. *Conversion time* is defined as the time it takes the ADC to convert the analog input to a digital (binary) number. The conversion time is dictated by the clock source connected to the ADC in addition to the method used for data conversion and technology used in the fabrication of the ADC chip such as MOS or TTL technology.

V_{ref}

V_{ref} is an input voltage used for the reference voltage. The voltage connected to this pin, along with the resolution of the ADC chip, dictate the step size. For an 8-bit ADC, the step size is V_{ref}/256 because it is an 8-bit ADC, and 2 to the power of 8 gives us 256 steps. See Table 13-1. For example, if the analog input range needs to be 0 to 4 volts, V_{ref} is connected to 4 volts. That gives 4 V/256 = 15.62 mV for the step size of an 8-bit ADC. In another case, if we need a step size

Table 13-2: V_{ref} Relation to V_{in} Range for an 8-bit ADC

V_{ref} (V)	V_{in} Range (V)	Step Size (mV)
5.00	0 to 5	5/256 = 19.53
4.0	0 to 4	4/256 = 15.62
3.0	0 to 3	3/256 = 11.71
2.56	0 to 2.56	2.56/256 = 10
2.0	0 to 2	2/256 = 7.81
1.28	0 to 1.28	1.28/256 = 5
1	0 to 1	1/256 = 3.90

Step size is V_{ref} / 256

Table 13-3: V_{ref} Relation to V_{in} Range for an 10-bit ADC

V_{ref} (V)	V_{in} (V)	Step Size (mV)
5.00	0 to 5	5/1024 = 4.88
4.096	0 to 4.096	4.096/1024 = 4
3.0	0 to 3	3/1024 = 2.93
2.56	0 to 2.56	2.56/1024 = 2.5
2.048	0 to 2.048	2.048/1024 = 2
1.28	0 to 1.28	1/1024 = 1.25
1.024	0 to 1.024	1.024/1024 = 1

of 10 mV for an 8-bit ADC, then V_{ref} = 2.56 V, because 2.56 V/256 = 10 mV. For the 10-bit ADC, if the V_{ref} = 5V, then the step size is 4.88 mV as shown in Table 13-1. Tables 13-2 and 13-3 show the relationship between the V_{ref} and step size for the 8- and 10-bit ADCs, respectively. In some applications, we need the differential reference voltage where V_{ref} = V_{ref} (+) − V_{ref} (−). Often the V_{ref} (−) pin is connected to ground and the V_{ref} (+) pin is used as the V_{ref}.

Digital data output

In an 8-bit ADC we have an 8-bit digital data output of D0–D7, while in the 10-bit ADC the data output is D0–D9. To calculate the output voltage, we use the following formula:

$$D_{out} = \frac{V_{in}}{step\ size}$$

where D_{out} = digital data output (in decimal), V_{in} = analog input voltage, and step size (resolution) is the smallest change, which is V_{ref}/256 for an 8-bit ADC. See Example 13-1. This data is brought out of the ADC chip either one bit at a time (serially), or in one chunk, using a parallel line of outputs. This is discussed next.

Example 13-1

For an 8-bit ADC, we have V_{ref} = 2.56 V. Calculate the D0–D7 output if the analog input is: (a) 1.7 V, and (b) 2.1 V.
Solution:

Because the step size is 2.56/256 = 10 mV, we have the following:
(a) D_{out} = 1.7 V/10 mV=170 in decimal, which gives us 10101010 in binary for D7–D0.
(b) D_{out}= 2.1 V/10 mV = 210 in decimal, which gives us 11010010 in binary for D7–D0.

Parallel versus serial ADC

The ADC chips are either parallel or serial. In parallel ADC, we have 8 or more pins dedicated to bringing out the binary data, but in serial ADC we have only one pin for data out. That means that inside the serial ADC, there is a parallel-in-serial-out shift register responsible for sending out the binary data one bit at a time. The D0–D7 data pins of the 8-bit ADC provide an 8-bit parallel data path between the ADC chip and the CPU. In the case of the 16-bit parallel ADC chip, we need 16 pins for the data path. In order to save pins, many 12- and 16-bit ADCs use pins D0–D7 to send out the upper and lower bytes of the binary data. In recent years, for many applications where space is a critical issue, using such a large number of pins for data is not feasible. For this reason, serial devices such as the serial ADC are becoming widely used. While the serial ADCs use fewer pins and their smaller packages take much less space on the printed circuit board, more CPU time is needed to get the converted data from the ADC because the CPU must get data one bit at a time, instead of in one single read operation as with the parallel ADC. ADC848 is an example of a parallel ADC with 8 pins for the data output, while the MAX1112 is an example of a serial ADC with a single pin for D_{out}. Figures 13-3 and 13-4 show the block diagram for ADC848 and MAX1112.

Analog input channels

Many data acquisition applications need more than one ADC. For this reason, we see ADC chips with 2, 4, 8, or even 16 channels on a single chip. Multiplexing of analog inputs is widely used as shown in the ADC848 and MAX1112. In these chips, we have 8 channels of analog inputs, allowing us to monitor multiple quantities such as temperature, pressure, heat, and so on. AVR microcontroller chips come with up to 16 ADC channels.

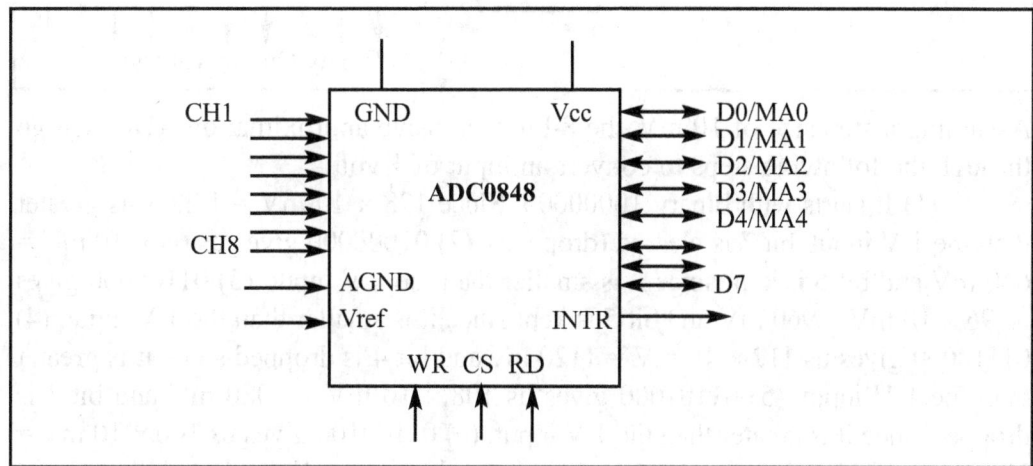

Figure 13-3. ADC0848 Parallel ADC Block Diagram

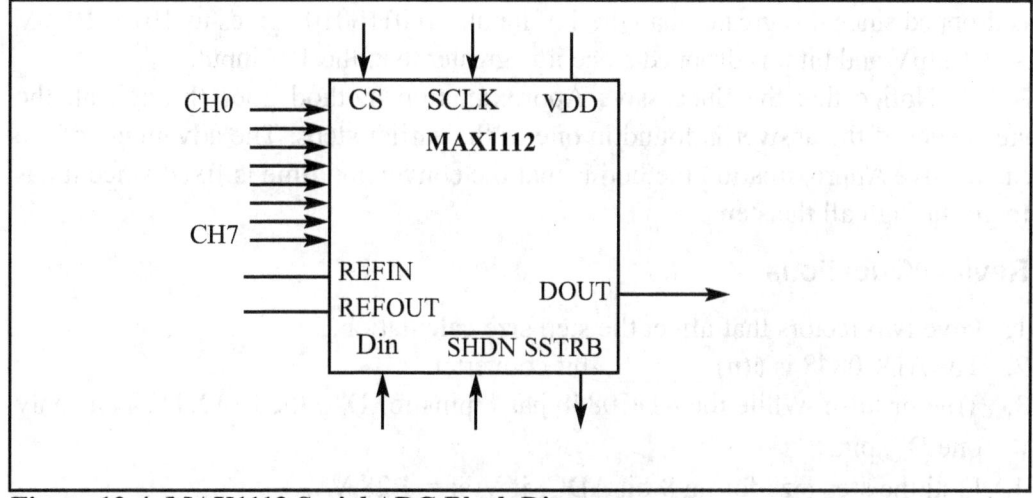

Figure 13-4. MAX1112 Serial ADC Block Diagram

Start conversion and end-of-conversion signals

The fact that we have multiple analog input channels and a single digital output register creats the need for start conversion (SC) and end-of-conversion (EOC) signals. When SC is activated, the ADC starts converting the analog input value of Vin to an n-bit digital number. The amount of time it takes to convert varies depending on the conversion method as was explained earlier. When the data conversion is complete, the end-of-conversion signal notifies the CPU that the converted data is ready to be picked up.

Successive Approximation ADC

Successive Approximation is a widely used method of converting an analog input to digital output. It has three main components: (a) successive approximation register (SAR), (b) comparator, and (c) control unit. See the figure below.

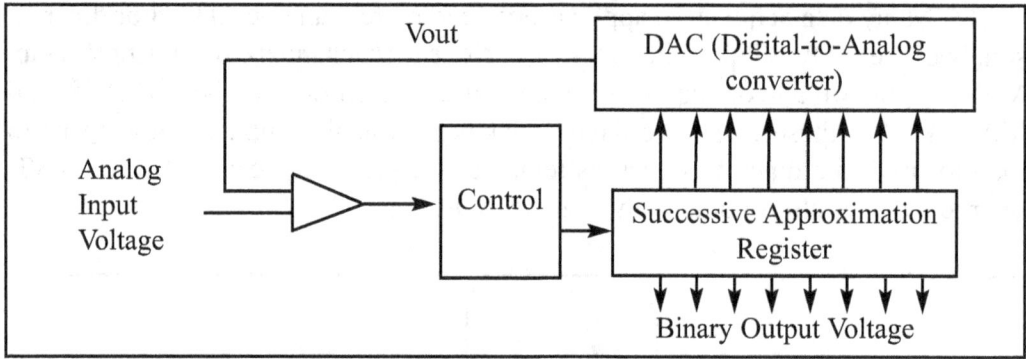

Assuming a step size of 10 mV, the 8-bit successive approximation ADC will go through the following steps to convert an input of 1 volt:

(1) It starts with binary 10000000. Since 128×10 mV = 1.28 V is greater than the 1 V input, bit 7 is cleared (dropped). (2) 01000000 gives us 64×10 mV = 640 mV and bit 6 is kept since it is smaller than the 1 V input. (3) 01100000 gives us 96×10 mV = 960 mV and bit 5 is kept since it is smaller than the 1 V input, (4) 01110000 gives us 112×10 mV = 1120 mv and bit 4 is dropped since it is greater than the 1 V input. (5) 01101000 gives us 108×10 mV = 1080 mV and bit 3 is dropped since it is greater than the 1 V input. (6) 01100100 gives us 100×10 mV = 1000 mV = 1 V and bit 2 is kept since it is equal to input. Even though the answer is found it does not stop. (7) 011000110 gives us 102×10 mV = 1020 mV and bit 1 is dropped since it is greater than the 1 V input. (8) 01100101 gives us 101×10 mV = 1010 mV and bit 0 is dropped since it is greater than the 1 V input.

Notice that the Successive Approximation method goes through all the steps even if the answer is found in one of the earlier steps. The advantage of the Successive Approximation method is that the conversion time is fixed since it has to go through all the steps.

Review Questions

1. Give two factors that affect the step size calculation.
2. The ADC0848 is a(n) _____-bit converter.
3. True or false. While the ADC0848 has 8 pins for D_{out}, the MAX1112 has only one D_{out} pin.
4. Find the step size for an 8-bit ADC, if Vref = 1.28 V.
5. For question 4, calculate the output if the analog input is: (a) 0.7 V, and (b) 1 V.

SECTION 13.2: ADC PROGRAMMING IN THE AVR

Because the ADC is widely used in data acquisition, in recent years an increasing number of microcontrollers have had an on-chip ADC peripheral, just like timers and USART. An on-chip ADC eliminates the need for an external ADC connection, which leaves more pins for other I/O activities. The vast majority of the AVR chips come with ADC. In this section we discuss the ADC feature of the ATmega32 and show how it is programmed in both Assembly and C.

ATmega32 ADC features

The ADC peripheral of the ATmega32 has the following characteristics:
(a) It is a 10-bit ADC.
(b) It has 8 analog input channels, 7 differential input channels, and 2 differential input channels with optional gain of 10x and 200x.
(c) The converted output binary data is held by two special function registers called ADCL (A/D Result Low) and ADCH (A/D Result High).
(d) Because the ADCH:ADCL registers give us 16 bits and the ADC data out is only 10 bits wide, 6 bits of the 16 are unused. We have the option of making either the upper 6 bits or the lower 6 bits unused.
(e) We have three options for V_{ref}. V_{ref} can be connected to AVCC (Analog V_{cc}), internal 2.56 V reference, or external AREF pin.
(f) The conversion time is dictated by the crystal frequency connected to the XTAL pins (Fosc) and ADPS0:2 bits.

AVR ADC hardware considerations

For digital logic signals a small variation in voltage level has no effect on the output. For example, 0.2 V is considered LOW, since in TTL logic, anything less than 0.5 V will be detected as LOW logic. That is not the case when we are dealing with analog voltage. See Example 13-2.

We can use many techniques to reduce the impact of ADC supply voltage and V_{ref} variation on the accuracy of ADC output. Next, we examine two of the most widely used techniques in the AVR.

Example 13-2

For an 10-bit ADC, we have V_{ref} = 2.56 V. Calculate the D0–D9 output if the analog input is: (a) 0.2 V, and (b) 0 V. How much is the variation between (a) and (b)?
Solution:

Because the step size is 2.56/1024 = 2.5 mV, we have the following:
(a) D_{out} = 0.2 V/2.5 mV = 80 in decimal, which gives us 1010000 in binary.

(b) D_{out} = 0 V/2.5 mV = 0 in decimal, which gives us 0 in binary.

The difference is 1010000, which is 7 bits!

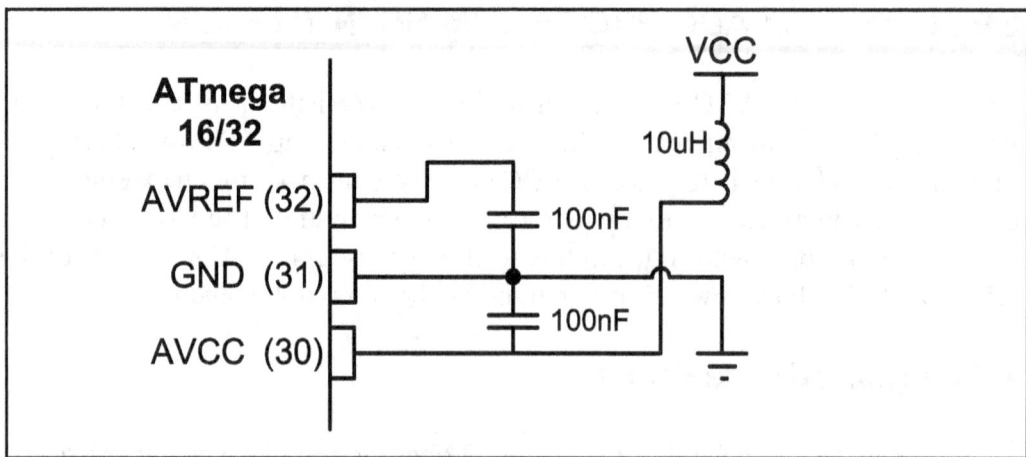

Figure 13-5. ADC Recommended Connection

Decoupling AVCC from VCC

As we mentioned in Chapter 8, the AVCC pin provides the supply for analog ADC circuitry. To get a better accuracy of AVR ADC we must provide a stable voltage source to the AVCC pin. Figure 13-5 shows how to use an inductor and a capacitor to achieve this.

Connecting a capacitor between V_{ref} and GND

By connecting a capacitor between the AVREF pin and GND you can make the V_{ref} voltage more stable and increase the precision of ADC. See Figure 13-5.

AVR programming in Assembly and C

In the AVR microcontroller five major registers are associated with the ADC that we deal with in this chapter. They are ADCH (high data), ADCL (low data), ADCSRA (ADC Control and Status Register), ADMUX (ADC multiplexer selection register), and SPIOR (Special Function I/O Register). We examine each of them in this section.

REFS1	REFS0	ADLAR	MUX4	MUX3	MUX2	MUX1	MUX0

REFS1:0 Bit 7:6 Reference Selection Bits
These bits select the reference voltage for the ADC.

ADLAR Bit 5 ADC Left Adjust Results
This bit dictates either the left bits or the right bits of the result registers ADCH:ADCL that are used to store the result. If we write a one to ADLAR, the result will be left adjusted; otherwise, the result is right adjusted.

MUX4:0 Bit 4:0 Analog Channel and Gain Selection Bits
The value of these bits selects the gain for the differential channels and also selects which combination of analog inputs are connected to the ADC.

Figure 13-6. ADMUX Register

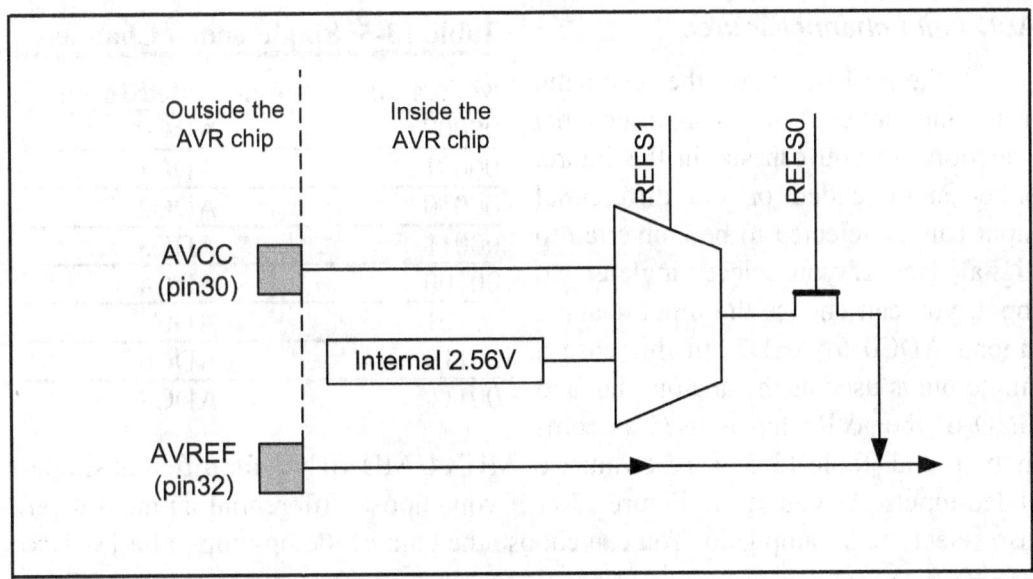

Figure 13-7. ADC Reference Source Selection

ADMUX register

Figure 13-6 shows the bits of ADMUX registers and their usage. In this section we will focus more on the function of these bits.

V_{ref} *source*

Figure 13-7 shows the block diagram of internal circuitry of V_{ref} selection. As you can see we have three options: (a) AREF pin, (b) AVCC pin, or (c) internal 2.56 V. Table 13-4 shows how the REFS1 and REFS0 bits of the ADMUX register can be used to select the V_{ref} source.

Table 13-4: V_{ref} Source Selection Table for AVR

REFS1	REFS0	V_{ref}	
0	0	AREF pin	Set externally
0	1	AVCC pin	Same as VCC
1	0	Reserved	----
1	1	Internal 2.56 V	Fixed regardless of VCC value

Notice that if you connect the VREF pin to an external fixed voltage you will not be able to use the other reference voltage options in the application, as they will be shorted with the external voltage.

Another important point to note is the fact that connecting a 100 nF external capacitor between the VREF pin and GND will increase the precision and stability of ADC, especially when you want to use internal 2.56 V. Refer to Figure 13-5 to see how to connect an external capacitor to the VREF pin of the ATmega32.

If you choose 2.56 V as the V_{ref}, the step size of ADC will be 2.56 / 1024 = 10/4 = 2.5 mV. Such a round step size will reduce the calculations in software.

ADC input channel source

Figure 13-8 shows the schematic of the internal circuitry of input channel selection. As you can see in the figure, either single-ended or the differential input can be selected to be converted to digital data. If you select single-ended input, you can choose the input channel among ADC0 to ACD7. In this case a single pin is used as the analog line, and GND of the AVR chip is used as common ground. Table 13-5 lists the values of MUX4–MUX0 bits for different single-ended inputs. As you see in Figure 13-8, if you choose differential input, you can also select the op-amp gain. You can choose the gain of the op-amp to be 1x, 10x,

Table 13-5: Single-ended Channels

MUX4...0	Single-ended Input
00000	ADC0
00001	ADC1
00010	ADC2
00011	ADC3
00100	ADC4
00101	ADC5
00110	ADC6
00111	ADC7

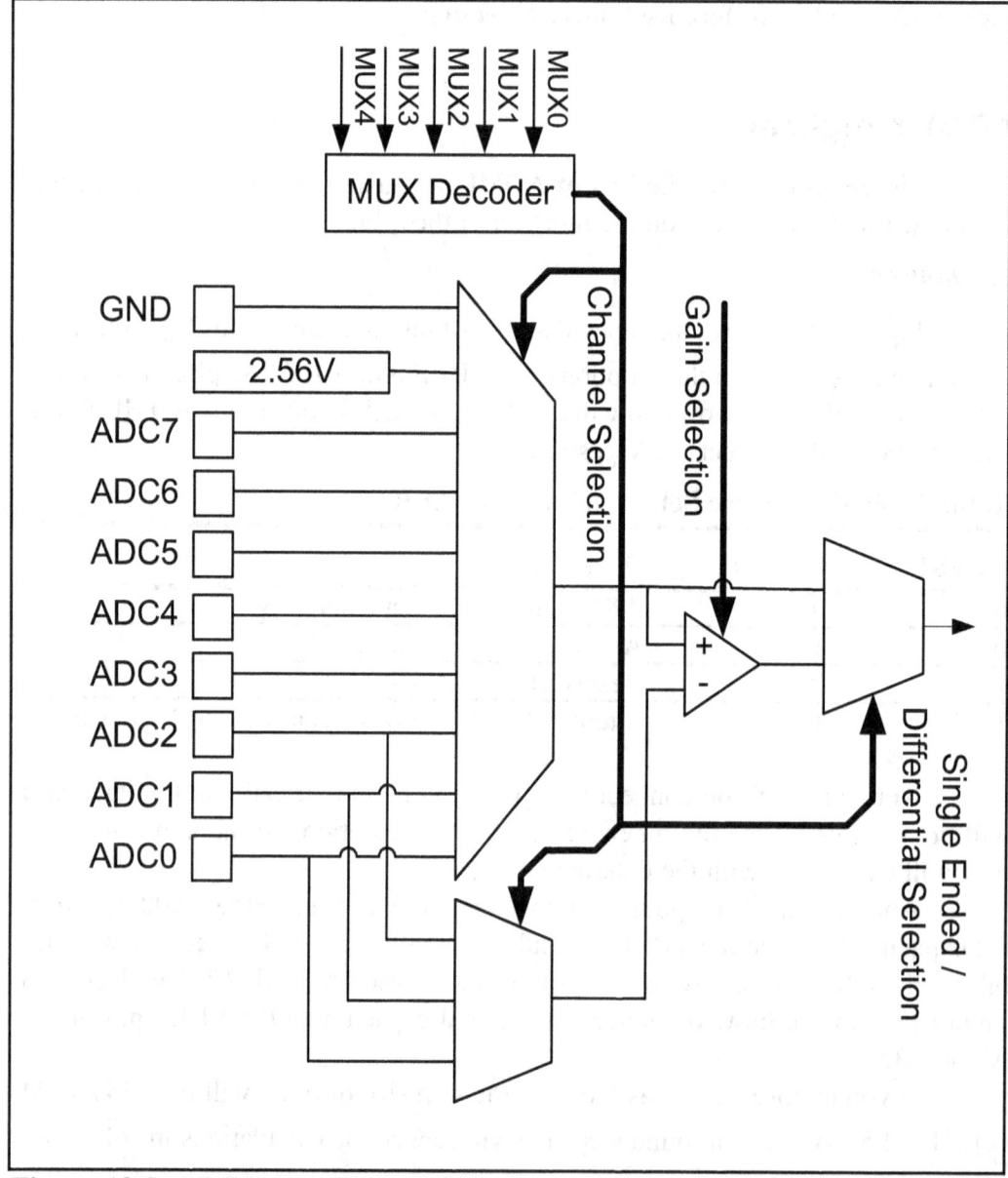

Figure 13-8. ADC Input Channel Selection

Table 13-6: V_{ref} Source Selection Table

MUX4...0	+ Differential Input	– Differential Input	Gain
01000 *	ADC0	ADC0	10x
01001	ADC1	ADC0	10x
01010 *	ADC0	ADC0	200x
01011	ADC1	ADC0	200x
01100 *	ADC2	ADC2	10x
01101	ADC3	ADC2	10x
01110 *	ADC2	ADC2	200x
01111	ADC3	ADC2	200x
10000	ADC0	ADC1	1x
10001 *	ADC1	ADC1	1x
10010	ADC2	ADC1	1x
10011	ADC3	ADC1	1x
10100	ADC4	ADC1	1x
10101	ADC5	ADC1	1x
10110	ADC6	ADC1	1x
10111	ADC7	ADC1	1x
11000	ADC0	ADC2	1x
11001	ADC1	ADC2	1x
11010 *	ADC2	ADC2	1x
11011	ADC3	ADC2	1x
11100	ADC4	ADC2	1x
11101	ADC5	ADC2	1x

*Note: The rows with * are not applicable.*

or 200x. You can select the positive input of the op-amp to be one of the pins ADC0 to ADC7, and the negative input of the op-amp can be any of ADC0, ADC1, or ADC2 pins. See Table 13-6.

ADLAR bit operation

The AVRs have a 10-bit ADC, which means that the result is 10 bits long and cannot be stored in a single byte. In AVR two 8-bit registers are dedicated to the ADC result, but only 10 of the 16 bits are used and 6 bits are unused. You can select the position of used bits in the bytes. If you set the ADLAR bit in ADMUX register, the result bits will be left-justified; otherwise, the result bits will be right-justified. See Figure 13-9. Notice that changing the ADLAR bit will affect the ADC data register immediately.

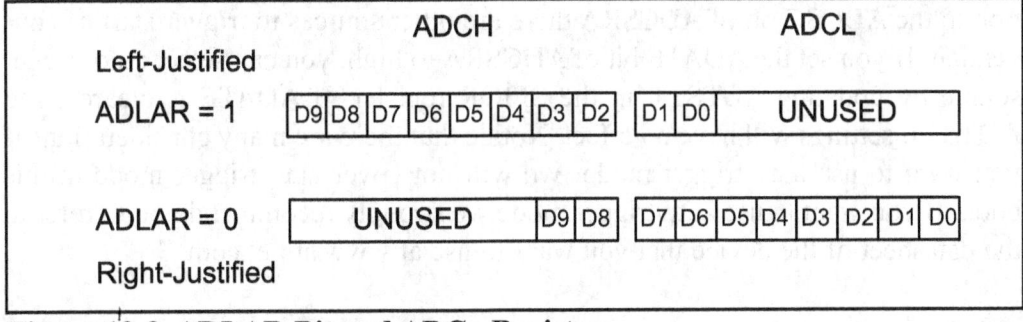

Figure 13-9. ADLAR Bit and ADCx Registers

ADCH: ADCL registers

After the A/D conversion is complete, the result sits in registers ADCL (A/D Result Low Byte) and ACDH (A/D Result High Byte). As we mentioned before, the ADLAR bit of the ADMUX is used for making it right-justified or left-justified because we need only 10 of the 16 bits.

ADCSRA register

The ADCSRA register is the status and control register of ADC. Bits of this register control or monitor the operation of the ADC. In Figure 13-10 you can see a description of each bit of the ADCSRA register. We will examine some of these bits in more detail.

ADEN	ADSC	ADATE	ADIF	ADIE	ADPS2	ADPS1	ADPS0

ADEN Bit 7 ADC Enable
This bit enables or disables the ADC. Setting this bit to one will enable the ADC, and clearing this bit to zero will disable it even while a conversion is in progress.

ADSC Bit 6 ADC Start Conversion
To start each conversion you have to set this bit to one.

ADATE Bit 5 ADC Auto Trigger Enable
Auto triggering of the ADC is enabled when you set this bit to one.

ADIF Bit 4 ADC Interrupt Flag
This bit is set when an ADC conversion completes and the data registers are updated.

ADIE Bit 3 ADC Interrupt Enable
Setting this bit to one enables the ADC conversion complete interrupt.

ADPS2:0 Bit 2:0 ADC Prescaler Select Bits
These bits determine the division factor between the XTAL frequency and the input clock to the ADC.

Figure 13-10. ADCSRA (A/D Control and Status Register A)

ADC Start Conversion bit

As we stated before, an ADC has a Start Conversion input. The AVR chip has a special circuit to trigger start conversion. As you see in Figure 13-11, in addition to the ADCSC bit of ADCSRA there are other sources to trigger start of conversion. If you set the ADATE bit of ADCSRA to high, you can select auto trigger source by updating ADTS2:0 in the SFIOR register. If ADATE is cleared, the ADTS2:0 settings will have no effect. Notice that there are many considerations if you want to use auto trigger mode. We will not cover auto trigger mode in this book. If you want to use auto trigger mode we strongly recommend you to refer to the datasheet of the device that you want to use at www.atmel.com.

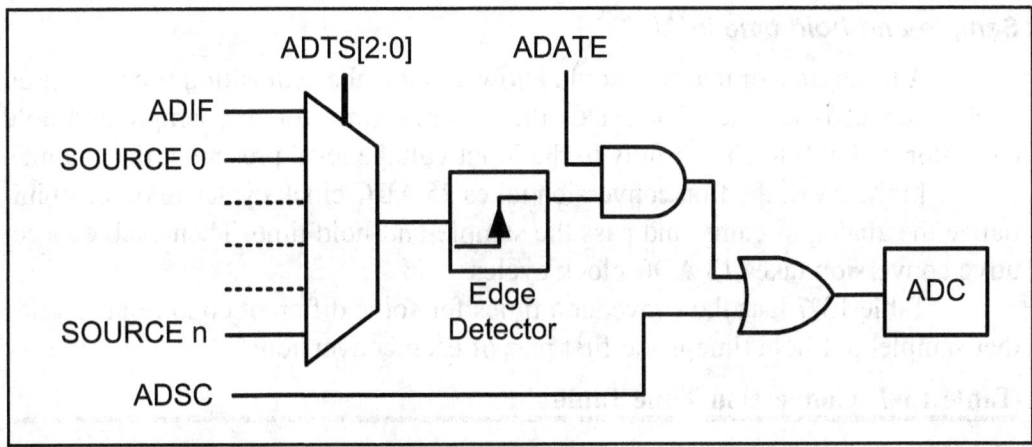

Figure 13-11. AVR ADC Trigger Source

A/D conversion time

As you see in Figure 13-12, by using the ADPS2:0 bits of the ADCSRA register we can set the A/D conversion time. To select the conversion time, we can select any of Fosc/2, Fosc/4, Fosc/8, Fosc/16, Fosc/32, Fosc/64, or Fosc/128 for ADC clock, where Fosc is the speed of the crystal frequency connected to the AVR chip. Notice that the multiplexer has 7 inputs since the option ADPS2:0 = 000 is reserved. For the AVR, the ADC requires an input clock frequency less than 200 kHz for the maximum accuracy. Look at Example 13-3 for clarification.

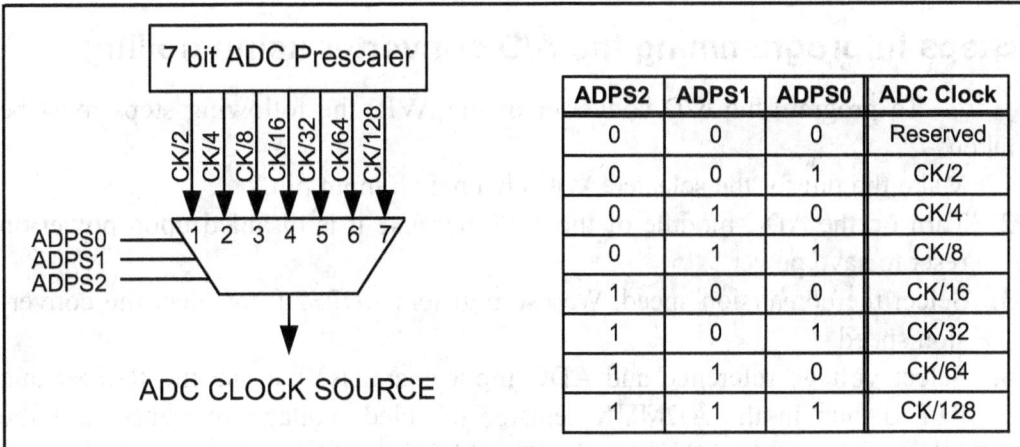

ADPS2	ADPS1	ADPS0	ADC Clock
0	0	0	Reserved
0	0	1	CK/2
0	1	0	CK/4
0	1	1	CK/8
1	0	0	CK/16
1	0	1	CK/32
1	1	0	CK/64
1	1	1	CK/128

Figure 13-12. AVR ADC Clock Selection

Example 13-3

An AVR is connected to the 8 MHz crystal oscillator. Calculate the ADC frequency for
(a) ADPS2:0 = 001 (b) ADPS2:0 = 100 (c) ADPS2:0 = 111
Solution:

(a) Because ADPS2:0 = 001 (1 decimal), the ck/2 input will be activated; we have
        8 MHz / 2 = 4 MHz (greater than 200 kHz and not valid)
(b) Because ADPS2:0 = 100 (4 decimal), the ck/8 input will be activated; we have
        8 MHz / 16 = 500 kHz (greater than 200 kHz and not valid)
(c) Because ADPS2:0 = 111 (7 decimal), the ck/128 input will be activated; we have
        8 MHz / = 62 kHz (a valid option since it is less than 200 kHz)

A timing factor that we should know about is the acquisition time. After an ADC channel is selected, the ADC allows some time for the sample-and-hold capacitor (C hold) to charge fully to the input voltage level present at the channel.

In the AVR, the first conversion takes 25 ADC clock cycles in order to initialize the analog circuitry and pass the sample-and-hold time. Then each consecutive conversion takes 13 ADC clock cycles.

Table 13-7 lists the conversion times for some different conditions. Notice that sample-and-hold time is the first part of each conversion.

Table 13-7: Conversion Time Table

Condition	Sample and Hold Time (Cycles)	Total Conversion Time (Cycles)
First Conversion	14.5	25
Normal Conversion, Single-ended	1.5	13
Normal Conversion, Differential	2	13.5
Auto trigger conversion	1.5 / 2.5	13/14

If the conversion time is not critical in your application and you do not want to deal with calculation of ADPS2:0 you can use ADPS2:0 = 111 to get the maximum accuracy of ADC.

Steps in programming the A/D converter using polling

To program the A/D converter of the AVR, the following steps must be taken:

1. Make the pin for the selected ADC channel an input pin.
2. Turn on the ADC module of the AVR because it is disabled upon power-on reset to save power.
3. Select the conversion speed. We use registers ADPS2:0 to select the conversion speed.
4. Select voltage reference and ADC input channels. We use the REFS0 and REFS1 bits in the ADMUX register to select voltage reference and the MUX4:0 bits in ADMUX to select the ADC input channel.
5. Activate the start conversion bit by writing a one to the ADSC bit of ADCSRA.
6. Wait for the conversion to be completed by polling the ADIF bit in the ADC-SRA register.
7. After the ADIF bit has gone HIGH, read the ADCL and ADCH registers to get the digital data output. Notice that you have to read ADCL before ADCH; otherwise, the result will not be valid.
8. If you want to read the selected channel again, go back to step 5.
9. If you want to select another V_{ref} source or input channel, go back to step 4.

Programming AVR ADC in Assembly and C

The Assembly language Program 13-1 illustrates the steps for ADC conversion shown above. Figure 13-13 shows the hardware connection of Program 13-1.

```
;Program 13-1: This program gets data from channel 0 (ADC0) of
;ADC and displays the result on Port C and Port D. This is done
;forever.
;**************** Program 13-1 ************************

.INCLUDE "M32DEF.INC"
        LDI   R16,0xFF
        OUT   DDRB, R16              ;make Port B an output
        OUT   DDRD, R16              ;make Port D an output
        LDI   R16,0
        OUT   DDRA, R16              ;make Port A an input for ADC
        LDI   R16,0x87               ;enable ADC and select ck/128
        OUT   ADCSRA, R16
        LDI   R16,0xC0               ;2.56V Vref, ADC0 single ended
        OUT   ADMUX, R16             ;input, right-justified data
READ_ADC:
        SBI   ADCSRA,ADSC            ;start conversion
KEEP_POLING:                        ;wait for end of conversion
        SBIS  ADCSRA,ADIF            ;is it end of conversion yet?
        RJMP  KEEP_POLING            ;keep polling end of conversion
        SBI   ADCSRA,ADIF            ;write 1 to clear ADIF flag
        IN    R16,ADCL               ;YOU HAVE TO READ ADCL FIRST
        OUT   PORTD,R16              ;give the low byte to PORTD
        IN    R16,ADCH               ;READ ADCH AFTER ADCL
        OUT   PORTB,R16              ;give the high byte to PORTB
        RJMP  READ_ADC               ;keep repeating it
```

Program 13-1: Reading ADC Using Polling Method in Assembly

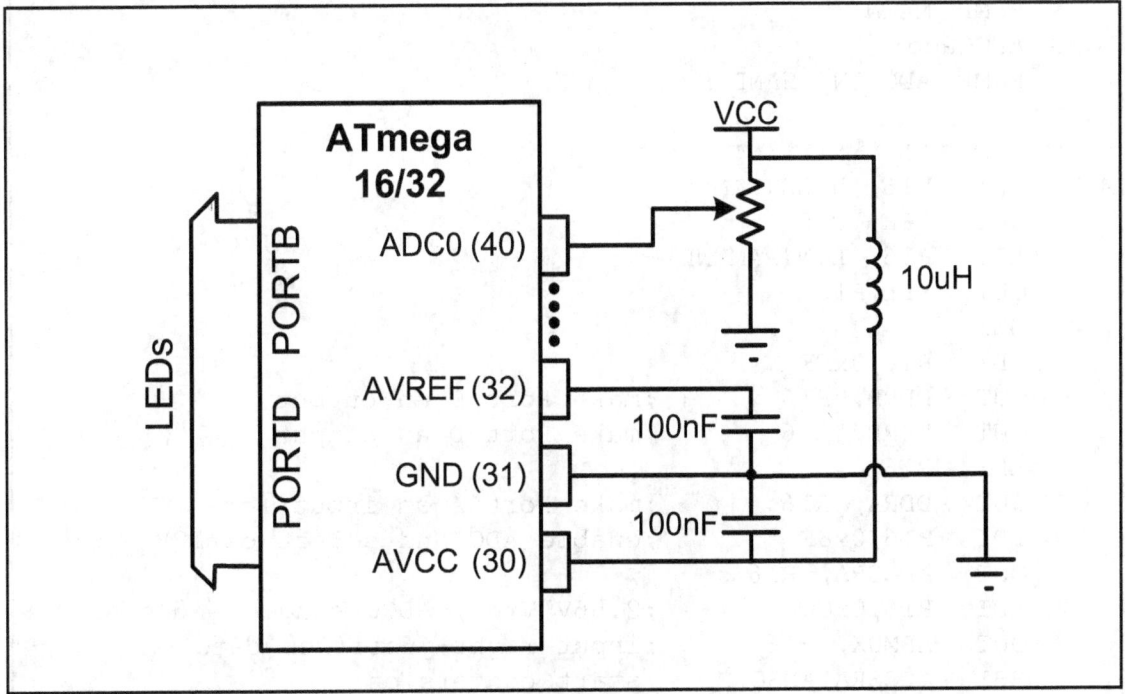

Figure 13-13. ADC Connection for Program 13-1

Program 13-1C is the C version of the ADC conversion for Program 13-1.

```c
#include <avr/io.h>              //standard AVR header
int main (void)
{
  DDRB = 0xFF;                   //make Port B an output
  DDRD = 0xFF;                   //make Port D an output
  DDRA = 0;                      //make Port A an input for ADC input
  ADCSRA= 0x87;                  //make ADC enable and select ck/128
  ADMUX= 0xC0;                   //2.56V Vref, ADC0 single ended input
                                 //data will be right-justified
  while (1){
    ADCSRA|=(1<<ADSC);           //start conversion
    while((ADCSRA&(1<<ADIF))==0);//wait for conversion to finish
    PORTD = ADCL;                //give the low byte to PORTD
    PORTB = ADCH;                //give the high byte to PORTB
  }
  return 0;
}
```

Program 13-1C: Reading ADC Using Polling Method in C

Programming A/D converter using interrupts

In Chapter 10, we showed how to use interrupts instead of polling to avoid tying down the microcontroller. To program the A/D using the interrupt method, we need to set HIGH the ADIE (A/D interrupt enable) flag. Upon completion of conversion, the ADIF (A/D interrupt flag) changes to HIGH; if ADIE = 1, it will force the CPU to jump to the ADC interrupt handler. Programs 13-2 and 13-2C show how to read ADC using interrupts.

```asm
.INCLUDE "M32DEF.INC"
.CSEG
      RJMP  MAIN
.ORG ADCCaddr
      RJMP  ADC_INT_HANDLER
.ORG 40
;***************************
MAIN: LDI   R16, HIGH(RAMEND)
      OUT   SPH, R16
      LDI   R16, LOW(RAMEND)
      OUT   SPL,R16
      SEI
      LDI   R16,0xFF
      OUT   DDRB, R16         ;make Port B an output
      OUT   DDRD, R16         ;make Port D an output
      LDI   R16,0
      OUT   DDRA, R16         ;make Port A an input for ADC
      LDI   R16,0x8F          ;enable ADC and select ck/128
      OUT   ADCSRA, R16
      LDI   R16,0xC0          ;2.56V Vref, ADC0 single ended
      OUT   ADMUX, R16        ;input right-justified data
      SBI   ADCSRA,ADSC       ;start conversion
```

Program 13-2: Reading ADC Using Interrupts in Assembly *(continued on next page)*

```
WAIT_HERE:
     RJMP  WAIT_HERE          ;keep repeating it
;*****************************
ADC_INT_HANDLER:
     IN    R16,ADCL           ;YOU HAVE TO READ ADCL FIRST
     OUT   PORTD,R16          ;give the low byte to PORTD
     IN    R16,ADCH           ;READ ADCH AFTER ADCL
     OUT   PORTB,R16          ;give the high byte to PORTB
     SBI   ADCSRA,ADSC        ;start conversion again
     RETI
```

Program 13-2: Reading ADC Using Interrupts in Assembly *(continued from previous page)*

Program 13-2C is the C version of Program 13-2. Notice that this program is checked under WinAVR (20080610). If you use another compiler you may need to read the documentation of your compiler to know how to deal with interrupts in your compiler.

```
#include <avr\io.h>
#include <avr\interrupt.h>
ISR(ADC_vect){
  PORTD = ADCL;            //give the low byte to PORTD
  PORTB = ADCH;            //give the high byte to PORTB
  ADCSRA|=(1<<ADSC);       //start conversion
}
int main (void){
  DDRB = 0xFF;             //make Port B an output
  DDRD = 0xFF;             //make Port D an output
  DDRA = 0;                //make Port A an input for ADC input
  sei();                   //enable interrupts
  ADCSRA= 0x8F;            //enable and interrupt select ck/128
  ADMUX= 0xC0;             //2.56V Vref and ADC0 single-ended
                           //input right-justified data
  ADCSRA|=(1<<ADSC);       //start conversion
  while (1);               //wait forever
  return 0;
}
```

Program 13-2C: Reading ADC Using Interrupts in C

Review Questions

1. What is the internal V_{ref} of the ATmega32?
2. The A/D of AVR is a(n) _____-bit converter.
3. True or false. The A/D of AVR has pins for D_{OUT}.
4. True or false. A/D in the AVR is an off-chip module.
5. Find the step size for an AVR ADC, if $V_{ref} = 2.56$ V.
6. For problem 5, calculate the D0–D9 output if the analog input is: (a) 0.7 V, and (b) 1 V.
7. How many single-ended inputs are available in the ATmega32 ADC?
8. Calculate the first conversion time for ADPS0–2 = 111 and Fosc = 4 MHz.
9. In AVR, the ADC requires an input clock frequency less than _____ .
10. Which bit is used to poll for the end of conversion?

CHAPTER 13: ADC, DAC, AND SENSOR INTERFACING   479

SECTION 13.3: SENSOR INTERFACING AND SIGNAL CONDITIONING

This section will show how to interface sensors to the microcontroller. We examine some popular temperature sensors and then discuss the issue of signal conditioning. Although we concentrate on temperature sensors, the principles discussed in this section are the same for other types of sensors such as light and pressure sensors.

Temperature sensors

Transducers convert physical data such as temperature, light intensity, flow, and speed to electrical signals. Depending on the transducer, the output produced is in the form of voltage, current, resistance, or capacitance. For example, temperature is converted to electrical signals using a transducer called a *thermistor*.

A thermistor responds to temperature change by changing resistance, but its response is not linear, as seen in Table 13-8.

The complexity associated with writing software for such non-linear devices has led many manufacturers to market a linear temperature sensor. Simple and widely used linear temperature sensors include the LM34 and LM35 series from National Semiconductor Corp. They are discussed next.

Table 13-8: Thermistor Resistance vs. Temperature

Temperature (C)	Tf (K ohms)
0	29.490
25	10.000
50	3.893
75	1.700
100	0.817

From William Kleitz, Digital Electronics

LM34 and LM35 temperature sensors

The sensors of the LM34 series are precision integrated-circuit temperature sensors whose output voltage is linearly proportional to the Fahrenheit temperature. See Table 13-9. The LM34 requires no external calibration because it is internally calibrated. It outputs 10 mV for each degree of Fahrenheit temperature. Table 13-9 is a selection guide for the LM34.

Table 13-9: LM34 Temperature Sensor Series Selection Guide

Part Scale	Temperature Range	Accuracy	Output
LM34A	−50 F to +300 F	+2.0 F	10 mV/F
LM34	−50 F to +300 F	+3.0 F	10 mV/F
LM34CA	−40 F to +230 F	+2.0 F	10 mV/F
LM34C	−40 F to +230 F	+3.0 F	10 mV/F
LM34D	−32 F to +212 F	+4.0 F	10 mV/F

Note: Temperature range is in degrees Fahrenheit.

Table 13-10: LM35 Temperature Sensor Series Selection Guide

Part	Temperature Range	Accuracy	Output Scale
LM35A	−55 C to +150 C	+1.0 C	10 mV/C
LM35	−55 C to +150 C	+1.5 C	10 mV/C
LM35CA	−40 C to +110 C	+1.0 C	10 mV/C
LM35C	−40 C to +110 C	+1.5 C	10 mV/C
LM35D	0 C to +100 C	+2.0 C	10 mV/C

Note: Temperature range is in degrees Celsius.

The LM35 series sensors are precision integrated-circuit temperature sensors whose output voltage is linearly proportional to the Celsius (centigrade) temperature. The LM35 requires no external calibration because it is internally calibrated. It outputs 10 mV for each degree of centigrade temperature. Table 13-10 is the selection guide for the LM35. (For further information see http://www.national.com.)

Signal conditioning

Signal conditioning is widely used in the world of data acquisition. The most common transducers produce an output in the form of voltage, current, charge, capacitance, and resistance. We need to convert these signals to voltage, however, in order to send input to an A-to-D converter. This conversion (modification) is commonly called *signal conditioning*. See Figure 13-14. Signal conditioning can be current-to-voltage conversion or signal amplification. For example, the thermistor changes resistance with temperature. The change of resistance must be translated into voltages to be of any use to an ADC. We now look at the case of connecting an LM34 (or LM35) to an ADC of the ATmega32.

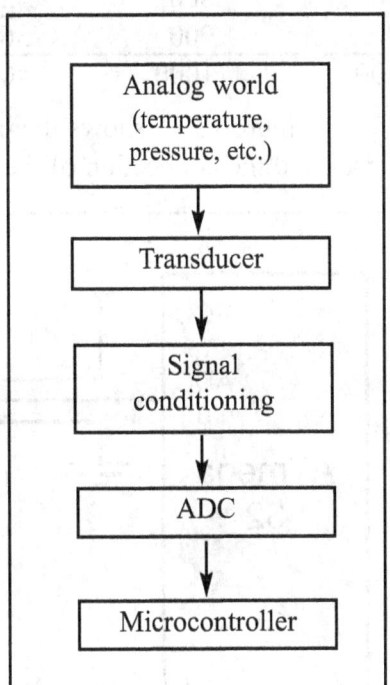

Figure 13-14. Getting Data from the Analog World

Interfacing the LM34 to the AVR

The A/D has 10-bit resolution with a maximum of 1024 steps, and the LM34 (or LM35) produces 10 mV for every degree of temperature change. Now, if we use the step size of 10 mV, the V_{out} will be 10,240 mV (10.24 V) for full-scale output. This is not acceptable even though the maximum temperature sensed by the LM34 is 300 degrees F, and the highest output we will get for the A/D is 3000 mV (3.00 V) .

Now if we use the internal 2.56 V reference voltage, the step size would be 2.56 V/1024 = 2.5 mV. This makes the binary output number for the ADC four times the real temperature because the sensor produces 10 mV for each degree of temperature change and the step size is 2.5 mV (10 mV/2.5 mV = 4). We can scale it by dividing it by 4 to get the real number for temperature. See Table 13-11.

Table 13-11: Temperature vs. V_{out} for AVR with $V_{ref} = 2.56$ V

Temp. (F)	V_{in} (mV)	# of steps	Binary V_{out} (b9–b0)	Temp. in Binary
0	0	0	00 00000000	00000000
1	10	4	00 00000100	00000001
2	20	8	00 00001000	00000010
3	30	12	00 00001100	00000011
10	100	20	00 00101000	00001010
20	200	80	00 01010000	00010100
30	300	120	00 01111000	00011110
40	400	160	00 10100000	00101000
50	500	200	00 11001000	00110010
60	600	240	00 11110000	00111100
70	700	300	01 00011000	01000110
80	800	320	01 01000000	01010000
90	900	360	01 01101000	01011010
100	1000	400	01 10010000	01100100

Figure 13-15 shows the pin configuration of the LM34/LM35 temperature sensor and the connection of the temperature sensor to the ATmega32.

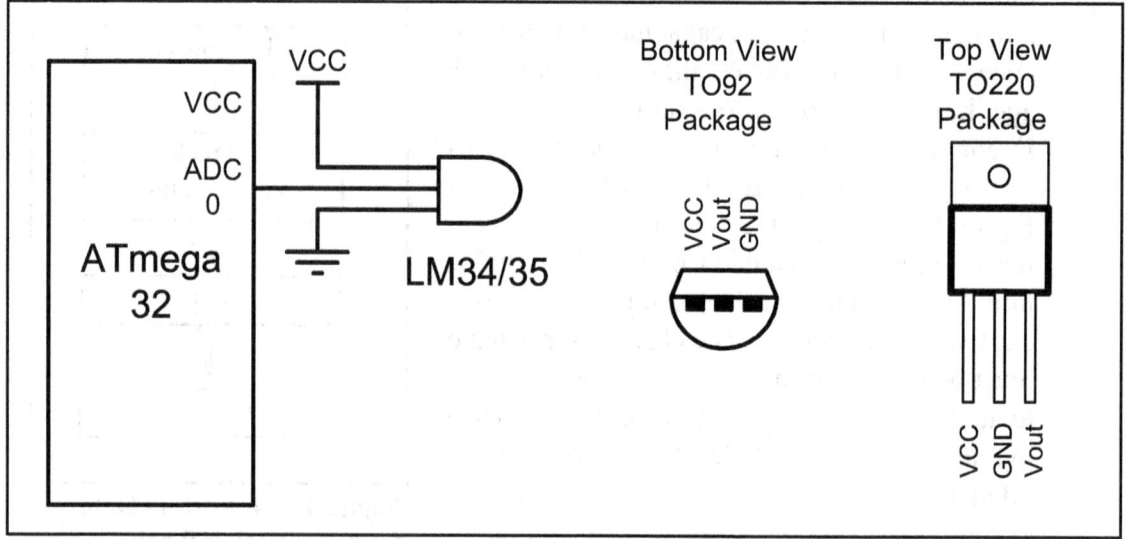

Figure 13-15. LM34/35 Connection to AVR and Its Pin Configuration

Reading and displaying temperature

Programs 13-4 and 13-4C show code for reading and displaying temperature in both Assembly and C, respectively.

The programs correspond to Figure 13-15. Regarding these two programs, the following points must be noted:

(1) The LM34 (or LM35) is connected to channel 0 (ADC0 pin).

(2) The 10-bit output of the A/D is divided by 4 to get the real temperature.

(3) To divide the 10-bit output of the A/D by 4 we choose the left-justified option and only read the ADCH register. It is same as shifting the result two bits right. See Example 13-4.

```
;this program reads the sensor and displays it on Port D
.INCLUDE "M32DEF.INC"
     LDI   R16,0xFF
     OUT   DDRD, R16          ;make Port D an output
     LDI   R16,0
     OUT   DDRA, R16          ;make Port A an input for ADC
     LDI   R16,0x87           ;enable ADC and select ck/128
     OUT   ADCSRA, R16
     LDI   R16,0xE0           ;2.56 V Vref, ADC0 single-ended
     OUT   ADMUX, R16         ;left-justified data
READ_ADC:
     SBI   ADCSRA,ADSC        ;start conversion
KEEP_POLING:                  ;wait for end of conversion
     SBIS  ADCSRA,ADIF        ;is it end of conversion?
     RJMP  KEEP_POLING        ;keep polling end of conversion
     SBI   ADCSRA,ADIF        ;write 1 to clear ADIF flag
     IN    R16,ADCH           ;read only ADCH for 8 MSB of
     OUT   PORTD,R16          ;result and give it to PORTD
     RJMP  READ_ADC           ;keep repeating
```
Program 13-3: Reading Temperature Sensor in Assembly

```
;this program reads the sensor and displays it on Port D
#include <avr/io.h>          //standard AVR header
int main (void)
{
  DDRD = 0xFF;               //make Port D an output
  DDRA = 0;                  //make Port A an input for ADC input
  ADCSRA = 0x87;             //make ADC enable and select ck/128
  ADMUX = 0xE0;              //2.56 V Vref and ADC0 single-ended
                             //data will be left-justified
  while (1){
     ADCSRA |= (1<<ADSC);  //start conversion
     while((ADCSRA&(1<<ADIF))==0); //wait for end of conversion
     PORTB = ADCH;           //give the high byte to PORTB
  }
  return 0;
}
```
Program 13-3C: Reading Temperature Sensor in C

Example 13-4

In Table 13-11, verify the AVR output for a temperature of 70 degrees. Find values in the AVR A/D registers of ADCH and ADCL for left-justified.
Solution:
The step size is 2.56/1024 = 2.5 mV because Vref = 2.56 V.
For the 70 degrees temperature we have 700 mV output because the LM34 provides 10 mV output for every degree. Now, the number of steps are 700 mV/2.5 mV = 280 in decimal. Now 280 = 0100011000 in binary and the AVR A/D output registers have ADCH = 01000110 and ADCL = 00000000 for left-justified. To get the proper result we must divide the result by 4. To do that, we simply read the ADCH register, which has the value 70 (01000110) in it.

CHAPTER 13: ADC, DAC, AND SENSOR INTERFACING

Review Questions

1. True or false. The transducer must be connected to signal conditioning circuitry before its signal is sent to the ADC.
2. The LM35 provides _____ mV for each degree of _____ (Fahrenheit, Celsius) temperature.
3. The LM34 provides ____ mV for each degree of ____ (Fahrenheit, Celsius) temperature.
4. Why do we set the V_{ref} of the AVR to 2.56 V if the analog input is connected to the LM35?
5. In Question 4, what is the temperature if the ADC output is 0011 1001?

SECTION 13.4: DAC INTERFACING

This section will show how to interface a DAC (digital-to-analog converter) to the AVR. Then we demonstrate how to generate a stair-step ramp on the scope using the DAC.

Digital-to-analog converter (DAC)

The digital-to-analog converter (DAC) is a device widely used to convert digital pulses to analog signals. In this section we discuss the basics of interfacing a DAC to the AVR.

Recall from your digital electronics course the two methods of creating a DAC: binary weighted and R/2R ladder. The vast majority of integrated circuit DACs, including the MC1408 (DAC0808) used in this section, use the R/2R method because it can achieve a much higher degree of precision. The first criterion for judging a DAC is its resolution, which is a function of the number of binary inputs. The common ones are 8, 10, and 12 bits. The number of data bit inputs decides the resolution of the DAC because the number of analog output levels is equal to 2^n, where n is the number of data bit inputs. Therefore, an 8-input DAC such as the DAC0808 provides 256 discrete voltage (or current) levels of output. See Figure 13-16. Similarly, the 12-bit DAC provides 4096 discrete voltage levels. There are also 16-bit DACs, but they are more expensive.

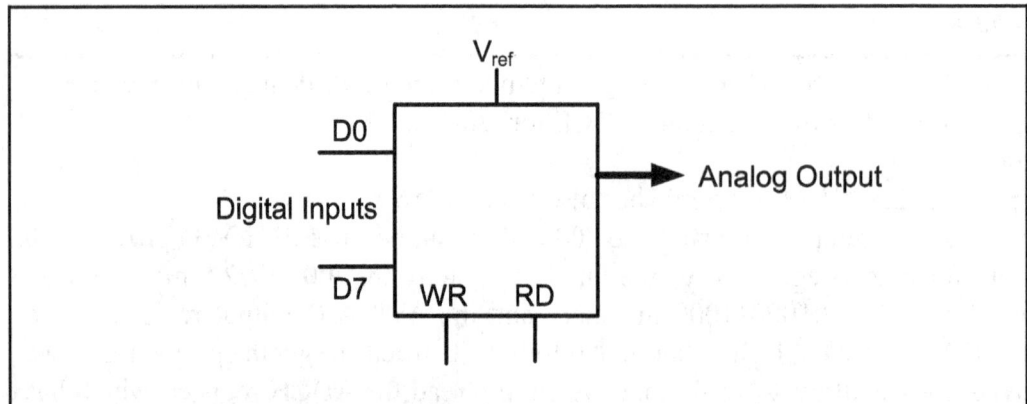

Figure 13-16. DAC Block Diagram

MC1408 DAC (or DAC0808)

In the MC1408 (DAC0808), the digital inputs are converted to current (I_{out}), and by connecting a resistor to the I_{out} pin, we convert the result to voltage. The total current provided by the I_{out} pin is a function of the binary numbers at the D0–D7 inputs of the DAC0808 and the reference current (I_{ref}), and is as follows:

$$I_{out} = I_{ref} \left(\frac{D7}{2} + \frac{D6}{4} + \frac{D5}{8} + \frac{D4}{16} + \frac{D3}{32} + \frac{D2}{64} + \frac{D1}{128} + \frac{D0}{256} \right)$$

where D0 is the LSB, D7 is the MSB for the inputs, and I_{ref} is the input current that must be applied to pin 14. The I_{ref} current is generally set to 2.0 mA. Figure 13-17 shows the generation of current reference (setting I_{ref} = 2 mA) by using the standard 5 V power supply. Now assuming that I_{ref} = 2 mA, if all the inputs to the DAC are high, the maximum output current is 1.99 mA (verify this for yourself).

Converting I_{out} to voltage in DAC0808

Ideally we connect the output pin I_{out} to a resistor, convert this current to voltage, and monitor the output on the scope. In real life, however, this can cause inaccuracy because the input resistance of the load where it is connected will also affect the output voltage. For this reason, the I_{ref} current output is isolated by connecting it to an op-amp such as the 741 with R_f = 5 kilohms for the feedback resistor. Assuming that R = 5 kilohms, by changing the binary input, the output voltage changes as shown in Example 13-5.

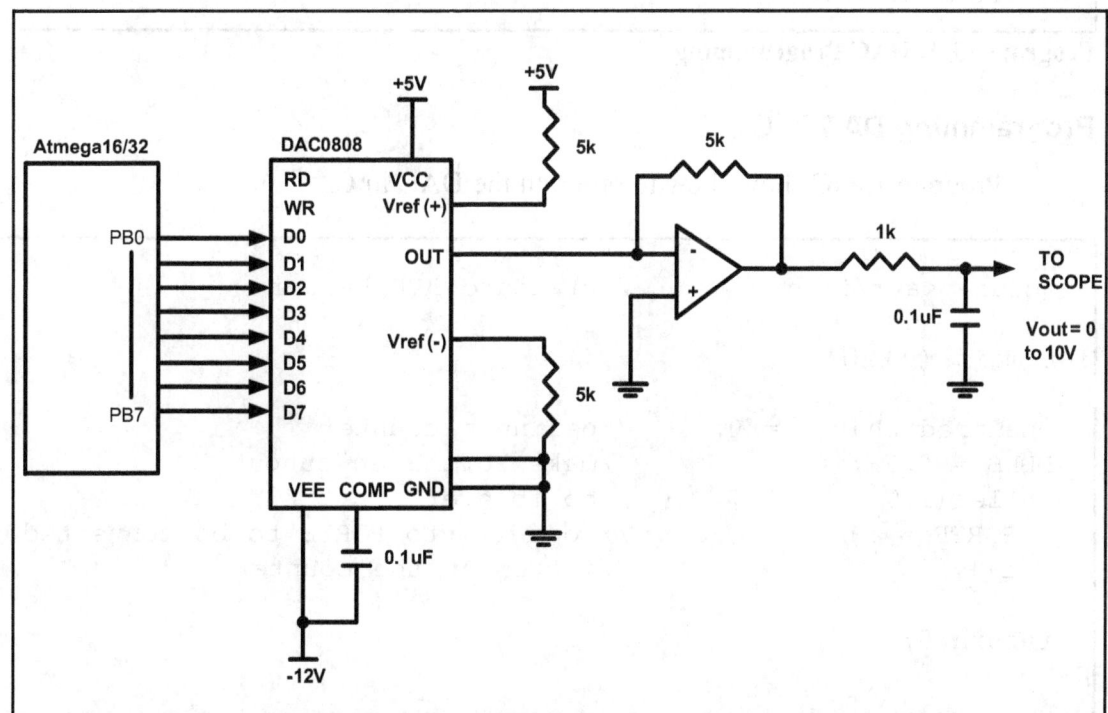

Figure 13-17. AVR Connection to DAC0808

Example 13-5

Assuming that R = 5 kilohms and I_{ref} = 2 mA, calculate V_{out} for the following binary inputs:

(a) 10011001 binary (99H)
(b) 11001000 (C8H)

Solution:

(a) I_{out} = 2 mA (153/256) = 1.195 mA and V_{out} = 1.195 mA × 5K = 5.975 V
(b) I_{out} = 2 mA (200/256) = 1.562 mA and V_{out} = 1.562 mA × 5K = 7.8125 V

Generating a stair-step ramp

In order to generate a stair-step ramp, you can set up the circuit in Figure 13-17 and load Program 13-4 on the AVR chip. To see the result wave, connect the output to an oscilloscope. Figure 13-18 shows the output.

```
LDI   R16,0xFF
      OUT   DDRB, R16          ;make Port B an output
AGAIN:
      INC   R16                ;increment R16
      OUT   PORTB,R16          ;sent R16 to PORTB
      NOP                      ;let DAC recover
      NOP
      RJMP  AGAIN
```

Program 13-4: DAC Programming

Programming DAC in C

Program 13-4C shows how to program the DAC in C.

```
#include <avr/io.h>         //standard AVR header

int main (void)
{
  unsigned char i = 0;      //define a counter
  DDRB = 0xFF;              //make Port B an output
  while (1){                //do forever
    PORTB = i;              //copy i into PORTB to be converted
    i++;                    //increment the counter
  }
  return 0;
}
```

Program 13-4C: DAC Programming in C

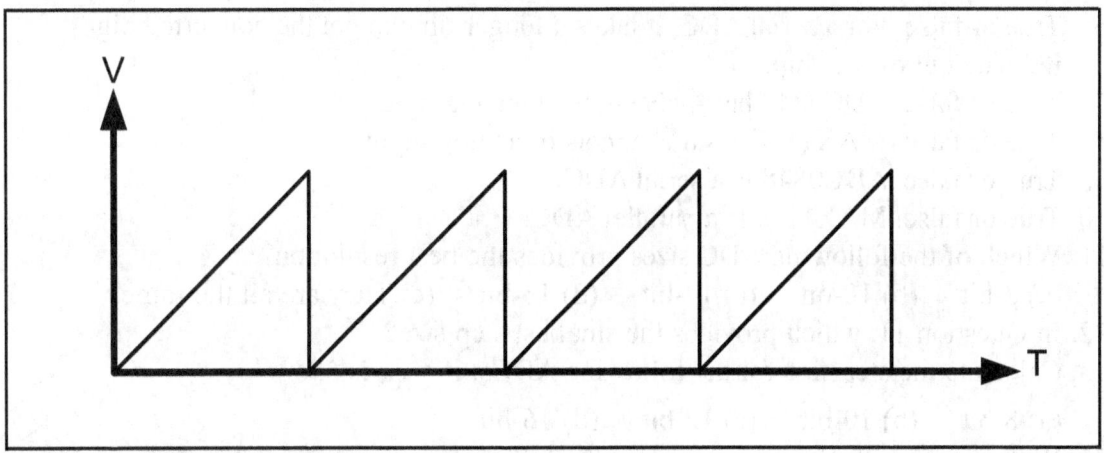

Figure 13-18. Stair Step Ramp Output

Review Questions

1. In a DAC, input is _____ (digital, analog) and output is _____ (digital, analog).
2. In an ADC, input is _____ (digital, analog) and output is _____ (digital, analog).
3. DAC0808 is a(n) ____-bit D-to-A converter.
4. (a) The output of DAC0808 is in _____ (current, voltage).
   (b) True or false. The output of DAC0808 is ideal to drive a motor.

SUMMARY

This chapter showed how to interface real-world devices such as DAC chips, ADC chips, and sensors to the AVR. First, we discussed both parallel and serial ADC chips, then described how the ADC module inside the AVR works and explained how to program it in both Assembly and C. Next we explored sensors. We also discussed the relation between the analog world and a digital device, and described signal conditioning, an essential feature of data acquisition systems. In the last section we studied the DAC chip, and showed how to interface it to the AVR.

PROBLEMS

SECTION 13.1: ADC CHARACTERISTICS

1. True or false. The output of most sensors is analog.
2. True or false. A 10-bit ADC has 10-bit digital output.
3. True or false. ADC0848 is an 8-bit ADC.
4. True or false. MAX1112 is a 10-bit ADC.
5. True or false. An ADC with 8 channels of analog input must have 8 pins, one for each analog input.

6. True or false. For a serial ADC, it takes a longer time to get the converted digital data out of the chip.
7. True or false. ADC0848 has 4 channels of analog input.
8. True or false. MAX1112 has 8 channels of analog input.
9. True or false. ADC0848 is a serial ADC.
10. True or false. MAX1112 is a parallel ADC.
11. Which of the following ADC sizes provides the best resolution?
    (a) 8-bit   (b) 10-bit   (c) 12-bit   (d) 16-bit   (e) They are all the same.
12. In Question 11, which provides the smallest step size?
13. Calculate the step size for the following ADCs, if V_{ref} is 5 V:

    (a) 8-bit   (b) 10-bit   (c) 12-bit   (d) 16-bit
14. With $V_{ref} = 1.28$ V, find the V_{in} for the following outputs:

    (a) D7–D0 = 11111111   (b) D7–D0 = 10011001   (c) D7–D0 = 1101100
15. In the ADC0848, what should the V_{ref} value be if we want a step size of 5 mV?
16. With $V_{ref} = 2.56$ V, find the V_{in} for the following outputs:

    (a) D7–D0 = 11111111   (b) D7–D0 = 10011001   (c) D7–D0 = 01101100

SECTION 13.2: ADC PROGRAMMING IN THE AVR

17. True or false. The ATmega32 has an on-chip A/D converter.
18. True or false. A/D of the ATmega32 is an 8-bit ADC.
19. True or false. ATmega32 has 8 channels of analog input.
20. True or false. The unused analog pins of the ATmega32 can be used for I/O pins.
21. True or false. The A/D conversion speed in the ATmega32 depends on the crystal frequency.
22. True or false. Upon power-on reset, the A/D module of the ATmega32 is turned on and ready to go.
23. True or false. The A/D module of the ATmega32 has an external pin for the start-conversion signal.
24. True or false. The A/D module of the ATmega32 can convert only one channel at a time.
25. True or false. The A/D module of the ATmega32 can have multiple external $V_{ref}+$ at any given time.
26. True or false. The A/D module of the ATmega32 can use the V_{cc} for V_{ref}.
27. In the A/D of ATmega32, what happens to the converted analog data? How do we know that the ADC is ready to provide us the data?
28. In the A/D of ATmega32, what happens to the old data if we start conversion again before we pick up the last data?
29. For the A/D of ATmega32, find the step size for each of the following V_{ref}:

    (a) $V_{ref} = 1.024$ V   (b) $V_{ref} = 2.048$ V   (c) $V_{ref} = 2.56$ V
30. In the ATmega32, what should the V_{ref} value be if we want a step size of 2 mV?
31. In the ATmega32, what should the V_{ref} value be if we want a step size of 3 mV?

32. With a step size of 1 mV, what is the analog input voltage if all outputs are 1?

33. With V_{ref} = 1.024 V, find the V_{in} for the following outputs:

    (a) D9–D0 = 0011111111 (b) D9–D0 = 0010011000 (c) D9–D0 = 0011010000

34. In the A/D of ATmega32, what should the V_{ref} value be if we want a step size of 4 mV?

35. With V_{ref} = 2.56 V, find the V_{in} for the following outputs:

    (a) D9–D0 = 1111111111 (b) D9–D0 = 1000000001 (c) D9–D0 = 1100110000

36. Find the first conversion times for the following cases if XTAL = 8 MHz. Are they acceptable?

    (a) Fosc/2   (b) Fosc/4   (c) Fosc/8   (d) Fosc/16   (e) Fosc/32

37. Find the first conversion times for the following cases if XTAL = 4 MHz. Are they acceptable?

    (a) Fosc/8   (b) Fosc/16   (c) Fosc/32   (d) Fosc/64

38. How do we start conversion in the ATmega32?

39. How do we recognize the end of conversion in the ATmega32?

40. Which bits of which register of the ATmega32 are used to select the A/D's conversion speed?

41. Which bits of which register of the ATmega32 are used to select the analog channel to be converted?

42. Give the names of the interrupt flags for the A/D of the ATmega32. State to which register they belong.

43. Upon power-on reset, the A/D of the ATmega32 is given (on, off).

SECTION 13.3: SENSOR INTERFACING AND SIGNAL CONDITIONING

44. What does it mean when a given sensor is said to have a linear output?

45. The LM34 sensor produces _____ mV for each degree of temperature.

46. What is signal conditioning?

SECTION 13.4: DAC INTERFACING

47. True or false. DAC0808 is the same as DAC1408.

48. Find the number of discrete voltages provided by the *n*-bit DAC for the following:

    (a) *n* = 8 (b) *n* = 10 (c) *n* = 12

49. For DAC1408, if I_{ref} = 2 mA, show how to get the I_{out} of 1.99 when all inputs are HIGH.

50. Find the I_{out} for the following inputs. Assume I_{ref} = 2 mA for DAC0808.

    (a) 10011001   (b) 11001100   (c) 11101110
    (d) 00100010   (e) 00001001   (f) 10001000

51. To get a smaller step, we need a DAC with _____ (more, fewer) digital inputs.

52. To get full-scale output, what should be the inputs for DAC?

ANSWERS TO REVIEW QUESTIONS

SECTION 13.1: ADC CHARACTERISTICS

1. Number of steps and V_{ref} voltage
2. 8
3. True
4. 1.28 V/256 = 5 mV
5. (a) 0.7 V/ 5 mV = 140 in decimal and D7–D0 = 10001100 in binary.
   (b) 1 V/ 5 mV = 200 in decimal and D7–D0 = 11001000 in binary.

SECTION 13.2: ADC PROGRAMMING IN THE AVR

1. 2.56 V
2. 10
3. False
4. False
5. 2.56/1024 = 2.5 mV
6. (a) 700 mV/2.5 mV = 280 (100011000), (b) 1000 mV/ 2.5 mV = 400 (110010000)
7. 8 channels
8. (1/(4 MHz/128)) × 25 = 800 microseconds
9. 200 kHz
10. ADIF bit of the ADCSRA register

SECTION 13.3: SENSOR INTERFACING AND SIGNAL CONDITIONING

1. True
2. 10, Celsius
3. 10, Fahrenheit
4. Using the 8-bit part of the 10-bit ADC, it gives us 256 steps, and 2.56 V/256 = 10 mV. The LM35 produces 10 mV for each degree of temperature, which matches the ADC's step size.
5. 00111001 = 57, which indicates it is 57 degrees.

SECTION 13.4: DAC INTERFACING

1. Digital, analog
2. Analog, digital
3. 8
4. (a) current   (b) true

CHAPTER 14

RELAY, OPTOISOLATOR, AND STEPPER MOTOR INTERFACING WITH AVR

OBJECTIVES

Upon completion of this chapter, you will be able to:

>> Describe the basic operation of a relay
>> Interface the AVR with a relay
>> Describe the basic operation of an optoisolator
>> Interface the AVR with an optoisolator
>> Describe the basic operation of a stepper motor
>> Interface the AVR with a stepper motor
>> Code AVR programs to control and operate a stepper motor
>> Define stepper motor operation in terms of step angle, steps
     per revolution, tooth pitch, rotation speed, and RPM

Microcontrollers are widely used in motor control. We also use relays and optoisolators in motor control. This chapter discusses motor control and shows AVR interfacing with relays, optoisolators, and stepper motors. We use both Assembly and C in our programming examples.

SECTION 14.1: RELAYS AND OPTOISOLATORS

This section begins with an overview of the basic operations of electro-mechanical relays, solid-state relays, reed switches, and optoisolators. Then we describe how to interface them to the AVR. We use both Assembly and C language programs to demonstrate their control.

Electromechanical relays

A *relay* is an electrically controllable switch widely used in industrial controls, automobiles, and appliances. It allows the isolation of two separate sections of a system with two different voltage sources. For example, a +5 V system can be isolated from a 120 V system by placing a relay between them. One such relay is called an *electromechanical* (or *electromagnetic*) *relay* (EMR) as shown in Figure 14-1. The EMRs have three components: the coil, spring, and contacts. In Figure 14-1, a digital +5 V on the left side can control a 12 V motor on the right side without any physical contact between them. When current flows through the coil, a magnetic field is created around the coil (the coil is energized), which causes the armature to be attracted to the coil. The armature's contact acts like a switch and closes or opens the circuit. When the coil is not energized, a spring pulls the armature to its normal state of open or closed. In the block diagram for electromechanical relays (EMR) we do not show the spring, but it does exist internally. There are all types of relays for all kinds of applications. In choosing a relay the following characteristics need to be considered:

1. The contacts can be normally open (NO) or normally closed (NC). In the NC type, the contacts are closed when the coil is not energized. In the NO type, the contacts are open when the coil is unenergized.
2. There can one or more contacts. For example, we can have SPST (single pole, single throw), SPDT (single pole, double throw), and DPDT (double pole, double throw) relays.
3. The voltage and current needed to energize the coil. The voltage can vary from a few volts to 50 volts, while the current can be from a few mA to 20 mA. The relay has a minimum voltage, below which the coil will not be energized. This minimum voltage is called the "pull-in" voltage. In the datasheets for relays we might not see current, but rather coil resistance. The V/R will give you the pull-in current. For example, if the coil voltage is 5 V, and the coil resistance is 500 ohms, we need a minimum of 10 mA (5 V/500 ohms = 10 mA) pull-in current.
4. The maximum DC/AC voltage and current that can be handled by the contacts. This is in the range of a few volts to hundreds of volts, while the current can be from a few amps to 40 A or more, depending on the relay. Notice the difference between this voltage/current specification and the voltage/current needed for energizing the coil. The fact that one can use such a small amount of volt-

age/current on one side to handle a large amount of voltage/current on the other side is what makes relays so widely used in industrial controls. Examine Table 14-1 for some relay characteristics.

Table 14-1: Selected DIP Relay Characteristics (www.Jameco.com)

Part No.	Contact Form	Coil Volts	Coil Ohms	Contact Volts-Current
106462CP	SPST-NO	5 VDC	500	100 VDC-0.5 A
138430CP	SPST-NO	5 VDC	500	100 VDC-0.5 A
106471CP	SPST-NO	12 VDC	1000	100 VDC-0.5 A
138448CP	SPST-NO	12 VDC	1000	100 VDC-0.5 A
129875CP	DPDT	5 VDC	62.5	30 VDC-1 A

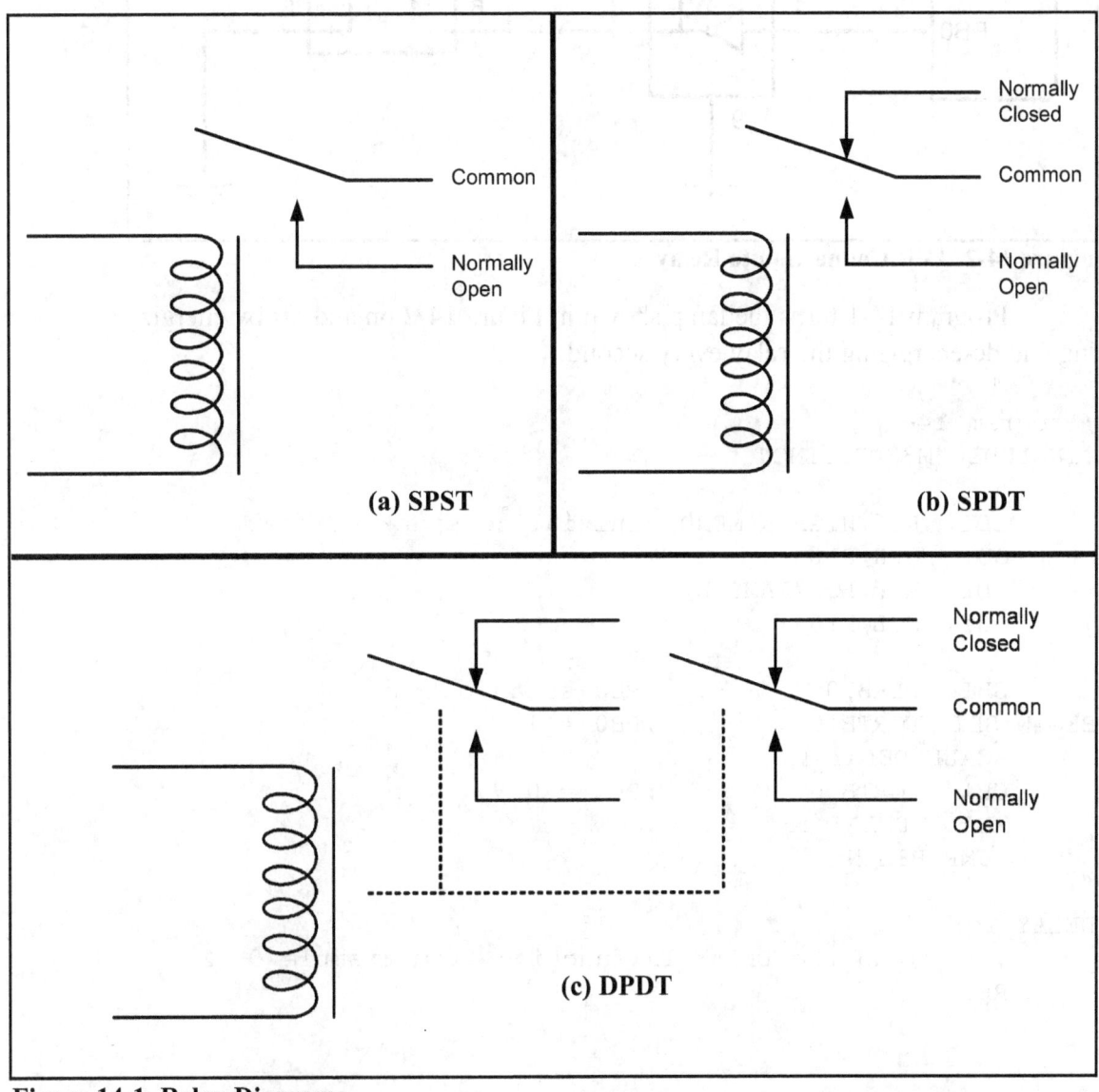

Figure 14-1. Relay Diagrams

Driving a relay

Digital systems and microcontroller pins lack sufficient current to drive the relay. While the relay's coil needs around 10 mA to be energized, the microcontroller's pin can provide a maximum of 1–2 mA current. For this reason, we place a driver, such as the ULN2803, or a power transistor between the microcontroller and the relay as shown in Figure 14-2.

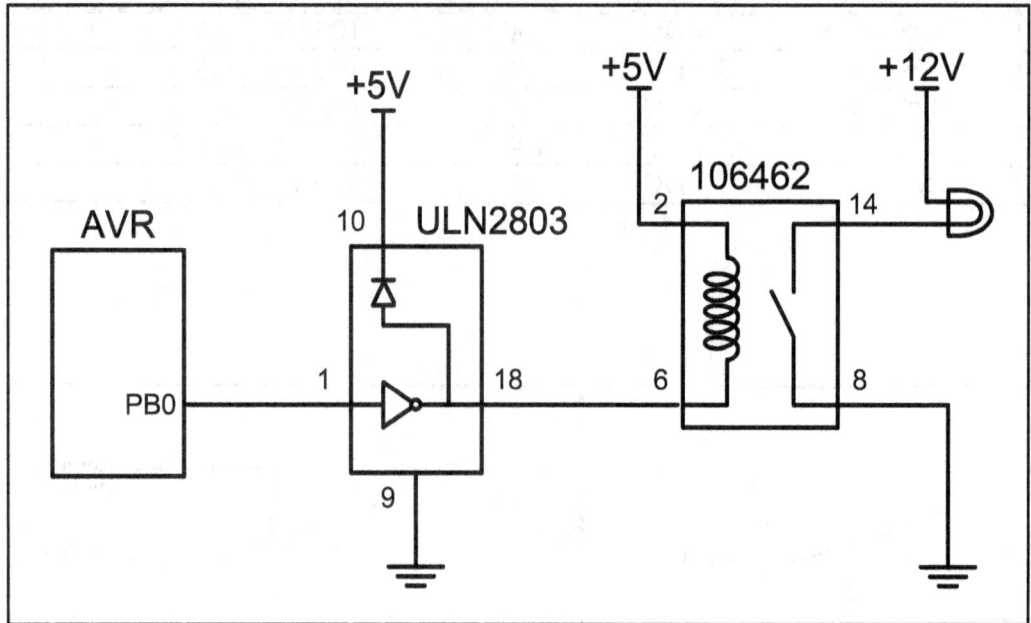

Figure 14-2. AVR Connection to Relay

Program 14-1 turns the lamp shown in Figure 14-2 on and off by energizing and de-energizing the relay every second.

```
;Program 14-1
.INCLUDE "M32DEF.INC"

        LDI    R16,HIGH(RAMEND)  ;initialize stack pointer
        OUT    SPH,R16
        LDI    R16,LOW(RAMEND)
        OUT    SPL,R16

        SBI    DDRB,0             ;PB0 as an output
BEGIN:SBI      PORTB,0            ;PB0 = 1
        RCALL  DELAY_1s
        CBI    PORTB,0            ;PB0 = 0
        RCALL  DELAY_1s
        RJMP   BEGIN

DELAY_1s:
        ...    ;add the DELAY_1s function from Example 9-32
        RET
```

Solid-state relay

Another widely used relay is the solid-state relay. See Table 14-2. In this relay, there is no coil, spring, or mechanical contact switch. The entire relay is made out of semiconductor materials. Because no mechanical parts are involved in solid-state relays, their switching response time is much faster than that of electromechanical relays. Another advantage of the solid-state relay is its greater life expectancy. The life cycle for the electromechanical relay can vary from a few hundred thousand to a few million operations. Wear and tear on the contact points can cause the relay to malfunction after a while. Solid-state relays, however, have no such limitations. Extremely low input current and small packaging make solid-state relays ideal for microcontroller and logic control switching. They are widely used in controlling pumps, solenoids, alarms, and other power applications. Some solid-state relays have a phase control option, which is ideal for motor-speed control and light-dimming applications. Figure 14-3 shows control of a fan using a solid-state relay (SSR).

Table 14-2: Selected Solid-State Relay Characteristics (www.Jameco.com)

Part No.	Contact Style	Control Volts	Contact Volts	Contact Current
143058CP	SPST	4–32 VDC	240 VAC	3 A
139053CP	SPST	3–32 VDC	240 VAC	25 A
162341CP	SPST	3–32 VDC	240 VAC	10 A
172591CP	SPST	3–32 VDC	60 VDC	2 A
175222CP	SPST	3–32 VDC	60 VDC	4 A
176647CP	SPST	3–32 VDC	120 VDC	5 A

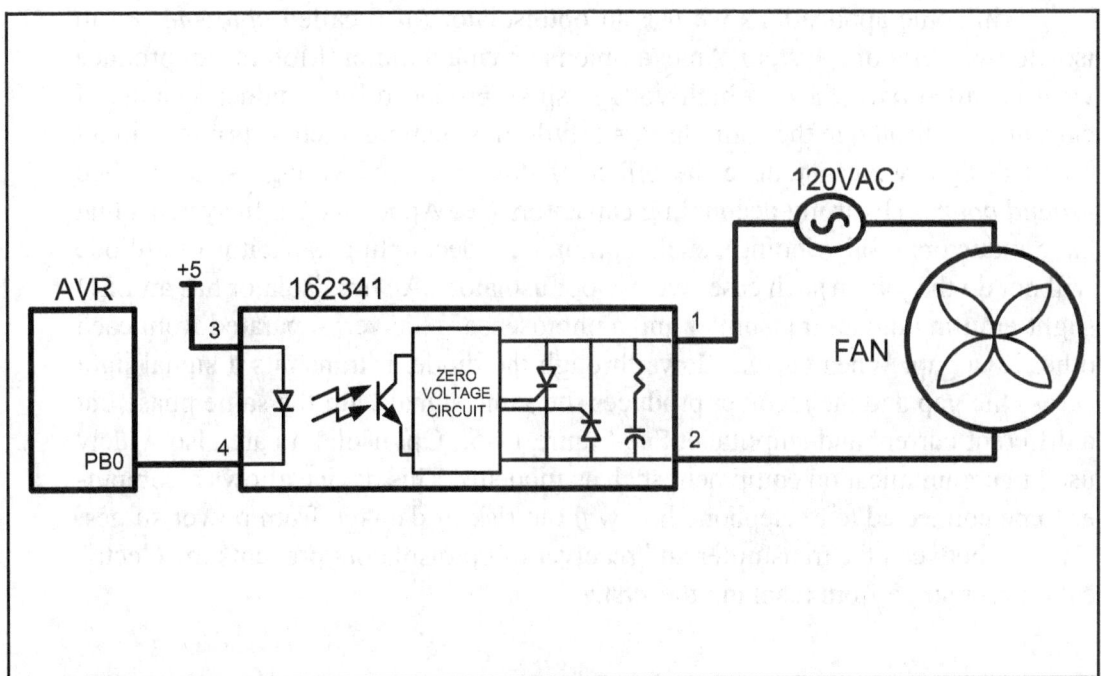

Figure 14-3. AVR Connection to a Solid-State Relay

Reed switch

Another popular switch is the reed switch. When the reed switch is placed in a magnetic field, the contact is closed. When the magnetic field is removed, the contact is forced open by its spring. See Figure 14-4. The reed switch is ideal for moist and marine environments where it can be submerged in fuel or water. Reed switches are also widely used in dirty and dusty atmospheres because they are tightly sealed.

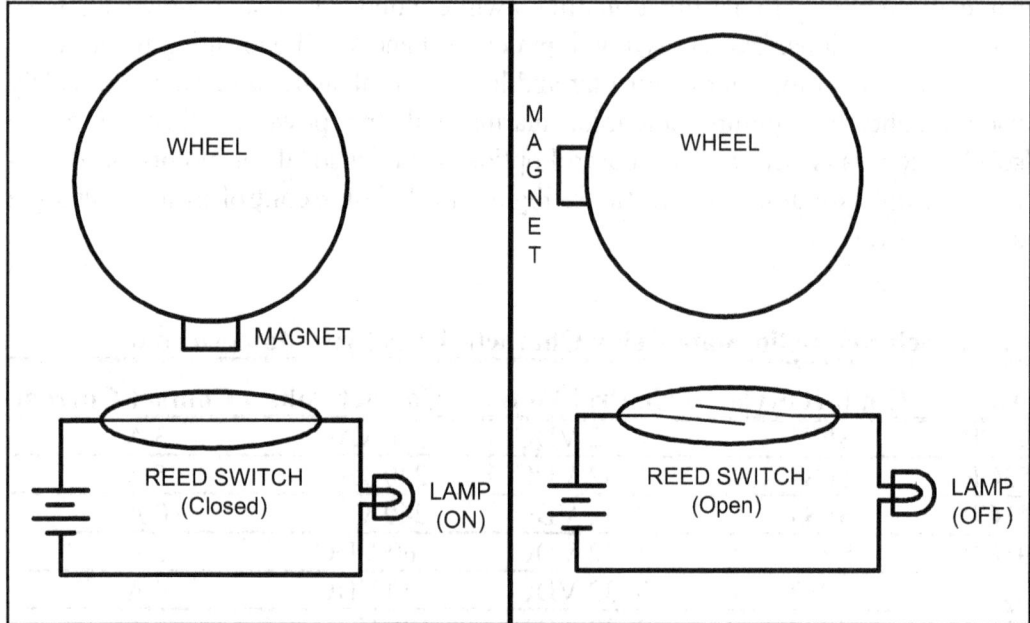

Figure 14-4. Reed Switch and Magnet Combination

Optoisolator

In some applications we use an optoisolator (also called *optocoupler*) to isolate two parts of a system. An example is driving a motor. Motors can produce what is called *back EMF*, a high-voltage spike produced by a sudden change of current as indicated in the formula V = Ldi/dt. In situations such as printed circuit board design, we can reduce the effect of this unwanted voltage spike (called *ground bounce*) by using decoupling capacitors (see Appendix C). In systems that have inductors (coil winding), such as motors, a decoupling capacitor or a diode will not do the job. In such cases we use optoisolators. An optoisolator has an LED (light-emitting diode) transmitter and a photosensor receiver, separated from each other by a gap. When current flows through the diode, it transmits a signal light across the gap and the receiver produces the same signal with the same phase but a different current and amplitude. See Figure 14-5. Optoisolators are also widely used in communication equipment such as modems. This device allows a computer to be connected to a telephone line without risk of damage from power surges. The gap between the transmitter and receiver of optoisolators prevents the electrical current surge from reaching the system.

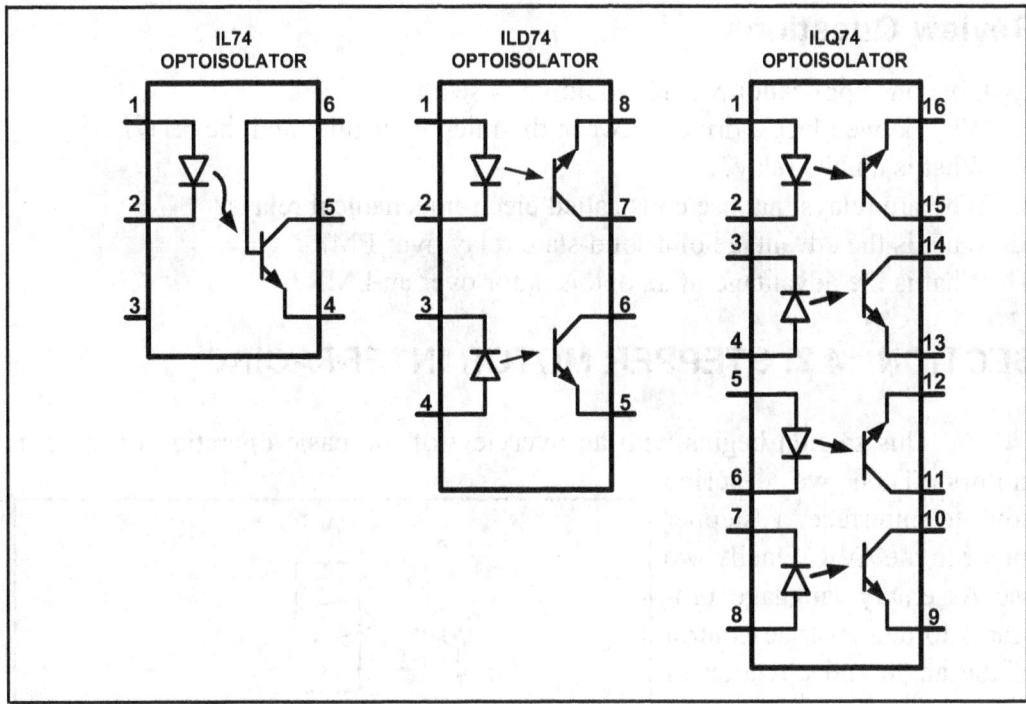

Figure 14-5. Optoisolator Package Examples

Interfacing an optoisolator

The optoisolator comes in a small IC package with four or more pins. There are also packages that contain more than one optoisolator. When placing an optoisolator between two circuits, we must use two separate voltage sources, one for each side, as shown in Figure 14-6. Unlike relays, no drivers need to be placed between the microcontroller/digital output and the optoisolators.

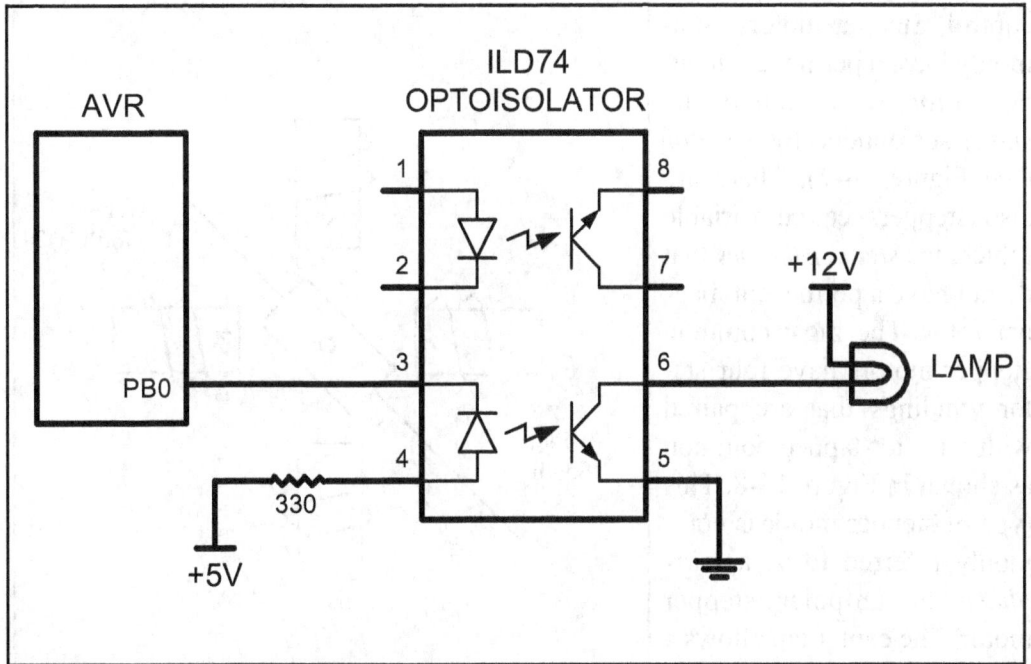

Figure 14-6. Controlling a Lamp via an Optoisolator

Review Questions

1. Give one application where would you use a relay.
2. Why do we place a driver between the microcontroller and the relay?
3. What is an NC relay?
4. Why are relays that use coils called electromechanical relays?
5. What is the advantage of a solid-state relay over EMR?
6. What is the advantage of an optoisolator over an EMR?

SECTION 14.2: STEPPER MOTOR INTERFACING

This section begins with an overview of the basic operation of stepper motors. Then we describe how to interface a stepper motor to the AVR. Finally, we use Assembly language programs to demonstrate control of the angle and direction of stepper motor rotation.

Stepper motors

A *stepper motor* is a widely used device that translates electrical pulses into mechanical movement. In applications such as disk drives, dot matrix printers, and robotics, the stepper motor is used for position control. Stepper motors commonly have a permanent magnet *rotor* (also called the *shaft*) surrounded by a *stator* (see Figure 14-7). There are also steppers called variable reluctance *stepper motors* that do not have a permanent magnet rotor. The most common stepper motors have four stator windings that are paired with a center-tapped common as shown in Figure 14-8. This type of stepper motor is commonly referred to as a *four-phase* or unipolar stepper motor. The center tap allows a change of current direction in

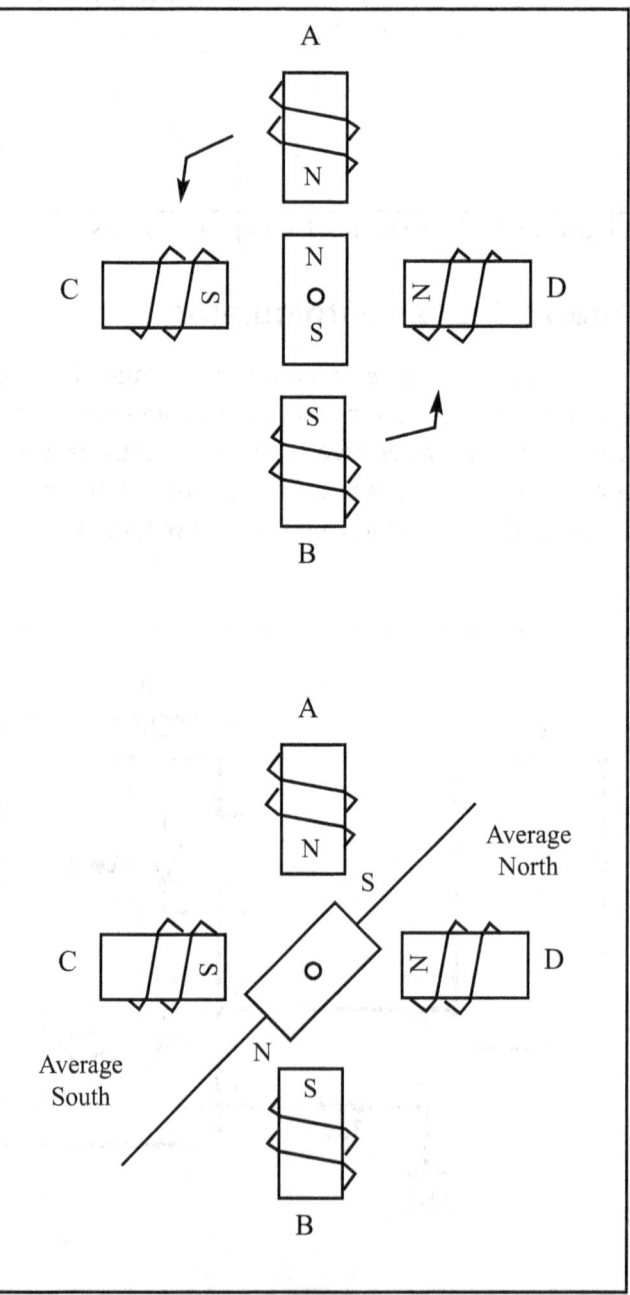

Figure 14-7. Rotor Alignment

498

each of two coils when a winding is grounded, thereby resulting in a polarity change of the stator. Notice that while a conventional motor shaft runs freely, the stepper motor shaft moves in a fixed repeatable increment, which allows it to move to a precise position. This repeatable fixed movement is possible as a result of basic magnetic theory where poles of the same polarity repel and opposite poles attract. The direction of the rotation is dictated by the stator

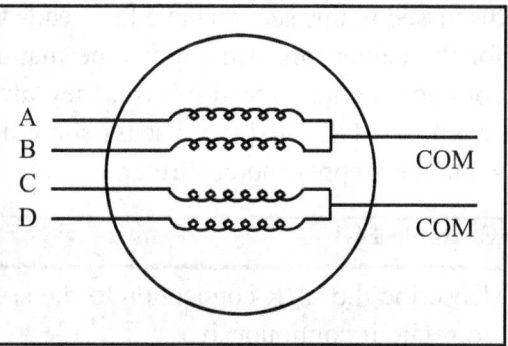

Figure 14-8. Stator Winding Configuration

poles. The stator poles are determined by the current sent through the wire coils. As the direction of the current is changed, the polarity is also changed causing the reverse motion of the rotor. The stepper motor discussed here has a total of six leads: four leads representing the four stator windings and two commons for the center-tapped leads. As the sequence of power is applied to each stator winding, the rotor will rotate. There are several widely used sequences, each of which has a different degree of precision. Table 14-3 shows a two-phase, four-step stepping sequence.

Note that although we can start with any of the sequences in Table 14-3, once we start we must continue in the proper order. For example, if we start with step 3 (0110), we must continue in the sequence of steps 4, 1, 2, and so on.

Table 14-3: Normal Four-Step Sequence

Clockwise	Step #	Winding A	Winding B	Winding C	Winding D	Counter-clockwise
	1	1	0	0	1	
	2	1	1	0	0	
	3	0	1	1	0	
	4	0	0	1	1	

Step angle

How much movement is associated with a single step? This depends on the internal construction of the motor, in particular the number of teeth on the stator and the rotor. The *step angle* is the minimum degree of rotation associated with a single step. Various motors have different step angles. Table 14-4 shows some step angles for various motors. In Table 14-4, notice the term *steps per revolution*. This is the total number of steps needed to rotate one complete rotation or 360 degrees (e.g., 180 steps × 2 degrees = 360).

It must be noted that perhaps contrary to one's initial impression, a stepper motor does not need more terminal leads for the stator to achieve smaller steps. All the stepper motors

Table 14-4: Stepper Motor Step Angles

Step Angle	Steps per Revolution
0.72	500
1.8	200
2.0	180
2.5	144
5.0	72
7.5	48
15	24

discussed in this section have four leads for the stator winding and two COM wires for the center tap. Although some manufacturers set aside only one lead for the common signal instead of two, they always have four leads for the stators. See Example 14-1. Next we discuss some associated terminology in order to understand the stepper motor further.

Example 14-1

Describe the AVR connection to the stepper motor of Figure 14-9 and code a program to rotate it continuously.

Solution:

The following steps show the AVR connection to the stepper motor and its programming:

1. Use an ohmmeter to measure the resistance of the leads. This should identify which COM leads are connected to which winding leads.
2. The common wire(s) are connected to the positive side of the motor's power supply. In many motors, +5 V is sufficient.
3. The four leads of the stator winding are controlled by four bits of the AVR port (PB0–PB3). Because the AVR lacks sufficient current to drive the stepper motor windings, we must use a driver such as the ULN2003 (or ULN2803) to energize the stator. Instead of the ULN2003, we could have used transistors as drivers, as shown in Figure 14-11. However, notice that if transistors are used as drivers, we must also use diodes to take care of inductive current generated when the coil is turned off. One reason that using the ULN2003 is preferable to the use of transistors as drivers is that the ULN2003 has an internal diode to take care of back EMF.

```
.INCLUDE  "M32DEF.INC"
      LDI    R20,HIGH(RAMEND) ;initialize stack pointer
      OUT    SPH,R20
      LDI    R20,LOW(RAMEND)
      OUT    SPL,R20
      LDI    R20,0xFF          ;Port B as output
      OUT    DDRB,R20
      LDI    R20,0x06          ;load step sequence
L1:   OUT    PORTB,R20         ;PORTB = R20
      LSR    R20               ;shift right
      BRCC   L2                ;if not carry skip next
      ORI    R20,0x8
L2:   RCALL  DELAY             ;wait
      RJMP   L1
DELAY:LDI    R16,0x50
D_L1: NOP
      NOP
      DEC    R16
      BRNE   D_L1
      RET
```

Change the value of DELAY to set the speed of rotation.

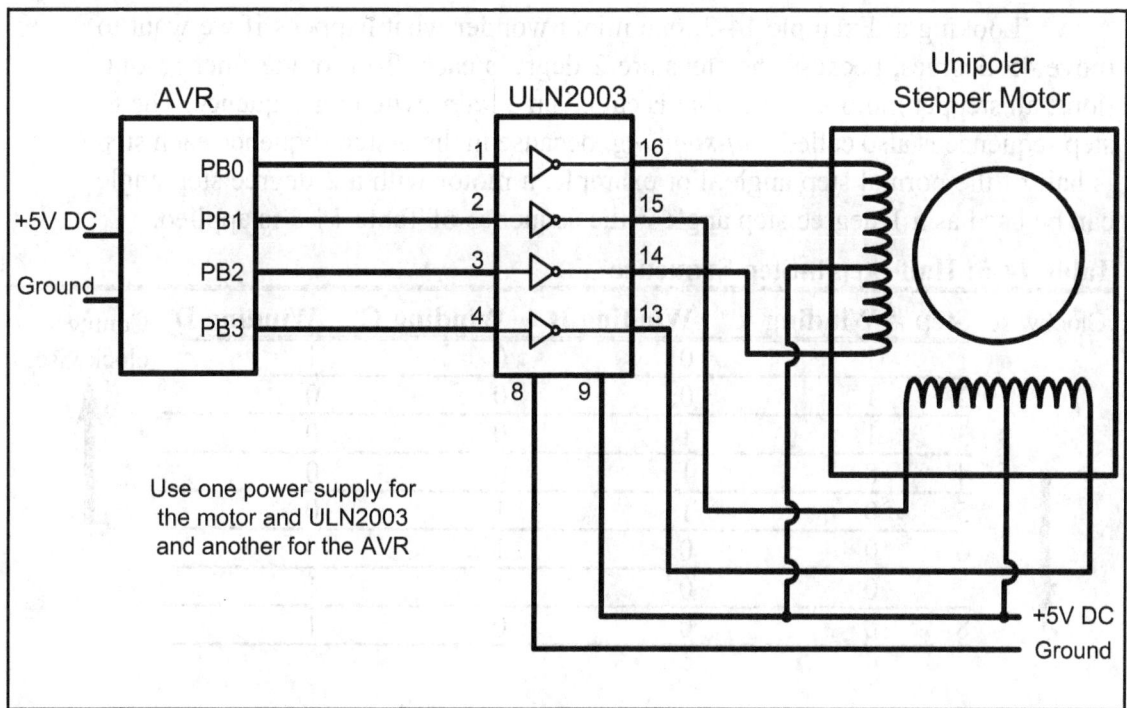

Figure 14-9. AVR Connection to Stepper Motor

Steps per second and RPM relation

The relation between RPM (revolutions per minute), steps per revolution, and steps per second is as follows.

$$Steps\ per\ second = \frac{RPM \times Steps\ per\ revolution}{60}$$

The 4-step sequence and number of teeth on rotor

The switching sequence shown earlier in Table 14-3 is called the 4-step switching sequence because after four steps the same two windings will be "ON". How much movement is associated with these four steps? After completing every four steps, the rotor moves only one tooth pitch. Therefore, in a stepper motor with 200 steps per revolution, the rotor has 50 teeth because $4 \times 50 = 200$ steps are needed to complete one revolution. This leads to the conclusion that the minimum step angle is always a function of the number of teeth on the rotor. In other words, the smaller the step angle, the more teeth the rotor has. See Example 14-2.

Example 14-2

Give the number of times the four-step sequence in Table 14-3 must be applied to a stepper motor to make an 80-degree move if the motor has a 2-degree step angle.

Solution:

A motor with a 2-degree step angle has the following characteristics:

Step angle:	2 degrees	Steps per revolution:	180
Number of rotor teeth:	45	Movement per 4-step sequence:	8 degrees

To move the rotor 80 degrees, we need to send 10 consecutive 4-step sequences, because 10×4 steps $\times$ 2 degrees = 80 degrees.

Looking at Example 14-2, one might wonder what happens if we want to move 45 degrees, because the steps are 2 degrees each. To provide finer resolutions, all stepper motors allow what is called an *8-step* switching sequence. The 8-step sequence is also called *half-stepping,* because in the 8-step sequence each step is half of the normal step angle. For example, a motor with a 2-degree step angle can be used as a 1-degree step angle if the sequence of Table 14-5 is applied.

Table 14-5: Half-Step 8-Step Sequence

Clockwise	Step #	Winding A	Winding B	Winding C	Winding D	Counter-clockwise
	1	1	0	0	1	
	2	1	0	0	0	
	3	1	1	0	0	
	4	0	1	0	0	
	5	0	1	1	0	
	6	0	0	1	0	
	7	0	0	1	1	
	8	0	0	0	1	

Motor speed

The motor speed, measured in steps per second (steps/s), is a function of the switching rate. Notice in Example 14-1 that by changing the length of the time delay loop, we can achieve various rotation speeds.

Holding torque

The following is a definition of holding torque: "With the motor shaft at standstill or zero rpm condition, the amount of torque, from an external source, required to break away the shaft from its holding position. This is measured with rated voltage and current applied to the motor." The unit of torque is ounce-inch (or kg-cm).

Wave drive 4-step sequence

In addition to the 8-step and the 4-step sequences discussed earlier, there is another sequence called the *wave drive 4-step sequence*. It is shown in Table 14-6. Notice that the 8-step sequence of Table 14-5 is simply the combination of the wave drive 4-step and normal 4-step normal sequences shown in Tables 14-6 and 14-3, respectively. Experimenting with the wave drive 4-step sequence is left to the reader.

Table 14-6: Wave Drive 4-Step Sequence

Clockwise	Step #	Winding A	Winding B	Winding C	Winding D	Counter-clockwise
	1	1	0	0	0	
	2	0	1	0	0	
	3	0	0	1	0	
	4	0	0	0	1	

Table 14-7: Selected Stepper Motor Characteristics (www.Jameco.com)

Part No.	Step Angle	Drive System	Volts	Phase Resistance	Current
151861CP	7.5	unipolar	5 V	9 ohms	550 mA
171601CP	3.6	unipolar	7 V	20 ohms	350 mA
164056CP	7.5	bipolar	5 V	6 ohms	800 mA

Unipolar versus bipolar stepper motor interface

There are three common types of stepper motor interfacing: universal, unipolar, and bipolar. They can be identified by the number of connections to the motor. A universal stepper motor has eight, while the unipolar has six and the bipolar has four. The universal stepper motor can be configured for all three modes, while the unipolar can be either unipolar or bipolar. Obviously the bipolar cannot be configured for universal nor unipolar mode. Table 14-7 shows selected stepper motor characteristics. Figure 14-10 shows the basic internal connections of all three type of configurations.

Unipolar stepper motors can be controlled using the basic interfacing shown in Figure 14-11, whereas the bipolar stepper requires H-Bridge circuitry. Bipolar stepper motors require a higher operational current than the unipolar; the advantage of this is a higher holding torque.

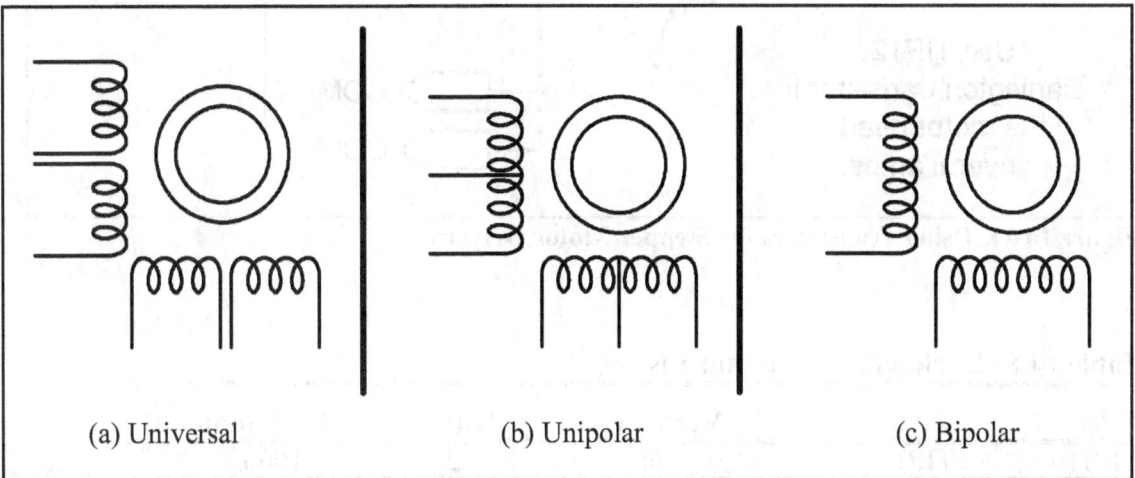

(a) Universal   (b) Unipolar   (c) Bipolar

Figure 14-10. Common Stepper Motor Types

Using transistors as drivers

Figure 14-11 shows an interface to a unipolar stepper motor using transistors. Diodes are used to reduce the back EMF spike created when the coils are energized and de-energized, similar to the electromechanical relays discussed earlier. TIP transistors can be used to supply higher current to the motor. Table 14-8 lists the common industrial Darlington transistors. These transistors can accommodate higher voltages and currents.

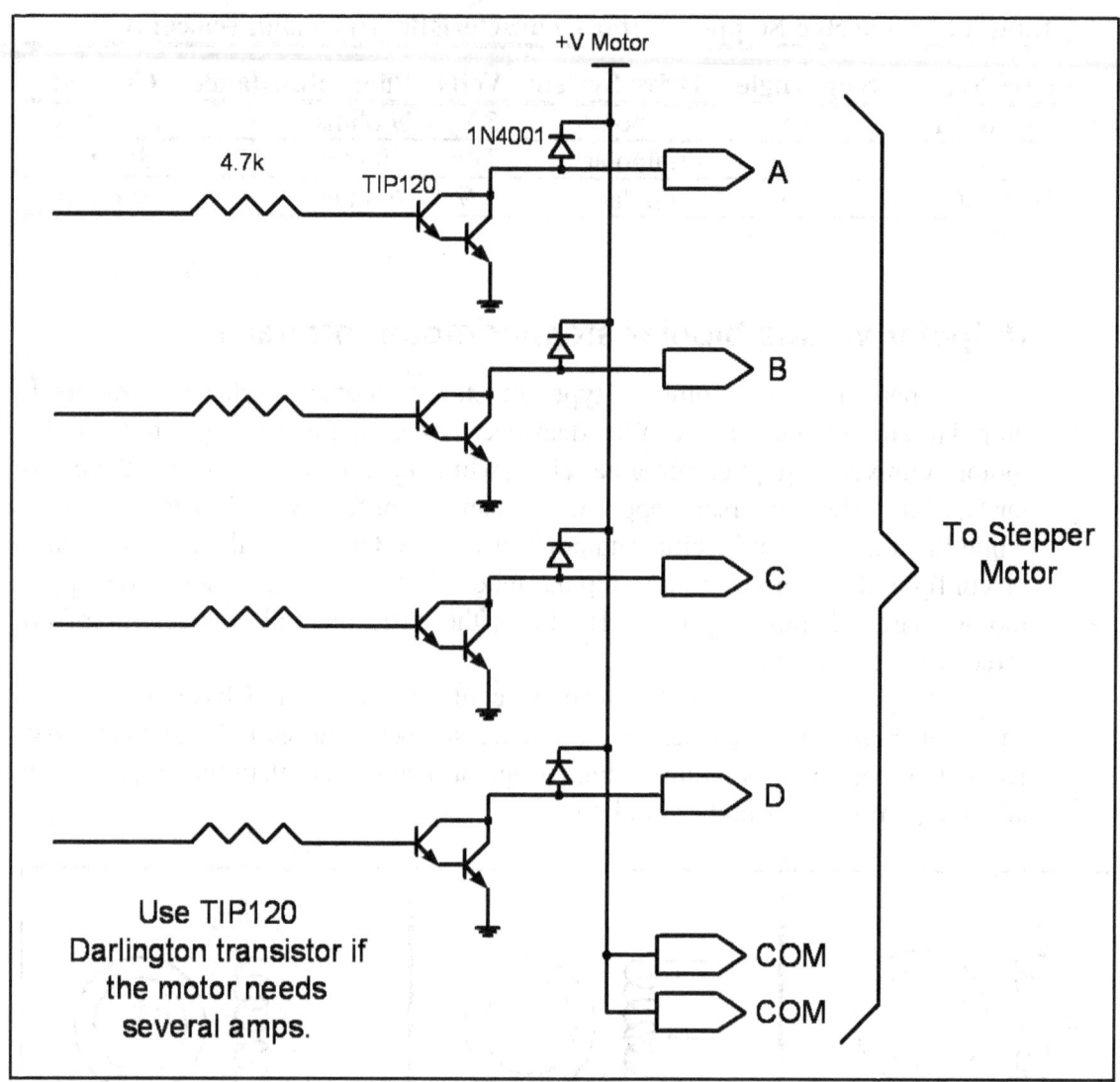

Figure 14-11. Using Transistors for Stepper Motor Driver

Table 14-8: Darlington Transistor Listing

NPN	PNP	Vceo (volts)	Ic (amps)	hfe (common)
TIP110	TIP115	60	2	1000
TIP111	TIP116	80	2	1000
TIP112	TIP117	100	2	1000
TIP120	TIP125	60	5	1000
TIP121	TIP126	80	5	1000
TIP122	TIP127	100	5	1000
TIP140	TIP145	60	10	1000
TIP141	TIP146	80	10	1000
TIP142	TIP147	100	10	1000

Controlling stepper motor via optoisolator

In the first section of this chapter we examined the optoisolator and its use. Optoisolators are widely used to isolate the stepper motor's EMF voltage and keep it from damaging the digital/microcontroller system. This is shown in Figure 14-12. See Examples 14-3 and 14-4.

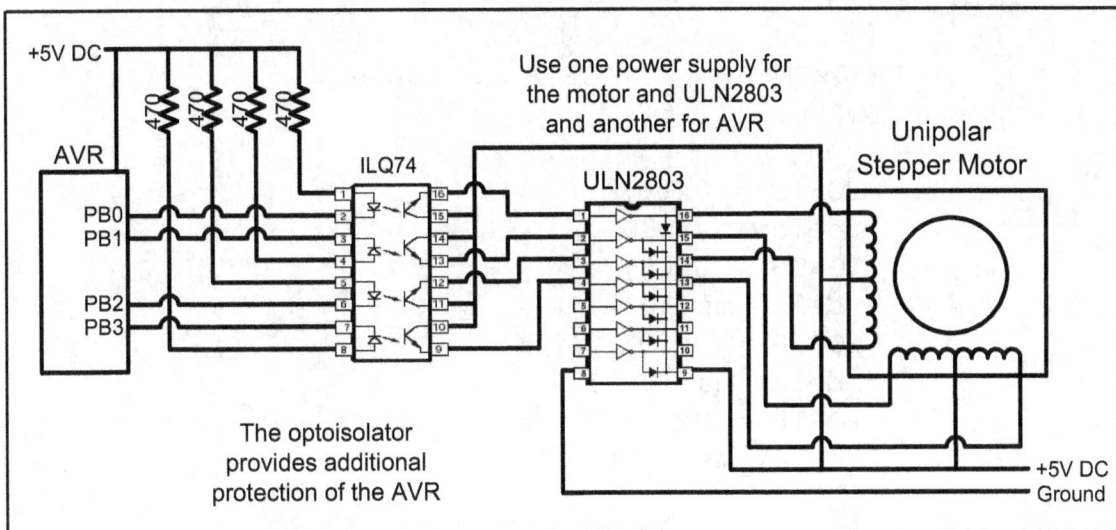

Figure 14-12. Controlling Stepper Motor via Optoisolator

Example 14-3

A switch is connected to pin PA7 (PORTA.7). Write a program to monitor the status of SW and perform the following:

(a) If SW = 0, the stepper motor moves clockwise.

(b) If SW = 1, the stepper motor moves counterclockwise.

Solution:
```
.INCLUDE "M32DEF.INC"
        LDI    R20,HIGH(RAMEND) ;initialize stack pointer
        OUT    SPH,R20
        LDI    R20,LOW(RAMEND)
        OUT    SPL,R20
        LDI    R20,0xFF         ;Port B as output
        OUT    DDRB,R20
        CBI    DDRA,7           ;make PA7 an input
        LDI    R20,0x66         ;starting phase value
L1:     OUT    PORTB,R20        ;PORTB = R20
        IN     R16,PINA
        BST    R16,7            ;T=PINA.7
        BRTS   CW
        LSR    R20              ;shift right
        BRCC   OV1              ;if not carry skip next
        ORI    R20,0x80
OV1:    RCALL  DELAY            ;wait
        RJMP   L1               ;repeat
CW:     LSL    R20              ;shift left
        BRCC   OV2              ;if not carry skip next
        ORI    R20,0x01
OV2:    RCALL  DELAY            ;wait
        RJMP   L1               ;repeat
```

Stepper motor control with AVR C

The AVR C version of the stepper motor control is given below. In this program we could have used << (shift left) and >> (shift right) as was shown in Chapter 7.

```
#include "avr/io.h"
void main()
  {
     DDRB=0xFF;        //PORTB as output
     while(1)
       {
          PORTB = 0x66;
          _delay_ms(100);
          PORTB = 0xCC;
          _delay_ms(100);
          PORTB = 0x99;
          _delay_ms(100);
          PORTB = 0x33;
          _delay_ms(100);
       }
  }
```

Example 14-4

A switch is connected to pin PA7. Write a C program to monitor the status of SW and perform the following:
(a) If SW = 0, the stepper motor moves clockwise.
(b) If SW = 1, the stepper motor moves counterclockwise.

Solution:

```
#define F_CPU    8000000UL       //XTAL = 8 MHz
#include "avr/io.h"
#include "util/delay.h"

int main ()
{
     DDRA = 0x00;
     DDRB = 0xFF;
     while (1)
     {
          if( (PINA&0x80) == 0)
          {
               PORTB = 0x66;
               _delay_ms (100);
               PORTB = 0xCC;
               _delay_ms (100);
               PORTB = 0x99;
               _delay_ms (100);
               PORTB = 0x33;
               _delay_ms (100);
          }
          else
          {
```

Example 14-4 Cont.

```
            PORTB = 0x66;
            _delay_ms (100);
            PORTB = 0x33;
            _delay_ms (100);
            PORTB = 0x99;
            _delay_ms (100);
            PORTB = 0xCC;
            _delay_ms (100);
        }
    }
}
```

Review Questions

1. Give the 4-step sequence of a stepper motor if we start with 0110.
2. A stepper motor with a step angle of 5 degrees has _____ steps per revolution.
3. Why do we put a driver between the microcontroller and the stepper motor?

PROBLEMS

SECTION 14.1: RELAYS AND OPTOISOLATORS

1. True or false. The minimum voltage needed to energize a relay is the same for all relays.
2. True or false. The minimum current needed to energize a relay depends on the coil resistance.
3. Give the advantages of a solid-state relay over an EMR.
4. True or false. In relays, the energizing voltage is the same as the contact voltage.
5. Find the current needed to energize a relay if the coil resistance is 1200 ohms and the coil voltage is 5 V.
6. Give two applications for an optoisolator.
7. Give the advantages of an optoisolator over an EMR.
8. Of the EMR and solid-state relay, which has the problem of back EMF?
9. True or false. The greater the coil inductance, the worse the back EMF voltage.
10. True or false. We should use the same voltage sources for both the coil voltage and the contact voltage.

SECTION 14.2: STEPPER MOTOR INTERFACING

11. If a motor takes 90 steps to make one complete revolution, what is the step angle for this motor?
12. Calculate the number of steps per revolution for a step angle of 7.5 degrees.
13. Finish the normal 4-step sequence clockwise if the first step is 0011 (binary).
14. Finish the normal 4-step sequence clockwise if the first step is 1100 (binary).
15. Finish the normal 4-step sequence counterclockwise if the first step is 1001 (binary).

---

16. Finish the normal 4-step sequence counterclockwise if the first step is 0110 (binary).
17. What is the purpose of the ULN2003 placed between the AVR and the stepper motor? Can we use that for 3A motors?
18. Which of the following cannot be a sequence in the normal 4-step sequence for a stepper motor?

    (a) $CC   (b) $DD   (c) $99   (d) $33
19. What is the effect of a time delay between issuing each step?
20. In Question 19, how can we make a stepper motor go faster?

ANSWERS TO REVIEW QUESTIONS

SECTION 14.1: RELAYS AND OPTOISOLATORS

1. With a relay we can use a 5 V digital system to control 12 V–120 V devices such as horns and appliances.
2. Because microcontroller/digital outputs lack sufficient current to energize the relay, we need a driver.
3. When the coil is not energized, the contact is closed.
4. When current flows through the coil, a magnetic field is created around the coil, which causes the armature to be attracted to the coil.
5. It is faster and needs less current to get energized.
6. It is smaller and can be connected to the microcontroller directly without a driver.

SECTION 14.2: STEPPER MOTOR INTERFACING

1. 0110, 0011, 1001, 1100 for clockwise; and 0110, 1100, 1001, 0011 for counterclockwise
2. 72
3. The microcontroller pins do not provide sufficient current to drive the stepper motor.

CHAPTER 15

INPUT CAPTURE AND WAVE GENERATION IN AVR

OBJECTIVES

Upon completion of this chapter, you will be able to:

>> Understand the compare and capture features of the AVR
>> Generate pulses with different frequencies
>> Explain how the wave generators of timers work
>> Explain the different operation modes of Timer0 and Timer1
>> Explain how the capture feature of Timer1 works
>> Code programs for the capture feature in Assembly and C

In Chapter 9, you learned how to use AVR timers to generate delay and count external events. AVR timers have other features as well. They can be used for generating different square waves or capturing events and measuring the frequency and duty cycle of waves. These usages are discussed in this chapter and Chapter 16. In Sections 15.1 and 15.2 you learn to generate waves using 8-bit and 16-bit timers, respectively. In Section 15.3 you learn to capture events and measure the frequency and duty cycle of waves. You can find the C versions of the programs in Section 15.4.

SECTION 15.1: WAVE GENERATION USING 8-BIT TIMERS

Examine Figure 15-1. As mentioned in Chapter 9, for each timer there is, at least, an OCRn register (like OCR0 for Timer0). The value of this register is constantly compared with the TCNTn register, and when a match occurs, the OCFn flag will be set to high.

As shown in Figures 15-1 and 15-2, in each AVR timer there is a waveform generator. The waveform generator can generate waves on the OCn pin. The WGMn and COMn bits of the TCCR register determine how the waveform generator works. When the TCNTn register reaches Top or Bottom or compare match

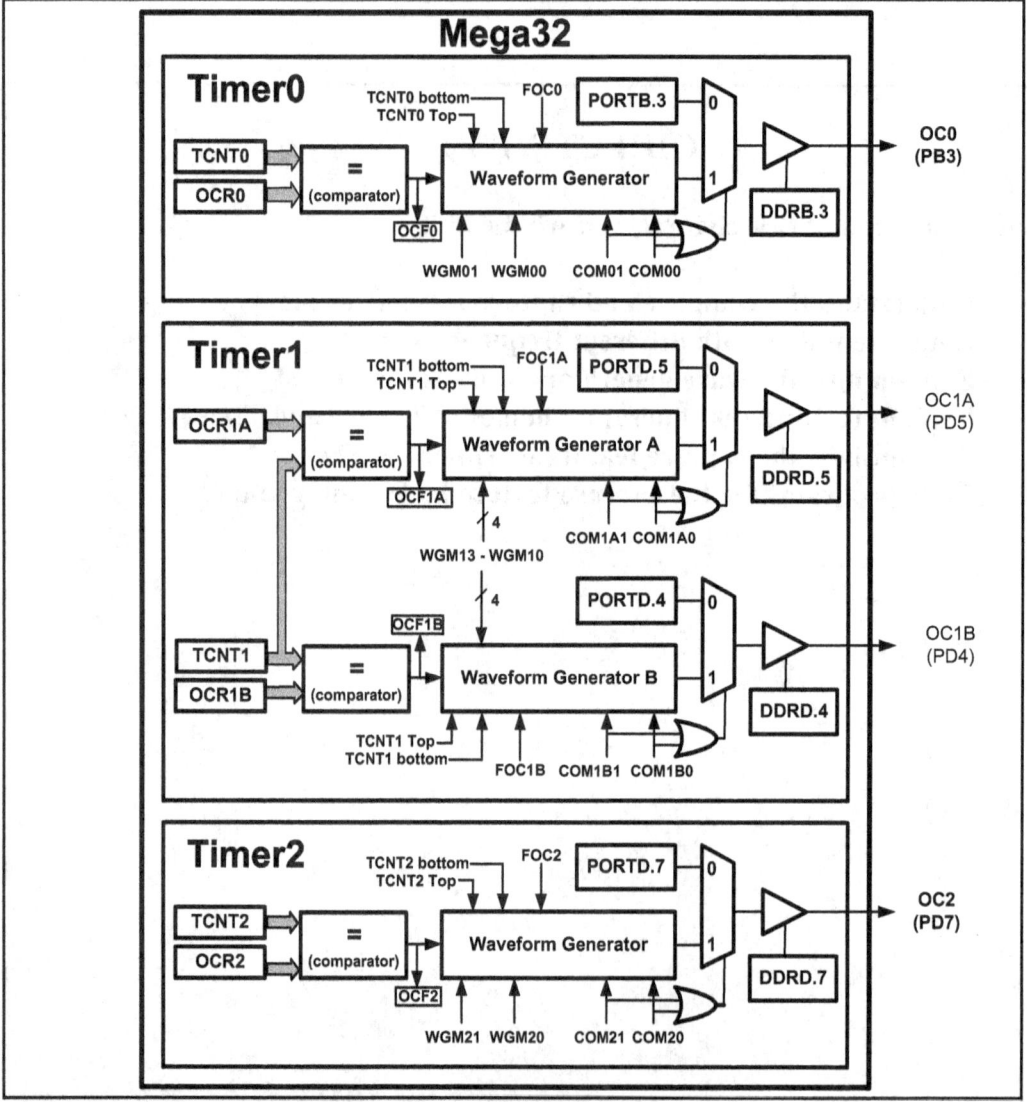

Figure 15-1. Waveform Generators in ATmega32

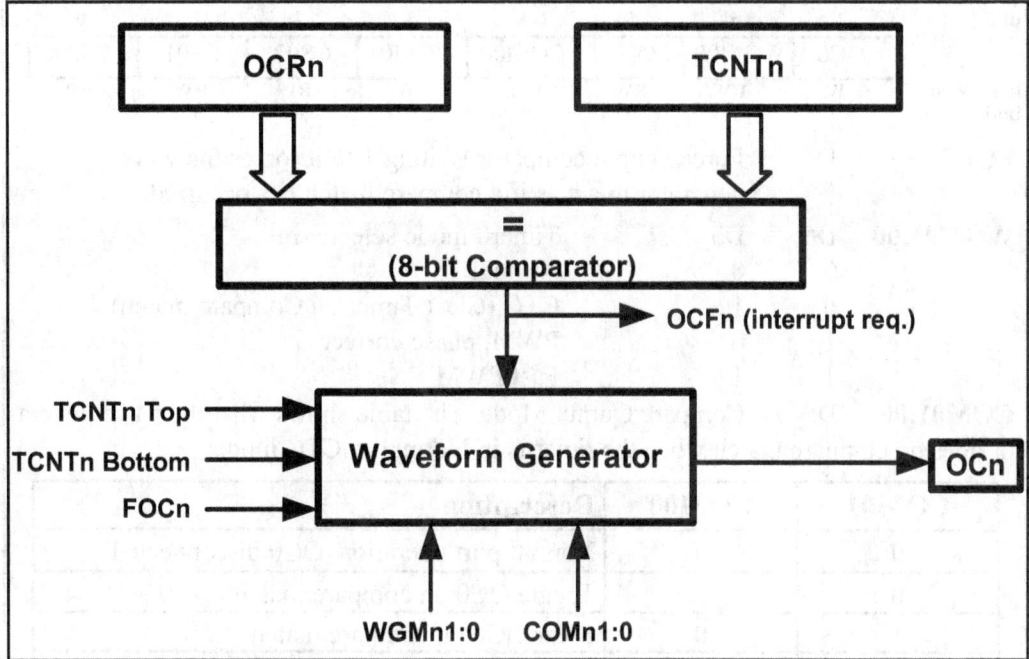

Figure 15-2. Waveform Generator

occurs, the waveform generator is informed. Then the waveform generator changes the state of the OC0 pin according to the mode of the timer (WGM01:00 bits of the TCCR0 register) and the COM01 (Compare Output Mode) and COM00 bits. See Figure 15-4.

In ATmega32/ATmega16, OC0 is the alternative function of PB3. In other words, the PB3 functions as an I/O port when both COM01 and COM00 are zero. Otherwise, the pin acts as a wave generator pin controlled by a waveform generator. See Figures 15-1 and 15-3. Notice that, since the DDR register represents the direction of the I/O pin, we should set the OC0 pin as an output pin when we want to use it for generating waves.

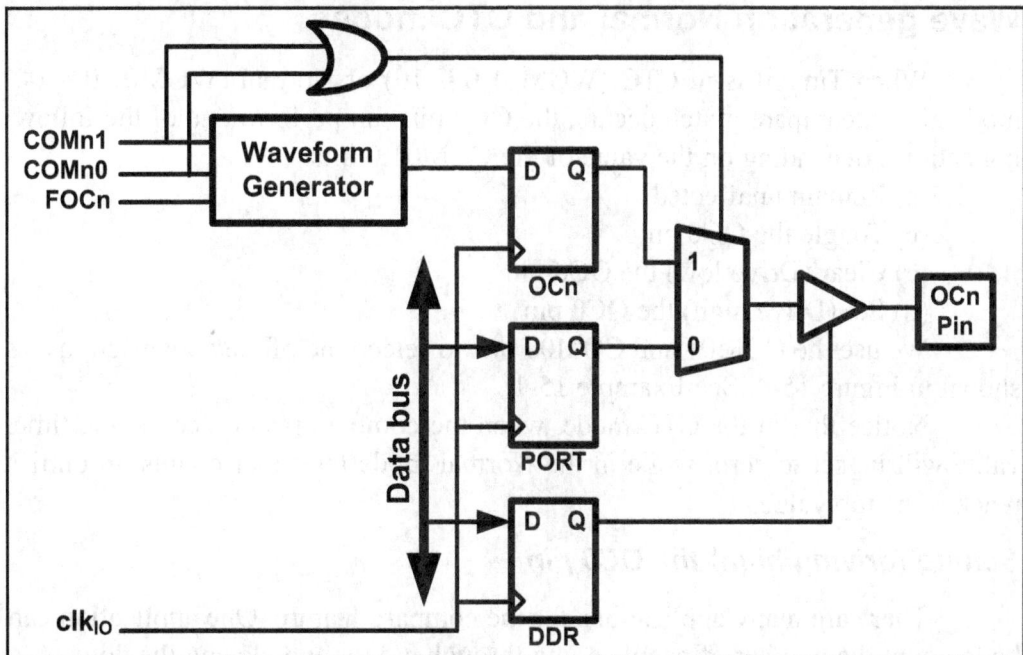

Figure 15-3. DDR Register and Waveform Generator

Bit	7	6	5	4	3	2	1	0
	FOC0	WGM00	COM01	COM00	WGM01	CS02	CS01	CS00
Read/Write	W	RW	RW	RW	RW	RW	RW	RW
Initial Value	0	0	0	0	0	0	0	0

FOC0   D7   Force Output compare: Writing 1 to it forces the wave generator to act as if a compare match has occurred.

WGM01:00   D6   D3   Timer0 mode selector bits

0	0	Normal
0	1	CTC (Clear Timer on Compare match)
1	0	PWM, phase correct
1	1	Fast PWM

COM01:00   D5 D4   Compare Output Mode; The table shows what the wave generator does on compare match when the timer is in Normal or CTC mode:

COM01	COM00	Description
0	0	Normal port operation, OC0 disconnected
0	1	Toggle OC0 on compare match
1	0	Clear OC0 on compare match
1	1	Set OC0 on compare match

CS02:00   D2D1D0   Timer0 clock selector

0	0	0	No clock source (Timer/Counter stopped)
0	0	1	clk (no prescaling)
0	1	0	clk / 8
0	1	1	clk / 64
1	0	0	clk / 256
1	0	1	clk / 1024
1	1	0	External clock source on T0 pin. Clock on falling edge
1	1	1	External clock source on T0 pin. Clock on rising edge

Figure 15-4. TCCR0 (Timer/Counter Control Register) Register

Wave generation Normal and CTC modes

When Timer0 is in CTC (WGM01:0 = 10) or Normal (WGM01:0 = 00) mode after a compare match occurs, the OC0 pin can perform one of the following actions, depending on the value of the COM01:0 bits:

(a) Remain unaffected

(b) Toggle the OC0 pin

(c) Clear (Drive low) the OC0 pin

(d) Set (Drive high) the OC0 pin

We use the COM01 and COM00 bits to select one of the above actions; as shown in Figure 15-4. See Example 15-1.

Notice that in the CTC mode, when the compare match occurs, the timer value will be set to zero, while in the Normal mode the timer counts up until it reaches the top value.

Setting (driving high) the OC0 pin

There are many applications for the compare feature. One application can be to count the number of people going through a door and closing the door when a certain number is reached. See Example 15-2.

Example 15-1

Using Figure 15-4, find the TCCR0 register value to:
(a) Set high the OC0 pin upon match. Use external clock, falling edge, and Normal mode.
(b) Toggle the OC0 pin upon match. Use external clock, falling edge, and CTC mode.

Solution:

(a)   TCCR0 =

0	0	1	1	0	1	1	0
FOC0	WGM00	COM01	COM00	WGM01	CS02	CS01	CS00

(b)   TCCR0 =

0	0	0	1	1	1	1	0
FOC0	WGM00	COM01	COM00	WGM01	CS02	CS01	CS00

Example 15-2

Write a program that (a) after 4 external clocks turns on an LED connected to the OC0 pin, (b) toggles the OC0 pin every 4 pulses.

Solution:

(a)
```
.INCLUDE  "M32DEF.INC"
      CBI    DDRB,0             ;PB0(T0) pin as input
      SBI    DDRB,3             ;PB3(OC0) pin as output
      LDI    R20, 3
      OUT    OCR0,R20           ;OCR0 = 3   the final count
      LDI    R20, 0
      OUT    TCNT0,R20          ;TCNT0 = 0
      LDI    R20,0x36           ;external clk, Normal mode, set OC0
      OUT    TCCR0,R20          ;load TCCR0 and start counting
HERE: RJMP   HERE
```

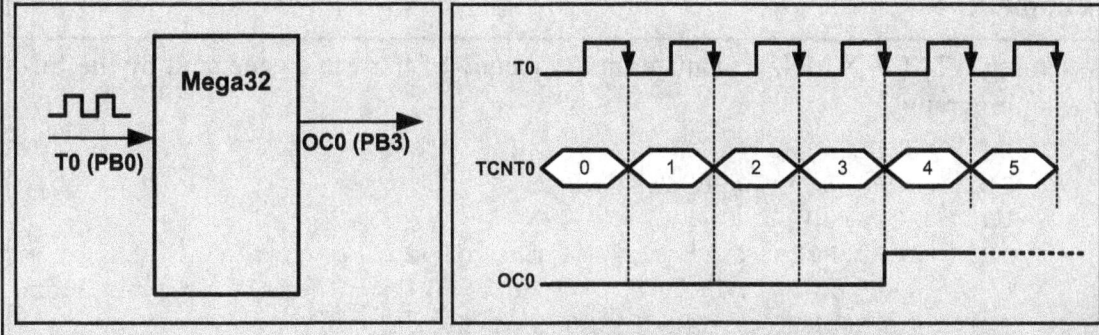

(b)
```
.INCLUDE  "M32DEF.INC"
      CBI    DDRB,0
      SBI    DDRB,3
      LDI    R20, 3
      OUT    OCR0,R20
      LDI    R20, 0
      OUT    TCNT0,R20
      LDI    R20,0x1E           ;external clk, CTC mode, toggle OC0
      OUT    TCCR0,R20          ;load TCCR0 and start counting
HERE: RJMP   HERE
```

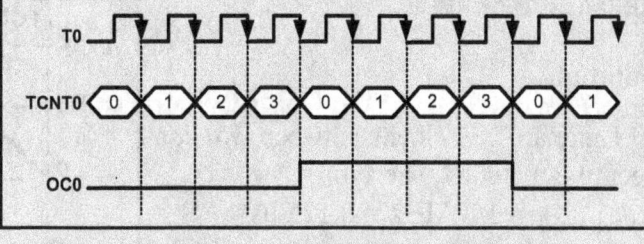

Notice that there is no need to monitor the OCF0 flag, which means the AVR can do other tasks.

Generating square waves

To generate square waves we can set the timer to Normal mode or CTC mode and set the COM bits to the toggle mode (COM01:00 = 01). The OC0 pin will be toggled on each compare match and a square wave will be generated. See Figure 15-5. See Examples 15-3 and 15-4.

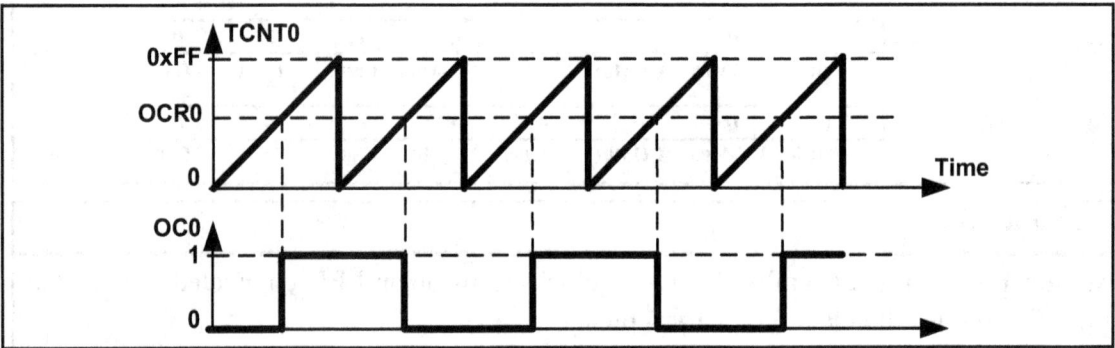

Figure 15-5. Generating Square Wave Using Normal

Example 15-3

Find the value for TCCR0 if we want to program Timer0 as a Normal mode square wave generator and no prescaler.

Solution:

TCCR0 =

0	0	0	1	0	0	0	1
FOC0	WGM00	COM01	COM00	WGM01	CS02	CS01	CS00

Example 15-4

Assuming XTAL = 8 MHz, calculate the frequency of the wave generated by the following program:

```
.INCLUDE "M32DEF.INC"
     SBI    DDRB,3     ;PB3 as output
     LDI    R22,100
     OUT    OCR0,R22   ;set the match value
     LDI    R22,0x11   ;COM01:00 = Toggle, Mode = Normal, no prescaler
     OUT    TCCR0,R22  ;load TCCR0 and start counting
HERE: RJMP  HERE
```

Solution:

There are 256 clocks between two consecutive matches. Therefore

$T_{timer\ clock}$ = 1/8 MHz = 0.125 μs

T_{wave} = 2 × 256 × 0.125 μs = 64 μs

F_{wave} = 1/64 μs = 15,625 Hz = 15.625 kHz

Note: In Normal mode, when match occurs, the OC0 pin toggles and the timer continues to count up until it reaches the top value.

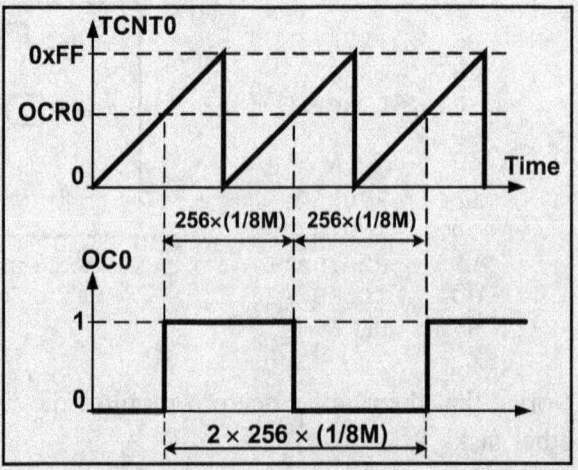

514

Generating square waves using CTC

The CTC mode is better than Normal mode for generating square waves, since the frequency of the wave can be easily adjusted using the OCR0 register. See Figure 15-6. In CTC mode, when OCR0 has a lower value, compare match occurs earlier and the period of the generated wave is smaller (higher frequency). When the OCR0 has a higher value, compare match occurs later and the period of the wave is longer (lower frequency).

Notice that in the CTC mode, when the compare match occurs, the timer value will be set to zero, while in the Normal mode the timer counts up until it reaches the top value. See Examples 15-5 through 15-7.

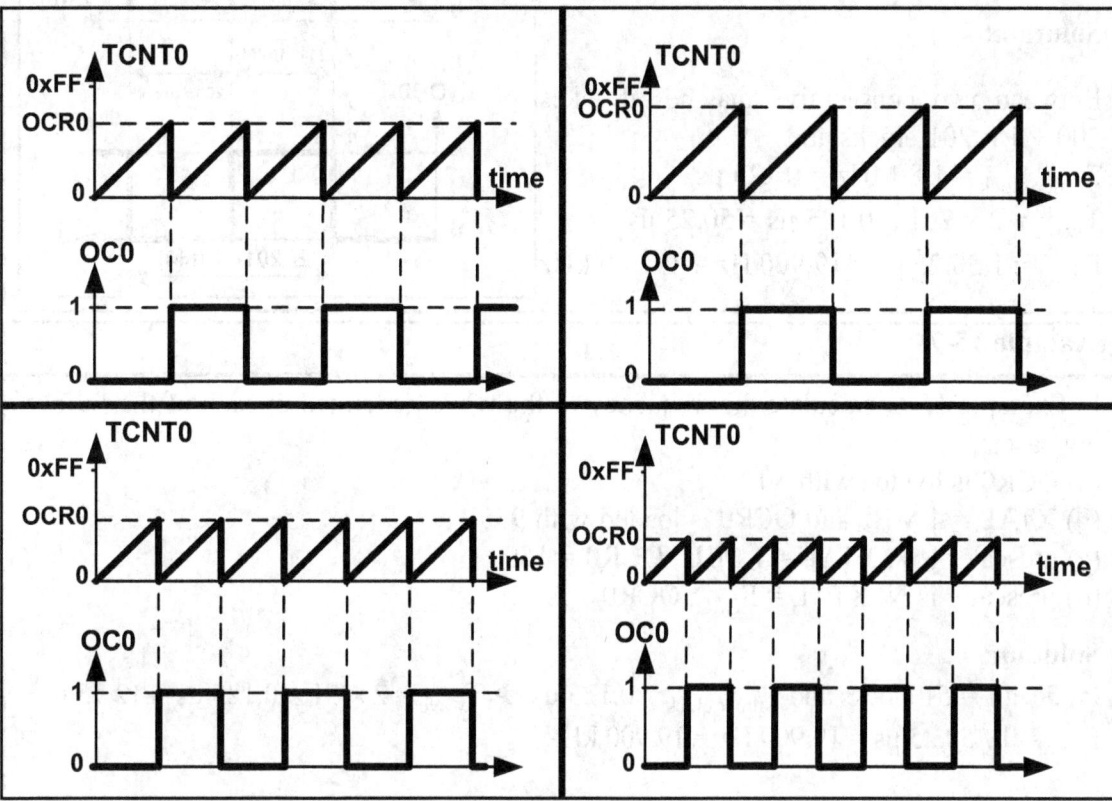

Figure 15-6. Generating Square Wave Using CTC Mode

Example 15-5

Find the value for TCCR0 if we want to program Timer0 as a CTC mode square wave generator and no prescaler.

Solution:

WGM01:00 = 10 = CTC mode
COM01:00 = 01 = Toggle
CS02:00 = 001 = No prescaler
FOC0 = 0

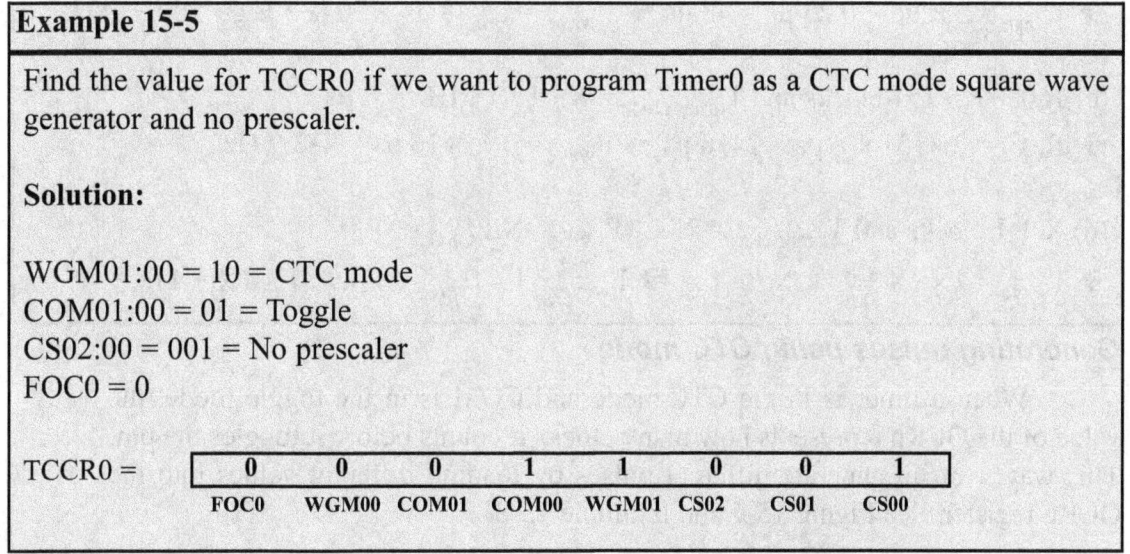

TCCR0 =	0	0	0	1	1	0	0	1
	FOC0	WGM00	COM01	COM00	WGM01	CS02	CS01	CS00

Example 15-6

Assuming XTAL = 8 MHz, calculate the frequency of the wave generated by the following program:

```
.INCLUDE "M32DEF.INC"
    SBI    DDRB,3
    LDI    R20,0x19   ;COM01:00 = Toggle, Mode = CTC, no prescaler
    OUT    TCCR0,R20
    LDI    R22,200
    OUT    OCR0,R22
HERE: RJMP  HERE
```

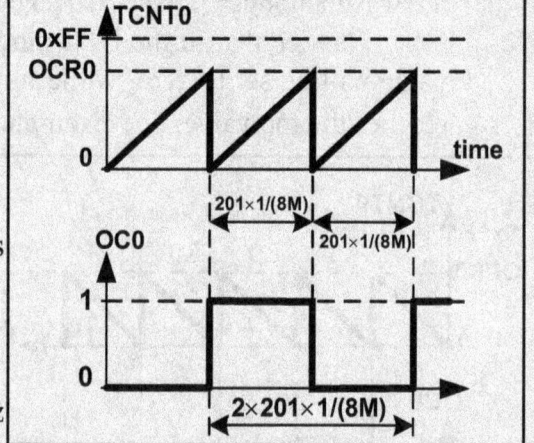

Solution:

Between two consecutive matches it takes $200 + 1 = 201$ clocks and

$T_{\text{timer clock}} = 1/8$ MHz $= 0.125$ μs

$T_{\text{wave}} = 2 \times 201 \times 0.125$ μs $= 50.25$ μs

$F_{\text{wave}} = 1/50.25$ μs $= 19{,}900$ Hz $= 19.900$ kHz

Example 15-7

In Example 15-6, calculate the frequency of the wave generated in each of the following cases:

(a) OCR0 is loaded with 50
(b) XTAL = 4 MHz and OCR0 is loaded with 95
(c) prescaler is 8, XTAL = 1 MHz, OCR0 = 150
(d) prescaler is N, XTAL = F_{OSC}, OCR0 = X

Solution:

(a) $50 + 1 = 51$ clocks and $T_{\text{timer clock}} = 0.125$ μs ➜ $T_{\text{wave}} = 2 \times 51 \times 0.125$ μs $= 12.75$ μs
$F_{\text{wave}} = 1/50.25$ μs $= 19{,}900$ Hz $= 19.900$ kHz

(b) $95 + 1 = 96$ clocks and $T_{\text{timer clock}} = 1/4$ MHz $= 0.25$ μs

➜ $T_{\text{wave}} = 2 \times 96 \times 0.25$ μs $= 48$ μs ➜ $F_{\text{wave}} = 1/48$ μs $= 20{,}833$ Hz $= 20.833$ kHz

(c) $150 + 1 = 151$ clocks and $T_{\text{timer clock}} = 8 \times 1/1$ MHz $= 8$ μs

➜ $T_{\text{wave}} = 2 \times 151 \times 8$ μs $= 2416$ μs ➜ $F_{\text{wave}} = 1/2416$ μs $= 413.9$ Hz

(d) $X + 1$ clocks and $T_{\text{timer clock}} = N \times 1/F_{\text{OSC}} = N/F_{\text{OSC}}$

➜ $T_{\text{wave}} = 2 \times (X + 1) \times N/F_{\text{OSC}}$ ➜ $F_{\text{wave}} = 1/T_{\text{wave}} = F_{\text{OSC}}/[2N(X + 1)]$

Generating pulses using CTC mode

When a timer is in the CTC mode and COM is in the toggle mode, the value of the OCRn represents how many clocks it counts before it toggles the pin. This way, we can generate different pulses by loading different values into the OCRn register. See Figure 15-7 and Example 15-8.

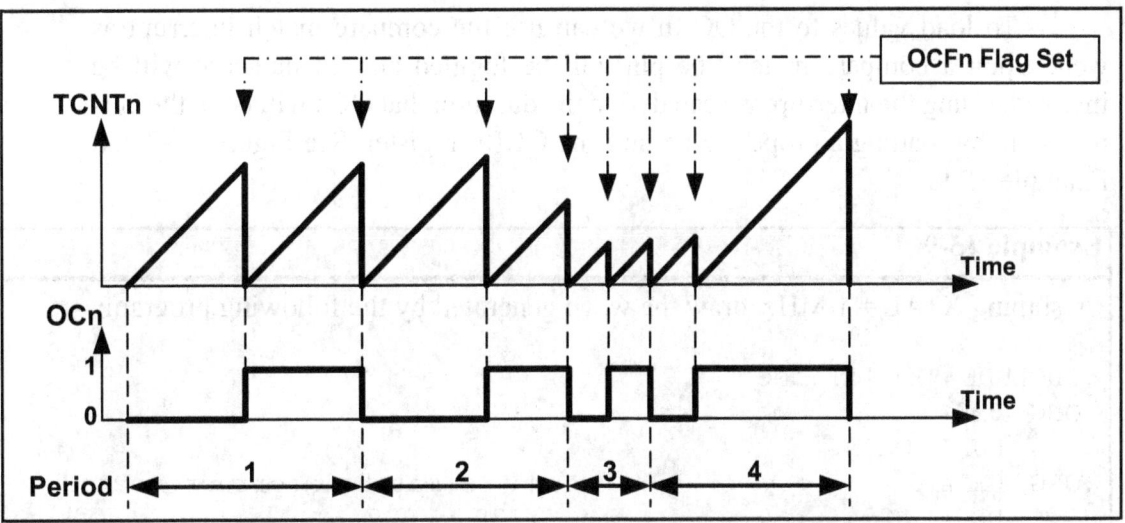

Figure 15-7. Generating Different Pulses Using CTC and Toggle Modes

---

Example 15-8

Assuming XTAL = 1 MHz, draw the wave generated by the following program:

```
.INCLUDE "M32DEF.INC"
       SBI    DDRB,3
BEGIN:LDI    R20,69
       OUT    OCR0,R20     ;OCR0 = 69
       LDI    R20,0x19
       OUT    TCCR0,R20    ;CTC, no prescaler, set on match
L1:    IN     R20,TIFR
       SBRS   R20,OCF0     ;skip next instruction if OCF0 = 1
       RJMP   L1
       LDI    R16,1<<OCF0
       OUT    TIFR,R16     ;clear OCF0
       LDI    R20,99
       OUT    OCR0,R20     ;OCR0 = 99
       LDI    R20,0x29
       OUT    TCCR0,R20    ;CTC, no prescaler, clear on match
L2:    IN     R20,TIFR
       SBRS   R20,OCF0     ;skip next instruction if OCF0 = 1
       RJMP   L2
       LDI    R16,1<<OCF0  ;clear OCF0
       OUT    TIFR,R16
       RJMP   BEGIN
```

Solution:

$T_{\text{timer clock}} = 1 / 1 \text{ MHz} = 1 \text{ μs}$

$T_0 = 70 \times 1 \text{ μs} = 70 \text{ μs}$

$T_1 = 100 \times 1 \text{ μs} = 100 \text{ μs}$

$T_{\text{wave}} = 70 \text{ μs} + 100 \text{ μs} = 170 \text{ μs}$

$F_{\text{wave}} = 1 / 170 \text{ μs} = 5882 \text{ Hz}$

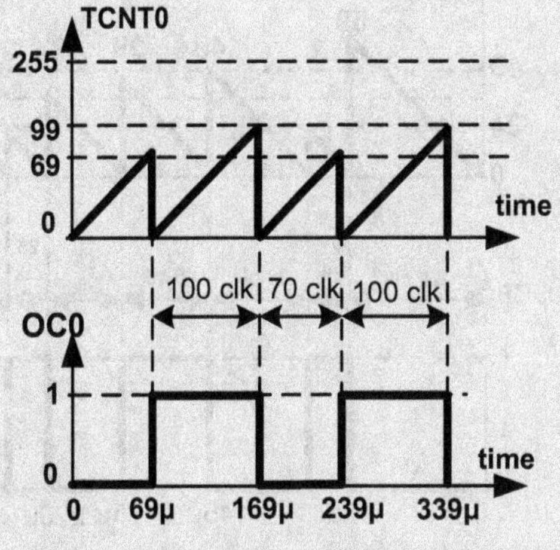

To load values to the OCRn we can use the compare match interrupt as well. Upon a compare match, the pin will be toggled and an interrupt will be invoked. Using the interrupt we can define the duration that OCn will be in the current state by loading a proper value into the OCR0 register. See Figure 15-7 and Example 15-9.

Example 15-9

Assuming XTAL = 1 MHz, draw the wave generated by the following program:

```
.INCLUDE "M32DEF.INC"
.ORG  0x0
      RJMP  MAIN
.ORG  0x14                        ;compare match interrupt vector
      DEC   R29                   ;R29 = R29 - 1
      BRPL  L1                    ;if (R29 >= 0) go to L1
      LDI   R30,WAVE_TABLE<<1     ;Z points to WAVE_TABLE
      LDI   R29,3                 ;R29 = 3
L1:   LPM   R28,Z+               ;R28 = [Z], Z = Z + 1
      OUT   OCR0,R28              ;OCR0 = 99
      RETI                       ;return from interrupt
WAVE_TABLE:      .DB   24,49,39,34
MAIN: LDI   R20,HIGH(RAMEND)
      OUT   SPH,R20
      LDI   R20,LOW(RAMEND)
      OUT   SPL,R20              ;initialize stack
      SBI   DDRB,3               ;PB3 as output
      LDI   R20,69
      OUT   OCR0,R20             ;OCR0 = 69
BEGIN:LDI   R20,0x19
      OUT   TCCR0,R20            ;CTC, no prescaler, toggle on match
      LDI   R20,1<<OCIE0
      OUT   TIMSK,R20            ;activate compare match interrupt
      SEI
HERE: RJMP  HERE
```

Solution:

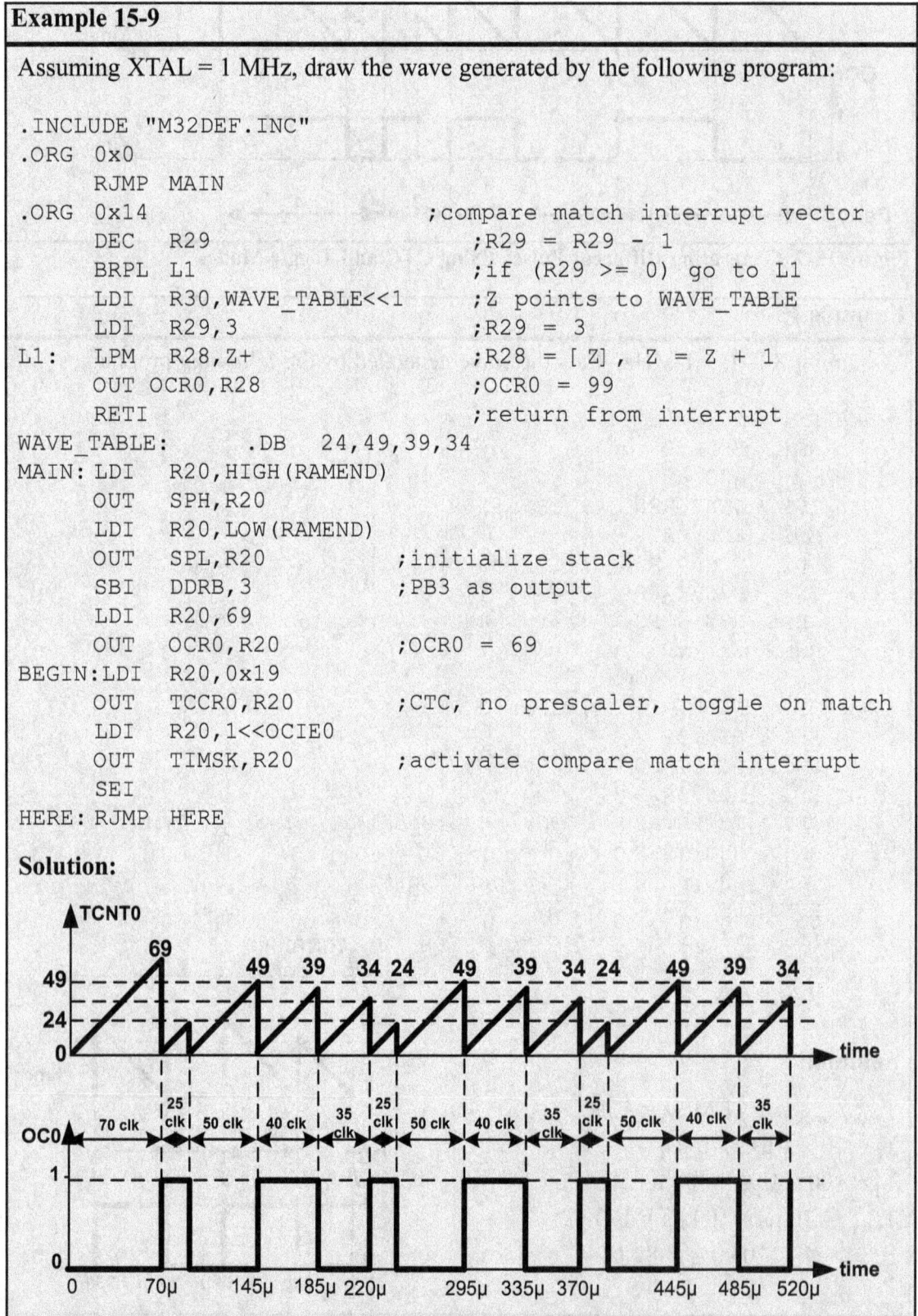

518

FOC0 (Force Output Compare) flag

Sometimes you might need to force the waveform generator to act as if a compare match has occurred. This can be done by setting the FOC0 bit of the TCCR0 register. See Example 15-10.

Example 15-10

Assuming XTAL = 1 MHz, draw the wave generated by the following program:

```
.INCLUDE "M32DEF.INC"
    SBI   DDRB,3
    LDI   R20,0x98
BEGIN:OUT  TCCR0,R20  ;CTC,timer stopped,toggle on match,FOC0=1
    RJMP  BEGIN
```

Solution:

The wave generator is in toggle mode. So, it toggles on compare match. Setting the FOC0 bit causes the wave generator to act as if a real compare match has occurred. The execution of instructions "OUT TCCR0,R20" and "RJMP BEGIN" takes 1 and 2 clocks, respectively. So, toggle occurs after 1 + 2 = 3 clocks.

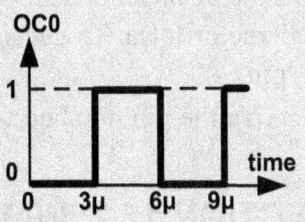

Generating waves using Timer2

We can generate waves using Timer2 or any other 8-bit timer the same way as we did using Timer0. We should simply use the proper registers and monitor the associated flag.

As the prescaler values are different in Timer2 we should be careful to load TCCR2 with proper value. For example, if we load 0x14 into TCCR0 the prescaler is 256, whereas loading TCCR2 with 0x14 means a prescaler of 64. See Examples 15-11 and 15-12.

Example 15-11

Rewrite the program of Example 15-4 using Timer2.

Solution:

```
.INCLUDE "M32DEF.INC"
    SBI   DDRD,7    ;OC2 (PD7) as output
    LDI   R22,100
    OUT   OCR2,R22  ;set the match value
    LDI   R22,0x11  ;COM21:20=Toggle, Mode=Normal, no prescaler
    OUT   TCCR2,R22 ;load TCCR2 and start counting
HERE: RJMP  HERE
```

Example 15-12

Rewrite the program of Example 15-6 using Timer2.

Solution:

```
.INCLUDE "M32DEF.INC"
      SBI    DDRD,7       ;OC2 (PD7) as output
      LDI    R22,0x19
      OUT    TCCR2,R22    ;COM21:20 = Toggle, Mode = CTC, no prescaler
      LDI    R22,200
      OUT    OCR2,R22     ;OCR2 = 200
HERE: RJMP HERE
```

Review Questions

1. True or false. In ATmega32, Timer0 has a wave generator.
2. True or false. CTC mode can be used to generate square waves.
3. True or false. To generate waves the OC0 pin must be configured as an input pin.
4. Give the pin number used by the wave generator of Timer0 in ATmega32.

SECTION 15.2: WAVE GENERATION USING TIMER1

In Chapter 9, we discussed Timer1. In this section we first discuss the different modes of Timer1 in more detail and then show how to generate waves using Timer1.

The different modes of Timer1

The WGM13, WGM12, WGM11, and WGM10 bits define the mode of Timer1, as shown in Figure 15-8. Timer1 has 16 different modes. Of these 16 modes, mode 13 is reserved (not implemented). These modes can be categorized into five groups: Normal, CTC, Fast PWM, Phase Correct PWM, and Phase and Frequency Correct PWM. We learned about the operation of the first two categories in Chapter 9; the operation of the other categories will be discussed in this part. Before discussing the operation of the different modes we should define the meaning of Top.

Top in Timer1

Top is the highest value that the TCNT register reaches while counting. In 8-bit timers (e.g., Timer0) the top value is 0xFF except for the CTC mode, whose top can be defined by OCRn. See Figure 15-8. In 16-bit timers such as Timer1 the top values are as follows:

- In Normal mode (mode 0) the top value is 0xFFFF.
- In some modes the top value is fixed and is other than the maximum; the top value can be 0xFF, 0x1FF, or 0x3FF.
- In some other modes the top can be defined by either the OCR1A register or the ICR1 register. See Figure 15-8.

Bit	7	6	5	4	3	2	1	0	
	ICNC1	ICES1	-	WGM13	WGM12	CS12	CS11	CS10	**TCCR1B**
Read/Write	R/W	R/W	R	R/W	R/W	R/W	R/W	R/W	
Initial Value	0	0	0	0	0	0	0	0	

ICNC1   D7   Input Capture Noise Canceller
                           0 = Input Capture Noise Canceller is disabled
                           1 = Input Capture Noise Canceller is enabled

ICES1   D6   Input Capture Edge Select
                           0 = Capture on the falling (negative) edge
                           1 = Capture on the rising (positive) edge

                   D5   Not used

WGM13:WGM12   D4 D3   Timer1 mode

Mode	WGM13	WGM12	WGM11	WGM10	Timer/Counter Mode of Operation	Top	Update of OCR1x	TOV1 Flag Set on
0	0	0	0	0	Normal	0xFFFF	Immediate	MAX
1	0	0	0	1	PWM, Phase Correct, 8-bit	0x00FF	TOP	BOTTOM
2	0	0	1	0	PWM, Phase Correct, 9-bit	0x01FF	TOP	BOTTOM
3	0	0	1	1	PWM, Phase Correct, 10-bit	0x03FF	TOP	BOTTOM
4	0	1	0	0	CTC	OCR1A	Immediate	MAX
5	0	1	0	1	Fast PWM, 8-bit	0x00FF	TOP	TOP
6	0	1	1	0	Fast PWM, 9-bit	0x01FF	TOP	TOP
7	0	1	1	1	Fast PWM, 10-bit	0x03FF	TOP	TOP
8	1	0	0	0	PWM, Phase and Frequency Correct	ICR1	BOTTOM	BOTTOM
9	1	0	0	1	PWM, Phase and Frequency Correct	OCR1A	BOTTOM	BOTTOM
10	1	0	1	0	PWM, Phase Correct	ICR1	TOP	BOTTOM
11	1	0	1	1	PWM, Phase Correct	OCR1A	TOP	BOTTOM
12	1	1	0	0	CTC	ICR1	Immediate	MAX
13	1	1	0	1	Reserved	-	-	-
14	1	1	1	0	Fast PWM	ICR1	TOP	TOP
15	1	1	1	1	Fast PWM	OCR1A	TOP	TOP

CS12:CS10   D2D1D0   Timer1 clock selector
                   0 0 0   No clock source (Timer/Counter stopped)
                   0 0 1   clk (no prescaling)
                   0 1 0   clk / 8
                   0 1 1   clk / 64
                   1 0 0   clk / 256
                   1 0 1   clk / 1024
                   1 1 0   External clock source on the T1 pin. Clock on falling edge
                   1 1 1   External clock source on the T1 pin. Clock on rising edge

Figure 15-8. TCCR1B (Timer 1 Control) Register

CTC mode

As shown in Figure 15-8, modes 4 and 12 operate in the CTC mode. They are almost the same. The only difference between them is that in mode 4, the top value is defined by OCR1A, whereas in mode 12, ICR specifies the top.

As mentioned in Chapter 9, in mode 4, the timer counts up until it reaches OCR1A; then the timer will be cleared and the OCF1A flag will be set as a result of compare match. See Figure 15-9.

In mode 12, the timer counts up until it reaches ICR; then the timer will be cleared and the ICF1 flag will be set, as shown in Figure 15-10. So, in mode 12, the timer works almost the same way as mode 4. See Example 15-13 and compare it with Example 9-22.

In other words, in Normal, CTC, and Fast PWM, the timer counts up until it reaches the top and then rolls over to zero. But the top value is different in the different modes and as a result, different flags are set when the timer rolls over. When the top value is a fixed value, the TOV1 flag is set; when the OCR1A register defines the top, the OCF1 flag will be set; and when the top is defined by the ICR1 register, the ICF1 flag will be set. See Figures 15-9 through 15-11.

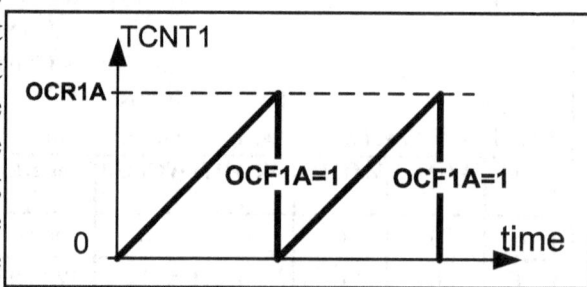

Figure 15-9. Modes 4 and 15

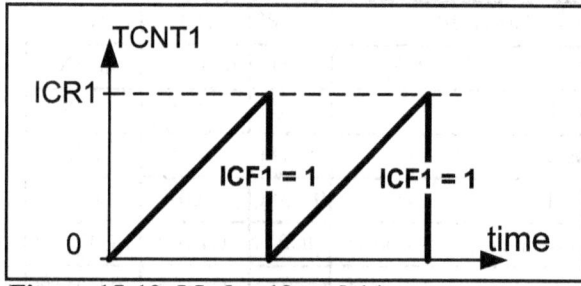

Figure 15-10. Modes 12 and 14

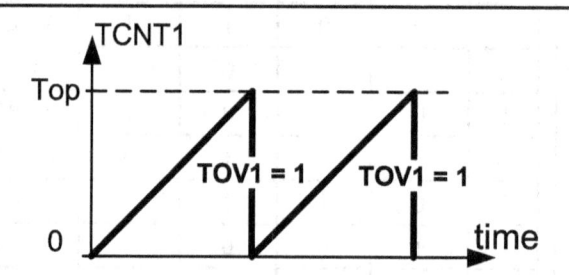

Figure 15-11. TOV1 in Modes 0, 5, 6, and 7

You might find the contents of these two pages confusing. There is no need to memorize the details. All you need to know is how the timer counts in each of the five categories of operations (Normal, CTC, etc.) and how to use the information mentioned in Figure 15-8. The following is a summary:

Counting:

In Normal, CTC, and Fast PWM modes the timer counts up until it reaches the top value. Then the timer rolls over to zero and a flag is set:
- If the top is a fixed value, TOV1 will be set.
- If the OCR1A register represents the top, the OCF1A will be set.
- If the ICR1 register defines the top, the ICF1 will be set.

Highlights of Figure 15-8:
- Column 6 (Timer/Counter Mode of Operation): mentions which of the five operation modes (Normal, CTC, Fast PWM, etc.) it belongs to.
- Column 7 (Top): represents the highest value that the timer reaches while counting; in some modes the top is a fixed value such as 0xFF, 0x1FF, 0x3FFF, and 0xFFFF, while in the others the top value can be determined by the OCR1A or ICR1 register.
- Column 8 is discussed in Chapter 16.

Example 15-13

Rewrite Example 9-27 using the ICR1 flag.

Solution:

To wait 10,000 clocks we should load the ICR1 flag with $10,000 - 1 = 9999 = 0x270F$ and use mode 14.

```
.INCLUDE "M32DEF.INC"

        LDI    R16,HIGH(RAMEND)   ;initialize stack pointer
        OUT    SPH,R16
        LDI    R16,LOW(RAMEND)
        OUT    SPL,R16
        SBI    DDRB,5             ;PB5 as an output
BEGIN:SBI    PORTB,5              ;PB5 = 1
        RCALL  DELAY_1ms
        CBI    PORTB,5            ;PB5 = 0
        RCALL  DELAY_1ms
        RJMP   BEGIN

DELAY_1ms:
        LDI    R20,HIGH(9999)
        OUT    ICR1H,R20          ;TEMP = 0x27
        LDI    R20,LOW(9999)
        OUT    ICR1L,R20          ;ICR1L = 0x0F, ICR1H = TEMP
        LDI    R20,0
        OUT    TCNT1H,R20         ;TEMP = 0x0
        OUT    TCNT1L,R20         ;TCNT1L = 0x0, TCNT1H = TEMP
        LDI    R20,0x02
        OUT    TCCR1A,R20         ;WGM11:10 = 10
        LDI    R20,0x19
        OUT    TCCR1B,R20         ;WGM13:12 = 11, CS = CLK, mode = 14
AGAIN:IN     R20,TIFR             ;read TIFR
        SBRS   R20,ICF1           ;if ICF1 is set skip next instruction
        RJMP   AGAIN
        LDI    R20,1<<ICF1
        OUT    TIFR,R20           ;clear ICF1 flag
        LDI    R19,0
        OUT    TCCR1B,R19         ;stop timer
        OUT    TCCR1A,R19
        RET
```

Waveform generators in Timer1

In examining Figures 15-12 and 15-13 we see that Timer1 has two independent waveform generators: Waveform Generator A and Waveform Generator B.

The compare match between OCR1A and TCNT1 affects Waveform Generator A, and the wave generated by Waveform Generator A shows up on the OC1A pin.

The compare match between OCR1B and TCNT1 affects Waveform Generator B, and the wave generated by Waveform Generator B shows up on the OC1B pin.

The COM1A1 and COM1A0 bits have control over Waveform Generator A; whereas COM1B1 and COM1B0 control Waveform Generator B. All of the COM bits are in the TCCR1A register, as shown in Figure 15-14.

The operation mode of Timer1 (WGM13, WGM12, WGM11, and WGM10 bits of TCCR1A and TCCR1B) affect both generators, as shown in Figures 15-12 and 15-13.

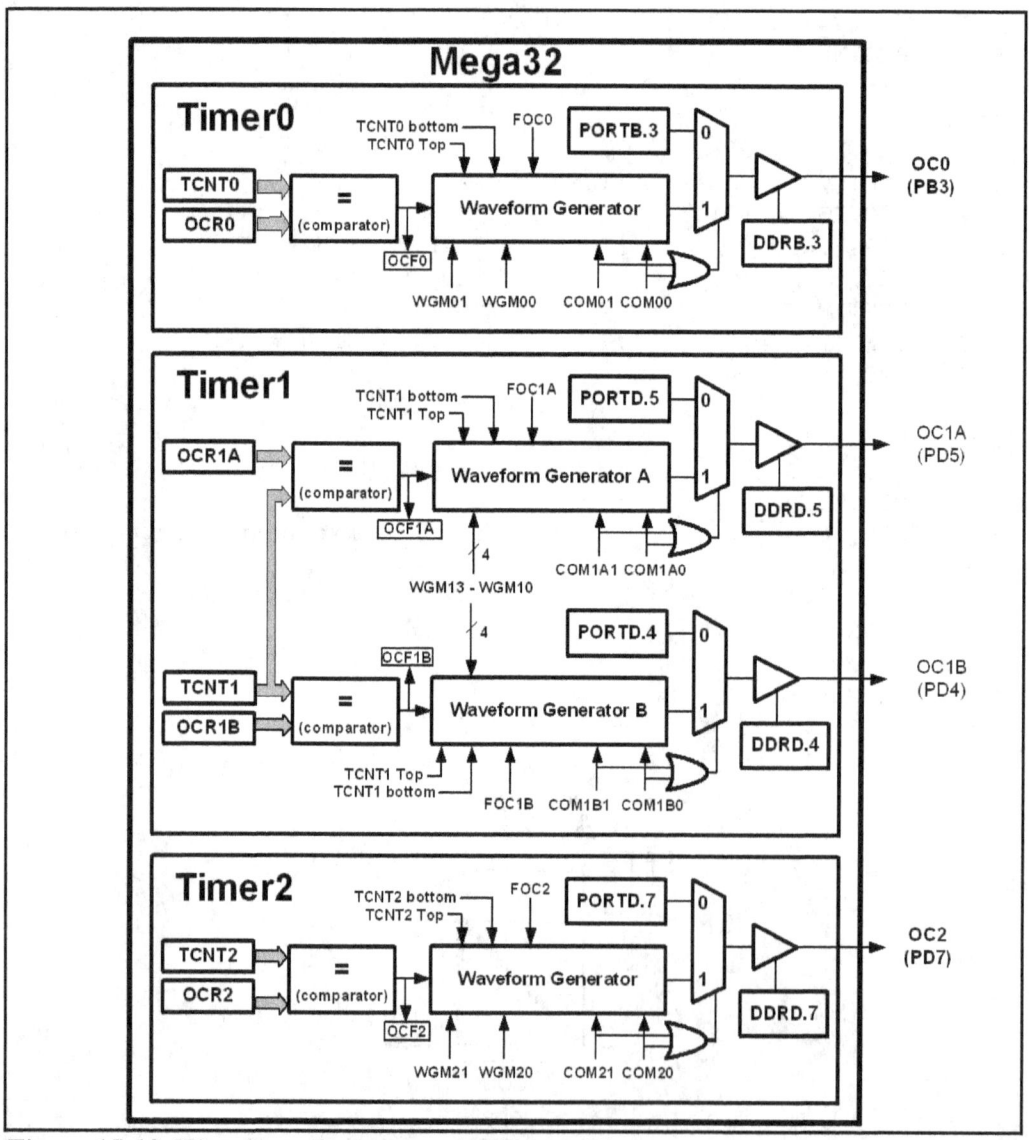

Figure 15-12. Waveform Generators in ATmega32

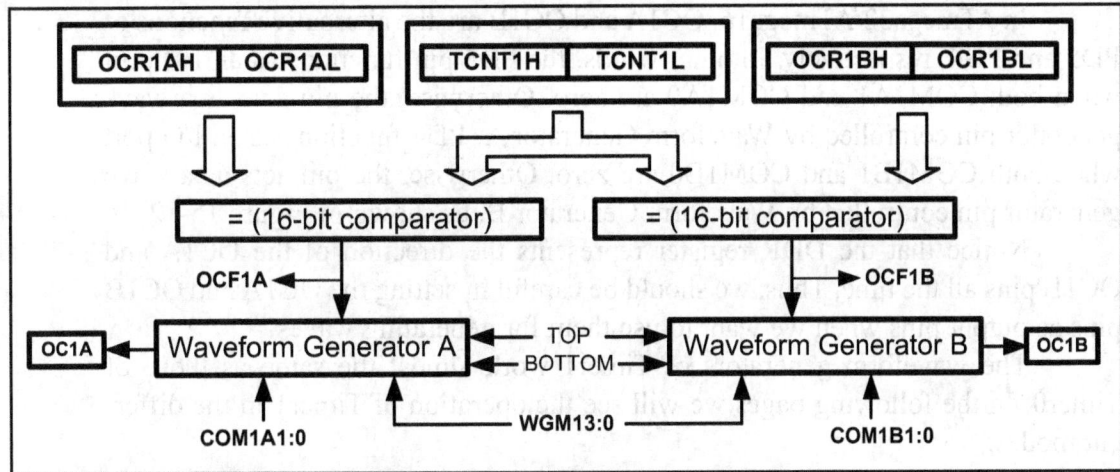

Figure 15-13. Simplified Waveform Generator Block Diagram

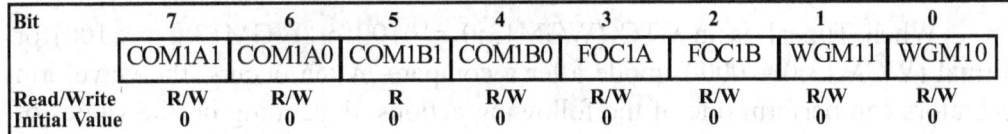

Bit	7	6	5	4	3	2	1	0
	COM1A1	COM1A0	COM1B1	COM1B0	FOC1A	FOC1B	WGM11	WGM10
Read/Write	R/W	R/W	R	R/W	R/W	R/W	R/W	R/W
Initial Value	0	0	0	0	0	0	0	0

COM1A1:COM1A0   D7 D6   Compare Output Mode for Channel A

COM1A1	COM1A0	Description
0	0	Normal port operation, OC1A disconnected
0	1	Toggle OC1A on compare match
1	0	Clear OC1A on compare match
1	1	Set OC1A on compare match

COM1B1:COM1B0   D5 D4   Compare Output Mode for Channel B

COM1B1	COM1B0	Description
0	0	Normal port operation, OC1B disconnected
0	1	Toggle OC1B on compare match
1	0	Clear OC1B on compare match
1	1	Set OC1B on compare match

FOC1A   D3   Force Output Compare for Channel A

FOC1B   D2   Force Output Compare for Channel B

WGM11:10   D1 D0 Timer1 mode (discussed in Figure 15-8)

Figure 15-14. TCCR1A (Timer 1 Control) Register

In ATmega32/ATmega16, OC1A and OC1B are the alternative functions of PD5 and PD4, respectively. In other words, the PD5 pin functions as an I/O port when both COM1A1 and COM1A0 are zero. Otherwise, the pin acts as a wave generator pin controlled by Waveform Generator A. PD4 functions as an I/O port when both COM1B1 and COM1B0 are zero. Otherwise, the pin acts as a wave generator pin controlled by Waveform Generator B, as shown in Figure 15-12.

Notice that the DDR register represents the direction of the OC1A and OC1B pins all the time. Thus, we should be careful in setting the OC1A and OC1B pins as output pins when we want to use them for generating waves.

The waveform generators of Timer1 work almost the same as those of Timer0. In the following pages we will see the operation of Timer1 in the different modes.

Wave generation in Normal and CTC modes

When Timer1 is in CTC (WGM13:0 = 0100 or WGM13:0 = 1100) or Normal (WGM13:0 = 0000) mode after a compare match occurs, the waveform generators can perform one of the following actions, depending on the values of COM1A1:0 and COM1B1:0 bits, respectively:

(a) Remain unaffected
(b) Toggle the OC1x pin (OC1A or OC1B)
(c) Clear (drive low) the OC1x pin
(d) Set (drive high) the OC1x pin

The COM1A1 and COM1A0 bits select the operation of OC1A, while COM1B1 and COM1B0 select the operation of OC1B, as shown in Figure 15-14. See Example 15-14.

Example 15-14

Using Figures 15-8 and 15-14, find the values of the TCCR1A and TCCR1B registers if we want to clear the OC1A pin upon match, with no prescaler, internal clock, and Normal mode.

Solution:

WGM13:10 = 0000 = Normal mode
COM1A1:0 = 10 = Clear
CS12:10 = 001 = No prescaler

TCCR1A =	1	0	0	0	0	0	0	0
	COM1A1	COM1A0	COM1B1	COM1B0	FOC1A	FOC1B	WGM11	WGM10

TCCR1B =	0	0	0	0	0	0	0	1
	ICNC1	ICES1	-	WGM13	WGM12	CS12	CS11	CS10

Generating square waves

To generate square waves we can set the timer to Normal or CTC mode and set the COM1x1 and COM1x0 bits of one of the Waveform Generators to the toggle mode (COM1A1:0 = 01 to generate waves with Waveform Generator A or COM1B1:0 = 01 for generating waves using Waveform Generator B).

The OC1x pin will be toggled on each compare match and a square wave will be generated, as shown in Figure 15-15. See Examples 15-15 and 15-16.

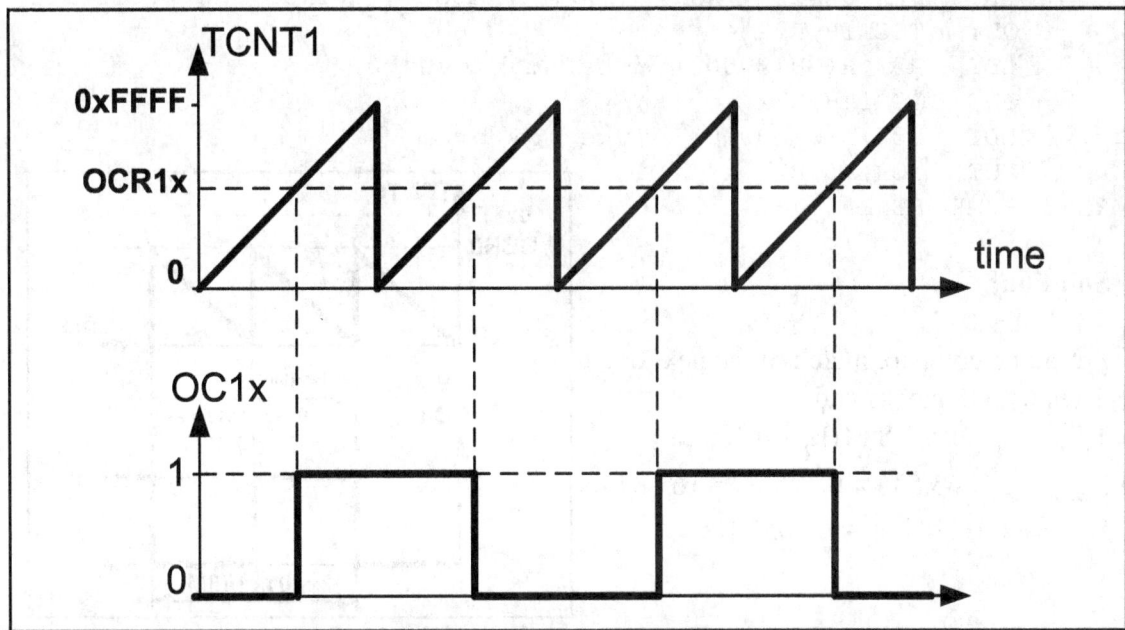

Figure 15-15. Generating Square Wave Using Normal Mode and Toggle Mode

Example 15-15

Find the value for TCCR1A and TCCR1B to program Timer1 as Normal mode and the OC1A generator as square wave generator and no prescaler.

Solution:

WGM13:10 = 0000 = Normal mode
COM1A1:0 = 01 = Toggle
CS12:10 = 001 = No prescaler
FOC1A = 1
FOC1B = 1

TCCR1A =	0	1	0	0	0	0	0	0
	COM1A1	COM1A0	COM1B1	COM1B0	FOC1A	FOC1B	WGM11	WGM10

TCCR1B =	0	0	0	0	0	0	0	1
	ICNC1	ICES1	-	WGM13	WGM12	CS12	CS11	CS10

Example 15-16

Assuming XTAL = 8 MHz, calculate the frequency of the wave generated by the following program:

```
.INCLUDE "M32DEF.INC"
     SBI   DDRD,5
     LDI   R22,0x40    ;COM1A = Toggle
     OUT   TCCR1A,R22
     LDI   R22,0x01    ;WGM = Toggle, Mode = Normal, no prescaler
     OUT   TCCR1B,R22
     LDI   R22,HIGH(30000)  ;the high byte
     OUT   OCR1AH,R22
     LDI   R22,LOW(30000)   ;the low byte
     OUT   OCR1AL,R22
HERE: RJMP  HERE
```

Solution:

From one compare match to the next one it takes 65,536 clocks and

$T_{timer\ clock}$ = 1/8 MHz = 0.125 μs

T_{wave} = 2 × 65,536 × 0.125 μs = 16,384 μs

F_{wave} = 1/16,384 μs = 61.035 Hz

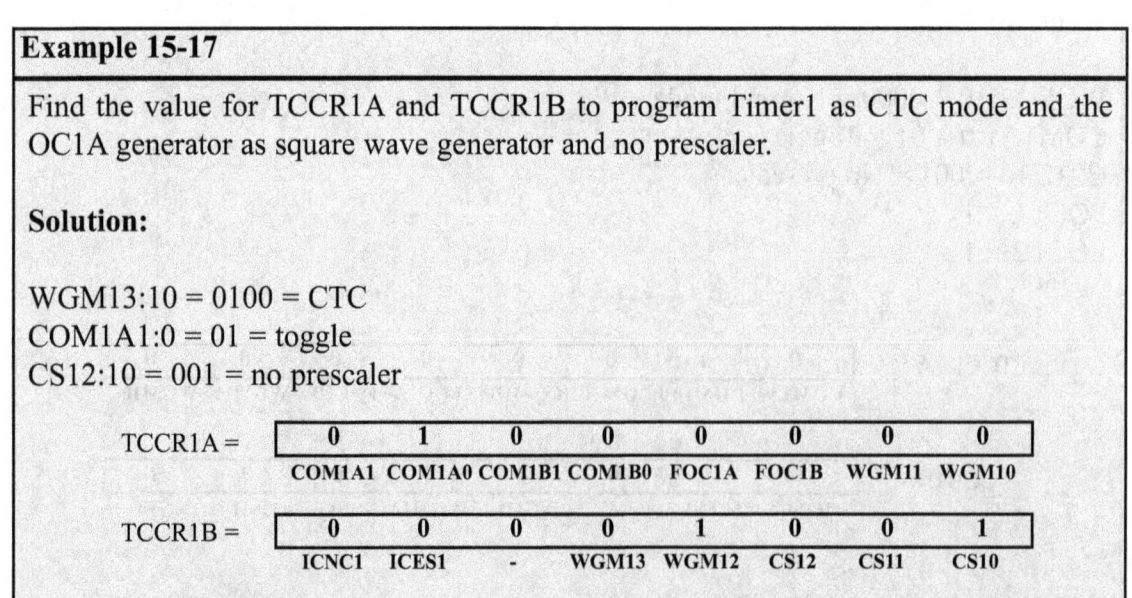

CTC mode is better than Normal mode for generating square waves, as the frequency of the wave can be easily adjusted by changing the top value (the value of the OCR1x register in mode 4, and ICR1 in mode12). See Figure 15-16. In CTC mode, when OCR1x (or ICR1 in mode 12) has a lower value, compare match occurs earlier and the period of the generated wave is smaller (higher frequency). When the OCR0 has a higher value, compare match occurs later and the period of the wave is longer (lower frequency). See Examples 15-17 through 15-19.

Example 15-17

Find the value for TCCR1A and TCCR1B to program Timer1 as CTC mode and the OC1A generator as square wave generator and no prescaler.

Solution:

WGM13:10 = 0100 = CTC
COM1A1:0 = 01 = toggle
CS12:10 = 001 = no prescaler

TCCR1A =	0	1	0	0	0	0	0	0
	COM1A1	COM1A0	COM1B1	COM1B0	FOC1A	FOC1B	WGM11	WGM10

TCCR1B =	0	0	0	0	1	0	0	1
	ICNC1	ICES1	-	WGM13	WGM12	CS12	CS11	CS10

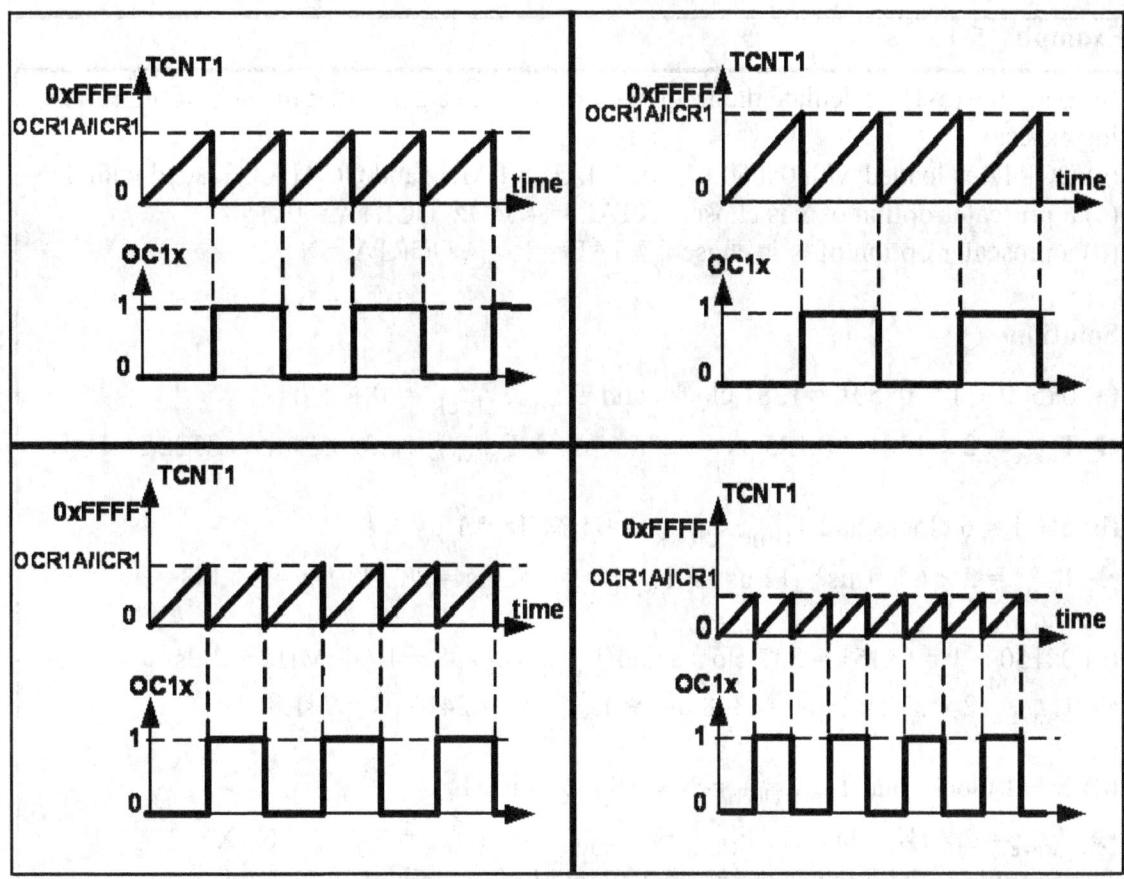

Figure 15-16. Generating Square Wave Using CTC Mode and Toggle Mode

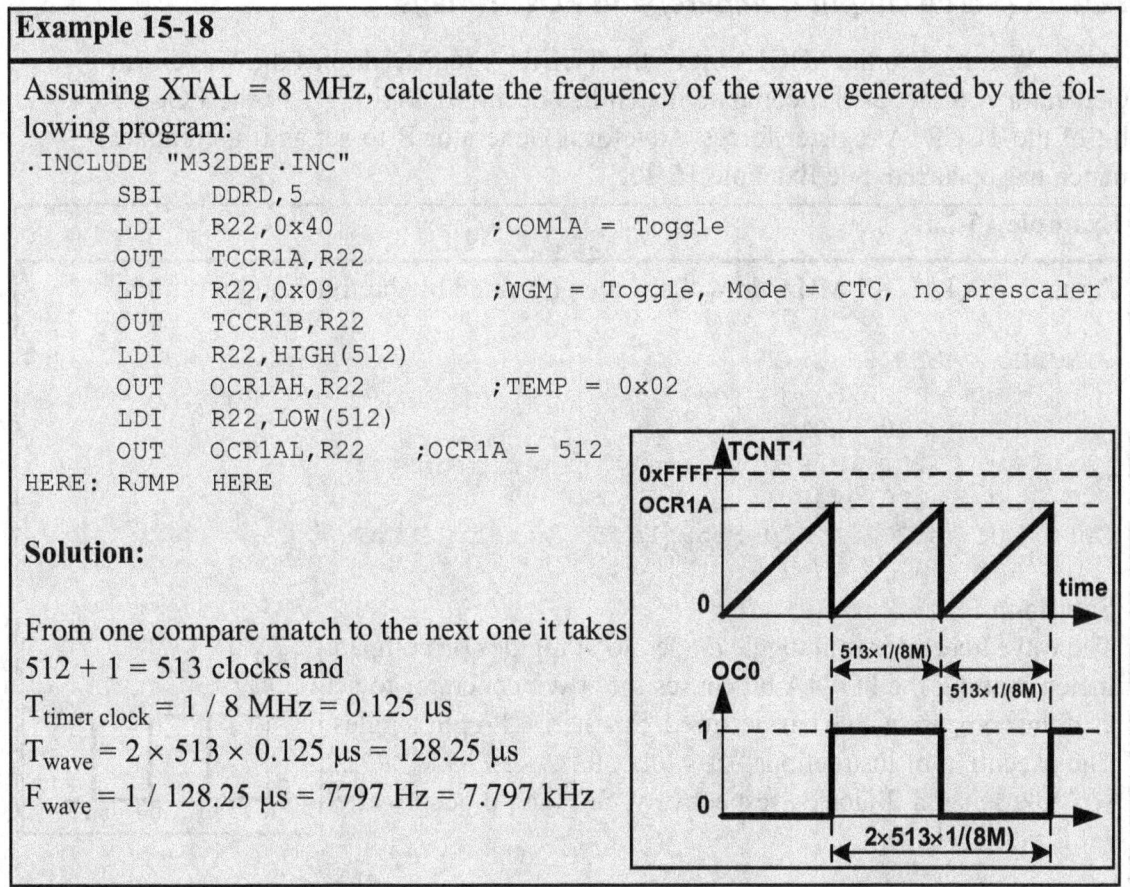

Example 15-18

Assuming XTAL = 8 MHz, calculate the frequency of the wave generated by the following program:

```
.INCLUDE "M32DEF.INC"
    SBI    DDRD,5
    LDI    R22,0x40           ;COM1A = Toggle
    OUT    TCCR1A,R22
    LDI    R22,0x09           ;WGM = Toggle, Mode = CTC, no prescaler
    OUT    TCCR1B,R22
    LDI    R22,HIGH(512)
    OUT    OCR1AH,R22         ;TEMP = 0x02
    LDI    R22,LOW(512)
    OUT    OCR1AL,R22    ;OCR1A = 512
HERE: RJMP   HERE
```

Solution:

From one compare match to the next one it takes
$512 + 1 = 513$ clocks and
$T_{timer\ clock} = 1\ /\ 8\ MHz = 0.125\ \mu s$
$T_{wave} = 2 \times 513 \times 0.125\ \mu s = 128.25\ \mu s$
$F_{wave} = 1\ /\ 128.25\ \mu s = 7797\ Hz = 7.797\ kHz$

Example 15-19

In Example 15-18, calculate the frequency of the wave generated in each of the following cases:
(a) OCR1A is loaded with 0x0500   (b) XTAL = 1 MHz and OCR1A is loaded with 0x5
(c) a prescaler option of 8 is chosen, XTAL = 4 MHz, OCR1A = 0x150
(d) a prescaler option of N is chosen, XTAL = F_{OSC}, OCR1A = X

Solution:

(a) 0x500 + 1 = 0x501 = 1281 clocks and $T_{timer\ clock}$ = 0.125 μs

➜ T_{wave} = 2 × 1281 × 0.125 μs = 320.25 μs ➜ F_{wave} = 1 / 320.25 μs = 3122.56 Hz

(b) 5 + 1 = 6 clocks and $T_{timer\ clock}$ = 1/1 MHz = 1 μs

➜ T_{wave} = 2 × 6 × 1 μs = 12 μs ➜ F_{wave} = 1 / 12 μs = 83,333 Hz = 83.333 kHz

(c) 0x150 + 1 = 0x151 = 337 clocks and $T_{timer\ clock}$ = 8 × 1 / 4 MHz = 2 μs

➜ T_{wave} = 2 × 337 × 2 μs = 1348 μs ➜ F_{wave} = 1 / 2416 μs = 741.8 Hz

(d) X + 1 clocks and $T_{timer\ clock}$ = N × 1/F_{OSC} = N / F_{OSC}

➜ T_{wave} = 2 × (X + 1) × N / F_{OSC} ➜ F_{wave} = 1 / T_{wave} = F_{OSC} / [2N (X + 1)]
The formula is the same as the one calculated in Example 15-7 d.

FOC1A (Force Output Compare) and FOC1B flags

Writing 1 to the FOC1A bit of the TCCR1A register forces the Waveform Generator A to act as if a compare match has occurred. Writing 1 to the FOC1B bit of the TCCR1A register forces Waveform Generator B to act as if a compare match has occurred. See Example 15-20.

Example 15-20

Assuming XTAL = 1 MHz, draw the wave generated by the following program:

```
.INCLUDE "M32DEF.INC"
     SBI   DDRD,5
     LDI   R20,0x01
     OUT   TCCR1B,R20  ;Normal, timer stopped
     LDI   R20,0x48
L1:  OUT   TCCR1A,R20  ;toggle on match, FOC1A = 1
     RJMP  L1
```

Solution:

The wave generator is in toggle mode. So, it toggles on compare match. Setting the FOC1A bit causes the wave generator to act as if the compare match has occurred. So, the OC1A pin toggles. The execution of instructions "OUT TCCR1A,R20" and "RJMP L1" takes 1 and 2 clocks, respectively. So, toggle occurs after 1 + 2 = 3 clocks

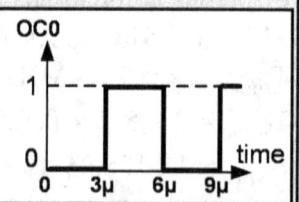

Review Questions

1. True or false. In ATmega32, Timer1 has three waveform generators.
2. True or false. In CTC modes the TOP value is determined by OCR1A or ICR1.
3. True or false. We can associate each of the pins with each of the waveform generators.
4. True or false. In CTC modes we cannot change the frequency of the generated wave.

SECTION 15.3: INPUT CAPTURE PROGRAMMING

The Input Capture function is widely used for many applications. Among them are (a) recording the arrival time of an event, (b) pulse width measurement, and (c) period measurement. In ATmega32, Timer1 can be used as the Input Capture to detect and measure the events happening outside the chip. Upon detection of an event, the TCNT value is loaded into the ICR1 register, and the ICF1 flag is set.

As shown in Figure 15-17, there are two event sources: (1) the ICP1 pin, which is PORTD.6 in ATmega32, and (2) the output of the analog comparator. We can use the ACIC flag to select the event source. ACIC is a bit of the ACSR register, as shown in Figure 15-18.

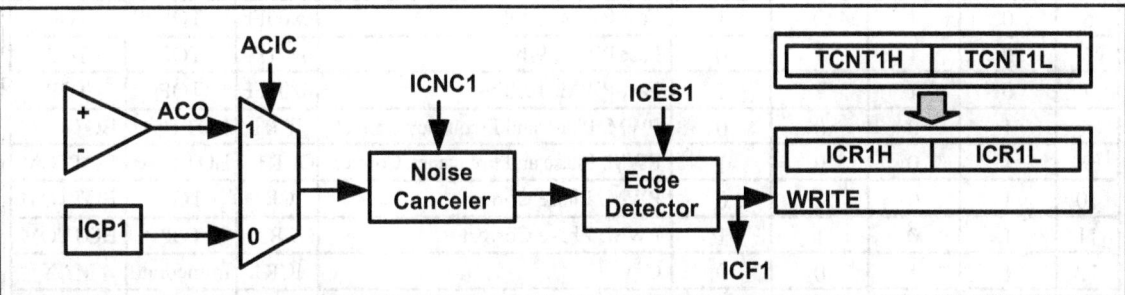

Figure 15-17. Capturing Circuit

ACD	ACBG	ACO	ACI	ACIE	ACIC	ACIS1	ACIS0

ACD (Analog Comparator Disable) When the bit is one, the power to the Analog Comparator is switched off, which reduces power consumption.

ACBG (Analog Comparator Bandgap Select) See the datasheet.

ACO (Analog Comparator Output) The output of the analog comparator is connected to the bit. ACO is read only. See Figure 15-17.

ACI (Analog Comparator Interrupt Flag)

ACIE (Analog Comparator Interrupt Enable)

ACIC (Analog Comparator Input Capture Enable) When the bit is one, the input capture is triggered by the Analog Comparator; otherwise, the ICP1 pin (PD6 in ATmega32) provides the capturing signal. See Figure 15-17.

ACIS1, ACIS0 (Analog Comparator Interrupt Mode Select) See the datasheet.

Figure 15-18. TCCR1B (Timer/Counter Control Register) Register, ICNC1, ICES1

ICNC1	ICES1	-	WGM13	WGM12	CS12	CS11	CS10

ICNC1 (Input Capture Noise Canceller) Setting the bit activates the noise canceller. When the noise canceller is activated, each change is considered only if it persists for at least 4 successive system clocks. Notice that although activating the noise canceller prevents the detection of noises as signals, it causes 4 clocks of delay from the event occurrence to the load of the ICR1 register.

ICES1 (Input Capture Edge Select) Selects edge detection for the input capture function. When an edge is detected, the TCNT is loaded into the ICRx register. It also raises the ICFn (input capture flag) flag in the TIFR register.

0	Capture on falling edge
1	Capture on rising edge

WGM13:WGM12 D4 D3 Timer1 mode

Mode	WGM13	WGM12	WGM11	WGM10	Timer/Counter Mode of Operation	Top	Update of OCR1x	TOV1 Flag Set on
0	0	0	0	0	Normal	0xFFFF	Immediate	MAX
1	0	0	0	1	PWM, Phase Correct, 8-bit	0x00FF	TOP	BOTTOM
2	0	0	1	0	PWM, Phase Correct, 9-bit	0x01FF	TOP	BOTTOM
3	0	0	1	1	PWM, Phase Correct, 10-bit	0x03FF	TOP	BOTTOM
4	0	1	0	0	CTC	OCR1A	Immediate	MAX
5	0	1	0	1	Fast PWM, 8-bit	0x00FF	TOP	TOP
6	0	1	1	0	Fast PWM, 9-bit	0x01FF	TOP	TOP
7	0	1	1	1	Fast PWM, 10-bit	0x03FF	TOP	TOP
8	1	0	0	0	PWM, Phase and Frequency Correct	ICR1	BOTTOM	BOTTOM
9	1	0	0	1	PWM, Phase and Frequency Correct	OCR1A	BOTTOM	BOTTOM
10	1	0	1	0	PWM, Phase Correct	ICR1	TOP	BOTTOM
11	1	0	1	1	PWM, Phase Correct	OCR1A	TOP	BOTTOM
12	1	1	0	0	CTC	ICR1	Immediate	MAX
13	1	1	0	1	Reserved	-	-	-
14	1	1	1	0	Fast PWM	ICR1	TOP	TOP
15	1	1	1	1	Fast PWM	OCR1A	TOP	TOP

CS12:CS10	D2D1D0	Timer1 clock selector
	0 0 0	No clock source (Timer/Counter stopped)
	0 0 1	clk (no prescaling)
	0 1 0	clk / 8
	0 1 1	clk / 64
	1 0 0	clk / 256
	1 0 1	clk / 1024
	1 1 0	External clock source on T1 pin. Clock on falling edge
	1 1 1	External clock source on T1 pin. Clock on rising edge

Figure 15-19. TCCR1B (Timer/Counter Control Register) Register, ICNC1, ICES1

As shown in Figures 15-17 and 15-19, we use the TCCR1B register to select the type of edge detection and activate/deactivate the noise canceller unit.

Notice that the input capture unit does not work in the timer modes for which the ICR1 defines the top value (modes 8, 10, 12, 14). See Example 15-21.

Example 15-21

Using Figures 15-12 and 15-19, find TCCR1A and TCCR1B, for capturing on rising edge, no noise canceller, no prescaler, and timer mode = Normal.

Solution:

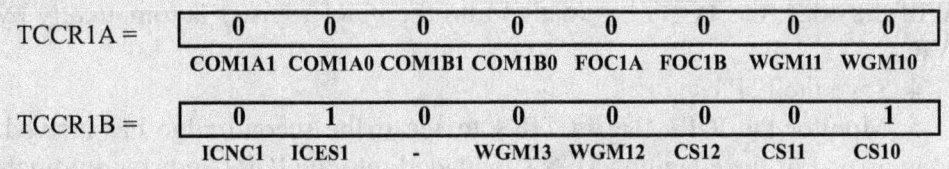

TCCR1A =	0	0	0	0	0	0	0	0
	COM1A1	COM1A0	COM1B1	COM1B0	FOC1A	FOC1B	WGM11	WGM10

TCCR1B =	0	1	0	0	0	0	0	1
	ICNC1	ICES1	-		WGM13	WGM12	CS12	CS11	CS10

Steps to program the Input Capture function

We use the following steps to measure the edge arrival time for the Input Capture function.

1. Initialize the TCCR1A and TCCR1B for a proper timer mode (any mode other than modes 8, 10, 12, and 14), enable or disable the noise canceller, and select the edge (positive or negative) we want to measure the arrival time for.

2. Initialize the ACSR to select the desired event source.

3. Monitor the ICF1 flag in TIFR to see if the edge has arrived. Upon the arrival of the edge, the TCNT1 value is loaded into the ICR1 register automatically by the AVR. Example 15-22 shows how the Input Capture function works. The Input Capture function is widely used to measure the period or the pulse width of an incoming signal.

Example 15-22

Assuming that clock pulses are fed into pin ICP1, write a program to read the TCNT1 value on every rising edge. Place the result on PORTA and PORTB.

Solution:

```
.INCLUDE "M32DEF.INC"
      LDI    R16,0xFF
      OUT    DDRA,R16     ;PORTA as output
      OUT    DDRB,R16     ;PORTB as output
      OUT    PORTD,R16    ;activate pull-up
BEGIN:LDI    R20,0x00
      OUT    TCCR1A,R20   ;timer mode = Normal
      LDI    R20,0x41
      OUT    TCCR1B,R20   ;rising edge, no prescaler, no noise canceller
L1:   IN     R21,TIFR
      SBRS   R21,ICF1     ;skip next if ICF1 flag is set
      RJMP   L1           ;jump L1
      OUT    TIFR,R21     ;clear ICF1
      IN     R22,ICR1L    ;TEMP = ICR1H, R22 = ICR1L
      OUT    PORTA,R22    ;PORTA = R22
      IN     R22,ICR1H    ;R22 = TEMP = ICR1H
      OUT    PORTB,R22    ;PORTB = R22
      RJMP   BEGIN        ;jump begin
```

Note: Upon the detection of each rising edge, the TCNT1 value is loaded into ICR1. Also notice that we clear the ICF1 flag bit.

Measuring period

We can use the following steps to measure the period of a wave.

1. Initialize the TCCR1A and TCCR1B.

2. Initialize the ACSR to select the desired event source.

3. Monitor the ICF1 flag in TIFR to see if the edge has arrived. Upon the arrival of the edge, the TCNT1 is loaded into the ICR1 register automatically by the AVR.

4. Save the ICR1.

5. Monitor the ICF1 flag in TIFR to see if the second edge has arrived. Upon the arrival of the edge, the TCNT is loaded into the ICR1 register automatically by the AVR.

6. Save the ICR1 for the second edge. By subtracting the second edge value from the first edge value we get the time. See Examples 15-23 and 15-24. Also see Figure 15-20.

Example 15-23

Assuming that clock pulses are fed into pin PORTD.6, write a program to measure the period of the pulses. Place the binary result on PORTA and PORTB.

Solution:

```
.INCLUDE "M32DEF.INC"
        LDI    R16,0xFF
        OUT    DDRA,R16     ;PORTA as output
        OUT    DDRB,R16     ;PORTB as output
        OUT    PORTD,R16
BEGIN:LDI      R20,0x00
        OUT    TCCR1A,R20   ;timer mode = Normal
        LDI    R20,0x41
        OUT    TCCR1B,R20   ;rising edge, no prescaler, no noise canceller
L1:     IN     R21,TIFR
        SBRS   R21,ICF1     ;skip next instruction if ICF1 flag is set
        RJMP   L1           ;jump L1
        IN     R23,ICR1L    ;R23 = ICR1L, TEMP = ICR1H (first edge value)
        IN     R24,ICR1H    ;R24 = ICR1H
        OUT    TIFR,R21     ;ICF1 = 0
L2:     IN     R21,TIFR
        SBRS   R21,ICF1     ;skip next if ICF1 flag is set
        RJMP   L2
        OUT    TIFR,R21     ;clear ICF1
        IN     R22,ICR1L    ;R22 = ICR1L, TEMP = ICR1H (second edge value)
        SUB    R22,R23      ;Period = Second edge - First edge
        OUT    PORTA,R22    ;PORTA = R22
        IN     R22,ICR1H    ;R22 = TEMP
        SBC    R22,R24      ;R22 = R22 - R24 - C
        OUT    PORTB,R22    ;PORTB = R22
L3:     RJMP   L3           ;wait forever
```

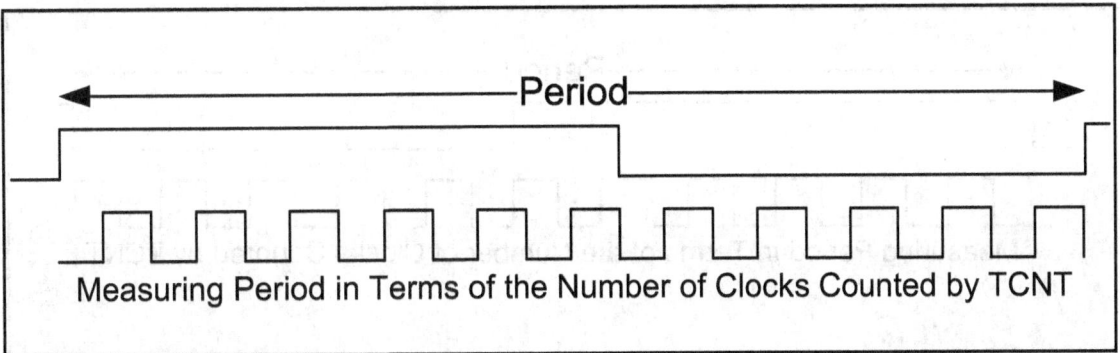

Measuring Period in Terms of the Number of Clocks Counted by TCNT

Figure 15-20. Using Input Capture to Measure Period

Example 15-24

The frequency of a pulse is between 50 Hz and 60 Hz. Assume that a pulse is connected to ICP1 (pin PD6). Write a program to measure its period and display it on PORTB. Use the prescaler value that gives the result in a single byte. Assume XTAL = 8 MHz.

Solution:

8 MHz × 1/1024 = 7812.5 Hz due to prescaler and T = 1/7812.5 Hz = 128 μs.
The frequency of 50 Hz gives us the period of 1/50 Hz = 20 ms. So, the output is
20 ms/128 μs = 156.
The frequency of 60 Hz gives us the period of 1/60 Hz = 16.6 ms. So, the output is
16.6 ms/128 μs = 130.

```
.INCLUDE "M32DEF.INC"
        LDI    R16,0xFF
        OUT    DDRB,R16       ;PORTB as output
        OUT    PORTD,R16
BEGIN:LDI      R20,0x00
        OUT    TCCR1A,R20     ;timer mode = Normal
        LDI    R20,0x45
        OUT    TCCR1B,R20     ;rising edge, prescaler = 1024, no noise canc.
L1:     IN     R21,TIFR
        SBRS   R21,ICF1       ;skip next instruction if ICF1 flag is set
        RJMP   L1             ;jump L1
        IN     R16,ICR1L      ;R16 = ICR1L (first edge value)
        OUT    TIFR,R21       ;ICF1 = 0
L2:     IN     R21,TIFR
        SBRS   R21,ICF1       ;skip next if ICF1 flag is set
        RJMP   L2
        IN     R22,ICR1L      ;R22 = ICR1L, TEMP = ICR1H (second edge value)
        SUB    R22,R16        ;period = second edge - first edge
        OUT    PORTB,R22      ;PORTB = R22
        OUT    TIFR,R21       ;clear ICF1
L3:     RJMP   L3             ;wait forever
```

AVR

PB

to
LEDs

PD6

60/50 Hz clock

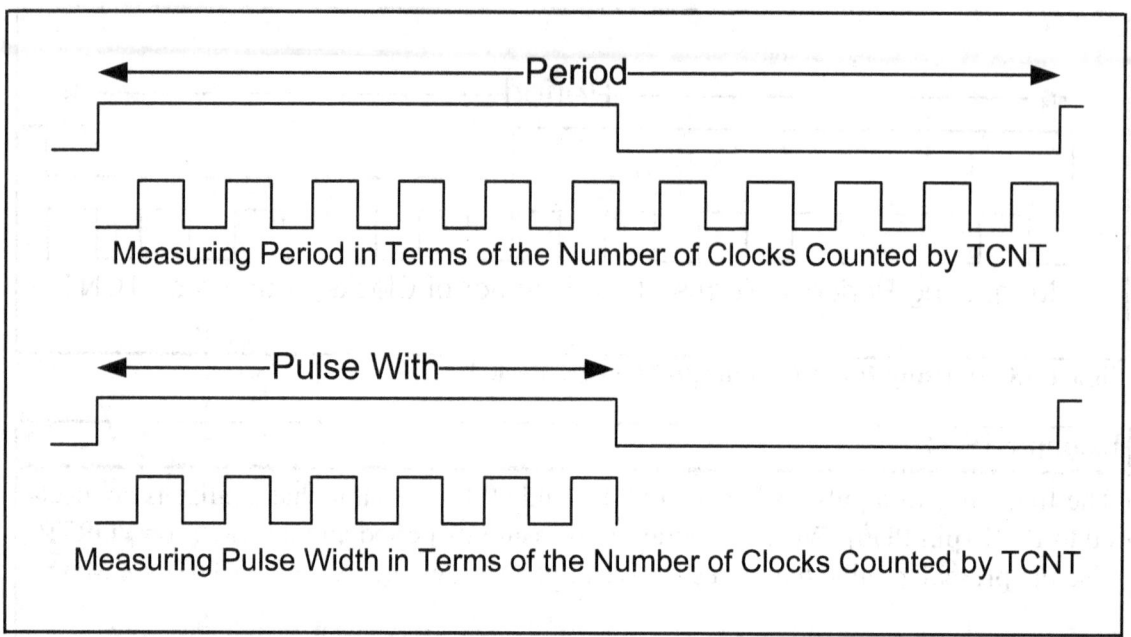

Measuring Period in Terms of the Number of Clocks Counted by TCNT

Measuring Pulse Width in Terms of the Number of Clocks Counted by TCNT

Figure 15-21. Using Input Capture to Measure Period and Pulse Width

Measuring pulse width

We can use the following steps to measure the pulse width of a wave.

1. Initialize TCCR1A and TCCR1B, and select capturing on rising edge.

2. Initialize ACSR to select the desired event source.

3. Monitor the ICF1 flag in TIFR to see if the edge has arrived. Upon the arrival of the edge, the TCNT1 value is loaded into the ICR1 register automatically by the AVR.

4. Save the ICR1 and change the capturing edge to the falling edge.

5. Monitor the ICF1 flag in TIFR to see if the second edge has arrived. Upon the arrival of the edge, the TCNT value is loaded into the ICR1 register automatically by the AVR.

6. Save the ICR1 for the second edge. Subtract the second edge value from the first edge value to get the time.

See Figure 15-21 and Examples 15-25 through 15-27 to see how it is done.

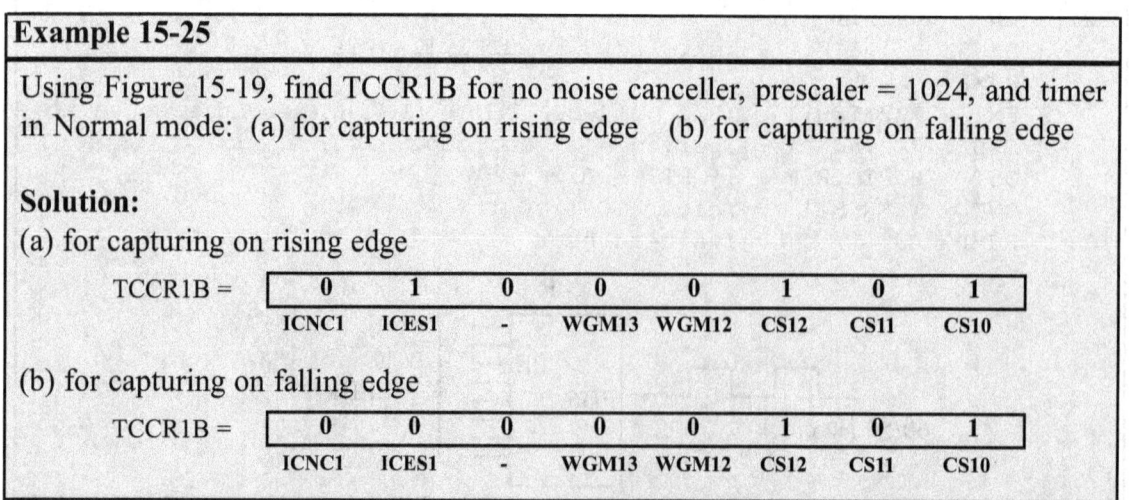

Example 15-25

Using Figure 15-19, find TCCR1B for no noise canceller, prescaler = 1024, and timer in Normal mode: (a) for capturing on rising edge   (b) for capturing on falling edge

Solution:
(a) for capturing on rising edge

0	1	0	0	0	1	0	1
ICNC1	ICES1	-	WGM13	WGM12	CS12	CS11	CS10

TCCR1B =

(b) for capturing on falling edge

0	0	0	0	0	1	0	1
ICNC1	ICES1	-	WGM13	WGM12	CS12	CS11	CS10

TCCR1B =

Example 15-26

Assume that a 60-Hz frequency pulse is connected to ICP1 (pin PD6). Write a program to measure its pulse width. Use the prescaler value that gives the result in a single byte. Display the result on PORTB. Assume XTAL = 8 MHz.

Solution:

The frequency of 60 Hz gives us the period of 1/60 Hz = 16.6 ms.
Now, 8 MHz × 1/1024 = 7812.5 Hz due to prescaler and T = 1/7812.5 Hz = 128 μs for TCNT. That means we get the value of 130 (1000 0010 binary) for the period since 16.6 ms / 128 μs = 130. Now the pulse width can be anywhere between 1 to 129.

```
.INCLUDE "M32DEF.INC"
        LDI     R16,0xFF
        OUT     DDRB,R16        ;PORTB as output
        OUT     PORTD,R16
BEGIN:LDI       R20,0x00
        OUT     TCCR1A,R20      ;timer mode = Normal
        LDI     R20,0x45
        OUT     TCCR1B,R20      ;rising edge, prescaler = 1024, no noise canc.
L1:     IN      R21,TIFR
        SBRS    R21,ICF1        ;skip next instruction if ICF1 flag is set
        RJMP    L1              ;jump L1
        IN      R16,ICR1L       ;R16 = ICR1L (rising edge value)
        OUT     TIFR,R21        ;ICF1 = 0 (for next round)
        LDI     R20,0x05
        OUT     TCCR1B,R20      ;falling edge, prescaler = 1024, no noise canc.
L2:     IN      R21,TIFR
        SBRS    R21,ICF1        ;skip next if ICF1 flag is set
        RJMP    L2
        IN      R22,ICR1L       ;R22 = ICR1L, TEMP = ICR1H (falling edge value)
        SUB     R22,R16         ;pulse width = falling edge - rising edge
        OUT     PORTB,R22       ;PORTB = R22
        OUT     TIFR,R21        ;clear ICF1 (for next round)
L3:     RJMP    L3              ;wait forever
```

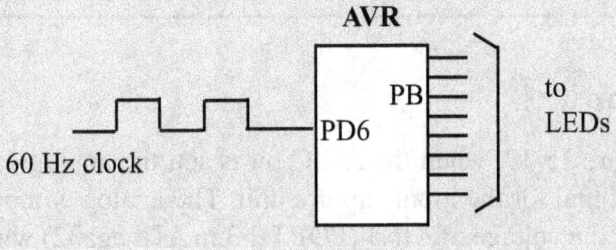

Example 15-27

Assume that a temperature sensor is connected to pin PD6. The temperature provided by the sensor is proportional to pulse width and is in the range of 1 μs to 250 μs. Write a program to measure the temperature if 1 μs is equal to 1 degree. Use the prescaler value that gives the result in a single byte. Display the result on PORTB. Assume XTAL = 8 MHz.

Solution:

8 MHz × 1 / 8 = 1 MHz = 1,000,000 Hz due to prescaler and T = 1/1,000,000 Hz = 1 μs for TCNT. That means we get the values between 1 and 65,536 μs for the TCNT, but since the pulse width never goes beyond 250 μs we should be able to display the temperature value on PORTB.

```
.INCLUDE "M32DEF.INC"
        LDI     R16,0xFF
        OUT     DDRB,R16      ;PORTB as output
        OUT     PORTD,R16
BEGIN:LDI       R20,0x00
        OUT     TCCR1A,R20    ;timer mode = Normal
        LDI     R20,0x42
        OUT     TCCR1B,R20    ;rising edge, prescaler 8, no noise canceller
L1:     IN      R21,TIFR      ;stay here for ICP rising
        SBRS    R21,ICF1      ;skip next instruction if ICF1 flag is set
        RJMP    L1            ;jump L1
        IN      R16,ICR1L     ;R16 = ICR1L
        OUT     TIFR,R21      ;ICF1 = 0
        LDI     R20,0x02
        OUT     TCCR1B,R20    ;falling edge, prescaler 8, no noise canceller
L2:     IN      R21,TIFR      ;stay here for ICP falling edge
        SBRS    R21,ICF1      ;skip next if ICF1 flag is set
        RJMP    L2
        IN      R22,ICR1L     ;R22 = ICR1L, TEMP = ICR1H
        SUB     R22,R16       ;period = falling edge - rising edge
        OUT     PORTB,R22     ;PORTB = R22
        OUT     TIFR,R21      ;clear ICF1
L3:     RJMP    L3            ;wait forever
```

Analog comparator

As shown in Figure 15-17, when the ACIC bit is set, the analog comparator provides the trigger signal for the input capture unit. The analog comparator is an op-amp that compares the voltage of AIN1 (PORTB.3 in ATmega32) with AIN0 (PORTB.2 in ATmega32). If the voltage of AIN1 is higher than AIN0, the comparator's output is 1; otherwise, its output is 0. For more information, see the datasheet of the ATmega32.

Review Questions

1. True or false. In the ATmega32, only Timer1 has the Input Capture function.
2. True or false. TCNT1 is also used by the Input Capture function.
3. True or false. Activating the noise canceller causes the capturing to occur instantly when an event rises.
4. Indicate the registers used by the Input Capture function.
5. True or false. The Input Capture function can capture the timing of an incoming pulse on the rising edge only.

SECTION 15.4: C PROGRAMMING

Examples 15-28 through 15-42 show the C versions of the earlier programs.

Example 15-28 (C version of Example 15-2)

Write a program that (a) after 4 external clocks turns on an LED connected to the OC0 pin, and (b) toggles the OC0 pin every 4 pulses.

Solution:

```
(a)
#include "avr/io.h"
int main ( )
{
     DDRB &= ~(1<<0);          //PB0(T0) pin as input
     DDRB = DDRB|(1<<3);       //PB3(OC0) pin as output
     OCR0 = 3;
     TCNT0 = 0;                //load timer with 0
     TCCR0 = 0x36;             //external clock, Normal mode, set OC0
     while (1);
     return 0;
}


(b)
#include "avr/io.h"
int main ( )
{
     DDRB &= ~(1<<0);          //PB0(T0) pin as input
     DDRB = DDRB|(1<<3);       //PB3(OC0) pin as output
     OCR0 = 3;
     TCNT0 = 0;                //load timer with 0
     TCCR0 = 0x1E;             //external clock, CTC mode, set OC0
     while (1);
     return 0;
}
```

Example 15-29 (C version of Example 15-4)

Rewrite the program of Example 15-4 using C.

Solution:

```c
#include "avr/io.h"
int main ( )
{
     DDRB = DDRB|(1<<3);   //PB3(OC0) = output
     TCCR0 = 0x11;   //COM01:00=Toggle, Mode=Normal, no prescaler
     OCR0 = 100;
     while (1);
     return 0;
}
```

Example 15-30 (C version of Example 15-6)

Rewrite the program of Example 15-6 using C.

Solution:

```c
#include "avr/io.h"
int main ( )
{
     DDRB = DDRB|(1<<3);   //PB3(OC0) = output
     TCCR0 = 0x19;   //COM01:00=Toggle, Mode=CTC, no prescaler
     OCR0 = 200;
     while (1);
}
```

Example 15-31 (C version of Example 15-8)

Rewrite the program of Example 15-8 using C.

Solution:

```c
#include "avr/io.h"
int main ( )
{
     DDRB |= (1<<3);            //PB3 = output
     while (1)
     {
         OCR0 = 99;
         TCCR0 = 0x19;       //CTC, no prescaler, set on match
         while ((TIFR&(1<<OCF0)) == 0);
         TIFR = (1<<OCF0);     //clear OCF0
         OCR0 = 69;
         TCCR0 = 0x39;       //CTC, no prescaler, set on match
         while ((TIFR&(1<<OCF0)) == 0);
         TIFR = (1<<OCF0);     //clear OCF0
     }
     return 0;
}
```

Example 15-32 (C version of Example 15-9)

Rewrite the program of Example 15-9 using C.

Solution:

```c
#include "avr/io.h"
#include "avr/interrupt.h"

int main ( )
{
    DDRB = DDRB |(1<<3);  //PB3 = output
    OCR0 = 69;
    TCCR0 = 0x19;       //CTC, no prescaler, toggle on match
    TIMSK = (1<<OCIE0);  //enable compare match interrupt
    sei();               //enable interrupts
    while(1);

    return 0;
}

ISR(TIMER0_COMP_vect)
{
    const unsigned char waveTable [] = { 24,49,39,34};
    static unsigned char index = 0;

    OCR0 = waveTable[ index] ;
    index ++;

    if(index >= 4)
        index = 0;
}
```

Example 15-33 (C version of Example 15-10)

Rewrite the program of Example 15-10 using C.

Solution:

```c
#include "avr/io.h"

int main ( )
{
    DDRB = DDRB |(1<<3);          //PB3 = output

    while(1)
      TCCR0 = 0x98; //CTC, timer stopped, toggle on match, FOC0=1

    return 0;
}
```

Example 15-34 (C version of Example 15-12)

Rewrite the program of Example 15-12 using C.

Solution:

```c
#include "avr/io.h"

int main ( )
{
    DDRD = DDRD |(1<<7);  //PD7(OC2) = output
    TCCR2 = 0x19;    //COM21:20=Toggle, Mode=CTC, no prescaler
    OCR2 = 200;

    while (1);
}
```

Example 15-35 (C version of Example 15-13)

Rewrite the program of Example 15-13 using C.

Solution:

```c
#include "avr/io.h"

void delay_1ms ( );

int main ()
{
    DDRB = (1<<5);
    while (1)
    {
        PORTB = PORTB ^ (1<<5);
        delay_1ms ( );
    }
    return 0;
}
void delay_1ms ( )
{
    ICR1H = 0x27;
    ICR1L = 0x0F;    //ICR1L = 0x0F, ICR1H = TEMP
    TCNT1H = 0;
    TCNT1L = 0;
    TCCR1A = 0x02;   //WGM11:10 = 10

    TCCR1B = 0x19;   //WGM13:12 = 11, CS = CLK, mode = 14
    while((TIFR&(1<<ICF1)) == 0);
    TIFR = (1<<ICF1);
    TCCR1B = 0;
    TCCR1A = 0;      //stop timer
}
```

Example 15-36 (C version of Example 15-18)

Rewrite the program of Example 15-18 using C.

Solution:
```c
#include "avr/io.h"
int main ()
{
    DDRD = (1<<5);
    TCCR1A = 0x40;   //COM1A = Toggle
    TCCR1B = 0x09;   //WGM = Toggle, Mode = CTC, no prescaler
    OCR1AH = 0x02;   //TEMP = 0x02
    OCR1AL = 0x00;   //OCR1A = 0x200 = 512
    while (1);
    return 0;
}
```

Example 15-37 (C version of Example 15-20)

Rewrite the program of Example 15-20 using C.

Solution:
```c
#include "avr/io.h"
int main ()
{
    DDRD = DDRD |(1<<5);
    TCCR1B = 0x01;   //Normal, timer stopped
    while (1)
      TCCR1A = 0x48;        //toggle on match, FOC1A = 1
}
```

Example 15-38 (C version of Example 15-22)

Assuming that clock pulses are fed into pin ICP1, write a program to read the TCNT1 value on every rising edge. Place the result on PORTA and PORTB.

Solution:
```c
#include "avr/io.h"
int main ( )
{
    DDRA = 0xFF;     //port A as output
    DDRB = 0xFF;     //port B as output
    PORTD = 0xFF;    //activate pull-up
    while(1) {
      TCCR1A = 0;    //Mode = Normal
      TCCR1B = 0x41;//rising edge, no scaler, no noise canceller
      while ((TIFR&(1<<ICF1)) == 0);
      TIFR = (1<<ICF1);   //clear ICF1
      PORTA = ICR1L;
      PORTB = ICR1H;
    }
    return 0;
}
```

Example 15-39 (C version of Example 15-23)

Assuming that clock pulses are fed into pin PORTD.6, write a program to measure the period of the pulses. Place the binary result on PORTA and PORTB.

Solution:

```c
#include "avr/io.h"
int main ( )
{
    unsigned int t;
    DDRA = 0xFF;      //PORTA as output
    DDRB = 0xFF;      //PORTB as output
    PORTD = 0xFF;     //activate pull-up
    TCCR1A = 0;       //Mode = Normal
    TCCR1B = 0x41;  //rising edge, no scaler, no noise canceller
    while ((TIFR&(1<<ICF1)) == 0);
    t = ICR1;
    TIFR = (1<<ICF1);      //clear ICF1
    while ((TIFR&(1<<ICF1)) == 0);
    t = ICR1 - t;
    PORTA = t;        //the low byte
    PORTB = t>>8;     //the high byte
    while (1);
    return 0;
}
```

Example 15-40 (C version of Example 15-24)

The frequency of a pulse is either 50 Hz or 60 Hz. Assume that a the pulse is connected to ICP1 (pin PD6). Write a program to measure its period and display it on PORTB. Use the prescaler value that gives the result in a single byte. Assume XTAL = 8 MHz.

Solution:

```c
#include "avr/io.h"
int main ( )
{
    unsigned char t1;
    DDRB = 0xFF;      //PORTB as output
    PORTD = 0xFF;
    TCCR1A = 0;       //Timer Mode = Normal
    TCCR1B = 0x45;  //rising edge, prescaler=1024, no noise canc.

    TIFR = (1<<ICF1);      //clear ICF1
    while ((TIFR&(1<<ICF1)) == 0); //wait while ICF1 is clear
    t1 = ICR1L;            //first edge value
    TIFR = (1<<ICF1);      //clear ICF1
    while ((TIFR&(1<<ICF1)) == 0); //wait while ICF1 is clear
    PORTB = ICR1L - t1;  //period = second edge - first edge
    TIFR = (1<<ICF1);      //clear ICF1
    while (1);             //wait forever
}
```

Example 15-41 (C version of Example 15-26)

Assume that a 60-Hz frequency pulse is connected to ICP1 (pin PD6). Write a program to measure its pulse width. Use the prescaler value that gives the result in a single byte. Display the result on PORTB. Assume XTAL = 8 MHz.

Solution:

```c
#include "avr/io.h"
int main ( )
{
    unsigned char t1;
    DDRB = 0xFF;            //Port B as output
    PORTD = 0xFF;
    TCCR1A = 0;             //Timer Mode = Normal
    TCCR1B = 0x45; //rising edge, prescaler=1024, no noise canc.
    while ((TIFR&(1<<ICF1)) == 0);
    t1 = ICR1L;             //first edge value
    TIFR = (1<<ICF1);       //clear ICF1 flag
    TCCR1B = 0x05;          //falling edge
    while ((TIFR&(1<<ICF1)) == 0);
    PORTB = ICR1L - t1;     //pulse width = falling - rising
    TIFR = (1<<ICF1);       //clear ICF1 flag
    while (1);              //wait forever
    return 0;
}
```

Example 15-42 (C version of Example 15-27)

Assume that a temperature sensor is connected to pin PD6. The temperature provided by the sensor is proportional to pulse width and is in the range of 1 μs to 250 μs. Write a program to measure the temperature if 1 μs is equal to 1 degree. Use the prescaler value that gives the result on PORTB. Assume XTAL = 8 MHz.

Solution:

```c
#include "avr/io.h"
int main ( )
{
    unsigned char t1;
    DDRB = 0xFF;            //Port B as output
    PORTD = 0xFF;
    TCCR1A = 0;             //Timer Mode = Normal
    TCCR1B = 0x42;   //rising edge, prescaler = 8, no noise canc.
    while ((TIFR&(1<<ICF1)) == 0);
    t1 = ICR1L;
    TIFR = (1<<ICF1);       //clear ICF1 flag
    TCCR1B = 0x02;          //falling edge
    while ((TIFR&(1<<ICF1)) == 0);
    PORTB = ICR1L - t1;     //pulse width = falling - rising
    TIFR = (1<<ICF1);       //clear ICF1 flag
    while (1);              //wait forever
    return 0;
}
```

SUMMARY

This chapter began by describing the pulse wave generating features of the AVR family. We discussed how to generate square waves and pulses using CTC mode. We discussed how to generate waves using Timer0 as an 8-bit timer and Timer1 as a 16-bit timer. We also described the input capture feature. We used the input capture feature of AVR to measure the pulse width and period of incoming pulses.

PROBLEMS

SECTION 15.1: WAVE GENERATION USING 8-BIT TIMERS

1. True or false. The ATmega32 has only one 8-bit timer.
2. True or false. In the ATmega32, Timer0 has a 16-bit register accessible as TCNT0L and TCNT0H.
3. True or false. Each waveform generator has a single pin.
4. Give the pin used for Timer2 waveform generator in the ATmega32.
5. Using Timer0, no prescaler, and CTC mode, write a program that generates a square wave with a frequency of 80 kHz. Assume XTAL = 8 MHz.
6. Using Timer0, no prescaler, and CTC mode, write a program that generates a square wave with a frequency of 5 kHz. Assume XTAL = 1 MHz.
7. Using Timer0 and CTC mode, write a program that generates a square wave with a frequency of 625 Hz. Assume XTAL = 8 MHz.
8. Using Timer0 and CTC mode, write a program that generates a square wave with a frequency of 3125 Hz. Assume XTAL = 16 MHz.

SECTION 15.2: WAVE GENERATION USING TIMER1

9. True or false. In the ATmega32, Timer1 has two waveform generator channels.
10. Give the number of waveform generators in the ATmega32.
11. Using Timer1, no prescaler, and CTC mode, write a program that generates a square wave with a frequency of 1 kHz. Assume XTAL = 8 MHz.
12. Using Timer1, no prescaler, and CTC mode, write a program that generates a square wave with a frequency of 5 kHz. Assume XTAL = 8 MHz.
13. Using Timer1 and CTC mode, write a program that generates a square wave with a frequency of 50 Hz. Assume XTAL = 8 MHz.
14. Using Timer1 and CTC mode, write a program that generates a square wave with a frequency of 20 Hz. Assume XTAL = 16 MHz.

SECTION 15.3: INPUT CAPTURE PROGRAMMING

15. What is the use of capturing?
16. True or false. In the ATmega32, all of the timers have the capturing capability.
17. True or false. To use capture mode, we must make the ICP pin an output pin.
18. Which timers can be used for the capture mode?
19. Find the value for the TCCR1B register in capture mode if we want to capture

on the falling edge.

20. Find the value for the TCCR1B register in capture mode if we want to capture on the rising edge while the noise canceller is active.

SECTION 15.4: C PROGRAMMING

21. Using Timer0, no prescaler, and CTC mode, write a program that generates a square wave with a frequency of 50 kHz. Assume XTAL = 8 MHz.
22. Using Timer2, no prescaler, and CTC mode, write a program that generates a square wave with a frequency of 20 kHz. Assume XTAL = 1 MHz.
23. Using Timer2, prescaler = 256, and CTC mode, write a program that generates a square wave with a frequency of 100 Hz. Assume XTAL = 8 MHz.
24. Using Timer0, prescaler = 64, and CTC mode, write a program that generates a square wave with a frequency of 95 Hz. Assume XTAL = 1 MHz.
25. As shown in the Figure, a switch is connected to PB1. Using CTC mode and prescaler = 1024, write a program in which, if the switch is closed, the waveform generator creates a 60 Hz wave; otherwise, it generates a wave with a frequency of 50 Hz. Assume XTAL = 8 MHz.

26. Using Timer1, no prescaler, and CTC mode, write a program that generates a square wave with a frequency of 3 kHz. Assume XTAL = 8 MHz.
27. Using Timer1, no prescaler, and CTC mode, write a program that generates a square wave with a frequency of 44 kHz. Assume XTAL = 8 MHz.

ANSWERS TO REVIEW QUESTIONS

SECTION 15.1: WAVE GENERATION USING 8-BIT TIMERS

1. True
2. True
3. False
4. PB3 (PORTB.3)

SECTION 15.2: WAVE GENERATION USING TIMER1

1. False
2. True
3. False
4. False

SECTION 15.3: INPUT CAPTURE PROGRAMMING

1. True
2. True
3. False
4. ICR, TCNT
5. False

CHAPTER 16

PWM PROGRAMMING AND DC MOTOR CONTROL IN AVR

OBJECTIVES

Upon completion of this chapter, you will be able to:

>> Describe the basic operation of a DC motor
>> Code AVR programs to control and operate a DC motor
>> Describe how PWM is used to control motor speed
>> Generate waves with different duty cycles using 8-bit and 16-bit timers
>> Code PWM programs to control and operate a DC motor

This chapter discusses the topic of PWM (pulse width modulation) and shows AVR interfacing with DC motors. The characteristics of DC motors are discussed along with their interfacing to the AVR. We use both Assembly and C programming examples to create PWM pulses.

SECTION 16.1: DC MOTOR INTERFACING AND PWM

This section begins with an overview of the basic operation of DC motors. Then we describe how to interface a DC motor to the AVR. Finally, we use Assembly and C language programs to demonstrate the concept of pulse width modulation (PWM) and show how to control the speed and direction of a DC motor.

DC motors

A direct current (DC) motor is a widely used device that translates electrical pulses into mechanical movement. In the DC motor we have only + and − leads. Connecting them to a DC voltage source moves the motor in one direction. By reversing the polarity, the DC motor will move in the opposite direction. One can easily experiment with the DC motor. For example, the small fans used in many motherboards to cool the CPU are run by DC motors. When the leads are connected to the + and − voltage source, the DC motor moves. While a stepper motor moves in steps of 1 to 15 degrees, the DC motor moves continuously. In a stepper motor, if we know the starting position we can easily count the number of steps the motor has moved and calculate the final position of the motor. This is not possible with a DC motor. The maximum speed of a DC motor is indicated in rpm and is given in the data sheet. The DC motor has two rpms: no-load and loaded. The manufacturer's datasheet gives the no-load rpm. The no-load rpm can be from a few thousand to tens of thousands. The rpm is reduced when moving a load and it decreases as the load is increased. For example, a drill turning a screw has a much lower rpm speed than when it is in the no-load situation. DC motors also have voltage and current ratings. The nominal voltage is the voltage for that motor under normal conditions, and can be from 1 to 150 V, depending on the motor. As we increase the voltage, the rpm goes up. The current rating is the current consumption when the nominal voltage is applied with no load, and can be from 25 mA to a few amps. As the load increases, the rpm is decreased, unless the current or voltage provided to the motor is increased, which in turn increases the torque. With a fixed voltage, as the load increases, the current (power) consumption of a DC motor is increased. If we overload the motor it will stall, and that can damage the motor due to the heat generated by high current consumption.

Unidirectional control

Figure 16-1 shows the DC motor rotation for clockwise (CW) and counterclockwise (CCW) rotations. See Table 16-1 for selected DC motors.

Bidirectional control

With the help of relays or some specially designed chips we can change the direction of the DC motor rotation. Figures 16-2 through 16-4 show the basic concepts of H-bridge control of DC motors.

Table 16-1: Selected DC Motor Characteristics (www.Jameco.com)

Part No.	Nominal Volts	Volt Range	Current	RPM	Torque
154915CP	3 V	1.5–3 V	0.070 A	5,200	4.0 g-cm
154923CP	3 V	1.5–3 V	0.240 A	16,000	8.3 g-cm
177498CP	4.5 V	3–14 V	0.150 A	10,300	33.3 g-cm
181411CP	5 V	3–14 V	0.470 A	10,000	18.8 g-cm

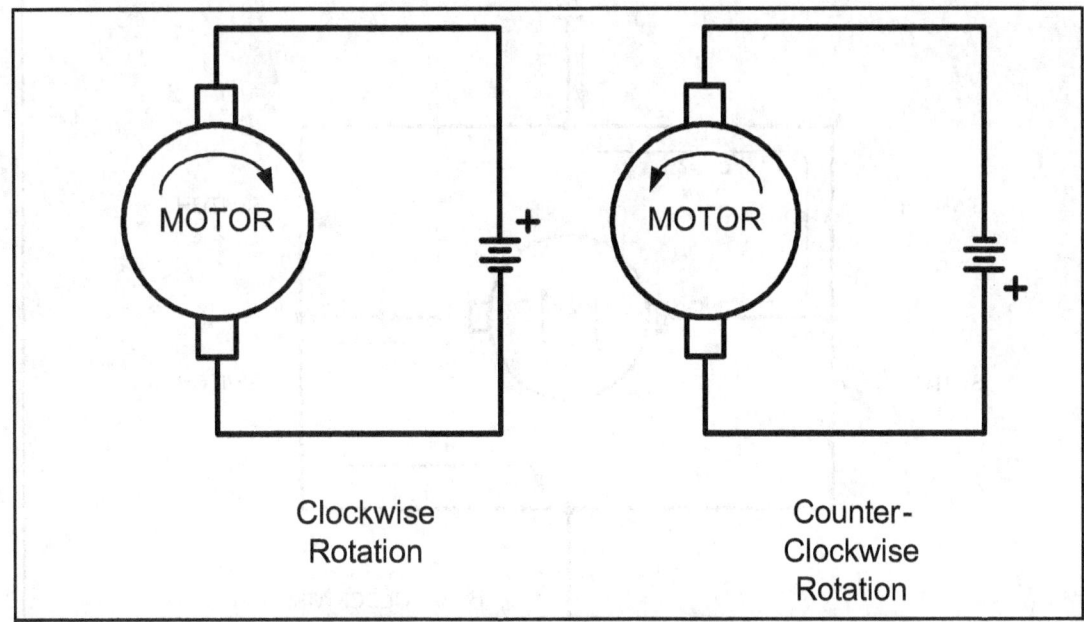

Figure 16-1. DC Motor Rotation (Permanent Magnet Field)

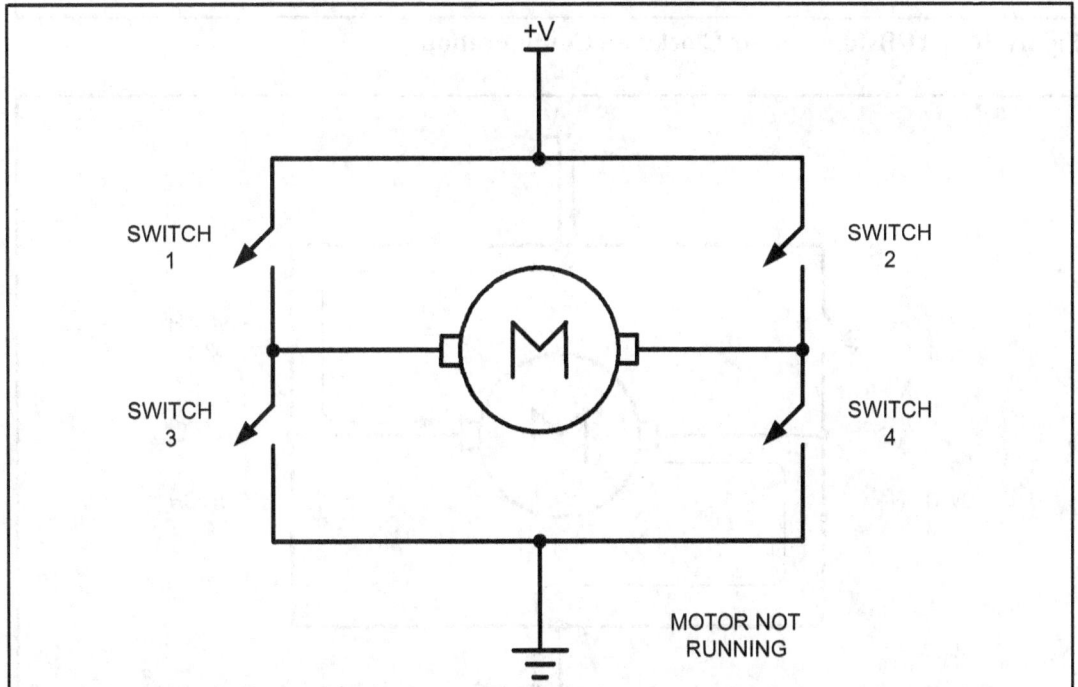

Figure 16-2. H-Bridge Motor Configuration

Figure 16-2 shows the connection of an H-bridge using simple switches. All the switches are open, which does not allow the motor to turn.

Figure 16-3 shows the switch configuration for turning the motor in one direction. When switches 1 and 4 are closed, current is allowed to pass through the motor.

Figure 16-4 shows the switch configuration for turning the motor in the opposite direction from the configuration of Figure 16-3. When switches 2 and 3 are closed, current is allowed to pass through the motor.

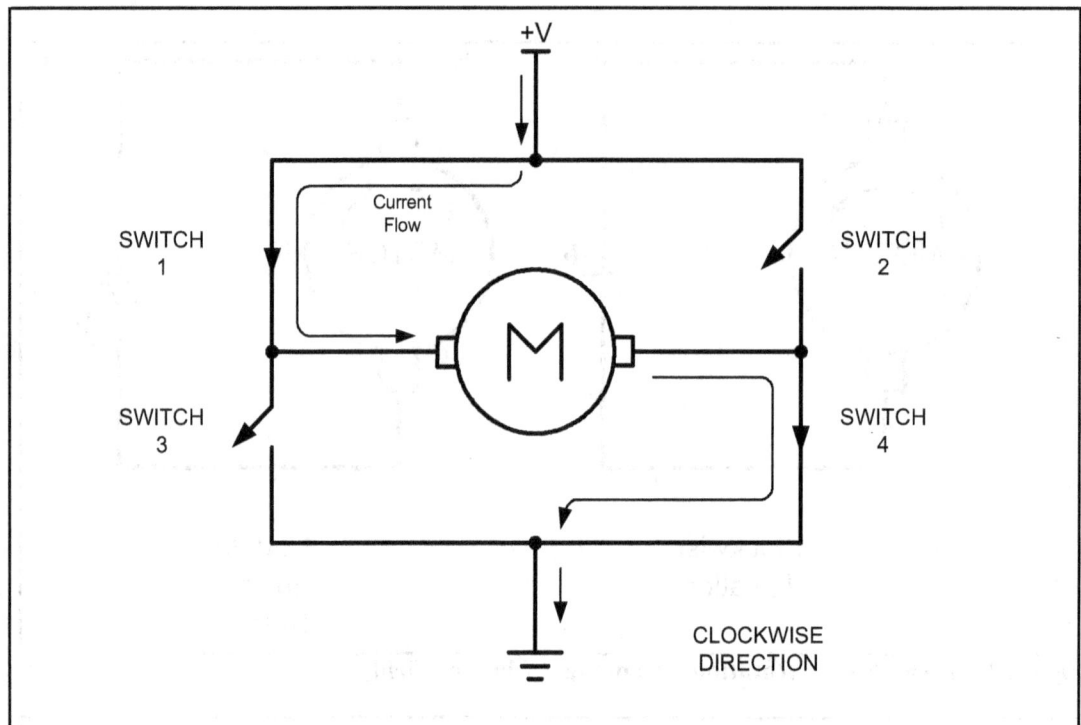

Figure 16-3. H-Bridge Motor Clockwise Configuration

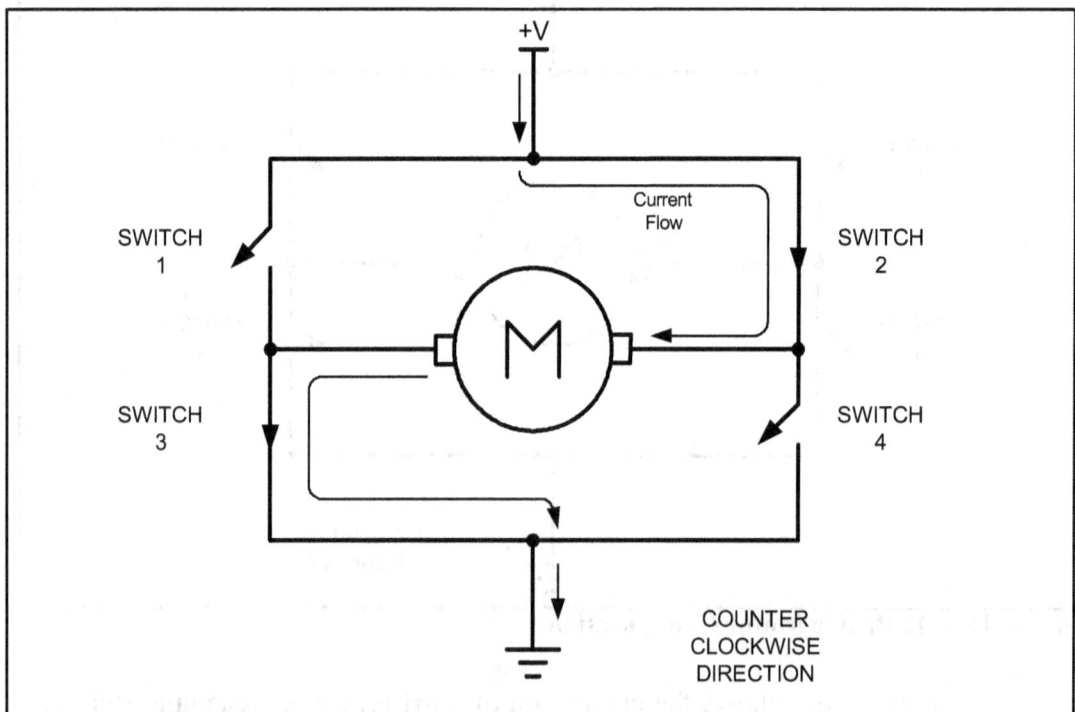

Figure 16-4. H-Bridge Motor Counterclockwise Configuration

Figure 16-5 shows an invalid configuration. Current flows directly to ground, creating a short circuit. The same effect occurs when switches 1 and 3 are closed or switches 2 and 4 are closed.

Table 16-2: Some H-Bridge Logic Configurations for Figure 16-2

Motor Operation	SW1	SW2	SW3	SW4
Off	Open	Open	Open	Open
Clockwise	Closed	Open	Open	Closed
Counterclockwise	Open	Closed	Closed	Open
Invalid	Closed	Closed	Closed	Closed

Table 16-2 shows some of the logic configurations for the H-bridge design.
H-bridge control can be created using relays, transistors, or a single IC solution such as the L298. When using relays and transistors, you must ensure that invalid configurations do not occur.

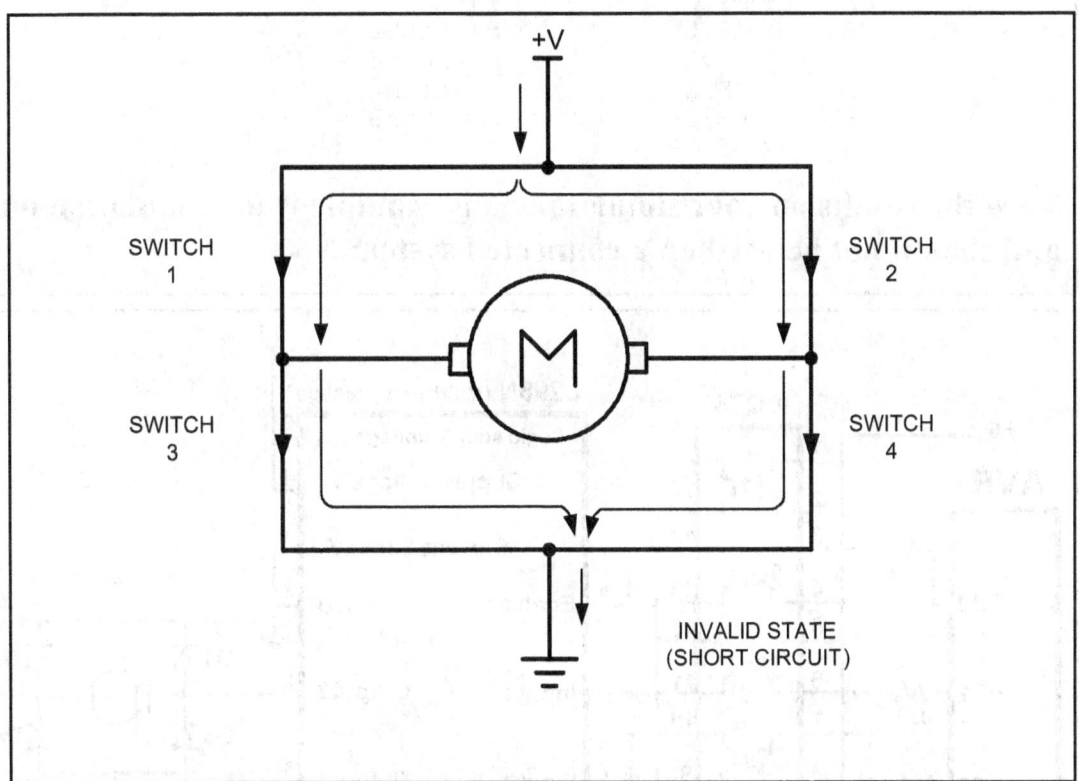

Figure 16-5. H-Bridge in an Invalid Configuration

Although we do not show the relay control of an H-bridge, Example 16-1 shows a simple program to operate a basic H-bridge.
Figure 16-6 shows the connection of the L298 to an AVR. Be aware that the L298 will generate heat during operation. For sustained operation of the motor, use a heat sink. Example 16-2 shows control of the L298.

Example 16-1

A switch is connected to pin PA7 (PORTA.7). Using a simulator, write a program to simulate the H-bridge in Table 16-2. We must perform the following:
(a) If PA7 = 0, the DC motor moves clockwise.
(b) If PA7 = 1, the DC motor moves counterclockwise.

Solution:

```
.INCLUDE "M32DEF.INC"
          SBI    DDRB,0          ;make PB0 an output (switch1)
          SBI    DDRB,1          ;make PB1 an output (switch2)
          SBI    DDRB,2          ;make PB2 an output (switch3)
          SBI    DDRB,3          ;make PB3 an output (switch4)
          CBI    DDRA,7          ;make PA7 an input
MONITOR:  SBIS   PINA,7          ;skip next if PINA.7 is set
          RJMP   CLKWISE         ;if PA7 = 0 go to CLKWISE
          CBI    DDRB,1          ;switch2 = 0
          CBI    DDRB,2          ;switch3 = 0
          SBI    DDRB,0          ;switch1 = 1
          SBI    DDRB,3          ;switch4 = 1
          JMP    MONITOR
CLKWISE:  CBI    DDRB,0          ;switch1 = 0
          CBI    DDRB,3          ;switch4 = 0
          SBI    DDRB,1          ;switch2 = 1
          SBI    DDRB,2          ;switch3 = 1
          JMP    MONITOR
```

View the results on your simulator. This example is for simulation only and should not be used on a connected system.

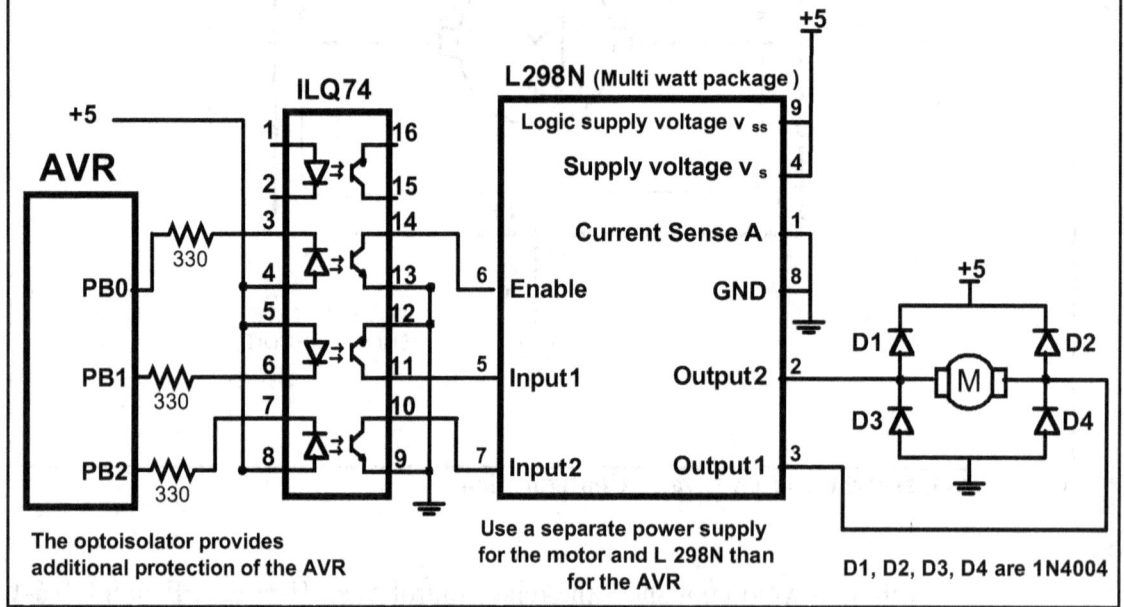

Figure 16-6. Bidirectional Motor Control Using an L298 Chip

DC motor control with optoisolator

As we discussed in Chapter 14, the optoisolator is indispensable in many motor control applications. Figures 16-6 through 16-8 show the connections to a

Example 16-2

Figure 16-6 shows the connection of an L298. Add a switch to pin PA7 (PORTA.7). Write a program to monitor the status of SW and perform the following:
(a) If SW = 0, the DC motor moves clockwise.
(b) If SW = 1, the DC motor moves counterclockwise.

Solution:

```
.INCLUDE "M32DEF.INC"
          SBI   DDRB,0          ;make PB0 an output (Enable)
          SBI   DDRB,1          ;make PB1 an output (clock)
          SBI   DDRB,2          ;make PB2 an output (counter)
          SBI   PORTB,0         ;Enable = 1
          CBI   DDRA,7          ;make PA7 an input
          SBI   PORTA,7
MONITOR:  SBIS  PINA,7          ;skip next if PINA.7 is set
          RJMP  CLKWISE         ;if PA7 = 0 go to CLKWISE
          CBI   PORTB,1         ;switch1 = 0
          SBI   PORTB,2         ;switch2 = 1
          JMP   MONITOR
CLKWISE:  SBI   PORTB,1         ;switch1 = 0
          CBI   PORTB,2         ;switch2 = 1
          JMP   MONITOR
```

simple DC motor using an optoisolator. Notice that the AVR is protected from EMI created by motor brushes by using an optoisolator and a separate power supply.

Figures 16-7 and 16-8 show optoisolators for single directional motor control, and the same principle should be used for most motor applications. Separating the power supplies of the motor and logic will reduce the possibility of damage to the control circuit.

Figure 16-7 shows the connection of a bipolar transistor to a motor. Protection of the control circuit is provided by the optoisolator. The motor and

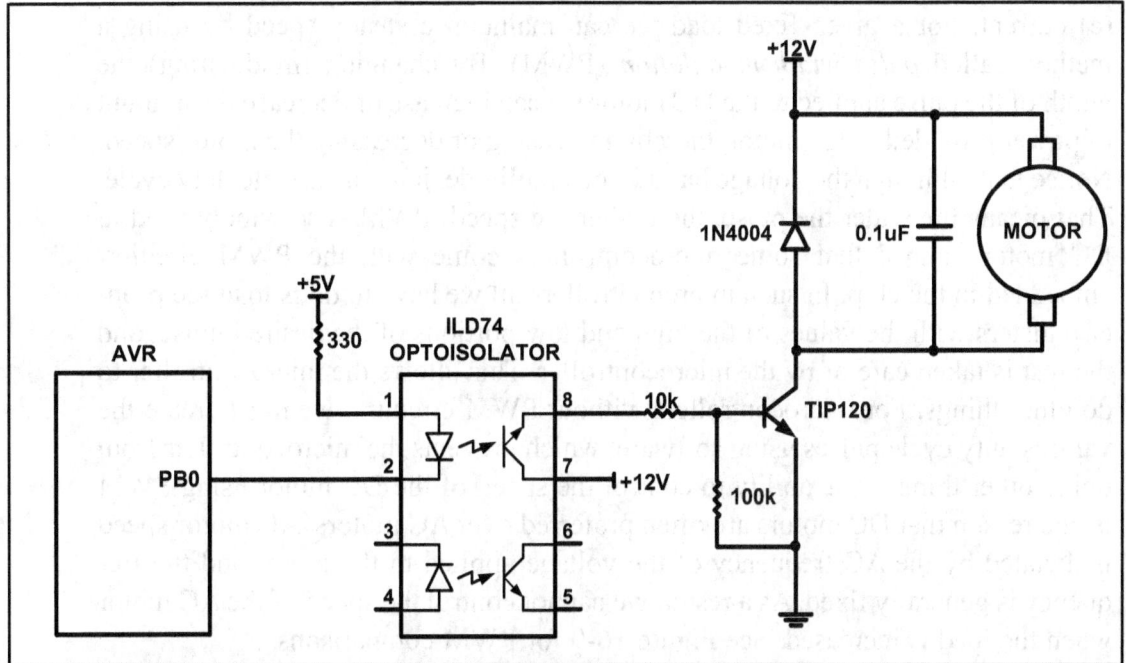

Figure 16-7. DC Motor Connection Using a Darlington Transistor

AVR use separate power supplies. The separation of power supplies also allows the use of high-voltage motors. Notice that we use a decoupling capacitor across the motor; this helps reduce the EMI created by the motor. The motor is switched on by clearing bit PB0.

Figure 16-8 shows the connection of a MOSFET transistor. The optoisolator protects the AVR from EMI. The zener diode is required for the transistor to reduce gate voltage below the rated maximum value. See Example 16-3.

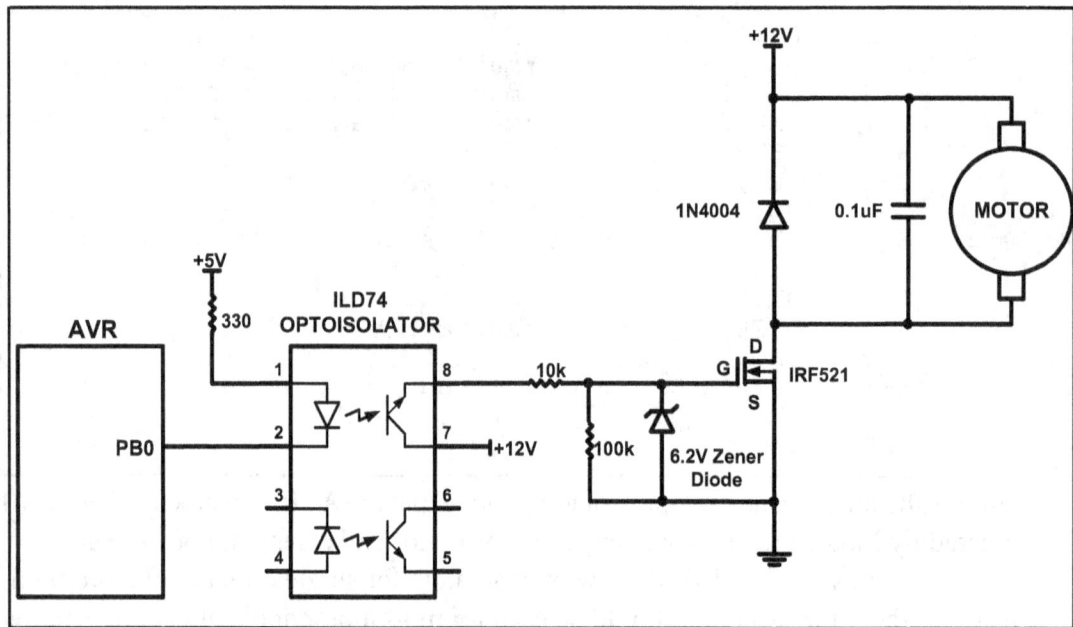

Figure 16-8. DC Motor Connection Using a MOSFET Transistor

Pulse width modulation (PWM)

The speed of the motor depends on three factors: (a) load, (b) voltage, and (c) current. For a given fixed load we can maintain a steady speed by using a method called *pulse width modulation* (PWM). By changing (modulating) the width of the pulse applied to the DC motor we can increase or decrease the amount of power provided to the motor, thereby increasing or decreasing the motor speed. Notice that, although the voltage has a fixed amplitude, it has a variable duty cycle. That means the wider the pulse, the higher the speed. PWM is so widely used in DC motor control that some microcontrollers come with the PWM circuitry embedded in the chip. In such microcontrollers all we have to do is load the proper registers with the values of the high and low portions of the desired pulse, and the rest is taken care of by the microcontroller. This allows the microcontroller to do other things. For microcontrollers without PWM circuitry, we must create the various duty cycle pulses using software, which prevents the microcontroller from doing other things. The ability to control the speed of the DC motor using PWM is one reason that DC motors are often preferred over AC motors. AC motor speed is dictated by the AC frequency of the voltage applied to the motor and the frequency is generally fixed. As a result, we cannot control the speed of the AC motor when the load is increased. See Figure 16-9 for PWM comparisons.

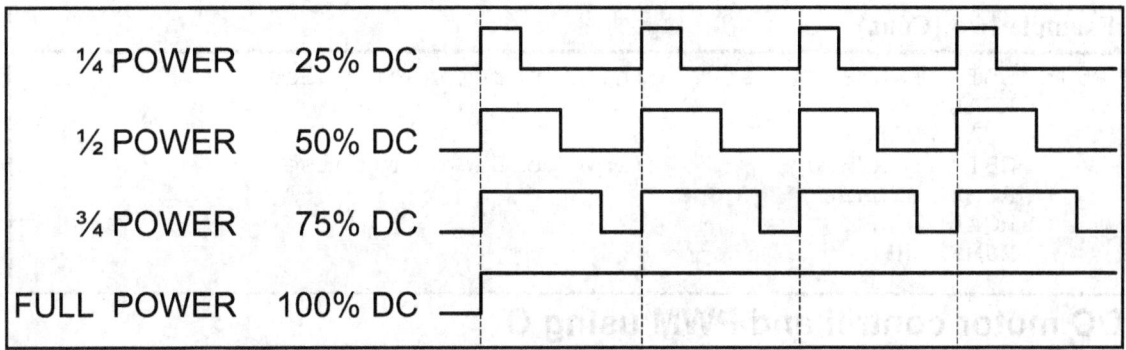

Figure 16-9. Pulse Width Modulation Comparisons

Example 16-3

Refer to the figure in this example. Write a program to monitor the status of the switch and perform the following:

(a) If PORTA.7 = 1, the DC motor moves with 25% duty cycle pulse.

(b) If PORTA.7 = 0, the DC motor moves with 50% duty cycle pulse.

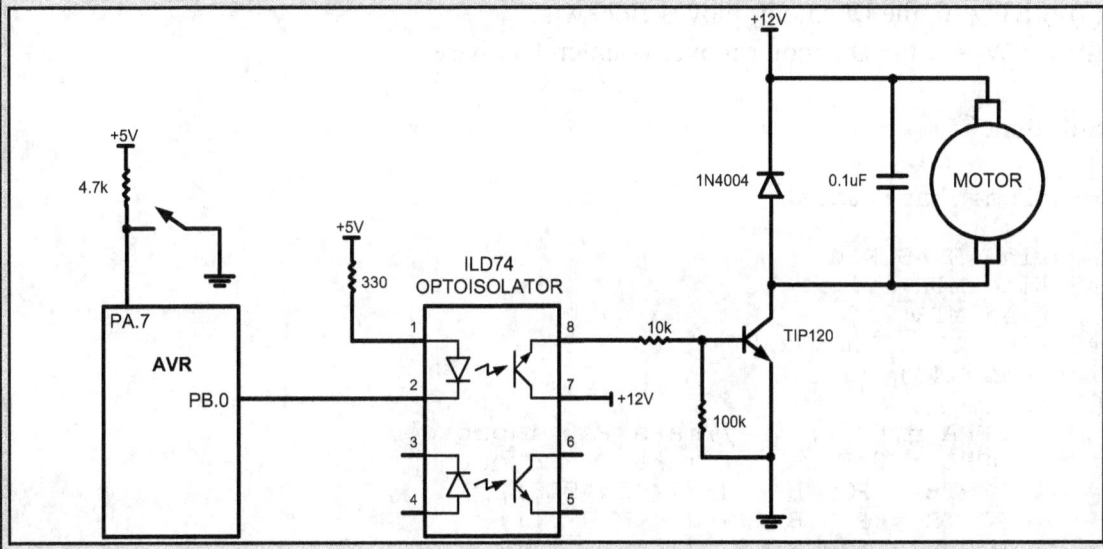

Solution:

```
.INCLUDE  "M32DEF.INC"
      LDI    R16,HIGH(RAMEND)
      OUT    SPH,R16
      LDI    R16,LOW(RAMEND)
      OUT    SPL,R16            ;initialize stack pointer
      SBI    DDRB,0             ;PORTB.0 as output
      CBI    DDRA,7             ;PORTA.7 as input
      SBI    PORTA,7            ;enable pull-up
      CBI    PORTB,0            ;turn off motor
CHK:  SBIC   PINA,7
      RJMP   P50
      SBI    PORTB,0            ;high portion of pulse
      RCALL DELAY
      RCALL DELAY
      RCALL DELAY
      CBI    PORTB,0            ;low portion of pulse
      RCALL DELAY
      RJMP   CHK
```

Example 16-3 (Cont.)

```
P50:  SBI   PORTB,0            ;high portion of pulse
      RCALL DELAY
      RCALL DELAY
      CBI   PORTB,0            ;low portion of pulse
      RCALL DELAY
      RCALL DELAY
      RJMP  CHK
```

DC motor control and PWM using C

Examples 16-4 through 16-5 show the C versions of the earlier programs controlling the DC motor.

Example 16-4 (C version of Example 16-2)

Refer to Figure 16-6 for connection of the motor. A switch is connected to pin PA7. Write a C program to monitor the status of SW and perform the following:
(a) If SW = 0, the DC motor moves clockwise.
(b) If SW = 1, the DC motor moves counterclockwise.

Solution:

```c
#include "avr/io.h"

#define ENABLE 0
#define MTR_1 1
#define MTR_2 2
#define SW (PINA&0x80)
int main ( )
{
      DDRA = 0x7F;    //make PA7 input pin
      DDRB = 0xFF;    //make PORTB output pin
      PORTB = PORTB & (~(1<<ENABLE));
      PORTB = PORTB & (~(1<<MTR_1));
      PORTB = PORTB & (~(1<<MTR_2));

      while (1)
      {
          PORTB = PORTB | (1<<ENABLE);

          if(SW == 1)
          {
              PORTB = PORTB | (1<<MTR_1);      //MTR_1 = 1
              PORTB = PORTB & (~(1<<MTR_2));   //MTR_2 = 0
          }
          else{
              PORTB = PORTB & (~(1<<MTR_1)); //MTR_1 = 0
              PORTB = PORTB | (1<<MTR_2);      //MTR_2 = 1
          }
      }
      return 0;
}
```

Example 16-5 (C version of Example 16-3)

Refer to the figure in this example. Write a C program to monitor the status of SW and perform the following:

(a) If SW = 0, the DC motor moves with 50% duty cycle pulse.
(b) If SW = 1, the DC motor moves with 25% duty cycle pulse.

Solution:

```
#define F_CPU   8000000UL         //XTAL = 8 MHz
#define SW      (PORTA&(1<<7))

#include "avr/io.h"
#include "util/delay.h"

void main()
{
    DDRA=0x7F;          //make PA7 input pin
    DDRB=0x01;          //make PB0 output pin
    while(1)
    {
        if(SW == 1)
        {
            PORTB = PORTB | (1<<0);
            _delay_ms(75);
            PORTB = PORTB & (~(1<<0));
            _delay_ms(25);
        }
        else
        {
            PORTB = PORTB | (1<<0);
            _delay_ms(50);
            PORTB = PORTB & (~(1<<0));
            _delay_ms(50);
        }
    }
}
```

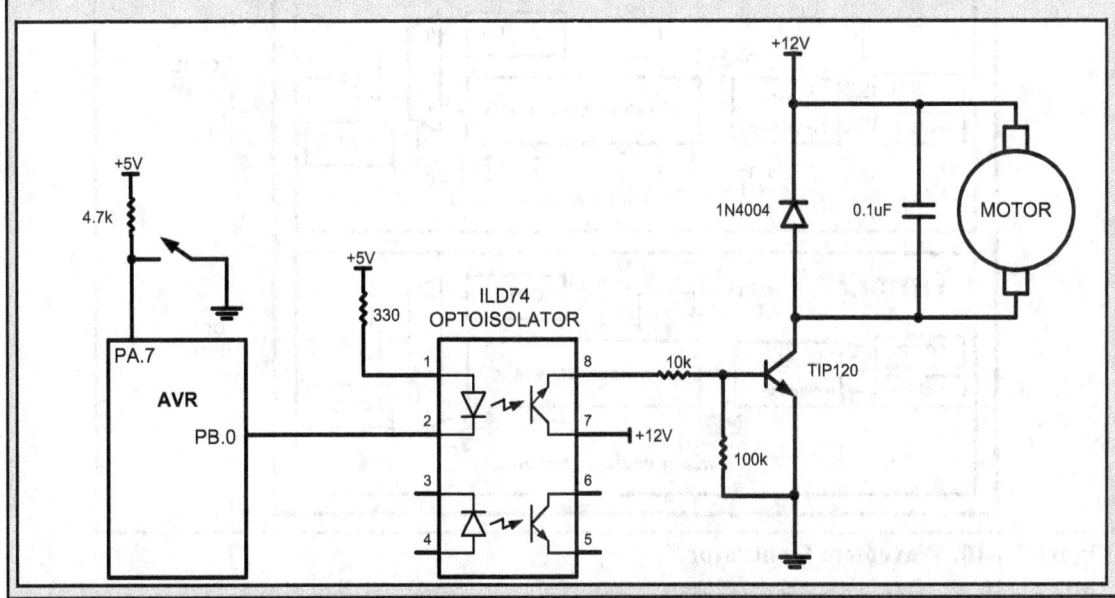

Review Questions

1. True or false. The permanent magnet field DC motor has only two leads for + and − voltages.
2. True or false. Just like a stepper motor, one can control the exact angle of a DC motor's move.
3. Why do we put a driver between the microcontroller and the DC motor?
4. How do we change a DC motor's rotation direction?
5. What is stall in a DC motor?
6. The RPM rating given for the DC motor is for _____ (no-load, loaded).

SECTION 16.2: PWM MODES IN 8-BIT TIMERS

This section and the next section discuss the PWM feature of the AVR. The ATmega32 comes with three timers, which can be used as wave generators, as shown in Figure 16-10. In the first section of this chapter we showed how to use the

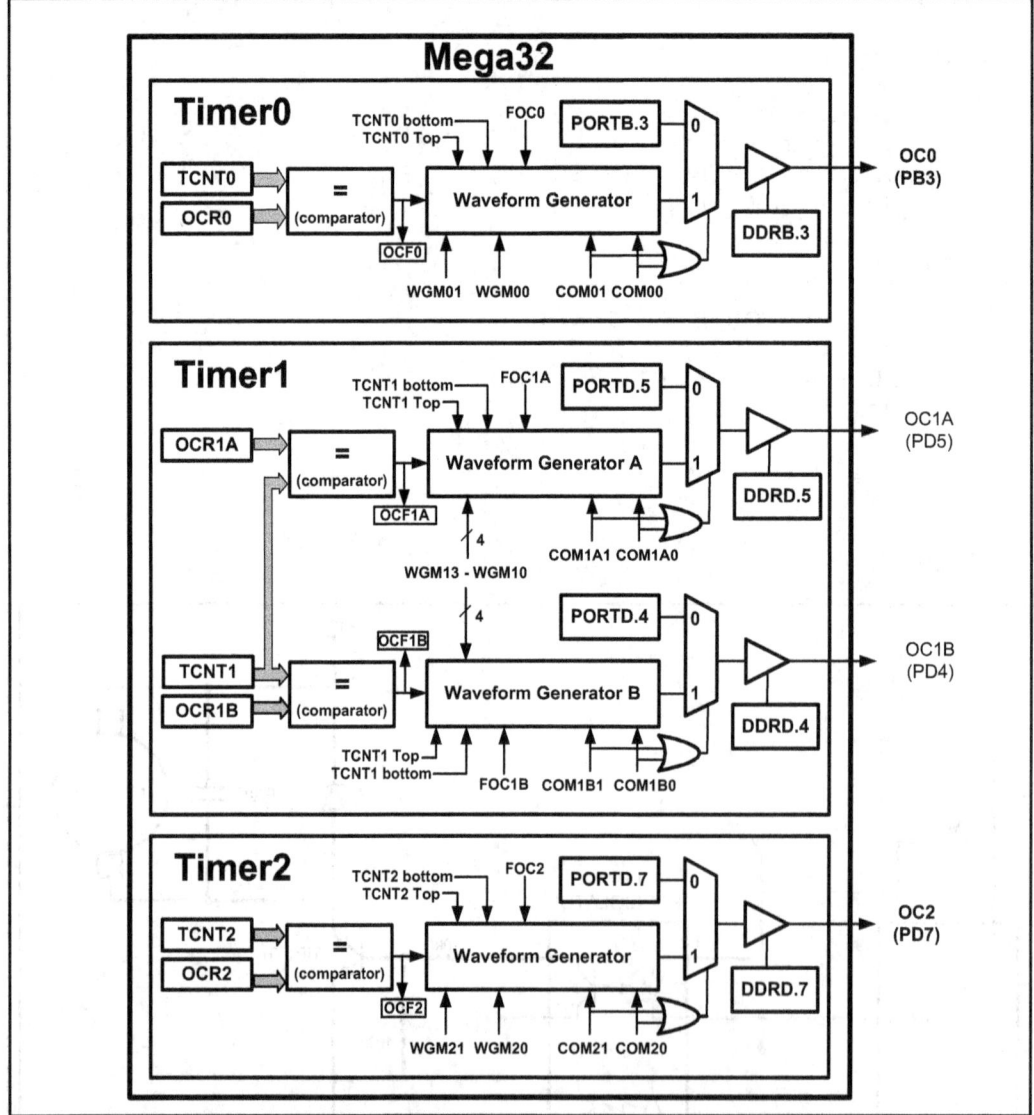

Figure 16-10. Waveform Generator

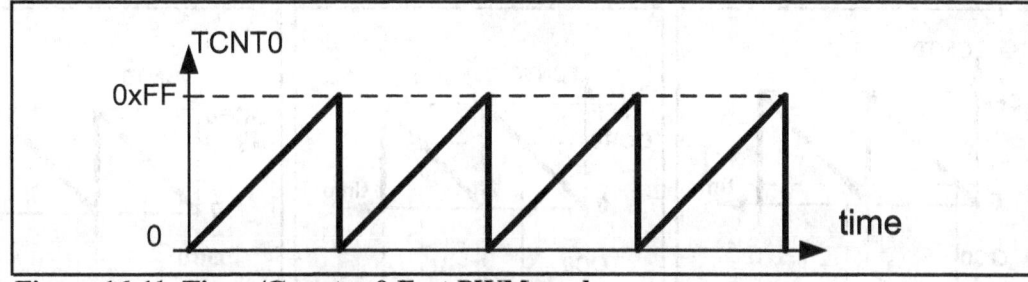

Figure 16-11. Timer/Counter 0 Fast PWM mode

CPU itself to create the equivalent of PWM outputs. The advantage of using the built-in PWM feature of the AVR is that it gives us the option of programming the period and duty cycle, therefore relieving the CPU to do other important things.

Fast PWM mode

In the Fast PWM, the counter counts like it does in the Normal mode. After the timer is started, it starts to count up. It counts up until it reaches its limit of 0xFF. When it rolls over from 0xFF to 00, it sets HIGH the TOV0 flag. See Figure 16-11.

In Figure 16-12 you see the reaction of the waveform generator when compare match occurs while the timer is in Fast PWM mode.

Bit	7	6	5	4	3	2	1	0
	FOC0	WGM00	COM01	COM00	WGM01	CS02	CS01	CS00
Read/Write	W	RW	RW	RW	RW	RW	RW	RW
Initial Value	0	0	0	0	0	0	0	0

FOC0 D7 Force compare match: it is a write-only bit, which can be used while generating a wave. Writing 1 to it causes the wave generator to act as if a compare match has occurred (see Chapter 15).

WGM01:00 D3D6 Timer0 mode selector bit
- 0 0 Normal
- 0 1 PWM, Phase correct
- 1 0 CTC (Clear Timer on Compare match)
- 1 1 Fast PWM

COM01:00 D5 D4 Compare Output Mode when Timer0 is in Fast PWM mode

COM01	COM00	Mode Name	Description
0	0	Disconnected	Normal port operation, OC0 disconnected
0	1	Reserved	Reserved
1	0	Non-inverted	Clear OC0 on compare match, set OC0 at TOP
1	1	Inverted PWM	Set OC0 on compare match, clear OC0 at TOP

CS02:00 D2D1D0 Timer0 clock selector
- 0 0 0 No clock source (Timer/Counter stopped)
- 0 0 1 clk (no prescaling)
- 0 1 0 clk / 8
- 0 1 1 clk / 64
- 1 0 0 clk / 256
- 1 0 1 clk / 1024
- 1 1 0 External clock source on T0 pin. Clock on falling edge
- 1 1 1 External clock source on T0 pin. Clock on rising edge

Figure 16-12. TCCR0 (Timer/Counter Control Register) Register

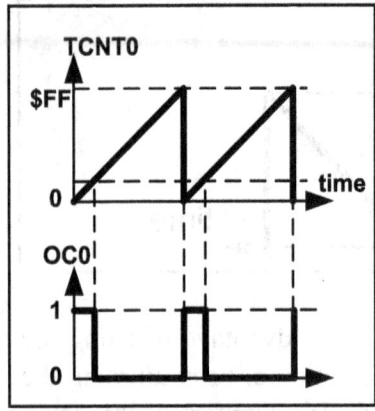

Figure 16-13A. Non-inverted

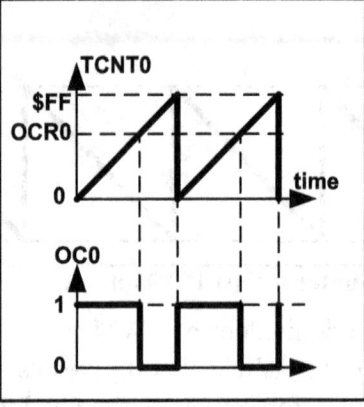

Figure 16-13B. Non-inverted

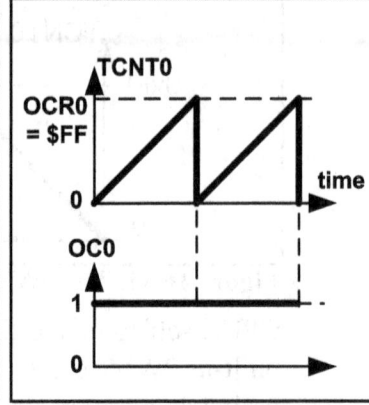

Figure 16-13C. Non-inverted

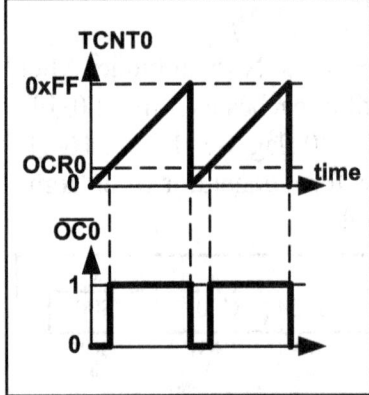

Figure 16-14A. Inverted

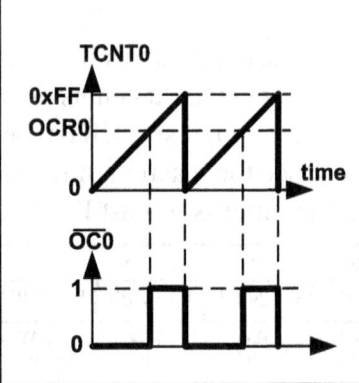

Figure 16-14B. Inverted

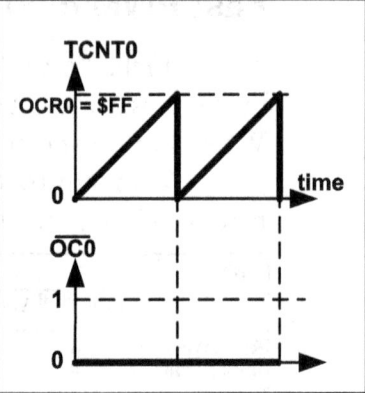

Figure 16-14C. Inverted

When COM01:00 = 00 the OC0 pin operates as an I/O port. When COM01:00 = 10, the waveform generator clears the OC0 pin whenever compare match occurs, and sets it at top. This mode is called *non-inverted PWM*. See Figures 16-13A through 16-13C. As you see from these figures, in the non-inverted PWM, the duty cycle of the generated wave increases when the value of OCR0 increases.

When COM01:00 = 11, the waveform generator sets the OC0 pin whenever compare match occurs, and clears it at top. This mode is referred as inverted PWM mode. See Figures 16-14A through 16-14C. As you see, in the inverted PWM mode when the value of OCR0 increases, the duty cycle of the generated wave decreases.

Frequency of the generated wave in Fast PWM mode

In Fast PWM mode, the timer counts from 0 to top (0xFF in 8-bit counters) and then rolls over. So, the frequency of the generated wave is 1/256 of the frequency of timer clock. As you saw in Section 9-1, the frequency of the timer clock can be selected using the prescaler. So, in 8-bit timers the frequency of the generated wave can be calculated as follows (N is determined by the prescaler):

$$\left. \begin{array}{l} F_{\text{generated wave}} = \dfrac{F_{\text{timer clock}}}{256} \\[2em] F_{\text{timer clock}} = \dfrac{F_{\text{oscillator}}}{N} \end{array} \right\} \Longrightarrow F_{\text{generated wave}} = \dfrac{F_{\text{oscillator}}}{256 \times N}$$

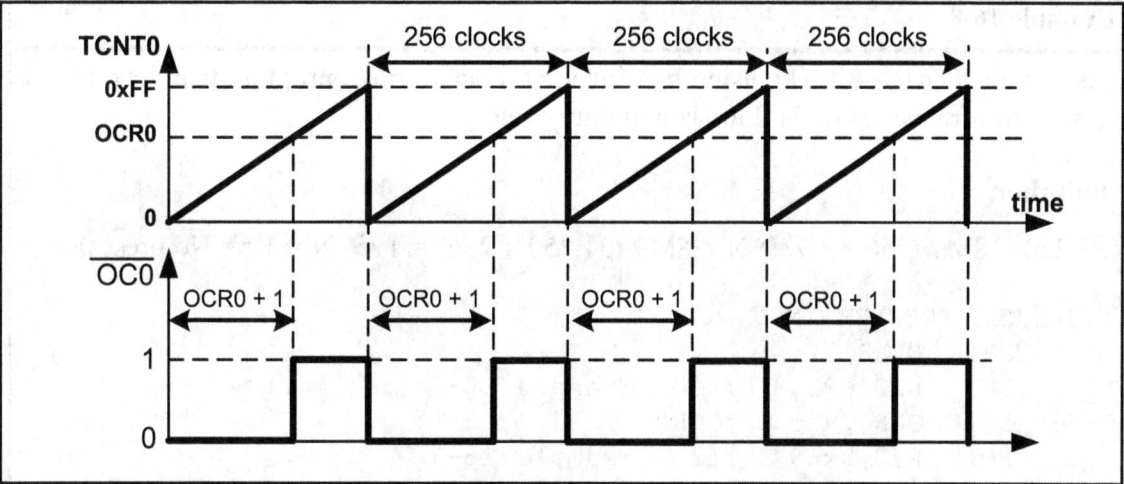

Figure 16-15. Timer/Counter 0 Fast PWM mode

Duty cycle of the generated wave in Fast PWM mode

The duty cycle of the generated mode can be determined using the OCR0 register. When COM01:00 = 10 (in non-inverted mode), the bigger OCR0 value results in a bigger duty cycle; When OCR0 = 255, the OC0 is 256 clocks out of 256 clocks, which means always high (duty cycle = 100%). Generally speaking, the OC0 is high, for a total of OCR0 + 1 clocks. See Figure 16-15. So, the duty cycle can be calculated using the following formula in non-inverted mode:

$$\text{Duty Cycle} = \frac{\text{OCR0} + 1}{256} \times 100$$

Similarly, the duty cycle formula for inverted mode is as follows:

$$\text{Duty Cycle} = \frac{255 - \text{OCR0}}{256} \times 100$$

Examine Figures 16-13 and 16-14 once again, and then examine Examples 16-6 through 16-10.

Example 16-6

To generate a wave with duty cycle of 75% in non-inverted mode, calculate the OCR0.

Solution:

75 = (OCR0 + 1) × 100 / 256 ➜ OCR0 + 1 = 75 × 256 / 100 = 192 ➜ OCR0 = 191

Example 16-7

Find the value for TCCR0 to initialize Timer0 for Fast PWM mode, non-inverted PWM wave generator, and no prescaler.

Solution:

WGM01:00 = 11 = Fast PWM mode   CS02:00 = 001 = No prescaler
COM01:00 = 10 = Non-inverted PWM

TCCR0 =	0	1	1	0	1	0	0	1
	FOC0	WGM00	COM01	COM00	WGM01	CS02	CS01	CS00

Example 16-8

Assuming XTAL = 8 MHz, using non-inverted mode, write a program that generates a wave with frequency of 31,250 Hz and duty cycle of 75%.

Solution:

31,250 = 8M / (256 × N) ➔ N = 8M / (31,250 × 256) = 1 ➔ N = 1 ➔ No prescaler

```
.INCLUDE "M32DEF.INC"
     SBI   DDRB,3
     LDI   R20,191    ;from Example 16-6
     OUT   OCR0,R20   ;OCR0 = 191
     LDI   R20,0x69   ;from Example 16-7
     OUT   TCCR0,R20  ;Fast PWM, no prescaler, non-inverted
HERE: RJMP  HERE       ;infinite loop
```

Notice that instead of the infinite loop we can use the CPU to perform other things.

Example 16-9

Assuming XTAL = 8 MHz, using non-inverted mode, write a program that generates a wave with frequency of 3906.25 Hz and duty cycle of 37.5%.

Solution:

3906.25 = 8M / (256 × N) ➔ N = 8M / (3906.25 × 256) = 8 ➔ the prescaler value = 8
37.5 = 100 × (OCR0 + 1) / 256 ➔ OCR0 + 1 = (256 × 37.5) / 100 = 96 ➔ OCR0 = 95

```
.INCLUDE "M32DEF.INC"
     SBI   DDRB,3
     LDI   R20,95
     OUT   OCR0,R20   ;OCR0 = 95
     LDI   R20,0x6A
     OUT   TCCR0,R20  ;Fast PWM, N = 8, non-inverted
HERE: RJMP  HERE
```

Example 16-10

Rewrite Example 16-9 using inverted mode.

Solution:

37.5 = 100 × (255 − OCR0)/256 ➔ 255 − OCR0 = (256 × 37.5)/100 = 96 ➔ OCR0 = 159

```
.INCLUDE "M32DEF.INC"
     SBI   DDRB,3
     LDI   R20,159
     OUT   OCR0,R20   ;OCR0 = 159
     LDI   R20,0x7A
     OUT   TCCR0,R20  ;Fast PWM, N = 8, inverted
HERE: RJMP  HERE
```

Loading values into the OCRx register in PWM modes

In the non-PWM modes (CTC mode and Normal mode), when we load a value into the OCR0 register, the value will be loaded instantly into the OCR0 register, but in the PWM modes (Fast PWM and Phase correct PWM), there is a buffer between us and the OCR0 register. When we read/write a value from/into the OCR0 we are dealing with the buffer. The contents of the buffer will be loaded into the OCR0 register only when the TCNT0 reaches to its topmost value. The top value is 0xFF in the 8-bit timers. See Figure 16-16 and Example 16-11.

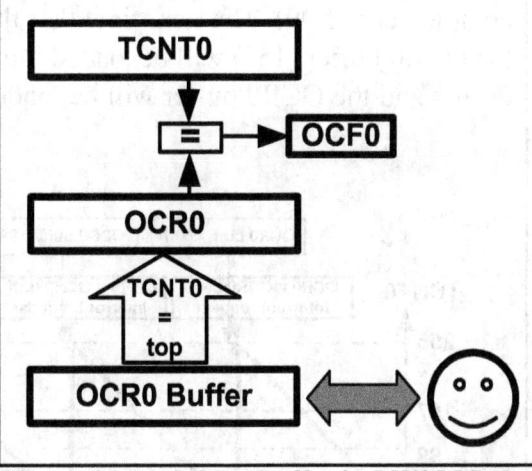

Figure 16-16. OCRn Buffer in PWM Modes

Example 16-11

Draw the wave generated by the following program. Assume XTAL = 1 MHz.

```
        .INCLUDE "M32DEF.INC"
        RJMP  MAIN
        .ORG  0x16                ;Timer0 overflow interrupt vector
        NEG   R20                 ;Negative R20
        OUT   OCR0,R20            ;OCR0 = R20
        RETI                      ;return interrupt
MAIN:
        LDI   R16,HIGH(RAMEND)
        OUT   SPH,R16
        LDI   R16,LOW(RAMEND)
        OUT   SPL,R16             ;initialize stack
        SBI   DDRB,3              ;OC0 as output
        LDI   R20,99              ;R20 = 99
        OUT   OCR0,R20            ;OCR0 = 99
        LDI   R16,0x69    ;Fast PWM mode, non-inverted, no prescaler
        OUT   TCCR0,R16
        OUT   OCR0,R20            ;OCR0 buffer = 99
        LDI   R16,(1<<TOIE0)      ;enable overflow interrupt
        OUT   TIMSK,R16
        SEI                       ;enable interrupt
HERE: RJMP  HERE                  ;wait here
```

Solution:

The wave generator is in non-inverted Fast PWM mode, which means that on compare match the OC0 pin will be set high. The OCR0 register is loaded with 99; so compare match occurs when TCNT0 reaches 99. When the timer reaches the top value and over-flows, the interrupt request occurs, and the OCR0 buffer is loaded with 157 (the two's

---

Example 16-11 (Cont.)

complement of 99). The next time that the timer reaches the top value, the contents of the OCR0 buffer (157) will be loaded into the OCR0 register. Then the second interrupt occurs and the OCR0 buffer will be loaded with 99 (the two's complement of 157).

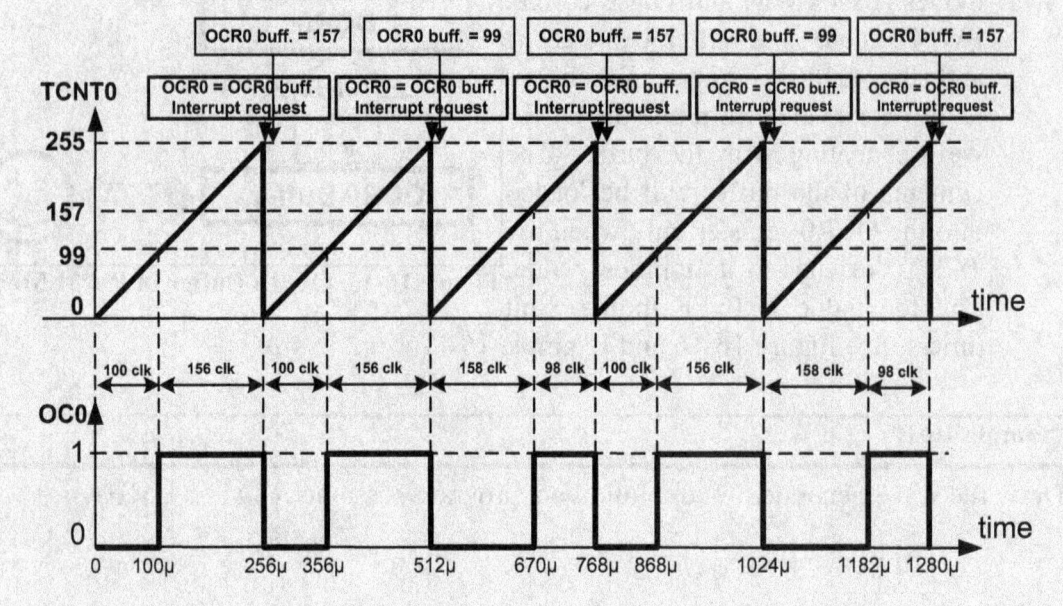

Phase correct PWM mode programming of Timer0

In the Phase correct PWM, the TCNT0 goes up and down like a yo-yo! First it counts up until it reaches the top value. Then it counts down until it reaches zero. The TOV0 flag is set whenever it reaches zero. See Figure 16-17.

Phase correct PWM mode

In Figure 16-18 you see the reaction of the waveform generator when compare match occurs in Phase correct PWM mode. When COM01:00 = 00 the OC0 pin operates as an I/O port. When COM01:00 = 10, the waveform generator clears the OC0 pin on compare match when counting up, and sets it on compare match when counting down. This mode is called *non-inverted Phase correct PWM*. See Figures 16-19A through 16-19C.

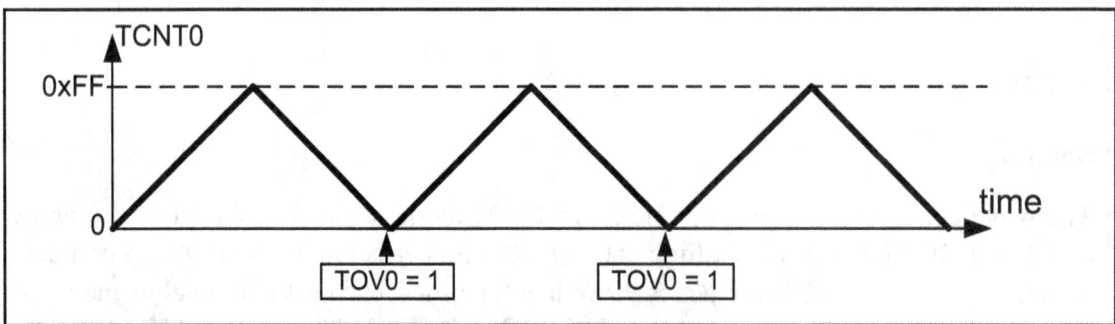

Figure 16-17. Timer/Counter 0 Phase Correct PWM Mode

Bit	7	6	5	4	3	2	1	0
	FOC0	WGM00	COM01	COM00	WGM01	CS02	CS01	CS00
Read/Write	W	RW	RW	RW	RW	RW	RW	RW
Initial Value	0	0	0	0	0	0	0	0

FOC0   D7   Force compare match: This is a write-only bit, which can be used while generating a wave. Writing 1 to it causes the wave generator to act as if a compare match had occurred (Chapter 11).

WGM01:00   D3D6   Timer0 mode selector bit

       0 0   Normal
       0 1   PWM, Phase correct
       1 0   CTC (Clear Timer on Compare match)
       1 1   Fast PWM

COM01:00   D5 D4   Compare Output Mode when Timer0 is in Phase correct PWM mode:

COM01	COM00	Description
0	0	Normal port operation, OC0 disconnected
0	1	Reserved
1	0	Clear OC0 on compare match when up-counting. Set OC0 on compare match when down-counting.
1	1	Set OC0 on compare match when up-counting. Clear OC0 on compare match when down-counting.

CS02:00   D2D1D0   Timer0 clock selector

       0 0 0   No clock source (Timer/Counter stopped)
       0 0 1   clk (no prescaling)
       0 1 0   clk / 8
       0 1 1   clk / 64
       1 0 0   clk / 256
       1 0 1   clk / 1024
       1 1 0   External clock source on T0 pin. Clock on falling edge
       1 1 1   External clock source on T0 pin. Clock on rising edge

Figure 16-18. TCCR0 (Timer/Counter Control Register) Register

When COM01:00 = 11, the waveform generator sets the OC0 pin on compare match when counting up, and clears it on compare match when counting down. This mode is referred as *inverted Phase correct PWM mode*. See Figures 16-20A through 16-20C.

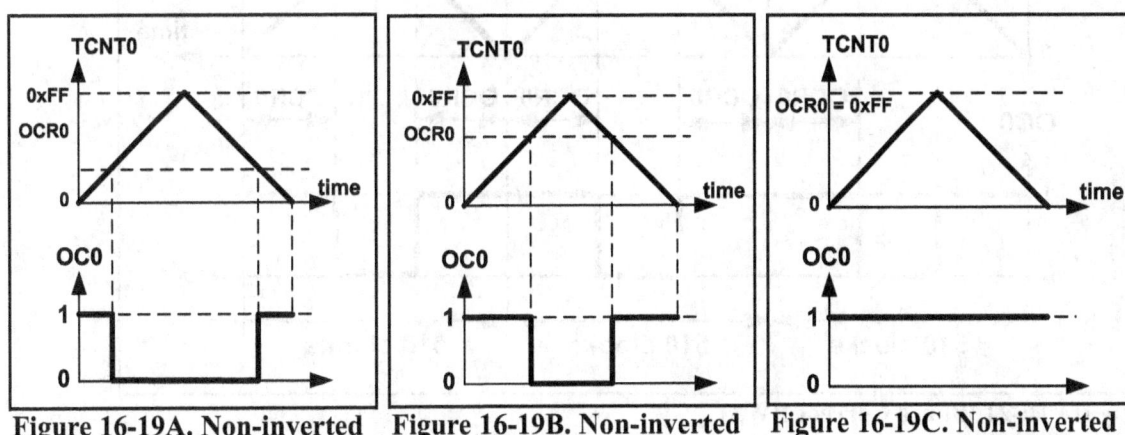

Figure 16-19A. Non-inverted   **Figure 16-19B. Non-inverted**   **Figure 16-19C. Non-inverted**

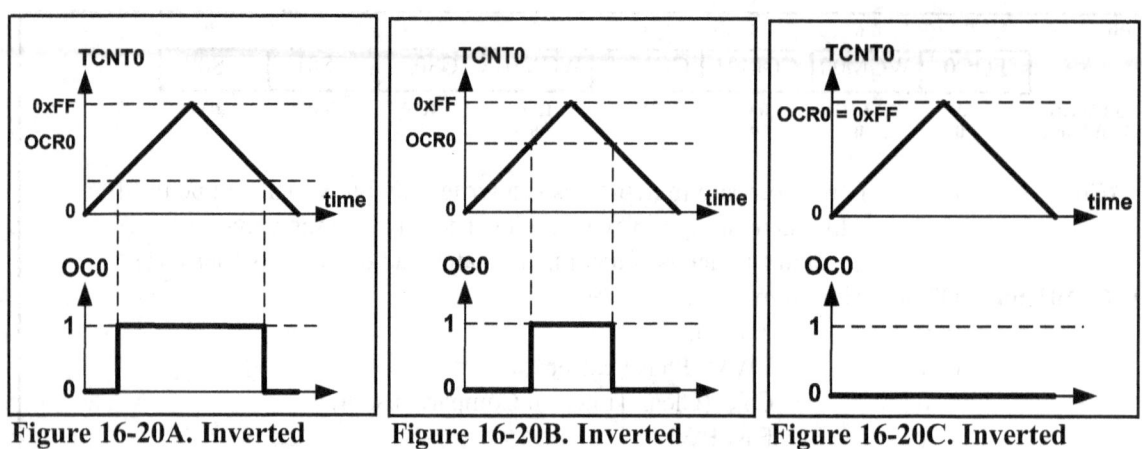

Figure 16-20A. Inverted **Figure 16-20B. Inverted** **Figure 16-20C. Inverted**

Frequency of the generated wave in Phase correct PWM mode

As you see in Figure 16-21, the frequency of the generated wave is 1/510 of the frequency of timer clock. As you saw in Section 9-1, the frequency of timer clock can be selected using the prescaler. So, in 8-bit timers the frequency of the generated wave can be calculated as follows:

$$
\left.
\begin{array}{c}
F_{generated\ wave} = \dfrac{F_{timer\ clock}}{510} \\[2em]
F_{timer\ clock} = \dfrac{F_{oscillator}}{N}
\end{array}
\right\} \implies F_{generated\ wave} = \dfrac{F_{oscillator}}{510 \times N}
$$

Duty cycle of the generated wave in Phase correct PWM mode

The duty cycle of the generated mode can be determined using the OCR0 register. When COM01:00 = 10 (in non-inverted mode), the bigger OCR0 value results in a bigger duty cycle. When OCR0 = 255, the OC0 is high, 510 clocks out of 510 clocks, which means always (duty cycle = 100%). Generally speaking, the OC0 is high for a total of 2 × OCR0 clocks. See Figure 16-21. So, the duty cycle

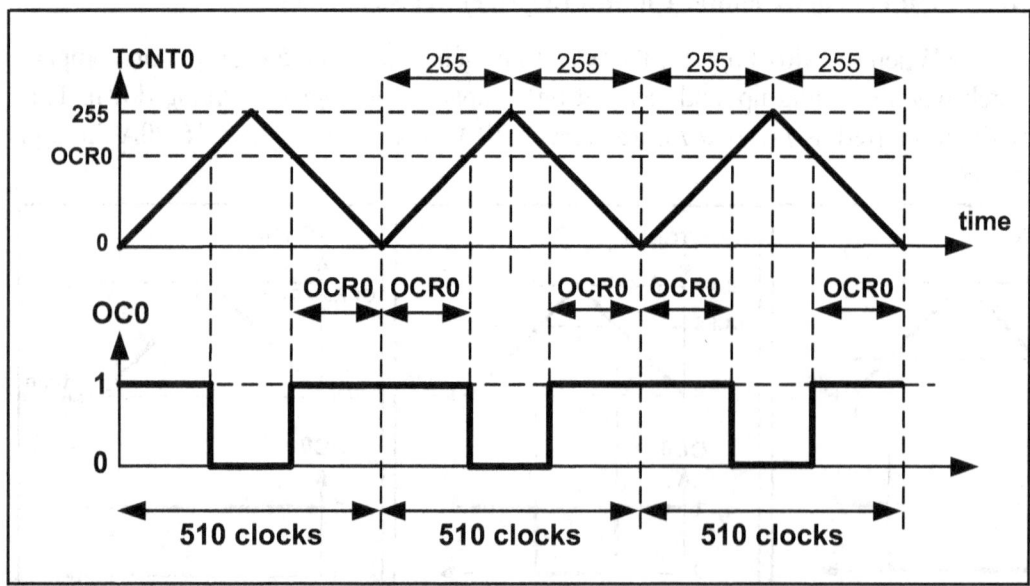

Figure 16-21. Phase Correct PWM

can be calculated using the following formula in non-inverted mode:

$$\text{Duty Cycle} = \frac{2 \times OCR0}{510} \times 100 \quad \Longrightarrow \quad \text{Duty Cycle} = \frac{OCR0}{255} \times 100$$

Similarly, the duty cycle formula for inverted mode is as follows:

$$\text{Duty Cycle} = \frac{510 - 2 \times OCR0}{510} \times 100 \quad \Longrightarrow \quad \text{Duty Cycle} = \frac{255 - OCR0}{255} \times 100$$

See Examples 16-12 through 16-15.

Example 16-12

Find the value for TCCR0 for Phase correct PWM, non-inverted PWM wave generator, and no prescaler.

Solution:

WGM01:00 = 01 = Phase correct PWM mode
COM01:00 = 10 = Non-inverted PWM
CS02:00 = 001 = No prescaler

TCCR0 =	0	1	1	0	0	0	0	1
	FOC0	WGM00	COM01	COM00	WGM01	CS02	CS01	CS00

Example 16-13

Assuming XTAL = 8 MHz, using non-inverted mode, write a program that generates a wave with frequency of 15,686 Hz and duty cycle of 75%.

Solution:

15,686 = 8M / (510 × N) ➔ N = 8M / (15,626 × 510) = 1 ➔ No prescaler
75 = OCR0 × 100 / 255 ➔ OCR0 = 75 × 255 / 100 = 191 ➔ OCR0 = 191

```
.INCLUDE "M32DEF.INC"
     SBI   DDRB,3
     LDI   R20,191
     OUT   OCR0,R20   ;OCR0 = 191
     LDI   R20,0x61
     OUT   TCCR0,R20  ;Phase c. PWM, no prescaler, non-inverted
HERE: RJMP  HERE
```

Comparing the program with the program in Example 16-8, you see that they are almost the same. The only difference is that the TCCR0 is loaded with 0x61 instead of 0x69.

Example 16-14

Find the value for TCCR0 for Phase correct PWM, inverted PWM wave generator, and prescaler = 256.

Solution:

WGM01:00 = 01 = Phase correct PWM mode
COM01:00 = 11 = Inverted PWM
CS02:00 = 100 = Scale 256

TCCR0 =

0	1	1	1	0	1	0	0
FOC0	WGM00	COM01	COM00	WGM01	CS02	CS01	CS00

Example 16-15

Assuming XTAL = 8 MHz, using inverted mode, write a program that generates a wave with frequency of 61 Hz and duty cycle of 87.5%.

Solution:

$61 = 8M/(510 \times N) \rightarrow N = 8M/(61 \times 510) = 256$
$87.5 = 100 \times (255 - OCR0)/255 \rightarrow 255 - OCR0 = (255 \times 87.5)/100 = 223 \rightarrow OCR0 = 32$

```
.INCLUDE "M32DEF.INC"
    SBI   DDRB,3      ;OC0 as output
    LDI   R20,32
    OUT   OCR0,R20    ;OCR0 = 32
    LDI   R20,0x74    ;from Example 16-14
    OUT   TCCR0,R20   ;Phase c. PWM, N = 256, inverted
HERE: RJMP  HERE
```

Difference between the wave generated by Phase correct PWM and Fast PWM

As you see in Figure 16-22, in Fast PWM, the phase of the wave is different for different duty cycles, while it remains unchanged in the Phase correct PWM as shown in Figure 16-23.

In non-inverted Fast PWM, the duty cycle of the generated wave is (OCR0 + 1)/256. Becuase the value of OCR0 is between 0 and 255, the duty cycle of the wave can be changed between 1/256 and 256/256. Therefore, in non-inverted Fast PWM the duty cycle of wave cannot be 0% (unless we turn off the waveform generator). Similarly, in inverted Fast PWM, the duty cycle changes between 0/256 and 255/256; thus, the duty cycle cannot be 100%. But in Phase correct PWM, the duty cycle changes between 0/255 and 255/255. Therefore, the wave can change between 0% (completely off) and 100% (completely on).

For driving motors, it is preferable to use Phase correct PWM rather than Fast PWM. In Fast PWM the frequency of the generated wave is twice that of the Phase correct mode. Thus, Fast PWM mode is preferable when we need to generate waves with high frequencies.

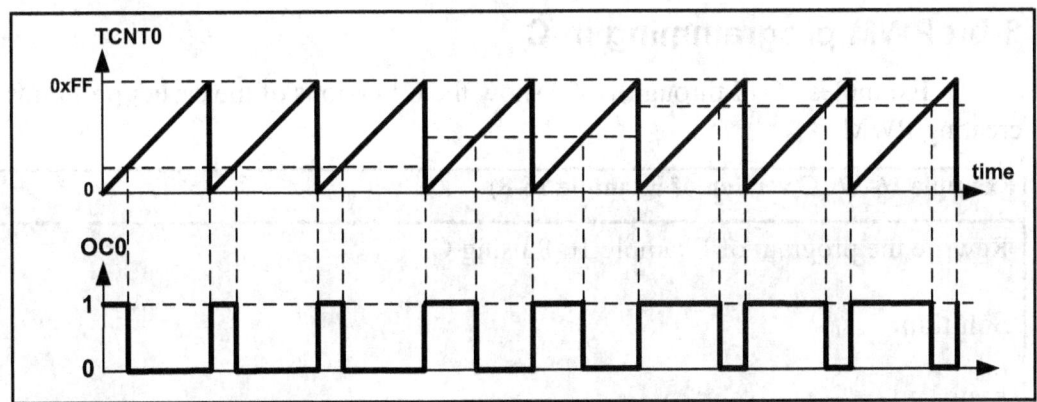

Figure 16-22. Fast PWM

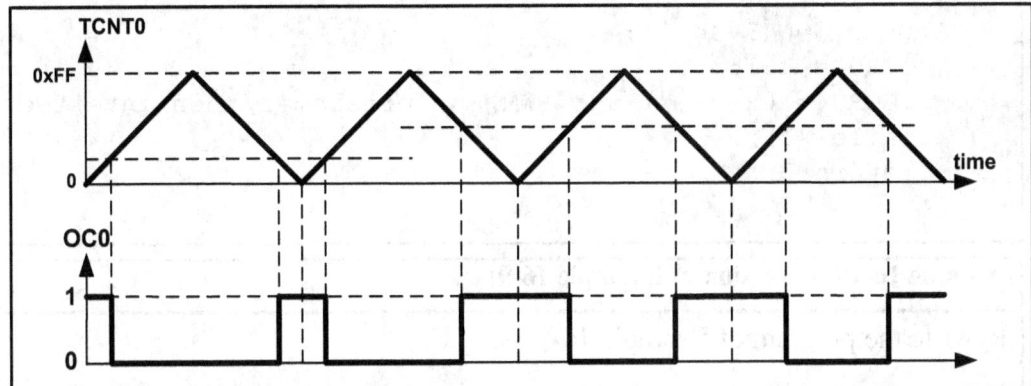

Figure 16-23. Phase Correct PWM

Generating waves using Timer2

Timer2 is an 8-bit timer. Therefore, it works similar to Timer0. The differences are register names, output port, and the prescaler values of TCCRn register. See Example 16-16.

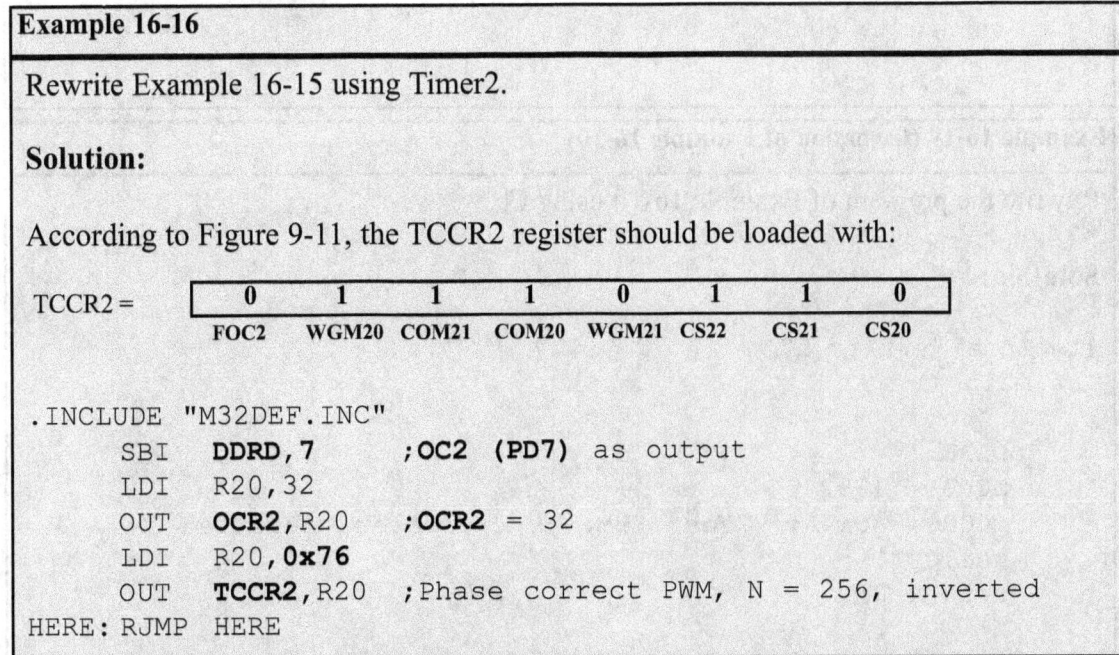

Example 16-16

Rewrite Example 16-15 using Timer2.

Solution:

According to Figure 9-11, the TCCR2 register should be loaded with:

TCCR2 =	0	1	1	1	0	1	1	0
	FOC2	WGM20	COM21	COM20	WGM21	CS22	CS21	CS20

```
.INCLUDE "M32DEF.INC"
     SBI   DDRD,7      ;OC2 (PD7) as output
     LDI   R20,32
     OUT   OCR2,R20    ;OCR2 = 32
     LDI   R20,0x76
     OUT   TCCR2,R20   ;Phase correct PWM, N = 256, inverted
HERE: RJMP  HERE
```

8-bit PWM programming in C

Examples 16-17 through 16-22 show the C versions of the earlier programs creating PWM.

Example 16-17 (C version of Example 16-8)

Rewrite the program of Example 16-8 using C.

Solution:

```c
#include "avr/io.h"
int main ()
{
    DDRB |= (1 << 3);
    OCR0 = 191;
    TCCR0 = 0x69; //Fast PWM, no prescaler, non-inverted
    while (1);
    return 0;
}
```

Example 16-18 (C version of Example 16-9)

Rewrite the program of Example 16-9 using C.

Solution:

```c
#include "avr/io.h"
int main ()
{
    DDRB |= (1 << 3);
    OCR0 = 95;
    TCCR0 = 0x6A; //Fast PWM, no prescaler, non-inverted
    while (1);
    return 0;
}
```

Example 16-19 (C version of Example 16-10)

Rewrite the program of Example 16-10 using C.

Solution:

```c
#include "avr/io.h"
int main ()
{
    DDRB |= (1 << 3);
    OCR0 = 159;
    TCCR0 = 0x7A; //Fast PWM, no prescaler, inverted
    while (1);
    return 0;
}
```

Example 16-20 (C version of Example 16-13)

Rewrite the program of Example 16-13 using C.

Solution:

```c
#include "avr/io.h"

int main ()
{
    DDRB |= (1 << 3);
    OCR0 = 191;
    TCCR0 = 0x61; //Phase c. PWM, no prescaler, non-inverted
    while (1);
    return 0;
}
```

Example 16-21 (C version of Example 16-15)

Rewrite the program of Example 16-15 using C.

Solution:

```c
#include "avr/io.h"

int main ()
{
    DDRB |= (1 << 3);
    OCR0 = 32;
    TCCR0 = 0x74; //Phase correct PWM, N = 256, inverted
    while (1);
    return 0;
}
```

Example 16-22 (C version of Example 16-16)

Rewrite the program of Example 16-16 using C.

Solution:

```c
#include "avr/io.h"

int main ()
{
    DDRD |= (1 << 7);
    OCR2 = 32;
    TCCR2 = 0x76; //Phase correct PWM, N = 256, inverted
    while (1);
    return 0;
}
```

Review Questions

1. True or false. In Fast PWM and Phase correct PWM modes, we can change the duty cycle.
2. True or false. In Fast PWM, we cannot change the frequency of the wave.
3. True or false. For 8-bit timers, in Phase correct PWM mode, the period is 510 clocks.
4. True or false. In Fast PWM, phase does not change when the duty cycle is changed.
5. Which of the PWM modes is preferable for controlling motors?

SECTION 16.3: PWM MODES IN TIMER1

Fast PWM mode

In the Fast PWM, the counter counts like it does in the Normal mode. After the timer is started, it starts to count up. It counts up until it reaches its top limit. See Figure 16-24.

From Figure 16-30 we see that we have five Fast PWM modes in Timer1: modes 5, 6, 7, 14, and 15. In modes 5, 6, and 7 the top value is fixed at 0xFF, 0x1FF, and 0x3FF; while in modes 14 and 15, the ICR1 and OCR1A registers represent the top value, respectively. See Figures 16-25 through 16-29.

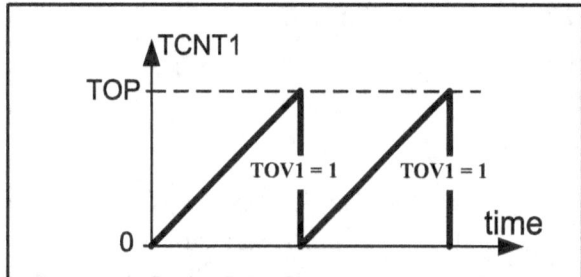

Figure 16-24. Fast PWM Mode

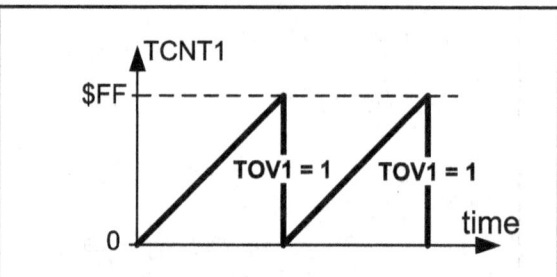

Figure 16-25. TOV in Mode 5

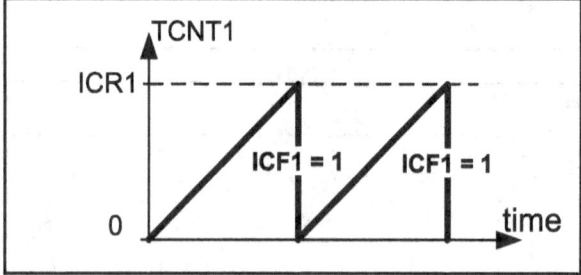

Figure 16-26. Mode 14

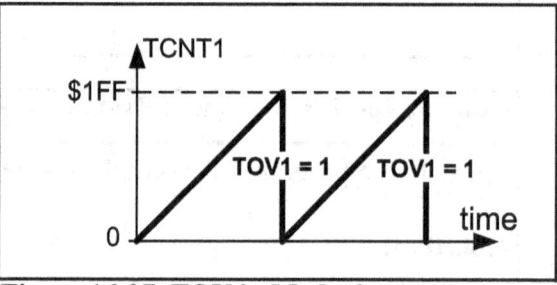

Figure 16-27. TOV in Mode 6

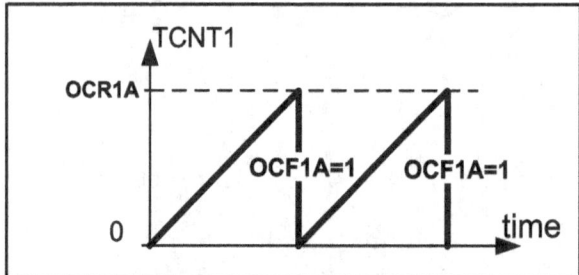

Figure 16-28. Mode 15

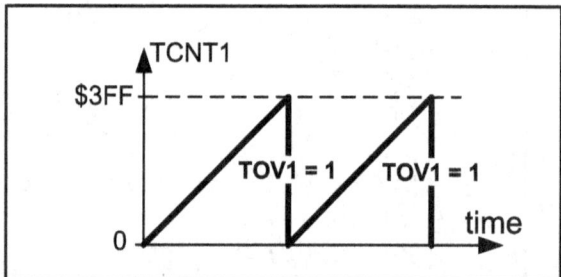

Figure 16-29. TOV in Mode 7

Bit	7	6	5	4	3	2	1	0	
	ICNC1	ICES1	–	WGM13	WGM12	CS12	CS11	CS10	TCCR1B
Read/Write	R/W	R/W	R	R/W	R/W	R/W	R/W	R/W	
Initial Value	0	0	0	0	0	0	0	0	

ICNC1   D7   Input Capture Noise Canceler
                   0 = Input Capture Noise Canceler is disabled.
                   1 = Input Capture Noise Canceler is enabled.

ICES1   D6   Input Capture Edge Select
                   0 = Capture on the falling (negative) edge
                   1 = Capture on the rising (positive) edge

               D5   Not used

WGM13:WGM12   D4 D3   Timer1 mode

Mode	WGM13	WGM12	WGM11	WGM10	Timer/Counter Mode of Operation	Top	Update of OCR1x	TOV1 Flag Set on
0	0	0	0	0	Normal	0xFFFF	Immediate	MAX
1	0	0	0	1	PWM, Phase Correct, 8-bit	0x00FF	TOP	BOTTOM
2	0	0	1	0	PWM, Phase Correct, 9-bit	0x01FF	TOP	BOTTOM
3	0	0	1	1	PWM, Phase Correct, 10-bit	0x03FF	TOP	BOTTOM
4	0	1	0	0	CTC	OCR1A	Immediate	MAX
5	0	1	0	1	Fast PWM, 8-bit	0x00FF	TOP	TOP
6	0	1	1	0	Fast PWM, 9-bit	0x01FF	TOP	TOP
7	0	1	1	1	Fast PWM, 10-bit	0x03FF	TOP	TOP
8	1	0	0	0	PWM, Phase and Frequency Correct	ICR1	BOTTOM	BOTTOM
9	1	0	0	1	PWM, Phase and Frequency Correct	OCR1A	BOTTOM	BOTTOM
10	1	0	1	0	PWM, Phase Correct	ICR1	TOP	BOTTOM
11	1	0	1	1	PWM, Phase Correct	OCR1A	TOP	BOTTOM
12	1	1	0	0	CTC	ICR1	Immediate	MAX
13	1	1	0	1	Reserved	–	–	–
14	1	1	1	0	Fast PWM	ICR1	TOP	TOP
15	1	1	1	1	Fast PWM	OCR1A	TOP	TOP

CS12:CS10   D2D1D0   Timer1 clock selector
           0   0   0   No clock source (Timer/Counter stopped)
           0   0   1   clk (no prescaling)
           0   1   0   clk / 8
           0   1   1   clk / 64
           1   0   0   clk / 256
           1   0   1   clk / 1024
           1   1   0   External clock source on T1 pin. Clock on falling edge
           1   1   1   External clock source on T1 pin. Clock on rising edge

Figure 16-30. TCCR1B (Timer1 Control) Register

         In Modes 5, 6, and 7, which have fixed top values, the TOV1 flag will be set when the timer rolls over. See Figures 16-25, 16-27, and 16-29.
         In Mode 14, whose top value is represented by ICR1, the ICF1 flag will be set when the timer rolls over, as shown in Figure 16-26.
         In Mode 15, when the timer rolls over, the OCF1A flag will be set. See Figure 16-28.

Bit	7	6	5	4	3	2	1	0	
	COM1A1	COM1A0	COM1B1	COM1B0	FOC1A	FOC1B	WGM11	WGM10	TCCR1A
Read/Write	R/W	R/W	R	R/W	R/W	R/W	R/W	R/W	
Initial Value	0	0	0	0	0	0	0	0	

COM1A1:COM1A0 D7 D6 Compare Output Mode for Channel A

COM1A1	COM1A0	Description
0	0	Normal port operation, OC1A disconnected
0	1	In mode 15, toggle OC1A on compare match. In other modes OC1A disconnected (Normal I/O port)
1	0	Clear OC1A on compare match. Set OC1A at Top.
1	1	Set OC1A on compare match. Clear OC1A at Top.

COM1B1:COM1B0 D5 D4 Compare Output Mode for Channel B

COM1B1	COM1B0	Description
0	0	Normal port operation, OC1B disconnected
0	1	Normal port operation, OC1B disconnected
1	0	Clear OC1B on compare match. Set OC1B at Top.
1	1	Set OC1B on compare match. Clear OC1B at Top.

FOC1A   D3   Force Output Compare for Channel A

FOC1B   D2   Force Output Compare for Channel B

WGM11:10   D1 D0 Timer1 mode (discussed in Figure 16-30)

Figure 16-31. TCCR1A (Timer1 Control) Register

In Figure 16-31 you see the reaction of the waveform generator when compare match occurs while the timer is in Fast PWM mode. When COM1A1:0 = 00 the OC1A pin operates as an I/O port. When COM1A1:0 = 10, the waveform generator clears the OC1A pin whenever compare match occurs, and sets it at the top value. This mode is called *non-inverted PWM*. See Figures 16-32A through 16-32C. As you see, in non-inverted PWM, the duty cycle of the generated wave increases when the value of OCR1A increases.

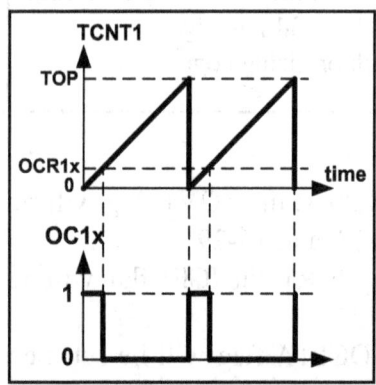

Figure 16-32A. Non-inverted

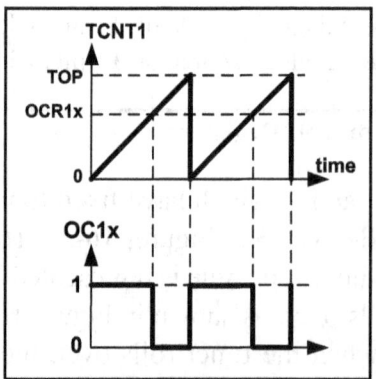

Figure 16-32B. Non-inverted

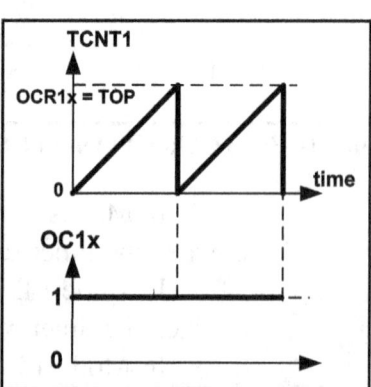

Figure 16-32C. Non-inverted

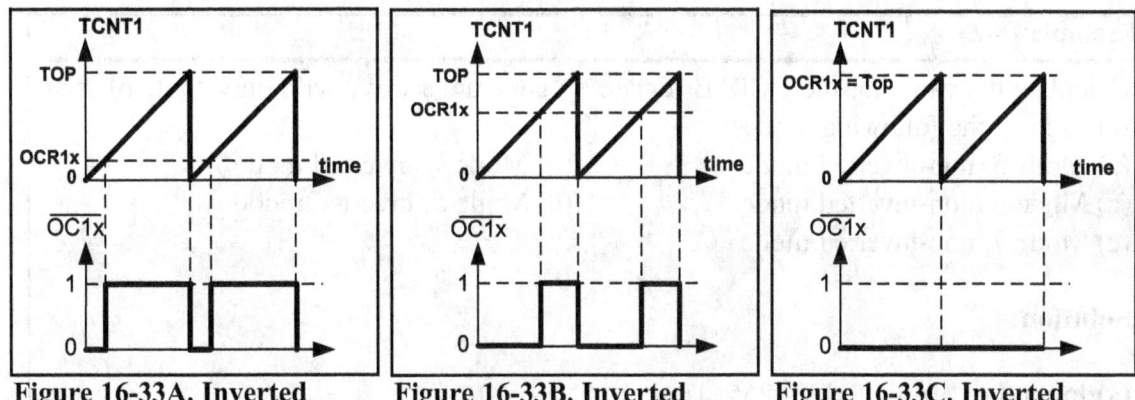

Figure 16-33A. Inverted   **Figure 16-33B. Inverted**   **Figure 16-33C. Inverted**

When COM1A1:0 = 11, the waveform generator sets the OC1A pin whenever compare match occurs, and clears it at the top value. This mode is referred to as *inverted PWM mode*. See Figures 16-33A through 16-33C. As you see, in inverted PWM, the duty cycle of the generated wave decreases when the value of OCR1A increases.

The same thing is true about the OCR1B register and COM1B1:10 bits.

Frequency of the generated wave in Fast PWM mode

In Fast PWM mode, timer counts from 0 to top value and then rolls over. Thus, the frequency of the generated wave is 1/(Top + 1) of the frequency of timer clock. As you saw in Section 9-1, the frequency of the timer clock can be selected using the prescaler. Therefore, the frequency of the generated wave can be calculated as follows (N is determined by the prescaler):

$$F_{generated\ wave} = \frac{F_{timer\ clock}}{Top + 1}$$
$$F_{timer\ clock} = \frac{F_{oscillator}}{N}$$
$$\Longrightarrow F_{generated\ wave} = \frac{F_{oscillator}}{(Top + 1) \times N}$$

Duty cycle of the generated wave in Fast PWM mode

The duty cycle of the generated mode can be determined using the OCR1x register. When COM1x1:0 = 10 (in non-inverted mode), the bigger OCR1x value results in a bigger duty cycle. When OCR1x = Top, the OC1 is always high (duty cycle = 100%). Generally speaking, the OC1x is high for a total of OCR1x + 1 clocks. So, the duty cycle can be calculated using the following formula in non-inverted mode:

$$Duty\ Cycle = \frac{OCR1x + 1}{Top + 1} \times 100$$

In inverted mode, the duty cycle can be calculated using the following formula:

$$Duty\ Cycle = \frac{Top - OCR1x}{Top + 1} \times 100$$

See Examples 16-23 through 16-28.

Example 16-23

Calculate the value for the OCR1B register to generate a wave with duty cycle of 75% for each of the following modes:
(a) Mode 5, non-inverted mode   (b) Mode 7, inverted mode
(c) Mode 6, non-inverted mode   (d) Mode 5, inverted mode
(e) Mode 7, non-inverted mode

Solution:

(a) In mode 5, Top = 0xFF= 255. Thus,
$75 = (OCR1x + 1) \times 100 / (Top + 1)$ ➔ $OCR1x + 1 = 75 \times 256 / 100 = 192$
➔ OCR1B = 191

(b) In mode 7, Top = 0x3FF = 1023. Thus,
$75 = (Top - OCR1x) \times 100 / (Top + 1)$ ➔ $1023 - OCR1x = 75 \times 1024 / 100 = 768$
➔ OCR1B = 255

(c) In mode 6, Top = 0x1FF = 511. Thus,
$75 = (OCR1x + 1) \times 100 /(Top + 1)$ ➔ $OCR1x + 1 = 75 \times 512 / 100 = 384$
➔ OCR1A = 383

(d) In mode 5, Top = 0xFF = 255. Thus,
$75 = (Top - OCR1x) \times 100 / (Top + 1)$ ➔ $75 = (255 - OCR1x) \times 100 / 256$
➔ $255 - OCR1x = 75 \times 256 / 100 = 192$ ➔ OCR1B = 255 − 192 = 63

(e) In mode 7, Top = 0x3FF = 1023. Thus,
$75 = (OCR1x + 1) \times 100 / (Top + 1)$ ➔ $OCR1x + 1 = 75 \times 1024 / 100 = 768$
➔ OCR1B = 767

Example 16-24

Find the values for TCCR1A and TCCR1B to initialize Timer1 for mode 5 (Fast PWM mode, top = 0xFF), non-inverted PWM wave generator, and no prescaler, using waveform generator A.

Solution:

WGM13:10 = 0101 = Fast PWM mode   CS02:00 = 001 = No prescaler
COM01:00 = 10 = Non-inverted PWM

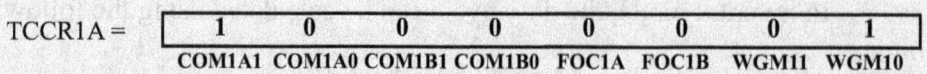

TCCR1A =	1	0	0	0	0	0	0	1
	COM1A1	COM1A0	COM1B1	COM1B0	FOC1A	FOC1B	WGM11	WGM10

TCCR1B =	0	0	0	0	1	0	0	1
	ICNC1	ICES1	–	WGM13	WGM12	CS12	CS11	CS10

Example 16-25

Assuming XTAL = 8 MHz, using non-inverted mode, and mode 5, write a program that generates a wave with frequency of 31,250 Hz and duty cycle of 75%.

Solution:

31,250 = 8M / (256 × N) ➔ N = 8M / (31,250 × 256) = 1 ➔ No prescaler

```
.INCLUDE "M32DEF.INC"
      SBI   DDRD,5              ;PD5 = output
      LDI   R16,HIGH(191)       ;from Example 16-23
      OUT   OCR1AH,R16          ;Temp = 0x00
      LDI   R16,LOW(191)        ;R16 = 191
      OUT   OCR1AL,R16          ;OCR1A = 191
      LDI   R16,0x81            ;from Example 16-24
      OUT   TCCR1A,R16          ;COM1A = non-inverted
      LDI   R16,0x09
      OUT   TCCR1B,R16          ;WGM = mode 5, clock = no scaler
HERE: RJMP  HERE
```

Example 16-26

Assuming XTAL = 8 MHz, using non-inverted mode and mode 7, write a program that generates a wave with frequency of 7812.5 Hz and duty cycle of 75%.

Solution:

7812.5 = 8M/(1024 × N) ➔ N = 8M/(7812.5 × 1024) = 1 ➔ No prescaler

```
.INCLUDE "M32DEF.INC"
      SBI   DDRD,5              ;PD5 = output
      LDI   R16,HIGH(767)       ;R16 = the high byte
      OUT   OCR1AH,R16
      LDI   R16,LOW(767)        ;R16 = the low byte
      OUT   OCR1AL,R16          ;OCR1A = 767 (from Example 16-23)
      LDI   R16,0x83
      OUT   TCCR1A,R16          ;COM1A = non-inverted
      LDI   R16,0x09
      OUT   TCCR1B,R16          ;WGM = mode 7, clock = no scaler
HERE: RJMP  HERE                ;wait here forever
```

Example 16-27

Assuming XTAL = 8 MHz, using non-inverted mode and mode 6, write a program that generates a wave with frequency of 1,953 Hz and duty cycle of 60%.

Solution:

In mode 6, Top = 0x1FF= 511. Thus,
$60 = (OCR1x + 1) \times 100 / (Top + 1)$ → $OCR1x + 1 = 60 \times 512 / 100 = 307$
→ $OCR1B = \mathbf{306}$
$1953 = 8M / (512 \times N)$ → $N = 8M / (1953 \times 512) = 8$ → prescaler = 1:8 → CS12:0 = 010

TCCR1A =	1	0	0	0	0	0	1	0
	COM1A1	COM1A0	COM1B1	COM1B0	FOC1A	FOC1B	WGM11	WGM10

TCCR1B =	0	0	0	0	1	0	1	0
	ICNC1	ICES1	–	WGM13	WGM12	CS12	CS11	CS10

```
.INCLUDE  "M32DEF.INC"
     SBI   DDRD,5            ;PD5 = output
     LDI   R16,HIGH(306)     ;R16 = the high byte
     OUT   OCR1AH,R16        ;Temp = R16
     LDI   R16,LOW(306)      ;R16 = the low byte
     OUT   OCR1AL,R16        ;OCR1A = 306
     LDI   R16,0x82
     OUT   TCCR1A,R16        ;COM1A = non-inverted.
     LDI   R16,0x0A
     OUT   TCCR1B,R16        ;WGM = mode 6, clock = no scaler
HERE: RJMP  HERE             ;wait here forever
```

Example 16-28

Rewrite Example 16-27 using inverted mode.

Solution:
$60 = (Top - OCR1x) \times 100 / (Top + 1)$ → $511 - OCR1x = 60 \times 512 / 100 = 307$
→ $OCR1B = 511 - 307 = \mathbf{204}$

TCCR1A =	1	1	0	0	0	0	1	0
	COM1A1	COM1A0	COM1B1	COM1B0	FOC1A	FOC1B	WGM11	WGM10

TCCR1B =	0	0	0	0	1	0	1	0
	ICNC1	ICES1	–	WGM13	WGM12	CS12	CS11	CS10

```
.INCLUDE  "M32DEF.INC"
     SBI   DDRD,5            ;PD5 = output
     LDI   R16,HIGH(204)
     OUT   OCR1AH,R16        ;Temp = the high byte
     LDI   R16,LOW(204)
     OUT   OCR1AL,R16        ;OCR1A = 204
     LDI   R16,0xB2
     OUT   TCCR1A,R16        ;COM1A = inverted
     LDI   R16,0x0A
     OUT   TCCR1B,R16        ;WGM = mode 6, clock = no scaler
HERE: RJMP  HERE             ;wait here forever
```

Loading values into the OCR1A and OCR1B registers in PWM modes

In the non-PWM modes (CTC mode and Normal mode), when we load a value into the OCR1x register, the value will be loaded instantly, but in the PWM modes (Fast PWM, Phase correct PWM, and phase and frequency correct PWM mode), there is a buffer between us and the OCR1A and OCR1B registers. When we read/write a value from/into the OCR1A or OCR1B register we are dealing with the buffer. The contents of the buffer will be loaded into the OCR1A/OCR1B registers only when the TCNT1 reaches its topmost value. See Figure 16-34 and Example 16-29.

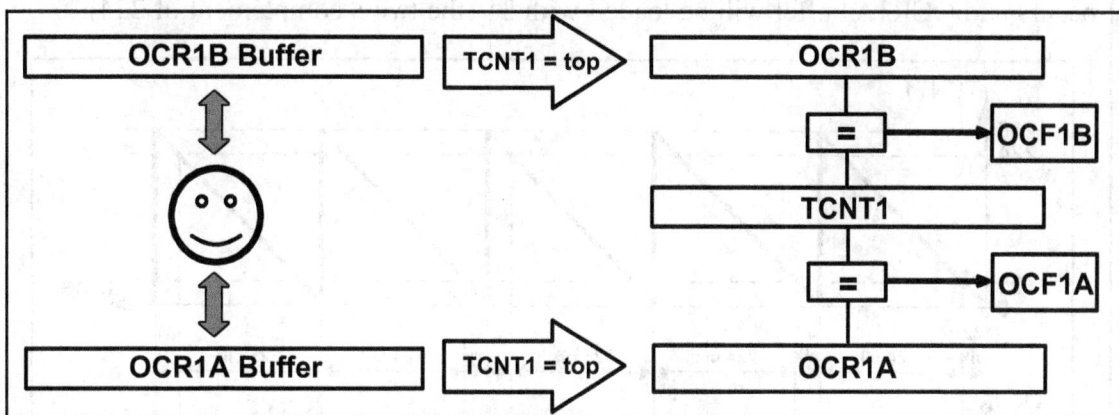

Figure 16-34. OCRnx Buffer in PWM Modes

Example 16-29

Draw the wave generated by the following program. Assume XTAL = 1 MHz.

```
.INCLUDE    "M32DEF.INC"
      RJMP  MAIN
.ORG  0x12                   ;Timer1 overflow interrupt vector
      OUT   OCR1AH,R19        ;OCR1AH = R19 = 0
      NEG   R20
      OUT   OCR1AL,R20        ;OCR1A = R20
      RETI                   ;return from interrupt
MAIN: LDI   R16,LOW(RAMEND)
      OUT   SPL,R16
      LDI   R16,HIGH(RAMEND)
      OUT   SPH,R16           ;initialize stack pointer
      SBI   DDRD,5            ;PD5 = output
      LDI   R19,0
      OUT   OCR1AH,R19        ;Temp = 0x00
      LDI   R20,31
      OUT   OCR1AL,R20        ;OCR1A = 31
      LDI   R16,0x81
      OUT   TCCR1A,R16        ;COM1A = non-inverted.
      LDI   R16,0x0A
      OUT   TCCR1B,R16        ;WGM = mode 5, clock = no scaler
      LDI   R16,(1<<TOIE1)
      OUT   TIMSK,R16         ;enable timer interrupt
      SEI
HERE: RJMP  HERE             ;wait here forever
```

Example 16-29 (Cont.)

Solution:

The wave generator is in non-inverted Fast PWM mode, which means that on compare match the OC1A pin will be set high. The OCR1A register is loaded with 31, so compare match occurs when TCNT1 reaches 31. When the timer reaches top and overflows, the interrupt request occurs, and OCR1A buffer is loaded with 224 (the two's complement of 31). The next time that the timer reaches the top value the contents of the OCR1A buffer (224) will be loaded into the OCR1A register. Then the second interrupt occurs and OCR1A buffer will be loaded with 31 (the two's complement of 224).

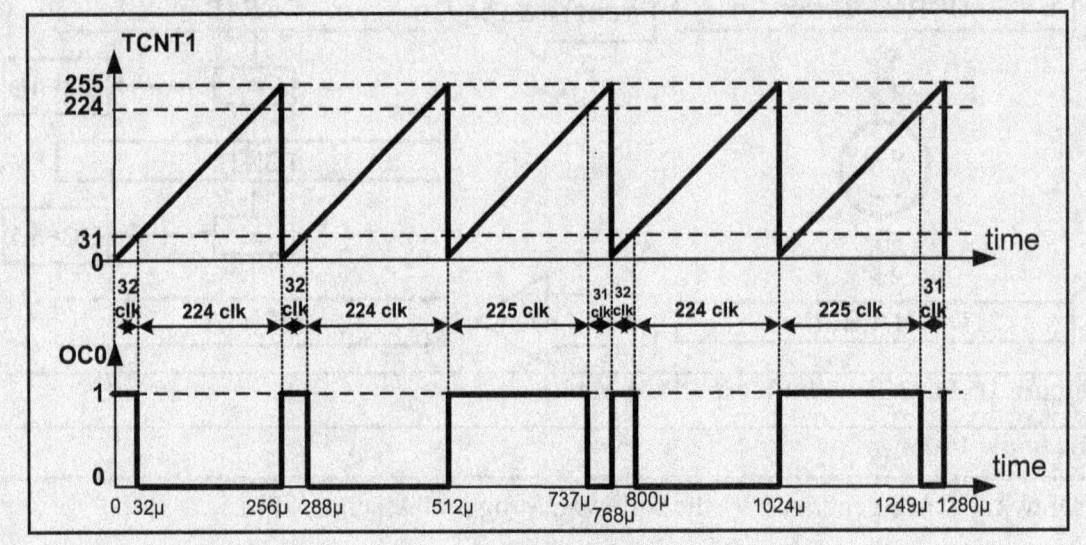

Generating waves with different frequencies (case study)

As we mentioned earlier, the frequency of the generated wave is equal to $F_{Oscillator}/[N \times (Top + 1)]$. In modes 5, 6, and 7, the Top value is fixed. Therefore, in these modes the only way to change the frequency of the generated wave is to change N (the prescaler). In Figure 16-35, you see the different frequencies that can be generated using modes 5, 6, and 7.

Prescaler	1	1:8	1:64	1:256	1:1024
Mode = 5	$\dfrac{F_{oscillator}}{1 \times 256}$	$\dfrac{F_{oscillator}}{8 \times 256}$	$\dfrac{F_{oscillator}}{64 \times 256}$	$\dfrac{F_{oscillator}}{256 \times 256}$	$\dfrac{F_{oscillator}}{1024 \times 256}$
Mode = 6	$\dfrac{F_{oscillator}}{1 \times 512}$	$\dfrac{F_{oscillator}}{8 \times 512}$	$\dfrac{F_{oscillator}}{64 \times 512}$	$\dfrac{F_{oscillator}}{256 \times 512}$	$\dfrac{F_{oscillator}}{1024 \times 512}$
Mode = 7	$\dfrac{F_{oscillator}}{1 \times 1024}$	$\dfrac{F_{oscillator}}{8 \times 1024}$	$\dfrac{F_{oscillator}}{64 \times 1024}$	$\dfrac{F_{oscillator}}{256 \times 1024}$	$\dfrac{F_{oscillator}}{1024 \times 1024}$

Figure 16-35. Different Frequencies Can Be Made Using Modes 5, 6, and 7

Thus, in these modes we can make a very limited number of frequencies. What if we want to make some other frequencies? In modes 14 and 15, the Top value can be specified by ICR1 and the OCR1A registers. Thus, we can change the frequency by loading proper values to ICR1 and OCR1A. See Examples 16-30 through 16-32.

Example 16-30

Assuming XTAL = 8 MHz, find TCCR1A and TCCR1B to generate a wave with frequency of 80 kHz using mode 14.

Solution:

$80K = 8M / [N \times (Top + 1)]$ ➔ $N \times (Top + 1) = 8M / 80K = 100$
➔ $N \times (Top + 1) = 100$ ➔ $N = 1$; $Top + 1 = 100$
$Top = 99$ ➔ $ICR1 = 99$
$N = 1$ ➔ $CS12{:}0 = 001$

TCCR1A =	1	0	0	0	0	0	1	0
	COM1A1	COM1A0	COM1B1	COM1B0	FOC1A	FOC1B	WGM11	WGM10

TCCR1B =	0	0	0	1	1	0	0	1
	ICNC1	ICES1	–	WGM13	WGM12	CS12	CS11	CS10

Example 16-31

Calculate the OCR1B to generate a wave with duty cycle of 20% in each of the following modes:
(a) mode 14, inverted mode, ICR1 = 45, (b) mode 15, non-inverted mode, OCR1A = 124, and (c) mode 14, non-inverted mode, ICR1 = 99.

Solution:

(a) In mode 14, Top = ICR1 = 45. Thus,
$20 = (Top - OCR1x) \times 100 / (Top + 1)$ ➔ $45 - OCR1x = 20 \times 46 / 100 = 9$ ➔ $OCR1A = 36$

(b) In mode 15, Top = OCR1A = 124. Thus,
$20 = (OCR1x + 1) \times 100 / (124 + 1)$ ➔ $OCR1x + 1 = 20 \times 125 / 100 = 25$ ➔ $OCR1x = 24$

(c) In mode 14, Top = ICR1 = 99. Therefore,
$20 = (OCR1x + 1) \times 100 / (99 + 1)$ ➔ $OCR1x + 1 = 20$ ➔ $OCR1x = 19$

Example 16-32

Assume XTAL = 8 MHz. Using mode 14 write a program that generates a wave with duty cycle of 20% and frequency of 80 kHz.

Solution:

```
.INCLUDE  "M32DEF.INC"
      LDI   R16,LOW(RAMEND)
      OUT   SPL,R16
      LDI   R16,HIGH(RAMEND)
      OUT   SPH,R16           ;initialize stack pointer
      SBI   DDRD,5            ;PD5 = output
      LDI   R16,HIGH(99)
      OUT   ICR1H,R16         ;Temp = 0
      LDI   R16,LOW(99)
      OUT   ICR1L,R16         ;ICR1 = 99
      LDI   R16,HIGH(19)
      OUT   OCR1AH,R16        ;Temp = 0
      LDI   R16,LOW(19)
      OUT   OCR1AL,R16        ;OCR1A = 19 (from Example 16-31)
      LDI   R16,0x82          ;from Example 16-30
      OUT   TCCR1A,R16        ;COM1A = non-inverted
      LDI   R16,0x19          ;from Example 16-30
      OUT   TCCR1B,R16        ;WGM = mode 14, clock = no scaler
HERE: RJMP  HERE             ;wait here forever
```

If we use mode 15 instead of mode 14, OCR1A is buffered, and the contents of the buffer will be loaded into OCR1A when the timer reaches its top value. In mode 15 we can only use the OC1B wave generator and not the OC1A wave generator since the OCR1A register is used for defining the top value. See Examples 16-33 and 16-34.

Example 16-33

Assuming XTAL = 8 MHz, find TCCR1A and TCCR1B to generate a wave with frequency of 64 kHz using mode 15.

Solution:

64k = 8M / [N × (Top + 1)] ➔ N × (Top + 1) = 8M / 64k = 125 ➔ N = 1; Top + 1 = 125 ➔ Top = 124 ➔ OCR1A = 124
N = 1 ➔ CS12:0 = 001

TCCR1A =	0	0	1	0	0	0	1	1
	COM1A1	COM1A0	COM1B1	COM1B0	FOC1A	FOC1B	WGM11	WGM10

TCCR1B =	0	0	0	1	1	0	0	1
	ICNC1	ICES1	–	WGM13	WGM12	CS12	CS11	CS10

Example 16-34

Assume XTAL = 8 MHz. Using mode 15 write a program that generates a wave with duty cycle of 20% and frequency of 64 kHz.

Solution:

```
.INCLUDE "M32DEF.INC"
     LDI   R16,LOW(RAMEND)
     OUT   SPL,R16
     LDI   R16,HIGH(RAMEND)
     OUT   SPH,R16            ;initialize stack pointer

     SBI   DDRD,4             ;PD4 = output
     LDI   R16,HIGH(124)      ;R16 = the high byte
     OUT   OCR1AH,R16
     LDI   R16,LOW(124)       ;R16 = the low byte
     OUT   OCR1AL,R16         ;OCR1A = 124

     LDI   R16,HIGH(24)       ;R16 = the high byte
     OUT   OCR1BH,R16
     LDI   R16,LOW(24)        ;R16 = the low byte
     OUT   OCR1BL,R16         ;OCR1B = 24

     LDI   R16,0x23           ;from Example 16-33
     OUT   TCCR1A,R16         ;COM1B = non-inverted
     LDI   R16,0x19           ;from Example 16-33
     OUT   TCCR1B,R16         ;WGM = mode 15, clock = no scaler
HERE: RJMP  HERE             ;wait here forever
```

Phase correct PWM mode

In the Phase correct PWM, the timer counts up until it reaches the top value then counts down until it reaches zero. The TOV1 flag will be set when the timer returns to zero, as shown in Figure 16-36.

There are five Phase correct PWM modes: modes 1, 2, 3, 10, and 11. See Figure 16-37.

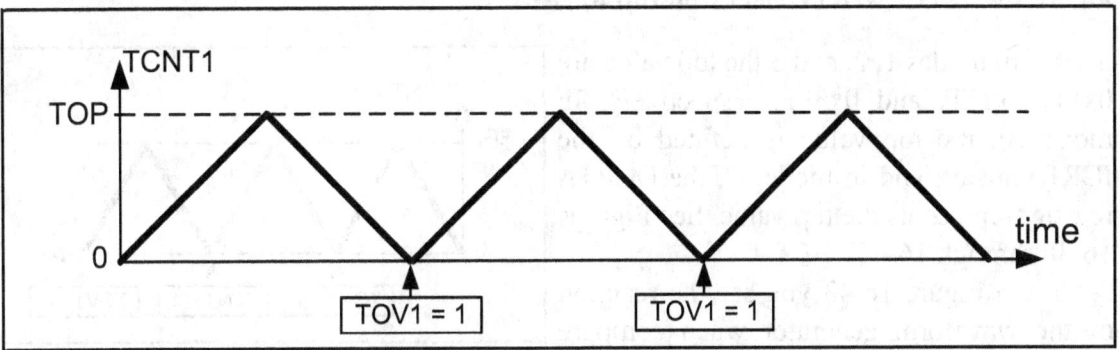

Figure 16-36. Timer/Counter 1 Phase Correct PWM Mode

Bit	7	6	5	4	3	2	1	0	
	ICNC1	ICES1	–	WGM13	WGM12	CS12	CS11	CS10	**TCCR1B**
Read/Write	R/W	R/W	R	R/W	R/W	R/W	R/W	R/W	
Initial Value	0	0	0	0	0	0	0	0	

ICNC1   D7   Input Capture Noise Canceler
                                      0 = Input Capture is disabled.
                                      1 = Input Capture is enabled.

ICES1   D6   Input Capture Edge Select
                                      0 = Capture on the falling (negative) edge
                                      1 = Capture on the rising (positive) edge

WGM13:WGM12   D4 D3   Timer1 mode

Mode	WGM13	WGM12	WGM11	WGM10	Timer/Counter Mode of Operation	Top	Update of OCR1x	TOV1 Flag Set on
0	0	0	0	0	Normal	0xFFFF	Immediate	MAX
1	0	0	0	1	PWM, Phase Correct, 8-bit	0x00FF	TOP	BOTTOM
2	0	0	1	0	PWM, Phase Correct, 9-bit	0x01FF	TOP	BOTTOM
3	0	0	1	1	PWM, Phase Correct, 10-bit	0x03FF	TOP	BOTTOM
4	0	1	0	0	CTC	OCR1A	Immediate	MAX
5	0	1	0	1	Fast PWM, 8-bit	0x00FF	TOP	TOP
6	0	1	1	0	Fast PWM, 9-bit	0x01FF	TOP	TOP
7	0	1	1	1	Fast PWM, 10-bit	0x03FF	TOP	TOP
8	1	0	0	0	PWM, Phase and Frequency Correct	ICR1	BOTTOM	BOTTOM
9	1	0	0	1	PWM, Phase and Frequency Correct	OCR1A	BOTTOM	BOTTOM
10	1	0	1	0	PWM, Phase Correct	ICR1	TOP	BOTTOM
11	1	0	1	1	PWM, Phase Correct	OCR1A	TOP	BOTTOM
12	1	1	0	0	CTC	ICR1	Immediate	MAX
13	1	1	0	1	Reserved	–	–	–
14	1	1	1	0	Fast PWM	ICR1	TOP	TOP
15	1	1	1	1	Fast PWM	OCR1A	TOP	TOP

CS12:CS10   D2D1D0   Timer1 clock selector
                         0 0 0   No clock source (Timer/Counter stopped)
                         0 0 1   clk (no prescaling)
                         0 1 0   clk / 8
                         0 1 1   clk / 64
                         1 0 0   clk / 256
                         1 0 1   clk / 1024
                         1 1 0   External clock source on T1 pin. Clock on falling edge
                         1 1 1   External clock source on T1 pin. Clock on rising edge

Figure 16-37. TCCR1B (Timer1 Control) Register

In modes 1, 2, and 3 the top value are 0xFF, 0x1FF, and 0x3FF, respectively. In mode 10, the top value is defined by the ICR1 register; and in mode 11, the OCR1A register represents the top value. See Figures 16-38 through 16-42.

In Figure 16-43 you see the reaction of the waveform generator when compare match occurs in Phase correct PWM mode.

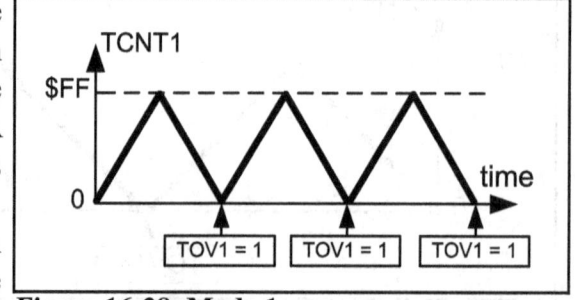

Figure 16-38. Mode 1

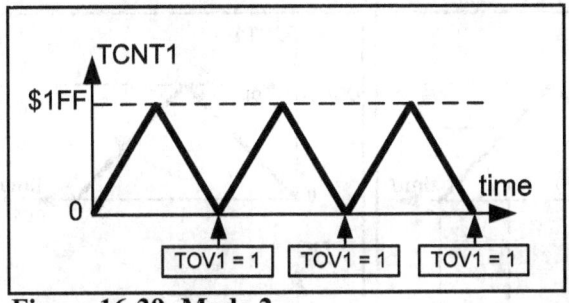

Figure 16-39. Mode 2

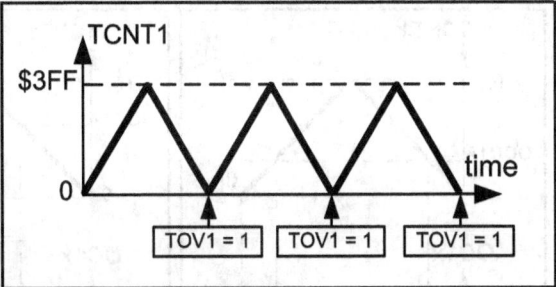

Figure 16-40. Mode 3

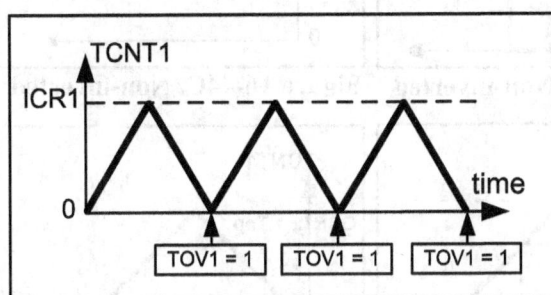

Figure 16-41. Mode 10

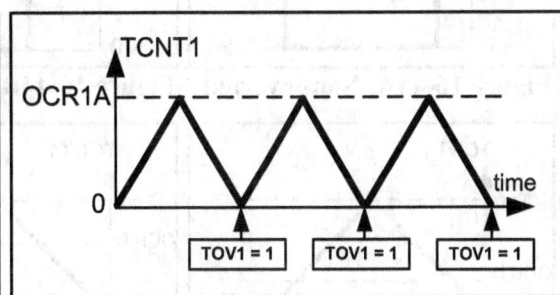

Figure 16-42. Mode 11

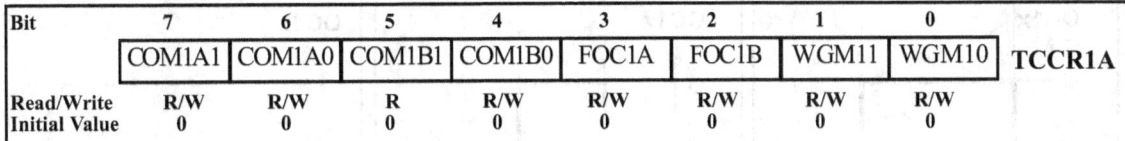

Bit	7	6	5	4	3	2	1	0	
	COM1A1	COM1A0	COM1B1	COM1B0	FOC1A	FOC1B	WGM11	WGM10	**TCCR1A**
Read/Write	R/W	R/W	R	R/W	R/W	R/W	R/W	R/W	
Initial Value	0	0	0	0	0	0	0	0	

COM1A1:COM1A0   D7 D6 Compare Output Mode for Channel A

COM1A1	COM1A0	Description
0	0	Normal port operation, OC1A disconnected
0	1	In mode 9 or 14 toggles on compare match. In other modes OC1A is disconnected (Normal I/O port).
1	0	Clear OC1A on compare match when up-counting. Set OC1A on compare match when down-counting.
1	1	Set OC1A on compare match when up-counting. Clear OC1A on compare match when down-counting.

COM1B1:COM1B0   D5 D4 Compare Output Mode for Channel B

COM1B1	COM1B0	Description
0	0	Normal port operation, OC1B disconnected
0	1	Normal port operation, OC1B disconnected
1	0	Clear OC1B on compare match when up-counting. Set OC1B on compare match when down-counting.
1	1	Set OC1B on compare match when up-counting. Clear OC1B on compare match when down-counting.

FOC1A   D3   Force Output Compare for Channel A

FOC1B   D2   Force Output Compare for Channel B

WGM11:10   D1 D0 Timer1 mode (discussed in Figure 16-37)

Figure 16-43. TCCR1A (Timer1 Control) Register

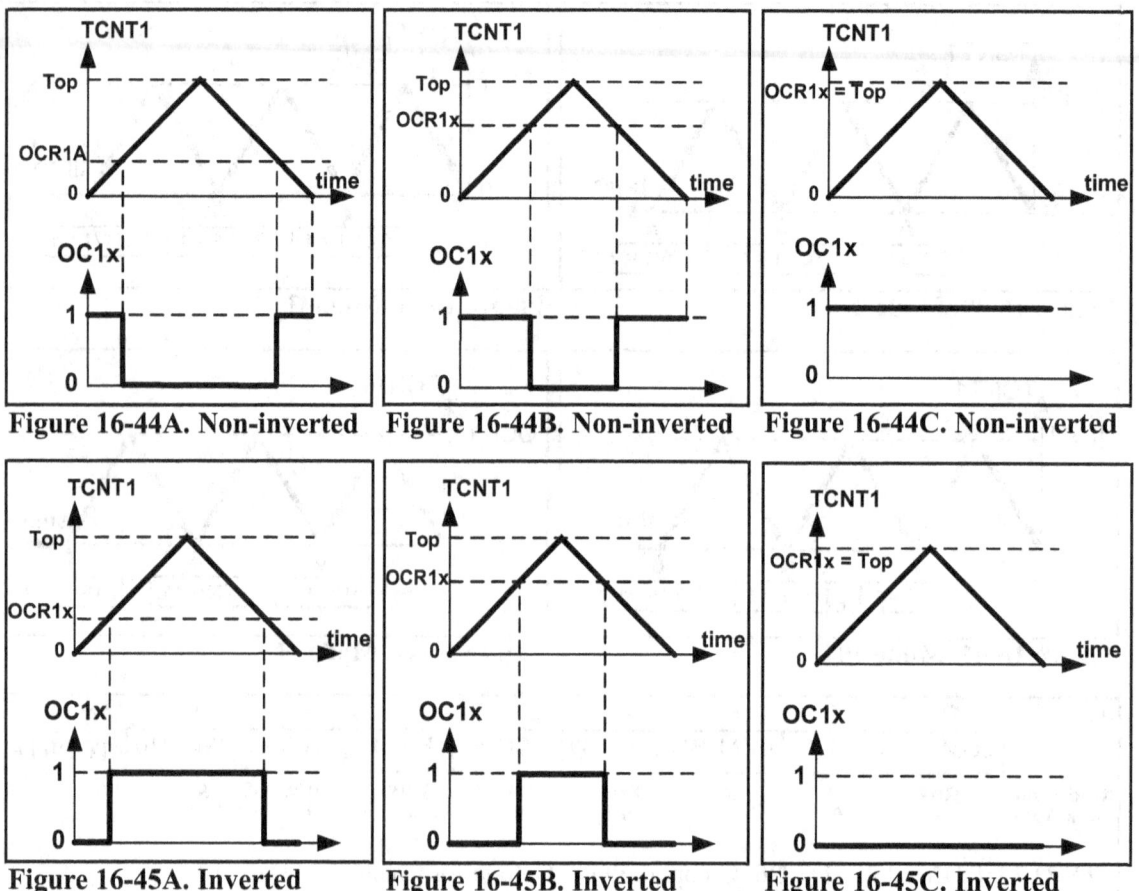

Figure 16-44A. Non-inverted   Figure 16-44B. Non-inverted   Figure 16-44C. Non-inverted

Figure 16-45A. Inverted   Figure 16-45B. Inverted   Figure 16-45C. Inverted

When COM1A1:10 = 00 the OC1A pin operates as an I/O port. When COM1A1:00 = 10, the waveform generator clears the OC1A pin on compare match when up-counting, and sets it on compare match when down-counting. This mode is called *non-inverted*.

See Figures 16-44A through 16-44C. As you see, in the non-inverted mode, the duty cycle of the generated wave increases when the value of OCR1A increases.

When COM1A1:00 = 11, the waveform generator sets the OC1A pin on compare match when up-counting, and clears it on compare match when down-counting. This mode is referred to as *inverted mode*. As you can see from Figures 16-45A through 16-45C, in inverted PWM, the duty cycle of the generated wave decreases when the value of OCR1A increases.

The same thing is true about the OCR1B register and COM1B1:10 bits.

Frequency of the generated wave in Phase correct PWM mode

As you see in Figure 16-46, the frequency of the generated wave is 1/2 TOP of the frequency of timer clock. As you saw in Section 9-1, the frequency of the timer clock can be selected using the prescaler. Therefore, in 8-bit timers the frequency of the generated wave can be calculated as follows:

$$F_{generated\ wave} = \frac{F_{timer\ clock}}{2 \times Top}$$

$$F_{timer\ clock} = \frac{F_{oscillator}}{N}$$

$$\Longrightarrow \quad F_{generated\ wave} = \frac{F_{oscillator}}{2 \times N \times Top}$$

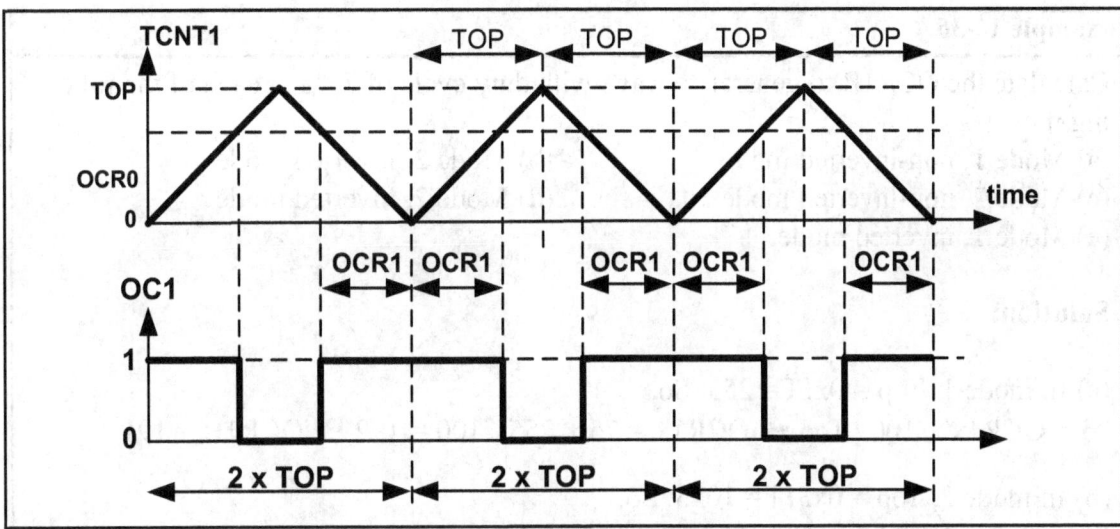

Figure 16-46. Timer/Counter 1 Phase Correct PWM Mode

Duty cycle of the generated wave in Phase correct PWM mode

The duty cycle of the generated mode can be determined using the OCR1x register. When COM1x1:0 = 10 (in non-inverted mode), the bigger OCR1x value results in a bigger duty cycle. When OCR1x = Top, the OC1x is always high (duty cycle = 100%). Generally speaking, OC1x is high for a total of OCR1x clocks. See Figure 16-46. So, the duty cycle can be calculated using the following formula in non-inverted mode:

$$\text{Duty Cycle} = \frac{2 \times OCR1A}{2 \times Top} \times 100 \quad \Longrightarrow \quad \text{Duty Cycle} = \frac{OCR1A}{Top} \times 100$$

Similarly, the duty cycle formula for inverted mode is as follows:

$$\text{Duty Cycle} = \frac{2 \times Top - 2 \times OCR1A}{2 \times Top} \times 100 \quad \Longrightarrow \quad \text{Duty Cycle} = \frac{Top - OCR1A}{Top} \times 100$$

See Examples 16-35 through 16-37.

Example 16-35

Find the values for TCCR1A and TCCR1B to initialize Timer1 for mode 1 (Phase correct PWM mode, top = 0xFF), non-inverted PWM wave generator, and no prescaler.

Solution:

WGM13:10 = 0001 = Phase correct PWM mode   CS02:00 = 001 = No prescaler
COM01:00 = 10 = Non-inverted PWM

TCCR1A =	1	0	0	0	0	0	0	1
	COM1A1	COM1A0	COM1B1	COM1B0	FOC1A	FOC1B	WGM11	WGM10

TCCR1B =	0	0	0	0	0	0	0	1
	ICNC1	ICES1	–	WGM13	WGM12	CS12	CS11	CS10

Example 16-36

Calculate the OCR1B to generate a wave with duty cycle of 75% in each of the following modes:

(a) Mode 1, non-inverted mode   (b) Mode 3, inverted mode

(c) Mode 2, non-inverted mode   (d) Mode 2, inverted mode

(e) Mode 1, inverted mode

Solution:

(a) In mode 1, Top = 0xFF = 255. So,

$75 = \text{OCR1x} \times 100 / \text{Top}$ ➜ $\text{OCR1x} = 75 \times 255 / 100 = 192$ ➜ $\text{OCR1B} = 191$

(b) In mode 3, Top = 0x3FF = 1023. So,

$75 = (\text{Top} - \text{OCR1x}) \times 100 / \text{Top}$ ➜ $1023 - \text{OCR1x} = 75 \times 1023 / 100 = 767$

➜ $\text{OCR1B} = 255$

(c) In mode 2, Top = 0x1FF = 511. So,

$75 = \text{OCR1x} \times 100 / \text{Top}$ ➜ $\text{OCR1x} = 75 \times 511 / 100 = 383$ ➜ $\text{OCR1B} = 383$

(d) In mode 2, Top = 0x1FF = 511. So,

$75 = (\text{Top} - \text{OCR1x}) \times 100 / \text{Top}$ ➜ $75 = (511 - \text{OCR1x}) \times 100 / 511$

➜ $511 - \text{OCR1x} = 75 \times 511 / 100 = 383$ ➜ $\text{OCR1B} = 511 - 383 = 128$

(e) In mode 1, Top = 0xFF = 255. So,

$75 = (\text{Top} - \text{OCR1x}) \times 100 / \text{Top}$ ➜ $75 = (255 - \text{OCR1x}) \times 100 / 255$

➜ $255 - \text{OCR1x} = 75 \times 255 / 100 = 191$ ➜ $\text{OCR1B} = 255 - 191 = 64$

Example 16-37

Assuming XTAL = 8 MHz, using non-inverted mode and mode 1, write a program that generates a wave with frequency of 15,686 Hz and duty cycle of 75%.

Solution:

$15{,}686 = 8M / (510 \times N)$ ➜ $N = 8M / (15{,}686 \times 510) = 1$ ➜ No prescaler

```
.INCLUDE  "M32DEF.INC"
    SBI   DDRD,5           ;PD5 = output
    LDI   R16,HIGH(191)    ;from Example 16-36
    OUT   OCR1AH,R16        ;Temp = 0x00
    LDI   R16,LOW(191)     ;R16 = 191
    OUT   OCR1AL,R16        ;OCR1A = 191
    LDI   R16,0x81         ;from Example 16-35
    OUT   TCCR1A,R16        ;COM1A = non-inverted
    LDI   R16,0x01
    OUT   TCCR1B,R16        ;WGM = mode 1, clock = no scaler
HERE: RJMP  HERE
```

Generating waves with different frequencies (case study)

As we mentioned earlier, the frequency of the generated wave is equal to $F_{oscillator}/(2N \times Top)$. In modes 1, 2, and 3, the Top value is fixed. Therefore, in these modes the only way to change the frequency of the generated wave is to change N (the prescaler). In Figure 16-47, you see the different frequencies that can be generated using modes 1, 2, and 3.

Prescaler	1	1:8	1:64	1:256	1:1024
Mode = 1	$\dfrac{F_{oscillator}}{510}$	$\dfrac{F_{oscillator}}{8*510}$	$\dfrac{F_{oscillator}}{64*510}$	$\dfrac{F_{oscillator}}{256*510}$	$\dfrac{F_{oscillator}}{1024*510}$
Mode = 2	$\dfrac{F_{oscillator}}{1*1022}$	$\dfrac{F_{oscillator}}{8*1022}$	$\dfrac{F_{oscillator}}{64*1022}$	$\dfrac{F_{oscillator}}{256*1022}$	$\dfrac{F_{oscillator}}{1024*1022}$
Mode = 3	$\dfrac{F_{oscillator}}{1*2046}$	$\dfrac{F_{oscillator}}{8*2046}$	$\dfrac{F_{oscillator}}{64*2046}$	$\dfrac{F_{oscillator}}{256*2046}$	$\dfrac{F_{oscillator}}{1024*2046}$

Figure 16-47. Different Frequencies Can Be Made Using Modes 1, 2, and 3

So, in these modes we can make a very limited number of frequencies. What if we want to make some other frequencies? In modes 10 and 11, the Top value can be specified by ICR1 and the OCR1A registers. Thus, we can change the frequency by loading proper values to ICR1 and OCR1A. See Examples 16-38 through 16-40.

Example 16-38

Assuming XTAL = 8 MHz, find TCCR1A and TCCR1B to generate two waves with frequency of 125 Hz on OC1A and OC1B using mode 10, non-inverted mode, and prescaler = 1:256.

Solution:

$125 = 8M / (2N \times Top)$ ➔ $2N \times Top = 8M / 125 = 64,000$ ➔ $Top = 64,000 / 512 = 250$
Top = 250 ➔ ICR1 = 250
N = 256 ➔ CS12:0 = 100
Mode = 10 ➔ WGM12:10 = 1010
OC1A in non-inverted mode ➔ COM1A1:COM1A0 = 10
OC1B in non-inverted mode ➔ COM1B1:COM1B0 = 10

TCCR1A =	1	0	1	0	0	0	1	0
	COM1A1	COM1A0	COM1B1	COM1B0	FOC1A	FOC1B	WGM11	WGM10

TCCR1B =	0	0	0	1	0	1	0	0
	ICNC1	ICES1	–	WGM13	WGM12	CS12	CS11	CS10

Example 16-39

Calculate the OCR1x to generate the following waves in each of the following modes:
(a) Mode 11, inverted mode, OCR1A=50, duty cycle = 30%
(b) Mode 10, non-inverted mode, ICR1 = 250, duty cycle = 30%
(c) Mode 10, non-inverted mode, ICR1 = 250, duty cycle = 60%

Solution:

(a) In mode 11, Top = OCR1A = 50. So,
$30 = (Top - OCR1B) \times 100 / Top$ ➜ $50 - OCR1B = 50 \times 30 / 100 = 15$ ➜ OCR1B = 35

(b) In mode 10, Top = ICR1 = 250. So,
$30 = OCR1x \times 100 / Top$ ➜ $OCR1x = 30 \times 250 / 100 = 75$ ➜ OCR1x = 75

(c) In mode 10, Top = ICR1 = 250. So,
$60 = OCR1x \times 100 / Top$ ➜ $OCR1x = 60 \times 250 / 100 = 150$ ➜ OCR1x = 150

Example 16-40

Assume XTAL = 8 MHz. Using mode 10 write a program that generates waves with duty cycles of 30% and 60% on the OC1A and OC1B pins, respectively. Frequency of the generated waves should be 125 Hz.

Solution:

```
.INCLUDE "M32DEF.INC"
     LDI    R16,LOW(RAMEND)
     OUT    SPL,R16
     LDI    R16,HIGH(RAMEND)
     OUT    SPH,R16            ;initialize stack pointer

     SBI    DDRD,5             ;PD5 (OC1A) = output
     SBI    DDRD,4             ;PD4 (OC1B) = output
     LDI    R16,0
     OUT    OCR1AH,R16         ;Temp = 0
     LDI    R16,75             ;from Example 16-39
     OUT    OCR1AL,R16         ;OCR1AL = 75, OCR1AH = Temp = 0
     LDI    R16,150            ;from Example 16-39
     OUT    OCR1BL,R16         ;OCR1BL = 150, OCR1BH = Temp = 0
     LDI    R16,250
     OUT    ICR1L,R16          ;ICR1L = 250, ICR1H = Temp = 0
     LDI    R16,0xA2           ;from Example 16-38
     OUT    TCCR1A,R16    ;COM1A = non-inverted, COM1B = non-inv.
     LDI    R16,0x14           ;from Example 16-38
     OUT    TCCR1B,R16         ;WGM = mode 10, clock = no scaler

HERE: RJMP  HERE               ;wait here forever
```

If we use mode 11 instead of mode 10, OCR1A is buffered, and the contents of the buffer will be loaded into OCR1A, when the timer reaches its top value. In mode 11 we can only use the OC1B wave generator and we cannot use the OC1A wave generator since the OCR1A register is used for defining the Top value.

16-bit PWM programming in C

Examples 16-41 through 16-49 show the C versions of the earlier programs.

Example 16-41 (C version of Example 16-25)

Assuming XTAL = 8 MHz, using non-inverted mode and mode 5, write a program that generates a wave with frequency of 31,250 Hz and duty cycle of 75%.

Solution:

```
#include "avr/io.h"
int main ( )
{
     DDRD |= (1<<5); //PD5 = output
     OCR1AH = 0;       //Temp = 0
     OCR1AL = 191;     //OCR1A = 191
     TCCR1A = 0x81;    //COM1A = non-inverted
     TCCR1B = 0x09;    //WGM = mode 5, clock = no scaler

     while (1);
     return 0;
}
```

Example 16-42 (C version of Example 16-26)

Assuming XTAL = 8 MHz, using non-inverted mode and mode 7, write a program that generates a wave with frequency of 7812.5 Hz and duty cycle of 75%.

Solution:

```
#include "avr/io.h"
int main ( )
{
     DDRD |= (1<<5); //PD5 = output
     OCR1AH = 767>>8;//OCR1AH = HIGH (767)
     OCR1AL = 767;    //OCR1AL = LOW (767)
     TCCR1A = 0x83;   //COM1A = non-inverted
     TCCR1B = 0x09;   //WGM = mode 7, clock = no scaler
     while (1);
     return 0;
}
```

Example 16-43 (C version of Example 16-27)

Assuming XTAL = 8 MHz, using non-inverted mode and mode 6, write a program that generates a wave with frequency of 1953 Hz and duty cycle of 60%.

Solution:
```c
#include "avr/io.h"
int main ( )
{
    DDRD |= (1<<5); //PD5 as output
    OCR1AH = 306>>8;//OCR1AH = HIGH (306)
    OCR1AL = 306;   //OCR1AL = LOW (306)
    TCCR1A = 0x82;  //COM1A = non-inverted
    TCCR1B = 0x0A;  //WGM = mode 6, clock = no prescaler
    while (1);
    return 0;
}
```

Example 16-44 (C version of Example 16-28)

Rewrite Example 16-43 using inverted mode.

Solution:
```c
#include "avr/io.h"
int main ( )
{
    DDRD |= (1<<5); //PD5 as output
    OCR1AH = 204>>8;//OCR1AH = HIGH(204) = 0
    OCR1AL = 204;   //OCR1AL = LOW(204) = 204
    TCCR1A = 0xB2;  //COM1A = inverted
    TCCR1B = 0x0A;  //WGM = mode 6, clock = no scaler
    while (1);
    return 0;
}
```

Example 16-45 (C version of Example 16-29)

Rewrite the program of Example 16-29 using C.

Solution:

```c
#include "avr/io.h"
#include "avr/interrupt.h"
ISR (TIMER1_OVF_vect)
{
    OCR1AH = 0;
    OCR1AL = ~OCR1AL;
}
int main ( )
{
    DDRD |= (1<<5); //PD5 as output
```

Example 16-45 (Cont.)

```
    OCR1AH = 0;       //Temp = 0
    OCR1AL = 31;      //OCR1A = 31
    TCCR1A = 0x81;    //COM1A = non-inverted
    TCCR1B = 0x0A;    //WGM = mode 5, clock = no scaler
    TIMSK = (1<<TOIE1);
    sei ( );
    while (1);
    return 0;
}
```

Example 16-46 (C version of Example 16-32)

Assume XTAL = 8 MHz. Using mode 14 write a program that generates a wave with duty cycle of 20% and frequency of 80 kHz.

Solution:
```
#include "avr/io.h"
int main ( )
{
    DDRD |= (1<<5); //PD5 as output
    ICR1H = 0x00;     //Temp = 0x00
    ICR1L = 99;       //ICR1 = 99
    OCR1AH = 0;       //OCR1AH = 0
    OCR1AL = 19;      //OCR1A = 19
    TCCR1A = 0x82;    //COM1A = non-inverted
    TCCR1B = 0x19;    //WGM = mode 14, clock = no scaler
    while (1);
    return 0;
}
```

Example 16-47 (C version of Example 16-34)

Assume XTAL = 8 MHz. Using mode 15 write a program that generates a wave with duty cycle of 20% and frequency of 64 kHz.

Solution:
```
#include "avr/io.h"
int main ( )
{
    DDRD |= (1<<4); //PD4 as output
    OCR1AH = 0;       //Temp = 0
    OCR1AL = 124;     //OCR1A = 124
    OCR1BH = 0;       //Temp = 0
    OCR1BL = 24;      //OCR1B = 24
    TCCR1A = 0x23;    //COM1B = non-inverted
    TCCR1B = 0x19;    //WGM = mode 15, clock = no scaler
    while (1);
    return 0;
}
```

Example 16-48 (C version of Example 16-37)

Assuming XTAL = 8 MHz, using non-inverted mode and mode 1, write a program that generates a wave with frequency of 15,686 Hz and duty cycle of 75%.

Solution:

```
#include "avr/io.h"
int main ( )
{
    DDRD |= (1<<5);  //PD5 as output
    OCR1AH = 0;      //Temp = 0
    OCR1AL = 191;    //OCR1A = 191
    TCCR1A = 0x23;   //COM1A = non-inverted
    TCCR1B = 0x01;   //WGM = mode 1, clock = no prescaler
    while (1);
    return 0;
}
```

Example 16-49 (C version of Example 16-40)

Assume XTAL = 8 MHz. Using mode 10 write a program that generates waves with duty cycles of 30% and 60% on the OC1A and OC1B pins, respectively. The frequency of the generated waves should be 125 Hz.

Solution:

```
#include "avr/io.h"
int main ( )
{
    DDRD = DDRD|(1<<5)|(1<<4);  //PD4 and PD5 as output
    OCR1AH = 0x00;   //Temp = 0
    OCR1AL = 75;     //OCR1A = 75
    OCR1BL = 150;    //OCR1B = 150
    ICR1L = 250;     //ICR1 = 250
    TCCR1A = 0xA2;   //COM1A = non-inverted, COM1B = non-inv.
    TCCR1B = 0x14;   //WGM = mode 10, clock = no prescaler
    while (1);
    return 0;
}
```

Review Questions

1. True or false. We can associate each of the pins with each of the waveform generators.
2. True or false. In PWM modes (Fast PWM and Phase correct PWM) we can change the duty cycle of the generated wave.
3. True or false. In inverted Phase correct PWM mode, the duty cycle increases when the OCR1A value increases.

SECTION 16.4: DC MOTOR CONTROL USING PWM

To generate the PWM waves for controlling the DC motor we can use the PWM features of AVR. See Examples 16-50 and 16-51.

Example 16-50 (Example 16-3 using AVR PWM features)

Refer to the figure in this example. Write a program to monitor the status of the switch and perform the following:

(a) If PORTA.7 = 1, the DC motor moves with 25% duty cycle pulse.
(b) If PORTA.7 = 0, the DC motor moves with 50% duty cycle pulse.

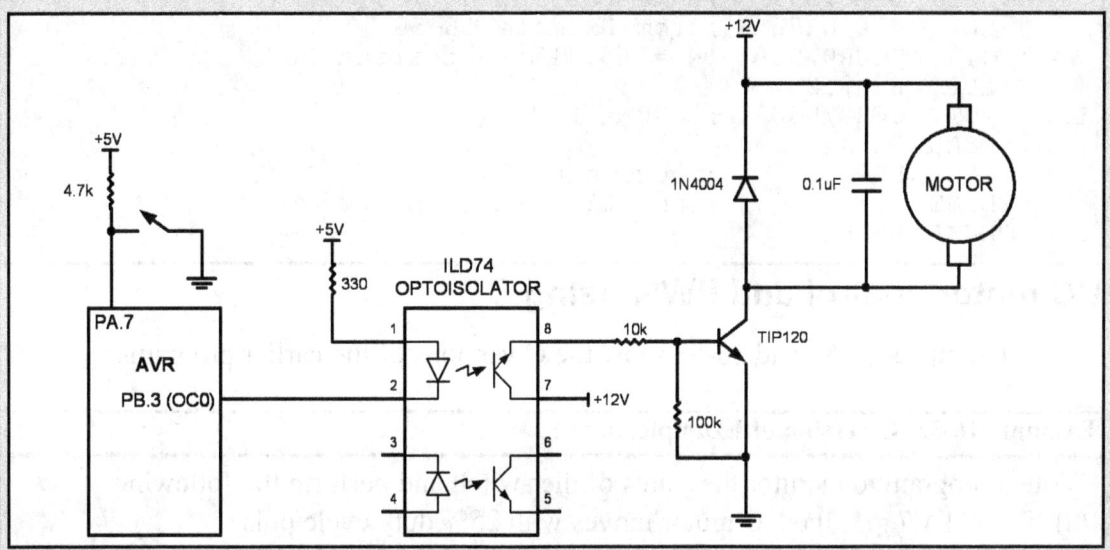

Solution:

For driving motors it is preferable to use the Phase correct PWM mode.
OCR0 / 255 = duty cycle / 100 ➜ OCR0 = 255 × duty cycle / 100
For duty cycle = 25% ➜ OCR0 = 255 × 25 / 100 = 64
For duty cycle = 50% ➜ OCR0 = 255 × 50 / 100 = 127

In this example we generate waves with frequency of 245 Hz. To do so,
245 = 8M / (510 × N) ➜ N = 8M / (245 × 510) = 64 ➜ Prescaler = 64

TCCR0 =	0	1	1	1	0	0	1	1
	FOC0	WGM00	COM01	COM00	WGM01	CS02	CS01	CS00

```
.INCLUDE "M32DEF.INC"
      SBI    DDRB,3        ;make PB3 output
      SBI    PORTA,7       ;activate pull-up of PA7
      LDI    R16,0x73
      OUT    TCCR0,R16     ;N = 64, Phase correct PWM, inverted
L1:   SBIC   PINA,7        ;skip next instruct if PINA.7 is zero
      LDI    R16,64        ;if PINA.7 is one then R16 = 64
      SBIS   PINA,7        ;skip next instruct if PINA.7 is one
      LDI    R16,127       ;if PINA.7 is zero then R16 = 127
      OUT    OCR0,R16      ;OCR0 = R16
      RJMP   L1            ;jump L1
```

Example 16-51

Write a program that gradually changes the speed of a DC motor from 50% to 100%. Use information given in Example 16-50.

Solution:

```
.INCLUDE "M32DEF.INC"
      LDI   R16,HIGH(RAMEND)
      OUT   SPH,R16
      LDI   R16,LOW(RAMEND)
      OUT   SPL,R16     ;initialize stack pointer

      SBI   DDRB,3      ;make PB3 output
      LDI   R16,0x73    ;from Example 16-50
      OUT   TCCR0,R16   ;N = 64, Phase correct PWM, inverted
      LDI   R20,127
L1:   OUT   OCR0,R20    ;OCR0 = R17
      RCALL DELAY
      INC   R20         ;increment R20
      BRNE  L1          ;jump L1 if R20 is not zero
HERE: RJMP  HERE
```

DC motor control and PWM using C

Examples 16-52 and 16-53 show the C versions of the earlier programs.

Example 16-52 (C version of Example 16-50)

Write a program to monitor the status of the switch and perform the following:
(a) If PORTA.7 = 1, the DC motor moves with 25% duty cycle pulse.
(b) If PORTA.7 = 0, the DC motor moves with 50% duty cycle pulse.

Solution:

```
#include "avr/io.h"

int main ( )
{
      DDRB = 0x08;      //PB3 as output
      PORTA = 0x80;     //pull-up resistor
      TCCR0 = 0x73;     //Phase correct PWM, inverted, N = 64

      while (1)
      {
            switch ((PINA&0x80))
            {
                  case 0: OCR0 = 64; break;     //25%
                  case 1: OCR0 = 127; break;    //50%
            }
      }

      return 0;
}
```

Example 16-53 (C version of Example 16-51)

Write a program that gradually changes the speed of a DC motor from 50% to 100%.

Solution:

```
#define  F_CPU  8000000UL  //XTAL = 8 MHz
#include "avr/io.h"
#include "util/delay.h"
int main ( )
{
    unsigned char i;

    DDRB = 0x08;      //PB3 as output
    i = 127;
    OCR0 = 127;       //duty cycle = 50%
    TCCR0 = 0x73;     //Phase correct PWM, inverted, N = 64

    while (i != 0)
    {
        OCR0 = i;
        _delay_ms(25);   //use AVR Studio library delay
        i++;
    }
    while (1);
    return 0;
}
```

SUMMARY

In the first section, The AVR was interfaced with DC motors. A typical DC motor will take electronic pulses and convert them to mechanical motion. This chapter showed how to interface the AVR with a DC motor. Then, simple Assembly and C programs were written to show the concept of PWM.

We discussed the PWM features of AVR timers in sections two and three, and in the last section we used the PWM feature of AVR to control DC motors.

PROBLEMS

SECTION 16.1: DC MOTOR INTERFACING AND PWM

1. Which motor is best for moving a wheel exactly 90 degrees?
2. True or false. Current dissipation of a DC motor is proportional to the load.
3. True or false. The RPM of a DC motor is the same for no-load and loaded.
4. The RPM given in data sheets is for _____ (no-load, loaded).
5. What is the advantage of DC motors over AC motors?
6. What is the advantage of stepper motors over DC motors?
7. True or false. Higher load on a DC motor slows it down if the current and voltage supplied to the motor are fixed.
8. What is PWM, and how is it used in DC motor control?
9. A DC motor is moving a load. How do we keep the RPM constant?
10. What is the advantage of placing an optoisolator between the motor and the microcontroller?

SECTION 16.2: PWM MODES IN 8-BIT TIMERS

11. Using Timer0 and non-inverted Fast PWM mode, write a program that generates a wave with frequency of 62.5 kHz and duty cycle of 60%. Assume XTAL = 16 MHz.

12. Using Timer0 and inverted Fast PWM mode, write a program that generates a wave with frequency of 46.875 kHz and duty cycle of 70%. Assume XTAL = 12 MHz.

13. Using Timer0 and inverted Fast PWM mode, write a program that generates a wave with frequency of 1953 Hz and duty cycle of 20%. Assume XTAL = 4 MHz.

14. Using Timer0 and non-inverted Fast PWM mode, write a program that generates a wave with frequency of 15.25 Hz and duty cycle of 10%. Assume XTAL = 1 MHz.

15. Using Timer0 and inverted Phase correct PWM mode, write a program that generates a wave with frequency of 1960 Hz and duty cycle of 20%. Assume XTAL = 1 MHz.

16. Using Timer0 and inverted Phase correct PWM mode, write a program that generates a wave with frequency of 1.96 kHz and duty cycle of 95%. Assume XTAL = 1 MHz.

17. Using Timer2 and non-inverted Phase correct PWM mode, write a program that generates a wave with frequency of 61.3 Hz and duty cycle of 19%. Assume XTAL = 8 MHz.

18. Using Timer2 and inverted Phase correct PWM mode, write a program that generates a wave with frequency of 245 Hz and duty cycle of 82%. Assume XTAL = 1 MHz.

SECTION 16.3: PWM MODES IN TIMER1

19. Using mode 6 of Timer1 and non-inverted Fast PWM, write a program that generates a wave with frequency of 15,625 Hz and duty cycle of 40%. Assume XTAL = 8 MHz.

20. Using mode 7 of Timer1 and inverted Fast PWM, write a program that generates a wave with frequency of 3906 Hz and duty cycle of 45%. Assume XTAL = 4 MHz.

21. Using mode 7 of Timer1 and inverted Fast PWM, write a program that generates a wave with frequency of 1953 Hz and duty cycle of 35%. Assume XTAL = 16 MHz.

22. Using mode 6 of Timer1 and non-inverted Fast PWM, write a program that generates a wave with frequency of 1953 Hz and duty cycle of 50%. Assume XTAL = 8 MHz.

23. Using mode 1 of Timer1 and inverted Phase correct PWM, write a program that generates a wave with frequency of 976.5 Hz and duty cycle of 35%. Assume XTAL = 4 MHz.

24. Using mode 2 of Timer1 and non-inverted Phase correct PWM, write a program that generates a wave with frequency of 30.5 Hz and duty cycle of 25%. Assume XTAL = 8 MHz.

25. Using mode 1 of Timer1 and non-inverted Phase correct PWM, write a program that generates a wave with frequency of 245 Hz and duty cycle of 19%. Assume XTAL = 8 MHz.

26. As shown in the figure, a switch is connected to PB0. Using mode 2 of Timer1 and non-inverted Phase correct PWM, write a program that generates a wave with frequency of 978 Hz. When the switch is closed the duty cycle is 20%, and when it is open the duty cycle is 85%. Assume XTAL = 8 MHz.

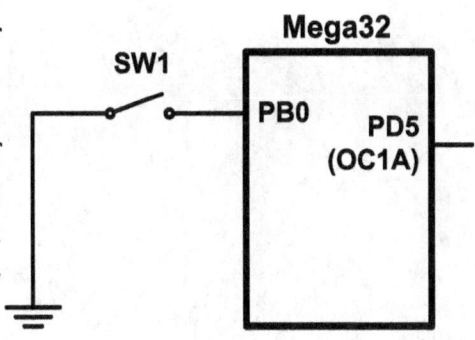

ANSWERS TO REVIEW QUESTIONS

SECTION 16.1: DC MOTOR INTERFACING AND PWM

1. True
2. False
3. Because microcontroller/digital outputs lack sufficient current to drive the DC motor, we need a driver.
4. By reversing the polarity of voltages connected to the leads
5. The DC motor is stalled if the load is beyond what it can handle.
6. No-load

SECTION 16.2: PWM MODES IN 8-BIT TIMERS

1. True
2. False
3. True
4. False
5. Phase correct PWM

SECTION 16.3: PWM MODES IN TIMER1

1. False
2. True
3. False

CHAPTER 17

SPI PROTOCOL AND
MAX7221 DISPLAY
INTERFACING

OBJECTIVES

Upon completion of this chapter, you will be able to:

>>   Understand the Serial Peripheral Interfacing (SPI) protocol
>>   Explain how the SPI read and write operations work
>>   Examine the SPI pins SDO, SDI, CE, and SCLK
>>   Code programs in Assembly and C for SPI
>>   Explain how 7-segment displays work
>>   Explain the function of the MAX7221 pins
>>   Explain the function of the MAX7221 registers
>>   Understand the interfacing of the MAX7221 to the AVR
>>   Code programs to display numbers in Assembly and C

This chapter discusses the SPI bus. In Section 17.1 we examine the different pins of SPI protocol and then focus on the concept of clock polarity. We distinguish differences between single-byte read/write and multibyte burst read/write. Section 17.2 discusses SPI programming in AVR using both Assembly and C. We will examine how to interface MAX7221 to the AVR in Section 17.3.

SECTION 17.1: SPI BUS PROTOCOL

The SPI (serial peripheral interface) is a bus interface connection incorporated into many devices such as ADC, DAC, and EEPROM. In this section we examine the pins of the SPI bus and show how the read and write operations in the SPI work.

The SPI bus was originally started by Motorola Corp. (now Freescale), but in recent years has become a widely used standard adapted by many semiconductor chip companies. SPI devices use only 2 pins for data transfer, called SDI (Din) and SDO (Dout), instead of the 8 or more pins used in traditional buses. This reduction of data pins reduces the package size and power consumption drastically, making them ideal for many applications in which space is a major concern. The SPI bus has the SCLK (shift clock) pin to synchronize the data transfer between two chips. The last pin of the SPI bus is CE (chip enable), which is used to initiate and terminate the data transfer. These four pins, SDI, SDO, SCLK, and CE, make the SPI a 4-wire interface. See Figure 17-1. In many chips the SDI, SDO, SCLK, and CE signals are alternatively named as MOSI, MISO, SCK, and SS as shown in Figure 17-2 (compare with Figure 17-1). There is also a widely used standard called a *3-wire interface bus*. In a 3-wire interface bus, we have SCLK and CE, and only a single pin for data transfer. The SPI 4-wire bus can become a 3-wire interface when the SDI and SDO data pins are tied together. However, there are some major differences between the SPI and 3-wire devices in the data transfer protocol. For that reason, a device must support the 3-wire protocol internally in order to be used as a 3-wire device. Many devices such as the DS1306 RTC (real-time clock) support both SPI and 3-wire protocols.

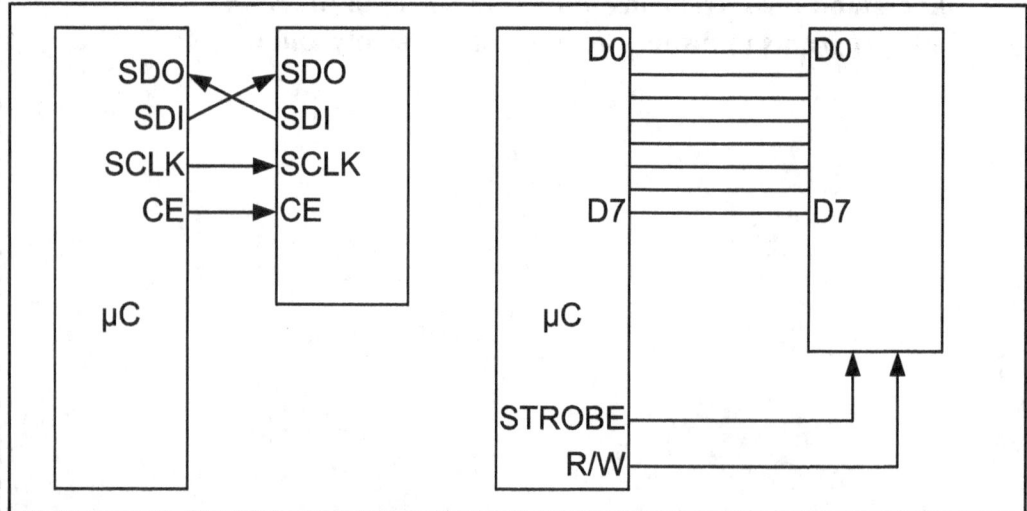

Figure 17-1. SPI Bus vs. Traditional Parallel Bus Connection to Microcontroller

How SPI works

SPI consists of two shift registers, one in the master and the other in the slave side. Also, there is a clock generator in the master side that generates the clock for the shift registers.

As you can see in Figure 17-2, the serial-out pin of the master shift register is connected to the serial-in pin of the slave shift register by MOSI (Master Out Slave In), and the serial-in pin of the master shift register is connected to the serial-out pin of the slave shift register by MISO (Master In Slave Out). The master clock generator provides clock to the shift registers in both the master and slave. The clock input of the shift registers can be falling- or rising-edge triggered. This will be discussed shortly.

In SPI, the shift registers are 8 bits long. It means that after 8 clock pulses, the contents of the two shift registers are interchanged. When the master wants to send a byte of data, it places the byte in its shift register and generates 8 clock pulses. After 8 clock pulses the byte is transmitted to the other shift register. When the master wants to receive a byte of data, the slave side should place the byte in its shift register, and after 8 clock pulses the data will be received by the master shift register. It must be noted that SPI is full duplex, meaning that it sends and receives data at the same time.

SPI read and write

In connecting a device with an SPI bus to a microcontroller, we use the microcontroller as the master while the SPI device acts as a slave. This means that the microcontroller generates the SCLK, which is fed to the SCLK pin of the SPI

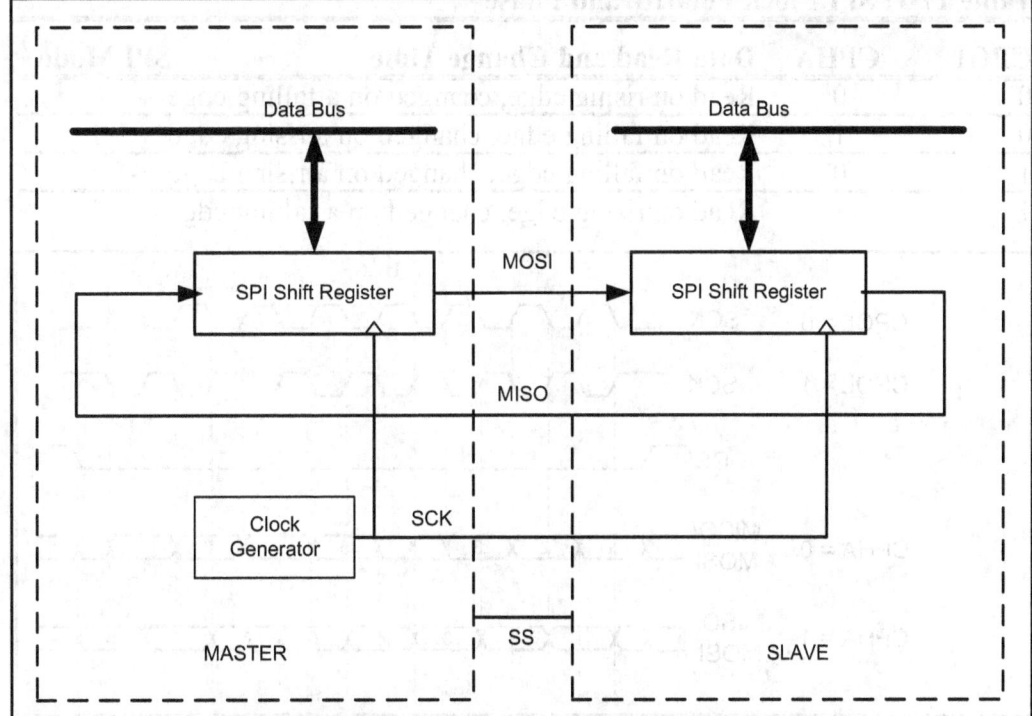

Figure 17-2. SPI Architecture

device. The SPI protocol uses SCLK to synchronize the transfer of information one bit at a time, where the most-significant bit (MSB) goes in first. During the transfer, the CE must stay HIGH. The information (address and data) is transferred between the microcontroller and the SPI device in groups of 8 bits, where the address byte is followed immediately by the data byte. To distinguish between the read and write operations, the D7 bit of the address byte is always 1 for write, while for the read, the D7 bit is LOW, as we will see next.

Clock polarity and phase in SPI device

As we mentioned before in USART communication, transmitter and receiver must agree on a clock frequency. In SPI communication, the master and slave(s) must agree on the clock polarity and phase with respect to the data. Freescale names these two options as CPOL (clock polarity) and CPHA (clock phase), respectively, and most companies like Atmel have adopted that convention. At CPOL= 0 the base value of the clock is zero, while at CPOL = 1 the base value of the clock is one. CPHA = 0 means sample on the leading (first) clock edge, while CPHA = 1 means sample on the trailing (second) clock. Notice that if the base value of the clock is zero, the leading (first) clock edge, is the rising edge but if the base value of the clock is one, the leading (first) clock edge is falling edge. See Table 17-1 and Figure 17-3.

Steps for writing data to an SPI device

In accessing SPI devices, we have two modes of operation: single-byte and multibyte. We will explain each one separately.

Table 17-1: SPI Clock Polarity and Phase

CPOL	CPHA	Data Read and Change Time	SPI Mode
0	0	Read on rising edge, changed on a falling edge	0
0	1	Read on falling edge, changed on a rising edge	1
1	0	Read on falling edge, changed on a rising edge	2
1	1	Read on rising edge, changed on a falling edge	3

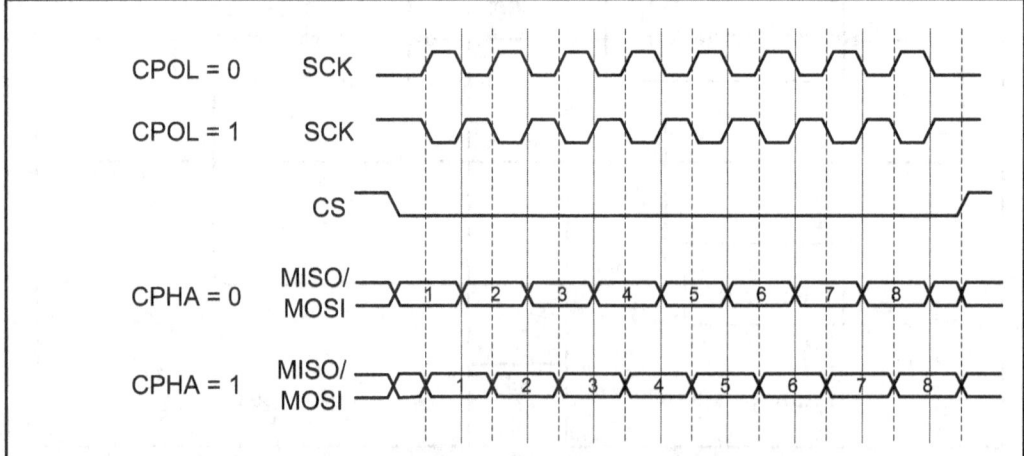

Figure 17-3. SPI Clock Polarity and Phase

Single-byte write

The following steps are used to send (write) data in single-byte mode for SPI devices, as shown in Figure 17-4:

1. Make CE = 0 to begin writing.
2. The 8-bit address is shifted in, one bit at a time, with each edge of SCLK. Notice that A7 = 1 for the write operation, and the A7 bit goes in first.
3. After all 8 bits of the address are sent in, the SPI device expects to receive the data belonging to that address location immediately.
4. The 8-bit data is shifted in one bit at a time, with each edge of the SCLK.
5. Make CE = 1 to indicate the end of the write cycle.

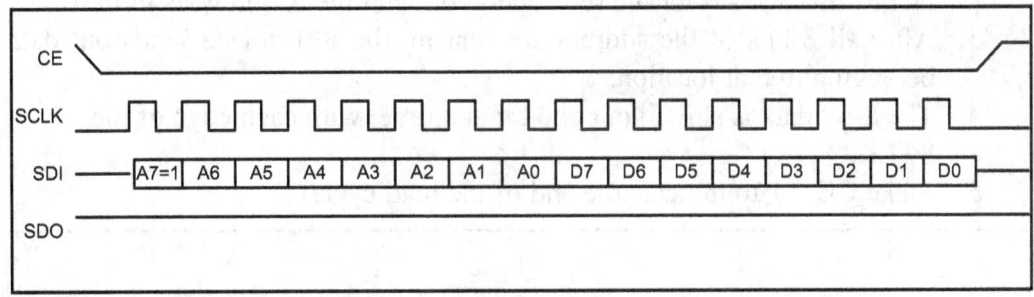

Figure 17-4. SPI Single-Byte Write Timing (Notice A7 = 1)

Multibyte burst write

Burst mode writing is an effective means of loading consecutive locations. In burst mode, we provide the address of the first location, followed by the data for that location. From then on, while CE = 0, consecutive bytes are written to consecutive memory locations. In this mode, the SPI device internally increments the address location as long as CE is LOW. The following steps are used to send (write) multiple bytes of data in burst mode for SPI devices as shown in Figure 17-5:

1. Make CE = 0 to begin writing.
2. The 8-bit address of the first location is provided and shifted in, one bit at a time, with each edge of SCLK. Notice that A7 = 1 for the write operation and the A7 bit goes in first.
3. The 8-bit data for the first location is provided and shifted in, one bit at a time, with each edge of the SCLK. From then on, we simply provide consecutive bytes of data to be placed in consecutive memory locations. In the process, CE must stay low to indicate that this is a burst mode multibyte write operation.
4. Make CE = 1 to end writing.

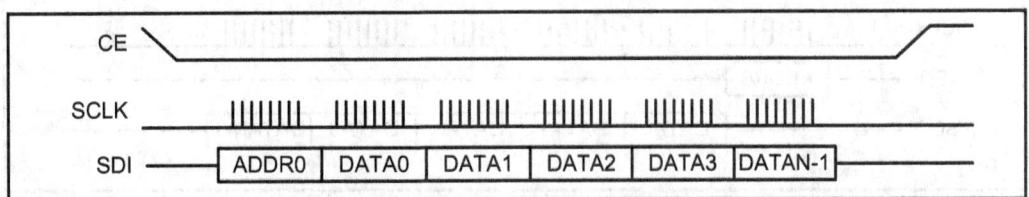

Figure 17-5. SPI Burst (Multibyte) Mode Writing

Steps for reading data from an SPI device

In reading SPI devices, we also have two modes of operation: single-byte and multibyte. We will explain each one separately.

Single-byte read

The following steps are used to get (read) data in single-byte mode from SPI devices, as shown in Figure 17-6:

1. Make CE = 0 to begin reading.
2. The 8-bit address is shifted in one bit at a time, with each edge of SCLK. Notice that A7 = 0 for the read operation, and the A7 bit goes in first.
3. After all 8 bits of the address are sent in, the SPI device sends out data belonging to that location.
4. The 8-bit data is shifted out one bit at a time, with each edge of the SCLK.
5. Make CE = 1 to indicate the end of the read cycle.

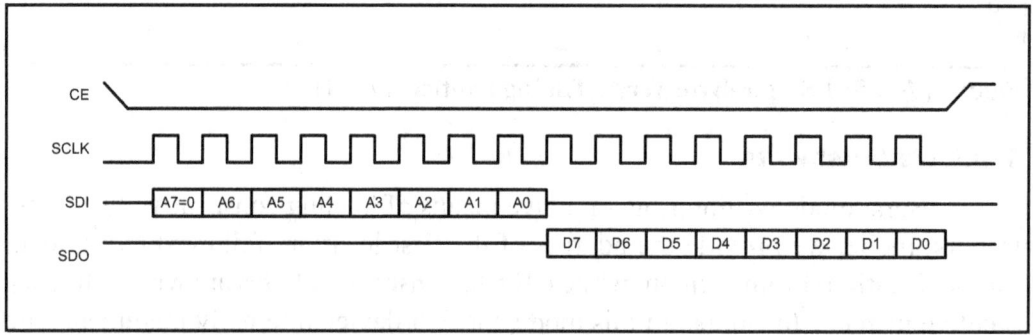

Figure 17-6. SPI Single-Byte Read Timing (Notice A7 = 0)

Multibyte burst read

Burst mode reading is an effective means of bringing out the contents of consecutive locations. In burst mode, we provide the address of the first location only. From then on, while CE = 0, consecutive bytes are brought out from consecutive memory locations. In this mode, the SPI device internally increments the address location as long as CE is LOW. The following steps are used to get (read) multiple bytes of data in burst mode for SPI devices, as shown in Figure 17-7:

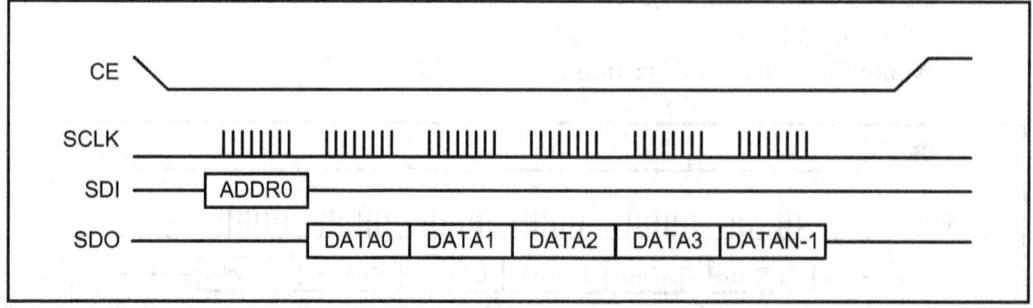

Figure 17-7. SPI Burst (Multibyte) Mode Reading

608

1. Make CE = 0 to begin reading.
2. The 8-bit address of the first location is provided and shifted in, one bit at a time, with each edge of SCLK. Notice that A7 = 0 for the read operation, and the A7 bit goes in first.
3. The 8-bit data for the first location is shifted out, one bit at a time, with each edge of the SCLK. From then on, we simply keep getting consecutive bytes of data belonging to consecutive memory locations. In the process, CE must stay LOW to indicate that this is a burst mode multibyte read operation.
4. Make CE = 1 to end reading.

Review Questions

1. True or false. The SPI protocol writes and reads information in 8-bit chunks.
2. True or false. In SPI, the address is immediately followed by the data.
3. True or false. In an SPI write cycle, bit A7 of the address is LOW.
4. True or false. In an SPI write, the LSB goes in first.
5. State the difference between the single-byte and burst modes in terms of the CE signal.

SECTION 17.2: SPI PROGRAMMING IN AVR

Most AVRs, including ATmega family members, support SPI protocols. In AVR three registers are associated with SPI. They are SPSR (SPI Status Register), SPCR (SPI Control Register), and SPDR (SPI Data Register). In this section we will focus on these registers.

SPSR (SPI Status Register)

Figure 17-8 shows the bits of the SPSR register used for SPI.

SPIF	WCOL	-	-	-	-	-	SPI2X

Bit 7 – SPIF (SPI Interrupt Flag)
In master mode, this bit is set in two situations: when a serial transfer is completed, or when SS pin is an input and is driven low by an external device. Setting the SPIF flag to one will cause an interrupt if SPIE in SPCR is set and global interrupts are enabled.

Bit 6 – WCOL (Write COLlision Flag)
The WCOL bit is set if you write on SPDR during a data transfer.

Bit 0 – SPI2X (Double SPI Speed)
When the SPI is in master mode, setting this bit to one doubles the SPI speed.

Notice that both the WCOL bit and the SPIF bit are cleared when you read the SPI Status Register and then access the SPI Data Register. Alternatively, the SPIF bit is cleared by hardware when executing the corresponding interrupt handler.

Figure 17-8. SPI Status Register *(Note: The portion shown is used for SPI.)*

SPCR (SPI Control Register)

Figure 17-9 shows details of each bit in SPCR.

SPIE	SPE	DORD	MSTR	CPOL	CPHA	SPR1	SPR0

Bit 7 – SPIE: SPI Interrupt Enable
Setting this bit to one enables the SPI interrupt.

Bit 6 – SPE: SPI Enable
Setting this bit to one enables the SPI.

Bit 5 – DORD: Data Order
This bit lets you choose to either transmit MSB and then LSB or vice versa. The LSB is transmitted first if DORD is one; otherwise, the MSB is transmitted first.

Bit 4 – MSTR: Master/Slave Select
If you want to work in master mode then set this bit to one; otherwise, slave mode is selected. Notice that if the SS pin is configured as an input and is driven low while MSTR is set, MSTR will be cleared, and SPIF will become set.

Bit 3 – CPOL: Clock Polarity
This bit set the base value of clock when it is idle. At CPOL = 0 the base value of the clock is zero while at CPOL = 1 the base value of the clock is one.

Bit 2 – CPHA: Clock Phase
CPHA = 0 means sample on the leading (first) clock edge, while CPHA = 1 means sample on the trailing (second) clock.

Bits 1, 0 – SPR1, SPR0: SPI Clock Rate Select 1 and 0
These two bits control the SCK rate of the device in master mode. See Table 17-2.

Figure 17-9. SPI Control Register

In Table 17-2 you see how SPI2X, SPR1, and SPR0 are combined to make different clock frequencies for master. As you see in Table 17-2, by setting SPI2X to one, the SCK frequency is doubled.

Table 17-2: SCK Frequency

SPI2X	SPR1	SPR0	SCK Frequency
0	0	0	Fosc/4
0	0	1	Fosc/16
0	1	0	Fosc/64
0	1	1	Fosc/128
1	0	0	Fosc/2 (Not recommended!)
1	0	1	Fosc/8
1	1	0	Fosc/32
1	1	1	Fosc/64

SPDR (The SPI Data Register)

The SPI Data Register is a read/write register. To write into SPI shift register, data must be written to SPDR. To read from the SPI shift register, you should read from SPDR. Writing to the SPDR register initiates data transmission. Notice that you cannot write to SPDR before the last byte is transmitted completely, otherwise a collision will happen. You can read the received data before another byte of data is received completely.

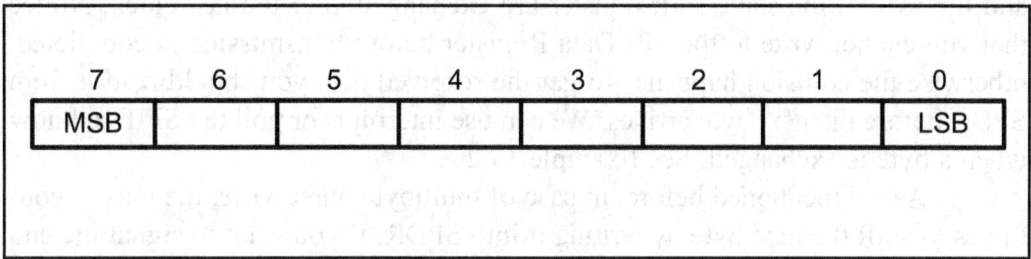

Figure 17-10. SPI Data Register

SS pin in AVR

As we mentioned before, the Slave Select (SS) pin of the SPI bus is used to initiate and terminate the data transfer. In AVR, there are some points regarding this pin that you should pay attention to.

When you are in master mode, you can choose to make this pin either input or output. If you make it output, the SPI circuit of AVR will not control the SS pin and you can make it one or zero by software. When you make the SS pin an input, it will control the function of SPI. In this case you should externally make SS pin high to ensure master SPI operation. If an external device makes the SS pin low, the SPI module stops working in master mode and switches to slave mode by clearing the MSTR bit in SPCR, and then sets the SPIF bit in SPSR. It is highly recommended to make the SS pin output if you do not want to be interrupted when you are working in master mode.

When you are in slave mode, the SS pin is always input and you cannot control it by software. You should hold it externally low to activate the SPI. When SS is driven high, SPI is disabled and all pins of SPI are input. Also the SPI module will immediately clear any partially received data in the shift register. As we mentioned before, it can be used in packet synchronizing by initiating and terminating the data transfer.

Notice that when you are working in slave mode and the SS pin is driven high by an external device, the SPI module is reset but not disabled and it is not necessary to enable it again.

SPI programming in AVR

Before you start data transmission, you should set SPI Mode (Clock Polarity and Clock Phase) by setting the values of the CPOL and CPHA bits in SPCR. See Table 17-1. In AVR you can operate in either master or slave modes. Next, we will discuss each mode in detail.

If you want to work in master mode, you should set the MSTR bit to one. Also you should set SCK frequency by setting the values of SPI2X, SPR1, and SPR2 according to Table 17-2. Then you should enable SPI by setting the SPIE bit to one before you start data transmission.

Writing a byte to the SPI Data Register (SPDR) starts data exchange by starting the SPI clock generator. After shifting the last (8th) bit, the SPI clock generator stops and the SPIF flag changes to one. The byte in the master shift register and the byte in the slave shift register are exchanged after the last clock. Notice that you cannot write to the SPI Data Register before transmission is completed, otherwise the collision happens. To get the received data you should read it from SPDR before the next byte arrives. We can use interrupts or poll the SPIF to know when a byte is exchanged. See Example 17-1.

As we mentioned before, in case of multibyte burst write, the master continues to shift the next byte by writing it into SPDR. If you want to signal the end of the packet, you should pull high the SS pin.

Example 17-1

Write an AVR program to initialize the SPI for master, mode 0, with CLCK frequency = Fosc/16, and then transmit 'G' via SPI repeatedly. The received data should be displayed on Port A.

Solution:

```
.INCLUDE "M32DEF.INC"
.equ MOSI = 5                      ;for ATmega32
.equ SCK = 7
.equ SS = 4

      LDI  R17,0xFF                ;Port A is output
      OUT  DDRA,R17

      LDI  R17, (1<<MOSI)|(1<<SCK)|(1<<SS)
      OUT  DDRB,R17                ;MOSI,SCK, and SS output

      LDI  R17, (1<<SPE)|(1<<MSTR)|(1<<SPR0) ;enable SPI
      OUT  SPCR,R17                ;master, CLK = fck/16
Transmit:
      CBI  PORTB,SS                ;enable slave device
      LDI  R17,'G'                 ;move G letter to R17
      OUT  SPDR,R17                ;start transmission of G
Wait:
      SBIS SPSR,SPIF               ;wait for transmission
      RJMP Wait                    ;to complete
      IN   R18,SPDR                ;read received data into R18
      OUT  PORTA,R18               ;move R18 to PORTA

      SBI  PORTB,SS                ;disable slave device
      RJMP Transmit                ;do it again
```

When AVR is configured as a master, the SPI will not control the SS pin. If you want to make SS high or low, you have to do it by writing 1 or 0, respectively, to the SS bit of Port B.

Slave operating mode

When AVR is configured as a slave, the function of the SPI interface depends on the SS pin. If the SS is driven high, MISO is tri-stated and the SPI interface sleeps. Only the contents of SPDR may be updated in this state. When SS is driven low, the data will be shifted by incoming clock pulses on the SCK pin. SPIF changes to one when the last bit of a byte has been shifted completely. Notice that the slave can place new data to be sent into SPDR before reading the incoming data; this is because in AVR there are two one-byte buffers to store received data.

In slave mode there is no need to set SCK frequency because the SCK is generated by the master, but you must select the SPI mode (Clock Phase and Clock Polarity) and Data Order to match with SPI mode and Data Order of the other side (master device). Finally you should enable the SPI by setting the SPIE bit of SPCR to one. See Example 17-2. Notice that Example 17-2 is the slave version of Example 17-1.

Example 17-2

Write an AVR program to initialize the SPI for slave, mode 0, with CLCK frequency = fck/16, and then transmit 'G' via SPI repeatedly. The received data should be displayed on Port A.

Solution:

```
.INCLUDE "M32DEF.INC"
.equ MISO = 6

      LDI   R17,0xFF              ;Port A is output
      OUT   DDRA,R17
      LDI   R17,(1<<MISO)         ;MISO is output
      OUT   DDRB,R17

      LDI   R17,(1<<SPE)          ;enable SPI slave mode 0
      OUT   SPCR,R17              ;
Again:
      LDI   R17,'G'               ;move letter G to R17
      OUT   SPDR,R17              ;send data to SPDR to be
                                  ;transmitted
Wait:
      SBIS  SPSR,SPIF             ;skip next instruction if IF=1
      RJMP  Wait                  ;otherwise jump wait
      IN    R18,SPDR              ;read received data into R18
      OUT   PORTA,R18             ;send R18 to PORTA
      RJMP  Again                 ;do it again
;
;It must be noted that slave will not start transfer or
;receive until it senses the clock from master
```

SPI programming in C for AVR

Examples 17-3 and 17-4 are C versions of the last two examples.

Example 17-3 (C version of 17-1)

Rewrite Example 17-1 in C.
Solution:

```
#include <avr/io.h>                      //standard AVR header
#define MOSI 5
#define SCK 7
int main (void)
{
  DDRB = (1<<MOSI)|(1<<SCK);             //MOSI and SCK are output
  DDRA = 0xFF;                           //Port A is output
  SPCR = (1<<SPE)|(1<<MSTR)|(1<<SPR0);   //enable SPI as master
  while(1){                              //do for ever
    SPDR = 'G';                          //start transmission
    while(!(SPSR & (1<<SPIF)));          //wait transfer finish
    PORTA = SPDR;                        //move received data to
  }                                      //Port A
  return 0;
}
```

Example 17-4 (C version of 17-2)

Rewrite Example 17-2 in C.
Solution:

```
#include <avr/io.h>                //standard AVR header
#define MISO 6
int main (void)
{
  DDRA = 0xFF ;                    //Port A is output
  DDRB = (1<<MISO);                //MISO is output
  SPCR = (1<<SPE);                 //enable SPI as slave
  while(1){
    SPDR = 'G';
    while(!(SPSR &(1<<SPIF)));      //wait for transfer finish
    PORTA = SPDR;                   //move received data to PORTA
  }
  return 0;
}
```

Review Questions

1. Which registers in the AVR are dedicated to SPI?
2. How do we set the SPI to operate in master mode 1?
3. How do we set the SPI clock frequency to be Fosc/120?
4. True or false. SPI is half duplex.
5. What is the maximum recommended frequency of the SPI clock?

SECTION 17.3: MAX7221 INTERFACING AND PROGRAMMING

In this section we first give a brief description of 7-segments and then show how to interface the MAX7221 chip to AVR.

What is a 7-segment display?

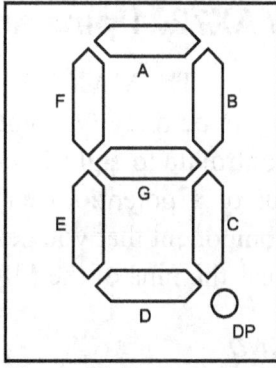

Figure 17-11. 7-segment LEDs

In many applications, when you want to display numbers, 7-segments are the best choice. These displays are made of 7 LEDs to show different numbers plus another LED to display the decimal point. See Figure 17-11. Some characters like A, b, c, d, E, and H are also displayed by 7-segments. Figure 17-12 shows how to display digits.

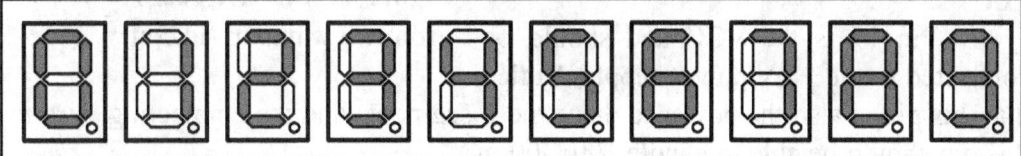

Figure 17-12. 7-Segment Display

There are two types of 7-segments, common anode and common cathode. The MAX7221 supports common cathode only. See Figure 17-13.

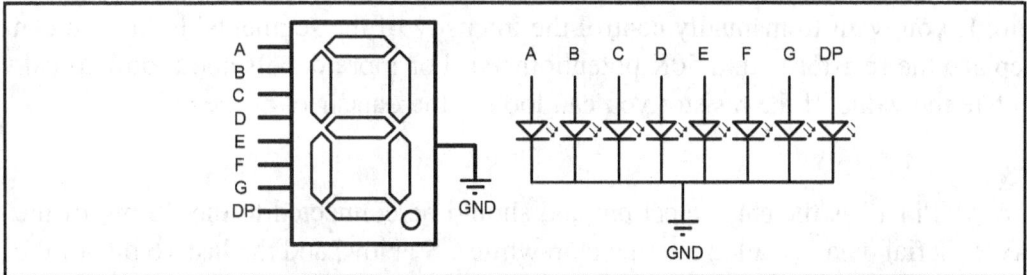

Figure 17-13. Common Cathode Connections in a 7-Segment Display

MAX7221

In many applications you need to connect two or more 7-segment LEDs to a microcontroller. For example, if you want to connect four 7-segment LEDs directly to a microcontroller you need $4 \times 8 = 32$ pins. This is not feasible. The MAX7221 IC is an ideal chip for such applications since it supports up to eight 7-segment LEDs. We can connect the MAX7221 to the AVR chip using SPI protocol and control up to eight 7-segment LEDs. The MAX7221 contains an internal decoder that can be used to convert binary numbers to 7-segment codes. That means we do not need to refresh the 7-segment LEDs. All you need to do is to send a binary number to the MAX7221, and the chip decodes the binary data and displays the number. The device includes analog and digital brightness control, an 8×8 static RAM that stores each digit, and a test mode that forces all LEDs on. Next, we will show how to interface an MAX7221 to the AVR and program it using SPI protocol.

MAX7221 pins and connections

The MAX7221 is a 24-pin DIP chip. It can be directly connected to the AVR and control up to eight 7-segment LEDs. A resistor or a potentiometer is the only external component that you need. Next, we will discuss the pins of the MAX7221.

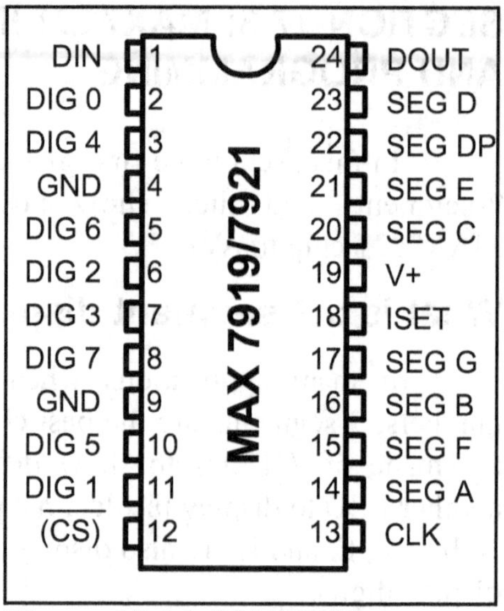

GND

Pin 4 and pin 9 are the ground. Notice that both of the ground pins should be connected to system ground and you cannot leave any of them unconnected.

```
            _____
DIN   [ 1   \_/    24 ]  DOUT
DIG 0 [ 2          23 ]  SEG D
DIG 4 [ 3          22 ]  SEG DP
GND   [ 4          21 ]  SEG E
DIG 6 [ 5          20 ]  SEG C
DIG 2 [ 6          19 ]  V+
DIG 3 [ 7          18 ]  ISET
DIG 7 [ 8          17 ]  SEG G
GND   [ 9          16 ]  SEG B
DIG 5 [ 10         15 ]  SEG F
DIG 1 [ 11         14 ]  SEG A
(CS)  [ 12         13 ]  CLK
```
(MAX 7919/7921)

Figure 17-14. MAX7221
From www.maxim-ic.com

VCC

Pin 19 is the VCC and should be connected to the +5 V power supply. Notice that this pin is also the power to drive the 7-segments and the connecting wire to this pin should be able to handle 100–300 mA.

ISET

Pin 18 is ISET and sets the maximum segment current. This pin should be connected to VCC through a resistor. A 10 kilohm resistor can be connected to this pin. If you want to manually control the intensity of the segments' light, you can replace the resistor with a 50K potentiometer. For more details about how to calculate the value of the resistor you can look at the datasheet of the chip.

CS

Pin 12 is the chip select pin and should be connected to the SS pin of the AVR. Serial data is loaded into the chip while CS is low, and the last 16 bits of the serial data are latched on CS's rising edge.

DIN

Pin 1 is the serial data input and should be connected to the MOSI pin of the AVR. On CLK's rising edge, data on this pin is loaded into the internal shift register. Notice that the MAX7221 uses the SPI Mode 0, that is, read on rising edge and change on falling edge as shown in Table 17-1.

CLK

Pin 13 is the serial clock input and should be connected to the SCK pin of the AVR. On MAX7221 the clock input is inactive when CS is high.

DOUT

Pin 24 is the serial data output and is used to connect more than one MAX7221 to a single SPI bus.

DIG0–DIG7

The DIG pins are the 7-segment selector pins and should be connected to the 7-segments' common cathode pin. The MAX7221 chip can control up to eight 7-segment LEDs. These eight 7-segment dispalys are designated as DIG0 to DIG7.

SEGA–SEGG and DP

These pins select each segment and should be connected to segments of each 7-segment accordingly. Figure 17-15 shows the connection for two 7-segments. You can connect up to eight 7-segments to MAX7221.

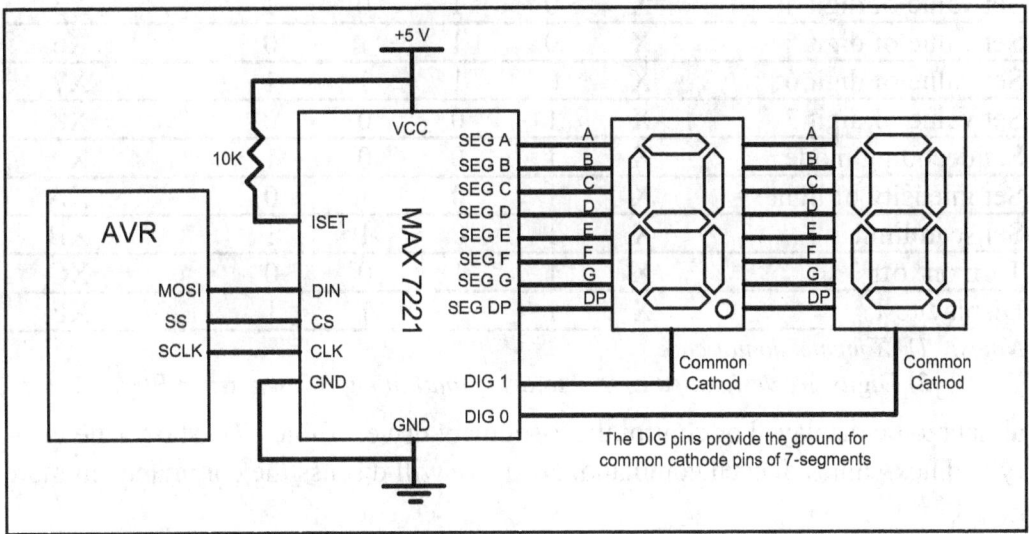

Figure 17-15. MAX7221 Connections to the AVR

MAX7221 data packet format

In MAX7221, data packets are 16 bits long (two bytes). You should first make CS low before transmitting; then you transmit two bytes of data and terminate the transmission by making CS high.

The first byte (MSBs) of each packet contains the command control bits, and the second byte is the data to be displayed. See Figure 17-16. The upper four bits (D15–D12) of the command byte are don't care and the lower four bits (D11–D8) are used to identify the meaning of the data byte to be followed. The second byte (D7–D0) of the two-byte packet is called the data byte and is the actu-

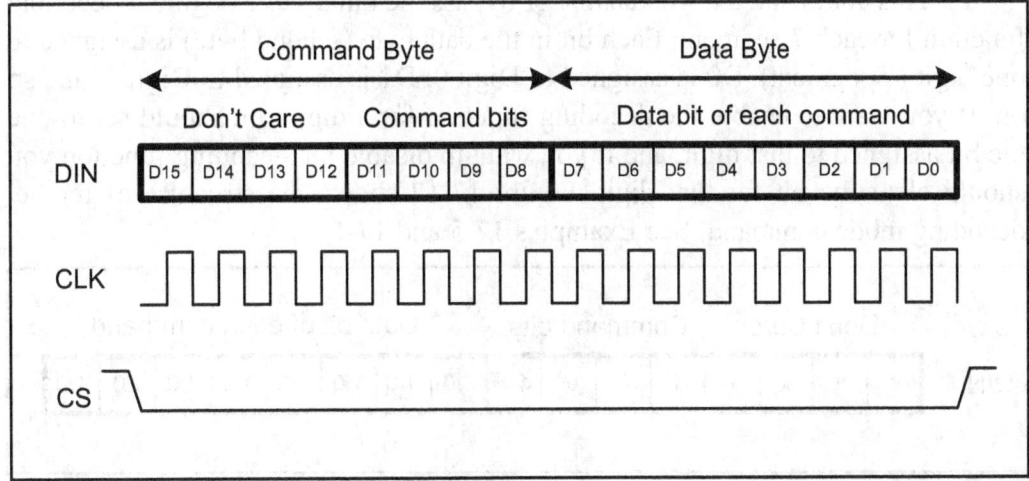

Figure 17-16. MAX7221 Packet Format

Table 17-3: List of Commands in MAX7221

Command	D15-12	D11	D10	D9	D8	Hex Code
No operation	X	0	0	0	0	X0
Set value of digit 0	X	0	0	0	1	X1
Set value of digit 1	X	0	0	1	0	X2
Set value of digit 2	X	0	0	1	1	X3
Set value of digit 3	X	0	1	0	0	X4
Set value of digit 4	X	0	1	0	1	X5
Set value of digit 5	X	0	1	1	0	X6
Set value of digit 6	X	0	1	1	1	X7
Set value of digit 7	X	1	0	0	0	X8
Set decoding mode	X	1	0	0	1	X9
Set intensity of light	X	1	0	1	0	XA
Set scan limit	X	1	0	1	1	XB
Turn on/ off	X	1	1	0	0	XC
Display test	X	1	1	1	1	XF

Notes: 1) *X means do not care.*
2) *Digits are designated as 0–7 to drive total of eight 7-segment LEDs.*

al data to be displayed or control the 7-segment driver. Table 17-3 shows the binary and hex values of each command. Next, we will discuss the commands in more detail.

Set value of digit 0–digit 7 (commands X1–X8)

These commands set what is to be displayed on each 7-segment. You can either send a binary number to the chip decoder and let it turn on/off the segments accordingly, or you may decide to turn on/off each segment of the 7-segment by yourself. The first way is useful when you do not want to deal with converting a binary number to 7-segment codes. The second way is useful when you want to show a character or any other thing that is not predefined. For example, if you want to show U, you should use the second way and turn on/off segments yourself. Next, you will see how to enable or bypass the decoder for each 7-segment.

Set decoding mode (command X9)

This command lets you enable or bypass the binary to 7-segment decoding function for each 7-segment. Each bit in the data byte (second byte) is assigned to one digit (7-segment). D0 is assigned to Digit 0, D1 is assigned to Digit 1, and so on. If you want to enable the decoding function for a digit you should set to one the bit assigned to that digit, and if you want to disable the decoding function you should clear the bit for that digit. Figure 17-17 shows the structure of the set decoding mode command. See Examples 17-5 and 17-6.

DIN		Don't Care			Command bits				Data bit of each command							
	x	x	x	x	1	0	0	1	1/0	1/0	1/0	1/0	1/0	1/0	1/0	1/0

Figure 17-17. Set Decoding Mode Command Format

Example 17-5

What sequence of bytes should be sent to the MAX7221 in order to enable the decoding function for digit 0 and digit 2, and disable the decoding function for other digits?

Solution:

The first byte should be xxxx 1001 (X9 hex) to execute the "Set decoding mode" command, and the second byte (argument of the command) should be 0000 0101 to enable the decoding function for digit 0 and digit 2.

Don't Care	Command 9	Data bits
x x x x	1 0 0 1	0 0 0 0 0 1 0 1

Example 17-6

After running Example 17-5, what sequence of numbers should be sent to the MAX7221 in order to write 5 on digit 2?

Solution:

The first byte should be xxxx 0011 (X3 hex) to execute the "Set value of digit 2" command, and the second byte (argument of the command) should be 0000 0101 (05 hex) to write 5 on digit 2. Notice that the decoding function for digit 2 has been enabled before.

Don't Care	Command 3	Data bits
x x x x	0 0 1 1	0 0 0 0 0 1 0 1

If you want to turn on/off each segment by yourself to display a specific letter on a 7-segment, you should bypass the decoding function and then use the "Set value of digit x" command to turn on/off each bit of a segment. As you see in Figure 17-18, each bit of the data bits is assigned to a segment of the 7-segment. For example, D0 is assigned to the G segment, D1 is assigned to the F segment, and so on. If you want to turn on a segment, you should write one to its bit, and if you want to turn off a segment, you should write zero to its bit. Figure 17-18 shows

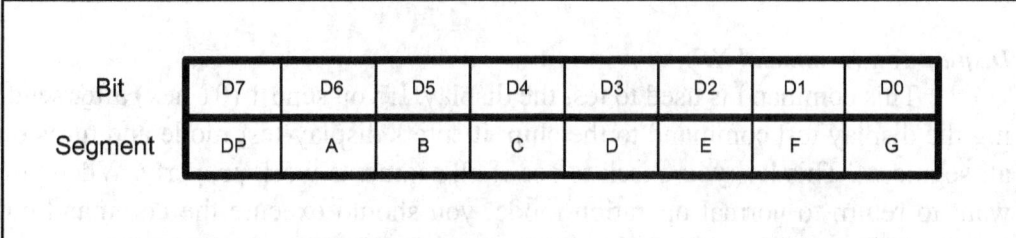

Bit	D7	D6	D5	D4	D3	D2	D1	D0
Segment	DP	A	B	C	D	E	F	G

Figure 17-18. Bits Assigned to Segments

which bits are assigned to which segments. See Example 17-7.

Example 17-7

After running Example 17-5, what sequence of numbers should be sent to the MAX7221 in order to write U on digit 1?

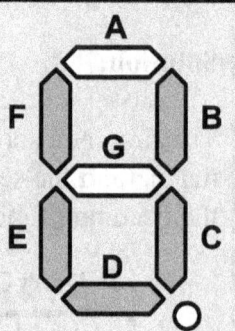

Solution:

The decoding function for digit 1 has been disabled before in Example 17-5, and we have to turn on/off each segment manually. As you see in the figure, segments B, C, D, E, and F should be turned on. To turn on these segments of digit 1, we should send the first byte xxxx 0010 (X2 hex) to execute the "Set value of digit 1" command and then we should send 0011 1110 (3E hex) to write U on digit 1. Notice that the decoding function for digit 1 has been enabled before. The figure below shows the bits.

Don't Care				Command 3				Data bits							
x	x	x	x	0	0	1	0	0	0	1	1	1	1	1	0
								DP	A	B	C	D	E	F	G

Set Intensity of Light (command XA)

This command sets the light intensity of the segments. The intensity can be any value between 0 and 16 (0F hex). 0 is the minimum value of intensity, and 16 is the maximum value of intensity. Notice that 0 does not mean off but it is the minimum intensity. As we mentioned before, you can also change the light intensity of segments by changing the resistor that connects the ISET pin to VCC.

Set Scan Limit (command XB)

This command sets the number of 7-segments that are connected to the chip. This number can vary from 1 to 8.

Turn On/ Off (command XC)

This command turns the display on or off. 1 (01 hex) turns the display on, while 0 (00 hex) turns off the display. This command is useful when you want to reduce the power consumption of your device.

Display Test (command XF)

This command is used to test the display. If you send 1 (01 hex) after sending the display test command to the chip, it enters display-test mode and turns on all segments. This lets you check to see if all segments work properly. When you want to return to normal operation mode, you should execute the command but send 0 (00 hex) as data to the chip.

MAX7221 programming in the AVR

To program MAX7221 in the AVR you should do the following steps. Notice that step 4 is optional and can be ignored:

1. Initialize the SPI to operate in master mode 0.
2. Enable or disable decoding mode by executing command 9 (x9 hex).
3. Set the scan limit.
4. Set the intensity of light (optional).
5. Turn on the display.
6. Set the value of each digit.

See Programs 17-1 and 17-2. Program 17-1 shows how to display 57 on the 7-segment display of Figure 17-15 by use of the decoding function. Program 17-2 shows how to display 2U on the 7-segment of Figure 17-15 without using the decoding function.

```
.INCLUDE "M32DEF.INC"

.equ MOSI = 5
.equ SCK = 7
.equ SS = 4

        LDI    R21,HIGH(RAMEND) ;set the high byte of stack
        OUT    SPH,R21          ;pointer
        LDI    R21,LOW(RAMEND)  ;set the low byte of stack
        OUT    SPL,R21          ;pointer

        LDI    R17,(1<<MOSI)|(1<<SCK)|(1<<SS)
        OUT    DDRB,R17         ;MOSI, SCK, and SS are output

        LDI    R17,(1<<SPE)|(1<<MSTR)|(1<<SPR0) ;enable SPI
        OUT    SPCR,R17         ;master mode 0, CLK = fck/16

        LDI    R17,0x09         ;set decoding mode command
        LDI    R18,0b00000011   ;enable decoding for digit 0,1
        CALL   RunCMD           ;send CMD and DATA to the chip

        LDI    R17,0x0B         ;set scan limit command
        LDI    R18,0x02         ;scan two 7-segments
        CALL   RunCMD           ;send CMD and DATA to the chip

        LDI    R17,0x0C         ;turn on/off command
        LDI    R18,0x01         ;turn on the chip
        CALL   RunCMD           ;send CMD and DATA to the chip

        LDI    R17,0x01         ;select digit 0
        LDI    R18,0x07         ;display value 7
```

Program 17-1: Display 57 on 7-Segment LEDs *(continued on next page)*

```
            CALL  RunCMD                ;send CMD and DATA to the chip

        LDI   R17,0x02             ;select digit 1
        LDI   R18,0x05             ;display value 5
        CALL  RunCMD               ;send CMD and DATA to the chip

H:      RJMP  H                    ;stop here
;------------------------------------------------------------
;this function sends a command and its argument (data) to SPI
;command should be in R17 and data should be in R18 before
;the function is invoked
;------------------------------------------------------------

RunCMD:
        CBI  PORTB,SS              ;CS = 0 to start packet
        OUT  SPDR,R17              ;transmit the command in R17
Wait1:
        SBIS SPSR,SPIF             ;skip next instruction if IF is set
        RJMP Wait1                 ;otherwise jump to wait1

        OUT  SPDR,R18              ;transmit the data in R18
Wait2:
        SBIS SPSR,SPIF             ;skip next instruction if IF is set
        RJMP Wait2                 ;otherwise jump to wait2

        SBI  PORTB,SS              ;CS = 1 to terminate packet
        RET                        ;return
```

Program 17-1: Display 57 on 7-Segment LEDs *(continued from previous page)*

Program 17-2 shows how to display 2U on the 7-segments of Figure 17-15 without using the decoding function.

```
.INCLUDE "M32DEF.INC"

.equ  MOSI = 5
.equ  SCK = 7
.equ  SS = 4

        LDI   R21,HIGH(RAMEND) ;set the high byte of stack
        OUT   SPH,R21          ;pointer
        LDI   R21,LOW(RAMEND)  ;set the low byte of stack
        OUT   SPL,R21          ;pointer

        LDI   R17,(1<<MOSI)|(1<<SCK)|(1<<SS)
        OUT   DDRB,R17               ;MOSI, SCK, and SS are output

        LDI   R17,(1<<SPE)|(1<<MSTR)|(1<<SPR0);enable SPI
```

Program 17-2: Display 2U on the 7-Segment LEDs *(continued on next page)*

```
        OUT    SPCR,R17            ;master mode 0, CLK = fck/16

        LDI    R17,0x09            ;set decoding mode command
        LDI    R18,0b00000010      ;enable decoding for digit 1
        CALL RunCMD                ;send CMD and DATA to the chip

        LDI    R17,0x0B            ;set scan limit command
        LDI    R18,0x02            ;scan two 7-segments
        CALL RunCMD                ;send CMD and DATA to the chip

        LDI    R17,0x0C            ;turn on/off command
        LDI    R18,0x01            ;turn on the chip
        CALL RunCMD                ;send CMD and DATA to the chip

        LDI    R17,0x01            ;select digit 0
        LDI    R18,0x3E            ;display U (see Example 17-7)
        CALL RunCMD                ;send CMD and DATA to the chip

        LDI    R17,0x02            ;select digit 1
        LDI    R18,0x02            ;display 2
        CALL RunCMD                ;send CMD and DATA to the chip

H:      RJMP H                     ;stop here


;------------------------------------------------------------
;this function sends a command and its argument (data) to SPI
;command should be in R17 and data should be in R18 before
;the function is invoked
;------------------------------------------------------------


RunCMD:
        CBI PORTB,SS               ;CS = 0 to start packet
        OUT SPDR,R17               ;transmit the command in R17
Wait1:
        SBIS SPSR,SPIF             ;skip next instruction if IF is set
        RJMP Wait1                 ;otherwise jump to wait1

        OUT SPDR,R18               ;transmit the data in R18
Wait2:
        SBIS SPSR,SPIF             ;skip next instruction if IF is set
        RJMP Wait2                 ;otherwise jump to wait2

        SBI PORTB,SS               ;CS = 1 to terminate packet
        RET                        ;return
```

Program 17-2: Display 2U on 7-Segment LEDs *(continued from previous page)*

MAX7221 programming in C

Example 17-8 and Example 17-9 are C versions of Programs 17-1 and 17-2, respectively.

Example 17-8

Write an AVR C program to display 57 on the 7-segments of Figure 17-15.

Solution:

```c
#include <avr/io.h>                        //standard AVR header
#define MOSI 5
#define SCK 7
#define SS 4

void execute( unsigned char cmd , unsigned char data)
{
  PORTB &= ~(1<<SS);                       //initializing the packet
                                           //by pulling SS low
  SPDR = cmd;                              //start CMD transmission
  while(!(SPSR & (1<<SPIF)));              //wait transfer finish

  SPDR = data;                             //start DATA transmission
  while(!(SPSR & (1<<SPIF)));              //wait transfer finish

  PORTB |= (1<<SS);                        //terminate the packet by
}                                          //pulling SS high


int main (void)
{
  DDRB = (1<<MOSI)|(1<<SCK)|(1<<SS);  //MOSI and SCK are output
  SPCR = (1<<SPE)|(1<<MSTR)|(1<<SPR0);//enable SPI as master

  execute(0x09,0b00000011);                //enable decoding for
                                           //digits 1,2
  execute(0x0B,0x02);                      //scan two 7-segments
  execute(0x0C,0x01);                      //turn on the chip
  execute(0x01,0x07);                      //display 7
  execute(0x02,0x05);                      //display 5

  while(1);
  return 0;
}
```

Example 17-9

Write an AVR C program to display 2U on the 7-segments of Figure 17-15.

Solution:

```c
#include <avr/io.h>                          //standard AVR header
#define MOSI 5
#define SCK 7
#define SS 4

void execute( unsigned char cmd , unsigned char data)
{
  PORTB &= ~(1<<SS);                         //initializing the packet
                                             //by pulling SS low

  SPDR = cmd;                                //start CMD transmission
  while(!(SPSR & (1<<SPIF)));                //wait transfer finish

  SPDR = data;                               //start DATA transmission
  while(!(SPSR & (1<<SPIF)));                //wait transfer finish

  PORTB |= (1<<SS);                          //terminate the packet by
}                                            //pulling SS high


int main (void)
{
  DDRB = (1<<MOSI)|(1<<SCK)|(1<<SS);  //MOSI and SCK are output
  SPCR = (1<<SPE)|(1<<MSTR)|(1<<SPR0);//enable SPI as master

  execute(0x09,0b00000010);          //enable decoding for digit 1
  execute(0x0B,0x02);                //scan two 7-segments
  execute(0x0C,0x01);                //turn on the chip
  execute(0x01,0x3E);                //display U (see Example 17-7)
  execute(0x02,0x02);                //display 2

  while(1);
  return 0;
}
```

Review Questions

1. How many 7-segments can be controlled by MAX7221?
2. What would happen if you do not set the scan limit?
3. True or False. If you want to show P on a 7-segment you can use the decoding function.
4. Which segments should be on to display P on a 7-segment?
5. What is the recommended value of the ISET resistor?

SUMMARY

This chapter began by describing the SPI bus connection and protocol. We also discussed the function of each pin of the MAX7221 chip. The MAX7221 can be used to drive up to eight 7-segments. Various features of the MAX7221 were explained, and numerous programming examples were given.

PROBLEMS

SECTION 17.1: SPI BUS PROTOCOL

1. True or false. The SPI bus needs an external clock.
2. True or false. The SPI CE is active-LOW.
3. True or false. The SPI bus has a single Din pin.
4. True or false. The SPI bus has multiple Dout pins.
5. True or false. When the SPI device is used as a slave, the SCLK is an input pin.
6. True or false. In SPI devices, data is transferred in 8-bit chunks.
7. True or false. In SPI devices, each bit of information (data, address) is transferred with a single clock pulse.
8. True or false. In SPI devices, the 8-bit data is followed by an 8-bit address.
9. In terms of data pins, what is the difference between the SPI and 3-wire connections?
10. How does the SPI protocol distinguish between the read and write cycles?

SECTION 17.2: SPI PROGRAMMING IN AVR

11. True or false. The ATmega family does not support SPI.
12. How many registers in the AVR are dedicated to SPI?
13. How do we set the SPI to operate in master mode 2?
14. True or false. The SPI module does not control the SS pin in master mode.
15. True or false. The SS pin should be taken low externally in order to enable the SPI module in slave mode.

SECTION 17.3: MAX7221 INTERFACING AND PROGRAMMING

16. The MAX7221 DIP package is a(n) _____-pin package.
17. Which pin is assigned as V_{cc}?
18. How much is the maximum current of the V_{cc} pin?
19. True or false. The MAX7221 has a pin for controlling the intensity of light of the segments.
20. What is the recommended resistor value for light intensity?
21. How many 7-segments can be interfaced by a single MAX7221?
22. What is the first byte in a 16-bit packet in the MAX7221?
23. What is the second byte in a 16-bit packet in the MAX7221?
24. True or false. The decoding function should be enabled to write L on a 7-segment.

ANSWERS TO REVIEW QUESTIONS

SECTION 17.1: SPI BUS PROTOCOL

1. True
2. True
3. False
4. False
5. In single-byte mode, after each byte, the CE pin must go HIGH before the next cycle. In burst mode, the CE pin stays LOW for the duration of the burst (multibyte) transfer.

SECTION 17.2: SPI PROGRAMMING IN AVR

1. SPSR (SPI Status Register), SPCR (SPI Control Register), and SPDR (SPI Data Register)
2. We set the MSTR bit in SPCR to one, clear CPOL to zero, and set CPHA to one.
3. We set SPI2X, SPR1, and SPR0 to 0, 1, and 1, respectively.
4. False
5. Fosc/4

SECTION 17.3: MAX7221 INTERFACING AND PROGRAMMING

1. 8
2. The scan limit would be 0 and nothing would be shown on the 7-segment.
3. False
4. A, B, E, F, G
5. 10 kilohms

CHAPTER 18

I2C PROTOCOL AND DS1307 RTC INTERFACING

OBJECTIVES

Upon completion of this chapter, you will be able to:

>> Understand the Inter-Integrated Circuit (I2C) protocol
>> Explain how the I2C read and write operations work
>> Examine the I2C pins SCK and SCL
>> Explain the function of I2C (TWI) registers in AVR
>> Code programs in Assembly and C for I2C (TWI)
>> Explain how the real-time clock (RTC) chip works
>> Explain the function of the DS1307 RTC pins
>> Explain the function of the DS1307 RTC registers
>> Understand the interfacing of the DS1307 RTC to the AVR
>> Code programs to display time and date in Assembly and C

This chapter discusses the I2C bus and shows the interfacing of the DS1307 real-time clock (RTC), an I2C chip. In Section 18.1, we describe the I2C bus and focus on I2C terminology and protocols. In Section 18.2, we describe the registers of AVR associated with I2C. In Section 18.3, we show how to write a simple program in Assembly and C to use the I2C features of AVR. In Section 18.4, we describe the DS1307 RTC's pin functions and show its interfacing and programming with the AVR. Advanced programming of the I2C (TWI) is discussed in Section 18.5.

SECTION 18.1: I2C BUS PROTOCOL

The IIC (Inter-Integrated Circuit) is a bus interface connection incorporated into many devices such as sensors, RTC, and EEPROM. The IIC is also referred to as I2C (I^2C) or I square C in many technical literatures. In this section we examine the pins of the I2C bus and focus on I2C terminology and protocols.

I2C bus

The I2C bus was originally started by Philips, but in recent years has become a widely used standard adapted by many semiconductor chip companies. I2C is ideal for attaching low-speed peripherals to a motherboard or embedded system or anywhere that a reliable communication over a short distance is required. As we will see in this chapter, I2C provides a connection-oriented communication with acknowledge. I2C devices use only 2 pins for data transfer, instead of the 8 or more pins used in traditional buses. They are called SCL (Serial Clock), which synchronize the data transfer between two chips, and SDA (Serial Data). This reduction of communication pins reduces the package size and power consumption drastically, making them ideal for many applications in which space is a major concern. These two pins, SDA and SCK, make the I2C a 2-wire interface. In many application notes, including AVR datasheets, I2C is referred to as *Two-Wire Serial Interface (TWI)*. In this chapter we use I2C and TWI interchangeably.

I2C line electrical characteristics

I2C devices use only 2 bidirectional open-drain pins for data communication. To implement I2C, only a 4.7 kilohm pull-up resistor for each of bus lines is needed (see Figure 18-1). This implements a wired-AND, which is needed to implement I2C protocols. This means that if one or more devices pull the line to

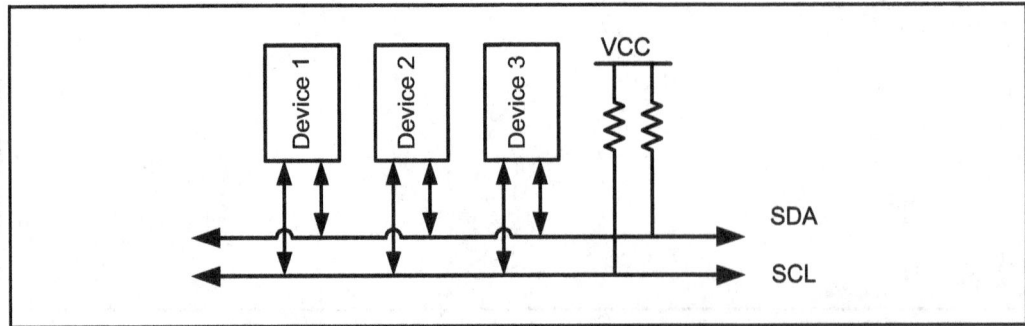

Figure 18-1. I2C Bus

low (zero) level, the line state is zero and the level of line will be 1 only if none of devices pull the line to low level.

I2C nodes

In the AVR up to 120 different devices can share an I2C bus. Each of these devices is called a *node*. In I2C terminology, each node can operate as either master or slave. Master is a device that generates the clock for the system; it also initiates and terminates a transmission. Slave is the node that receives the clock and is addressed by the master. In I2C, both master and slave can receive or transmit data, so there are four modes of operation. They are master transmitter, master receiver, slave transmitter, and slave receiver. Notice that each node can have more than one mode of operation at different times, but it has only one mode of operation at a given time. See Example 18-1.

Example 18-1

Give an example to show how a device (node) can have more than one mode of operation.
Solution:

If you connect the AVR to an EEPROM with I2C, the AVR does a master transmit operation to write to EEPROM. The AVR also does master receive operations to read from EEPROM. In the following sections, you will see that a node can do the operations of master and slave at different times.

Bit format

I2C is a synchronous serial protocol; each data bit transferred on the SDA line is synchronized by a high-to-low pulse of clock on the SCL line. According to I2C protocols the data line cannot change when the clock line is high; it can change only when the clock line is low. See Figure 18-2. The STOP and START conditions are the only exceptions to this rule.

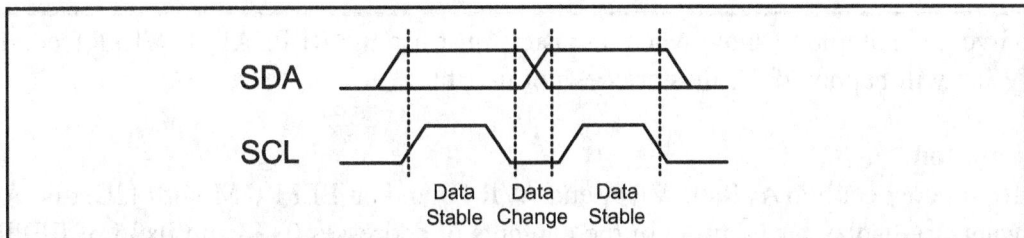

Figure 18-2. I2C Bit Format

START and STOP conditions

As we mentioned before, I2C is a connection-oriented communication protocol. This means that each transmission is initiated by a START condition and is terminated by a STOP condition. Remember that the START and STOP conditions are generated by the master.

STOP and START conditions must be distinguished from bits of address or data. That is why they do not obey the bit format rule that we mentioned before.

START and STOP conditions are generated by keeping the level of the SCL line high and then changing the level of the SDA line. The START condition is generated by a high-to-low change in the SDA line when SCL is high. The STOP condition is generated by a low-to-high change in the SDA line when SCL is low. See Figure 18-3.

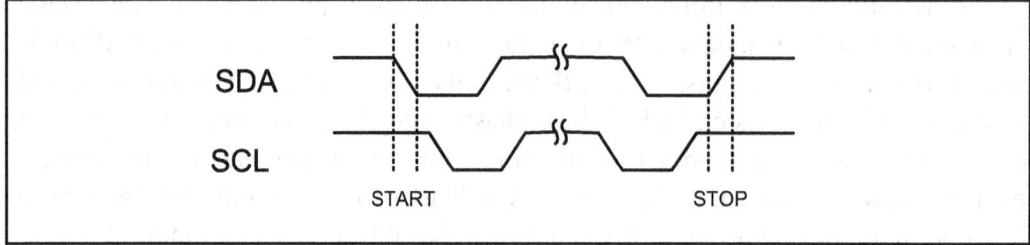

Figure 18-3. START and STOP Conditions

The bus is considered busy between each pair of START and STOP conditions, and no other master tries to take control of the bus when it is busy. If a master, which has the control of the bus, wishes to initiate a new transfer and does not want to release the bus before starting the new transfer, it issues a new START condition between a pair of START and STOP conditions. It is called the REPEATED START condition. See Figure 18-4.

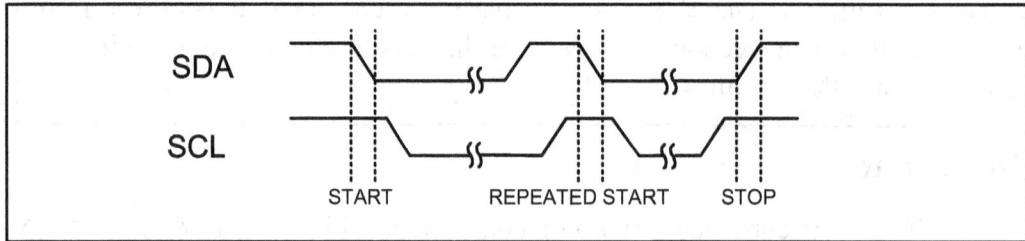

Figure 18-4. REPEATED START Condition

Example 18-2 shows why the REPEATED START condition is necessary.

Example 18-2

Give an example to show when a master must use the REPEATED START condition. What will happen if the master does not use it?

Solution:

If you connect two AVRs (AVR A and AVR B) and an EEPROM with I2C, and AVR A wants to display the addition of the contents of addresses 0x34 and 0x35 of EEPROM, it has to use the REPEATED START condition. Let's see what may happen if AVR A does not use the REPEATED START condition. AVR A transmits a START condition, reads the content of address 0x34 of EEPROM into R16, and transmits a STOP condition to release the bus. Before AVR A reads the contents of address 0x35 into R17, AVR B seizes the bus and changes the contents of addresses 0x34 and 0x35 of EEPROM. Then AVR A reads the content of address 0x35 into R17, adds it to R16, and displays the result on the LCD. The result on the LCD is neither the sum of the old values of addresses 0x34 and 0x35 nor the sum of the new values of addresses 0x34 and 0x35 of EEPROM!

Packet format in I2C

In I2C, each address or data to be transmitted must be framed in a packet. Each packet is 9 bits long. The first 8 bits are put on the SDA line by the transmitter, and the 9th bit is an acknowledge by the receiver or it may be NACK (not acknowledge). The clock is generated by the master, regardless of whether it is the transmitter or receiver. To get an acknowledge, the transmitter releases the SDA line during the ninth clock so that the receiver can pull the SDA line low to indicate an ACK. If the receiver doesn't pull the SDA line low, it is considered as NACK. See Figure 18-5.

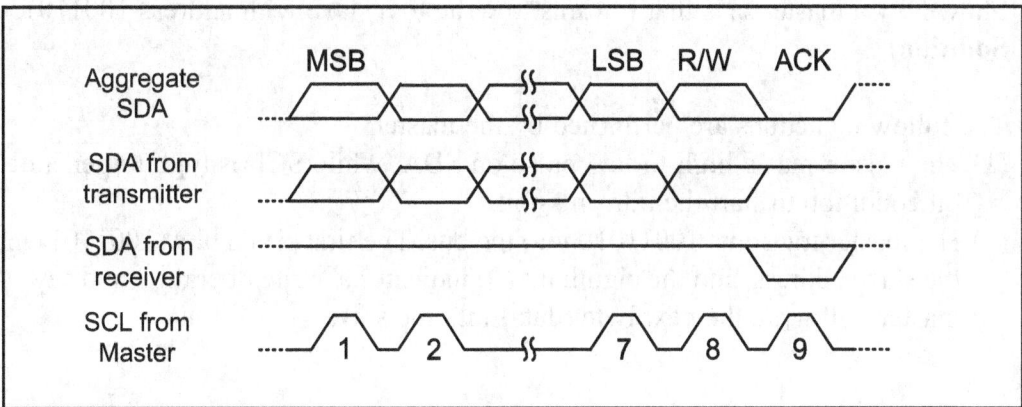

Figure 18-5. Packet Format in I2C

In I2C, each packet may contain either address or data. Also notice that START condition + address packet + one or more data packet + STOP condition together form a complete data transfer. Next we will study address and data packet formats and how to combine them to make a complete transmission.

Address packet format

Like any other packets, all address packets transmitted on the I2C bus are nine bits long. An address packet consists of seven address bits, one READ/WRITE control bit, and an acknowledge bit (see Figure 18-6).

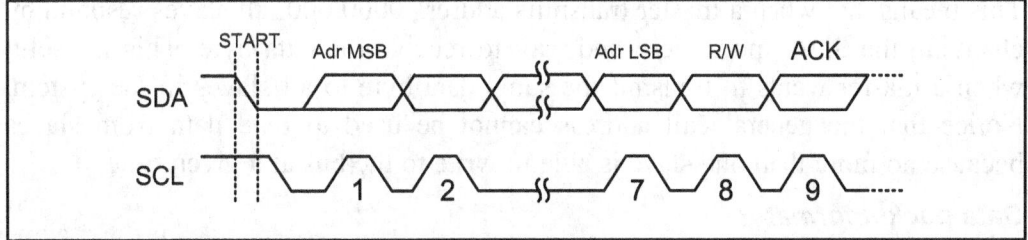

Figure 18-6. Address Packet Format in I2C

Address bits are used to address a specific slave device on the bus. The 7-bit address lets the master address a maximum of 128 slaves on the bus, although the address 0000 000 is reserved for general call and all addresses of the format 1111 xxx are reserved. That means 119 (128 − 1 − 8) devices can share an I2C bus. In the I2C bus the MSB of the address is transmitted first.

The eighth bit in the packet is the READ/WRITE control bit. If this bit is set, the master will read the next frame (Data) from the slave, otherwise, the mas-

---

CHAPTER 18: I2C PROTOCOL AND DS1307 RTC INTERFACING   633

ter will write the next frame (Data) on the bus to the slave. When a slave detects its address on the bus, it knows that it is being addressed and it should acknowledge in the ninth SCL (ACK) cycle by changing SDA to zero. If the addressed slave is not ready or for any reason does not want to service the master, it should leave the SDA line high in the ninth clock cycle. This is considered to be NACK. In case of NACK, the master can transmit a STOP condition to terminate the transmission, or a REPEATED START condition to initiate a new transmission.

Example 18-3 shows how a master says that it wants to write to a slave.

Example 18-3

Show how a master says that it wants to write to a slave with address 1001101.
Solution:

The following actions are performed by the master:
(1) The master puts a high-to-low pulse on SDA, while SCL is high to generate a start bit condition to start the transmission.
(2) The master transmits 10011010 into the bus. The first seven bits (1001101) indicates the slave address, and the eighth bit (0) indicates a Write operation and says that the master will write the next byte (data) into the slave.

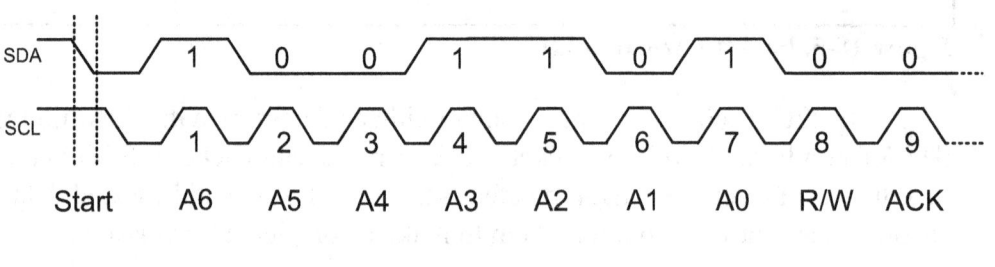

An address packet consisting of a slave address and a READ is called SLA+R, while an address packet consisting of a slave address and a WRITE is called SLA+W.

As we mentioned before, address 0000 000 is reserved for general call. This means that when a master transmits address 0000 000, all slaves respond by changing the SDA line to zero and wait to receive the data byte. This is useful when a master wants to transmit the same data byte to all slaves in the system. Notice that the general call address cannot be used to read data from slaves because no more than one slave is able to write to the bus at a given time.

Data packet format

Like other packets, data packets are 9 bits long too. The first 8 bits are a byte of data to be transmitted, and the 9th bit is ACK. If the receiver has received the last byte of data and there is no more data to be received, or the receiver cannot receive or process more data, it will signal a NACK by leaving the SDA line high. In data packets, like address packets, MSB is transmitted first.

Combining address and data packets into a transmission

In I2C, normally, a transmission is started by a START condition, followed by an address packet (SLA + R/W), one or more data packets, and finished by a

STOP condition. Figure 18-7 shows a typical data transmission. Try to understand each element in the figure (see Example 18-4).

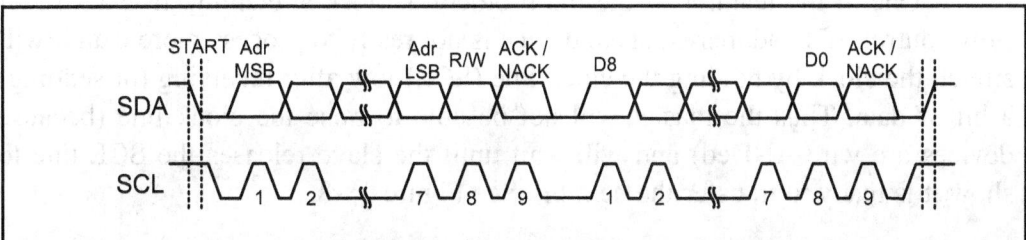

Figure 18-7. Typical Data Transmission

Example 18-4

Show how a master writes the value 11110000 to a slave with address 1001101.
Solution:

The following actions are performed by the master:
(1) The master puts a high-to-low pulse on SDA while, SCL is high to generate a START condition to start the transmission.
(2) The master transmits 10011010 into the bus. The first seven bits (1001101) indicate the slave address, and the eighth bit (0) indicates the Write operation stating that the master will write the next byte (data) into the slave.
(3) The slave pulls the SDA line low to signal an ACK to say that it is ready to receive the data byte.
(4) After receiving the ACK, the master will transmit the data byte (1111000) on the SDA line (MSB first).
(5) When the slave device receives the data it leaves the SDA line high to signal NACK. This informs the master that the slave received the last data byte and does not need any more data.
(6) After receiving the NACK, the master will know that no more data should be transmitted. The master changes the SDA line when the SCL line is high to transmit a STOP condition and then releases the bus.

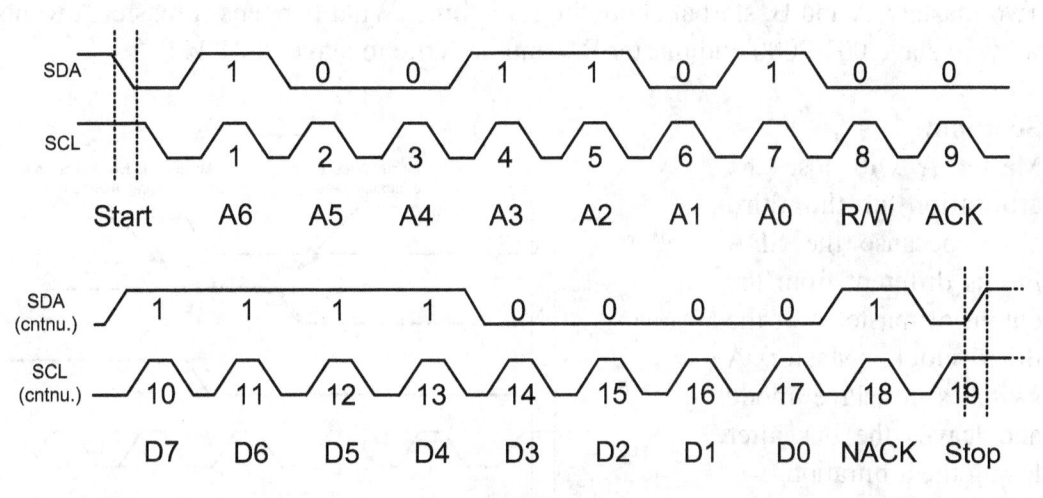

Clock stretching

One of the features of the I²C protocol is clock stretching. It is a kind of flow control. If an addressed slave device is not ready to process more data it will stretch the clock by holding the clock line (SCL) low after receiving (or sending) a bit of data. Thus the master will not be able to raise the clock line (because devices are wire-ANDed) and will wait until the slave releases the SCL line to show it is ready to transfer the next bit. See Figure 18-8.

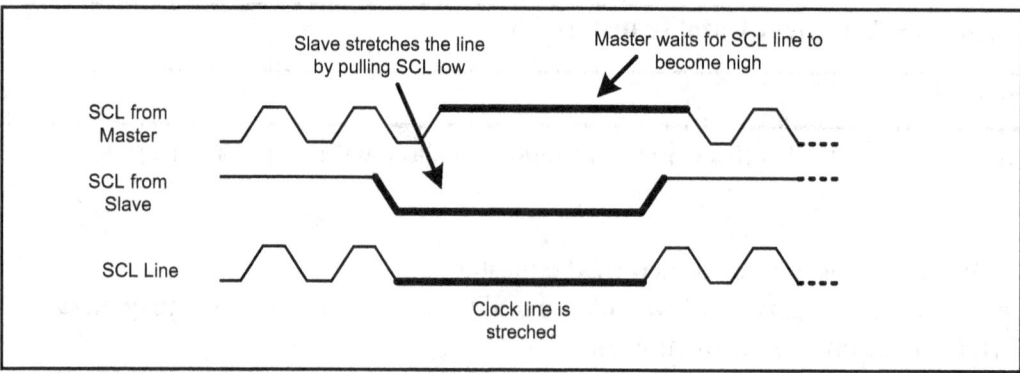

Figure 18-8. Clock Stretching

Arbitration

I2C protocol supports a multimaster bus system. This doesn't mean that more than one master can use the bus at the same time. Rather, each master waits for the current transmission to finish and then starts to use the bus. But it is possible that two or more masters initiate a transmission at about the same time. In this case the arbitration happens.

Each transmitter has to check the level of the bus and compare it with the level it expects; if it doesn't match, that transmitter has lost the arbitration, and will switch to slave mode. In the case of arbitration, the winning master will continue its job. Notice that neither the bus is corrupted nor the data is lost. See Example 18-5.

Example 18-5
Two masters, A and B, start at about the same time. What happens if master A wants to write to slave 0010 000 and master B wants to write to slave 0001 111?

Solution:

Master A will lose the arbitration in the third clock because the SDA line is different from the output of master A at the third clock. Master A switches to slave mode and leaves the bus after losing the arbitration.

Multibyte burst write

Burst mode writing is an effective means of loading consecutive locations. It is supported in I2C, SPI, and many other serial protocols. In burst mode, we provide the address of the first location, followed by the data for that location. From then on, consecutive bytes are written to consecutive memory locations. In this mode, the I2C device internally increments the address location as long as the STOP condition is not detected. The following steps are used to send (write) multiple bytes of data in burst mode for I2C devices.

1. Generate a START condition.
2. Transmit the slave address followed by zero (for write).
3. Transmit the address of the first location.
4. Transmit the data for the first location and from then on, simply provide consecutive bytes of data to be placed in consecutive memory locations.
5. Generate a STOP condition.

Figure 18-9 shows how to write 0x01, 0x02, and 0x03 to three consecutive locations starting from location 00001111 of slave 1111000.

Start	Slave address	Write	ACK	First location address	ACK	Data byte #1	ACK	Data byte #2	ACK	Data byte #3	ACK	Stop
S	1111000	0	A	00001111	A	00000001	A	00000010	A	00000011	A	P

Figure 18-9. Multibyte Burst Write

Multibyte burst read

Burst mode reading is an effective way of bringing out the contents of consecutive locations. In burst mode, we provide the address of the first location only. From then on, contents are brought out from consecutive memory locations. In this mode, the I2C device internally increments the address location as long as the STOP condition is not detected. The following steps are used to get (read) multiple bytes of data using burst mode for I2C devices.

1. Generate a START condition.
2. Transmit the slave address followed by zero (for address write).
3. Transmit the address of the first location.
4. Generate a START (REPEATED START) condition.
5. Transmit the slave address followed by one (for read).
6. Read the data from the first location and from then on, bring contents out from consecutive memory locations.
7. Generate a STOP condition.

Figure 18-10 shows how to read three consecutive locations starting from location 00001111 of slave number 1111000.

Start	Slave address	Write	ACK	First location address	ACK	Start	Slave address	Read	ACK	Data byte #1	ACK	Data byte #2	ACK	Data byte #3	ACK	Stop
S	1111000	0	A	00001111	A	S	1111000	1	A	xxxxxxxx	A	xxxxxxxx	A	xxxxxxxx	A	P

Figure 18-10. Multibyte Burst Read

Review Questions

1. True or false. I2C protocol is ideal for short distances.
2. How many bits are there in a frame? Which bit is for acknowledge?
3. True or false. START and STOP conditions are generated when the SDA is high.
4. What is the name of the flow control method in the I2C protocol?
5. What is the recommended value for the pull-up resistors in the I2C protocol?
6. True or false. After the arbitration of two masters, both of them must start transmission from the beginning.

SECTION 18.2: TWI (I2C) IN THE AVR

In many applications, including AVR datasheet, I2C is referred to as *Two-wire Serial Interface (TWI)*. From now on, in this book we use TWI to conform with the AVR data sheets. In this section we discuss the TWI module and registers of the AVR. Then we show how to program the AVR to address a slave device and send or receive data using TWI. The TWI module in the AVR is composed of four submodules: bit rate generation unit, bus interface unit, address match unit, and control unit. Figure 18-11 shows the TWI module. All registers drawn with a thick line are accessible through the AVR data bus.

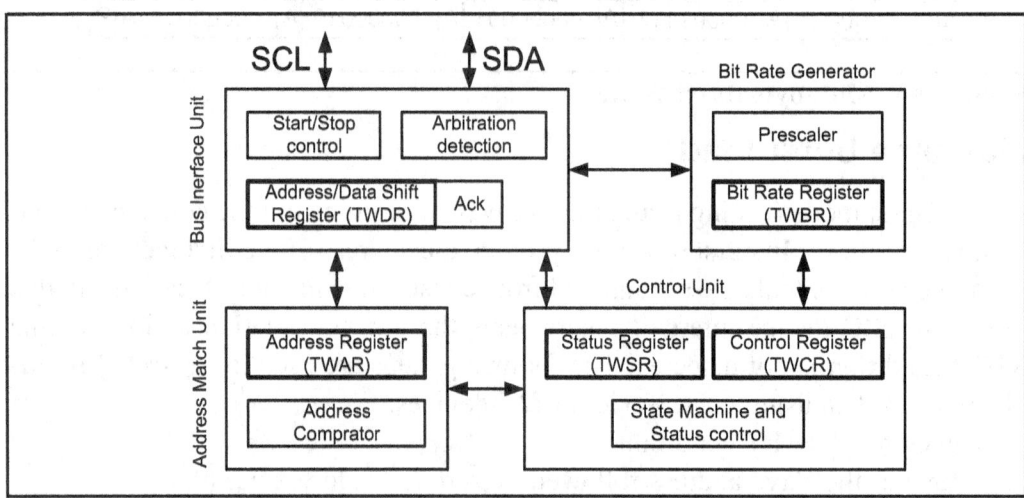

Figure 18-11. TWI (I2C) in AVR

The bit rate generation unit controls the frequency of the system clock (SCL) when operating in a master mode. The bus interface unit detects and generates START, REPEATED START and STOP conditions. It also detects arbitration, controls sending or receiving ACK, and also transfers packets of data or address. The address match unit compares the received address byte with the 7-bit address in TWI address register and informs the control unit upon an address match. The control unit controls the TWI module and generates responses according to settings in the TWI control register. It also sets the contents of the status register according to current state.

In the AVR microcontroller, five major registers are associated with the TWI. They are TWBR (TWI Bit rate Register), TWCR (TWI Control Register), TWSR (TWI Status Register), TWAR (TWI Address Register), and TWDR (TWI

Data Register). Next, we will focus on registers related to TWI and study each bit of them in detail.

TWI Bit Rate Register (TWBR)

The following figure shows the TWBR register and its bits.

TWBR7	TWBR6	TWBR5	TWBR4	TWBR3	TWBR2	TWBR1	TWBR0

TWBR selects the division factor to control the SCL clock frequency in master mode. The SCL frequency is controlled by settings in the TWBR and the prescaler bits in the TWSR (TWI status register is discussed next). The following equation demonstrates the relation between SCL frequency, TWBR, and TWPS bits in TWI status register:

$$SCL\ frequency = \frac{CPU\ Clock\ frequency}{16 + 2\ (TWBR) \times 4^{TWPS}}$$

Notice that the value of TWBR should be 10 or higher if the TWI operates in master mode. Example 18-6 shows how the frequency of SCL is calculated.

Example 18-6

Calculate the SCL frequency if the value of TWPS bits in TWSR is 01 (1 Dec) and the value of TWBR is 00100110 (38 Dec). Assume that CPU clock frequency is 8 MHz.

Solution:

The SCL frequency will be: 8 MHz / ((16 + 2 (38) × 4) = 25 kHz

---

TWI Status Register (TWSR)

As you see in Figure 18-12, five bits of TWSR are dedicated to show the status of the TWI logic and bus. Notice that if you read TWSR, you will read both the status bits and the prescaler value. To check the status bits, you should mask the two LSB bits (prescaler values) to zero. In this book we do not list all of the status codes and their meanings, but we will cover some of more common ones. To see the complete list of status register codes, you should refer to the data sheet of the chip. Next we will see how to use these bits when we want to program the AVR to use the TWI module.

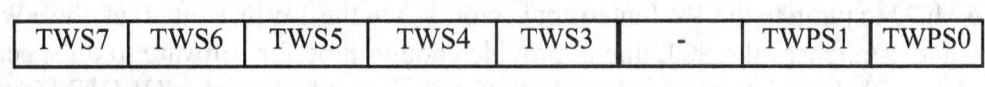

TWS7	TWS6	TWS5	TWS4	TWS3	-	TWPS1	TWPS0

Bits 7..3 – TWS: TWI Status
These five bits show the status of the TWI control and bus.

Bits 1..0 – TWPS: TWI Prescaler Bits
These bits control the bit rate prescaler.

Figure 18-12. TWSR: TWI Status Register

TWI Control Register (TWCR)

TWCR controls the operation of the TWI. In Figure 18-13 you see each bit of TWCR and a short description of it. Here we will describe some of these bits in more detail.

TWINT	TWEA	TWSTA	TWSTO	TWWC	TWEN	-	TWIE

Bit 7 – TWINT: TWI Interrupt
This bit is set by hardware when the TWI module has finished its current job. If the TWI and general interrupt are enabled, changing TWINT to one will cause the MCU to jump to the TWI interrupt vector. Clearing this flag starts the operation of the TWI. TWINT must be cleared by software.

Bit 6 – TWEA: TWI Enable Acknowledge
Making this bit HIGH will enable the generation of ACK when needed in slave or receiver mode.

Bit 5 – TWSTA: TWI START condition Bit
Making this bit HIGH will generate a START condition if the bus is free; otherwise, the TWI module waits for the bus to become free and then generates a START condition

Bit 4 – TWSTO: TWI STOP condition bit
In master mode, making this bit HIGH causes the TWI to generate a STOP condition. This bit is cleared by hardware when the STOP condition is transmitted.

Bit 3 – TWWC: TWI Write Collision Flag
This bit is set HIGH when we attempt to access the TWI Data Register when TWINT is low. This flag is cleared by writing to the TWDR register when TWINT is high.

Bit 2 – TWEN: TWI Enable
Making this bit HIGH enables the TWI module.

Bit 0 – TWIE: TWI Interrupt Enable
Making this bit HIGH enables the TWI interrupt if the general interrupt is enabled.

Figure 18-13. TWCR: TWI Control Register

TWI Interrupt (TWINT) flag

When the TWI hardware finishes its job, it sets the TWINT bit to one. If the TWI and general interrupts are enabled, changing TWINT to HIGH will cause the MCU to jump to the TWI interrupt vector. When the TWINT bit is set, the TWI module "stretches" the SCL line to provide enough time for software to do specified jobs. When the software finishes its job, it must clear the TWINT bit to resume the operation of the TWI module. Notice that all accesses to the TWI address, status, and data registers must be complete before clearing this flag. If you try to write to the TWI Data Register when TWINT is low, a collision will happen and the TWI collision flag (TWWC) will be set to HIGH by hardware. Software can monitor (poll) the TWI bit to know when the TWI module finishes its job and is ready for a new command.

TWI Enable Acknowledge (TWEA) bit

Making this bit HIGH will enable the generation of the ACK bit if any of the following conditions are met:

1. The TWI Address Match module detects that the TWI module is addressed by receiving its own slave address from the bus.
2. A general call has been received while the TWGCE bit in the TWAR is set to one (to enable accepting of global calls).
3. A data byte has been received in each of the receiving modes, master receiver or slave receiver mode.

If you clear the TWEA bit to zero, the device will not generate ACK and will be virtually disconnected from the TWI bus.

TWI Start bit and TWI Stop bit (TWSTA and TWSTO)

To generate START or STOP conditions, you have to set the TWSTA or TWSTO bit to one respectively and then clear the TWINT flag to zero by writing a one to it.

TWI Data Register (TWDR)

In Receive mode, the last received byte will be in the TWDR, and in Transmit mode, you should write the next byte into TWDR to be transmitted. As we mentioned before, you can access the TWDR only when the TWIE is set to one otherwise collision happens. This means the Data Register cannot be initialized by the user before the first interrupt occurs.

TWI Address Register (TWAR)

TWAR contains the 7-bit slave address to which the TWI will respond when working as slave. The eighth bit (LSB) of TWAR is TWGCE (TWI General Call Recognition Enable). It controls recognition of general call address (00). If this bit is set to one, receiving of a general call address will cause an interrupt request.

Review Questions

1. True or false. The AVR has an internal TWI module.
2. What are the TWI registers in AVR?
3. How do we generate START or STOP conditions in the AVR?
4. True or false. After reading status register we should mask the 2 MSB bits.
5. Which bit is polled to know if the TWI is ready now?
6. True or false. We can write to TWDR when the TWI module is busy.
7. Which bit controls the generation of ACK?

SECTION 18.3: AVR TWI PROGRAMMING IN ASSEMBLY AND C

In this section we discuss TWI programming in Assembly and C. Here we will focus on the simplest form of TWI programming without checking the status register. In most applications, if you are not dealing with critical systems and there is not more than one master on a single bus, you can use this method. If you want to deal with multimaster or critical designs you must check the value of the status flag. TWI programming with checking the value of the status flag is discussed in a later section.

In I2C protocol, a device can be either master or slave. In this section we will discuss the steps of programming in each mode.

Programming of the AVR TWI in master operating mode

To work in master operating mode, we must be able to initialize the TWI, transmit a START condition, send or receive data, and transmit a STOP condition. Next we will discuss each one in more detail.

Initialization

To initialize the TWI module to operate in master operating mode, we should do the following steps:
1. Set the TWI module clock frequency by setting the values of the TWBR register and the TWPS bits in the TWSR register.
2. Enable the TWI module by setting the TWEN bit in the TWCR register to one.

Transmit START condition

To start data transfer in master operating mode, we must transmit a START condition. This is done by setting the TWEN, TWSTA, and TWINT bits of TWCR to one. Setting the TWEN bit to one enables the TWI module. Setting the TWSTA bit to one tells the TWI to initiate a START condition when the bus is free, and setting the TWINT bit to one clears the interrupt flag to initiate operation of the TWI module to transmit the START condition. Then we should poll the TWINT flag in the TWCR register to see whether the START condition transmitted completely.

Send data

To send a byte of data, after transmitting the START condition, we should do the following steps:
1. Copy the data byte to the TWDR.
2. Set the TWEN and TWINT bits of the TWCR register to one to start sending the byte.
3. Poll the TWINT flag in the TWCR register to see whether the byte transmitted completely.

Notice that right after the START condition, we should transmit SLA + W (Slave Address + Write) or SLA + R (Slave Address + Read). As we mentioned in the first section, right after sending SLA+W we should write to the slave, and right after sending SLA+R we should read from it. To transmit SLA + R, SLA + W, and to write a byte of data to a slave we use a function called I2C_WIRITE.

Receive data

To receive a byte of data, after transmitting of SLA + R, we should do the following steps:

1. Set the TWEN and TWINT bits of the TWCR register to one to start receiving a byte. Notice that if you want to return ACK after receiving data you should also set the TWEA bit of the TWCR register to one.
2. Poll the TWINT flag in the TWCR register to see whether a byte has been received completely.
3. Copy the received byte from the TWDR to another register to save it.

Transmit STOP condition

To stop data transfer, we must transmit a STOP condition. This is done by setting the TWEN, TWSTO, and TWINT bits of the TWCR register to one. Notice that we cannot poll the TWINT flag after transmitting the STOP condition.

Program 18-1 shows how a master writes 11110000 to a slave with address 1101000.

```
;Tested OK-ok
.INCLUDE "M32DEF.INC"

        LDI    R21,HIGH(RAMEND)      ;set up stack
        OUT    SPH,R21
        LDI    R21,LOW(RAMEND)
        OUT    SPL,R21


        CALL I2C_INIT                ;initialize TWI module
        CALL I2C_START               ;transmit START condition
        LDI    R27, 0b11010000       ;SLA(1001100) + W (0)
        CALL I2C_WRITE               ;write R27 to I2C bus
        LDI    R27, 0b11110000       ;data to be transmitted
        CALL I2C_WRITE               ;write R27 to I2C bus
        CALL I2C_STOP                ;transmit STOP condition

HERE: RJMP  HERE                     ;wait here forever


;***********************************************************
I2C_INIT:
        LDI    R21, 0
        OUT    TWSR,R21              ;set prescaler bits to zero
        LDI    R21, 0x47             ;move 0x47 into R21
        OUT    TWBR,R21          ;set clock freq. to 50k (8 MHz XTAL)
        LDI    R21, (1<<TWEN)        ;move 0x04 into R21
        OUT    TWCR,R21              ;enable the TWI
        RET


;***********************************************************
I2C_START:
        LDI    R21, (1<<TWINT)|(1<<TWSTA)|(1<<TWEN)
        OUT    TWCR,R21                  ;transmit a START condition
```

Program 18-1: Writing a Byte in Master Mode *(continued on next page)*

```
WAIT1:
    IN    R21, TWCR              ;read control register into R21
    SBRS  R21, TWINT             ;skip next line if TWINT is 1
    RJMP  WAIT1                  ;jump to WAIT1 if TWINT is 1
    RET

;*********************************************************
I2C_WRITE:
    OUT   TWDR, R27              ;move the byte into TWDR
    LDI   R21, (1<<TWINT)|(1<<TWEN)
    OUT   TWCR, R21              ;configure TWCR to send TWDR
WAIT3:
    IN    R21, TWCR              ;read control register into R21
    SBRS  R21, TWINT             ;skip next line if TWINT is 1
    RJMP  WAIT3                  ;jump to WAIT3 if TWINT is 1
    RET

;*********************************************************
I2C_STOP:
    LDI   R21, (1<<TWINT)|(1<<TWSTO)|(1<<TWEN)
    OUT   TWCR, R21              ;transmit STOP condition
    RET
```

Program 18-1: Writing a Byte in Master Mode *(continued from previous page)*

Program 18-2 shows how to read a byte from a slave with address 1001100 and displays the result on Port A.

```
;Tested OK- ok
.INCLUDE "M32DEF.INC"

    LDI   R21,HIGH(RAMEND)       ;set up stack
    OUT   SPH,R21
    LDI   R21,LOW(RAMEND)
    OUT   SPL,R21

    LDI   R21,$FF                ;move $FF to R21
    OUT   DDRA, R21              ;Port A is output

    CALL  I2C_INIT               ;initialize TWI module
    CALL  I2C_START              ;transmit START condition
    LDI   R27, 0b11010001        ;SLA(1001100) + R (1)
    CALL  I2C_WRITE              ;write R27 to I2C bus
    CALL  I2C_READ               ;write R27 to I2C bus
    OUT   PORTA,R27              ;show received data on Port A
    CALL  I2C_STOP               ;transmit STOP condition

HERE:
    RJMP  HERE                   ;wait here forever
```

Program 18-2: Reading a Byte in Master Mode *(continued on next page)*

```
;*********************************************************
I2C_INIT:
     LDI   R21, 0
     OUT   TWSR,R21                ;set prescaler bits to zero
     LDI   R21, $47                ;move $47 into R21
     OUT   TWBR,R21                ;SCL freq. is 50k for 8MHz XTAL
     LDI   R21, (1<<TWEN)          ;move 0x04 into R21
     OUT   TWCR,R21                ;enable the TWI
     RET

;*********************************************************
I2C_START:
     LDI   R21, (1<<TWINT)|(1<<TWSTA)|(1<<TWEN)
     OUT   TWCR,R21                ;transmit a START condition
WAIT1:
     IN    R21, TWCR               ;read control register into R21
     SBRS  R21, TWINT              ;skip next line if TWINT is 1
     RJMP  WAIT1                   ;jump to WAIT1 if TWINT is 1
     RET

;*********************************************************
I2C_READ:
     LDI   R21,(1<<TWINT)|(1<<TWEN)
     OUT   TWCR, R21
WAIT2:
     IN    R21, TWCR               ;read control register into R21
     SBRS  R21, TWINT              ;skip next line if TWINT is 1
     RJMP  WAIT2                   ;jump to WAIT2 if TWINT is 0
     IN    R27, TWDR               ;read received data into R21
     RET

;*********************************************************
I2C_WRITE:
     OUT   TWDR, R27               ;move the byte into TWDR
     LDI   R21, (1<<TWINT)|(1<<TWEN)
     OUT   TWCR, R21               ;configure TWCR to send TWDR
WAIT3:
     IN    R21, TWCR               ;read control register into R21
     SBRS  R21, TWINT              ;skip next line if TWINT is 1
     RJMP  WAIT3                   ;jump to WAIT3 if TWINT is 1
     RET

;*********************************************************
I2C_STOP:
     LDI   R21, (1<<TWINT)|(1<<TWSTO)|(1<<TWEN)
     OUT   TWCR, R21               ;transmit STOP condition
     RET
```

Program 18-2: Reading a Byte in Master Mode *(continued from previous page)*

C programming of the AVR TWI in master operating mode

Program 18-3 shows how a master writes 11110000 to a slave with address 1101000. This program is the C version of Program 18-1.

Program 18-4 shows how a master reads from a slave with address 1101000 and displays the result on Port A. This program is the C version of Program 18-2.

```c
#include <avr/io.h>

void i2c_write(unsigned char data)
{
  TWDR = data ;
  TWCR = (1<< TWINT)|(1<<TWEN);
  while ((TWCR & (1 <<TWINT)) == 0);
}

//*******************************************************

void i2c_start(void)
{
  TWCR = (1 << TWINT) | (1 << TWSTA) | (1 << TWEN);
  while ((TWCR & (1 << TWINT)) == 0);
}

//*******************************************************

void i2c_stop()
{
  TWCR = (1<< TWINT)|(1<<TWEN)|(1<<TWSTO);
}

//*******************************************************

void i2c_init(void)
{
  TWSR=0x00;                 //set prescaler bits  to zero
  TWBR=0x47;                 //SCL frequency is 50K for XTAL = 8M
  TWCR=0x04;                 //enable the TWI module
}

//*******************************************************

int main (void)
{
  i2c_init();
  i2c_start();               //transmit START condition
  i2c_write(0b11010000);     //transmit SLA + W(0)
  i2c_write(0b11110000);     //transmit data
  i2c_stop();                //transmit STOP condition
  while(1);                  //stay here forever
  return 0 ;
}
```

Program 18-3: Writing a Byte in Master Mode in C

```c
#include <avr/io.h>

void i2c_init(void)
{
   TWSR=0x00;               //set prescaler bits to zero
   TWBR=0x47;               //SCL Frequency is 50K for XTAL=8M
   TWCR=0x04;               //enable the TWI module
}
//***********************************************************
void i2c_start(void)
{
   TWCR = (1 << TWINT) | (1 << TWSTA) | (1 << TWEN);
   while ((TWCR & (1 << TWINT)) == 0);
}
//***********************************************************
void i2c_write(unsigned char data)
{
   TWDR = data ;
   TWCR = (1<< TWINT)|(1<<TWEN);
   while ((TWCR & (1 <<TWINT)) == 0);
}
//***********************************************************
unsigned char i2c_read(unsigned char isLast)
{
   if (isLast == 0)         //if want to read more than 1 byte
      TWCR = (1<< TWINT)|(1<<TWEN)|(1<<TWEA);
   else                     //if want to read only one byte
      TWCR = (1<< TWINT)|(1<<TWEN);
   while ((TWCR & (1 <<TWINT)) == 0);
   return TWDR ;
}
//***********************************************************
void i2c_stop()
{
   TWCR = (1<< TWINT)|(1<<TWEN)|(1<<TWSTO);
}
//***********************************************************
int main (void)
{
   unsigned char i = 0 ;
   DDRA = 0xFF;             //Port A is output
   i2c_init();             //initialize TWI for master mode
   i2c_start();            //transmit START condition
   i2c_write(0b11010001);   //transmit SLA + R(1)
   i=i2c_read(1);          //read only one byte of data
   PORTA= i;               //show the byte on Port A
   i2c_stop();             //transmit STOP condition
   while(1);               //stay here forever
   return 0 ;
}
```

Program 18-4: Reading a Byte in Master Mode in C

Programming of the AVR TWI in slave operating mode

To work in slave operating mode, we must be able to initialize the TWI and we must also be able to send or receive data. In slave mode we cannot transmit START or STOP conditions. A slave device should listen to the bus and wait to be addressed by a master device or general call.

Initialization

To initialize the TWI module to operate in slave operating mode, we should do the following steps:
1. Set the slave address by setting the values for the TWAR registers. As we mentioned before, the upper seven bits of TWAR are the slave address, and the eighth bit is TWGCE. If you set this bit to one, the TWI will respond to the general call address ($00); otherwise, it will ignore the general call address.
2. Enable the TWI module by setting the TWEN bit in the TWCR register to one.
3. Set the TWEN, TWINT, and TWEA bits of TWCR to one to enable the TWI and acknowledge generation.

Notice that we cannot combine steps 2 and 3 into a single step. We have to enable the TWI module before doing the third step.

Listen to the bus

After initializing the TWI module, a slave device should listen to the bus to detect when it is addressed by a master device. When the TWI module detects its own address on the bus, it returns ACK and then sets the TWINT flag in the TWCR register to one. We should poll the TWINT flag to see when the slave is addressed by a master device.

Send data

After being addressed by a master device for read, we should do the following steps to send a byte of data:
1. Copy the data byte to the TWDR.
2. Set the TWEN, TWEA, and TWINT bits of the TWCR register to one to start sending the byte. Notice that if you expect not to receive ACK after receiving data you can leave the TWEA cleared. It will have no effect on generation of ACK by master and will only change the internal state of the TWI module. We recommend that you set the TWEA bit of the TWCR register to one anyway.
3. Poll the TWINT flag in the TWCR register to see when the byte is completely transmitted.

Receive data

After being addressed by a master device, we should do the following steps to receive a byte of data:
1. Set the TWEN and TWINT bits of the TWCR register to one to start receiving a byte. Notice that if you want to return ACK after receiving data you should also set the TWEA bit of the TWCR register to one.
2. Poll the TWINT flag in the TWCR register to see whether a byte has been received completely.
3. Copy the received byte from the TWDR to another register to save it.

Programs 18-5 and 18-6 show how to initialize the TWI module to operate in slave mode. In Program 18-5 the TWI module listens to the bus and waits to be addressed by a master device. Then it transmits the letter 'G' to the master device.

```
.INCLUDE "M32DEF.INC"

       LDI   R21,HIGH(RAMEND) ;set up stack
       OUT   SPH,R21
       LDI   R21,LOW(RAMEND)
       OUT   SPL,R21

       CALL  I2C_INIT          ;initialize the TWI module as slave
       CALL  I2C_LISTEN        ;listen to the bus to be addressed
       LDI   R21, 'G'          ;load 'G' into R21
       CALL  I2C_WRITE         ;write the byte to the bus
HERE:
       RJMP  HERE              ;wait here forever

;*********************************************************

I2C_INIT:
       LDI   R21, 0x10         ;load slave address 00010000 into R21
       OUT   TWAR,R21          ;load TWI Address Register
       LDI   R21, (1<<TWEN)    ;move 0x04 into R21
       OUT   TWCR,R21          ;enable the TWI
       LDI   R21, (1<<TWINT)|(1<<TWEN)|(1<<TWEA)
       OUT   TWCR,R21          ;enable TWI and ACK(can't be ignored)
       RET

;*********************************************************

I2C_LISTEN:
W1:
       IN    R21, TWCR         ;read control register into R21
       SBRS  R21, TWINT        ;skip next intruction if TWINT is 1
       RJMP  W1                ;jump to W1 if TWINT is 0
       RET

;*********************************************************

I2C_WRITE:

       OUT   TWDR, R21         ;move R21 to TWDR
       LDI   R21, (1<<TWINT)|(1<<TWEN)
       OUT   TWCR, R21         ;configure TWCR to send TWDR
W2:
       IN    R21, TWCR         ;read control register into R21
       SBRS  R21, TWINT        ;skip next line if TWINT is 1
       RJMP  W2                ;jump to W2 if TWINT is 0
       RET
```

Program 18-5: Writing a Byte in Slave Mode

In Program 18-6 the TWI module listens to the bus and waits to be addressed by a master device. Then it reads a byte of data from the master device and displays it on Port A.

```
.INCLUDE "M32DEF.INC"

        LDI   R21,HIGH(RAMEND);set up stack
        OUT   SPH,R21
        LDI   R21,LOW(RAMEND)
        OUT   SPL,R21

        LDI   R21, 0xFF          ;move 0xFF into R21
        OUT   DDRA,R21           ;set Port A as output

        CALL  I2C_INIT           ;initialize the TWI module as slave
        CALL  I2C_LISTEN         ;listen to the bus to be addressed
        CALL  I2C_READ           ;read a byte and copy it to R27
        OUT   PORTA,R27          ;copy R27 to PORTA
HERE:
        RJMP  HERE               ;wait here forever
;**********************************************************

I2C_INIT:
        LDI   R21, 0x10          ;load 00010000 into R21
        OUT   TWAR,R21           ;set address register
        LDI   R21, (1<<TWEN)     ;move 0x04 into R21
        OUT   TWCR,R21           ;enable the TWI
        LDI   R21, (1<<TWINT)|(1<<TWEN)|(1<<TWEA)
        OUT   TWCR,R21           ;enable TWI and ACK(can't be ignored)
        RET
;**********************************************************

I2C_LISTEN:
W1:
        IN    R21, TWCR          ;read control register into R21
        SBRS  R21, TWINT         ;skip next intruction if TWINT is 1
        RJMP  W1                 ;jump to W1 if TWINT is 0
        RET
;**********************************************************

I2C_READ:
        LDI   R21, (1<<TWINT)|(1<<TWEN)|(1<<TWEA)
        OUT   TWCR, R21          ;configure TWCR to receive TWDR
W2:     IN    R21, TWCR          ;read control register into R21
        SBRS  R21, TWINT         ;skip next line if TWINT is 1
        RJMP  W2                 ;jump to W2 if TWINT is 0
        IN    R27,TWDR           ;move received data into R27
        RET
```

Program 18-6: Reading a Byte in Slave Mode

C programming of the AVR TWI in slave operating mode

Program 18-7 is the C version of Program 18-5. Program 18-7 shows how to initialize the TWI module to operate in slave mode. In Program 18-7 the TWI module listens to the bus and waits to be addressed by a master device. Then it transmits the letter 'G' to the master device.

```c
#include <avr/io.h>                      //standard AVR header

void i2c_initSlave(unsigned char slaveAddress)
{
  TWCR = 0x04;                           //enable TWI module
  TWAR = slaveAddress;                   //set the slave address
  TWCR = (1<<TWINT)|(1<<TWEN)|(1<<TWEA);//init. TWI module
}

//*********************************************************

void i2c_send(unsigned char data)
{
  TWDR = data;                           //copy data to TWDR
  TWCR = (1<< TWINT)|(1<<TWEN);          //start transmission
  while ((TWCR & (1 <<TWINT))==0);       //wait to complete
}

//*********************************************************

void i2c_listen()
{
  while ((TWCR & (1 <<TWINT))==0);       //wait to be addressed
}

//*********************************************************

int main (void)
{
  i2c_initSlave(0x10);                   //init. TWI module as
                                         //slave with address
                                         //0b0001000 and do not
                                         //accept general call
  i2c_listen();                          //listen to be addressed
  i2c_send('G');                         //transmit letter 'G'
  while(1);                              //stay here forever
  return 0;
}
```

Program 18-7: Writing a Byte in Slave Mode in C

Program 18-8 is the C version of Program 18-6. In Program 18-8 the TWI module listens to the bus and waits to be addressed by a master device. Then it reads a byte of data from the master device and displays it on Port A.

```c
#include <avr/io.h>                     //standard AVR header

void i2c_initSlave(unsigned char slaveAddress)
{
  TWCR = 0x04;                          //enable TWI module
  TWAR = slaveAddress;                  //set the slave address
  TWCR = (1<<TWINT)|(1<<TWEN)|(1<<TWEA);//init. TWI module
}


//*********************************************************

unsigned char i2c_receive(unsigned char isLast)
{
  if (isLast == 0)        //if want to read more than 1 byte
    TWCR = (1<< TWINT)|(1<<TWEN)|(1<<TWEA);
  else                    //if want to read only one byte
    TWCR = (1<< TWINT)|(1<<TWEN);

  while ((TWCR & (1 <<TWINT))==0);    //wait to complete
  return (TWDR);
}


//*********************************************************

void i2c_listen()
{
  while ((TWCR & (1 <<TWINT))==0);    //wait to be addressed
}


//*********************************************************

int main (void)
{
  DDRA = 0xFF;
  i2c_initSlave(0x10);                //init. TWI module as
                                      //slave with address
                                      //0b0001000 and do not
                                      //accept general call
  i2c_listen();                       //listen to be addressed
  PORTA = i2c_receive(1);             //
  while(1);                           //stay here forever
  return 0;
}
```

Program 18-8: Reading a Byte in Slave Mode in C

Review Questions

1. True or false. We can ignore the status flag in multimaster systems.
2. Which of the following is not needed to initialize the TWI module to operate in master operating mode? (More than one choice can be true.)
   (a) Enable the TWI module.
   (b) Set the value of the prescaler bits.
   (c) Set the value of the TWBR register.
   (d) Set the value of the TWAR register.
3. Which of the following is not needed to initialize the TWI module to operate in slave operating mode? (More than one choice can be true.)
   (a) Enable the TWI module.
   (b) Set the value of the prescaler bits.
   (c) Set the value of the TWBR register.
   (d) Set the value of the TWAR register.
4. Which of the following instructions is used to transmit a STOP condition in master operating mode?
   (a) LDI   R21,   (1<<TWINT) | (1<<TWEN) | (1<<TWEA)
       OUT   TWCR,R21
   (b) LDI   R21,   (1<<TWINT) | (1<<TWEN)
       OUT   TWCR, R21
   (c) LDI   R21,   (1<<TWINT) | (1<<TWSTO) | (1<<TWEN)
       OUT   TWCR, R21
   (d) LDI   R21,   (1<<TWINT) | (1<<TWSTA) | (1<<TWEN)
       OUT   TWCR,R21
5. Which of the following instructions is used to transmit a STOP condition in master operating mode?
   (a) LDI   R21,   (1<<TWINT) | (1<<TWEN) | (1<<TWEA)
       OUT   TWCR,R21
   (b) LDI   R21,   (1<<TWINT) | (1<<TWEN)
       OUT   TWCR, R21
   (c) LDI   R21,   (1<<TWINT) | (1<<TWSTO) | (1<<TWEN)
       OUT   TWCR, R21
   (d) LDI   R21,   (1<<TWINT) | (1<<TWSTA) | (1<<TWEN)
       OUT   TWCR,R21

SECTION 18.4: DS1307 RTC INTERFACING AND PROGRAMMING

The real-time clock (RTC) is a widely used device that provides accurate time and date information for many applications. Many systems such as the x86 PC come with such a chip on the motherboard. The RTC chip in the x86 PC provides the time components of hour, minute, and second, in addition to the date/calendar components of year, month, and day. Many RTC chips use an internal battery, which keeps the time and date even when the power is off. Although some microcontrollers, such as the DS5000T and some of AVRs, come with the RTC already embedded into the chip, we have to interface the vast majority of them to an external RTC chip. One of the most widely used RTC chips is the DS12887 from Dallas Semiconductor/Maxim Corp. This chip is found in the vast majority of x86 PCs. The original IBM PC/AT used the MC14618B RTC from Motorola (now Freescale). The DS12887 is the replacement for that chip. It uses an internal lithium battery to keep operating for over 10 years in the absence of external power. The DS12887 is a parallel RTC with 8 pins for the data bus. The DS1307 is a serial RTC with an I2C bus. In this section, we interface and program the DS1307 RTC. According to the DS1307 data sheet from Maxim, "The clock/calendar provides seconds, minutes, hours, day, date, month, and year information. The end of the month date is automatically adjusted for months with fewer than 31 days, including corrections for leap year. The clock operates in either the 24-hour or 12-hour format with AM/PM indicator. The DS1307 has a built-in power-sense circuit that detects power failures and automatically switches to the battery supply." The DS1307 does not support the Daylight Savings Time option. Next, we describe the pins of the DS1307. See Figure 18-14.

X1–X2

These are input pins that allow the DS1307 connection to an external crystal oscillator to provide the clock source to the chip. We must use the standard 32.768 kHz quartz crystal. The accuracy of the clock depends on the quality of this crystal oscillator. Heat can cause a drift on the oscillator. To avoid this, we can use the DS32KHZ chip, which automatically adjusts for temperature variations.

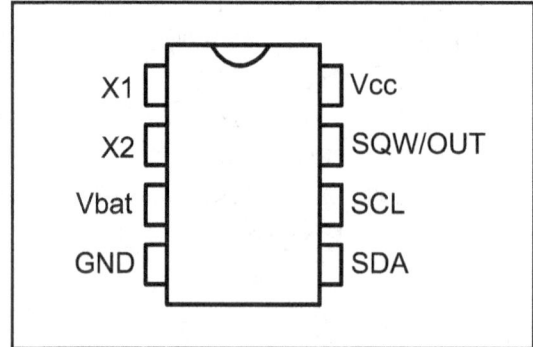

Figure 18-14. DS1307 Pin Out

Notice that when using the DS32KHZ or similar clock generators, we only need to connect X1 because the X2 loopback is not required.

V$_{bat}$

Pin 3 can be connected to an external +3 V lithium battery, thereby providing the power source to the chip when the external supply voltage is not available. We must connect this pin to ground if it is not used. A 48mAhr lithium battery can provide the power needed for more than 10 years to back up the chip.

GND

Pin 4 is the ground.

SDA (Serial Data)

Pin 5 is the SDA pin and must be connected to the SDA line of the I2C bus.

SCL (Serial Clock)

Pin 6 is the SCL pin and must be connected to the SCL line of the I2C bus.

SWQ/OUT

Pin 7 is an output pin providing 1 kHz, 4 kHz, 8 kHz, or 32 kHz frequency if enabled. This pin needs an external pull-up resistor to generate the frequency because it is open drain. If you do not want to use this pin you can omit the external pull-up resistor. We will see shortly how to control this pin.

V_{CC}

Pin 8 is used as the primary voltage supply to the chip. This primary voltage source is generally set to +5 V. When V_{cc} falls below the Vbat level, the DS1307 switches to Vbat and the external lithium battery provides power to the RTC. According to the DS1307 data sheet, "upon power-up, the device switches from Vbat to Vcc1 when Vcc1 is greater than V_{bat}+0.2 Volts." Also notice that the device is accessible only when V_{cc} is more than $1.25 \times V_{bat}$. Because we can connect the standard 3 V lithium battery to the V_{bat} pin, the Vcc voltage level must remain above 3.2 V in order for the Vcc to remain as the primary voltage source to the chip, and it must be more than 3.75 V if you want to access the chip.

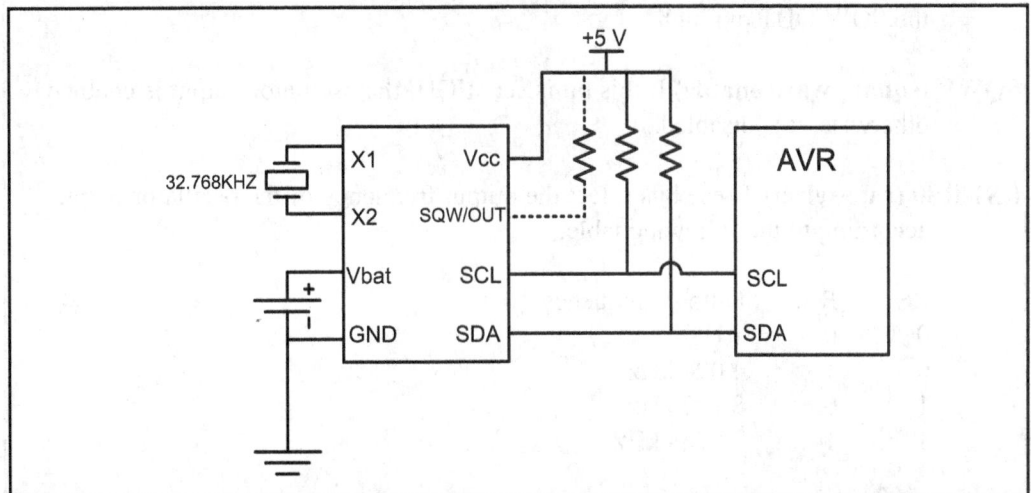

Figure 18-15. DS1307 Power Connection Options (Maxim/Dallas Semiconductor)

Address map of the DS1307

The DS1307 has a total of 64 bytes of RAM space with addresses 00–3FH. The first seven locations, 00–06, are set aside for RTC values of time and date. The next byte is used for the control register. It is located at address 07 in hex. That leaves 56 bytes, from addresses 07H to 3FH, available for general-purpose data storage. That means the entire 64 bytes of RAM are accessible directly for read or write. Figure 18-16 shows the address map of the DS1307. Next, we study the control register, and time and date access in DS1307.

ADDRESS	Bit7	Bit6	Bit5	Bit4	Bit3	Bit2	Bit1	Bit0	FUNCTION	RANGE
00H	CH	10 Seconds			Seconds				Seconds	00–59
01H	0	10 Minutes			Minutes				Minutes	00–59
02H	0	12	10 Hour	10 Hour	Hours				Hours	1–12 +AM/PM
		24	PM/AM							00–23
03H	0	0	0	0	0	DAY			Day	01–07
04H	0	0	10 Date		Date				Date	01–31
05H	0	0	0	10 Month	Month				Month	01–12
06H	10 Year				Year				Year	00–99
07H	OUT	0	0	SQWE	0	0	RS1	RS0	Control	—
08H–3FH									RAM 56 x 8	00H–FFH

Figure 18-16. Simplified Block Diagram of DS1307 (Maxim/Dallas Semiconductor)

The DS1307 control register

As shown in Figure 18-16, the control register has an address of 07H. In the DS1307 control register, the bits control the function of the SQW/OUT pin. In Figure 18-17 you see the function of each bit.

OUT	0	0	SQWE	0	0	RS1	RS0

OUT (output control) If the square wave output is disabled, setting the OUT bit to one will make the SQW/OUT pin low, and clearing the OUT bit to zero will make the SQW/OUT pin high.

SQWE (square wave enable) If this bit is set HIGH, the oscillator output is enabled; otherwise, it is disabled.

RS1-RS0 (rate select) These bits select the output frequency of the oscillator output according to the following table.

RS1	RS0	Output Frequency
0	0	1 Hz
0	1	4.096 kHz
1	0	8.192 kHz
1	1	32.768 kHz

Figure 18-17. DS1307 Control Register (Write location address is 8FH)

CH bit in address 00

One of the most important bits in the Seconds address location in the DS1307 is the CH (Clock Halt) bit. It is the seventh bit of address location 00. Setting the CH bit to one disables the oscillator, while setting CH to zero enables the oscillator. The CH bit is undefined upon reset. In order to enable the oscillator, we must clear the CH during initial configuration.

Time and date address locations and modes

The byte addresses 0–6 are set aside for the time and date, as shown in Figure 18-16. The DS1307 provides data in BCD format only. Notice the data range for the hour mode. We can select 12-hour or 24-hour mode with bit 6 of hour location 02. When D6 = 1, the 12-hour mode is selected, and D6 = 0 provides us the 24-hour mode. In the 12-hour mode, we decide the AM and PM with the bit 5. If D5 = 0, the AM is selected and D5 = 1 is for the PM. See Example 18-7.

Example 18-7

Find the values for address location $02 to set the hour to: (a) 21, (b) 11 AM, (c) 12 PM.

Solution:

(a) For 24-hour mode, we have D6 = 0. Therefore, we place 0010 0001 at location $02, which is 21 in BCD.
(b) For 12-hour mode, we have D6 = 1. Also, we have D5 = 0 for AM. Therefore, we place 0101 0001 at location $02, which is 51 in BCD.
(c) For 12-hour mode, we have D6 = 1. Also, we have D5 = 1 for PM. Therefore, we place 0111 0010 at location $02, which is 72 in BCD.

Register pointer

In DS1307 there is a register pointer that specifies the byte that will be accessed in the next read or write command. After each read or write operation, the content of the register pointer is automatically incremented. It is useful in multibyte read or write.

Writing to DS1307

To set the value of the register pointer and write one or more bytes of data to DS1307, you can use the following steps:
1. To access the DS1307 for a write operation, after sending a START condition, you should transmit the address of DS1307 (`1001101`) followed by 0 to indicate a write operation.
2. The first byte of data in the write operation will set the register pointer. For example, if you want to access the control register you should send 0x07.
3. If you want only to set the register pointer you should skip this step. If you want to write one or more bytes of data, you should transmit them one byte at a time. Remember that the register pointer is automatically incremented and you can simply transmit bytes of data to consecutive locations in a multibyte burst write.
4. Transmit a STOP bit condition.

Reading from DS1307

Notice that before reading a byte you should load the address of the byte to the register pointer by doing a write operation as mentioned before.
To read one or more bytes of data from the DS1307 you should do the fol-

---

lowing steps:

1. To access the DS1307 for a read operation, after sending a START condition, you should transmit the address of DS1307 (1001101) followed by 1 to indicate a read operation.
2. Now you can read one or more bytes of data. Remember that the register pointer indicates which address will be read. Also notice that the register pointer is automatically incremented and you can simply receive consecutive bytes of data in a multibyte burst read.
3. Transmit a STOP bit condition.

Setting the time in Assembly

Program 18-9 initializes the clock at 16:58:55 using the 24-hour clock mode. It uses the single-byte operation for writing into the control register of the DS1307 and multibyte burst mode for writing seconds, minutes, and hours. Notice that in this program we assume that there is only one master on the bus and we do not deal with checking the status register.

```
.INCLUDE  "M32DEF.INC"

    LDI   R21,HIGH(RAMEND)      ;set up stack
    OUT   SPH,R21
    LDI   R21,LOW(RAMEND)
    OUT   SPL,R21

    CALL  I2C_INIT              ;initialize the I2C module

    CALL  I2C_START             ;transmit a START condition
    LDI   R21, 0b11010000       ;SLA (1001101) + W(0)
    CALL  I2C_SEND              ;transmit R21 to I2C bus
    LDI   R21, 0x07             ;set register pointer to 07
    CALL  I2C_SEND              ;to access the control register
    LDI   R21, 0x00             ;set control register = 0
    CALL  I2C_SEND              ;transmit R21 to I2C bus
    CALL  I2C_STOP              ;transmit a STOP condition

    CALL  DELAY

    CALL  I2C_START             ;transmit a START condition
    LDI   R21, 0b11010000       ;SLA (1001101) + W(0)
    CALL  I2C_SEND              ;transmit R21 to I2C bus
    LDI   R21, 0x00             ;set register pointer to 0
    CALL  I2C_SEND              ;transmit R21 to I2C bus
    LDI   R21, 0x55             ;set seconds to 0x55 = 55 BCD
    CALL  I2C_SEND              ;transmit R21 to I2C bus
    LDI   R21, 0x58             ;set minutes to 0x58 = 58 BCD
    CALL  I2C_SEND              ;transmit R21 to I2C bus
    LDI   R21, 0b00010110       ;hour = 16 in 24 hours mode
    CALL  I2C_SEND              ;transmit R21 to I2C bus
```

Program 18-9: Setting the Time in Assembly

```
        CALL  I2C_STOP          ;transmit a STOP condition
HERE:
        RJMP  HERE              ;wait here forever


;****************************************************
I2C_INIT:
        LDI   R21, 0
        OUT   TWSR,R21          ;set prescaler bits to zero
        LDI   R21, 0x47         ;move 0x47 into r21
        OUT   TWBR,R21          ;SCL freq. is 50k for 8 MHz XTAL
        LDI   R21, (1<<TWEN)    ;move 0x04 into r21
        OUT   TWCR,R21          ;enable the TWI
        RET


;****************************************************
I2C_START:
        LDI   R21, (1<<TWINT)|(1<<TWSTA)|(1<<TWEN)
        OUT   TWCR,R21          ;transmit a START condition
W1:     IN    R21, TWCR         ;read control register into R21
        SBRS  R21, TWINT        ;mask the interrupt flag
        RJMP  W1                ;jump to W1 if TWINT is 1
        RET


;****************************************************
I2C_SEND:
        OUT   TWDR, R21         ;move SLA+W into TWDR
        LDI   R21, (1<<TWINT)|(1<<TWEN)
        OUT   TWCR, R21         ;configure TWCR to send TWDR
W2:     IN    R21, TWCR         ;read control register into R21
        SBRS  R21, TWINT        ;mask the interrupt flag
        RJMP  W2                ;jump to W2 if TWINT is 1
        RET


;****************************************************
I2C_STOP:
        LDI   R21, (1<<TWINT)|(1<<TWSTO)|(1<<TWEN)
        OUT   TWCR, R21         ;transmit STOP condition
W3:     IN    R21, TWCR         ;read control register into R21
        SBRS  R21, TWSTO        ;mask the interrupt flag
        RJMP  W3                ;jump to W3 if TWINT is 1
        RET


;****************************************************
DELAY:
        LDI   R22, 0xFF
A1:     DEC   R22               ;transmit STOP condition
        NOP
        BRNE  A1
        RET
```

Program 18-9: Setting the Time in Assembly *(continued from previous page)*

Setting the date in Assembly

Program 18-10 shows how to set the date to October 19th, 2009. It uses the single-byte operation for writing into the control register of the DS1307 and multibyte burst mode for writing day, month, and year. As you can see in the program, to access the location of the date, you should write 0x04 into the register pointer and then you can use multibyte burst write to write the values of month and year in the consecutive locations. Also, notice that in this code we assume that there is only one master on the bus and we do not deal with checking the status register.

```
.INCLUDE "M32DEF.INC"

        LDI   R21,HIGH(RAMEND)        ;set up stack
        OUT   SPH,R21
        LDI   R21,LOW(RAMEND)
        OUT   SPL,R21

        CALL  I2C_INIT                ;initialize the I2C module

        CALL  I2C_START               ;transmit a START condition
        LDI   R21, 0b11010000         ;SLA (1001101) + W(0)
        CALL  I2C_SEND                ;transmit R21 to I2C bus
        LDI   R21, 0x07               ;set register pointer to 07
        CALL  I2C_SEND                ;to access the control register
        LDI   R21, 0x00               ;set control register = 0
        CALL  I2C_SEND                ;transmit R21 to I2C bus
        CALL  I2C_STOP                ;transmit a STOP condition

        CALL  DELAY


        CALL  I2C_START               ;transmit a START condition
        LDI   R21, 0b11010000         ;SLA (1001101) + W(0)
        CALL  I2C_SEND                ;transmit R21 to I2C bus
        LDI   R21, 0x04               ;set register pointer to 4
        CALL  I2C_SEND                ;transmit R21 to I2C bus
        LDI   R21, 0x19               ;set day to 0x19 = 19 BCD
        CALL  I2C_SEND                ;transmit R21 to I2C bus
        LDI   R21, 0x10               ;set month to 0x10 = 10 BCD
        CALL  I2C_SEND                ;transmit R21 to I2C bus
        LDI   R21, 0x09               ;set year to 0x09 = 09 BCD
        CALL  I2C_SEND                ;transmit R21 to I2C bus
        CALL  I2C_STOP                ;transmit a STOP condition

HERE: RJMP  HERE                      ;wait here forever
```

Program 18-10: Setting the Date in Assembly

```
;*****************************************************
I2C_INIT:
      LDI   R21, 0
      OUT   TWSR,R21          ;set prescaler bits to zero
      LDI   R21, 0x47         ;move 0x47 into R21
      OUT   TWBR,R21          ;SCL freq. is 50k for 8 MHz XTAL
      LDI   R21, (1<<TWEN)    ;move 0x04 into R21
      OUT   TWCR,R21          ;enable the TWI
      RET


;*****************************************************
I2C_START:
      LDI   R21, (1<<TWINT)|(1<<TWSTA)|(1<<TWEN)
      OUT   TWCR,R21          ;transmit a START condition
W1:   IN    R21, TWCR         ;read control register into R21
      SBRS  R21, TWINT        ;mask the interrupt flag
      RJMP  W1                ;jump to W1 if TWINT is 1
      RET
;*****************************************************
I2C_SEND:
      OUT   TWDR, R21         ;move SLA+W into TWDR
      LDI   R21, (1<<TWINT)|(1<<TWEN)
      OUT   TWCR, R21         ;configure TWCR to send TWDR
W2:   IN    R21, TWCR         ;read control register into R21
      SBRS  R21, TWINT        ;mask the interrupt flag
      RJMP  W2                ;jump to W2 if TWINT is 1
      RET
;*****************************************************
I2C_STOP:
      LDI   R21, (1<<TWINT)|(1<<TWSTO)|(1<<TWEN)
      OUT   TWCR, R21         ;transmit STOP condition
W3:   IN    R21, TWCR         ;read control register into R21
      SBRS  R21, TWSTO        ;mask the interrupt flag
      RJMP  W3                ;jump to W3 if TWINT is 1
      RET
;*****************************************************
DELAY:
      LDI   R22, 0xFF
A1:   DEC   R22               ;transmit STOP condition
      NOP
      BRNE  A1
      RET
```

Program 18-10: Setting the Date in Assembly *(continued from previous page)*

Setting the time in C

Programs 18-11 and 18-12 are the C versions of the last two programs. Notice that you have to make the optimization level o0 (optimization 0); other-

```c
#include <avr/io.h>                //standard AVR header
void i2c_stop()
{    TWCR = (1<< TWINT)|(1<<TWEN)|(1<<TWSTO);
}
//************************************************************
void i2c_write(unsigned char data)
{

     TWDR = data ;
     TWCR = (1<< TWINT)|(1<<TWEN);
     while (!(TWCR & (1 <<TWINT)));

}
//************************************************************
void i2c_start(void)
{

     TWCR = (1 << TWINT) | (1 << TWSTA) | (1 << TWEN);
     while (!(TWCR & (1 << TWINT)));

}
//************************************************************
void i2c_init(void)
{

     TWSR=0x00;                //set prescaler bits  to zero
     TWBR=0x47;                //SCL freq. is 50k for XTAL=8M
     TWCR=0x04;                //enable TWI module

}
//************************************************************
int main (void)
{    i2c_init();              //initialize I2C module
     i2c_start();             //transmit START condition
     i2c_write(0b11010000);   //address DS1307 for write
     i2c_write(0x07);         //set register pointer to 7
     i2c_write(0x00);         //set value of location 7 to 0
     i2c_stop();              //transmit STOP condition

     for ( int k = 0 ; k<100 ; k++); //wait for a short time

     i2c_start();             //transmit START condition
     i2c_write(0b11010000);   //address DS1307 for write
     i2c_write(0);            //set register pointer to 7
     i2c_write(0x55);         //set seconds to 0x55 = 55 BCD
     i2c_write(0x58);         //set minutes to 0x58 = 58 BCD
     i2c_write(0b00010110);   //set hour=16 in 24 hours mode
     i2c_stop();              //transmit STOP condition

     while(1);                //stop here
     return 0;

}
```

Program 18-11: Setting the Time in C

wise, the compiler would omit the line "for (int k = 0 ; k<100 ; k++)"
and the program would not work correctly.

```c
#include <avr/io.h>                    //standard AVR header

void i2c_stop()
{     TWCR = (1<< TWINT)|(1<<TWEN)|(1<<TWSTO);
}
//************************************************************
void i2c_write(unsigned char data)
{     TWDR = data ;
      TWCR = (1<< TWINT)|(1<<TWEN);
      while (!(TWCR & (1 <<TWINT)));

}
//************************************************************
void i2c_start(void)
{

    TWCR = (1 << TWINT) | (1 << TWSTA) | (1 << TWEN);
    while (!(TWCR & (1 << TWINT)));

}
//************************************************************
void i2c_init(void)
{

    TWSR=0x00;                     //set prescaler bits  to zero
    TWBR=0x47;                     //SCL freq. is 50K for XTAL=8M
    TWCR=0x04;                     //enable TWI module

}
//************************************************************
int main (void)
{

    i2c_init();                    //initialize I2C module
    i2c_start();                   //transmit START condition
    i2c_write(0b11010000);         //address DS1307 for write
    i2c_write(0x07);               //set register pointer to 7
    i2c_write(0x00);               //set value of location 7 to 0
    i2c_stop();                    //transmit STOP condition

    for ( int k = 0 ; k<100 ; k++) ; //wait for a short time

    i2c_start();                   //transmit START condition
    i2c_write(0b11010000);         //address DS1307 for write
    i2c_write(0x04);               //set register pointer to 4
    i2c_write(0x19);               //set day to 0x19 = 19 BCD
    i2c_write(0x10);               //set month to 0x10 = 10 BCD
    i2c_write(0x09);               //set year to 0x09 = 09 BCD
    i2c_stop();                    //transmit STOP condition

    while(1);                      //stop here
    return 0;

}
```

Program 18-12: Setting the Date in C

Setting, reading, and displaying time and date in C

Program 18-13 is the complete C code for setting, reading, and displaying the time and date. The times and dates are sent to the IBM PC screen via the serial port after they are converted from packed BCD to ASCII.

```c
#include <avr/io.h>                    //standard AVR header

void usart_init(void)
{
     //initialize USART transmitter for 8-bit data no parity
     //and one stop bit
     UCSRB = (1<<TXEN) ;

     UCSRC = (1<< UCSZ1)|(1<<UCSZ0)|(1<<URSEL);
     UBRRL = 0x33 ;
}
//*********************************************************

void usart_send( unsigned char data )
{
     while (! (UCSRA & (1<<UDRE))); //wait until udr is empty
     UDR = data ;
}
//*********************************************************

void usart_send_packedBCD( unsigned char data )
{
     usart_send('0'+ (data>>4));
     usart_send('0'+ (data & 0x0F));
}

//*********************************************************

void i2c_init(void)
{
     TWSR=0x00;                //set prescaler bits  to zero
     TWBR=0x47;                //SCL frequency is 50K for XTAL = 8M
     TWCR=0x04;                //enable TWI module
}

//*********************************************************
void i2c_start(void)
{
    TWCR = (1<<TWINT)|(1<<TWSTA)|(1<<TWEN);
    while (!(TWCR & (1<<TWINT)));
}
//*********************************************************
```

Program 18-13: A Complete DS1307 Code Example in C

```c
void i2c_write(unsigned char data)
{
    TWDR = data ;
    TWCR = (1<< TWINT)|(1<<TWEN);
    while (!(TWCR & (1 <<TWINT)));
}


//***************************************************************

unsigned char i2c_read(unsigned char ackVal)
{
    TWCR = (1<< TWINT)|(1<<TWEN)|(ackVal<<TWEA);
    while (!(TWCR & (1 <<TWINT)));
    return TWDR  ;
}


//***************************************************************

void i2c_stop()
{
    TWCR = (1<< TWINT)|(1<<TWEN)|(1<<TWSTO);
    for ( int k = 0 ; k<100 ; k++) ; //wait for a short time
}


//***************************************************************

void rtc_init(void)
{
    i2c_init();             //initialize I2C module
    i2c_start();            //transmit START condition
    i2c_write(0xD0);        //address DS1307 for write
    i2c_write(0x07);        //set register pointer to 7
    i2c_write(0x00);        //set value of location 7 to 0
    i2c_stop();             //transmit STOP condition
}


//***************************************************************

void rtc_setTime(unsigned char h,unsigned char m,unsigned char
s)
{
    i2c_start();            //transmit START condition
    i2c_write(0xD0);        //address DS1307 for write
    i2c_write(0);           //set register pointer to 0
    i2c_write(s);           //set seconds
    i2c_write(m);           //set minutes
    i2c_write(h);           //set hour
    i2c_stop();             //transmit STOP condition
}
```

Program 18-13: A Complete DS1307 Code Example in C *(continued from previous page)*

```
//*********************************************************
void rtc_setDate(unsigned char y,unsigned char m,unsigned char
d)
{
        i2c_start();            //transmit START condition
        i2c_write(0xD0);        //address DS1307 for write
        i2c_write(0x04);        //set register pointer to 4
        i2c_write(d);           //set day
        i2c_write(m);           //set month
        i2c_write(y);           //set year
        i2c_stop();             //transmit STOP condition
}


//*********************************************************

void rtc_getTime(unsigned char *h,unsigned char *m,unsigned char
*s)
{
        i2c_start();            //transmit START condition
        i2c_write(0xD0);        //address DS1307 for write
        i2c_write(0);           //set register pointer to 0
        i2c_stop();             //transmit STOP condition

        i2c_start();            //transmit START condition
        i2c_write(0xD1);        //address DS1307 for read
        *s = i2c_read(1);       //read second, return ACK
        *m = i2c_read(1);       //read minute, return ACK
        *h = i2c_read(0);       //read hour, return NACK
        i2c_stop();             //transmit STOP condition


}


//*********************************************************

void rtc_getDate(unsigned char *y,unsigned char *m,unsigned char
*d)
{
        i2c_start();            //transmit START condition
        i2c_write(0xD0);        //address DS1307 for write
        i2c_write(0x04);        //set register pointer to 4
        i2c_stop();             //transmit STOP condition

        i2c_start();            //transmit START condition
        i2c_write(0xD1);        //address DS1307 for read
        *d = i2c_read(1);       //read day, return ACK
        *m = i2c_read(1);       //read month, return ACK
        *y = i2c_read(0);       //read year, return NACK
        i2c_stop();             //transmit STOP condition


}
```

Program 18-13: A Complete DS1307 Code Example in C *(continued from previous page)*

```
//*************************************************************
int main (void)
{
        unsigned char i,j,k;
        rtc_init();
        rtc_setTime(0x19,0x45,0x30);       //19:45:30 (hh:mm:ss)
        rtc_setDate(0x09,0x01,0x10);       //09:01:10 (yy:mm:dd)
        usart_init();
        rtc_getTime(&i,&j,&k);
        usart_send_packedBCD(i);
        usart_send_packedBCD(j);
        usart_send_packedBCD(k);
        rtc_getDate(&i,&j,&k);
        usart_send_packedBCD(i);
        usart_send_packedBCD(j);
        usart_send_packedBCD(k);


        while(1);                          //stop here
        return 0;

}
```

Program 18-13: A Complete DS1307 Code Example in C *(continued from previous page)*

Review Questions

1. True or false. All of the RAM contents of the DS1307 are nonvolatile.
2. How many bytes of RAM in the DS1307 are set aside for the clock and date?
    (a) 7 bytes
    (b) 8 bytes
    (c) 56 bytes
    (d) 64 bytes
3. How many bytes of RAM in the DS1307 are set aside for general-purpose applications?
    (a) 7 bytes
    (b) 8 bytes
    (c) 56 bytes
    (d) 64 bytes
4. True or false. The DS1307 has a single pin for data.
5. Which pin of the DS1307 is used for clock in I2C connection?
6. What is the common voltage for Vbat in the DS1307?
7. True or false. The value of the CH bit is zero at power-up time.
8. What is the address location for the control register?
    (a) 07H
    (b) 08H
    (c) 56H
    (d) 64H

SECTION 18.5: TWI PROGRAMMING WITH CHECKING STATUS REGISTER

In this section we discuss TWI programming with checking the value of status register. By checking the value of the status register you can monitor the TWI module current state and operation. This helps you to detect an error when it happens and resolve it at the same time. This is an advanced topic and used only if you are connecting I2C to multiple masters.

As we mentioned before, there are four modes of operation: master transmitter, master receiver, slave transmitter, and slave receiver. We will discuss each mode separately because each mode has its own special status codes. For each mode of operation there is a flowchart that shows the sequence of steps in each mode and also a figure that summarizes most of the status values for each mode in a single table.

Programming of the AVR TWI in master transmitter operating mode

Figure 18-18 shows the steps of programming the AVR TWI in master transmitter mode. Here we focus on each step in more detail:

Initialization

To initialize the TWI module to operate in master operating mode, we should do the following steps:
1. Set the TWI module clock frequency by setting the values of the TWBR register and the TWPS bits in the TWSR register.
2. Enable the TWI module by setting the TWEN bit in the TWCR register to one.

Transmit START condition

To start data transfer in master operating mode, we must transmit a START condition. To transmit a START condition we should do the following steps:
1. Set the TWEN, TWSTA, and TWINT bits of TWCR to one. Setting the TWEN bit to one enables the TWI module. Setting the TWSTA bit to one tells the TWI to initiate a START condition when the bus is free, and setting the TWINT bit to one clears the interrupt flag to initiate operation of the TWI module to transmit a START condition.
2. Poll the TWINT flag in the TWCR register to see when the START condition is completely transmitted.
3. When the TWINT flag is set to one, check the value of the status register to see if the START condition transmitted successfully. Notice that you have to mask the two LSB bits of the status register to get ride of prescalers. If the status value is 0x08 it indicates that the START condition has been transmitted successfully.

Send SLA + W

To send SLA + W, after transmitting the START condition, we should do the following steps:
1. Copy SLA + W to the TWDR.

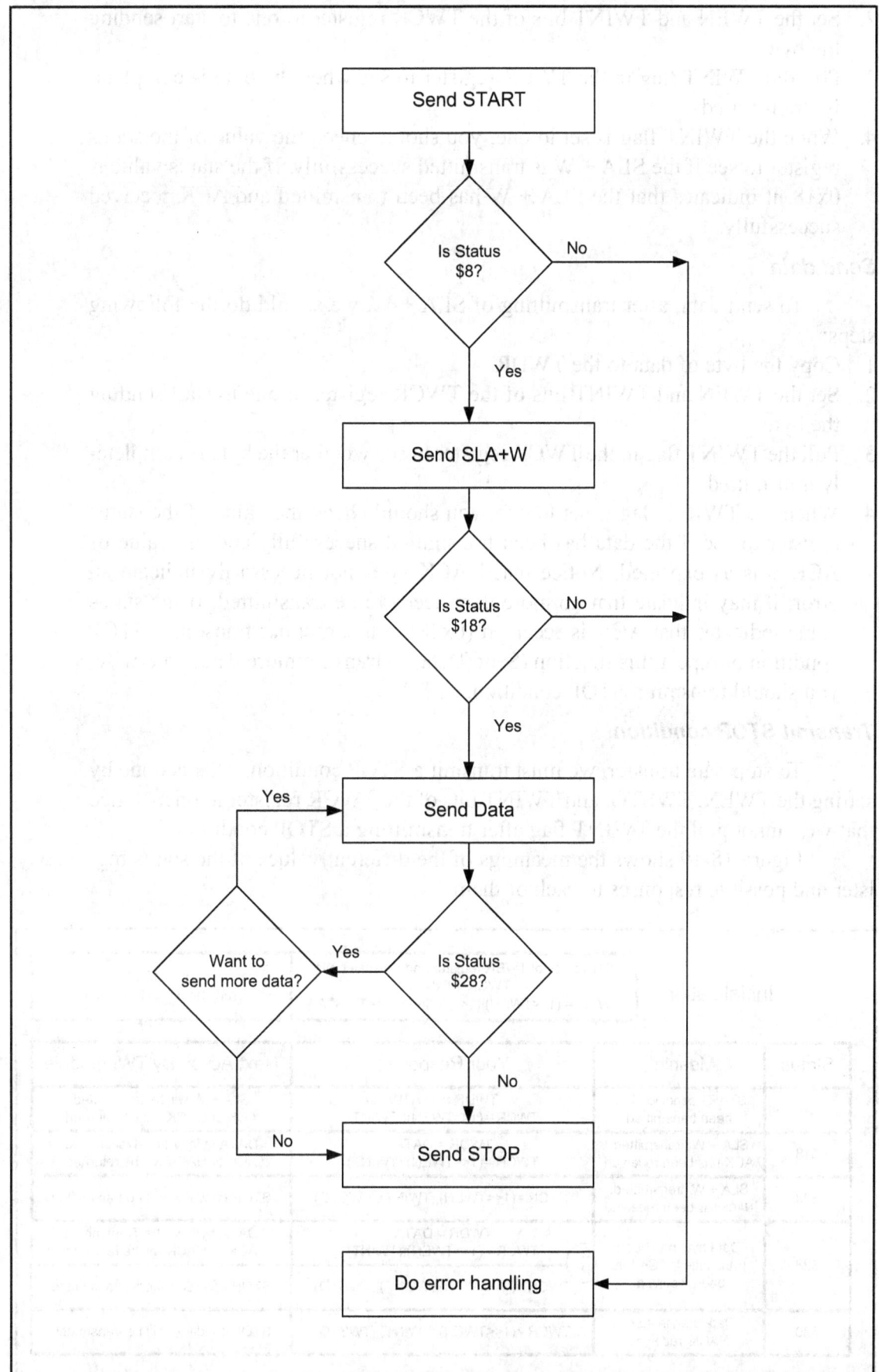

Figure 18-18. Programming Steps of Master Transmitter Mode with Checking of Flags

2. Set the TWEN and TWINT bits of the TWCR register to one to start sending the byte.
3. Poll the TWINT flag in the TWCR register to see when the byte is completely transmitted.
4. When the TWINT flag is set to one, you should check the value of the status register to see if the SLA + W is transmitted successfully. If the status value is 0x18, it indicates that the SLA + W has been transmitted and ACK received successfully.

Send data

To send data, after transmitting of SLA + W, we should do the following steps:
1. Copy the byte of data to the TWDR.
2. Set the TWEN and TWINT bits of the TWCR register to one to start sending the byte.
3. Poll the TWINT flag in the TWCR register to see whether the byte is completely transmitted.
4. When the TWINT flag is set to one, you should check the value of the status register to see if the data has been transmitted successfully and the value of ACK was as expected. Notice that NACK does not necessarily indicate an error; it may indicate that no more data needs to be transmitted. If the status value indicates that ACK is received (0x28) you can either transmit a STOP condition or repeat this function (Send Data) to transmit more data; otherwise, you should transmit a STOP condition.

Transmit STOP condition

To stop data transfer, we must transmit a STOP condition. This is done by setting the TWEN, TWSTO, and TWINT bits of the TWCR register to one. Notice that we cannot poll the TWINT flag after transmitting a STOP condition.

Figure 18-19 shows the meanings of the different values of the status register and possible responses to each of them.

Initialization:	Set values of TWBR register and prescaler bits TWCR = 0x04 TWCR = (1<<TWEN)|(1<<TWINT)|(1<<TWSTA)	Enable TWI Transmit START condition

Status	Meaning	Your Response	Next Action By TWI module
$8	START condition has been transmitted	TWDR = SLA+W TWCR =(1<<TWEN)|(TWINT)	SLA + W will be transmitted ACK or NACK will be returned
$18	SLA + W transmitted. ACK has been received	TWDR = DATA TWCR =(1<<TWEN)|(TWINT)	DATA byte will be Transmitted ACK or NACK will be returned
$20	SLA + W transmitted. NACK has been received	TWCR =(1<<TWEN)|(TWINT)|(TWSTO)	STOP condition will be transmitted
$28	Data byte has been transmitted. ACK has been received.	OR TWDR = DATA TWCR =(1<<TWEN)|(TWINT)	DATA byte will be Transmitted ACK or NACK will be returned
		TWCR =(1<<TWEN)|(TWINT)|(TWSTO)	STOP condition will be transmitted
$30	Data transmitted. NACK received	TWCR =(1<<TWEN)|(TWINT)|(TWSTO)	STOP condition will be transmitted

Figure 18-19. TWSR Register Values for Master Transmitter

Program 18-14 shows how a master writes 11110000 on a slave with address 1101000. The program checks the value of the status register in each step of the operation.

```
.INCLUDE "M32DEF.INC"


        LDI    R21,HIGH(RAMEND);set up stack
        OUT    SPH,R21
        LDI    R21,LOW(RAMEND)
        OUT    SPL,R21


        CALL   I2C_INIT          ;initialize TWI module
        CALL   I2C_START         ;transmit START condition
        CALL   I2C_READ_STATUS   ;read status register
        CPI    R26, 0x08         ;was START transmitted correctly?
        BRNE   ERROR             ;else jump to error function
        LDI    R27, 0b11010000   ;SLA (11010000) + W(0)
        CALL   I2C_WRITE         ;write R27 to I2C bus
        CALL   I2C_READ_STATUS   ;read status register
        CPI    R26, 0x18         ;was SLA+W transmitted, ACK received?
        BRNE   ERROR             ;else jump to error function
        LDI    R27, 0b11110000   ;data to be transmitted
        CALL   I2C_WRITE         ;write R27 to I2C bus
        CALL   I2C_READ_STATUS   ;read status register
        CPI    R26, 0x28         ;was data transmitted, ACK received?
        BRNE   ERROR             ;else jump to error function
        CALL   I2C_STOP          ;transmit STOP condition
HERE:   RJMP   HERE              ;wait here forever
ERROR:                          ;you can type error handler here
        LDI    R21,0xFF
        OUT    DDRA,R21          ;Port A is output
        OUT    PORTA,R26         ;send error code to Port A
        RJMP   HERE              ;some error code
;************************************************************
I2C_INIT:
        LDI    R21, 0
        OUT    TWSR,R21          ;set prescaler bits to zero
        LDI    R21, 0x47         ;move 0x47 into R21
        OUT    TWBR,R21          ;clock frequency is 50k (XTAL=50MHZ)
        LDI    R21, (1<<TWEN)    ;move 0x04 into R21
        OUT    TWCR,R21          ;enable the TWI
        RET


;************************************************************
I2C_START:
        LDI    R21, (1<<TWINT)|(1<<TWSTA)|(1<<TWEN)
        OUT    TWCR,R21          ;transmit a START condition
WAIT1:
        IN     R21, TWCR         ;read control register into R21
        SBRS   R21, TWINT        ;skip next line if TWINT is 1
```

Program 18-14: Writing a Byte in Master Mode with Status Checking

CHAPTER 18: I2C PROTOCOL AND DS1307 RTC INTERFACING

```
        RJMP  WAIT1              ;jump to WAIT1 if TWINT is 1
        RET
;************************************************************
I2C_WRITE:
        OUT   TWDR, R27          ;move the byte into TWDR
        LDI   R21, (1<<TWINT)|(1<<TWEN)
        OUT   TWCR, R21          ;configure TWCR to send TWDR
WAIT3:
        IN    R21, TWCR          ;read control register into R21
        SBRS  R21, TWINT         ;skip next line if TWINT is 1
        RJMP  WAIT3              ;jump to WAIT3 if TWINT is 1
        RET
;************************************************************
I2C_STOP:
        LDI   R21, (1<<TWINT)|(1<<TWSTO)|(1<<TWEN)
        OUT   TWCR, R21          ;transmit STOP condition
        RET
;************************************************************
I2C_READ_STATUS:
        IN    R26, TWSR          ;read status register into R21
        ANDI  R26, 0xF8          ;mask the prescaler bits
        RET
```

Program 18-14: Writing a Byte in Master Mode with Status Checking *(cont. from prev. page)*

Program 18-15 is the C version of Program 18-10 and shows how a master writes 11110000 to a slave with address 1101000. The program checks the value of the status register in each step of the operation.

```c
#include <avr/io.h>

void i2c_write(unsigned char data)
{
  TWDR = data ;
  TWCR = (1<< TWINT)|(1<<TWEN);
  while ((TWCR & (1 <<TWINT)) == 0);
}
//************************************************************
void i2c_start(void)
{
  TWCR = (1 << TWINT) | (1 << TWSTA) | (1 << TWEN);
  while ((TWCR & (1 << TWINT)) == 0);
}
//************************************************************
void i2c_showError(unsigned char er)
{
  DDRA = 0xFF;
  PORTA = er;
}
```

Program 18-15: Writing a Byte in Master Mode with Status Checking in C

```
//*************************************************
unsigned char i2c_readStatus(void)
{
  unsigned char i = 0;
  i = TWSR & 0xF8;
  return i;
}
//*************************************************
void i2c_stop()
{
  TWCR = (1<< TWINT)|(1<<TWEN)|(1<<TWSTO);
}
//*************************************************
void i2c_init(void)
{
  TWSR=0x00;                 //set prescaler bits  to zero
  TWBR=0x47;                 //SCL frequency is 50K for XTAL = 8M
  TWCR=0x04;                 //enable the TWI module
}
//*************************************************

int main (void)
{
  unsigned char s = 0;
  i2c_init();
  i2c_start();               //transmit START condition
  s = i2c_readStatus();
  if (s != 0x08)
  {
     i2c_showError(s);
     return 0;
  }
  i2c_write(0b11010000);     //transmit SLA + W(0)
  s = i2c_readStatus();
  if (s != 0x18)
  {
     i2c_showError(s);
     return 0;
  }
  i2c_write(0b11110000);     //transmit data
  s = i2c_readStatus();
  if (s != 0x28)
  {
     i2c_showError(s);
     return 0;
  }
  i2c_stop();                //transmit STOP condition
  while(1);                  //stay here forever
  return 0;
}
```

Program 18-15: Writing a Byte in Master Mode with Status Checking in C *(continued)*

Programming of the AVR TWI in master receiver operating mode

The steps to program the AVR TWI to operate in master receiver mode are somewhat similar to the steps for programming for master transmitter mode. Figure 18-20 shows the steps for programming of the AVR TWI in master receiver mode. Here we focus on each step in more detail:

Initialization

To initialize the TWI module to operate in master operating mode, we should do the following steps:
1. Set the TWI module clock frequency by setting the values of the TWBR register and the TWPS bits in the TWSR register.
2. Enable the TWI module by setting the TWEN bit in the TWCR register to one.

Transmit START condition

To start data transfer in master operating mode, we must transmit a START condition. To transmit a START condition we should do the following steps:
1. Set the TWEN, TWSTA, and TWINT bits of TWCR to one. Setting the TWEN bit to one enables the TWI module. Setting the TWSTA bit to one tells the TWI module to initiate a START condition when the bus is free, and setting the TWINT bit to one clears the interrupt flag to initiate operation of the TWI module to transmit a START condition.
2. Poll the TWINT flag in the TWCR register to see when the START condition is completely transmitted.
3. When the TWINT flag is set to one, check the value of the status register to see if the START condition was successfully transmitted. Notice that you have to mask the two LSB bits of the status register to get rid of prescalers. If the status value is 0x08 it indicates that the START condition was successfully transmitted.

Send SLA + R

To send SLA + R, after transmitting a START condition, we should do the following steps:
1. Copy SLA + R to the TWDR.
2. Set the TWEN and TWINT bits of the TWCR register to one to start sending the byte.
3. Poll the TWINT flag in the TWCR register to see whether the byte has completely transmitted.
4. When the TWINT flag is set to one, you should check the value of status register to see if the SLA + R transmitted successfully. 0x40 means that the SLA + R transmitted and ACK was successfully received.

Receive data return NACK

If we want to receive only one byte of data, we should receive data and return NACK by doing the following steps:
1. Set the TWEN and TWINT bits of the TWCR register to one to start receiving a byte.
2. Poll the TWINT flag in the TWCR register to see whether a byte was com-

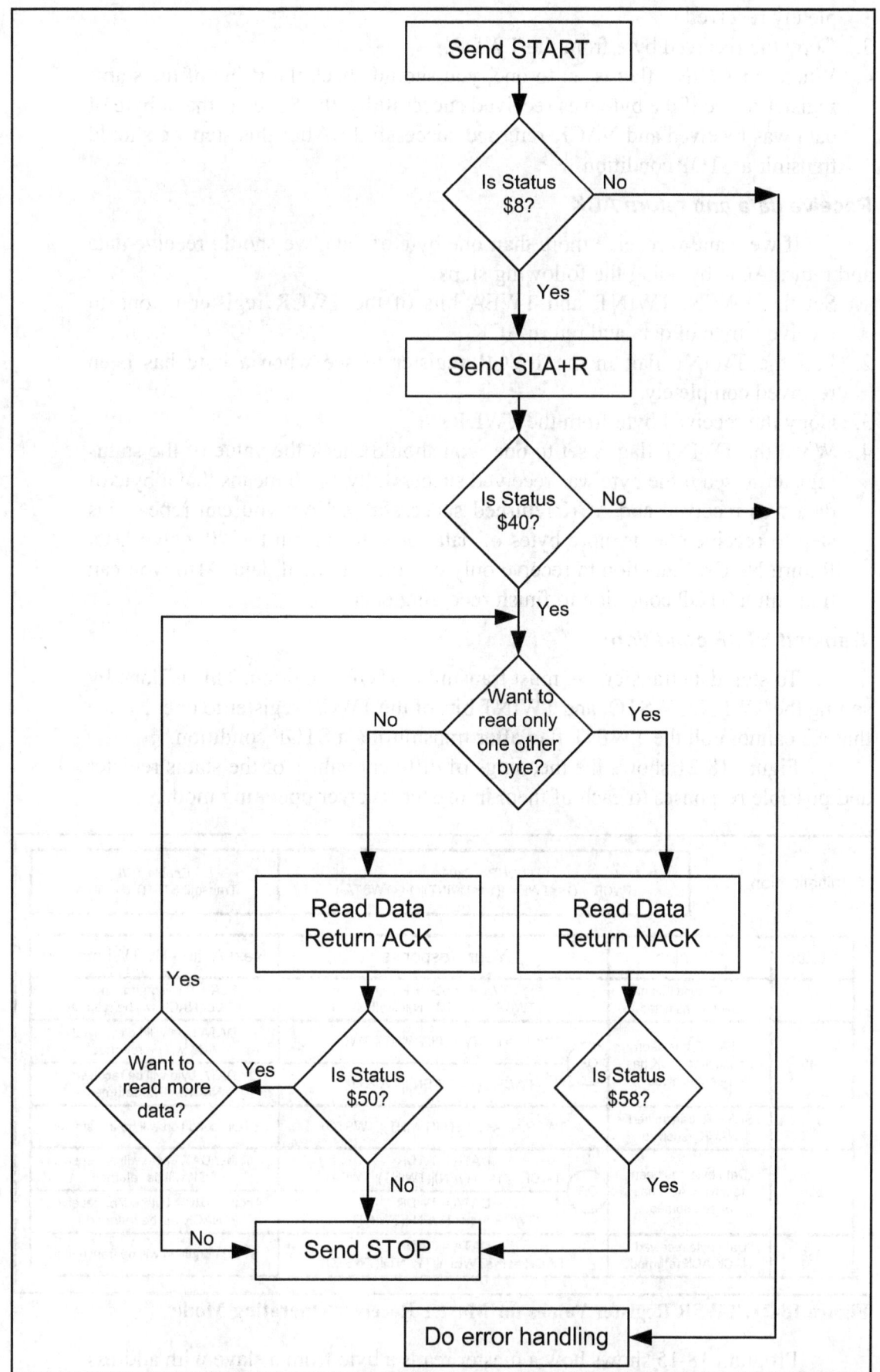

Figure 18-20. TWI Programming Steps of Master Receiver Mode with Checking of Flags

pletely received.

3. Copy the received byte from the TWDR.
4. When the TWINT flag is set to one, you should check the value of the status register to see if the byte was received successfully. 0x58 means that a byte of data was received and NACK returned successfully. After this step we should transmit a STOP condition.

Receive data and return ACK

If we want to receive more than one byte of data, we should receive data and return ACK by doing the following steps:

1. Set the TWEN, TWINT, and TWEA bits of the TWCR register to one to receive a byte of data and return ACK.
2. Poll the TWINT flag in the TWCR register to see when a byte has been received completely.
3. Copy the received byte from the TWDR.
4. When the TWINT flag is set to one, you should check the value of the status register to see if the byte was received successfully. 0x50 means that a byte of data was received and ACK returned successfully. Now you can repeat this step to receive one or more bytes of data, or you can run the "Receive Data Return NACK" function to receive only one other byte of data. Also, you can transmit a STOP condition to finish receiving data.

Transmit STOP condition

To stop data transfer, we must transmit a STOP condition. This is done by setting the TWEN, TWSTO, and TWINT bits of the TWCR register to one. Notice that we cannot poll the TWINT flag after transmitting a STOP condition.

Figure 18-21 shows the meanings of different values of the status register and possible responses to each of them in master receiver operating mode.

Initialization:	TWCR = 0x04 TWCR = (1<<TWEN)|(1<<TWINT)|(1<<TWSTA)		Enable TWI Transmit START condition.
Status	Meaning	Your Response	Next Action By TWI module
$8	START condition has been transmitted	TWDR = SLA + R (1) TWCR =(1<<TWEN)|(TWINT)	SLA + R will be transmitted ACK or NACK will be returned
$40	SLA + R has been transmitted. ACK has been received OR	TWCR =(1<<TWEN)|(TWINT)|(TWEA)	DATA byte will be received ACK will be returned
		TWCR =(1<<TWEN)|(TWINT)	DATA byte will be received NACK will be returned
$48	SLA + R transmitted. NACK received	TWCR =(1<<TWEN)|(TWINT)|(TWSTO)	STOP condition will be transmitted
$50	Data byte has been received. ACK has been returned OR	DATA = TWDR TWCR =(1<<TWEN)|(TWINT)|(TWEA)	Another DATA byte will be received ACK will be returned
		DATA = TWDR TWCR =(1<<TWEN)|(TWINT)	Another DATA byte will be received NACK will be returned
$58	Data byte received. NACK ACK returned.	DATA = TWDR TWCR =(1<<TWEN)|(TWINT)|(TWSTO)	STOP condition will be transmitted

Figure 18-21. TWSR Register Values for Master Receiver Operating Mode

Program 18-15 shows how a master reads a byte from a slave with address 1101000 and displays the result on Port A. The program checks the value of the status register in each step of the operation.

```
.INCLUDE "M32DEF.INC"
     LDI    R21,HIGH(RAMEND) ;set up stack
     OUT    SPH,R21
     LDI    R21,LOW(RAMEND)
     OUT    SPL,R21
     LDI    R21,0xFF
     OUT    DDRA,R21          ;Port A is output
     CALL   I2C_INIT          ;initialize TWI module
     CALL   I2C_START         ;transmit START condition
     CALL   I2C_READ_STATUS   ;read status register
     CPI    R26, 0x08         ;was start transmitted correctly?
     BRNE   ERROR             ;else jump to error function
     LDI    R27, 0b11010001   ;SLA (11010000) + R(1)
     CALL   I2C_WRITE         ;write R27 to I2C bus
     CALL   I2C_READ_STATUS   ;read status register
     CPI    R26, 0x40         ;was SLA+R transmitted, ACK received?
     BRNE   ERROR             ;else jump to error function
     CALL   I2C_READ
     CALL   I2C_READ_STATUS   ;read status register
     CPI    R26, 0x58         ;was data transmitted, ACK received?
     BRNE   ERROR             ;else jump to error function
     OUT    PORTA,R27
     CALL   I2C_STOP          ;transmit STOP condition
HERE: RJMP  HERE              ;wait here forever
ERROR:RJMP  HERE              ;you can type error handler here
;****************************************************
I2C_INIT:
     LDI    R21, 0
     OUT    TWSR,R21          ;set prescaler bits to zero
     LDI    R21, 0x47         ;move 0x47 into R21
     OUT    TWBR,R21          ;SCL freq. is 50k for 8 MHz XTAL
     LDI    R21, (1<<TWEN)    ;move 0x04 into R21
     OUT    TWCR,R21          ;enable the TWI
     RET
;****************************************************
I2C_START:
     LDI    R21, (1<<TWINT)|(1<<TWSTA)|(1<<TWEN)
     OUT    TWCR,R21          ;transmit a START condition
WAIT1:
     IN     R21, TWCR         ;read control register into R21
     SBRS   R21, TWINT        ;skip next line if TWINT is 1
     RJMP   WAIT1             ;jump to WAIT1 if TWINT is 1
     RET
;****************************************************
I2C_WRITE:
     OUT    TWDR, R27         ;move the byte into TWDR
     LDI    R21, (1<<TWINT)|(1<<TWEN)
     OUT    TWCR, R21         ;configure TWCR to send TWDR
```

Program 18-16: TWI Reading a Byte in Master Mode with Status Checking

```
W3:     IN    R21, TWCR               ;read control register into R21
        SBRS  R21, TWINT              ;skip next line if TWINT is 1
        RJMP  W3                      ;jump to W3 if TWINT is 1
        RET
;******************************************************************
I2C_READ:
        LDI   R21,(1<<TWINT)|(1<<TWEN)
        OUT   TWCR, R21
W2:     IN    R21, TWCR               ;read control register into R21
        SBRS  R21, TWINT              ;skip next line if TWINT is 1
        RJMP  W2                      ;jump to W2 if TWINT is 0
        IN    R27, TWDR               ;read received data into R21
        RET
;******************************************************************
I2C_STOP:
        LDI   R21, (1<<TWINT)|(1<<TWSTO)|(1<<TWEN)
        OUT   TWCR, R21               ;transmit STOP condition
        RET
;******************************************************************
I2C_READ_STATUS:
        IN    R26, TWSR               ;read status register into R21
        ANDI  R26, 0xF8               ;mask the prescaler bits
        RET
```

Program 18-16: TWI Reading a Byte in Master Mode with Status Checking *(continued)*

Program 18-17 is the C version of Program 18-16.

```
#include <avr/io.h>
void i2c_showError(unsigned char er)
{
   DDRA = 0xFF;
   PORTA = er;
} //*****************************************************************
unsigned char i2c_readStatus(void)
{
   unsigned char i = 0;
   i = TWSR & 0xF8;
   return i;
} //*****************************************************************
void i2c_init(void)
{
   TWSR=0x00;              //set prescaler bits to zero
   TWBR=0x47;              //SCL frequency is 50K for XTAL=8M
   TWCR=0x04;              //enable the TWI module
} //*****************************************************************
void i2c_start(void)
{
   TWCR = (1 << TWINT) | (1 << TWSTA) | (1 << TWEN);
   while ((TWCR & (1 << TWINT)) == 0);
}  //*****************************************************************
```

Program 18-17: TWI Reading a Byte in Master Mode with Status Checking in C

```
void i2c_write(unsigned char data)
{
  TWDR = data;
  TWCR = (1<< TWINT)|(1<<TWEN);
  while ((TWCR & (1 <<TWINT)) == 0);
} //********************************************************
unsigned char i2c_read(unsigned char isLast)
{
  if (isLast == 0)            //if want to read more than 1 byte
    TWCR = (1<< TWINT)|(1<<TWEN)|(1<<TWEA);
  else                        //if want to read only one byte
    TWCR = (1<< TWINT)|(1<<TWEN);
  while ((TWCR & (1 <<TWINT)) == 0);
  return TWDR;
} //********************************************************
void i2c_stop()
{
  TWCR = (1<< TWINT)|(1<<TWEN)|(1<<TWSTO);
} //********************************************************
int main (void)
{
  DDRA = 0xFF;              //Port A is output
  unsigned char s,i;
  i2c_init();
  i2c_start();             //transmit START condition
  s = i2c_readStatus();
  if (s != 0x08)
  {
    i2c_showError(s);
    return 0;
  }
  i2c_write(0b11010001);   //transmit SLA + R(1)
  s = i2c_readStatus();
  if (s != 0x40)
  {
    i2c_showError(s);
    return 0;
  }
  i=i2c_read(1);
  s = i2c_readStatus();
  if (s != 0x58)
  {
    i2c_showError(s);
    return 0;
  }
  PORTA= i;                //show the byte on Port A
  i2c_stop();              //transmit STOP condition
  while(1);                //stay here forever
  return 0;
}
```

Program 18-17: TWI Reading a Byte in Master Mode with Status Checking in C *(continued)*

Programming of the AVR TWI in slave transmitter operating mode

Before programming the AVR to operate in slave mode, there are some points that we must pay attention to. As we mentioned before, the slave device, regardless of whether it is receiver or transmitter, does not generate the clock pulse. To control the clock rate and let the software to complete its job, the slave device uses clock stretching. The slave device does not start or stop a transmission; it listens to the bus and replies when it is addressed by a master device.

In the slave transmitter mode, one or more bytes of data are transmitted from the slave to a master receiver. The following steps show the transmission of one or more bytes of data in slave transmitter mode.

Initialization

To initialize the TWI module to operate in slave operating mode, we should do the following steps:
1. Set the TWAR. As we mentioned before, the upper seven bits of TWAR are the slave address. It is the address to which the TWI will respond when addressed by a master. The eighth bit is TWGCE. If you set this bit to one, the TWI will respond to the general call address ($00); otherwise, it will ignore the general call address.
2. Enable the TWI module by setting the TWEN bit in the TWCR register to one.
3. Set the TWEN and TWEA bits of TWCR to one to enable the TWI and acknowledge generation.

Wait to be addressed for read

In slave mode, the TWI hardware waits until it is addressed by its own slave address (or the general call address, if enabled) followed by the R/W bit, and then sets the TWINT flag and updates the status register. If the R/W bit is zero (write), it means that the slave should operate in slave receiver mode; otherwise, the slave should operate in slave transmitter mode. Notice that you can not directly read the value of the R/W bit. Instead you should read the value of the status register. Next, we will show how to wait to be addressed by a master device.
1. Poll the TWINT flag in the TWCR register to see whether a byte has received completely.
2. When the TWINT flag is set to one, you should check the value of the status register to see if the SLA + R is received successfully. $A8 means that the SLA + R was received and ACK returned successfully.

Now if you want to transmit only one byte of data you should run the "Send Data and Wait for NACK" function. Otherwise, if you want to send more than one byte of data you should run the "Send Data and Wait for ACK" function. Next we will examine each function in detail.

Send data and wait for ACK

In slave transmitter mode, if you want to transmit more than one byte of data you should send a byte of data and wait for ACK by doing the following steps:
1. Copy the byte of data to the TWDR.
2. Set the TWEN, TWINT, and TWEA bits of the TWCR register to one to send

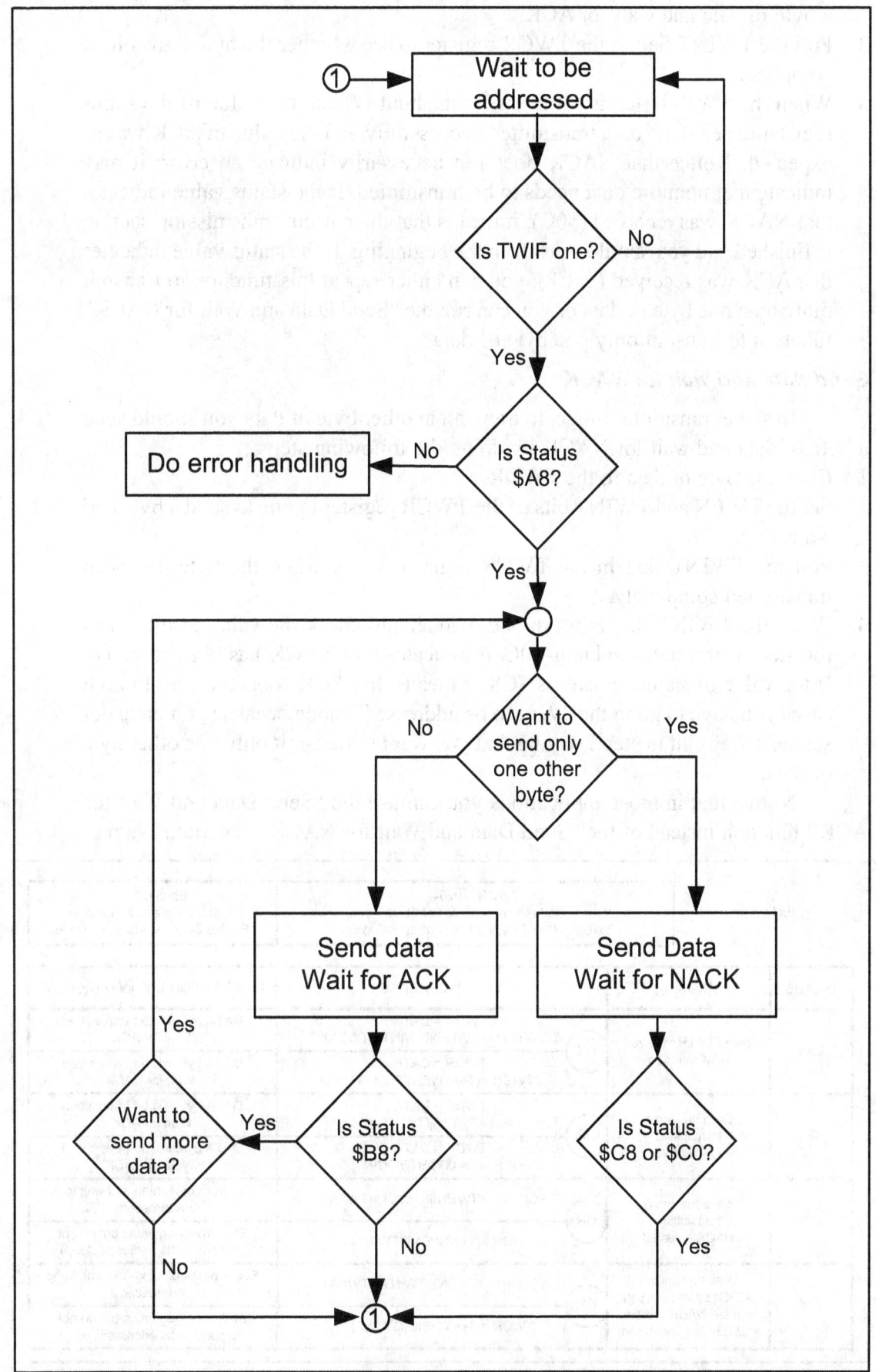

Figure 18-22. TWI Programming Steps of Slave Transmitter Mode with Checking of Flags

a byte of data and wait for ACK.

3. Poll the TWINT flag in the TWCR register to see whether the byte transmitted completely.

4. When the TWINT flag is set to one, you should check the value of the status register to see if the data transmitted successfully and the value of ACK was as expected. Notice that NACK does not necessarily indicate an error; it may indicate that no more data needs to be transmitted. If the status value indicates that NACK was received ($0C), it means that the current transmission section is finished and you should start from the beginning. If the status value indicates that ACK was received (0xC8), you can either repeat this function to transmit more than one byte of data or you can run the "Send Data and Wait for NACK" function to transmit only one byte of data.

Send data and wait for NACK

In slave transmitter mode, to transmit another byte of data you should send a byte of data and wait for NACK by doing the following steps:

1. Copy the byte of data to the TWDR.

2. Set the TWEN and TWINT bits of the TWCR register to one to send a byte and wait for NACK.

3. Poll the TWINT flag in the TWCR register to see when the byte has been transmitted completely.

4. When the TWINT flag is set to one, you should check the value of the status register. If the status value is $0C, it indicates that NACK has been received. If the value of status register is $C8, it means that ACK was received. In both cases you have to go to the "Wait to be addressed" mode because you have not set the TWEA bit in step 2 saying that you want to transmit only one other byte of data.

Notice that in most applications you can use the "Send Data and Wait for ACK" function instead of the "Send Data and Wait for NACK" function. We rec-

Initialization:		TWCR = 0x04 TWAR = the address of Slave TWCR = (1<<TWEN)|(1<<TWIF)|(1<<TWEA)		Enable TWI Set the slave address Enable Acknowledging by slave

Status	Meaning		Your Response	Next Action By TWI module
$A8	Own SLA+R received ACK returned	OR	TWDR = DATA TWCR =(1<<TWEN)|(TWINT)|(TWEA)	DATA byte will be transmitted Wait for ACK
			TWDR = DATA TWCR =(1<<TWEN)|(TWINT)	DATA byte will be transmitted Wait for NACK
$B8	Data has been transmitted ACK received	OR	TWDR = DATA TWCR =(1<<TWEN)|(TWINT)|(TWEA)	DATA byte will be transmitted Wait for ACK
			TWDR = DATA TWCR =(1<<TWEN)|(TWINT)	DATA byte will be transmitted Wait for NACK
$C0	Data has been transmitted NACK received	OR	TWCR =(1<<TWEN)|(TWINT)|(TWEA)	Start from beginning and wait to be addressed
			TWCR =(1<<TWEN)|(TWINT)	Start from beginning but do not respond to Its address (Sleep)
$C8	Data transmitted ACK received but you wanted NACK (TWEA was 0 in last command)	OR	TWCR =(1<<TWEN)|(TWINT)|(TWEA)	Start from beginning and wait to be addressed
			TWCR =(1<<TWEN)|(TWINT)	Start from beginning but do not respond to Its address (Sleep)

Figure 18-23. TWSR Register Values for Slave Transmitter Operating Mode

ommend that you use the first one.

Program 18-18 shows how to initialize the TWI module to operate in slave transmitter mode. In this program the TWI module listens to the bus and waits to be addressed by a master device. Then it transmits the letter 'G' to the master device.

```
.INCLUDE "M32DEF.INC"

        LDI    R21,HIGH(RAMEND) ;set up stack
        OUT    SPH,R21
        LDI    R21,LOW(RAMEND)
        OUT    SPL,R21

        CALL   I2C_INIT           ;initialize the TWI module as slave
        CALL   I2C_LISTEN         ;listen to the bus to be addressed
        CALL   I2C_READ_STATUS    ;read the status value into R26
        CPI    R26, 0xA8          ;addressed as slave tranmitter ?
        BRNE   ERROR              ;else jump to error function
        LDI    R27, 'G'           ;load 'G' into R21
        CALL   I2C_WRITE
        CALL   I2C_READ_STATUS    ;read the status value into R26
        CPI    R21, 0xc0          ;was data transmitted, NACK received?
        BRNE   ERROR              ;else jump to error function

HERE:
        RJMP   HERE               ;wait here forever
ERROR:                           ;you can type error handler here
        LDI    R21,0xFF
        OUT    DDRA,R21           ;Port A is output
        OUT    PORTA,R26
        RJMP   HERE
;***************************************************************

I2C_INIT:
        LDI    R21, 0x10          ;load 00010000 into R21
        OUT    TWAR,R21           ;set address register
        LDI    R21, (1<<TWEN)     ;move 0x04 into R21
        OUT    TWCR,R21           ;enable the TWI
        LDI    R21, (1<<TWINT)|(1<<TWEN)|(1<<TWEA)
        OUT    TWCR,R21           ;enable TWI and ACK(can't be ignored)
        RET
;***************************************************************

I2C_LISTEN:
W1:
        IN     R21, TWCR          ;read control register into R21
        SBRS   R21, TWINT         ;skip next intruction if TWINT is 1
        RJMP   W1                 ;jump to W1 if TWINT is 0
        RET
```

Program 18-18: Writing a Byte in Slave Mode with Status Checking

CHAPTER 18: I2C PROTOCOL AND DS1307 RTC INTERFACING   683

```
;**********************************************************
I2C_WRITE:

    OUT   TWDR, R27          ;move R21 to TWDR
    LDI   R21, (1<<TWINT)|(1<<TWEN)
    OUT   TWCR, R21          ;configure TWCR to send TWDR
W2:
    IN    R21, TWCR          ;read control register into R21
    SBRS  R21, TWINT ;skip next intruction if TWINT is 1
    RJMP  W2                 ;jump to W2 if TWINT is 0
    RET


;**********************************************************
I2C_READ_STATUS:
    IN    R26, TWSR          ;read status register into R21
    ANDI  R26, 0xF8          ;mask the prescaler bits
    RET
```

Program 18-18: Writing a Byte in Slave Mode with Status Checking *(cont. from prev. page)*

Program 18-19 is the C version of Program 18-18. Program 18-19 shows how to initialize the TWI module to operate in slave transmitter mode. In Program 18-19 the TWI module listens to the bus and waits to be addressed by a master device. Then it transmits the letter 'G' to the master device.

```c
#include <avr/io.h>                        //standard AVR header

void i2c_showError(unsigned char er)
{
  DDRA = 0xFF;
  PORTA = er;
}  //***********************************************************

unsigned char i2c_readStatus(void)
{
  unsigned char i = 0;
  i = TWSR & 0xF8;
  return i;
}  //***********************************************************

void i2c_initSlave(unsigned char slaveAddress)
{
  TWCR = 0x04;                             //enable TWI module
  TWAR = slaveAddress;                     //set the slave address
  TWCR = (1<<TWINT)|(1<<TWEN)|(1<<TWEA);//init TWI module
}
```

Program 18-19: Writing a Byte in Slave Mode with Status Checking in C

```
//**********************************************************

void i2c_send(unsigned char data)
{
  TWDR = data;                              //copy data to TWDR
  TWCR = (1<< TWINT)|(1<<TWEN);             //start transmission
  while ((TWCR & (1 <<TWINT))==0);          //wait to complete
}


//**********************************************************

void i2c_listen()
{
  while ((TWCR & (1 <<TWINT))==0);          //wait to be addressed
}


//**********************************************************

int main (void)
{
  i2c_initSlave(0x10);                      //init TWI module as
                                            //slave with address
                                            //0b0001000 and do not
                                            //accept general call
  i2c_listen();                             //listen to be addressed

  unsigned char s,i;
  s = i2c_readStatus();
  if (s != 0xA8)
  {
    i2c_showError(s);
    return 0;
  }
  i2c_send('G');
  s = i2c_readStatus();
  if (s != 0xC0)
  {
    i2c_showError(s);
    return 0;
  }

  while(1);                                 //stay here forever
  return 0;

}
```

Program 18-19: Writing a Byte in Slave Mode with Status Checking in C *(continued)*

Programming of the AVR TWI in slave receiver operating mode

In the slave receiver mode, one or more bytes of data are transmitted from a master transmitter to the slave receiver. The following steps show the functions needed to receive one or more bytes of data in slave receiver mode.

Initialization

To initialize the TWI module to operate in slave operating mode, we should do the following steps:

1. Set the TWAR. As we mentioned before, the upper seven bits of TWAR are the slave address. It is the address to which the Two-wire Serial Interface will respond when addressed by a master. The eighth bit is TWGCE. If you set this bit to one, the TWI will respond to the general call address ($00); otherwise, it will ignore the general call address.
2. Enable the TWI module by setting the TWEN bit in the TWCR register to one.
3. Set the TWEN and TWEA bits of TWCR to one to enable the TWI and acknowledge generation.

Wait to be addressed for write

In slave mode, we should do the following steps to wait to be addressed by a master for a write operation.

1. Poll the TWINT flag in the TWCR register to see when a byte has been received completely.
2. When the TWINT flag is set to one, we should check the value of the status register to see if the SLA + W was received successfully. $60 or $70 (for general call) means that the SLA + W was received and ACK returned successfully.

Now if you want to receive only one byte of data you should run the "Receive Data and Return NACK" function. Otherwise, if you want to send more than one byte of data you should run the "Receive Data and Return ACK" function. Next, we will examine each function in detail.

Receive data and Return ACK

In slave receiver mode, if you want to receive more than one byte of data you should receive a byte of data and return ACK by doing the following steps:

1. Set the TWEN, TWINT, and TWEA bits of the TWCR register to one to receive a byte and return ACK.
2. Poll the TWINT flag in the TWCR register to see when a byte has been received completely.
3. When the TWINT flag is set to one, you should check the value of the status register to see if the data was received successfully and ACK was returned. If the status value is $80 or $90 (for general call), it means that a byte of data has been received and ACK was returned. You can either repeat this function to receive more than one bytes of data or you can run the "Receive Data and Return NACK" function to receive only one byte of data.
4. Copy the received byte from the TWDR.

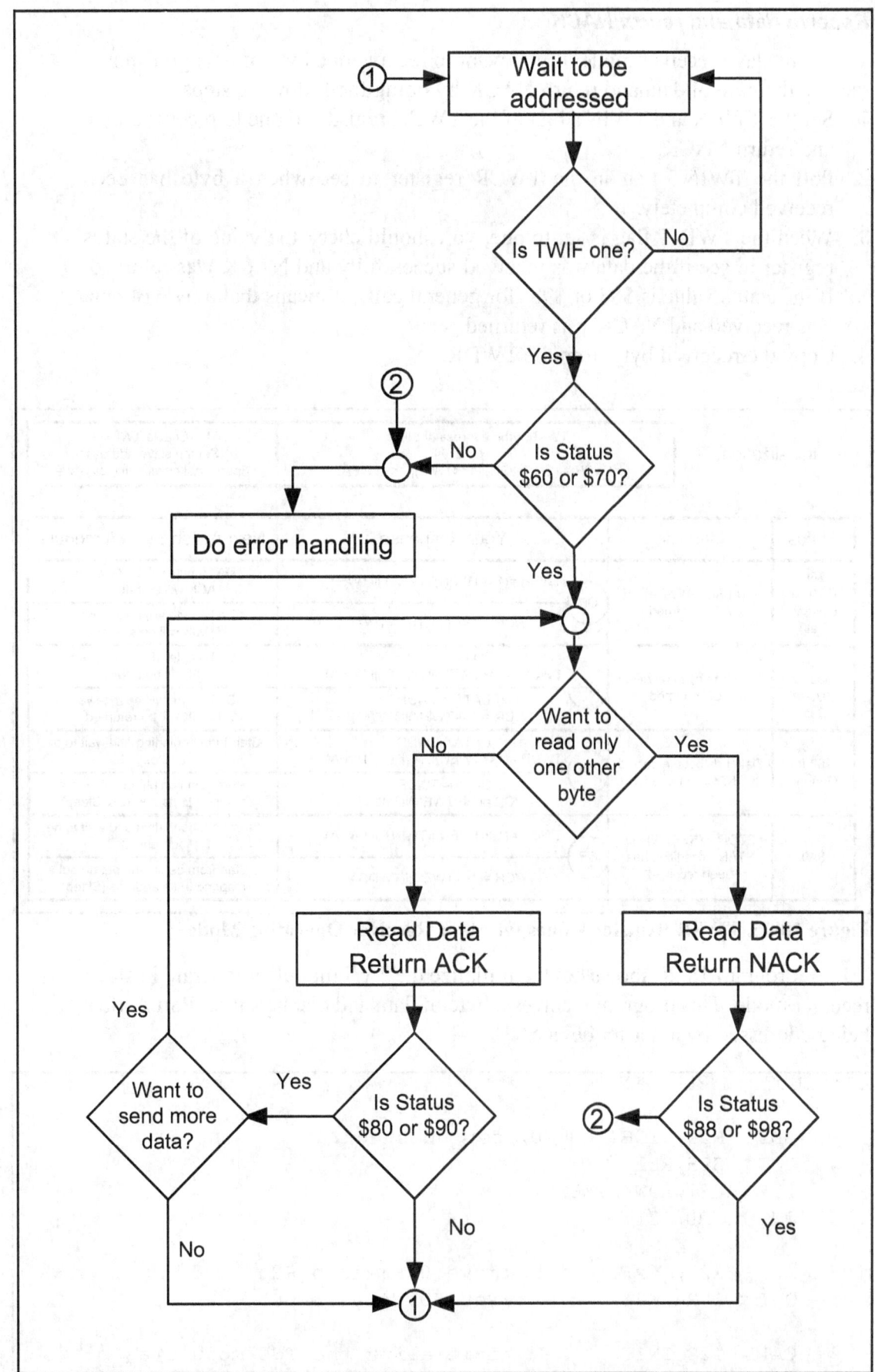

Figure 18-24. TWI Programming Steps of Slave Receiver Mode with Checking of Flags

Receive data and return NACK

In slave receiver mode, if you want to receive one byte of data you should receive the byte of data and return NACK by doing the following steps:

1. Set the TWEN and TWINT bits of the TWCR register to one to receive a byte and return NACK.
2. Poll the TWINT flag in the TWCR register to see when a byte has been received completely.
3. When the TWINT flag is set to one, you should check the value of the status register to see if the data was received successfully and NACK was returned. If the status value is $88 or $98 (for general call), it means that a byte of data was received and NACK was returned.
4. Copy the received byte from the TWDR.

Initialization:	TWAR = the address of Slave TWCR = 0x04 TWCR = (1<<TWEN)|(1<<TWIF)|(1<<TWEA)		Enable TWI Set the slave address Enable Acknowledging by slave
Status	**Meaning**	**Your Response**	**Next Action By TWI module**
$60 ($70 for General Call)	Own SLA+W received ACK returned	TWCR =(1<<TWEN)|(TWINT)|(TWEA) OR TWCR =(1<<TWEN)|(TWINT)	DATA byte will be received ACK will be returned DATA byte will be received NACK will be returned
$80 ($90 for General Call)	Data has been received ACK returned	DATA = TWDR TWCR =(1<<TWEN)|(TWINT)|(TWEA) OR DATA = TWDR TWCR =(1<<TWEN)|(TWINT)	DATA byte will be received ACK will be returned DATA byte will be received NACK will be returned
$88 ($98 for General Call)	Data has been received NACK returned	DATA = TWDR TWCR =(1<<TWEN)|(TWINT)|(TWEA) OR DATA = TWDR TWCR =(1<<TWEN)|(TWINT)	Start from beginning and wait to be addressed Start from beginning but do not respond to Its address (Sleep)
$A0	STOP or REPEATED START condition has been received	TWCR =(1<<TWEN)|(TWINT)|(TWEA) OR TWCR =(1<<TWEN)|(TWINT)	Start from beginning and wait to be addressed Start from beginning but do not respond to Its address (Sleep)

Figure 18-25. TWSR Register Values for Slave Receiver Operating Mode

Program 18-20 shows how to initialize the TWI module to operate in slave receiver mode. This program receives a byte of data and displays it on Port A after being addressed by a master device.

```
.INCLUDE "M32DEF.INC"

    LDI   R21,HIGH(RAMEND);set up stack
    OUT   SPH,R21
    LDI   R21,LOW(RAMEND)
    OUT   SPL,R21

    LDI   R21, 0xFF         ;move 0xFF into R21
    OUT   DDRA,R21          ;set PORTA as ouput

    CALL  I2C_INIT          ;initialize the TWI module as slave
```

Program 18-20: Reading a Byte in Slave Mode with Status Checking

```
        CALL  I2C_LISTEN          ;listen to the bus to be addressed
        CALL  I2C_READ_STATUS
        CPI   R26, 0x60           ;addressed as slave receiver?
        BRNE  ERROR               ;else jump to error function
        CALL  I2C_READ            ;read a byte and copy it to R27
        CALL  I2C_READ_STATUS
        CPI   R26, 0x80           ;addressed as slave receiver?
        BRNE  ERROR               ;else jump to error function
        OUT   PORTA,R27           ;copy R27 to PORTA

HERE:
        RJMP  HERE                ;wait here forever
ERROR:
        RJMP  HERE
;****************************************************************

I2C_INIT:
        LDI   R21, 0x10           ;load 00010000 into R21
        OUT   TWAR,R21            ;set address register
        LDI   R21, (1<<TWEN)      ;move 0x04 into R21
        OUT   TWCR,R21            ;enable the TWI
        LDI   R21, (1<<TWINT)|(1<<TWEN)|(1<<TWEA)
        OUT   TWCR,R21            ;enable TWI and ACK(can't be ignored)
        RET

;****************************************************************

I2C_LISTEN:
W1:
        IN    R21, TWCR           ;read control register into R21
        SBRS  R21, TWINT          ;skip next intruction if TWINT is 1
        RJMP  W1                  ;jump to W1 if TWINT is 0
        RET

;****************************************************************

I2C_READ:
        LDI   R21, (1<<TWINT)|(1<<TWEN)|(1<<TWEA)
        OUT   TWCR, R21           ;configure TWCR to receive TWDR
W2:     IN    R21, TWCR           ;read control register into R21
        SBRS  R21, TWINT          ;skip next line if TWINT is 1
        RJMP  W2                  ;jump to W2 if TWINT is 0
        IN    R27,TWDR            ;move received data into R21
        RET

;****************************************************************
I2C_READ_STATUS:
        IN    R26, TWSR           ;read status register into R21
        ANDI  R26, 0xF8           ;mask the prescaler bits
        RET
```

Program 18-20: Reading a Byte in Slave Mode with Status Checking *(cont. from prev. page)*

Program 18-21 is the C version of Program 18-20. This program receives a byte of data and displays it on Port A after being addressed by a master device.

```c
#include <avr/io.h>                    //standard AVR header

void i2c_showError(unsigned char er)
{
  DDRA = 0xFF;
  PORTA = er;
}


//*************************************************************

unsigned char i2c_readStatus(void)
{
  unsigned char i = 0;
  i = TWSR & 0xF8;
  return i;
}


//*************************************************************

void i2c_initSlave(unsigned char slaveAddress)
{
  TWCR = 0x04;                         //enable TWI module
  TWAR = slaveAddress;                 //set the slave address
  TWCR = (1<<TWINT)|(1<<TWEN)|(1<<TWEA);//init. TWI module
}
//*************************************************************

unsigned char i2c_receive(unsigned char isLast)
{
  if (isLast == 0)         //if want to read more than 1 byte
    TWCR = (1<< TWINT)|(1<<TWEN)|(1<<TWEA);
  else                     //if want to read only one byte
    TWCR = (1<< TWINT)|(1<<TWEN);

  while ((TWCR & (1 <<TWINT))==0);    //wait to complete
  return (TWDR);
}


//*************************************************************

void i2c_listen()
{
  while ((TWCR & (1 <<TWINT))==0);    //wait to be addressed
}

//*************************************************************
```

Program 18-21: Reading a Byte in Slave Mode with Status Checking in C

```
int main (void)
{
  DDRA = 0xFF;
  i2c_initSlave(0x10);                    //init. TWI module as
                                          //slave with address
                                          //0b0001000 and do not
                                          //accept general call
  i2c_listen();                           //listen to be addressed

  unsigned char s,i;
  s = i2c_readStatus();
  if (s != 0x60)
  {
     i2c_showError(s);
     return 0;
  }
  i=i2c_receive(0);
  s = i2c_readStatus();
  if (s != 0x80)
  {
     i2c_showError(s);
     return 0;
  }
  PORTA = i;
  while(1);                               //stay here forever
  return 0;
}
```

Program 18-21: Reading a Byte in Slave Mode with Status Checking in C *(continued)*

Review Questions

1. True or false. We can ignore checking the status register when there is more than one master on the bus.
2. True or false. We can enable the TWI module and generate aSTART condition at the same time.
3. How can a slave device read the value of the R/W bit when it is being addressed by a master device?
4. True or false. We can check the status register to see if a STOP condition has been transmitted successfully.
5. What is the value of the status register when SLA + W is received and ACK has been returned?
6. What is the value of the status register when SLA + W is transmitted and ACK has been received?
7. What is the value of the status register when SLA + R is received and ACK has been returned?
8. What is the value of the status register when SLA + W is transmitted and ACK has been received?

SUMMARY

This chapter began by describing the TWI bus connection and protocol. Then we focused on programming of TWI in the AVR. We also discussed the function of each pin of the DS1307 RTC chip. The DS1307 can be used to provide a real-time clock and dates for many applications. Various features of the RTC were explained, and numerous programming examples were given.

PROBLEMS

SECTION 18.1: I2C BUS PROTOCOL

1. True or false. The I2C bus needs an external clock.
2. True or false. The SDA pin is internally pulled up.
3. True or false. The I2C bus needs two wires to transfer data.
4. True or false. The SDA line is output for the master device.
5. True or false. When a device is used as a slave, the SCL is an input pin.
6. True or false. In I2C, the data frame is 8 bits long.
7. True or false. In I2C devices, each bit of information (data, address, ACK/NACK) is transferred with a single clock pulse.
8. True or false. In I2C devices, the 8-bit data is followed by an ACK/NACK.
9. In terms of data pins, what is the difference between the SPI and I2C connections?
10. How does the I2C protocol distinguish between the read and write cycles?

SECTION 18.2: TWI (I2C) IN THE AVR

11. True or false. The AVR uses the term TWI instead of I2C.
12. What are the TWI submodules in AVR?
13. Which unit generates START or STOP conditions in the AVR?
14. True or false. After reading the status register we should mask the 2 LSB bits.
15. Which bits of TWSR are used to specify the clock of the TWI module?
16. Which bit of TWCR enables generation of interrupts when the TWINT flag is set?
17. How can we virtually disconnect the TWI module from the bus?

SECTION 18.3: AVR TWI PROGRAMMING IN ASSEMBLY AND C

18. Write a program to read a byte from a slave with address 0110 100 and write the byte to a slave with address 0110 101.
19. Write a program to operate in slave mode and transmit "Y" to the master when the slave device is addressed. The slave address should be 0110 100.

SECTION 18.4: DS1307 RTC INTERFACING AND PROGRAMMING

20. The DS1307 DIP package is a(n) _____-pin package.
21. Which pin is assigned as GND?

22. Which pin is assigned as V_{cc}?

23. True or false. The DS1307 needs an external battery.

24. True or false. The DS1307 needs an external crystal oscillator.

25. True or false. The DS1307's crystal oscillator and heat affect the time-keeping accuracy.

26. What is the maximum year that the DS1307 can provide?

27. Describe the functions of the SQW/OUT pin.

28 X1 is an _____ (input, output) pin.

29. The SQW/OUT pin is controlled by _____ , _____, and _____ bits.

30. DS1307 has a total of _____bytes of RAM locations.

31. When does the DS1307 switch to a battery energy source?

32. What are the addresses assigned to the real-time clock (time) registers?

33. What are the addresses assigned to the calendar?

34. Which bit is used to set the AM/PM mode?

35. Which bit is used to set the 24-hour mode?

36. At what memory location does the DS1307 store the year 2009?

37. What is the address of the last location of RAM for the DS1307?

38. True or false. The DS1307 provides data in BCD format only.

39. Write a C program to set the time to 9:15:05 PM.

40. Write a C program to set the time to 22:47:19.

41. Write a C program to set the date to May 14, 2009.

42. Write a C program to get the hour and minute data and send it to Port B and Port D.

ANSWERS TO REVIEW QUESTIONS

SECTION 18.1: I2C BUS PROTOCOL

1. True
2. 9 bits. The ninth bit
3. True
4. Clock stretching
5. 4.7 kilohms
6. False

SECTION 18.2: TWI (I2C) IN THE AVR

1. True
2. TWDR, TWAR, TWBR, TWCR, and TWSR
3. By writing 1 to the TWSTA and TWSTO bits, respectively
4. False
5. TWINT
6. False
7. TWEA

SECTION 18.3: AVR TWI PROGRAMMING IN ASSEMBLY AND C

1. False
2. d
3. b and c

4. d
5. c

SECTION 18.4: DS1307 RTC INTERFACING AND PROGRAMMING

1. True
2. a
3. c (64 − 8 = 56 bytes)
4. True
5. SCL
6. 3V
7. False
8. a

SECTION 18.5: TWI PROGRAMMING WITH CHECKING STATUS REGISTER

1. False
2. False. We have to first enable the TWI module by writing one to the TWEN bit and then we can generate a START condition.
3. It should read the value of the status register.
4. False
5. $60
6. $18
7. $A8
8. $40

APPENDIX A

AVR INSTRUCTIONS EXPLAINED

OVERVIEW

In this appendix, we describe each intruction of the ATmega32. In many cases, a simple code example is given to clarify the instruction.

At the end there is a table that shows all the registers and their bits.

SECTION A.1: INSTRUCTION SUMMARY

DATA TRANSFER INSTRUCTIONS

Mnemonics	Operands	Description	Operation	Flags
MOV	Rd, Rr	Move Between Registers	Rd ← Rr	None
MOVW	Rd, Rr	Copy Register Word	Rd + 1:Rd ← Rr + 1:Rr	None
LDI	Rd, K	Load Immediate	Rd ← K	None
LD	Rd, X	Load Indirect	Rd ← (X)	None
LD	Rd, X+	Load Indirect and Post-Inc.	Rd ← (X), X ← X + 1	None
LD	Rd, −X	Load Indirect and Pre-Dec.	X ← X − 1, Rd ← (X)	None
LD	Rd, Y	Load Indirect	Rd ← (Y)	None
LD	Rd, Y+	Load Indirect and Post-Inc.	Rd ← (Y), Y ← Y + 1	None
LD	Rd, −Y	Load Indirect and Pre-Dec.	Y ← Y − 1, Rd ← (Y)	None
LDD	Rd,Y+q	Load Indirect with Displacement	Rd ← (Y + q)	None
LD	Rd, Z	Load Indirect	Rd ← (Z)	None
LD	Rd, Z+	Load Indirect and Post-Inc.	Rd ← (Z), Z ← Z+1	None
LD	Rd, −Z	Load Indirect and Pre-Dec.	Z ← Z − 1, Rd ← (Z)	None
LDD	Rd, Z + q	Load Indirect with Displacement	Rd ← (Z + q)	None
LDS	Rd, k	Load Direct from SRAM	Rd ← (k)	None
ST	X, Rr	Store Indirect	(X) ← Rr	None
ST	X+, Rr	Store Indirect and Post-Inc.	(X) ← Rr, X ← X + 1	None
ST	−X, Rr	Store Indirect and Pre-Dec.	X ← X − 1, (X) ← Rr	None
ST	Y, Rr	Store Indirect	(Y) ← Rr	None
ST	Y+, Rr	Store Indirect and Post-Inc.	(Y) ← Rr, Y ← Y + 1	None
ST	−Y, Rr	Store Indirect and Pre-Dec.	Y ← Y − 1, (Y) ← Rr	None
STD	Y + q, Rr	Store Indirect with Displacement	(Y + q) ← Rr	None
ST	Z, Rr	Store Indirect	(Z) ← Rr	None
ST	Z+, Rr	Store Indirect and Post-Inc.	(Z) ← Rr, Z ← Z + 1	None
ST	−Z, Rr	Store Indirect and Pre-Dec.	Z ← Z − 1, (Z) ← Rr	None
STD	Z + q, Rr	Store Indirect with Displacement	(Z + q) ← Rr	None
STS	k, Rr	Store Direct to SRAM	(k) ← Rr	None
LPM		Load Program Memory	R0 ← (Z)	None
LPM	Rd, Z	Load Program Memory	Rd ← (Z)	None
LPM	Rd, Z+	Load Program Memory and Post-Inc.	Rd ← (Z), Z ← Z+1	None
SPM		Store Program Memory	(Z) ← R1:R0	None
IN	Rd, P	In Port	Rd ← P	None
OUT	P, Rr	Out Port	P ← Rr	None
PUSH	Rr	Push Register on Stack	Stack ← Rr	None
POP	Rd	Pop Register from Stack	Rd ← Stack	None

BRANCH INSTRUCTIONS

Mnem.	Oper.	Description	Operation	Flags
RJMP	k	Relative Jump	PC ← PC + k + 1	None
IJMP		Indirect Jump to (Z)	PC ← Z	None
JMP	k	Direct Jump	PC ← k	None
RCALL	k	Relative Subroutine Call	PC ← PC + k + 1	None
ICALL		Indirect Call to (Z)	PC ← Z	None
CALL	k	Direct Subroutine Call	PC ← k	None
RET		Subroutine Return	PC ← Stack	None
RETI		Interrupt Return	PC ← Stack	I
CPSE	Rd,Rr	Compare, Skip if Equal	if (Rd = Rr) PC ← PC + 2 or 3	None
CP	Rd,Rr	Compare	Rd − Rr	Z,N,V,C,H
CPC	Rd,Rr	Compare with Carry	Rd − Rr − C	Z,N,V,C,H
CPI	Rd,K	Compare Register with Immediate	Rd − K	Z,N,V,C,H
SBRC	Rr, b	Skip if Bit in Register Cleared	if (Rr(b)=0) PC ← PC + 2 or 3	None
SBRS	Rr, b	Skip if Bit in Register is Set	if (Rr(b)=1) PC ← PC + 2 or 3	None
SBIC	P, b	Skip if Bit in I/O Register Cleared	if (P(b)=0) PC ← PC + 2 or 3	None
SBIS	P, b	Skip if Bit in I/O Register is Set	if (P(b)=1) PC ← PC + 2 or 3	None
BRBS	s, k	Branch if Status Flag Set	if (SREG(s)=1) then PC←PC+k+1	None
BRBC	s, k	Branch if Status Flag Cleared	if (SREG(s)=0) then PC←PC+k+1	None
BREQ	k	Branch if Equal	if (Z = 1) then PC ← PC + k + 1	None
BRNE	k	Branch if Not Equal	if (Z = 0) then PC ← PC + k + 1	None
BRCS	k	Branch if Carry Set	if (C = 1) then PC ← PC + k + 1	None
BRCC	k	Branch if Carry Cleared	if (C = 0) then PC ← PC + k + 1	None
BRSH	k	Branch if Same or Higher	if (C = 0) then PC ← PC + k + 1	None
BRLO	k	Branch if Lower	if (C = 1) then PC ← PC + k + 1	None
BRMI	k	Branch if Minus	if (N = 1) then PC ← PC + k + 1	None
BRPL	k	Branch if Plus	if (N = 0) then PC ← PC + k + 1	None
BRGE	k	Branch if Greater or Equal,Signed	if (N and V= 0) then PC←PC + k +1	None
BRLT	k	Branch if Less Than Zero, Signed	if (N and V= 1) then PC←PC + k +1	None
BRHS	k	Branch if Half Carry Flag Set	if (H = 1) then PC ← PC + k + 1	None
BRHC	k	Branch if Half Carry Flag Cleared	if (H = 0) then PC ← PC + k + 1	None
BRTS	k	Branch if T Flag Set	if (T = 1) then PC ← PC + k + 1	None
BRTC	k	Branch if T Flag Cleared	if (T = 0) then PC ← PC + k + 1	None
BRVS	k	Branch if Overflow Flag is Set	if (V = 1) then PC ← PC + k + 1	None
BRVC	k	Branch if Overflow Flag is Cleared	if (V = 0) then PC ← PC + k + 1	None
BRIE	k	Branch if Interrupt Enabled	if (I = 1) then PC ← PC + k + 1	None
BRID	k	Branch if Interrupt Disabled	if (I = 0) then PC ← PC + k + 1	None

BIT AND BIT-TEST INSTRUCTIONS

Mnem.	Operan.	Description	Operation	Flags
SBI	P, b	Set Bit in I/O Register	I/O(P, b) ← 1	None
CBI	P, b	Clear Bit in I/O Register	I/O(P, b) ← 0	None
LSL	Rd	Logical Shift Left	Rd(n + 1) ← Rd(n), Rd(0) ← 0	Z,C,N,V
LSR	Rd	Logical Shift Right	Rd(n)←Rd(n+1), Rd(7)←0	Z,C,N,V
ROL	Rd	Rotate Left Through Carry	Rd(0)←C, Rd(n+1)←Rd(n), C←Rd(7)	Z,C,N,V
ROR	Rd	Rotate Right Through Carry	Rd(7) ← C, Rd(n) ← Rd(n + 1), C ← Rd(0)	Z,C,N,V
ASR	Rd	Arithmetic Shift Right	Rd(n) ← Rd(n + 1), n = 0..6	Z,C,N,V
SWAP	Rd	Swap Nibbles	Rd(3..0) ← Rd(7..4), Rd(7..4) ← Rd(3..0)	None
BSET	s	Flag Set	SREG(s) ← 1	SREG(s)
BCLR	s	Flag Clear	SREG(s) ← 0	SREG(s)
BST	Rr, b	Bit Store from Register to T	T ← Rr(b)	T
BLD	Rd, b	Bit load from T to Register	Rd(b) ← T	None
SEC		Set Carry	C ← 1	C
CLC		Clear Carry	C ← 0	C
SEN		Set Negative Flag	N ←1	N
CLN		Clear Negative Flag	N ← 0	N
SEZ		Set Zero Flag	Z ←1	Z
CLZ		Clear Zero Flag	Z ← 0	Z
SEI		Global Interrupt Enable	I ← 1	I
CLI		Global Interrupt Disable	I ← 0	I
SES		Set Signed Test Flag	S ← 1	S
CLS		Clear Signed Test Flag	S ← 0	S
SEV		Set Two's Complement Overflow	V ← 1	V
CLV		Clear Two's Complement Overflow	V ← 0	V
SET		Set T in SREG	T ← 1	T
CLT		Clear T in SREG	T ← 0	T
SEH		Set Half Carry Flag in SREG	H ←1	H
CLH		Clear Half Carry Flag in SREG	H ← 0	H

ARITHMETIC AND LOGIC INSTRUCTIONS

Mnem.	Operands	Description	Operation	Flags
ADD	Rd, Rr	Add two Registers	Rd ← Rd + Rr	Z,C,N,V,H
ADC	Rd, Rr	Add with Carry two Registers	Rd ← Rd + Rr + C	Z,C,N,V,H
ADIW	Rdl, K	Add Immediate to Word	Rdh:Rdl ← Rdh:Rdl + K	Z,C,N,V,S
SUB	Rd, Rr	Subtract two Registers	Rd ← Rd − Rr	Z,C,N,V,H
SUBI	Rd, K	Subtract Constant from Register	Rd ← Rd − K	Z,C,N,V,H
SBC	Rd, Rr	Subtract with Carry two Registers	Rd ← Rd − Rr − C	Z,C,N,V,H
SBCI	Rd, K	Subtract with Carry Constant from Reg.	Rd ← Rd − K − C	Z,C,N,V,H
SBIW	Rdl, K	Subtract Immediate from Word	Rdh:Rdl ← Rdh:Rdl − K	Z,C,N,V,S
AND	Rd, Rr	Logical AND Registers	Rd ← Rd • Rr	Z,N,V
ANDI	Rd, K	Logical AND Register and Constant	Rd ← Rd • K	Z,N,V
OR	Rd, Rr	Logical OR Registers	Rd ← Rd v Rr	Z,N,V
ORI	Rd, K	Logical OR Register and Constant	Rd ← Rd v K	Z,N,V
EOR	Rd, Rr	Exclusive OR Registers	Rd ← Rd Rr	Z,N,V
COM	Rd	One's Complement	Rd ← $FF − Rd	Z,C,N,V
NEG	Rd	Two's Complement	Rd ← $00 − Rd	Z,C,N,V,H
SBR	Rd, K	Set Bit(s) in Register	Rd ← Rd v K	Z,N,V
CBR	Rd, K	Clear Bit(s) in Register	Rd ← Rd • ($FF − K)	Z,N,V
INC	Rd	Increment	Rd ← Rd + 1	Z,N,V
DEC	Rd	Decrement	Rd ← Rd − 1	Z,N,V
TST	Rd	Test for Zero or Minus	Rd ← Rd • Rd	Z,N,V
CLR	Rd	Clear Register	Rd ← $00	Z,N,V
SER	Rd	Set Register	Rd ← $FF	None
MUL	Rd, Rr	Multiply Unsigned	R1:R0 ← Rd x Rr	Z,C
MULS	Rd, Rr	Multiply Signed	R1:R0 ← Rd x Rr	Z,C
MULSU	Rd, Rr	Multiply Signed with Unsigned	R1:R0 ← Rd x Rr	Z,C
FMUL	Rd, Rr	Fractional Multiply Unsigned	R1:R0 ← (Rd x Rr)<< 1	Z,C
FMULS	Rd, Rr	Fractional Multiply Signed	R1:R0 ← (Rd x Rr)<< 1	Z,C
FMULSU	Rd, Rr	Fractional Multiply Signed with Unsigned	R1:R0 ← (Rd x Rr)<< 1	Z,C

MCU CONTROL INSTRUCTIONS

Mnemonics	Operands	Description	Operation	Flags
NOP		No Operation		None
SLEEP		Sleep	(see specific descr. for Sleep function)	None
WDR		Watchdog Reset	(see specific descr. for WDR/timer)	None
BREAK		Break	For On-Chip Debug Only	None

SECTION A.2: AVR INSTRUCTIONS FORMAT

ADC Rd, Rr	; Add with carry
$0 \leq d \leq 31, 0 \leq r \leq 31$	; Rd ← Rd + Rr + C

Adds two registers and the contents of the C flag and places the result in the destination register Rd.

Flags: H, S, V, N, Z, C   Cycles: 1

Example:

```
                        ;Add R1:R0 to R3:R2
    add r2,r0           ;Add low byte
    adc r3,r1           ;Add with carry high byte
```

ADD Rd, Rr	; Add without carry
$0 \leq d \leq 31, 0 \leq r \leq 31$	; Rd ← Rd + Rr

Adds two registers without the C flag and places the result in the destination register Rd.

Flags: H, S, V, N, Z, C   Cycles: 1

Example:

```
    add r1,r2           ;Add r2 to r1 (r1=r1+r2)
    add r28,r28         ;Add r28 to itself (r28=r28+r28)
```

ADIW Rd+1:Rd, K	; Add Immediate to Word
$d \in \{24,26,28,30\}, 0 \leq K \leq 63$	; Rd + 1:Rd ← Rd + 1:Rd + K

Adds an immediate value (0–63) to a register pair and places the result in the register pair. This instruction operates on the upper four register pairs, and is well suited for operations on the pointer registers.

Flags: S, V, N, Z, C   Cycles: 2

Example:

```
    adiw r25:24,1       ;Add 1 to r25:r24
    adiw ZH:ZL,63       ;Add 63 to the Z-pointer (r31:r30)
```

AND Rd, Rr	; Logical AND
$0 \leq d \leq 31, 0 \leq r \leq 31$	; Rd ← Rd • Rr

Performs the logical AND between the contents of register Rd and register Rr and places the result in the destination register Rd.

Flags: S, V ← 0, N, Z   Cycles: 1

Example:

```
    and r2,r3           ;Bitwise and r2 and r3, result in r2
    ldi r16,1           ;Set bitmask 0000 0001 in r16
    and r2,r16          ;Isolate bit 0 in r2
```

ANDI Rd, K	; Logical AND with Immediate
$16 \leq d \leq 31, 0 \leq K \leq 255$	; Rd ← Rd • K

Performs the logical AND between the contents of register Rd and a constant and places the result in the destination register Rd.

Flags: S, V ← 0, N, Z   Cycles: 1

Example:
```
andi r17,$0F          ;Clear upper nibble of r17
andi r18,$10          ;Isolate bit 4 in r18
```

ASR   Rd   **; Arithmetic Shift Right**
0 ≤ d ≤ 31

Shifts all bits in Rd one place to the right. Bit 7
is held constant. Bit 0 is loaded into the C flag of the
SREG. This operation effectively divides a signed value
by two without changing its sign. The Carry flag can be used to round the result.

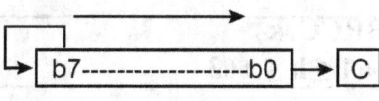

Flags: S, V, N, Z, C   Cycles: 1

Example:
```
ldi r16,$10           ;Load decimal 16 into r16
asr r16               ;r16=r16 / 2
ldi r17,$FC           ;Load -4 in r17
asr r17               ;r17=r17/2
```

BCLR s   **; Bit Clear in SREG**
0 ≤ s ≤7   **; SREG(s) ← 0**

Clears a single flag in SREG (Status Register).
Flags: I, T, H, S, V, N, Z, C Cycles: 1

Example:
```
bclr 0                ;Clear Carry flag
bclr 7                ;Disable interrupts
```

BLD   Rd, b   **; Bit Load from the T Flag in SREG to a Bit in Register**
0 ≤ d ≤ 31, 0 ≤ b ≤7   **; Rd(b) ← T**

Copies the T flag in the SREG (Status Register) to bit b in register Rd.
Flags: ---   Cycles: 1

Example:
```
bst r1,2              ;Store bit 2 of r1 in T flag
bld r0,4              ;Load T flag into bit 4 of r0
```

BRBC s, k   **; Branch if Bit in SREG is Cleared**
0 ≤ s ≤ 7, –64 ≤ k ≤ +63   **; If SREG(s) = 0 then PC ← PC + k + 1, else PC ← PC + 1**

Conditional relative branch. Tests a single bit in SREG (Status Register) and
branches relatively to PC if the bit is set.
Flags: ---   Cycles: 1or 2

Example:
```
     cpi r20,5             ;Compare r20 to the value 5
     brbc 1,noteq          ;Branch if Zero flag cleared
     ...
noteq:nop                  ;Branch destination (do nothing)
```

BRBS s, k   **; Branch if Bit in SREG is Set**
0 ≤ s ≤ 7, –64 ≤ k ≤ +63   **; If SREG(s) = 1 then PC ← PC + k + 1, else PC ← PC + 1**

Conditional relative branch. Tests a single bit in SREG (Status Register) and
branches relatively to PC if the bit is set.

Flags: ---   Cycles: 1 or 2

Example:
```
    bst r0,3              ;Load T bit with bit 3 of r0
    brbs 6,bitset         ;Branch T bit was set
    ...
    bitset: nop           ;Branch destination (do nothing)
```

BRCC k   **; Branch if Carry Cleared**
−64 ≤ k ≤ +63   **; If C = 0 then PC ← PC + k + 1, else PC ← PC + 1**

Conditional relative branch. Tests the Carry flag (C) and branches relatively to PC if C is cleared.

Flags: ---   Cycles: 1 or 2

Example:
```
    add r22,r23           ;Add r23 to r22
    brcc nocarry          ;Branch if carry cleared
    ...
nocarry:  nop             ;Branch destination (do nothing)
```

BRCS k   **; Branch if Carry Set**
−64 ≤ k ≤ +63   **; If C = 1 then PC ← PC + k + 1, else PC ← PC + 1**

Conditional relative branch. Tests the Carry flag (C) and branches relatively to PC if C is set.

Flags: ---   Cycles: 1 or 2

Example:
```
    cpi r26,$56           ;Compare r26 with $56
    brcs carry            ;Branch if carry set
    ...
carry:   nop              ;Branch destination (do nothing)
```

BREAK   **; Break**

The BREAK instruction is used by the on-chip debug system, and is normally not used in the application software. When the BREAK instruction is executed, the AVR CPU is set in the stopped mode. This gives the on-chip debugger access to internal resources.

Flags: ---   Cycles: 1

Example: ---

BREQ k   **; Branch if Equal**
−64 ≤ k ≤ +63   **; If Rd = Rr (Z = 1) then PC ← PC + k + 1, else PC ← PC + 1**

Conditional relative branch. Tests the Zero flag (Z) and branches relatively to PC if Z is set. If the instruction is executed immediately after any of the instructions CP, CPI, SUB, or SUBI, the branch will occur if and only if the unsigned or signed binary number represented in Rd was equal to the unsigned or signed binary number represented in Rr.

Flags: ---   Cycles: 1 or 2

Example:
```
    ccp r1,r0             ;Compare registers r1 and r0
    breq equal            ;Branch if registers equal
    ...
equal:   nop              ;Branch destination (do nothing)
```

BRGE k   **; Branch if Greater or Equal (Signed)**
−64 ≤ k ≤ +63   **; If Rd ≥ Rr (N⊕V = 0) then PC ← PC + k + 1, else PC ← PC + 1**

Conditional relative branch. Tests the Signed flag (S) and branches relatively to PC if S is cleared. If the instruction is executed immediately after any of the instructions CP, CPI, SUB, or SUBI, the branch will occur if and only if the signed binary number represented in Rd was greater than or equal to the signed binary number represented in Rr.

Flags: ---   Cycles: 1 or 2
Example:

```
          cp r11,r12        ;Compare registers r11 and r12
          brge greateq      ;Branch if r11 ≥ r12 (signed)
          ...
greateq:  nop               ;Branch destination (do nothing)
```

BRHC k   **; Branch if Half Carry Flag is Cleared**
−64 ≤ k ≤ +63   **; If H = 0 then PC ← PC + k + 1, else PC ← PC + 1**

Conditional relative branch. Tests the Half Carry flag (H) and branches relatively to PC if H is cleared.

Flags: ---   Cycles: 1 or 2
Example:

```
          brhc hclear       ;Branch if Half Carry flag cleared
          ...
hclear:   nop               ;Branch destination (do nothing)
```

BRHS k   **; Branch if Half Carry Flag is Set**
−64 ≤ k ≤ +63   **; If H = 1 then PC ← PC + k + 1, else PC ← PC + 1**

Conditional relative branch. Tests the Half Carry flag (H) and branches relatively to PC if H is set.

Flags: ---   Cycles: 1 or 2
Example:

```
          brhs hset         ;Branch if Half Carry flag set
          ...
hset:     nop               ;Branch destination (do nothing)
```

BRID k   **; Branch if Global Interrupt is Disabled**
−64 ≤ k ≤ +63   **; If I = 0 then PC←PC + k + 1, else PC←PC + 1**

Conditional relative branch. Tests the Global Interrupt flag (I) and branches relatively to PC if I is cleared.

Flags: ---   Cycles: 1 or 2
Example:

```
          brid intdis       ;Branch if interrupt disabled
          ...
intdis:   nop               ;Branch destination (do nothing)
```

BRIE k   **; Branch if Global Interrupt is Enabled**
−64 ≤ k ≤ +63   **; If I = 1 then PC ← PC + k + 1, else PC ← PC + 1**

Conditional relative branch. Tests the Global Interrupt flag (I) and branches relatively to PC if I is set.

Flags: ---   Cycles: 1 or 2

Example:

```
        brie inten        ;Branch if interrupt enabled
        . . .
inten:  nop               ;Branch destination (do nothing)
```

BRLO k   ; Branch if Lower (Unsigned)
$-64 \leq k \leq +63$; If Rd < Rr (C = 1) then PC $\leftarrow$ PC + k + 1, else PC $\leftarrow$ PC + 1

Conditional relative branch. Tests the Carry flag (C) and branches relatively to PC if C is set. If the instruction is executed immediately after any of the instructions CP, CPI, SUB, or SUBI, the branch will occur if and only if the unsigned binary number represented in Rd was smaller than the unsigned binary number represented in Rr.

Flags: ---   Cycles: 1 or 2

Example:

```
        eor r19,r19       ;Clear r19
loop:   inc r19           ;Increment r19
        . . .
        cpi r19,$10       ;Compare r19 with $10
        brlo loop         ;Branch if r19 < $10 (unsigned)
        nop               ;Exit from loop (do nothing)
```

BRLT k   ; Branch if Less Than (Signed)
$-64 \leq k \leq +63$; If Rd < Rr (N $\oplus$ V = 1) then PC$\leftarrow$ PC + k + 1, else PC $\leftarrow$ PC + 1

Conditional relative branch. Tests the Signed flag (S) and branches relatively to PC if S is set. If the instruction is executed immediately after any of the instructions CP, CPI, SUB, or SUBI, the branch will occur if and only if the signed binary number represented in Rd was less than the signed binary number represented in Rr.

Flags: ---   Cycles: 1 or 2

Example:

```
        bcp r16,r1        ;Compare r16 to r1
        brlt less         ;Branch if r16 < r1 (signed)
        . . .
less:   nop               ;Branch destination (do nothing)
```

BRMI k   ; Branch if Minus
$-64 \leq k \leq +63$; If N=1 then PC$\leftarrow$PC + k + 1, else PC$\leftarrow$PC + 1

Conditional relative branch. Tests the Negative flag (N) and branches relatively to PC if N is set.

Flags: ---   Cycles: 1 or 2

Example:

```
        subi r18,4        ;Subtract 4 from r18
        brmi negative     ;Branch if result negative
        . . .
negative:  nop            ;Branch destination (do nothing)
```

BRNE k   ; Branch if Not Equal
$-64 \leq k \leq +63$; If Rd $\neq$ Rr (Z = 0) then PC $\leftarrow$ PC + k + 1, else PC $\leftarrow$ PC + 1

Conditional relative branch. Tests the Zero flag (Z) and branches relatively to PC if Z is cleared. If the instruction is executed immediately after any of the instructions CP, CPI, SUB, or SUBI, the branch will occur if and only if the unsigned or signed binary number represented in Rd was not equal to the unsigned or signed binary number represented in Rr.

Flags: ---   Cycles: 1 or 2
Example:

```
        eor r27,r27     ;Clear r27
loop:   inc r27         ;Increment r27
        ...
        cpi r27,5       ;Compare r27 to 5
        brne loop       ;Branch if r27 not equal 5
        nop             ;Loop exit (do nothing)
```

BRPL k   **; Branch if Plus**
−64 ≤ k ≤ +63   **; If N = 0 then PC ← PC + k + 1, else PC ← PC + 1**

Conditional relative branch. Tests the Negative flag (N) and branches relatively to PC if N is cleared.

Flags: ---   Cycles: 1 or 2
Example:

```
        subi r26,$50    ;Subtract $50 from r26
        brpl positive   ;Branch if r26 positive
        ...
positive: nop           ;Branch destination (do nothing)
```

BRSH k   **; Branch if Same or Higher (Unsigned)**
−64 ≤ k ≤ +63   **; If Rd ≥Rr (C = 0) then PC ← PC + k + 1, else PC ← PC + 1**

Conditional relative branch. Tests the Carry flag (C) and branches relatively to PC if C is cleared. If the instruction is executed immediately after execution of any of the instructions CP, CPI, SUB, or SUBI, the branch will occur if and only if the unsigned binary number represented in Rd was greater than or equal to the unsigned binary number represented in Rr.

Flags: ---   Cycles: 1 or 2
Example:

```
        subi r19,4      ;Subtract 4 from r19
        brsh highsm     ;Branch if r19 >= 4 (unsigned)
        ...
highsm: nop             ;Branch destination (do nothing)
```

BRTC k   **; Branch if the T Flag is Cleared**
−64 ≤ k ≤ +63   **; If T = 0 then PC ← PC + k + 1, else PC ← PC + 1**

Conditional relative branch. Tests the T flag and branches relatively to PC if T is cleared.

Flags: ---   Cycles: 1 or 2
Example:

```
        bst r3,5        ;Store bit 5 of r3 in T flag
        brtc tclear     ;Branch if this bit was cleared
        ...
tclear: nop             ;Branch destination (do nothing)
```

BRTS k   ; **Branch if the T Flag is Set**
−64 ≤ k ≤ +63   ; **If T = 1 then PC←PC + k + 1, else PC ← PC + 1**

    Conditional relative branch. Tests the T flag and branches relatively to PC if T is set.

    Flags: ---   Cycles: 1 or 2

Example:

```
        bst   r3,5        ;Store bit 5 of r3 in T flag
        brts  tset        ;Branch if this bit was set
        ...
tset:   nop               ;Branch destination (do nothing)
```

BRVC k   ; **Branch if Overflow Cleared**
−64 ≤ k ≤ +63   ; **If V = 0 then PC ← PC + k + 1, else PC ← PC + 1**

    Conditional relative branch. Tests the Overflow flag (V) and branches relatively to PC if V is cleared.

    Flags: ---   Cycles: 1 or 2

Example:

```
        add   r3,r4       ;Add r4 to r3
        brvc  noover      ;Branch if no overflow
        ...
noover: nop               ;Branch destination (do nothing)
```

BRVS k   ; **Branch if Overflow Set**
−64 ≤ k ≤ +63   ; **If V=1 then PC←PC + k + 1, else PC←PC + 1**

    Conditional relative branch. Tests the Overflow flag (V) and branches relatively to PC if V is set.

    Flags: ---   Cycles: 1 or 2

Example:

```
        add   r3,r4       ;Add r4 to r3
        brvs  overfl      ;Branch if overflow
        ...
overfl: nop               ;Branch destination (do nothing)
```

BSET s   ; **Bit Set in SREG**
0 ≤ s ≤ 7   ; **SREG(s) ← 1**

    Sets a single flag or bit in SREG (Status Register).
    Flags: Any of the flags.   Cycles: 1

Example:

```
        bset  6           ;Set T flag
        bset  7           ;Enable interrupt
```

BST Rd,b   ; **Bit Store from Register to T Flag in SREG**
0 ≤ d ≤ 31, 0 ≤ b ≤ 7   ; **T ← Rd(b)**

    Stores bit b from Rd to the T flag in SREG (Status Register).
    Flags: T   Cycles: 1

Example:

```
                          ;Copy bit
        bst   r1,2        ;Store bit 2 of r1 in T flag
        bld   r0,4        ;Load T into bit 4 of r0t
```

CALL k   **; Long Call to a Subroutine**
0 ≤ k < 64K (Devices with 16 bits PC) or 0 ≤ k < 4M (Devices with 22 bits PC)

Calls to a subroutine within the entire program memory. The return address (to the instruction after the CALL) will be stored onto the stack. (See also RCALL.) The stack pointer uses a post-decrement scheme during CALL.

Flags: ---   Cycles: 4

Example:

```
        mov r16,r0          ;Copy r0 to r16
        call check          ;Call subroutine
        nop                 ;Continue (do nothing)
        ...
check:  cpi r16,$42         ;Check if r16 has a special value
        breq error          ;Branch if equal
        ret                 ;Return from subroutine
        ...
error:  rjmp error          ;Infinite loop
```

CBI A, b   **; Clear Bit in I/O Register**
0 ≤ A ≤ 31, 0 ≤ b ≤ 7   **; I/O(A,b) ← 0**

Clears a specified bit in an I/O Register. This instruction operates on the lower 32 I/O registers (addresses 0–31).

Flags: ---   Cycles: 2

Example:

```
        cbi $12,7           ;Clear bit 7 in Port D
```

CBR Rd, k   **; Clear Bits in Register**
16 ≤ d ≤ 31, 0 ≤ K ≤ 255   **; Rd ← Rd • ($FF – K)**

Clears the specified bits in register Rd. Performs the logical AND between the contents of register Rd and the complement of the constant mask K.

Flags: S, N, V ← 0, Z   Cycles: 1

Example:

```
        cbr r16,$F0         ;Clear upper nibble of r16
        cbr r18,1           ;Clear bit 0 in r18
```

CLC   **; Clear Carry Flag**
                                           ; C ← 0

Clears the Carry flag (C) in SREG (Status Register).

Flags: C ← 0.   Cycles: 1

Example:

```
        add r0,r0           ;Add r0 to itself
        clc                 ;Clear Carry flag
```

CLH   **; Clear Half Carry Flag**
                                           ; H ← 0

Clears the Half Carry flag (H) in SREG (Status Register).

Flags: H ← 0.   Cycles: 1

Example:

```
        clh                 ;Clear the Half Carry flag
```

CLI   **; Clear Global Interrupt Flag**
; I ← 0

Clears the Global Interrupt flag (I) in SREG (Status Register). The interrupts will be immediately disabled. No interrupt will be executed after the CLI instruction, even if it occurs simultaneously with the CLI instruction.

Flags: I ← 0.   Cycles: 1

Example:

```
        in temp, SREG       ;Store SREG value
                            ;(temp must be defined by user)
        cli                 ;Disable interrupts during timed sequence
        sbi EECR, EEMWE     ;Start EEPROM write
        sbi EECR, EEWE      ;
        out SREG, temp      ;Restore SREG value (I-flag)
```

CLN   **; Clear Negative Flag**
; N ← 0

Clears the Negative flag (N) in SREG (Status Register).

Flags: N ← 0.   Cycles: 1

Example:

```
        add r2,r3           ;Add r3 to r2
        cln                 ;Clear Negative flag
```

CLR Rd   **; Clear Register**
0 ≤ d ≤ 31   **; Rd ← Rd ⊕ Rd**

Clears a register. This instruction performs an Exclusive-OR between a register and itself. This will clear all bits in the register..

Flags: S ← 0 , N ← 0, V ← 0, Z ← 0   Cycles: 1

Example:

```
        clr r18             ;Clear r18
loop:   inc r18             ;Increment r18
        ...
        cpi r18,$50         ;Compare r18 to $50
        brne loop
```

CLS   **; Clear Signed Flag**
; S ← 0

Clears the Signed flag (S) in SREG (Status Register).

Flags: S ← 0.   Cycles: 1

Example:

```
        add r2,r3           ;Add r3 to r2
        cls                 ;Clear Signed flag
```

CLT   **; Clear T Flag**
; T ← 0

Clears the T flag in SREG (Status Register).

Flags: T ← 0.   Cycles: 1

Example:

```
        clt                 ;Clear T flag
```

CLV   ; **Clear Overflow Flag**
                                           ; **V ← 0**
_____

Clears the Overflow flag (V) in SREG (Status Register).

Flags: V ← 0.   Cycles: 1

Example:

```
        add r2,r3          ;Add r3 to r2
        clv                ;Clear Overflow flag
```

CLZ   ; **Clear Zero Flag**
                                           ; **Z ← 0**
_____

Clears the Zero flag (Z) in SREG (Status Register).

Flags: Z ← 0.   Cycles: 1

Example:

```
        clz                ;Clear zero
```

COM Rd   ; **One's Complement**
0 ≤ d ≤ 31   ; **Rd ← $FF – Rd**
_____

This instruction performs a one's complement of register Rd.

Flags: S, V ← 0, N , Z ← 1, C.   Cycles: 1

Example:

```
        com r4             ;Take one's complement of r4
        breq zero          ;Branch if zero
        . . .
zero:   nop                ;Branch destination (do nothing)
```

CP Rd,Rr   ; **Compare**
0 ≤ d ≤ 31, 0 ≤ r ≤ 31   ; **Rd – Rr**
_____

This instruction performs a compare between two registers, Rd and Rr. None of the registers are changed. All conditional branches can be used after this instruction.

Flags: H, S,V, N, Z, C.   Cycles: 1

Example:

```
        cp r4,r19          ;Compare r4 with r19
        brne noteq         ;Branch if r4 not equal r19
        . . .
noteq:  nop                ;Branch destination (do nothing)
```

CPC Rd,Rr   ; **Compare with Carry**
0 ≤ d ≤ 31, 0 ≤ r ≤ 31   ; **Rd – Rr – C**
_____

This instruction performs a compare between two registers, Rd and Rr, and also takes into account the previous carry. None of the registers are changed. All conditional branches can be used after this instruction.

Flags: H, S, V, N, Z, C.   Cycles: 1

Example:

```
                           ;Compare r3:r2 with r1:r0
        cp r2,r0           ;Compare low byte
        cpc r3,r1          ;Compare high byte
        brne noteq         ;Branch if not equal
        . . .
noteq:  nop                ;Branch destination (do nothing)
```
_____

CPI Rd,K   ; **Compare with Immediate**

16 ≤ d ≤ 31, 0 ≤ K ≤ 255   ; **Rd − K**

This instruction performs a compare between register Rd and a constant. The register is not changed. All conditional branches can be used after this instruction.

      Flags: H, S,V, N, Z, C.   Cycles: 1

Example:

```
            cpi r19,3        ;Compare r19 with 3
            brne error       ;Branch if r19 not equal 3
            ...
error:      nop              ;Branch destination (do nothing)
```

CPSE Rd,Rr   ; **Compare Skip if Equal**

0 ≤ d ≤ 31, 0 ≤ r ≤ 31   ; **If Rd = Rr then PC ← PC + 2 or 3 else PC ← PC + 1**

This instruction performs a compare between two registers Rd and Rr, and skips the next instruction if Rd = Rr.

      Flags:---   Cycles: 1, 2, or 3

Example:

```
            inc r4           ;Increment r4
            cpse r4,r0       ;Compare r4 to r0
            neg r4           ;Only executed if r4 not equal r0
            nop              ;Continue (do nothing)
```

DEC Rd   ; **Decrement**

0 ≤ d ≤ 31   ; **Rd ← Rd − 1**

Subtracts one from the contents of register Rd and places the result in the destination register Rd.

The C flag in SREG is not affected by the operation, thus allowing the DEC instruction to be used on a loop counter in multiple-precision computations.

When operating on unsigned values, only BREQ and BRNE branches can be expected to perform consistently. When operating on two's complement values, all signed branches are available.

      Flags: S,V, N, Z.   Cycles: 1

Example:

```
            ldi r17,$10      ;Load constant in r17
loop:       add r1,r2        ;Add r2 to r1
            dec r17          ;Decrement r17
            brne loop        ;Branch if r17 not equal 0
            nop              ;Continue (do nothing)
```

EOR Rd,Rr   ; **Exclusive OR**

0 ≤ d ≤ 31, 0 ≤ r ≤ 31   ; **Rd ← Rd ⊕ Rr**

Performs the logical Exclusive OR between the contents of register Rd and register Rr and places the result in the destination register Rd.

      Flags: S, V, Z ← 0, N, Z.   Cycles: 1

Example:

```
            eor r4,r4        ;Clear r4
            eor r0,r22       ;Bitwise XOR between r0 and r22
```

FMUL Rd,Rr   ; **Fractional Multiply Unsigned**
$16 \leq d \leq 23, 16 \leq r \leq 23$; **R1:R0 ← Rd × Rr (unsigned ← unsigned × unsigned)**

This instruction performs 8-bit × 8-bit → 16-bit unsigned multiplication and shifts the result one bit left.

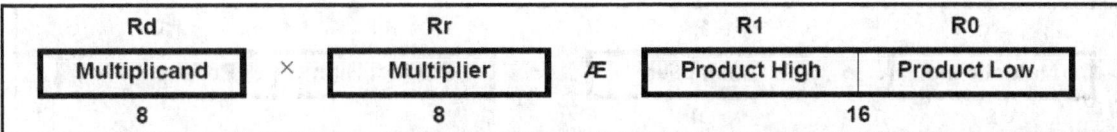

Let (N.Q) denote a fractional number with N binary digits left of the radix point, and Q binary digits right of the radix point. A multiplication between two numbers in the formats (N1.Q1) and (N2.Q2) results in the format ((N1 + N2).(Q1 + Q2)). For signal processing applications, the (1.7) format is widely used for the inputs, resulting in a (2.14) format for the product. A left shift is required for the high byte of the product to be in the same format as the inputs. The FMUL instruction incorporates the shift operation in the same number of cycles as MUL.

The (1.7) format is most commonly used with signed numbers, while FMUL performs an unsigned multiplication. This instruction is therefore most useful for calculating one of the partial products when performing a signed multiplication with 16-bit inputs in the (1.15) format, yielding a result in the (1.31) format. (Note: The result of the FMUL operation may suffer from a 2's complement overflow if interpreted as a number in the (1.15) format.) The MSB of the multiplication before shifting must be taken into account, and is found in the carry bit. See the following example.

The multiplicand Rd and the multiplier Rr are two registers containing unsigned fractional numbers where the implicit radix point lies between bit 6 and bit 7. The 16-bit unsigned fractional product with the implicit radix point between bit 14 and bit 15 is placed in R1 (high byte) and R0 (low byte).

Flags: Z, C.   Cycles: 2

Example:

```
;*****************************************************************
;* DESCRIPTION
;* Signed fractional multiply of two 16-bit numbers with 32-bit result.
;* r19:r18:r17:r16 = ( r23:r22 * r21:r20 ) << 1
;*****************************************************************
            fmuls 16x16_32:
            clr r2
            fmuls r23, r21          ;((signed)ah * (signed)bh) << 1
            movw r19:r18, r1:r0
            fmul r22, r20           ;(al * bl) << 1
            adc r18, r2
            movwr17:r16, r1:r0
            fmulsu r23, r20         ;((signed)ah * bl) << 1
            sbc r19, r2
            add r17, r0
            adc r18, r1
            adc r19, r2
            fmulsu r21, r22         ;((signed)bh * al) << 1
            sbc r19, r2
            add r17, r0
            adc r18, r1
            adc r19, r2
```

FMULS Rd,Rr   ; Fractional Multiply Signed
$16 \leq d \leq 23, 16 \leq r \leq 23$; R1:R0 ← Rd × Rr (signed ← signed × signed)

This instruction performs 8-bit × 8-bit → 16-bit signed multiplication and shifts the result one bit left.

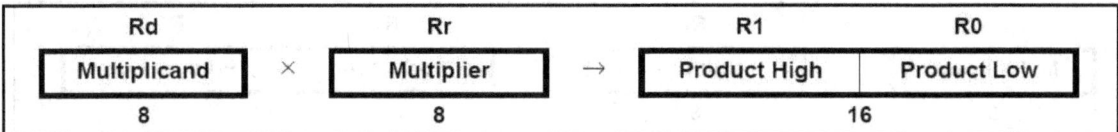

Let (N.Q) denote a fractional number with N binary digits left of the radix point, and Q binary digits right of the radix point. A multiplication between two numbers in the formats (N1.Q1) and (N2.Q2) results in the format ((N1 + N2).(Q1 + Q2)). For signal processing applications, the (1.7) format is widely used for the inputs, resulting in a (2.14) format for the product. A left shift is required for the high byte of the product to be in the same format as the inputs. The FMULS instruction incorporates the shift operation in the same number of cycles as MULS.

The multiplicand Rd and the multiplier Rr are two registers containing signed fractional numbers where the implicit radix point lies between bit 6 and bit 7. The 16-bit signed fractional product with the implicit radix point between bit 14 and bit 15 is placed in R1 (high byte) and R0 (low byte).

Note that when multiplying 0x80 (–1) with 0x80 (–1), the result of the shift operation is 0x8000 (–1). The shift operation thus gives a two's complement overflow. This must be checked and handled by software.

This instruction is not available in all devices. Refer to the device-specific instruction set summary.

Flags: Z, C.   Cycles: 2

Example:
```
fmuls r23,r22          ;Multiply signed r23 and r22 in
                       ;(1.7) format, result in (1.15) format
movw r23:r22,r1:r0     ;Copy result back in r23:r22
```

FMULSU Rd,Rr   ; Fractional Multiply Signed with Unsigned
$16 \leq d \leq 23, 16 \leq r \leq 23$; R1:R0 ← Rd × Rr

This instruction performs 8-bit × 8-bit → 16-bit signed multiplication and shifts the result one bit left.

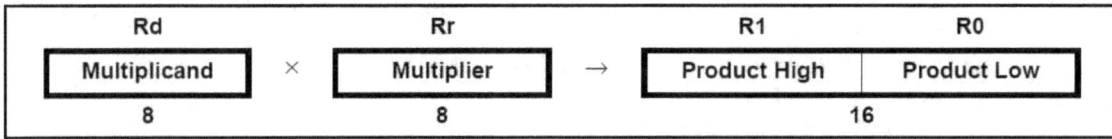

Let (N.Q) denote a fractional number with N binary digits left of the radix point, and Q binary digits right of the radix point. A multiplication between two numbers in the formats (N1.Q1) and (N2.Q2) results in the format ((N1 + N2).(Q1 + Q2)). For signal processing applications, the (1.7) format is widely used for the inputs, resulting in a (2.14) format for the product. A left shift is required for the high byte of the product to be in the same format as the inputs. The FMULSU instruction incorporates the shift operation in the same number of cycles as MULSU.

The (1.7) format is most commonly used with signed numbers, while FMULSU performs a multiplication with one unsigned and one signed input. This instruction is therefore most useful for calculating two of the partial products when performing a signed multiplication with 16-bit inputs in the (1.15) format, yielding a result in the (1.31) format. (Note: The result of the FMULSU operation may suffer from a 2's complement overflow if interpreted as a number in the (1.15) format.) The MSB of the multiplication before shifting must be taken into account, and is found in the carry bit. See the following example.

The multiplicand Rd and the multiplier Rr are two registers containing fractional numbers where the implicit radix point lies between bit 6 and bit 7. The multiplicand Rd is a signed fractional number, and the multiplier Rr is an unsigned fractional number. The 16-bit signed fractional product with the implicit radix point between bit 14 and bit 15 is placed in R1 (high byte) and R0 (low byte).

This instruction is not available in all devices. Refer to the device-specific instruction set summary.

Flags: Z, C.   Cycles: 2

Example:

```
;*********************************************************
;* DESCRIPTION
;* Signed fractional multiply of two 16-bit numbers with 32-bit result.
;* r19:r18:r17:r16 = ( r23:r22 * r21:r20 ) << 1
;*********************************************************
fmuls16x16_32:
        clrr2
        fmuls r23, r21    ;((signed)ah * (signed)bh) << 1
        movwr19:r18, r1:r0
        fmul r22, r20     ;(al * bl) << 1
        adc r18, r2
        movwr17:r16, r1:r0
        fmulsu r 23, r20  ;((signed)ah * bl) << 1
        sbc r19, r2
        add r17, r0
        adc r18, r1
        adc r19, r2
        fmulsu r21, r22   ;((signed)bh * al) << 1
        sbc r19, r2
        add r17, r0
        adc r18, r1
        adc r19, r2
```

ICALL   ; Indirect Call to Subroutine

Indirect call of a subroutine pointed to by the Z (16 bits) pointer register in the register file. The Z-pointer register is 16 bits wide and allows calls to a subroutine within the lowest 64K words (128K bytes) section in the program memory space. The stack pointer uses a post-decrement scheme during ICALL.

This instruction is not available in all devices. Refer to the device-specific instruction set summary.

Flags: ---   Cycles: 3

Example:

```
        mov r30,r0        ;Set offset to call table
        icall             ;Call routine pointed to by r31:r30
```

IJMP   **; Indirect Jump**

Indirect jump to the address pointed to by the Z (16 bits) pointer register in the register file. The Z-pointer register is 16 bits wide and allows jumps within the lowest 64K words (128K bytes) of the program memory.

This instruction is not available in all devices. Refer to the device-specific instruction set summary.

Flags:---   Cycles: 2

Example:

```
        mov r30,r0      ;Set offset to jump table
        ijmp            ;Jump to routine pointed to by r31:r30
```

IN Rd,A   **; Load an I/O Location to Register**
$0 \leq d \leq 31, 0 \leq A \leq 63$ **; Rd ← I/O(A)**

Loads data from the I/O space (ports, timers, configuration registers, etc.) into register Rd in the register file.

Flags:---   Cycles: 1

Example:

```
        in r25,$16      ;Read Port B
        cpi r25,4       ;Compare read value to constant
        breq exit       ;Branch if r25=4
        ...
exit:   nop             ;Branch destination (do nothing)
```

INC Rd   **; Increment**
$0 \leq d \leq 31$ **; Rd ← Rd + 1**

Adds one to the contents of register Rd and places the result in the destination register Rd.

The C flag in SREG is not affected by the operation, thus allowing the INC instruction to be used on a loop counter in multiple-precision computations.

When operating on unsigned numbers, only BREQ and BRNE branches can be expected to perform consistently. When operating on two's complement values, all signed branches are available.

Flags: S, V, N, Z.   Cycles: 1

Example:

```
        clr r22         ;Clear r22
loop:   inc r22         ;Increment r22
        ...
        cpi r22,$4F     ;Compare r22 to $4f
        brne loop       ;Branch if not equal
        nop             ;Continue (do nothing)
```

JMP k   **; Jump**
$0 \leq k < 4M$ **; PC ← k**

Jump to an address within the entire 4M (words) program memory. See also RJMP.

Flags:---   Cycles: 3

Example:

```
            mov r1,r0          ;Copy r0 to r1
            jmp farplc         ;Unconditional jump
            ...
farplc:     nop                ;Jump destination (do nothing)
```

LD	; **Load Indirect from Data Space to Register** ; **using Index X**

Loads one byte indirect from the data space to a register. For parts with SRAM, the data space consists of the register file, I/O memory, and internal SRAM (and external SRAM if applicable). For parts without SRAM, the data space consists of the register file only. The EEPROM has a separate address space.

The data location is pointed to by the X (16 bits) pointer register in the register file. Memory access is limited to the current data segment of 64K bytes. To access another data segment in devices with more than 64K bytes data space, the RAMPX in register in the I/O area has to be changed.

The X-pointer register can either be left unchanged by the operation, or it can be post-incremented or pre-decremented.

These features are especially suited for accessing arrays, tables, and stack pointer usage of the X-pointer register. Note that only the low byte of the X-pointer is updated in devices with no more than 256 bytes data space. For such devices, the high byte of the pointer is not used by this instruction and can be used for other purposes. The RAMPX register in the I/O area is updated in parts with more than 64K bytes data space or more than 64K bytes program memory, and the increment/ decrement is added to the entire 24-bit address on such devices.

Syntax:	Operation:	Comment:
(i) LD Rd, X	Rd ← (X)	X: Unchanged
(ii) LD Rd, X+	Rd ← (X) , X ← X + 1	X: Post-incremented
(iii) LD Rd, –X	X ← X – 1, Rd ← (X)	X: Pre-decremented

Flags:---   Cycles: 2

Example:

```
            clr r27            ;Clear X high byte
            ldi r26,$60        ;Set X low byte to $60
            ld r0,X+           ;Load r0 with data space loc. $60
                               ;X post inc)
            ld r1,X            ;Load r1 with data space loc. $61
            ldi r26,$63        ;Set X low byte to $63
            ld r2,X            ;Load r2 with data space loc. $63
            ld r3,-X           ;Load r3 with data space loc.
                               ;$62(X pre dec)
```

LD (LDD)	; **Load Indirect from Data Space to Register** ; **using Index Y**

Loads one byte indirect with or without displacement from the data space to a register. For parts with SRAM, the data space consists of the register file, I/O memory, and internal SRAM (and external SRAM if applicable). For parts without SRAM, the data space consists of the register file only. The EEPROM has a separate address space.

The data location is pointed to by the Y (16 bits) pointer register in the register file. Memory access is limited to the current data segment of 64K bytes. To access another data segment in devices with more than 64K bytes data space, the RAMPY in register in the I/O area has to be changed.

The Y-pointer register can either be left unchanged by the operation, or it can be post-incremented or pre-decremented. These features are especially suited for accessing arrays, tables, and stack pointer usage of the Y-pointer register. Note that only the low byte of the Y-pointer is updated in devices with no more than 256 bytes data space. For such devices, the high byte of the pointer is not used by this instruction and can be used for other purposes. The RAMPY register in the I/O area is updated in parts with more than 64K bytes data space or more than 64K bytes program memory, and the increment/ decrement/displacement is added to the entire 24-bit address on such devices.

Syntax:	Operation:	Comment:
(i) LD Rd, Y	Rd ← (Y)	Y: Unchanged
(ii) LD Rd, Y+	Rd ← (Y) ,Y ← Y + 1	Y: Postincremented
(iii) LD Rd, –Y	Y ← Y – 1, Rd ← (Y)	Y: Predecremented
(iiii) LDD Rd, Y + q	Rd ← (Y + q)	Y: Unchanged, q: Displacement

Flags:---   Cycles: 2

Example:
```
clr r29       ;Clear Y high byte
ldi r28,$60   ;Set Y low byte to $60
ld r0,Y+      ;Load r0 with data space loc. $60(Y post inc)
ld r1,Y       ;Load r1 with data space loc. $61
ldi r28,$63   ;Set Y low byte to $63
ld r2,Y       ;Load r2 with data space loc. $63
ld r3,-Y      ;Load r3 with data space loc. $62(Y pre dec)
ldd r4,Y+2    ;Load r4 with data space loc. $64
```

LD (LDD)   **; Load Indirect from Data Space to Register**
; using Index Z

Loads one byte indirect with or without displacement from the data space to a register. For parts with SRAM, the data space consists of the register file, I/O memory, and internal SRAM (and external SRAM if applicable). For parts without SRAM, the data space consists of the register file only. The EEPROM has a separate address space.

The data location is pointed to by the Z (16 bits) pointer register in the register file. Memory access is limited to the current data segment of 64K bytes. To access another data segment in devices with more than 64K bytes data space, the RAMPZ in register in the I/O area has to be changed.

The Z-pointer register can either be left unchanged by the operation, or it can be post-incremented or pre-decremented. These features are especially suited for stack pointer usage of the Z-pointer register, however because the Z-pointer register can be used for indirect subroutine calls, indirect jumps, and table lookup, it is often more convenient to use the X or Y-pointer as a dedicated stack pointer. Note that only the low byte of the Z-pointer is updated in devices with no more than 256 bytes data space. For such devices, the high byte of the pointer is not used by this instruction and can be used for other purposes. The RAMPZ register in the I/O area is updated in parts with more than 64K bytes data space or more than 64K bytes program memory, and the increment/decrement/displacement is added to the entire 24-bit address on such devices.

Syntax:	Operation:	Comment:
(i) LD Rd, Z	Rd ← (Z)	Z: Unchanged
(ii) LD Rd, Z+	Rd ← (Z) Z ← Z + 1	Z: Postincrement
(iii) LD Rd, –Z	Z ← Z – 1 Rd ← (Z)	Z: Predecrement
(iiii) LDD Rd, Z + q	Rd ← (Z + q)	Z: Unchanged, q: Displacement

Flags:---   Cycles: 2

Example:

```
clr r31        ;Clear Z high byte
ldi r30,$60    ;Set Z low byte to $60
ld r0,Z+       ;Load r0 with data space loc.$60(Z postinc.)
ld r1,Z        ;Load r1 with data space loc. $61
ldi r30,$63    ;Set Z low byte to $63
ld r2,Z        ;Load r2 with data space loc. $63
ld r3,-Z       ;Load r3 with data space loc. $62(Z predec.)
ldd r4,Z+2     ;Load r4 with data space loc. $64
```

LDI Rd,K	**; Load Immediate**
16 ≤ d ≤ 31, 0 ≤ K ≤ 255	**; Rd ← K**

Loads an 8-bit constant directly to registers 16 to 31.

Flags:---   Cycles: 1

Example:

```
clr r31        ;Clear Z high byte
ldi r30,$F0    ;Set Z low byte to $F0
lpm            ;Load constant from program
               ;memory pointed to by Z
```

LDS Rd,k	**; Load Direct from Data Space**
0 ≤ d ≤ 31, 0 ≤ k ≤ 65535	**; Rd ← (k)**

Loads one byte from the data space to a register. The data space consists of the register file, I/O memory, and SRAM.

Flags:---   Cycles: 2

Example:

```
lds r2,$FF00   ;Load r2 with the contents of
               ;data space location $FF00
add r2,r1      ;add r1 to r2
sts $FF00,r2   ;Write back
```

LPM	**; Load Program Memory**

Loads one byte pointed to by the Z-register into the destination register Rd. This instruction features a 100% space effective constant initialization or constant data fetch. The program memory is organized in 16-bit words while the Z-pointer is a byte address. Thus, the least significant bit of the Z-pointer selects either the low byte (ZLSB = 0) or the high byte (ZLSB = 1). This instruction can address the first 64K bytes (32K words) of program memory. The Z-pointer register can either be left unchanged by the operation, or it can be incremented. The incrementation does not apply to the RAMPZ register.

Devices with self-programming capability can use the LPM instruction to read the Fuse and Lock bit values. Refer to the device documentation for a detailed description.

```
Syntax:              Operation:              Comment:
(i) LPM              R0 ← (Z)                Z: Unchanged, R0 implied Rd
(ii) LPM Rd, Z       Rd ← (Z)                Z: Unchanged
(iii) LPM Rd, Z+     Rd ← (Z), Z ← Z + 1     Z: Postincremented
```

Flags:---   Cycles: 3

Example:
```
        ldi ZH, high(Table_1<<1);Initialize Z-pointer
        ldi ZL, low(Table_1<<1)
        lpm r16, Z               ;Load constant from program
                                 ;Memory pointed to by Z (r31:r30)
        ...
Table_1:
.dw 0x5876                       ;0x76 is addresses when ZLSB = 0
                                 ;0x58 is addresses when ZLSB = 1

        ...
```

LSL Rd   **; Logical Shift Left**
0 ≤ d ≤ 31

Shifts all bits in Rd one place to the left. Bit 0 is cleared. Bit 7 is loaded into the C flag of the SREG (Status Register). This operation effectively multiplies signed and unsigned values by two.

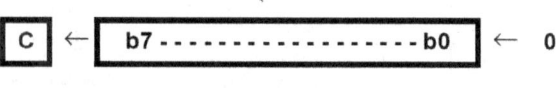

Flags: H, S, V, N, Z, C.   Cycles: 1

Example:
```
        add r0,r4                ;Add r4 to r0
        lsl r0                   ;Multiply r0 by 2
```

LSR Rd   **; Logical Shift Left**
0 ≤ d ≤ 31

Shifts all bits in Rd one place to the right. Bit 7 is cleared. Bit 0 is loaded into the C flag of the SREG. This operation effectively divides an unsigned value by two. The C flag can be used to round the result.

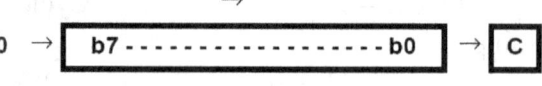

Flags: S, V, N ← 0, Z, C.   Cycles: 1

Example:
```
        add r0,r4                ;Add r4 to r0
        lsr r0                   ;Divide r0 by 2
```

MOV Rd,Rr   **; Copy Register**
0 ≤ d ≤ 31, 0 ≤ r ≤ 31   **; Rd ← Rr**

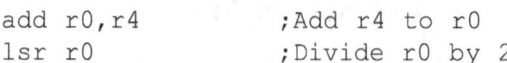

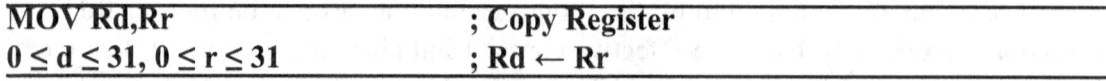

This instruction makes a copy of one register into another. The source register Rr is left unchanged, while the destination register Rd is loaded with a copy of Rr.

Flags: ---   Cycles: 1

Example:

```
            mov r16,r0        ;Copy r0 to r16
            call check        ;Call subroutine
            ...
check:      cpi r16,$11       ;Compare r16 to $11
            ...
            ret               ;Return from subroutine
```

MOVW Rd + 1:Rd,Rr + 1:Rrd   **; Copy RegisterWord**
d ∈ {0,2,...,30}, r ∈ {0,2,...,30}   **; Rd + 1:Rd ← Rr + 1:Rr**

This instruction makes a copy of one register pair into another register pair. The source register pair Rr + 1:Rr is left unchanged, while the destination register pair Rd + 1:Rd is loaded with a copy of Rr + 1:Rr.

Flags: ---   Cycles: 1

Example:

```
            movw r17:16,r1:r0 ;Copy r1:r0 to r17:r16
            call check        ;Call subroutine
            ...
check:      cpi r16,$11       ;Compare r16 to $11
            ...
            cpi r17,$32       ;Compare r17 to $32
            ...
            ret               ;Return from subroutine
```

MUL Rd,Rr   **; Multiply Unsigned**
0 ≤ d ≤ 31, 0 ≤ r ≤ 31   **; R1:R0 ← Rd × Rr(unsigned ← unsigned × unsigned)**

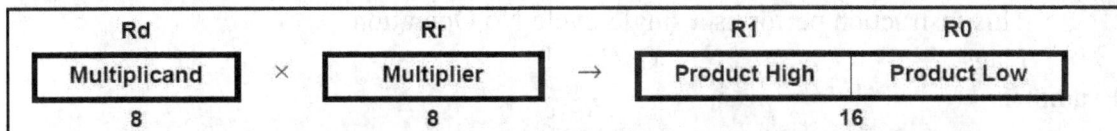

This instruction performs 8-bit × 8-bit → 16-bit unsigned multiplication.

The multiplicand Rd and the multiplier Rr are two registers containing unsigned numbers. The 16-bit unsigned product is placed in R1 (high byte) and R0 (low byte). Note that if the multiplicand or the multiplier is selected from R0 or R1 the result will overwrite those after multiplication.

Flags: Z, C.   Cycles: 2

Example:

```
        mul r5,r4         ;Multiply unsigned r5 and r4
        movw r4,r0        ;Copy result back in r5:r4
```

MULS Rd,Rr   **; Multiply Signed**
16 ≤ d ≤ 31, 16 ≤ r ≤ 31   **; R1:R0 ← Rd × Rr(signed ← signed × signed)**

This instruction performs 8-bit × 8-bit → 16-bit signed multiplication.

The multiplicand Rd and the multiplier Rr are two registers containing signed numbers. The 16-bit signed product is placed in R1 (high byte) and R0 (low byte).

Flags: Z, C.   Cycles: 2

Example:

```
        muls r21,r20      ;Multiply signed r21 and r20
        movw r20,r0       ;Copy result back in r21:r20
```

MULSU Rd,Rr   ; Multiply Signed with Unsigned
$16 \le d \le 31, 16 \le r \le 31$; R1:R0 ← Rd × Rr (signed ← signed × unsigned)

This instruction performs 8-bit × 8-bit → 16-bit multiplication of a signed and an unsigned number.

The multiplicand Rd and the multiplier Rr are two registers. The multiplicand Rd is a signed number, and the multiplier Rr is unsigned. The 16-bit signed product is placed in R1 (high byte) and R0 (low byte).

Flags: Z, C.   Cycles: 2

Example:---

NEG Rd   ; Two's Complement
$0 \le d \le 31$; Rd ← \$00 – Rd

Replaces the contents of register Rd with its two's complement; the value \$80 is left unchanged.

Flags: H, S, V, N, Z, C.   Cycles: 1

Example:

```
            sub r11,r0        ;Subtract r0 from r11
            brpl positive     ;Branch if result positive
            neg r11           ;Take two's complement of r11
positive:   nop               ;Branch destination (do nothing)
```

NOP   ; No Operation

This instruction performs a single-cycle No Operation.

Flags: ---.   Cycles: 1

Example:

```
            clr r16           ;Clear r16
            ser r17           ;Set r17
            out $18,r16       ;Write zeros to Port B
            nop               ;Wait (do nothing)
            out $18,r17       ;Write ones to Port B
```

OR Rd,Rr   ; Logical OR
$0 \le d \le 31, 0 \le r \le 31$; Rd ← Rd OR Rr

Performs the logical OR between the contents of register Rd and register Rr and places the result in the destination register Rd.

Flags: S, V ← 0, N, Z.   Cycles: 1

Example:

```
            or r15,r16        ;Do bitwise or between registers
            bst r15,6         ;Store bit 6 of r15 in T flag
            brts ok           ;Branch if T flag set
            ...
ok:         nop               ;Branch destination (do nothing)
```

ORI Rd,K   **; Logical OR with Immediate**
$16 \leq d \leq 31, 0 \leq K \leq 255$ **; Rd ← Rd OR K**

Performs the logical OR between the contents of register Rd and a constant and places the result in the destination register Rd.

     Flags: S, V ← 0, N, Z.   Cycles: 1

Example:

```
        ori r16,$F0     ;Set high nibble of r16
        ori r17,1       ;Set bit 0 of r17
```

OUT A,Rr   **; Store Register to I/O Location**
$0 \leq r \leq 31, 0 \leq A \leq 63$ **; I/O(A) ← Rr**

Stores data from register Rr in the register file to I/O space (ports, timers, configuration registers, etc.).

     Flags: ---.   Cycles: 1

Example:

```
        clr r16         ;Clear r16
        ser r17         ;Set r17
        out $18,r16     ;Write zeros to Port B
        nop             ;Wait (do nothing)
        out $18,r17     ;Write ones to Port B
```

POP Rd   **; Pop Register from Stack**
$0 \leq d \leq 31$ **; Rd ← STACK**

This instruction loads register Rd with a byte from the STACK. The stack pointer is pre-incremented by 1 before the POP.

     Flags: ---.   Cycles: 2

Example:

```
        call routine    ;Call subroutine
        ...
routine: push r14       ;Save r14 on the stack
        push r13        ;Save r13 on the stack
        ...
        pop r13         ;Restore r13
        pop r14         ;Restore r14
        ret             ;Return from subroutine
```

PUSH Rr   **; Push Register on Stack**
$0 \leq d \leq 31$ **; STACK ← Rr**

This instruction stores the contents of register Rr on the STACK. The stack pointer is post-decremented by 1 after the PUSH.

     Flags: ---.   Cycles: 2

Example:

```
        call routine    ;Call subroutine
        ...
routine: push r14       ;Save r14 on the stack
        push r13        ;Save r13 on the stack
        ...
        pop r13         ;Restore r13
        pop r14         ;Restore r14
        ret             ;Return from subroutine
```

RCALL k   ; Relative Call to Subroutine

−2K ≤ k < 2K   ; PC ← PC + k + 1

Relative call to an address within PC − 2K + 1 and PC + 2K (words). The return address (the instruction after the RCALL) is stored onto the stack. (See also CALL.) In the assembler, labels are used instead of relative operands. For AVR microcontrollers with program memory not exceeding 4K words (8K bytes) this instruction can address the entire memory from every address location. The stack pointer uses a post-decrement scheme during RCALL.

       Flags: ---.   Cycles: 3

Example:

```
          rcall routine   ;Call subroutine
          ...
routine:  push r14         ;Save r14 on the stack
          ...
          pop r14          ;Restore r14
          ret              ;Return from subroutine
```

RET   ; Return from Subroutine

Returns from subroutine. The return address is loaded from the stack. The stack pointer uses a pre-increment scheme during RET.

       Flags: ---.   Cycles: 4

Example:

```
          call routine    ;Call subroutine
          ...
routine:  push r14         ;Save r14 on the stack
          ...
          pop r14          ;Restore r14
          ret              ;Return from subroutine
```

RETI   ; Return from Interrupt

Returns from interrupt. The return address is loaded from the stack and the Global Interrupt flag is set.

Note that the Status Register is not automatically stored when entering an interrupt routine, and it is not restored when returning from an interrupt routine. This must be handled by the application program. The stack pointer uses a pre-increment scheme during RETI.

       Flags: ---.   Cycles: 4

Example:

```
          ...
extint:   push r0          ;Save r0 on the stack
          ...
          pop r0           ;Restore r0
          reti             ;Return and enable interrupts
```

RJMP k   ; **Relative Jump**
−2K ≤ k < 2K   ; **PC ← PC + k + 1**

Relative jump to an address within PC − 2K +1 and PC + 2K (words). In the assembler, labels are used instead of relative operands. For AVR microcontrollers with program memory not exceeding 4K words (8K bytes) this instruction can address the entire memory from every address location.

Flags: ---.   Cycles: 2

Example:

```
        cpi r16,$42     ;Compare r16 to $42
        brne error      ;Branch if r16 not equal $42
        rjmp ok         ;Unconditional branch
error:  add r16,r17     ;Add r17 to r16
        inc r16         ;Increment r16
ok:     nop             ;Destination for rjmp (do nothing)
```

ROL Rd   ; **Rotate Left through Carry**
0 ≤ d ≤ 31

Shifts all bits in Rd one place to the left. The C flag is shifted into bit 0 of Rd. Bit 7 is shifted into the C flag. This operation combined with LSL effectively multiplies multibyte signed and unsigned values by two.

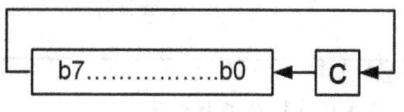

Flags: H, S, V, N, Z, C.   Cycles: 1

Example:

```
        lsl r18         ;Multiply r19:r18 by two
        rol r19         ;r19:r18 is a signed or unsigned word
        brcs oneenc     ;Branch if carry set
        ...
oneenc: nop             ;Branch destination (do nothing)
```

ROR Rd   ; **Rotate Right through Carry**
0 ≤ d ≤ 31

Shifts all bits in Rd one place to the right. The C flag is shifted into bit 7 of Rd. Bit 0 is shifted into the C flag. This operation combined with ASR effectively divides multibyte signed values by two.

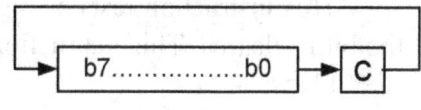

Combined with LSR, it effectively divides multibyte unsigned values by two. The Carry flag can be used to round the result.

Flags: S, V, N, Z, C.   Cycles: 1

Example:

```
        lsr r19         ;Divide r19:r18 by two
        ror r18         ;r19:r18 is an unsigned two-byte integer
        brcc zeroenc1   ;Branch if carry cleared
        asr r17         ;Divide r17:r16 by two
        ror r16         ;r17:r16 is a signed two-byte integer
        brcc zeroenc2   ;Branch if carry cleared
        ...
zeroenc1: nop           ;Branch destination (do nothing)
        ...
zeroenc2: nop           ;Branch destination (do nothing)
```

APPENDIX A: AVR INSTRUCTIONS EXPLAINED   **723**

SBC Rd,Rr   ; **Subtract with Carry**
$0 \le d \le 31, 0 \le r \le 31$; $Rd \leftarrow Rd - Rr - C$

Subtracts two registers and subtracts with the C flag and places the result in the destination register Rd.

     Flags: H, S, V, N, Z, C.   Cycles: 1

Example:

```
                              ;Subtract r1:r0 from r3:r2
           sub r2,r0          ;Subtract low byte
           sbc r3,r1          ;Subtract with carry high byte
```

SBCI Rd,K   ; **Subtract Immediate with Carry**
$0 \le d \le 31, 0 \le r \le 31$; $Rd \leftarrow Rd - K - C$

Subtracts a constant from a register and subtracts with the C flag and places the result in the destination register Rd.

     Flags: H, S, V, N, Z, C.   Cycles: 1

Example:

```
                              ;Subtract $4F23 from r17:r16
           subi r16,$23       ;Subtract low byte
           sbci r17,$4F       ;Subtract with carry high byte
```

SBI A,b   ; **Set Bit in I/O Register**
$0 \le A \le 31, 0 \le b \le 7$; $I/O(A,b) \leftarrow 1$

Sets a specified bit in an I/O register. This instruction operates on the lower 32 I/O registers.

     Flags: ---.   Cycles: 2

Example:

```
           out $1E,r0         ;Write EEPROM address
           sbi $1C,0          ;Set read bit in EECR
           in r1,$1D          ;Read EEPROM data
```

SBIC A,b   ; **Skip if Bit in I/O Register is Cleared**
$0 \le d \le 31, 0 \le r \le 31$; **If I/O(A,b) = 0 then PC ← PC + 2 (or 3) else PC ← PC + 1**

This instruction tests a single bit in an I/O register and skips the next instruction if the bit is cleared. This instruction operates on the lower 32 I/O registers.

     Flags:---.   Cycles: 1/2/3

Example:

```
e2wait:    sbic $1C,1         ;Skip next inst. if EEWE cleared
           rjmp e2wait        ;EEPROM write not finished
           nop                ;Continue (do nothing)
```

SBIS A,b   ; **Skip if Bit in I/O Register is Set**
$0 \le d \le 31, 0 \le r \le 31$; **If I/O(A,b) = 1 then PC ← PC + 2 (or 3) else PC ← PC + 1**

This instruction tests a single bit in an I/O register and skips the next instruction if the bit is set. This instruction operates on the lower 32 I/O registers.

     Flags: ---.   Cycles: 1/2/3

Example:

```
waitset:   sbis $10,0         ;Skip next inst. if bit 0 in Port D set
           rjmp waitset       ;Bit not set
           nop                ;Continue (do nothing)
```

SBIW Rd + 1:Rd,K	; Subtract Immediate from Word
d ∈ {24,26,28,30}, 0 ≤ K ≤ 63	; Rd + 1:Rd ← Rd + 1:Rd – K

Subtracts an immediate value (0–63) from a register pair and places the result in the register pair. This instruction operates on the upper four register pairs, and is well suited for operations on the pointer registers.

Flags: S, V, N, Z, C.   Cycles: 2

Example:

```
        sbiw r25:r24,1    ;Subtract 1 from r25:r24
        sbiw YH:YL,63     ;Subtract 63 from the Y-pointer
```

SBR Rd,K	; Set Bits in Register
16 ≤ d ≤ 31, 0 ≤ K ≤ 255	; Rd ← Rd OR K

Sets specified bits in register Rd. Performs the logical ORI between the contents of register Rd and a constant mask K and places the result in the destination register Rd.

Flags: S,V←0, N, Z.   Cycles: 1

Example:

```
        sbr r16,3     ;Set bits 0 and 1 in r16
        sbr r17,$F0   ;Set 4 MSB in r17
```

SBRC Rr,b	; Skip if Bit in Register is Cleared
0 ≤ r ≤ 31, 0 ≤ b ≤7	; If Rr(b) = 0 then PC ← PC + 2 or 3 else PC ← PC + 1

This instruction tests a single bit in an I/O register and skips the next instruction if the bit is set. This instruction operates on the lower 32 I/O registers.

Flags: ---   Cycles: 1/2/3

Example:

```
        sub r0,r1     ;Subtract r1 from r0
        sbrc r0,7     ;Skip if bit 7 in r0 cleared
        sub r0,r1     ;Only executed if bit7 in r0 not cleared
        nop           ;Continue (do nothing)
```

SBRS Rr,b	; Skip if Bit in Register is Set
0 ≤ r ≤ 31, 0 ≤ b ≤7	; If Rr(b) = 1 then PC ← PC + 2 or 3 else PC ← PC + 1

This instruction tests a single bit in a register and skips the next instruction if the bit is set.

Flags: H, S, V, N, Z, C.   Cycles: 1/2/3

Example:

```
        sub r0,r1     ;Subtract r1 from r0
        sbrs r0,7     ;Skip if bit 7 in r0 set
        neg r0        ;Only executed if bit 7 in r0 not set
        nop           ;Continue (do nothing)
```

SEC	; Set Carry Flag
	; C ← 1

Sets the Carry flag (C) in SREG (Status Register).

Flags: C ← 1.   Cycles: 1

Example:

```
        sec           ;Set Carry flag
        adc r0,r1     ;r0=r0+r1+1
```

SEH   **; Set Half Carry Flag**
                                                   ; H ← 1

Sets the Half Carry (H) in SREG (Status Register).

Flags: H ← 1.   Cycles: 1

Example:

```
        seh                 ;Set Half Carry flag
```

SEI   **; Set Global Interrupt Flag**
                                                   ; I ← 1

Sets the Global Interrupt flag (I) in SREG (Status Register). The instruction following SEI will be executed before any pending interrupts.

Flags: I ← 1.   Cycles: 1

Example:

```
        sei                 ;Set global interrupt enable
        sec                 ;Set Carry flag
        ;Note: will set Carry flag before any pending interrupt
```

SEN   **; Set Negative Flag**
                                                   ; N ← 1

Sets the Negative flag (N) in SREG (Status Register).

Flags: N ← 1.   Cycles: 1

Example:

```
        add r2,r19          ;Add r19 to r2
        sen                 ;Set Negative flag
```

SER Rd   **; Set all Bits in Register**
16 ≤ d ≤ 31   **; Rd ← $FF**

Loads $FF directly to register Rd.

Flags: ---.   Cycles: 1

Example:

```
        ser r17             ;Set r17
        out $18,r17         ;Write ones to Port B
```

SES   **; Set Signed Flag**
                                                   ; S ← 1

Sets the Signed flag (S) in SREG (Status Register).

Flags: S ← 1.   Cycles: 1

Example:

```
        add r2,r19          ;Add r19 to r2
        ses                 ;Set Negative flag
```

SET   **; Set T Flag**
                                                   ; T ← 1

Sets the T flag in SREG (Status Register).

Flags: T ← 1.   Cycles: 1

Example:

```
        set                 ;Set T flag
```

SEV   ; **Set Overflow Flag**
                                 ; $V \leftarrow 1$

Sets the Overflow flag (V) in SREG (Status Register).

Flags: $V \leftarrow 1$.   Cycles: 1

Example:

```
        sev               ;Set Overflow flag
```

SEZ   ; **Set Zero Flag**
                                 ; $Z \leftarrow 1$

Sets the Zero flag (Z) in SREG (Status Register).

Flags: $Z \leftarrow 1$.   Cycles: 1

Example:

```
        sez               ;Set Z flag
```

SLEEP

This instruction sets the circuit in sleep mode defined by the MCU control register.

Flags: ---.   Cycles: 1

Example:

```
        mov r0,r11        ;Copy r11 to r0
        ldi r16,(1<<SE)   ;Enable sleep mode
        out MCUCR, r16
        sleep             ;Put MCU in sleep mode
```

SPM   ; **Store Program Memory**

SPM can be used to erase a page in the program memory, to write a page in the program memory (that is already erased), and to set Boot Loader Lock bits. In some devices, the program memory can be written one word at a time, in other devices an entire page can be programmed simultaneously after first filling a temporary page buffer. In all cases, the program memory must be erased one page at a time. When erasing the program memory, the RAMPZ and Z-register are used as page address. When writing the program memory, the RAMPZ and Z-register are used as page or word address, and the R1:R0 register pair is used as data(1). When setting the Boot Loader Lock bits, the R1:R0 register pair is used as data.

Refer to the device documentation for detailed description of SPM usage. This instruction can address the entire program memory.

Flags: ---.   Cycles: depends on the operation

	Syntax:	Operation:	Comment:
(i)	SPM	(RAMPZ:Z) $\leftarrow$ \$ffff	Erase program memory page
(ii)	SPM	(RAMPZ:Z) $\leftarrow$ R1:R0	Write program memory word
(iii)	SPM	(RAMPZ:Z) $\leftarrow$ R1:R0	Write temporary page buffer
(iv)	SPM	(RAMPZ:Z) $\leftarrow$ TEMP	Write temporary page buffer to program memory
(v)	SPM	BLBITS $\leftarrow$ R1:R0	Set Boot Loader Lock bits

Stores one byte indirect from a register to data space. For parts with SRAM, the data space consists of the register file, I/O memory, and internal SRAM (and external SRAM if applicable). For parts without SRAM, the data space consists of the register file only. The EEPROM has a separate address space.

The data location is pointed to by the X (16 bits) pointer register in the register file. Memory access is limited to the current data segment of 64K bytes. To access another data segment in devices with more than 64K bytes data space, the RAMPX register in the I/O area has to be changed.

The X-pointer register can either be left unchanged by the operation, or it can be post-incremented or pre-decremented.These features are especially suited for accessing arrays, tables, and stack pointer usage of the X-pointer register. Note that only the low byte of the X-pointer is updated in devices with no more than 256 bytes data space. For such devices, the high byte of the pointer is not used by this instruction and can be used for other purposes. The RAMPX register in the I/O area is updated in parts with more than 64K bytes data space or more than 64K bytes program memory, and the increment/ decrement is added to the entire 24-bit address on such devices.

Flags: ---.   Cycles: 2

	Syntax:	Operation:	Comment:
(i)	ST X, Rr	(X) ← Rr	X: Unchanged
(ii)	ST X+, Rr	(X) ← Rr X ← X + 1	X: Postincremented
(iii)	ST −X, Rr	X ← X − 1 (X) ← Rr	X: Predecremented

Example:

```
    clr r27            ;Clear X high byte
    ldi r26,$60        ;Set X low byte to $60
    st X+,r0           ;Store r0 in data space loc. $60(X post inc)
    st X,r1            ;Store r1 in data space loc. $61
    ldi r26,$63        ;Set X low byte to $63
    st X,r2            ;Store r2 in data space loc. $63
    st -X,r3           ;Store r3 in data space loc. $62(X pre dec)
```

ST (STD)   ; Store Indirect From Register to Data Space
                                     ; using Index Y

Stores one byte indirect with or without displacement from a register to data space. For parts with SRAM, the data space consists of the register file, I/O memory, and internal SRAM (and external SRAM if applicable). For parts without SRAM, the data space consists of the register file only. The EEPROM has a separate address space.

The data location is pointed to by the Y (16 bits) pointer register in the register file. Memory access is limited to the current data segment of 64K bytes. To access another data segment in devices with more than 64K bytes data space, the RAMPY register in the I/O area has to be changed.

The Y-pointer register can either be left unchanged by the operation, or it can be post-incremented or pre-decremented. These features are especially suited for accessing arrays, tables, and stack pointer usage of the Y-pointer register. Note that only the low byte of the Y-pointer is updated in devices with no more than 256 bytes data space. For such devices, the high byte of the pointer is not used by this instruction and can be used for other purposes. The RAMPY register in the I/O area is updated in parts with more than 64K bytes data space or more than 64K bytes program memory, and the increment/ decrement/displacement is added to the entire 24-bit address on such devices.

Flags: ---.   Cycles:2

	Syntax:	Operation:	Comment:
(i)	ST Y, Rr	(Y) ← Rr	Y: Unchanged
(ii)	ST Y+, Rr	(Y) ← Rr Y ← Y + 1	Y: Postincremented
(iii)	ST –Y, Rr	Y ← Y – 1 (Y) ← Rr	Y: Predecremented
(iiii)	STD Y + q, Rr	(Y + q) ← Rr	Y: Unchanged
			q: Displacement

Example:

```
clr r29          ;Clear Y high byte
ldi r28,$60      ;Set Y low byte to $60
st Y+,r0         ;Store r0 in data space loc. $60 (Y postinc.)
st Y,r1          ;Store r1 in data space loc. $61
ldi r28,$63      ;Set Y low byte to $63
st Y,r2          ;Store r2 in data space loc. $63
st -Y,r3         ;Store r3 in data space loc. $62 (Y predec.)
std Y+2,r4       ;Store r4 in data space loc. $64
```

ST (STD)   **; Store Indirect From Register to Data Space using Index Z**

Stores one byte indirect with or without displacement from a register to data space. For parts with SRAM, the data space consists of the register file, I/O memory, and internal SRAM (and external SRAM if applicable). For parts without SRAM, the data space consists of the register file only. The EEPROM has a separate address space.

The data location is pointed to by the Z (16 bits) pointer register in the register file. Memory access is limited to the current data segment of 64K bytes. To access another data segment in devices with more than 64K bytes data space, the RAMPZ register in the I/O area has to be changed.

The Z-pointer register can either be left unchanged by the operation, or it can be post-incremented or pre-decremented. These features are especially suited for stack pointer usage of the Z-pointer register; however, because the Z-pointer register can be used for indirect subroutine calls, indirect jumps and table lookup, it is often more convenient to use the X or Y-pointer as a dedicated stack pointer. Note that only the low byte of the Z-pointer is updated in devices with no more than 256 bytes data space. For such devices, the high byte of the pointer is not used by this instruction and can be used for other purposes. The RAMPZ register in the I/O area is updated in parts with more than 64K bytes data space or more than 64K bytes program memory, and the increment/decrement/displacement is added to the entire 24-bit address on such devices.

Flags: ---.   Cycles: 2

	Syntax:	Operation:	Comment:
(i)	ST Z, Rr	(Z) ← Rr	Z: Unchanged
(ii)	ST Z+, Rr	(Z) ← Rr Z ← Z + 1	Z: Postincremented
(iii)	ST −Z, Rr	Z ← Z − 1 (Z) ← Rr	Z: Predecremented
(iiii)	STD Z + q, Rr	(Z + q) ← Rr	Z: Unchanged, q: Displacement

Example:

```
clr r31             ;Clear Z high byte
ldi r30,$60         ;Set Z low byte to $60
st Z+,r0            ;Store r0 in data space loc. $60 (Z postinc.)
st Z,r1             ;Store r1 in data space loc. $61
ldi r30,$63         ;Set Z low byte to $63
st Z,r2             ;Store r2 in data space loc. $63
st -Z,r3            ;Store r3 in data space loc. $62 (Z predec.)
std Z+2,r4          ;Store r4 in data space loc. $64
```

STS k,Rr   ; **Store Direct to Data Space**

$0 \leq r \leq 31, 0 \leq k \leq 65535$; (k) ← Rr

Stores one byte from a register to the data space. For parts with SRAM, the data space consists of the register file, I/O memory, and internal SRAM (and external SRAM if applicable). For parts without SRAM, the data space consists of the register file only. The EEPROM has a separate address space.

A 16-bit address must be supplied. Memory access is limited to the current data segment of 64K bytes. The STS instruction uses the RAMPD register to access memory above 64K bytes. To access another data segment in devices with more than 64K bytes data space, the RAMPD register in the I/O area has to be changed.

Flags:---.   Cycles: 2

Example:

```
lds r2,$FF00        ;Load r2 with the contents of location $FF00
add r2,r1           ;Add r1 to r2
sts $FF00,r2        ;Write back
```

SUB Rd,Rr   ; **Subtract without Carry**

$0 \leq d \leq 31, 0 \leq r \leq 31$; Rd ← Rd − Rr

Subtracts two registers and places the result in the destination register Rd.

Flags: H, S, V, N, Z, C.   Cycles: 1

Example:

```
        sub r13,r12     ;Subtract r12 from r13
        brne noteq      ;Branch if r12 not equal r13
        ...
noteq:  nop             ;Branch destination (do nothing)
```

SUBI Rd,K   ; **Subtract Immediate**

$16 \leq d \leq 31, 0 \leq K \leq 255$; Rd ← Rd − K

Subtracts a register and a constant and places the result in the destination register Rd. This instruction works on registers R16 to R31 and is very well suited for operations on the X, Y, and Z-pointers.

Flags: H, S, V, N, Z, C.   Cycles: 1

Example:

```
            subi r22,$11          ;Subtract $11 from r22
            brne noteq            ;Branch if r22 not equal $11
            ...
noteq:      nop                   ;Branch destination (do nothing)
```

SWAP Rd	**; Swap Nibbles**
0 ≤ d ≤ 31	**; R(7:4) ← Rd(3:0), R(3:0) ← Rd(7:4)**

Swaps high and low nibbles in a register.

Flags:---.   Cycles: 1

Example:

```
            inc r1                ;Increment r1
            swap r1               ;Swap high and low nibble of r1
            inc r1                ;Increment high nibble of r1
            swap r1               ;Swap back
```

TST Rd	**; Test for Zero or Minus**
0 ≤ d ≤ 31	**; Rd ← Rd • Rd**

Tests if a register is zero or negative. Performs a logical AND between a register and itself. The register will remain unchanged.

Flags: S, V ← 1, N, Z.   Cycles: 1

Example:

```
            tst r0                ;Test r0
            breq zero             ;Branch if r0=0
            ...
zero:       nop                   ;Branch destination (do nothing)
```

WDR	**; Watchdog Reset**

This instruction resets the watchdog timer. This instruction must be executed within a limited time given by the WD prescaler.

Flags:---.   Cycles: 1

Example:

```
            wdr                   ;Reset watchdog timer
```

SECTION A.3: AVR REGISTER SUMMARY

Address	Name	Bit 7	Bit 6	Bit 5	Bit 4	Bit 3	Bit 2	Bit 1	Bit 0
$3F ($5F)	SREG	I	T	H	S	V	N	Z	C
$3E ($5E)	SPH	–	–	–	–	SP11	SP10	SP9	SP8
$3D ($5D)	SPL	SP7	SP6	SP5	SP4	SP3	SP2	SP1	SP0
$3C ($5C)	OCR0	Timer/Counter0 Output Compare Register							
$3B ($5B)	GICR	INT1	INT0	INT2	–	–	–	IVSEL	IVCE
$3A ($5A)	GIFR	INTF1	INTF0	INTF2	–	–	–	–	–
$39 ($59)	TIMSK	OCIE2	TOIE2	TICIE1	OCIE1A	OCIE1B	TOIE1	OCIE0	TOIE0
$38 ($58)	TIFR	OCF2	TOV2	ICF1	OCF1A	OCF1B	TOV1	OCF0	TOV0
$37 ($57)	SPMCR	SPMIE	RWWSB	–	RWWSRE	BLBSET	PGWRT	PGERS	SPMEN
$36 ($56)	TWCR	TWINT	TWEA	TWSTA	TWSTO	TWWC	TWEN	–	TWIE
$35 ($55)	MCUCR	SE	SM2	SM1	SM0	ISC11	ISC10	ISC01	ISC00
$34 ($54)	MCUCSR	JTD	ISC2	–	JTRF	WDRF	BORF	EXTRF	PORF
$33 ($53)	TCCR0	FOC0	WGM00	COM01	COM00	WGM01	CS02	CS01	CS00
$32 ($52)	TCNT0	Timer/Counter0 (8 Bits)							
$31 ($51)	OSCCAL	Oscillator Calibration Register							
	OCDR	On-Chip Debug Register							
$30 ($50)	SFIOR	ADTS2	ADTS1	ADTS0	–	ACME	PUD	PSR2	PSR10
$2F ($4F)	TCCR1A	COM1A1	COM1A0	COM1B1	COM1B0	FOC1A	FOC1B	WGM11	WGM10
$2E ($4E)	TCCR1B	ICNC1	ICES1	–	WGM13	WGM12	CS12	CS11	CS10
$2D ($4D)	TCNT1H	Timer/Counter1 – Counter Register High Byte							
$2C ($4C)	TCNT1L	Timer/Counter1 – Counter Register Low Byte							
$2B ($4B)	OCR1AH	Timer/Counter1 – Output Compare Register A High Byte							
$2A ($4A)	OCR1AL	Timer/Counter1 – Output Compare Register A Low Byte							
$29 ($49)	OCR1BH	Timer/Counter1 – Output Compare Register B High Byte							
$28 ($48)	OCR1BL	Timer/Counter1 – Output Compare Register B Low Byte							
$27 ($47)	ICR1H	Timer/Counter1 – Input Capture Register High Byte							
$26 ($46)	ICR1L	Timer/Counter1 – Input Capture Register Low Byte							
$25 ($45)	TCCR2	FOC2	WGM20	COM21	COM20	WGM21	CS22	CS21	CS20
$24 ($44)	TCNT2	Timer/Counter2 (8 Bits)							
$23 ($43)	OCR2	Timer/Counter2 Output Compare Register							
$22 ($42)	ASSR	–	–	–	–	AS2	TCN2UB	OCR2UB	TCR2UB
$21 ($41)	WDTCR	–	–	–	WDTOE	WDE	WDP2	WDP1	WDP0
$20 ($40)	UBRRH	URSEL	–	–	–	UBRR[11:8]			
	UCSRC	URSEL	UMSEL	UPM1	UPM0	USBS	UCSZ1	UCSZ0	UCPOL
$1F ($3F)	EEARH	–	–	–	–	–	–	EEAR9	EEAR8
$1E ($3E)	EEARL	EEPROM Address Register Low Byte							
$1D ($3D)	EEDR	EEPROM Data Register							
$1C ($3C)	EECR	–	–	–	–	EERIE	EEMWE	EEWE	EERE
$1B ($3B)	PORTA	PORTA7	PORTA6	PORTA5	PORTA4	PORTA3	PORTA2	PORTA1	PORTA0
$1A ($3A)	DDRA	DDA7	DDA6	DDA5	DDA4	DDA3	DDA2	DDA1	DDA0
$19 ($39)	PINA	PINA7	PINA6	PINA5	PINA4	PINA3	PINA2	PINA1	PINA0
$18 ($38)	PORTB	PORTB7	PORTB6	PORTB5	PORTB4	PORTB3	PORTB2	PORTB1	PORTB0
$17 ($37)	DDRB	DDB7	DDB6	DDB5	DDB4	DDB3	DDB2	DDB1	DDB0
$16 ($36)	PINB	PINB7	PINB6	PINB5	PINB4	PINB3	PINB2	PINB1	PINB0
$15 ($35)	PORTC	PORTC7	PORTC6	PORTC5	PORTC4	PORTC3	PORTC2	PORTC1	PORTC0
$14 ($34)	DDRC	DDC7	DDC6	DDC5	DDC4	DDC3	DDC2	DDC1	DDC0
$13 ($33)	PINC	PINC7	PINC6	PINC5	PINC4	PINC3	PINC2	PINC1	PINC0
$12 ($32)	PORTD	PORTD7	PORTD6	PORTD5	PORTD4	PORTD3	PORTD2	PORTD1	PORTD0
$11 ($31)	DDRD	DDD7	DDD6	DDD5	DDD4	DDD3	DDD2	DDD1	DDD0
$10 ($30)	PIND	PIND7	PIND6	PIND5	PIND4	PIND3	PIND2	PIND1	PIND0
$0F ($2F)	SPDR	SPI Data Register							
$0E ($2E)	SPSR	SPIF	WCOL	–	–	–	–	–	SPI2X
$0D ($2D)	SPCR	SPIE	SPE	DORD	MSTR	CPOL	CPHA	SPR1	SPR0
$0C ($2C)	UDR	USART I/O Data Register							
$0B ($2B)	UCSRA	RXC	TXC	UDRE	FE	DOR	PE	U2X	MPCM
$0A ($2A)	UCSRB	RXCIE	TXCIE	UDRIE	RXEN	TXEN	UCSZ2	RXB8	TXB8
$09 ($29)	UBRRL	USART Baud Rate Register Low Byte							
$08 ($28)	ACSR	ACD	ACBG	ACO	ACI	ACIE	ACIC	ACIS1	ACIS0
$07 ($27)	ADMUX	REFS1	REFS0	ADLAR	MUX4	MUX3	MUX2	MUX1	MUX0
$06 ($26)	ADCSRA	ADEN	ADSC	ADATE	ADIF	ADIE	ADPS2	ADPS1	ADPS0
$05 ($25)	ADCH	ADC Data Register High Byte							
$04 ($24)	ADCL	ADC Data Register Low Byte							
$03 ($23)	TWDR	Two-wire Serial Interface Data Register							
$02 ($22)	TWSR	TWS7	TWS6	TWS5	TWS4	TWS3	TWA1	TWA0	TWGCE
$01 ($21)	TWAR	TWA6	TWA5	TWA4	TWA3	TWA2	–	TWPS1	TWPS0
$00 ($20)	TWBR	Two-wire Serial Interface Bit Rate Register							

APPENDIX B

BASICS OF
WIRE WRAPPING

OVERVIEW

This appendix shows the basics of wire wrapping.

BASICS OF WIRE WRAPPING

Note: For this tutorial appendix, you will need the following:
Wire-wrapping tool (Radio Shack part number 276-1570)
30-gauge (30-AWG) wire for wire wrapping
(Thanks to Shannon Looper and Greg Boyle for their assistance on this section.)

The following describes the basics of wire wrapping.

1. There are several different types of wire-wrap tools available. The best one is available from Radio Shack for less than $10. The part number for Radio Shack is 276-1570. This tool combines the wrap and unwrap functions in the same end of the tool and includes a separate stripper. We found this to be much easier to use than the tools that combined all these features on one two-ended shaft. There are also wire-wrap guns, which are, of course, more expensive.

2. Wire-wrapping wire is available prestripped in various lengths or in bulk on a spool. The prestripped wire is usually more expensive and you are restricted to the different wire lengths you can afford to buy. Bulk wire can be cut to any length you wish, which allows each wire to be custom fit.

3. Serveral different types of wire-wrap boards are available. These are usually called *perfboards* or *wire-wrap boards*. These types of boards are sold at many electronics stores (such as Radio Shack). The best type of board has plating around the holes on the bottom of the board. These boards are better because the sockets and pins can be soldered to the board, which makes the circuit more mechanically stable.

4. Choose a board that is large enough to accommodate all the parts in your design with room to spare so that the wiring does not become too cluttered. If you wish to expand your project in the future, you should be sure to include enough room on the original board for the complete circuit. Also, if possible, the layout of the IC on the board needs to be such that signals go from left to right just like the schematics.

5. To make the wiring easier and to keep pressure off the pins, install one stand-off on each corner of the board. You may also wish to put standoffs on the top of the board to add stability when the board is on its back.

6. For power hook-up, use some type of standard binding post. Solder a few single wire-wrap pins to each power post to make circuit connections (to at least one pin for each IC in the circuit).

7. To further reduce problems with power, each IC must have its own connection to the main power of the board. If your perfboard does not have built-in power buses, run a separate power and ground wire from each IC to the main power. In other words, DO NOT daisy chain (making a chip-to-chip connection is called *daisy chaining*) power connections, as each connection down the line will have more wire and more resistance to get power through. However, daisy chaining is acceptable for other connections such as data, address, and control buses.

8. You must use wire-wrap sockets. These sockets have long square pins whose edges will cut into the wire as it is wrapped around the pin.

9. Wire wrapping will not work on round legs. If you need to wrap to compo-

nents, such as capacitors, that have round legs, you must also solder these connections. The best way to connect single components is to install individual wire-wrap pins into the board and then solder the components to the pins. An alternate method is to use an empty IC socket to hold small components such as resistors and wrap them to the socket.

10. The wire should be stripped about 1 inch. This will allow 7 to 10 turns for each connection. The first turn or turn-and-a-half should be insulated. This prevents stripped wire from coming in contact with other pins. This can be accomplished by inserting the wire as far as it will go into the tool before making the connection.

11. Try to keep wire lengths to a minimum. This prevents the circuit from looking like a bird nest. Be neat and use color coding as much as possible. Use only red wires for V_{CC} and black wires for ground connections. Also use different colors for data, address, and control signal connections. These suggestions will make troubleshooting much easier.

12. It is standard practice to connect all power lines first and check them for continuity. This will eliminate trouble later on.

13. It's also a good idea to mark the pin orientation on the bottom of the board. Plastic templates are available with pin numbers preprinted on them specifically for this purpose or you can make your own from paper. Forgetting to reverse pin order when looking at the bottom of the board is a very common mistake when wire wrapping circuits.

14. To prevent damage to your circuit, place a diode (such as IN5338) in reverse bias across the power supply. If the power gets hooked up backwards, the diode will be forward biased and will act as a short, keeping the reversed voltage from your circuit.

15. In digital circuits, there can be a problem with current demand on the power supply. To filter the noise on the power supply, a 100 μF electrolytic capacitor and a 0.1 μF monolithic capacitor are connected from V_{CC} to ground, in parallel with each other, at the entry point of the power supply to the board. These two together will filter both the high- and the low-frequency noises. Instead of using two capacitors in parallel, you can use a single 20–100 μF tantalum capacitor. Remember that the long lead is the positive one.

16. To filter the transient current, use a 0.1 μF monolithic capacitor for each IC. Place the 0.1 μF monolithic capacitor between V_{CC} and ground of each IC. Make sure the leads are as short as possible.

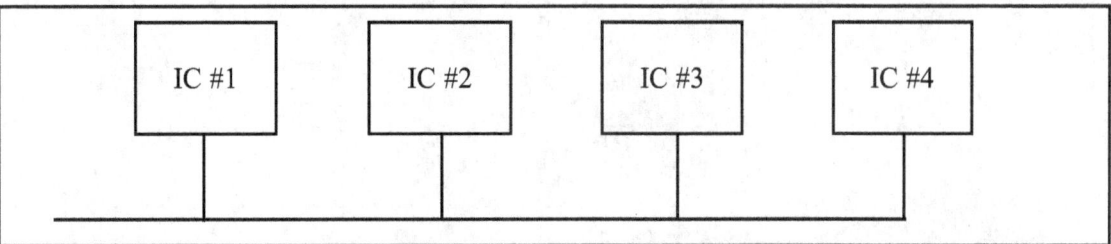

Figure B-1. Daisy Chain Connection (not recommended for power lines)

APPENDIX C

IC INTERFACING AND SYSTEM DESIGN ISSUES

OVERVIEW

This appendix provides an overview of IC technology and AVR interfacing. In addition, we look at the microcontroller-based system as a whole and examine some general issues in system design.

First, in Section C.1, we provide an overview of IC technology. Then, in Section C.2, the internal details of AVR I/O ports and interfacing are discussed. Section C.3 examines system design issues.

C.1: OVERVIEW OF IC TECHNOLOGY

In this section we examine IC technology and discuss some major developments in advanced logic families. Because this is an overview, it is assumed that the reader is familiar with logic families on the level presented in basic digital electronics books.

Transistors

The transistor was invented in 1947 by three scientists at Bell Laboratory. In the 1950s, transistors replaced vacuum tubes in many electronics systems, including computers. It was not until 1959 that the first integrated circuit was successfully fabricated and tested by Jack Kilby of Texas Instruments. Prior to the invention of the IC, the use of transistors, along with other discrete components such as capacitors and resistors, was common in computer design. Early transistors were made of germanium, which was later abandoned in favor of silicon. This was because the slightest rise in temperature resulted in massive current flows in germanium-based transistors. In semiconductor terms, it is because the band gap of germanium is much smaller than that of silicon, resulting in a massive flow of electrons from the valence band to the conduction band when the temperature rises even slightly. By the late 1960s and early 1970s, the use of the silicon-based IC was widespread in mainframes and minicomputers. Transistors and ICs at first were based on P-type materials. Later on, because the speed of electrons is much higher (about two-and-a-half times) than the speed of holes, N-type devices replaced P-type devices. By the mid-1970s, NPN and NMOS transistors had replaced the slower PNP and PMOS transistors in every sector of the electronics industry, including in the design of microprocessors and computers. Since the early 1980s, CMOS (complementary MOS) has become the dominant technology of IC design. Next we provide an overview of differences between MOS and bipolar transistors. See Figure C-1.

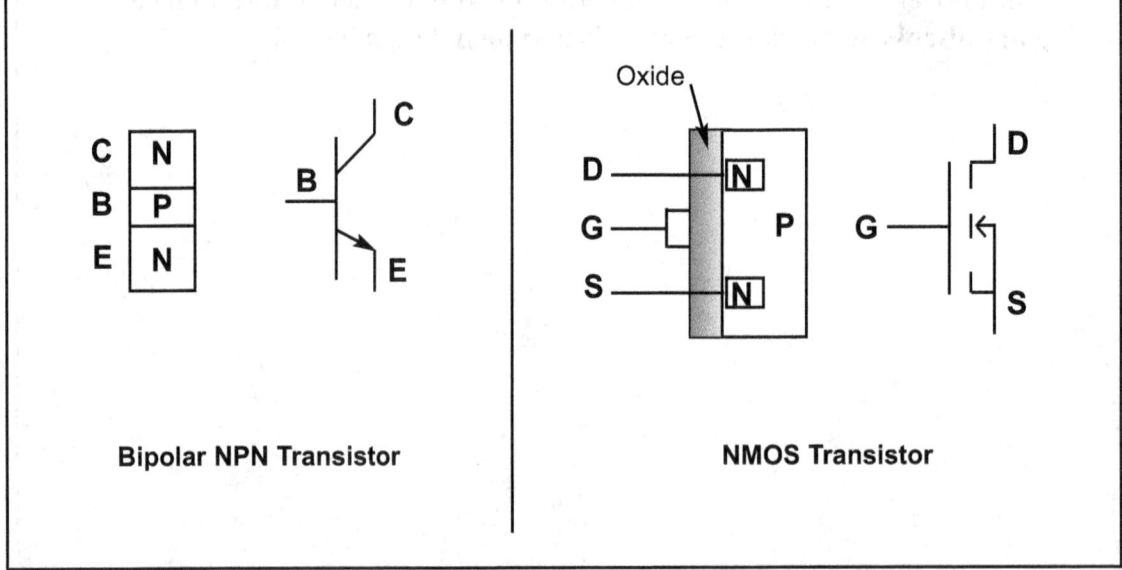

Figure C-1. Bipolar vs. MOS Transistors

MOS vs. bipolar transistors

There are two types of transistors: bipolar and MOS (metal-oxide semiconductor). Both have three leads. In bipolar transistors, the three leads are referred to as the *emitter*, *base*, and *collector*, while in MOS transistors they are named *source*, *gate*, and *drain*. In bipolar transistors, the carrier flows from the emitter to the collector, and the base is used as a flow controller. In MOS transistors, the carrier flows from the source to the drain, and the gate is used as a flow controller. In NPN-type bipolar transistors, the electron carrier leaving the emitter must overcome two voltage barriers before it reaches the collector (see Figure C-1). One is the N-P junction of the emitter-base and the other is the P-N junction of the base-collector. The voltage barrier of the base-collector is the most difficult one for the electrons to overcome (because it is reverse-biased) and it causes the most power dissipation. This led to the design of the unipolar type transistor called *MOS*. In N-channel MOS transistors, the electrons leave the source and reach the drain without going through any voltage barrier. The absence of any voltage barrier in the path of the carrier is one reason why MOS dissipates much less power than bipolar transistors. The low power dissipation of MOS allows millions of transistors to fit on a single IC chip. In today's technology, putting 10 million transistors into an IC is common, and it is all because of MOS technology. Without the MOS transistor, the advent of desktop personal computers would not have been possible, at least not so soon. The bipolar transistors in both the mainframes and minicomputers of the 1960s and 1970s were bulky and required expensive cooling systems and large rooms. MOS transistors do have one major drawback: They are slower than bipolar transistors. This is due partly to the gate capacitance of the MOS transistor. For a MOS to be turned on, the input capacitor of the gate takes time to charge up to the turn-on (threshold) voltage, leading to a longer propagation delay.

Overview of logic families

Logic families are judged according to (1) speed, (2) power dissipation, (3) noise immunity, (4) input/output interface compatibility, and (5) cost. Desirable qualities are high speed, low power dissipation, and high noise immunity (because it prevents the occurrence of false logic signals during switching transition). In interfacing logic families, the more inputs that can be driven by a single output, the better. This means that high-driving-capability outputs are desired. This, plus the fact that the input and output voltage levels of MOS and bipolar transistors are not compatible mean that one must be concerned with the ability of one logic family to drive the other one. In terms of the cost of a given logic family, it is high during the early years of its introduction but it declines as production and use rise.

The case of inverters

As an example of logic gates, we look at a simple inverter. In a one-transistor inverter, the transistor plays the role of a switch, and R is the pull-up resistor. See Figure C-2. For this inverter to work most effectively in digital circuits, however, the R value must be high when the transistor is "on" to limit the current flow from V_{CC} to ground in order to have low power dissipation (P = VI, where V

= 5 V). In other words, the lower the I, the lower the power dissipation. On the other hand, when the transistor is "off", R must be a small value to limit the voltage drop across R, thereby making sure that V_{OUT} is close to V_{CC}. This is a contradictory demand on R. This is one reason that logic gate designers use active components (transistors) instead of passive components (resistors) to implement the pull-up resistor R.

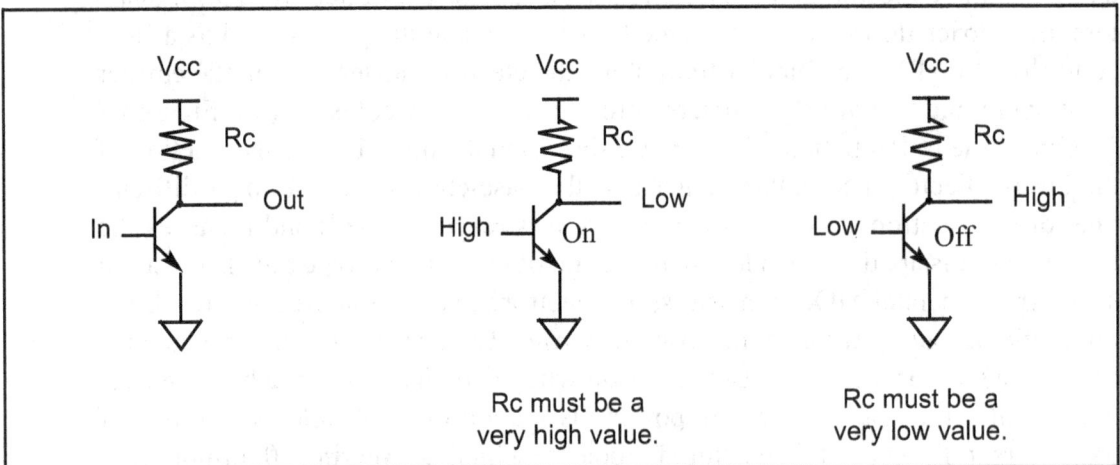

Figure C-2. One-Transistor Inverter with Pull-up Resistor

The case of a TTL inverter with totem-pole output is shown in Figure C-3. In Figure C-3, Q3 plays the role of a pull-up resistor.

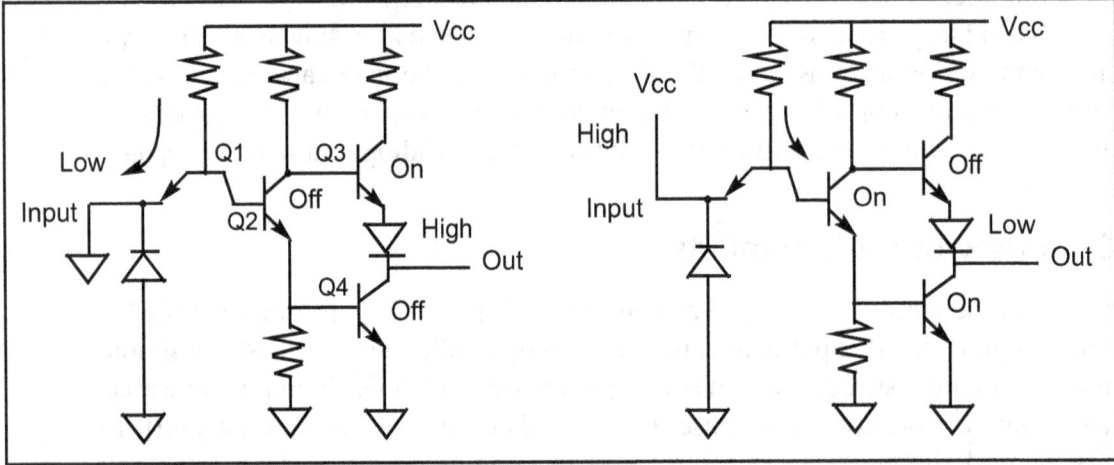

Figure C-3. TTL Inverter with Totem-Pole Output

CMOS inverter

In the case of CMOS-based logic gates, PMOS and NMOS are used to construct a CMOS (complementary MOS) inverter as shown in Figure C-4. In CMOS inverters, when the PMOS transistor is off, it provides a very high impedance path, making leakage current almost zero (about 10 nA); when the PMOS is on, it provides a low resistance on the path of V_{DD} to load. Because the speed of the hole is slower than that of the electron, the PMOS transistor is wider to compensate for this disparity; therefore, PMOS transistors take more space than NMOS transistors in the CMOS gates. At the end of this section we will see an open-collector gate in which the pull-up resistor is provided externally, thereby allowing system designers to choose the value of the pull-up resistor.

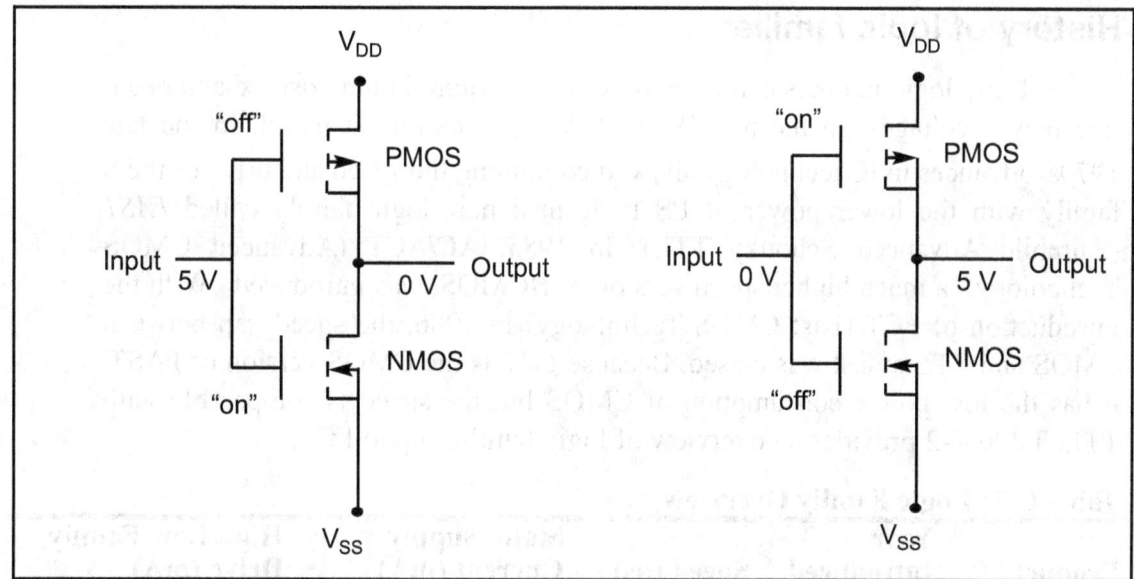

Figure C-4. CMOS Inverter

Input/output characteristics of some logic families

In 1968 the first logic family made of bipolar transistors was marketed. It was commonly referred to as the *standard TTL* (transistor-transistor logic) family. The first MOS-based logic family, the CD4000/74C series, was marketed in 1970. The addition of the Schottky diode to the base-collector of bipolar transistors in the early 1970s gave rise to the S family. The Schottky diode shortens the propagation delay of the TTL family by preventing the collector from going into what is called *deep saturation*. Table C-1 lists major characteristics of some logic families. In Table C-1, note that as the CMOS circuit's operating frequency rises, the power dissipation also increases. This is not the case for bipolar-based TTL.

Table C-1: Characteristics of Some Logic Families

Characteristic	STD TTL	LSTTL	ALSTTL	HCMOS
V_{CC}	5 V	5 V	5 V	5 V
V_{IH}	2.0 V	2.0 V	2.0 V	3.15 V
V_{IL}	0.8 V	0.8 V	0.8 V	1.1 V
V_{OH}	2.4 V	2.7 V	2.7 V	3.7 V
V_{OL}	0.4 V	0.5 V	0.4 V	0.4 V
I_{IL}	−1.6 mA	−0.36 mA	−0.2 mA	−1 μA
I_{IH}	40 μA	20 μA	20 μA	1 μA
I_{OL}	16 mA	8 mA	4 mA	4 mA
I_{OH}	−400 μA	−400 μA	−400 μA	4 mA
Propagation delay	10 ns	9.5 ns	4 ns	9 ns
Static power dissipation (f = 0)	10 mW	2 mW	1 mW	0.0025 nW
Dynamic power dissipation at f = 100 kHz	10 mW	2 mW	1 mW	0.17 mW

History of logic families

Early logic families and microprocessors required both positive and negative power voltages. In the mid-1970s, 5 V V_{CC} became standard. In the late 1970s, advances in IC technology allowed combining the speed and drive of the S family with the lower power of LS to form a new logic family called *FAST* (Fairchild Advanced Schottky TTL). In 1985, AC/ACT (Advanced CMOS Technology), a much higher speed version of HCMOS, was introduced. With the introduction of FCT (Fast CMOS Technology) in 1986, the speed gap between CMOS and TTL at last was closed. Because FCT is the CMOS version of FAST, it has the low power consumption of CMOS but the speed is comparable with TTL. Table C-2 provides an overview of logic families up to FCT.

Table C-2: Logic Family Overview

Product	Year Introduced	Speed (ns)	Static Supply Current (mA)	High/Low Family Drive (mA)
Std TTL	1968	40	30	–2/32
CD4K/74C	1970	70	0.3	–0.48/6.4
LS/S	1971	18	54	–15/24
HC/HCT	1977	25	0.08	–6/–6
FAST	1978	6.5	90	–15/64
AS	1980	6.2	90	–15/64
ALS	1980	10	27	–15/64
AC/ACT	1985	10	0.08	–24/24
FCT	1986	6.5	1.5	–15/64

Reprinted by permission of Electronic Design Magazine, c. 1991.

Recent advances in logic families

As the speed of high-performance microprocessors reached 25 MHz, it shortened the CPU's cycle time, leaving less time for the path delay. Designers normally allocate no more than 25% of a CPU's cycle time budget to path delay. Following this rule means that there must be a corresponding decline in the propagation delay of logic families used in the address and data path as the system frequency is increased. In recent years, many semiconductor manufacturers have responded to this need by providing logic families that have high speed, low noise, and high drive I/O. Table C-3 provides the characteristics of high-performance logic families introduced in recent years. ACQ/ACTQ are the second-generation advanced CMOS (ACMOS) with much lower noise. While ACQ has the CMOS input level, ACTQ is equipped with TTL-level input. The FCTx and FCTx-T are second-generation FCT with much higher speed. (The "x" in the FCTx and FCTx-T refers to various speed grades, such as A, B, and C, where A means low speed and C means high speed.) For designers who are well versed in using the FAST logic family, FASTr is an ideal choice because it is faster than FAST, has higher driving capability (I_{OL}, I_{OH}), and produces much lower noise than FAST. At the time of this writing, next to ECL and gallium arsenide logic gates, FASTr is the fastest logic family in the market (with the 5 V V_{CC}), but the power consumption is high relative to other logic families, as shown in Table C-3. The combining of high-speed bipolar TTL and the low power consumption of CMOS has given birth to what is called *BICMOS*. Although BICMOS seems to be the future trend in IC design, at this time it is expensive due to extra steps required in BICMOS IC fabrication, but in some cases there is no other choice. (For example, Intel's Pentium microprocessor, a BICMOS product, had to use high-speed bipolar transistors to speed up some of the internal functions.) Table C-3 provides advanced logic characteristics. The "x" is for different speeds designated as A, B, and C. A is the slowest one while C is the fastest one. The above data is for the 74244 buffer.

Table C-3: Advanced Logic General Characteristics

Family	Year	Number Suppliers	Tech Base	I/O Level	Speed (ns)	Static Current	I_{OH}/I_{OL}
ACQ	1989	2	CMOS	CMOS/CMOS	6.0	80 μA	−24/24 mA
ACTQ	1989	2	CMOS	TTL/CMOS	7.5	80 μA	−24/24 mA
FCTx	1987	3	CMOS	TTL/CMOS	4.1–4.8	1.5 mA	−15/64 mA
FCTxT	1990	2	CMOS	TTL/TTL	4.1–4.8	1.5 mA	−15/64 mA
FASTr	1990	1	Bipolar	TTL/TTL	3.9	50 mA	−15/64 mA
BCT	1987	2	BICMOS	TTL/TTL	5.5	10 mA	−15/64 mA

Reprinted by permission of Electronic Design Magazine, c. 1991.

Since the late 1970s, the use of a +5 V power supply has become standard in all microprocessors and microcontrollers. To reduce power consumption, 3.3 V V_{CC} is being embraced by many designers. The lowering of V_{CC} to 3.3 V has two

major advantages: (1) It lowers the power consumption, prolonging the life of the battery in systems using a battery, and (2) it allows a further reduction of line size (design rule) to submicron dimensions. This reduction results in putting more transistors in a given die size. As fabrication processes improve, the decline in the line size is reaching submicron level and transistor densities are approaching 1 billion transistors.

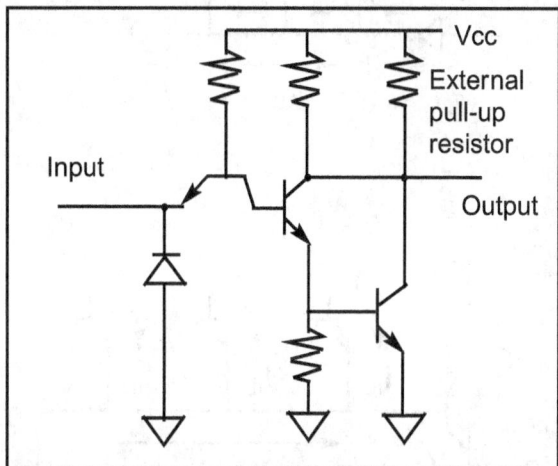

Figure C-5. Open Collector

Open-collector and open-drain gates

To allow multiple outputs to be connected together, we use open-collector logic gates. In such cases, an external resistor will serve as load. This is shown in Figures C-5 and C-6.

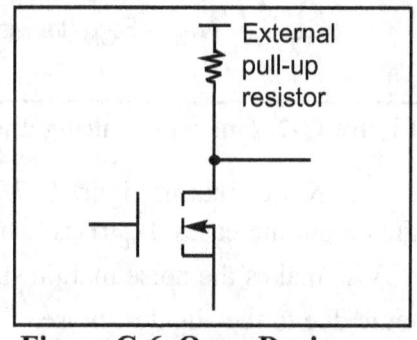

Figure C-6. Open Drain

SECTION C.2: AVR I/O PORT STRUCTURE AND INTERFACING

In interfacing the AVR microcontroller with other IC chips or devices, fan-out is the most important issue. To understand the AVR fan-out we must first understand the port structure of the AVR. This section provides a detailed discussion of the AVR port structure and its fan-out. It is very critical that we understand the I/O port structure of the AVR lest we damage it while trying to interface it with an external device.

IC fan-out

When connecting IC chips together, we need to find out how many input pins can be driven by a single output pin. This is a very important issue and involves the discussion of what is called *IC fan-out*. The IC fan-out must be addressed for both logic "0" and logic "1" outputs. See Example C-1. Fan-out for logic LOW and fan-out for logic HIGH are defined as follows:

$$\text{fan-out (of LOW)} = \frac{I_{OL}}{I_{IL}} \qquad\qquad \text{fan-out (of HIGH)} = \frac{I_{OH}}{I_{IH}}$$

Of the above two values, the lower number is used to ensure the proper noise margin. Figure C-7 shows the sinking and sourcing of current when ICs are connected together.

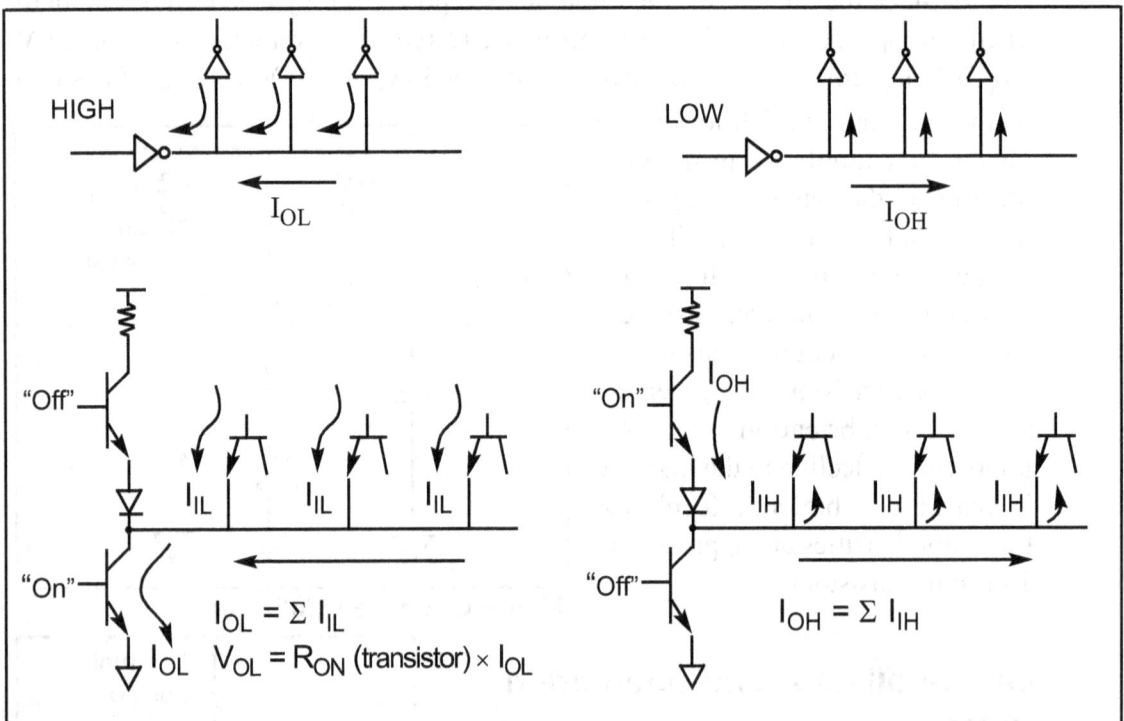

Figure C-7. Current Sinking and Sourcing in TTL

Notice that in Figure C-7, as the number of input pins connected to a single output increases, I_{OL} rises, which causes V_{OL} to rise. If this continues, the rise of V_{OL} makes the noise margin smaller, and this results in the occurrence of false logic due to the slightest noise.

Example C-1

Find how many unit loads (UL) can be driven by the output of the LS logic family.

Solution:

The unit load is defined as $I_{IL} = 1.6$ mA and $I_{IH} = 40$ μA. Table C-1 shows $I_{OH} = 400$ μA and $I_{OL} = 8$ mA for the LS family. Therefore, we have

$$\text{fan-out (LOW)} = \frac{I_{OL}}{I_{IL}} = \frac{8 \text{ mA}}{1.6 \text{ mA}} = 5$$

$$\text{fan-out (HIGH)} = \frac{I_{OH}}{I_{IH}} = \frac{400 \text{ μA}}{40 \text{ μA}} = 10$$

This means that the fan-out is 5. In other words, the LS output must not be connected to more than 5 inputs with unit load characteristics.

74LS244 and 74LS245 buffers/drivers

In cases where the receiver current requirements exceed the driver's capability, we must use buffers/drivers such as the 74LS245 and 74LS244. Figure C-8 shows the internal gates for the 74LS244 and 74LS245. The 74LS245 is used for bidirectional data buses, and the 74LS244 is used for unidirectional address buses.

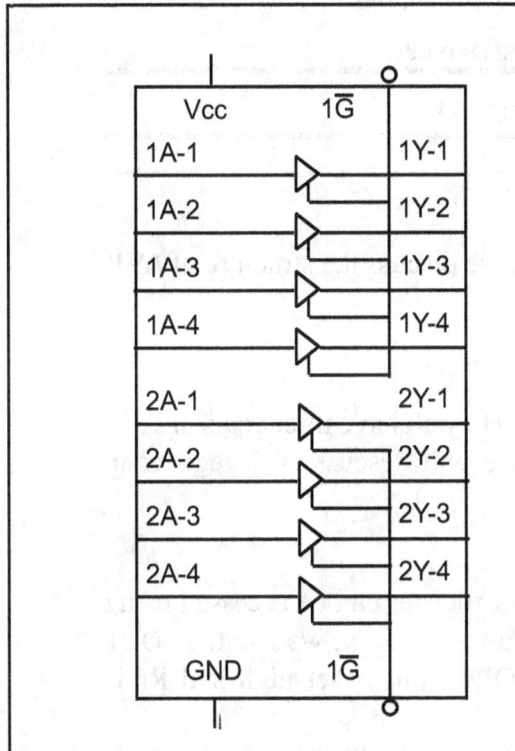

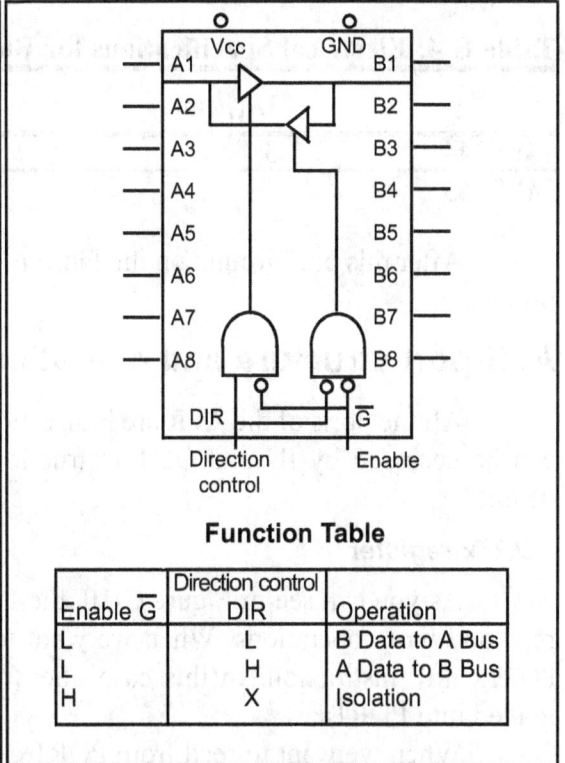

Function Table

Enable $\overline{G}$	Direction control DIR	Operation
L	L	B Data to A Bus
L	H	A Data to B Bus
H	X	Isolation

Figure C-8 (a). 74LS244 Octal Buffer
(Reprinted by permission of Texas Instruments, Copyright Texas Instruments, 1988)

Figure C-8 (b). 74LS245 Bidirectional Buffer
(Reprinted by permission of Texas Instruments, Copyright Texas Instruments, 1988)

Tri-state buffer

Notice that the 74LS244 is simply 8 tristate buffers in a single chip. As shown in Figure C-9 a tri-state buffer has a single input, a single output, and the enable control input. By activating the enable, data at the input is transferred to the output. The enable can be an active-LOW or an active-HIGH. Notice that the enable input for the 74LS244 is an active-LOW whereas the enable input pin for Figure C-9 is active-HIGH.

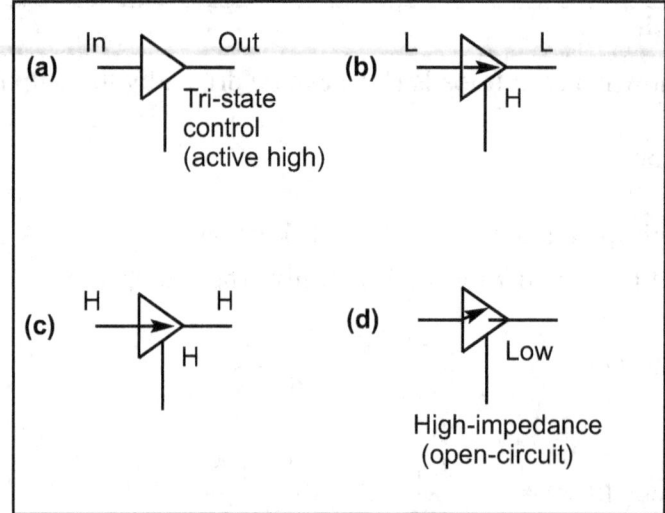

Figure C-9. Tri-State Buffer

74LS245 and 74LS244 fan-out

It must be noted that the output of the 74LS245 and 74LS244 can sink and source a much larger amount of current than that of other LS gates. See Table C-4. That is the reason we use these buffers for driver when a signal is travelling a long distance through a cable or it has to drive many inputs.

Table C-4: Electrical Specifications for Buffers/Drivers

	I_{OH} (mA)	I_{OL} (mA)
74LS244	3	12
74LS245	3	12

After this background on the fan-out, next we discuss the structure of AVR ports.

AVR port structure and operation

All the ports of the AVR are bidirectional. They all have three registers that can be accessed by IN and OUT instructions. We will descuss each register in detail.

PORTx register

As you can see in Figure C-10, the PORTx register can be accessed using read and write operations. When we want to write to PORTx, we use the "OUT PORTx, Rr" instruction. In this case, the WR-PORTx pin is set high and Rr is loaded into PORTx.

When we want to read from PORTx, we use "IN Rd, PORTx". In this case, the PRx pin is set to HIGH, which enables the buffer and makes it possible to read from PORTx.

The output of PORTx is either connected to the Px pin of the chip or con-

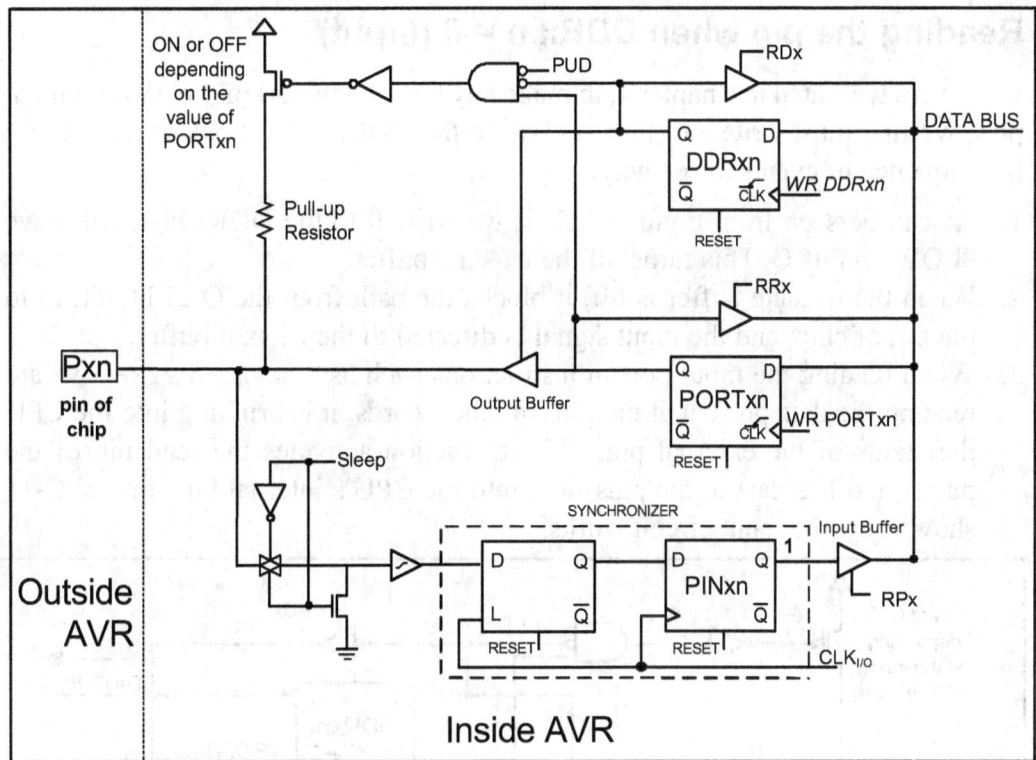

Figure C-10. The AVR Ports Structure

trols the pull-up resistor, as we will see next.

DDRx register

As shown in Figure C-10, the DDRx register can be accessed using read and write operations. When we want to write to DDRx, we use "OUT DDRx, Rr". In this case, the WR-DDRx pin is set to HIGH and enables writing to DDRx. When we want to read from DDRx, we use "IN Rd, DDRx". In this case, the RDx pin is set to LOW, which enables the buffer and makes it possible to read from DDRx.

The DDRx register controls the output buffer and the pull-up resistor. When the Q of DDRx is HIGH, it enables the output buffer and connects the Q of the PORTx register to the Px pin of the chip. In this case, the pin is configured as output. When the Q of DDRx is LOW, it disables the output buffer and configures the Px pin of the chip as input. In this case, assuming that the PUD bit is LOW, the Q of PORTx controls the pull-up resistor. When the Q of PORTx is HIGH, it enables the pull-up resistor, and when it is LOW, it disables the pull-up resistor.

PINx register

As you see in Figure C-10, when the AVR is not in sleep mode, the PINxn flip-flop is loaded with the value of the AVR pin on each machine cycle. Therefore, to read the current state of the Px pin of the chip, we should read the content of the PINx register. To do so, we use "IN Rd,PINx", which sets RPx high and enables the input buffer. In this case, the value of PINx passes through the internal data bus of AVR and will be loaded into the Rd register.

Reading the pin when DDRx.n = 0 (Input)

As we stated in Chapter 4, to make any bits of any port of the AVR an input port, we first must write a 0 (logic LOW) to the DDRx.n bit. Look at the following sequence of events to see why:

1. As can be seen from Figure C-11, if we write 0 to the DDRx.n, it will have "LOW" on its Q. This turns off the tri-state buffer.
2. When the tri-state buffer is off, it blocks the path from the Q of PORTx.n to the pin of chip, and the input signal is directed to the PINx.n buffer.
3. When reading the input port in instructions such as "IN R16, PINB" we are reading the data present at the pin. In other words, it is bringing into the CPU the status of the external pin. This instruction activates the read pin of the buffer and lets data at the pins flow into the CPU's internal bus. Figure C-11 shows how the input circuit works.

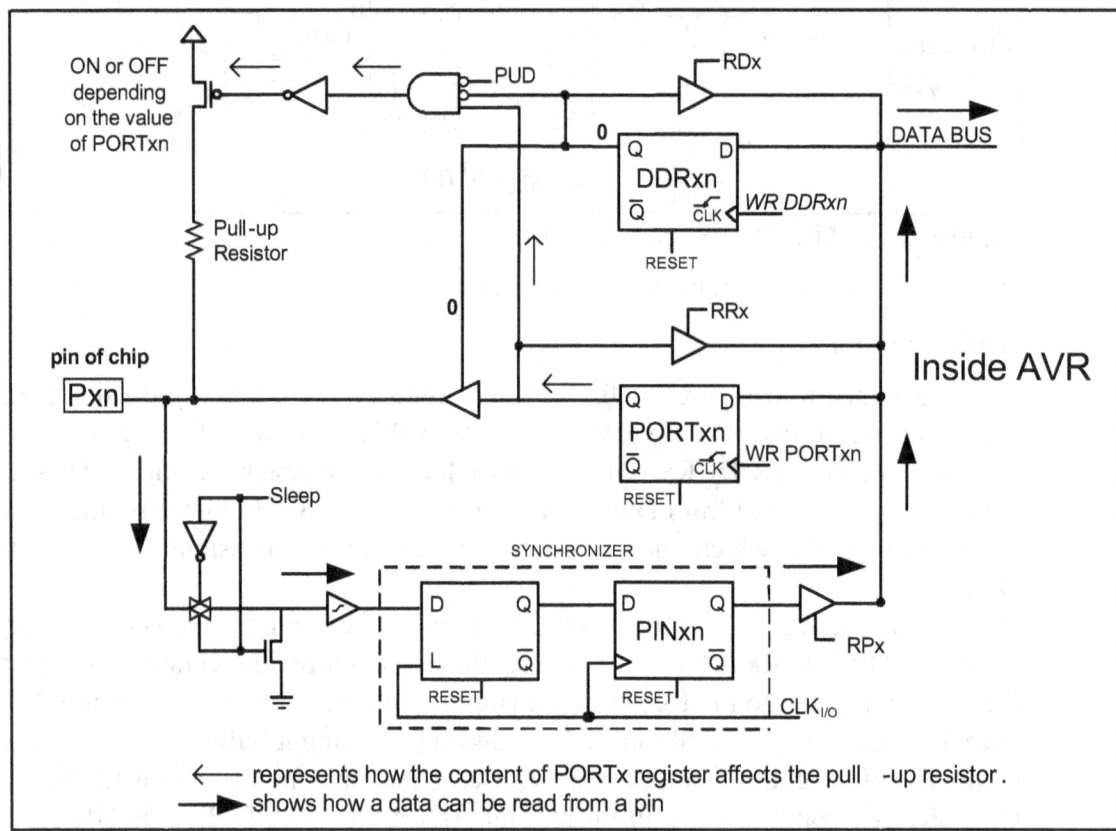

Figure C-11. Inputting (Reading) from a Pin via a PINx Register in the AVR

Writing to pin when DDRx.n = 1 (Output)

The above discussion showed why we must write a "LOW" to a port's DDRx.n bits in order to make it an input port. What happens if we write a "1" to DDRx.n that was configured as an input port? From Figure C-12 we see that when DDRx.n = 1, the DDRx.n latch has "HIGH" on its Q. This turns on the tri-state buffer, and the data of PORTx.n is transferred to the pin of chip.

From Figure C-12 we see that when DDRx.n = 1, if we write a 0 to the PORTx.n latch, then PORTx.n has "LOW" on its Q. This provides 0 to the pin of chip. Therefore, any attempt to read the input pin will always get the "LOW" ground signal. Figure C-13 shows what happens if we write "HIGH" to PORTx.n when DDRx.n = 1. Writing 1 to the PORTx.n makes Q = 1. As a result, a 1 is provided to the pin of the chip. Therefore, any attempt to read the input pin will always get the "HIGH" signal.

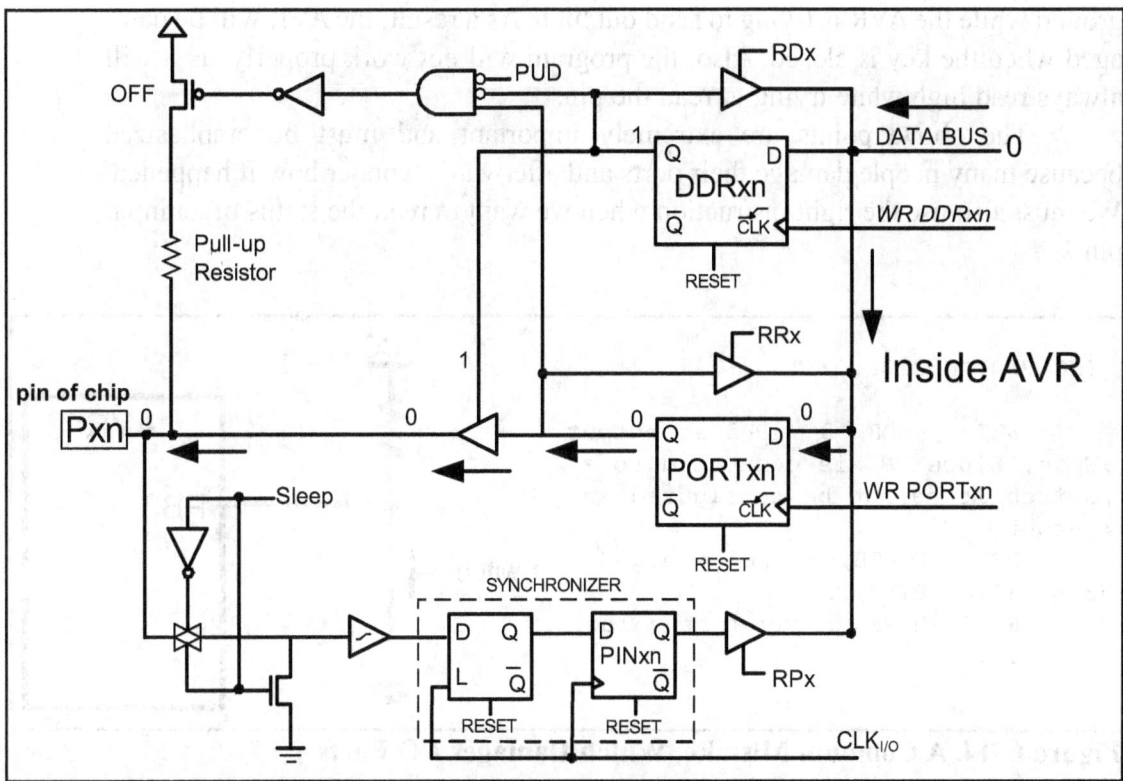

Figure C-12. Outputting (Writing) 0 to a Pin in the AVR

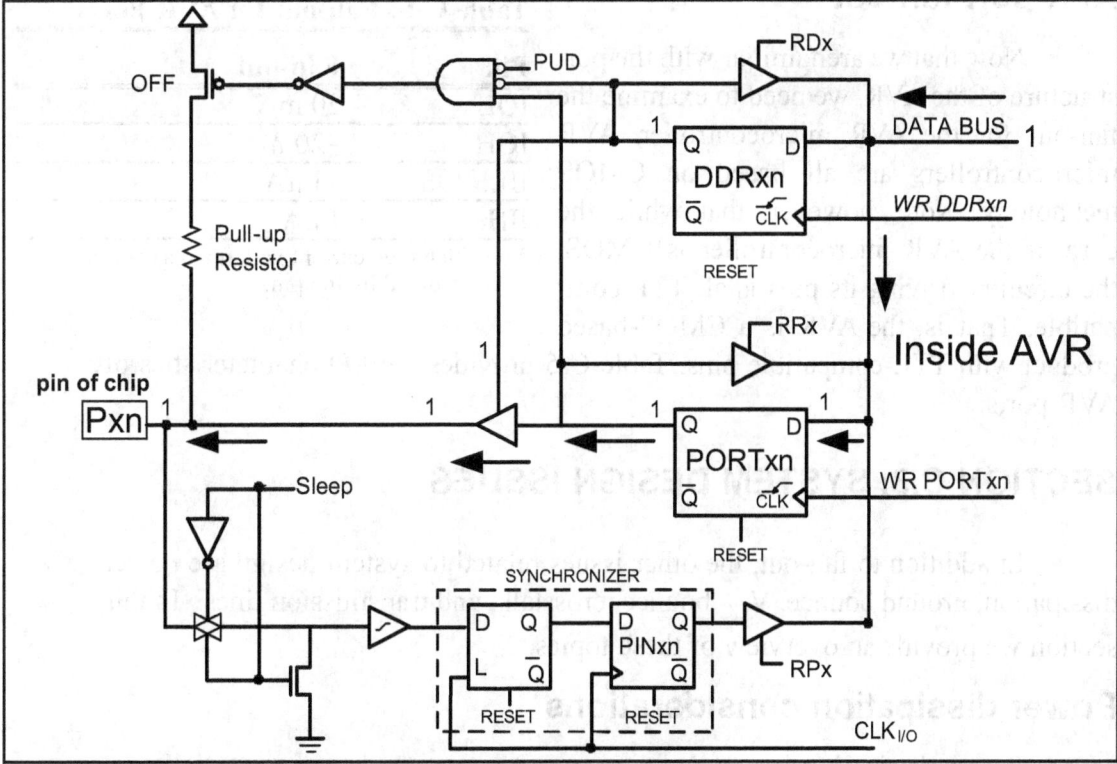

Figure C-13. Outputting (Writing) 1 to a Pin in the AVR

Notice that we should not make an I/O port output while it is externally connected to a voltage; otherwise, we might damage the ports.

For example, see Figure C-14. In this program, the PORTB.3 is mistakenly set as output. When the key is closed, the pin will be directly connected to ground while the AVR is trying to send out high. As a result, the AVR will be damaged when the key is closed. Also, the program will not work properly, as it will always read high while trying to read the pin.

The above points are extremely important and must be emphasized because many people damage their ports and afterwards wonder how it happened. We must also use the right instruction when we want to read the status of an input pin.

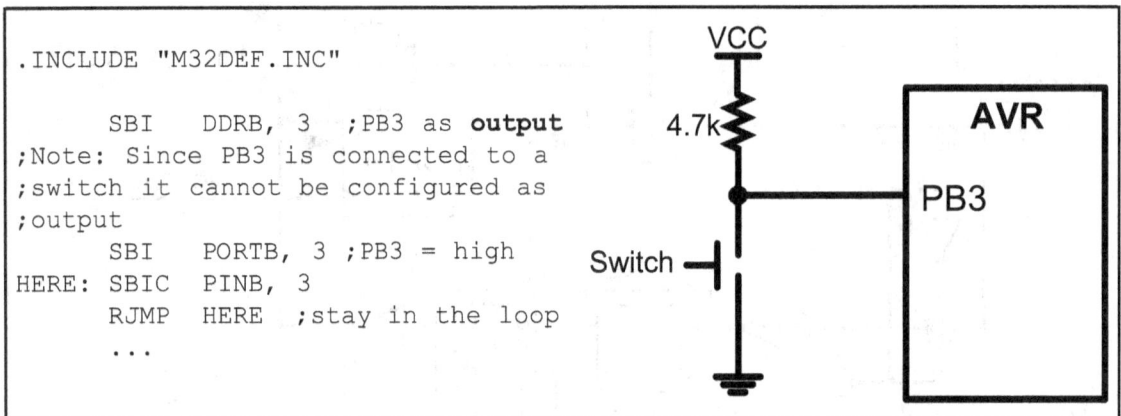

```
.INCLUDE "M32DEF.INC"

      SBI   DDRB, 3  ;PB3 as output
;Note: Since PB3 is connected to a
;switch it cannot be configured as
;output
      SBI   PORTB, 3 ;PB3 = high
HERE: SBIC  PINB, 3
      RJMP  HERE  ;stay in the loop
      ...
```

Figure C-14. A Common Mistake, Which Damages I/O Ports

AVR port fan-out

Now that we are familiar with the port structure of the AVR, we need to examine the fan-out for the AVR microcontroller. AVR microcontrollers are all based on CMOS technology. Note, however, that while the core of the AVR microcontroller is CMOS, the circuitry driving its pins is all TTL compatible. That is, the AVR is a CMOS-based product with TTL-compatible pins. Table C-5 provides the I/O characteristics of AVR ports.

Table C-5: Fan-out for AVR Ports

Pin	Fan-out
IOL	20 mA
IOH	−20 mA
IIL	−1 μA
IIH	1 μA

Note: Negative current is defined as current sourced by the pin.

SECTION C.3: SYSTEM DESIGN ISSUES

In addition to fan-out, the other issues related to system design are power dissipation, ground bounce, V_{CC} bounce, crosstalk, and transmission lines. In this section we provide an overview of these topics.

Power dissipation considerations

Power dissipation is a major concern of system designers, especially for laptop and hand-held systems in which batteries provide the power. Power dissipation is a function of frequency and voltage as shown below:

$$Q = CV$$

$$\frac{Q}{T} = \frac{CV}{T}$$

$$since \quad F = \frac{1}{T} \quad and \quad I = \frac{Q}{T}$$

$$I = CVF$$

$$now \quad P = VI = CV^2F$$

In the above equations, the effects of frequency and V_{CC} voltage should be noted. While the power dissipation goes up linearly with frequency, the impact of the power supply voltage is much more pronounced (squared). See Example C-2.

Example C-2

Compare the power consumption of two microcontroller-based systems. One uses 5 V and the other uses 3 V for V_{CC}.

Solution:
Because P = VI, by substituting I = V/R we have $P = V^2/R$. Assuming that R = 1, we have $P = 5^2 = 25$ W and $P = 3^2 = 9$ W. This results in using 16 W less power, which means power saving of 64% ($16/25 \times 100$) for systems using a 3 V power source.

Dynamic and static currents

Two major types of currents flow through an IC: dynamic and static. A dynamic current is I = CVF. It is a function of the frequency under which the component is working. This means that as the frequency goes up, the dynamic current and power dissipation go up. The static current, also called DC, is the current consumption of the component when it is inactive (not selected). The dynamic current dissipation is much higher than the static current consumption. To reduce power consumption, many microcontrollers, including the AVR, have power-saving modes. In the AVR, the power saving mode is called *sleep mode*. We describe the sleep mode next.

Sleep mode

In sleep mode the clocks of the CPU and some peripheral functions, such as serial ports, interrupts, and timers, are cut off. This brings power consumption down to an absolute minimum, while the contents of RAM and the SFR registers are saved and remain unchanged. The AVR provides six different sleeping modes, which enable you to choose which units will sleep. For more information see the AVR datasheets.

Ground bounce

One of the major issues that designers of high-frequency systems must grapple with is ground bounce. Before we define ground bounce, we will discuss lead inductance of IC pins. There is a certain amount of capacitance, resistance, and inductance associated with each pin of the IC. The size of these elements varies depending on many factors such as length, area, and so on.

The inductance of the pins is commonly referred to as *self-inductance* because there is also what is called *mutual inductance*, as we will show below. Of the three components of capacitor, resistor, and inductor, the property of self-inductance is the one that causes the most problems in high-frequency system design because it can result in ground bounce. Ground bounce occurs when a massive amount of current flows through the ground pin caused by many outputs changing from HIGH to LOW all at the same time. See Figure C-15 (a). The voltage is related to the inductance of the ground lead as follows:

$$V = L \frac{di}{dt}$$

As we increase the system frequency, the rate of dynamic current, di/dt, is also increased, resulting in an increase in the inductance voltage L (di/dt) of the ground pin. Because the LOW state (ground) has a small noise margin, any extra voltage due to the inductance can cause a false signal. To reduce the effect of ground bounce, the following steps must be taken where possible:

1. The V_{CC} and ground pins of the chip must be located in the middle rather than at opposite ends of the IC chip (the 14-pin TTL logic IC uses pins 14 and 7 for ground and V_{CC}). This is exactly what we see in high-performance logic gates such as Texas Instruments' advanced logic AC11000 and ACT11000 families. For example, the ACT11013 is a 14-pin DIP chip in which pin numbers 4 and 11 are used for the ground and V_{CC}, instead of 7 and 14 as in the traditional TTL family. We can also use the SOIC packages instead of DIP.
2. Another solution is to use as many pins for ground and V_{CC} as possible to reduce the lead length. This is exactly why all high-performance microprocessors and logic families use many pins for V_{CC} and ground instead of the traditional single pin for V_{CC} and single pin for GND. For example, in the case of Intel's Pentium processor there are over 50 pins for ground, and another 50 pins for V_{CC}.

The above discussion of ground bounce is also applicable to V_{CC} when a large number of outputs changes from the LOW to the HIGH state; this is referred to as V_{CC} *bounce*. However, the effect of V_{CC} bounce is not as severe as ground bounce because the HIGH ("1") state has a wider noise margin than the LOW ("0") state.

Filtering the transient currents using decoupling capacitors

In the TTL family, the change of the output from LOW to HIGH can cause what is called *transient current*. In a totem-pole output in which the output is LOW, Q4 is on and saturated, whereas Q3 is off. By changing the output from the

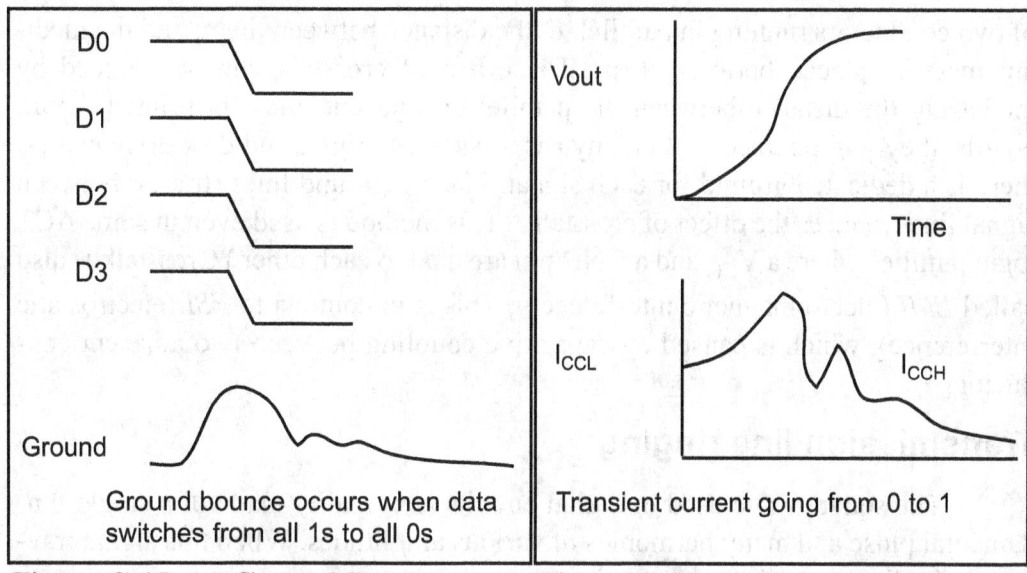

Figure C-15. (a) Ground Bounce

Ground bounce occurs when data switches from all 1s to all 0s

Figure C-15. (b) Transient Current

Transient current going from 0 to 1

LOW to the HIGH state, Q3 turns on and Q4 turns off. This means that there is a time when both transistors are on and drawing current from V_{CC}. The amount of current depends on the R_{ON} values of the two transistors, which in turn depend on the internal parameters of the transistors. The net effect of this, however, is a large amount of current in the form of a spike for the output current, as shown in Figure C-15 (b). To filter the transient current, a 0.01 μF or 0.1 μF ceramic disk capacitor can be placed between the V_{CC} and ground for each TTL IC. The lead for this capacitor, however, should be as small as possible because a long lead results in a large self-inductance, and that results in a spike on the V_{CC} line [V = L (di/dt)]. This spike is called V_{CC} *bounce*. The ceramic capacitor for each IC is referred to as a *decoupling capacitor*. There is also a bulk decoupling capacitor, as described next.

Bulk decoupling capacitor

If many IC chips change state at the same time, the combined currents drawn from the board's V_{CC} power supply can be massive and may cause a fluctuation of V_{CC} on the board where all the ICs are mounted. To eliminate this, a relatively large decoupling tantalum capacitor is placed between the V_{CC} and ground lines. The size and location of this tantalum capacitor vary depending on the number of ICs on the board and the amount of current drawn by each IC, but it is common to have a single 22 μF to 47 μF capacitor for each of the 16 devices, placed between the V_{CC} and ground lines.

Crosstalk

Crosstalk is due to mutual inductance. See Figure C-16. Previously, we discussed self-inductance, which is inherent in a piece of conductor. *Mutual inductance* is caused by two electric lines running parallel to each other. The mutual inductance is a function of l, the length

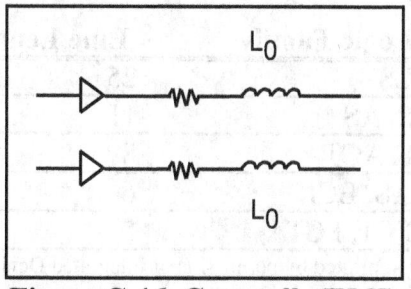

Figure C-16. Crosstalk (EMI)

of two conductors running in parallel; d, the distance between them; and the medium material placed between them. The effect of crosstalk can be reduced by increasing the distance between the parallel or adjacent lines (in printed circuit boards, they will be traces). In many cases, such as printer and disk drive cables, there is a dedicated ground for each signal. Placing ground lines (traces) between signal lines reduces the effect of crosstalk. (This method is used even in some ACT logic families where a V_{CC} and a GND pin are next to each other.) Crosstalk is also called *EMI* (electromagnetic interference). This is in contrast to *ESI* (electrostatic interference), which is caused by capacitive coupling between two adjacent conductors.

Transmission line ringing

The square wave used in digital circuits is in reality made of a single fundamental pulse and many harmonics of various amplitudes. When this signal travels on the line, not all the harmonics respond in the same way to the capacitance, inductance, and resistance of the line. This causes what is called *ringing*, which depends on the thickness and the length of the line driver, among other factors. To reduce the effect of ringing, the line drivers are terminated by putting a resistor at the end of the line. See Figure C-17. There are three major methods of line driver termination: parallel, serial, and Thevenin.

In serial termination, resistors of 30–50 ohms are used to terminate the line. The parallel and Thevenin methods are used in cases where there is a need to match the impedance of the line with the load impedance. This requires a detailed analysis of the signal traces and load impedance, which is beyond the scope of this book. In high-frequency systems, wire traces on the printed circuit board (PCB) behave like transmission lines, causing ringing. The severity of this ringing depends on the speed and the logic family used. Table C-6 provides the trace length, beyond which the traces must be looked at as transmission lines.

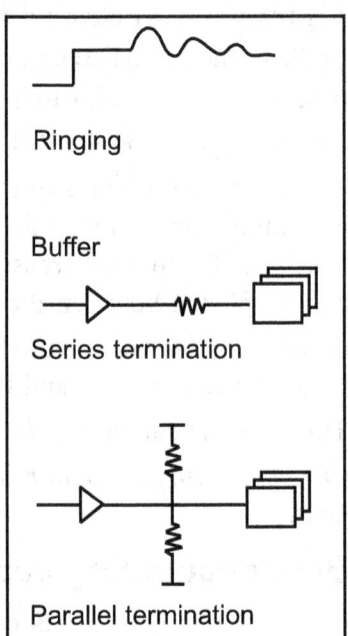

Figure C-17. Reducing Transmission Line Ringing

Table C-6: Line Length Beyond Which Traces Behave Like Transmission Lines

Logic Family	Line Length (in.)
LS	25
S, AS	11
F, ACT	8
AS, ECL	6
FCT, FCTA	5

(Reprinted by permission of Integrated Device Technology, copyright IDT 1991)

APPENDIX D

FLOWCHARTS AND PSEUDOCODE

OVERVIEW

This appendix provides an introduction to writing flowcharts and pseudocode.

Flowcharts

If you have taken any previous programming courses, you are probably familiar with flowcharting. Flowcharts use graphic symbols to represent different types of program operations. These symbols are connected together into a flowchart to show the flow of execution of a program. Figure D-1 shows some of the more commonly used symbols. Flowchart templates are available to help you draw the symbols quickly and neatly.

Pseudocode

Flowcharting has been standard practice in industry for decades. However, some find limitations in using flowcharts, such as the fact that you can't write much in the little boxes, and it is hard to get the "big picture" of what the program does without getting bogged down in the details. An alternative to using flowcharts is pseudocode, which involves writing brief descriptions of the flow of the code. Figures D-2 through D-6 show flowcharts and pseudocode for commonly used control structures.

Structured programming uses three basic types of program control structures: sequence, control, and itera-

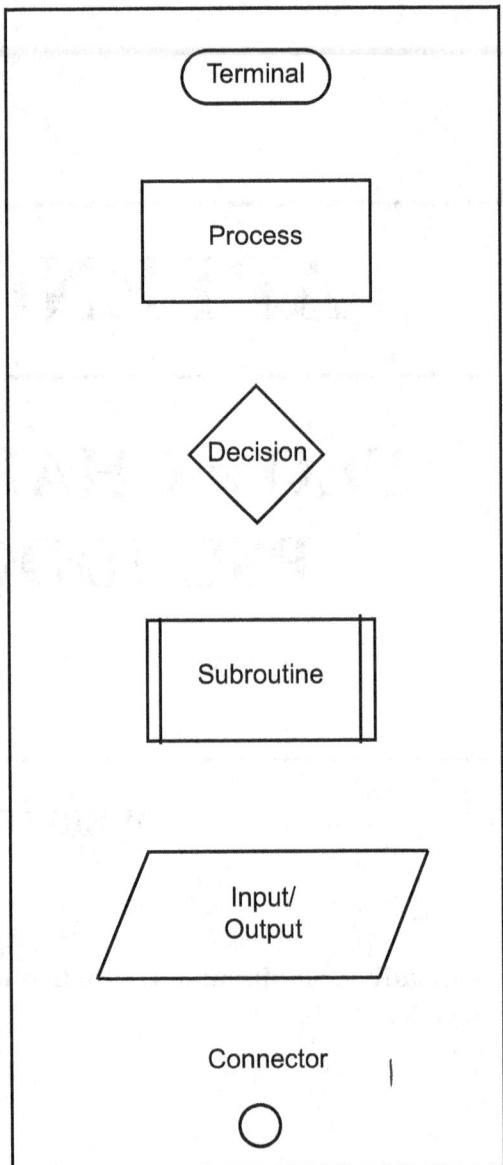

Figure D-1. Commonly Used Flowchart Symbols

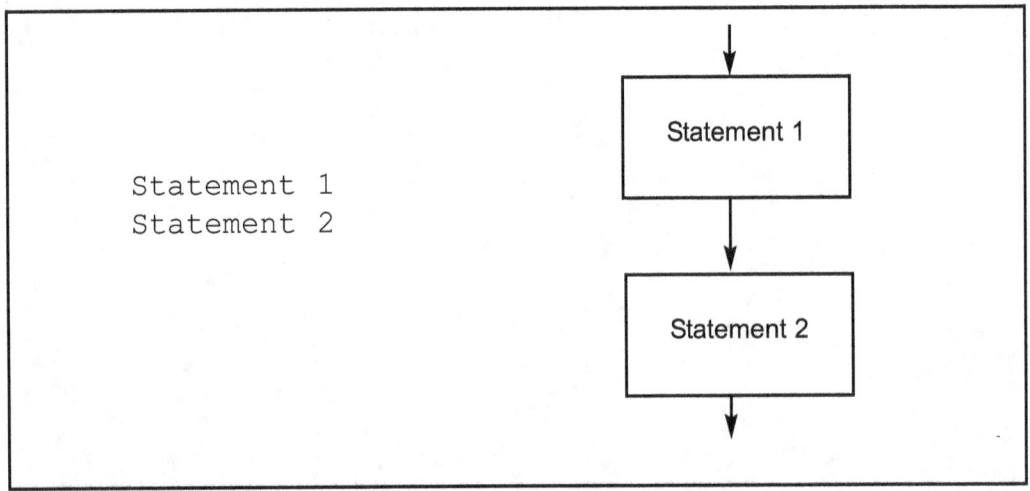

```
Statement 1
Statement 2
```

Figure D-2. SEQUENCE Pseudocode versus Flowchart

tion. Sequence is simply executing instructions one after another. Figure D-2 shows how sequence can be represented in pseudocode and flowcharts.

Figures D-3 and D-4 show two control programming structures: IF-THEN-ELSE and IF-THEN in both pseudocode and flowcharts.

Note in Figures D-2 through D-6 that "statement" can indicate one statement or a group of statements.

Figures D-5 and D-6 show two iteration control structures: REPEAT UNTIL and WHILE DO. Both structures execute a statement or group of statements repeatedly. The difference between them is that the REPEAT UNTIL structure always executes the statement(s) at least once, and checks the condition after each iteration, whereas the WHILE DO may not execute the statement(s) at all because the condition is checked at the beginning of each iteration.

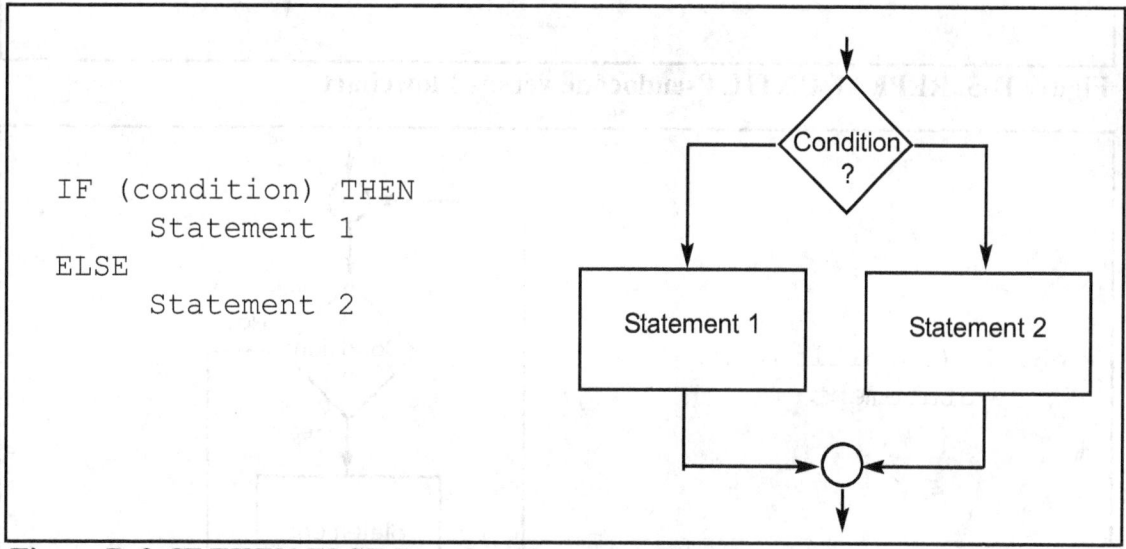

```
IF (condition) THEN
      Statement 1
ELSE
      Statement 2
```

Figure D-3. IF THEN ELSE Pseudocode versus Flowchart

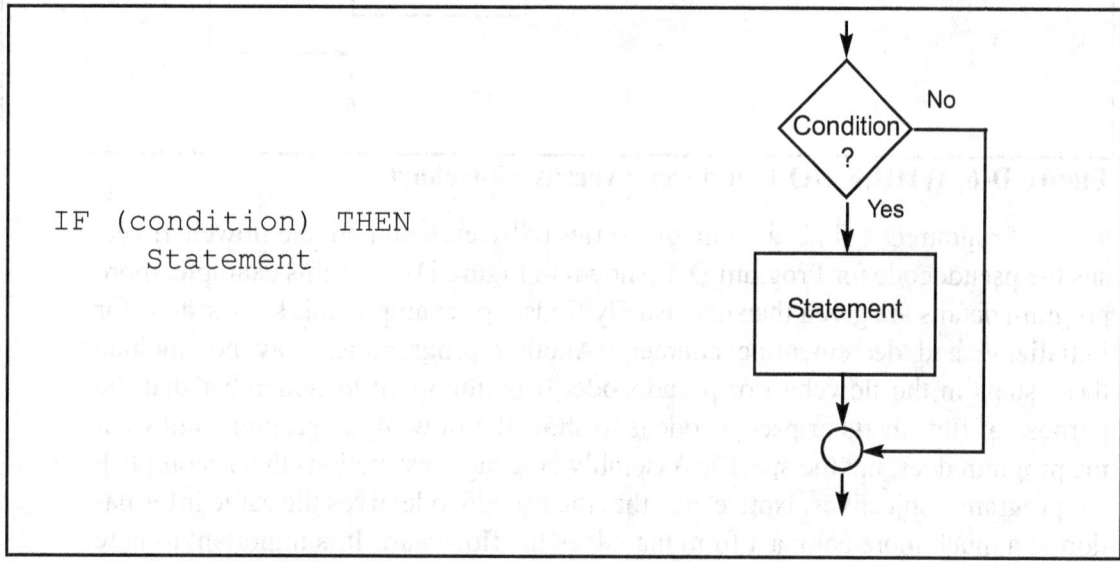

```
IF (condition) THEN
      Statement
```

Figure D-4. IF THEN Pseudocode versus Flowchart

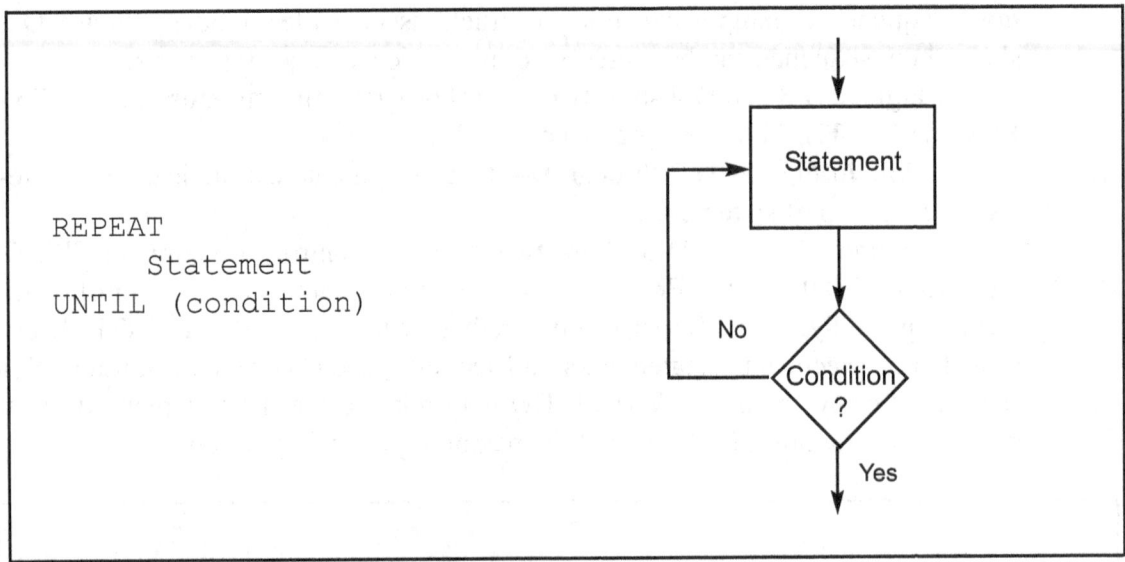

```
REPEAT
      Statement
UNTIL (condition)
```

Figure D-5. REPEAT UNTIL Pseudocode versus Flowchart

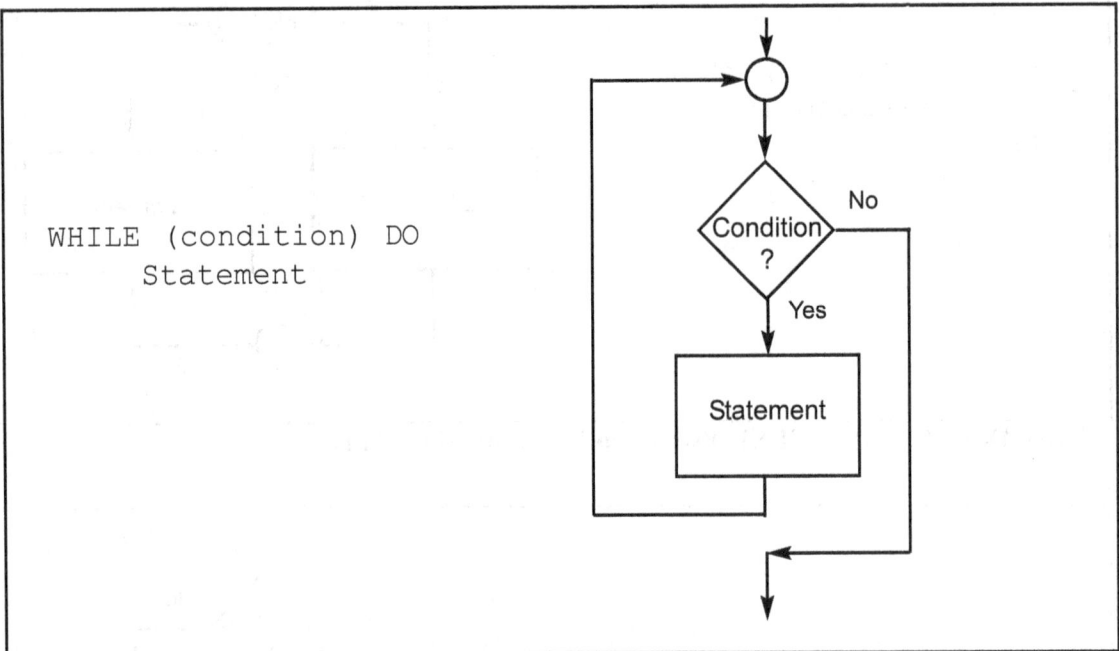

```
WHILE (condition) DO
      Statement
```

Figure D-6. WHILE DO Pseudocode versus Flowchart

Program D-1 finds the sum of a series of bytes. Compare the flowchart versus the pseudocode for Program D-1 (shown in Figure D-7). In this example, more program details are given than one usually finds. For example, this shows steps for initializing and decrementing counters. Another programmer may not include these steps in the flowchart or pseudocode. It is important to remember that the purpose of flowcharts or pseudocode is to show the flow of the program and what the program does, not the specific Assembly language instructions that accomplish the program's objectives. Notice also that the pseudocode gives the same information in a much more compact form than does the flowchart. It is important to note that sometimes pseudocode is written in layers, so that the outer level or layer shows the flow of the program and subsequent levels show more details of how the program accomplishes its assigned tasks.

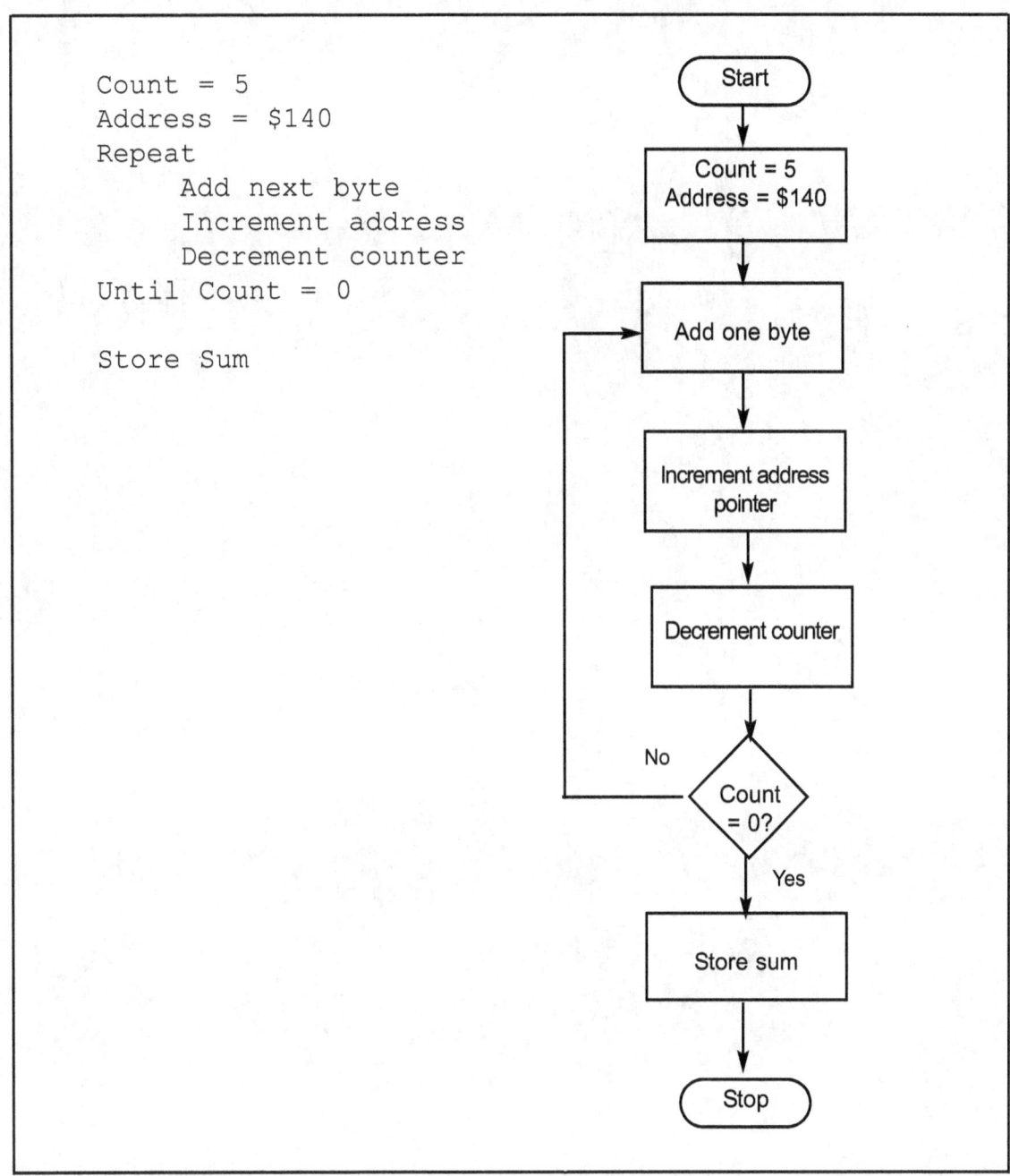

```
Count = 5
Address = $140
Repeat
      Add next byte
      Increment address
      Decrement counter
Until Count = 0

Store Sum
```

Figure D-7. Pseudocode versus Flowchart for Program D-1

```
#define     COUNTVAL      5        ;COUNT = 5
#define     COUNTER       R22
#define     SUM           R23
      LDI   COUNTER,COUNTVAL       ;R22 = 5
      CLR   SUM                    ;SUM = 0
      LDI   R26,LOW($140)          ;load pointer to RAM data address
      LDI   R27,HIGH($140)
L1:   LD    R24,x+                 ;copy RAM to R24 and increment pointer
      ADD   SUM,R24                ;add R24 to SUM
      DEC   COUNTER                ;decrement counter
      BRNE  L1                     ;loop until counter = zero
HERE: RJMP  HERE                   ;stay here forever
```

Program D-1

APPENDIX E

AVR PRIMER FOR 8051 PROGRAMMERS

	AVR	8051
8-bit registers:	32 general-purpose registers (R0 to R31)	A, B, R0, R1, R2, R3, R4, R5, R6, R7
16-bit (data pointer):	X, Y, Z	DPTR
Program Counter:	PC (up to 22-bit)	PC (16-bit)

Input:

```
        IN   Rn,PINx              MOV A,Pn ; (n = 0 - 3)
        (Use R0, R1, ..., R31.)
```

Output:

```
        OUT  PORTx,Rn             MOV Pn,A ; (n = 0 - 3)
```

Loop:

```
        DEC  Rn                   DJNZ R3,TARGET
        BRNE TARGET               (Using R0-R7)
```

Stack pointer:

```
        SP (16-bit)               SP (8-bit)
        As we PUSH data onto the  As we PUSH data onto the
        stack, it decrements the SP.  stack, it increments the SP.

        As we POP data from the stack,  As we POP data from the
        it increments the SP.     stack, it decrements the SP.
```

Data movement:
   From the code segment:

```
        LPM  Rn,Z                 MOVC A,@A+PC
        (Use Z only.)
```

   From RAM using indirect addressing:

```
        LD   Rn,X                 MOV A,@R0
        (Use X, Y, or Z.)         (Use R0 or R1 only.)
```

   From RAM using direct addressing:

```
        LDS  Rn,k                 MOV A,RAM_addr
```

   To RAM using indirect addressing mode:

```
        ST   X,Rn                 MOV @R0,A
        (Use X, Y, or Z.)
```

   To RAM using direct addressing mode:

```
        STS  k,X                  MOV RAM_addr,A
        (Use X, Y, or Z.)
```

APPENDIX F

ASCII CODES

Ctrl	Dec	Hex	Ch	Code	Dec	Hex	Ch	Dec	Hex	Ch	Dec	Hex	Ch
^@	0	00		NUL	32	20		64	40	@	96	60	`
^A	1	01	☺	SOH	33	21	!	65	41	A	97	61	a
^B	2	02	☻	STX	34	22	"	66	42	B	98	62	b
^C	3	03	♥	ETX	35	23	#	67	43	C	99	63	c
^D	4	04	♦	EOT	36	24	$	68	44	D	100	64	d
^E	5	05	♣	ENQ	37	25	%	69	45	E	101	65	e
^F	6	06	♠	ACK	38	26	&	70	46	F	102	66	f
^G	7	07	•	BEL	39	27	'	71	47	G	103	67	g
^H	8	08	◘	BS	40	28	(	72	48	H	104	68	h
^I	9	09	○	HT	41	29	)	73	49	I	105	69	i
^J	10	0A	◙	LF	42	2A	*	74	4A	J	106	6A	j
^K	11	0B	♂	VT	43	2B	+	75	4B	K	107	6B	k
^L	12	0C	♀	FF	44	2C	,	76	4C	L	108	6C	l
^M	13	0D	♪	CR	45	2D	–	77	4D	M	109	6D	m
^N	14	0E	♫	SO	46	2E	.	78	4E	N	110	6E	n
^O	15	0F	☼	SI	47	2F	/	79	4F	O	111	6F	o
^P	16	10	►	DLE	48	30	0	80	50	P	112	70	p
^Q	17	11	◄	DC1	49	31	1	81	51	Q	113	71	q
^R	18	12	↕	DC2	50	32	2	82	52	R	114	72	r
^S	19	13	‼	DC3	51	33	3	83	53	S	115	73	s
^T	20	14	¶	DC4	52	34	4	84	54	T	116	74	t
^U	21	15	§	NAK	53	35	5	85	55	U	117	75	u
^V	22	16	▬	SYN	54	36	6	86	56	V	118	76	v
^W	23	17	↨	ETB	55	37	7	87	57	W	119	77	w
^X	24	18	↑	CAN	56	38	8	88	58	X	120	78	x
^Y	25	19	↓	EM	57	39	9	89	59	Y	121	79	y
^Z	26	1A	→	SUB	58	3A	:	90	5A	Z	122	7A	z
^[	27	1B	←	ESC	59	3B	;	91	5B	[	123	7B	{
^\	28	1C	∟	FS	60	3C	<	92	5C	\	124	7C		
^]	29	1D	↔	GS	61	3D	=	93	5D	]	125	7D	}
^^	30	1E	▲	RS	62	3E	>	94	5E	^	126	7E	~
^_	31	1F	▼	US	63	3F	?	95	5F	_	127	7F	⌂

Dec	Hex	Ch	Dec	Hex	Ch	Dec	Hex	Ch	Dec	Hex	Ch
128	80	Ç	160	A0	á	192	C0	└	224	E0	α
129	81	ü	161	A1	í	193	C1	┴	225	E1	β
130	82	é	162	A2	ó	194	C2	┬	226	E2	Γ
131	83	â	163	A3	ú	195	C3	├	227	E3	π
132	84	ä	164	A4	ñ	196	C4	─	228	E4	Σ
133	85	à	165	A5	Ñ	197	C5	┼	229	E5	σ
134	86	å	166	A6	ª	198	C6	╞	230	E6	µ
135	87	ç	167	A7	º	199	C7	╟	231	E7	τ
136	88	ê	168	A8	¿	200	C8	╚	232	E8	Φ
137	89	ë	169	A9	⌐	201	C9	╔	233	E9	θ
138	8A	è	170	AA	¬	202	CA	╩	234	EA	Ω
139	8B	ï	171	AB	½	203	CB	╦	235	EB	δ
140	8C	î	172	AC	¼	204	CC	╠	236	EC	∞
141	8D	ì	173	AD	¡	205	CD	═	237	ED	ø
142	8E	Ä	174	AE	«	206	CE	╬	238	EE	∈
143	8F	Å	175	AF	»	207	CF	╧	239	EF	∩
144	90	É	176	B0	░	208	D0	╨	240	F0	≡
145	91	æ	177	B1	▒	209	D1	╤	241	F1	±
146	92	Æ	178	B2	▓	210	D2	╥	242	F2	≥
147	93	ô	179	B3	│	211	D3	╙	243	F3	≤
148	94	ö	180	B4	┤	212	D4	╘	244	F4	⌠
149	95	ò	181	B5	╡	213	D5	╒	245	F5	⌡
150	96	û	182	B6	╢	214	D6	╓	246	F6	÷
151	97	ù	183	B7	╖	215	D7	╫	247	F7	≈
152	98	ÿ	184	B8	╕	216	D8	╪	248	F8	°
153	99	Ö	185	B9	╣	217	D9	┘	249	F9	∙
154	9A	Ü	186	BA	║	218	DA	┌	250	FA	·
155	9B	¢	187	BB	╗	219	DB	█	251	FB	√
156	9C	£	188	BC	╝	220	DC	▄	252	FC	ⁿ
157	9D	¥	189	BD	╜	221	DD	▌	253	FD	²
158	9E	₧	190	BE	╛	222	DE	▐	254	FE	■
159	9F	ƒ	191	BF	┐	223	DF	▀	255	FF	

APPENDIX G

ASSEMBLERS, DEVELOPMENT RESOURCES, AND SUPPLIERS

This appendix provides various sources for AVR assemblers, compilers, and trainers. In addition, it lists some suppliers for chips and other hardware needs. While these are all established products from well-known companies, neither the author nor the publisher assumes responsibility for any problem that may arise with any of them. You are neither encouraged nor discouraged from purchasing any of the products mentioned; you must make your own judgment in evaluating the products. This list is simply provided as a service to the reader. It also must be noted that the list of products is by no means complete or exhaustive.

The AVR Studio from Atmel
http://www.atmel.com

MicroC from mikroElectronika
http://www.mikroe.com

CodeVision
http://www.hpinfotech.ro

ImageCraft
http://www.imagecraft.com

Micro IDE
http://www.micro-ide.com

Figure G-1. Suppliers of Assemblers and Compilers

AVR assemblers

The AVR assembler is provided by Atmel and other companies. Some of the companies provide shareware versions of their products, which you can download from their websites. However, the size of code for these shareware versions is limited to a few KB. Figure G-1 lists some suppliers of assemblers and compilers.

AVR trainers

There are many companies that produce and market AVR trainers. Figure G-2 provides a list of some of them.

MicroDigitalEd
http://www.MicroDigitalEd.com

Digilent
http://www.digilentinc.com

Atmel
http://www.atmel.com

Figure G-2. Trainer Suppliers

Parts suppliers

Figure G-3 provides a list of suppliers for many electronics parts.

RSR Electronics
Electronix Express
365 Blair Road
Avenel, NJ 07001
Fax: (732) 381-1572
Mail Order: 1-800-972-2225
In New Jersey: (732) 381-8020
http://www.elexp.com

Altex Electronics
11342 IH-35 North
San Antonio, TX 78233
Fax: (210) 637-3264
Mail Order: 1-800-531-5369
http://www.altex.com

Digi-Key
1-800-344-4539 (1-800-DIGI-KEY)
Fax: (218) 681-3380
http://www.digikey.com

Radio Shack
http://www.radioshack.com

JDR Microdevices
1850 South 10th St.
San Jose, CA 95112-4108
Sales 1-800-538-5000
(408) 494-1400
Fax: 1-800-538-5005
Fax: (408) 494-1420
http://www.jdr.com

Mouser Electronics
958 N. Main St.
Mansfield, TX 76063
1-800-346-6873
http://www.mouser.com

Jameco Electronic
1355 Shoreway Road
Belmont, CA 94002-4100
1-800-831-4242
(415) 592-8097
Fax: 1-800-237-6948
Fax: (415) 592-2503
http://www.jameco.com

B. G. Micro
P. O. Box 280298
Dallas, TX 75228
1-800-276-2206 (orders only)
(972) 271-5546
Fax: (972) 271-2462
This is an excellent source of LCDs, ICs,
keypads, etc.
http://www.bgmicro.com

Tanner Electronics
1100 Valwood Parkway, Suite #100
Carrollton, TX 75006
(972) 242-8702
http://www.tannerelectronics.com

Figure G-3. Electronics Suppliers

APPENDIX H

DATA SHEETS

27. Electrical Characteristics

27.1 Absolute Maximum Ratings*

Operating Temperature.................................. -55°C to +125°C	
Storage Temperature.................................... -65°C to +150°C	
Voltage on any Pin except RESET with respect to Ground ....-0.5V to V_{CC}+0.5V	
Voltage on RESET with respect to Ground......-0.5V to +13.0V	
Maximum Operating Voltage 6.0V	
DC Current per I/O Pin .... 40.0 mA	
DC Current V_{CC} and GND Pins........................ 200.0 mA and 400.0 mA TQFP/MLF	

*NOTICE: Stresses beyond those listed under "Absolute Maximum Ratings" may cause permanent damage to the device. This is a stress rating only and functional operation of the device at these or other conditions beyond those indicated in the operational sections of this specification is not implied. Exposure to absolute maximum rating conditions for extended periods may affect device reliability.

27.2 DC Characteristics

T_A = -40°C to 85°C, V_{CC} = 2.7V to 5.5V (Unless Otherwise Noted)

Symbol	Parameter	Condition	Min	Typ	Max	Units
V_{IL}	Input Low Voltage except XTAL1 and RESET pins	V_{CC}=2.7 - 5.5 V_{CC}=4.5 - 5.5	-0.5		0.2 V_{CC}[1]	V
V_{IH}	Input High Voltage except XTAL1 and RESET pins	V_{CC}=2.7 - 5.5 V_{CC}=4.5 - 5.5	0.6 V_{CC}[2]		V_{CC} + 0.5	V
V_{IL1}	Input Low Voltage XTAL1 pin	V_{CC}=2.7 - 5.5	-0.5		0.1 V_{CC}[1]	V
V_{IH1}	Input High Voltage XTAL1 pin	V_{CC}=2.7 - 5.5 V_{CC}=4.5 - 5.5	0.7 V_{CC}[2]		V_{CC} + 0.5	V
V_{IL2}	Input Low Voltage RESET pin	V_{CC}=2.7 - 5.5	-0.5		0.2 V_{CC}	V
V_{IH2}	Input High Voltage RESET pin	V_{CC}=2.7 - 5.5	0.9 V_{CC}[2]		V_{CC} + 0.5	V
V_{OL}	Output Low Voltage[3] (Ports A,B,C,D)	I_{OL} = 20 mA, V_{CC} = 5V I_{OL} = 10 mA, V_{CC} = 3V			0.7 0.5	V V
V_{OH}	Output High Voltage[4] (Ports A,B,C,D)	I_{OH} = -20 mA, V_{CC} = 5V I_{OH} = -10 mA, V_{CC} = 3V	4.2 2.2			V V
I_{IL}	Input Leakage Current I/O Pin	V_{CC} = 5.5V, pin low (absolute value)			1	μA
I_{IH}	Input Leakage Current I/O Pin	V_{CC} = 5.5V, pin high (absolute value)			1	μA
R_{RST}	Reset Pull-up Resistor		30	60	85	kΩ
R_{pu}	I/O Pin Pull-up Resistor		20		50	kΩ

8155A–AVR–06/08

27.3 Speed Grades

Figure 27-1. Maximum Frequency vs. V_{CC}.

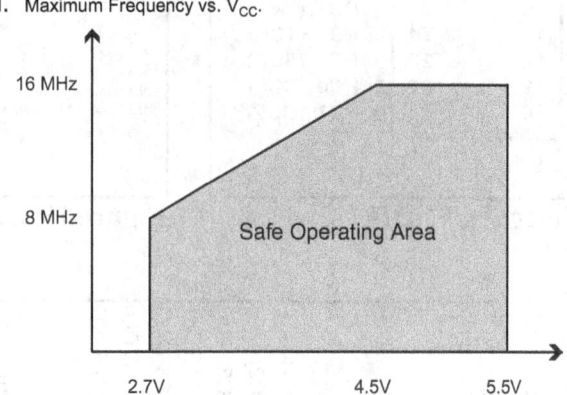

27.4 Clock Characteristics

27.4.1 External Clock Drive Waveforms

Figure 27-2. External Clock Drive Waveforms

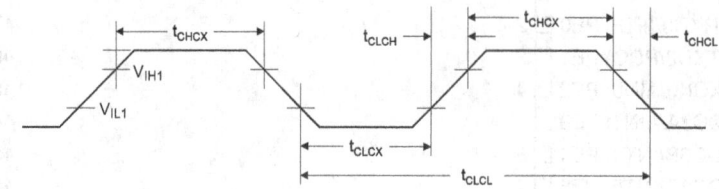

27.4.2 External Clock Drive

Figure 27-3. External Clock Drive

Symbol	Parameter	V_{CC} = 2.7V to 5.5V		V_{CC} = 4.5V to 5.5V		Units
		Min	Max	Min	Max	
$1/t_{CLCL}$	Oscillator Frequency	0	8	0	16	MHz
t_{CLCL}	Clock Period	125		62.5		ns
t_{CHCX}	High Time	50		25		ns
t_{CLCX}	Low Time	50		25		ns

298 **ATmega32A** ━━━━━━━━━━━━

8155A–AVR–06/08

All AVR data sheets are copyright of Atmel Semiconductor, Inc. 2009, used by permission.

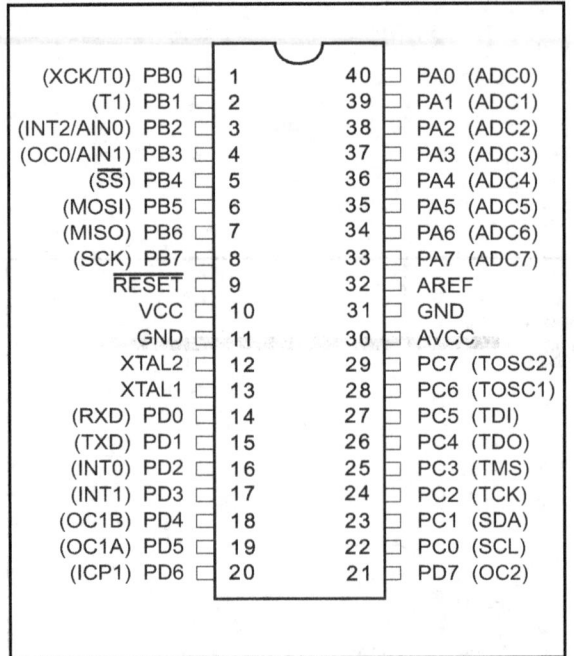

Figure H-1. ATmega16/32 DIP

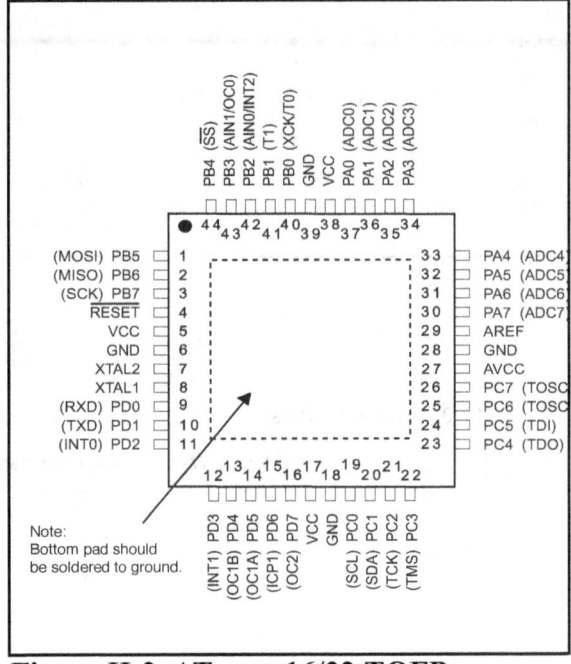

Figure H-2. ATmega16/32 TQFP

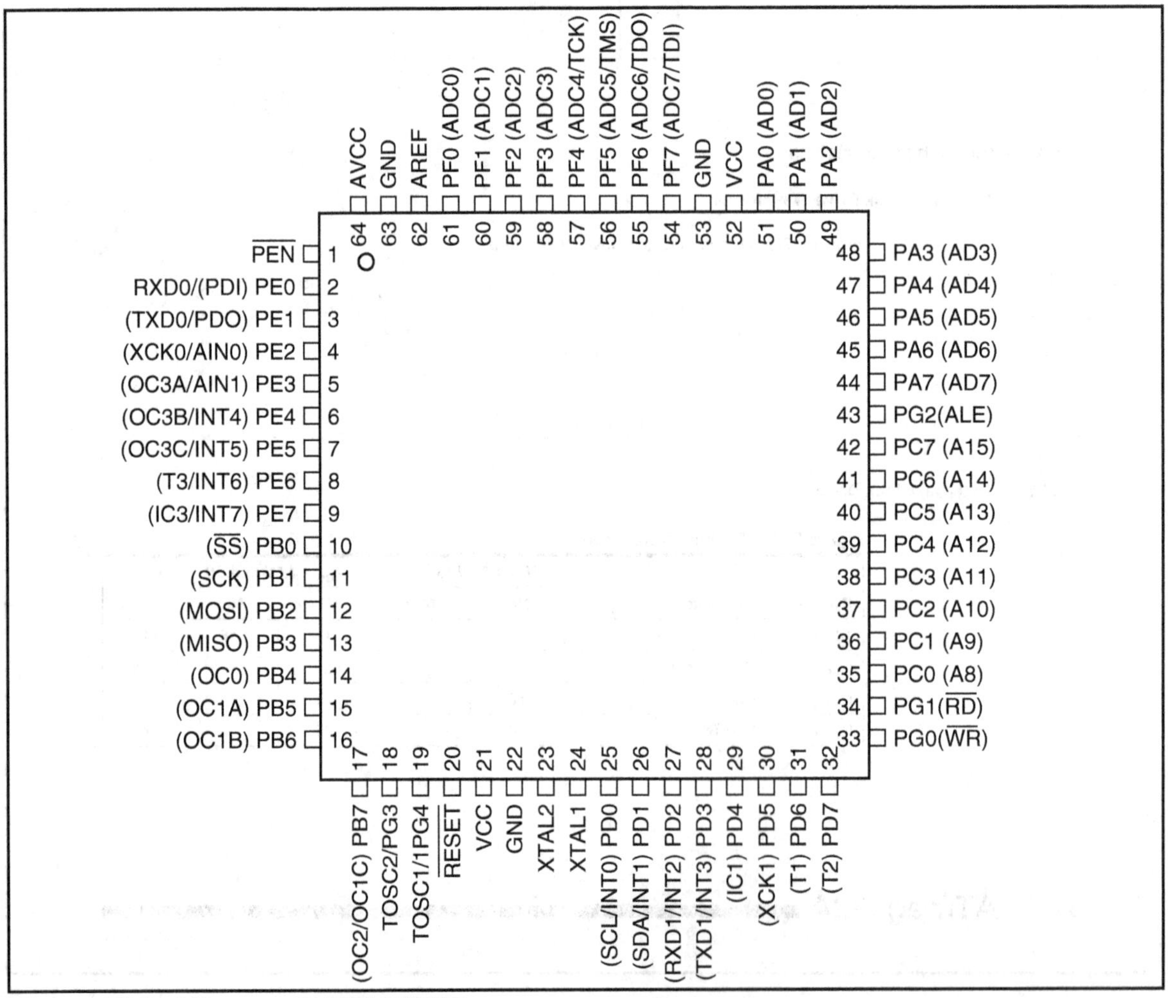

Figure H-3. ATmega 64/128 TQFP

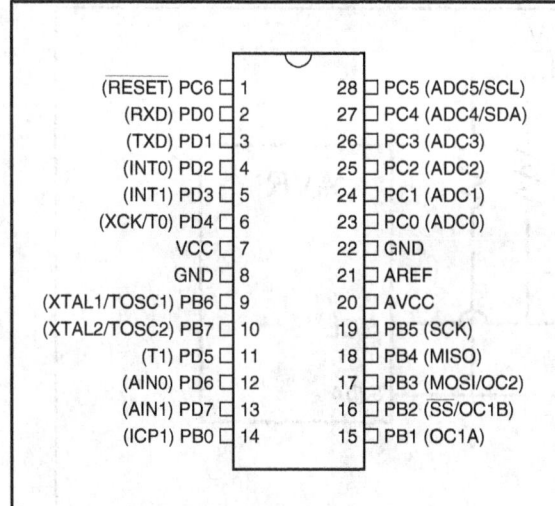

Figure H-4. ATmega8 DIP

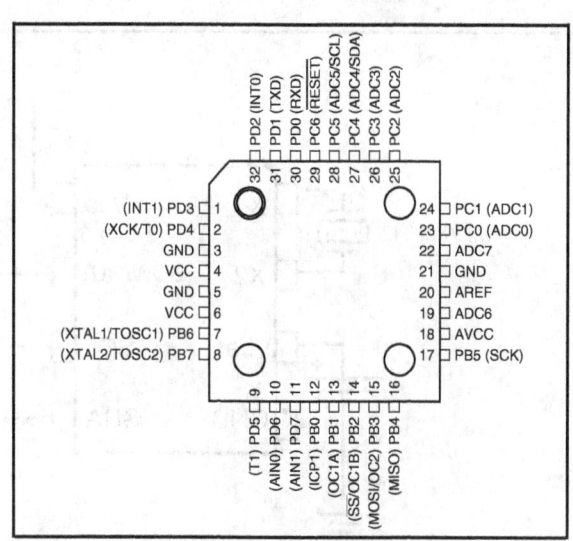

Figure H-5. ATmega8 TQFP

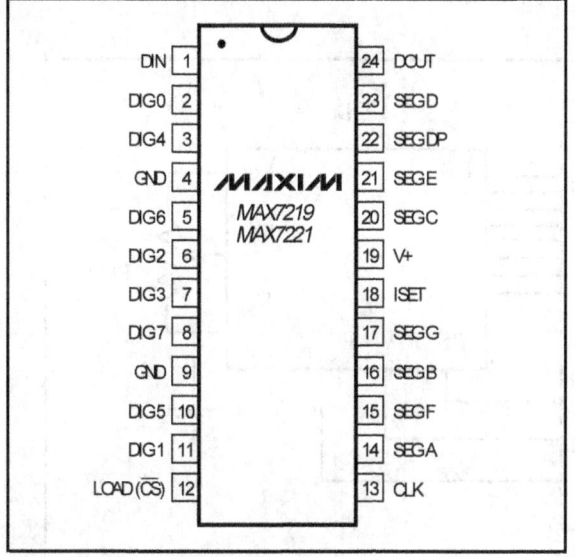

Figure H-6. MAX7221

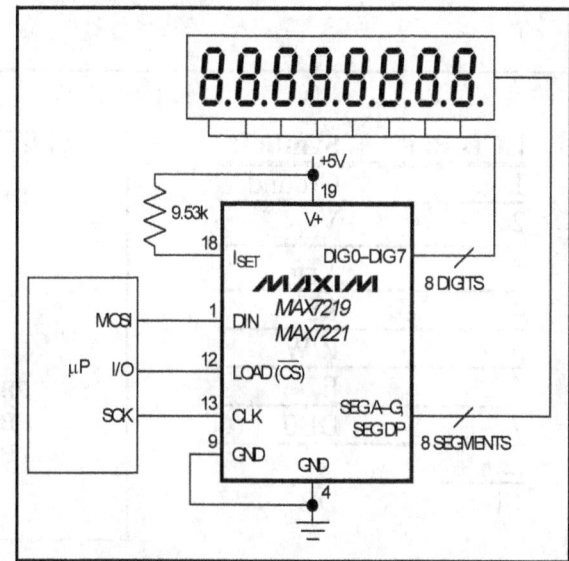

Figure H-7. MAX7221 Connections

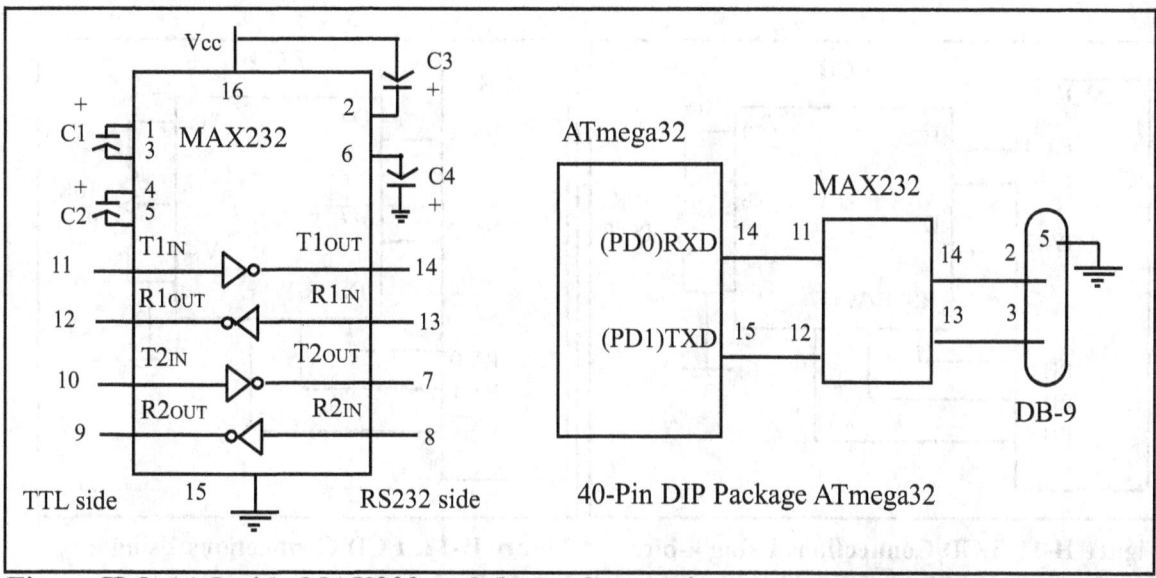

Figure H-8. (a) Inside MAX232 and (b) Its Connection to the ATmega32 (Null Modem)

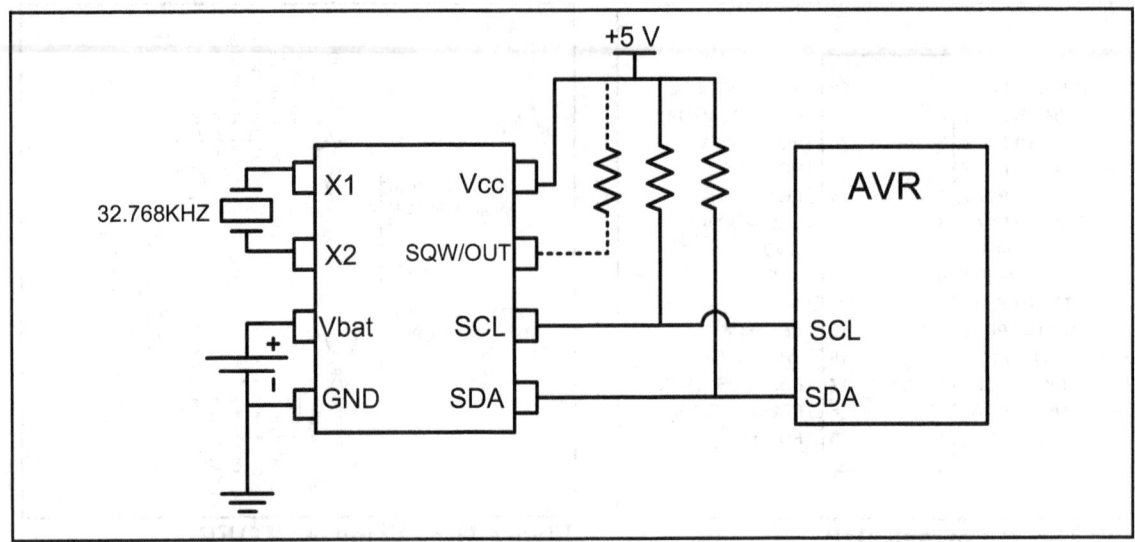

Figure H-9. DS1307 Power Connection Options (Maxim/Dallas Semiconductor)

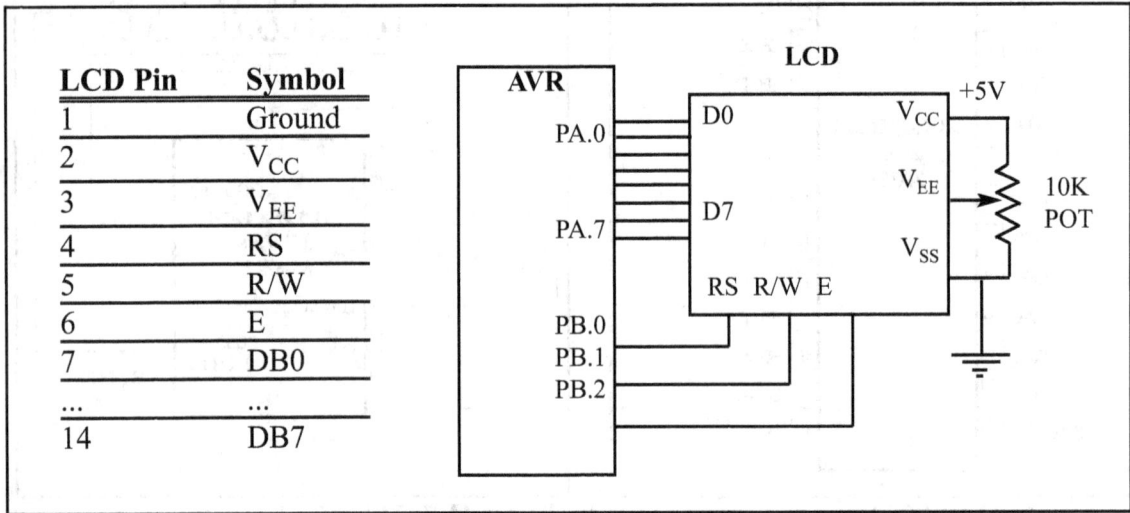

LCD Pin	Symbol
1	Ground
2	V_{CC}
3	V_{EE}
4	RS
5	R/W
6	E
7	DB0
...	...
14	DB7

Figure H-10. LCD Connections for 8-bit Data

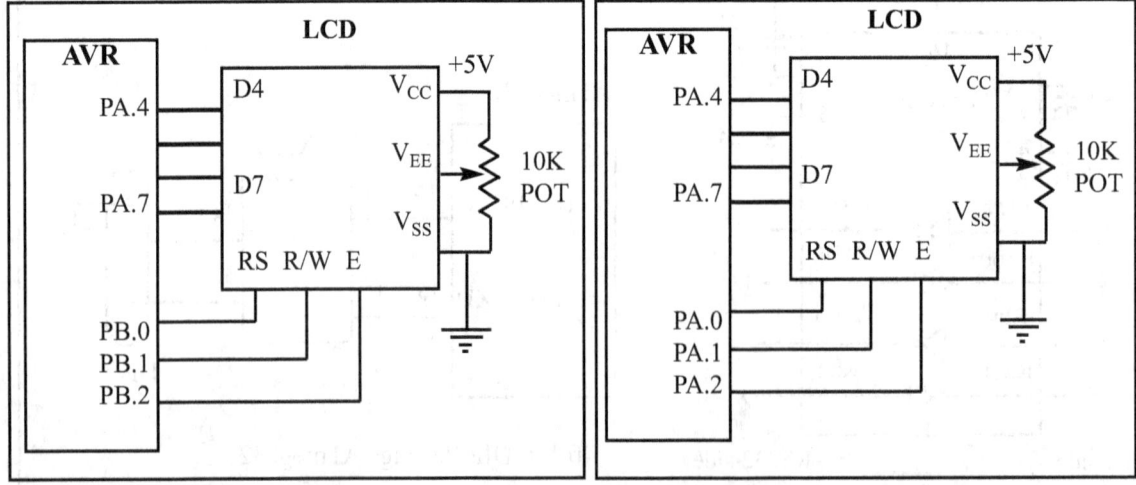

Figure H-11. LCD Connections Using 4-bit Data

Figure H-12. LCD Connections Using a Single Port

INDEX

A

ADC characteristics,   464
   analog input channels,   467
   conversion time,   465
   digital data output,   466
   end-of-conversion,   468
   parallel versus serial,   466
   resolution,   465
   start conversion,   468
   step size,   466
   V_{ref},   465
ADC hardware considerations,   469
ADC programming,   469
   A/D conversion time,   475
   ADC input channel source,   472
   ADC Start Conversion bit,   474
   ADCH: ADCL registers,   474
   ADCSRA register,   474
   ADLAR bit operation,   473
   ADMUX register,   471
   interrupts,   478
   polling,   476
   sample-and-hold time,   476
   V_{ref} source,   471
addition,   6
addition instruction,   162
   ADC,   163
   ADD,   162
addressing mode,   202
architectures,   32
   Harvard,   32
   von Neumann,   32
ASCII character,   76
ASCII code,   8
assembler,   80
assembler directive,   77
ATmega pins,   290
   AREF,   291
   AVCC,   291

   GND,   291
   RESET,   292
   VCC,   291
   XTAL1 and XTAL2,   291
AVR family,   49
   Classic AVR,   49
   Mega AVR,   49
   Special purpose AVR,   50
   Tiny AVR,   49
AVR fuse bits,   294
   AVR fuse bits,   297
   BODLEVEL,   297
   CKSEL0–CKSEL3,   295
   reset delay,   297
AVR programming,   305
   boot loader,   306
   in-circuit serial programming,   305
   parallel programming,   305
AVR serial port,   405
   doubling the baud rate,   416
   FE and PE flag bits,   412
   interrupt-based data receive,   422
   interrupt-based data transmit,   423
   monitoring the UDRE flag,   412
   receive data serially,   413
   transfer data serially,   412
   transmit and receive,   414
   UBRR register,   405
   UCSR registers,   408
   UDR registers,   407
AVR trainers,   306

B

back EMF,   496
baud rate error calculation,   416
big endian,   91
binary,   2
binary logic,   9
binary number,   76
bipolar,   503
bit,   13
bit-wise operators in C,   265
branch penalty,   130
BREQ,   112
BRNE,   108
brown-out detector,   297

bus, 14
  address bus, 14
  control bus, 14
  data bus, 14
byte, 13

C

C data types, 256
  signed int, 259
  unsigned char, 257
  unsigned int, 259
CALL, 118
carry flag, 71
choosing a microcontroller, 42
CISC, 94
compare instruction, 179
compare match, 328
compiler, 80
conditional flag, 71
context saving, 381
converting between binary and hex, 4
converting from binary to decimal, 3
converting from decimal to binary, 2
converting from decimal to hex, 4
converting from hex to decimal, 5
counter, 348
CPU architecture, 29
CTC, 328

D

DAC interfacing, 484
  converting I_{out} to voltage, 485
  DAC0808, 485
  generating a stair-step ramp, 486
  MC1408, 485
  programming DAC in C, 486
data memory, 59
DC motors, 550-553
  bidirectional control, 550-553
  pulse width modulation, 556
  unidirectional control, 550
decimal, 2
decimal number, 76
decoder, 12
direct addressing, 204

disabling interrupt, 366
division, 167
DPDT, 492
DS1307 interfacing, 654
  address map, 655
  control register, 656
  reading from DS1307, 657
  register pointer, 657
  setting the time in Assembly, 658
  setting the time in C, 662
  writing to DS1307, 657

E

edge-triggered, 378
EEPROM access in C, 284
electromagnetic relay, 492
electromechanical relay, 492
embedded system, 41
EMR, 492
enabling interrupt, 366
EQU, 77
extended I/O memory, 206
external hardware interrupt, 376

F

Fast PWM, 561,574
flag register, 71
flip-flop, 12

G

general purpose register, 56
generating pulses, 516
gigabyte, 13

H

half carry flag, 72
half-stepping, 502
Harvard architecture, 90
hex number, 75
hexadecimal, 4
HIGH( ), 201
holding torque, 502

I

I/O direct addressing,	205
I/O memory,	59
I/O port pins and their functions,	140
I/O register,	205
I2C bus protocol,	630
bit format,	631
clock stretching,	636
line electrical characteristics,	630
nodes,	631
packet format,	633
address packet format,	633
arbitration,	636
data packet format,	634
SLA+R,	634
SLA+W,	634
START and STOP conditions,	631
ICALL,	126
IJMP,	117
immediate addressing,	203
IN,	64
INCLUDE,	78
input capture,	531
instruction cycle time,	130
Intel Hex,	300
interrupt,	364
interrupt handler,	364
interrupt latency,	384
interrupt priority,	381
interrupt service routine,	364
interrupt vector table,	365
inverted,	567,577

J

JMP,	116

K

kilobyte,	13

L

LCD interfacing,	430
4-bit interfacing,	434
initializing the LCD,	432
LCD pin descriptions,	430
D0–D7,	431
E, enable,	430
R/W, read/write,	430
RS, register select,	430
VCC, VSS, and VEE,	430
sending commands to the LCD,	432
sending data to the LCD,	432
LDI,	57
LDS,	62
level-triggered,	378
little endian,	91
logic gates,	9
AND gate,	9
inverter,	10
NAND gate,	10
NOR gate,	10
OR gate,	9
tri-state buffer,	9
XOR gate,	10
logic instructions,	176
AND,	176
COM,	179
EX-OR,	177
NEG,	179
OR,	176
loop,	108
LOW( ),	201
lst file,	83

M

machine cycle,	130
machine language,	80
map file,	83
MAX232,	403
MAX233,	404
MAX7221 interfacing,	615
7-segment display,	615
data packet format,	617
MAX7221 pins,	616
CLK,	616
CS,	616
DIG0–DIG7,	616
DIN,	616
DOUT,	616
GND,	616

ISET,	616
SEGA–SEGG and DP,	617
VCC,	616
programming in Assembly,	621
programming in C,	624
measuring period,	534
measuring pulse width,	536
megabyte,	13
microcontroller,	40
microprocessor,	40
mnemonic,	80
MOV,	67
multiplication,	166
multistage execution,	129

N

NC,	492
negative flag,	71
nested loop,	110
nested loop delay,	133
nibble,	13
NO,	492
non-inverted,	566,576
Normal mode,	317
normally closed,	492
normally open,	492
numbering and coding systems,	2

O

operators in C,	271
bit-wise shift operation,	272
compound assignment,	271
optoisolator,	496
ORG,	78
oscillator clock source,	295
OUT,	67
overflow flag,	71

P

Phase correct PWM,	566, 585
pipelining,	128
polling,	364
port programming,	140
DDR for inputting data,	142
DDR for outputting data,	142
PIN register role in inputting data,	143
Port A,	144
Port B,	145
Port C,	146
Port D,	146
PORT register	143
SBIC (Skip if Bit in I/O register Cleared),	154
SBIS (Skip if Bit in I/O register Set),	153
synchronizer delay,	145
prescaler,	345
program counter,	85
PWM,	556,560

R

RAM,	20
DRAM,	23
NV-RAM,	22
SRAM,	21
RCALL,	124
reed switch,	496
relay,	492
RISC,	93
RJMP,	117
ROM,	16
EEPROM,	19
EPROM,	17
Flash memory,	19
Mask ROM,	20
PROM,	16
RPM,	501
RS232 pins,	399
CTS,	401
DCD,	401
DSR,	401
DTR,	401
RI,	401
RTS,	401

S

sensor interfacing,	480
interfacing the LM35,	481
LM34 and LM35,	480

reading temperature, 482
sensors, 480
transducers, 480
serial communication, 396
  framing, 398
  full-duplex, 397
  half-duplex, 397
  modem, 397
  space, 398
  transfer rate, 399
SET, 77
SFR, 60
short jump, 114
sign bit, 72
single-register addressing, 203
solid-state relay, 495
SPDT, 492
special function register, 60
SPI bus protocol, 604
  3-wire interface bus, 604
  clock polarity and phase, 606
    CPHA, 606
    CPOL, 606
  how SPI works, 605
  reading data, 608
    multibyte burst read, 608
    single-byte read, 608
  writing data, 606
    multibyte burst write, 607
    single-byte write, 607
SPI programming, 609
  master operating mode, 612
  programming in C, 614
  slave operating mode, 613
  SPCR (SPI Control Register), 610
  SPDR (SPI Data Register), 611
  SPSR (SPI Status Register), 609
  SS pin, 611
SPST, 492
stack, 119
  initializing the stack, 121
  POP, 119
  PUSH, 119
  stack pointer, 119
status register, 71
step angle, 499
stepper motor, 498

steps per revolution, 499
STS, 63
subtract instruction, 164
  SBC, 166
  SBIW, 165
  SUB, 164
  SUBI, 165
subtraction, 7
Successive Approximation, 468

T

task switching, 381
TIFR, 316
time delay in C, 261
timer interrupt, 369
Timer0, 315
Timer1, 335
Timer2, 332
TWI programming, 642
  checking status register, 668
    master receiver operating
    mode, 674
    master transmitter operating
    mode, 668
    slave receiver operating mode, 686
    slave transmitter operating
    mode, 680
  master operating mode, 642
  slave operating mode, 648
TWI registers, 638
  TWI Address Register (TWAR), 641
  TWI Bit Rate Register (TWBR), 639
  TWI Control Register (TWCR), 640
  TWI Data Register (TWDR), 641
  TWI Interrupt (TWINT) flag, 640
  TWI Status Register (TWSR), 639
two-register addressing, 203

U

unconditional branch, 116
unipolar, 503

V

von Neumann, 32

W

waveform generator,   510

Z

zero flag,   71